ADVANCED LECTURES IN MATHEMATICS

ADVANCED LECTURES IN MATHEMATICS

(*Executive Editors: Shing-Tung Yau, Kefeng Liu, Lizhen Ji*)

1. Superstring Theory (2007)
(Editors: Shing-Tung Yau, Kefeng Liu, Chongyuan Zhu)

2. Asymptotic Theory in Probability and Statistics with Applications (2007)
(Editors: Tze Leung Lai, Lianfen Qian, Qi-Man Shao)

3. Computational Conformal Geometry (2007)
(Authors: Xianfeng David Gu, Shing-Tung Yau)

4. Variational Principles for Discrete Surfaces (2007)
(Authors: Feng Luo, Xianfeng David Gu, Junfei Dai)

5. Proceedings of The 4th International Congress of Chinese Mathematicians Vol. I, II (2007)
(Editors: Lizhen Ji, Kefeng Liu, Lo Yang, Shing-Tung Yau)

6. Geometry, Analysis and Topology of Discrete Groups (2008)
(Editors: Lizhen Ji, Kefeng Liu, Lo Yang, Shing-Tung Yau)

7. Handbook of Geometric Analysis Vol. I (2008)
(Editors: Lizhen Ji, Peter Li, Richard Schoen, Leon Simon)

8. Recent Developments in Algebra and Related Areas (2009)
(Editors: Chongying Dong, Fu-An Li)

9. Automorphic Forms and the Langlands Program (2009)
(Editors: Lizhen Ji, Kefeng Liu, Shing-Tung Yau)

10. Trends in Partial Differential Equations (2009)
(Editors: Baojun Bian, Shenghong Li, Xu-Jia Wang)

11. Recent Advances in Geometric Analysis (2009)
(Editors: Yng-Ing Lee, Chang-Shou Lin, Mao-Pei Tsui)

12. Cohomology of Groups and Algebraic K-theory (2009)
(Editors: Lizhen Ji, Kefeng Liu, Shing-Tung Yau)

13–14. Handbook of Geometric Analysis Vol. I, II (2010)
(Editors: Lizhen Ji, Peter Li, Richard Schoen, Leon Simon)

15. An Introduction to Groups and Lattices (2010)
(Author: Robert L. Griess, Jr.)

16. Transformation Groups and Moduli Spaces of Curves (2010)
(Editors: Lizhen Ji, Shing-Tung Yau)

17–18. Geometry and Analysis Vol. I, II (2010)
(Editor: Lizhen Ji)

19. Arithmetic Geometry and Automorphic Forms (2011)
(Editors: James Cogdell, Jens Funke, Michael Rapoport, Tonghai Yang)

20. Surveys in Geometric Analysis and Relativity (2011)
(Editors: Hubert L. Bray, William P. Minicozzi II)

21. Advances in Geometric Analysis (2011)
(Editors: Stanisław Janeczko, Jun Li, Duong H. Phong)

22. Differential Geometry (2012)
(Editors: Yibing Shen, Zhongmin Shen, Shing-Tung Yau)

23. Recent Development in Geometry and Analysis (2012)
(Editors: Yuxin Dong, Jixiang Fu, Guozhen Lu, Weimin Sheng, Xiaohua Zhu)

24–26. Handbook of Moduli Vol. I, II, III (2012)
(Editors: Gavril Farkas, Ian Morrison)

ALM 25
Advanced Lectures in Mathematics

Handbook of Moduli

(Volume Ⅱ)

模手册（卷Ⅱ） Moshouce

Editors: Gavril Farkas · Ian Morrison

高等教育出版社·北京
HIGHER EDUCATION PRESS BEIJING

International Press

Higher Education Press Limited Company
4 Dewai Dajie, Beijing 100120, P. R. China, and
International Press
387 Somerville Ave, Somerville, MA, U. S. A.

The Handbook of Moduli was designed by Ian Morrison in LaTeX using a variant of the standard style files of the Higher Education Press. The text of the Handbook is set in ITC Giovanni and the mathematics in AMS Euler.

Giovanni was designed by Robert Slimbach in 1989 for ITC and was one of the early faces that earned him the Prix Charles Peignot, the Fields Medal of type design awarded "to a designer under the age of 35 who has made an outstanding contribution to type design". It combines the basic proportions of traditional oldstyle designs with the more even color and higher x-height of modern digital fonts to produce an inconspicuous but legible typeface.

Euler was designed in 1981 by Hermann Zapf, a major figure in 20th type design and a pioneer in digital typography, working in close cooperation with Donald Knuth, as an upright, cursive symbol font that would give the effect of mathematics handwritten on a blackboard. In 2008, Zapf reshaped many of the glyphs, with the assistance of Hans Hagen, Taco Hoekwater, and Volker RW Schaa, in order to harmonize the designs and bring them into line with contemporary standards of digital typography.

ADVANCED LECTURES IN MATHEMATICS

EXECUTIVE EDITORS

Shing-Tung Yau
Harvard University
Cambridge, MA. USA

Kefeng Liu
University of California, Los Angeles
Los Angeles, CA. USA

Lizhen Ji
University of Michigan
Ann Arbor, MI. USA

EXECUTIVE BOARD

Chongqing Cheng
Nanjing University
Nanjing, China

Tatsien Li
Fudan University
Shanghai, China

Zhong-Ci Shi
Institute of Computational Mathematics
Chinese Academy of Sciences (CAS)
Beijing, China

Zhiying Wen
Tsinghua University
Beijing, China

Lo Yang
Institute of Mathematics
Chinese Academy of Sciences (CAS)
Beijing, China

Zhouping Xin
The Chinese University of Hong Kong
Hong Kong, China

Weiping Zhang
Nankai University
Tianjin, China

Xiping Zhu
Sun Yat-sen University
Guangzhou, China

Xiangyu Zhou
Institute of Mathematics
Chinese Academy of Sciences (CAS)
Beijing, China

The Handbook of Moduli is dedicated to the memory of Eckart Viehweg, whose untimely death precluded a planned contribution, and to David Mumford, who first proposed the project, for all that they both did to nurture its subject; and to Angela Ortega and Jane Reynolds for everything that they do to sustain its editors.

Contents

Volume I

Volume II

Volume III

Parameter spaces of curves

Joe Harris

Abstract. In this article I will try to survey the state of our knowledge (and the much greater area of our ignorance) of the geometry of spaces parametrizing curves in projective space.

Contents

1. Introduction

Robin Hartshorne, in [6], describes the problem of classifying algebraic varieties as the guiding problem of algebraic geometry. I'd agree, for the most part; and, since you're currently reading a book entitled "Handbook of Moduli," presumably you would too.

But the question remains: what exactly are we classifying? To be specific, consider the problem of smooth, complete algebraic curves over $\mathbb{C}$. If you ask mathematicians today to describe the problem of classifying curves, they would naturally take "curve" to mean "abstract curve," in which case the answer to the problem, "classify all smooth complete curves" would consist of two parts. Algebraic

---

2000 *Mathematics Subject Classification.* Primary 14Dxx; Secondary 14Dxx.
Key words and phrases. moduli.

curves are classified first by their sole discrete numerical invariant, the genus, which can assume any value $g \in \mathbb{N}$; and the set of curves of a given genus g has naturally the structure of an irreducible quasi-projective variety M_g. Beyond this, the problem of classifying algebraic curves consists of studying the geometry of the variety M_g, and of relating properties of curves to the loci in M_g of curves with that property.

If you had posed the same question to an algebraic geometer of the 19th century, however, it would of necessity have been interpreted differently. Abstract curves didn't exist then (or, depending on your philosophical point of view, they hadn't been discovered); the word "curve" would have been taken to mean a subset of projective space defined by polynomial equations, smooth and irreducible of dimension 1. As such, a curve had not one but three numerical invariants: its degree d; the dimension r of the projective space in which it lay (or, more properly, the dimension of its span); and of course its genus. The problem of classifying all algebraic curves would thus amount to two things:

(1) To say for which triples g, r and d there exists a smooth, irreducible and nondegenerate curve of degree d and genus g in $\mathbb{P}^r$; and
(2) To describe, for each such triple (g, r, d), the geometry of the space $\mathcal{H}^\circ_{g,r,d}$ parametrizing such curves: its irreducible components, their dimensions and so on.

In this volume, there are many articles that address aspects of the problem of classification in its modern sense. But the classical version is still very much of interest, and has many fascinating aspects that are not fully understood: we haven't answered the first of the questions above; and we know the answer to the second only in an extremely limited range of cases. The goal of this article is to give a survey of what we do know about this problem, and likewise to suggest some of the numerous open problems.

The remainder of this paper will consist of three parts. In Section 2, we'll discuss the notion of parameter spaces of curves, and compare the two most commonly used such spaces, primarily the Hilbert scheme and the Kontsevich space. This may in a sense not be necessary if we're only concerned with smooth, irreducible curves in projective space, since the Hilbert scheme and the Kontsevich space have a common open subset parametrizing such curves (and indeed the reader can skip this section and go directly to the following ones). But for many purposes it's useful to have a compactification of the space of curves, and here the Hilbert scheme and the Kontsevich space differ dramatically, as we'll see.

In Section 3, we'll describe the conjectured answer to the Existence Problem, the first of the two questions listed above. This actually tells us a lot about curves of high genus: when g is more than roughly half the maximal possible genus of an irreducible, nondegenerate curve of degree d in $\mathbb{P}^r$, in addition to simply saying which triples (g, r, d) occur, we learn about the geometry of such curves, and the dimension and irreducible components of their families. But for g below this bound, all we can say is that such curves exist; we can't say much about the spaces parametrizing them

Finally, in Section 4, we address this issue. We can in fact give a pretty explicit description of the spaces of curves of low genus, using what we know about the moduli space of abstract curves and Brill-Noether theory. Again, our knowledge—even conjectured—drops off as we approach the middle range of possible genera; we'll try to indicate what are some of the main unresolved questions in this area.

2. Parameter spaces

First of all, some terminology. We propose to call a space whose points correspond naturally to isomorphism classes of varieties or schemes X of a given type a *moduli space*; we'll call a space whose points correspond naturally to subschemes $X \subset Z$ of a fixed scheme Z (not up to isomorphism) a *parameter space*. There is not always a clear line dividing the two—for example, the Kontsevich space parameterizing stable maps has elements of both—but it does reflect an important duality in how we view geometric objects. One of the fundamental ideas underlying much recent progress in the theory of curves, for example, is the fact that whenever we have a one-parameter family $\{C_t \subset \mathbb{P}^r\}_{t\in\Delta}$ of curves in projective space, with C_t smooth for $t \neq t_0$, we have two distinct notions of the "limit" $\lim_{t\to t_0} C_t$ of the curves C_t: the *flat limit*, which is a subscheme of $\mathbb{P}^r$ whose geometry can be pretty much arbitrarily messy; and the *stable limit*, which is the limit of the abstract curves C_t and has at worst nodes as singularities. (Other articles in this volume discuss alternative notions of stability, and correspondingly alternative definitions of the limit of the abstract curves C_t; as for the flat limit, we really don't have much of an alternative to that.)

That said, what should we take as the parameter space for curves of degree d and genus g in $\mathbb{P}^r$? There are principally three answers to this question: the *Chow variety*, the *Hilbert scheme* and the *Kontsevich space*. These agree on the common open subset $\mathcal{H}^{\circ}_{g,r,d}$ parametrizing smooth curves (at least if we ignore the scheme structure on these spaces), but give very different compactifications of $\mathcal{H}^{\circ}_{g,r,d}$. Now, the questions we raised earlier—when do there exist such curves $C \subset \mathbb{P}^r$, what are the irreducible components of $\mathcal{H}^{\circ}_{g,r,d}$ and what are their dimensions—really don't depend on the choice of compactification, as long as we restrict our attention to the closure of $\mathcal{H}^{\circ}_{g,r,d}$ in each. But for many other questions it is important to have a complete parameter space, and so we start with a brief discussion of the properties of each. Actually, we'll pretty much ignore the Chow variety—in many ways, it has all the drawbacks of the Hilbert scheme and the Kontsevich space, and none of the virtues—and focus primarily on the other two.

The following discussion is adapted from a forthcoming book, *3264 and All That: Intersection Theory in Algebraic Geometry*, by the author and David Eisenbud.

2.1. Hilbert schemes

The Hilbert scheme $\mathcal{H} = \mathcal{H}_{g,r,d}(\mathbb{P}^r)$ is a parameter space for subschemes of $\mathbb{P}^r$ with Hilbert polynomial $p(m) = md - g + 1$; in the case of curves (one-dimensional subschemes) this means all subschemes with fixed degree and arithmetic genus. The Hilbert scheme has many good properties. For example, there is a useful cohomological description of its tangent spaces, and, beyond that, a deformation theory that in some cases can describe its local structure. And, of course, associated to a point on the Hilbert scheme is all the rich structure of a homogenous ideal in the ring $K[x_0, \ldots, x_r]$ and its resolution.

There is one circumstance where the Hilbert scheme is particularly nice: the Hilbert scheme parametrizing plane curves of degree d (Hilbert polynomial $p(m) = md - \binom{d-1}{2} + 1$) is simply the projective space $\mathbb{P}^{\binom{d+2}{2}-1}$ of homogeneous forms of degree d. Beyond that, the geometry of the Hilbert scheme ranges from the mysterious to the pathological; we'll illustrate with some examples.

Beyond hypersurfaces, the simplest case is the Hilbert scheme containing twisted cubic curves in $\mathbb{P}^3$ (Hilbert polynomial $H(m) = 3m + 1$.) It has one component of dimension 12 whose general point corresponds to a twisted cubic curve, but it has a second component, whose general point corresponds to the union of a plane cubic $C \subset \mathbb{P}^2 \subset \mathbb{P}^3$ and a point $p \in \mathbb{P}^3$. Moreover, this second component has dimension 15 (the choice of plane has 3 degrees of freedom; the cubic inside the plane 9 more; and the point gives an additional 3.) These two components meet along the 11-dimensional subscheme of singular plane cubics C with an embedded point at the singularity, not contained in the plane spanned by C.

2.1.1. Report card for the Hilbert scheme The Hilbert scheme, as a compactification of the space of smooth curves, has drawbacks that sometimes make it difficult to use:

(1) **It has extraneous components, often of differing dimensions.** We see this phenomenon already in the case of twisted cubics, above. Of course we could take just the closure in the Hilbert scheme of the locus of smooth curves, but we would lose some of the nice properties, like the description of the tangent space. Thus while it is relatively easy to describe the singular locus of $\mathcal{H}$, we don't know how to describe singular locus of $\mathcal{H}^\circ$ along the locus where it intersects other components.

In fact, we don't know for curves of higher degree how many such extraneous components there are, or their dimensions: for $r \geqslant 3$ and large d the Hilbert scheme of zero-dimensional subschemes of degree d in $\mathbb{P}^r$ will have an unknown number of extraneous components of unknown dimensions, and this creates even more extraneous components in the Hilbert schemes of curves.

(2) **No one knows what's in the closure of the locus of smooth curves**. If we do choose to deal with the closure of the locus of smooth curves rather than the whole Hilbert scheme—as it seems we must—we face another problem: except in a few special cases, we can't tell if a given point in the Hilbert scheme is in this closure. That is, we don't know how to tell whether a given singular 1-dimensional scheme $C \subset \mathbb{P}^r$ is smoothable.

(3) **It has many singularities**. Vakil has shown that the singularities of the Hilbert scheme are, in a precise sense, arbitrarily bad: in[14] he proves that that the completion of every affine local K-algebra appears (up to adding variables) as the completion of a local ring on a Hilbert scheme of curves. But I want to focus here on a specific source of particularly bad singular points: points in the Hilbert scheme corresponding to curves that are flat limits of almost all other curves.

For example, start with an arbitrary smooth curve $C \subset \mathbb{P}^r$ of degree d and genus g, choose a general system of coordinates on $\mathbb{P}^r$ and apply the one-parameter group of automorphisms

$$A_t = \begin{pmatrix} t^{a_0} & 0 & 0 & \dots & 0 \\ 0 & t^{a_1} & 0 & \dots & 0 \\ 0 & 0 & t^{a_2} & \dots & 0 \\ \vdots & \vdots & \vdots & \ddots & \vdots \\ 0 & 0 & 0 & \dots & t^{a_n} \end{pmatrix}$$

with $a_0 \ll a_1 \ll \cdots \ll a_n$. Let C_0 and C_∞ be the flat limits of the curves $C_t = A_t(C)$ as $t \to 0$ or ∞—that is, the schemes defined by the *initial ideal* with Hilbert polynomial $dm - g + 1$ with respect to the lexicographical and reverse lexicographical orderings. The corresponding points $[C_0]$ and $[C_\infty]$ in the Hilbert scheme will then be in the closure of the orbit of a general point under the action of PGL_{r+1}; so that the local geometry of the Hilbert scheme at the point [C] necessarily encodes most of the global geometry of the Hilbert scheme itself, or at least its quotient by PGL_{r+1}.

2.2. The Kontsevich space

These drawbacks often make it difficult to study the global geometry of the Hilbert scheme. An alternative is the *Kontsevich space*; see [3] for a systematic treatment.

The Kontsevich space $\overline{M}_{g,0}(\mathbb{P}^r, d)$ parametrizes what are called *stable maps* of degree d and genus g to $\mathbb{P}^r$. These are morphisms

$$f : C \to \mathbb{P}^r$$

with C a connected curve of arithmetic genus g having only nodes as singularities, such that the image $f_*[C]$ of the fundamental class of C is equal to d times the class of a line in $A_1(\mathbb{P}^r)$, and satisfying the one additional condition that the automorphism group of the map f—that it, automorphisms ϕ of C such that $f \circ \phi = f$—is finite. (This last condition is automatically satisfied if the map f is finite; it's relevant only for maps that are constant on an irreducible component of C, and amounts to saying that any smooth, rational component C_0 of C on which f is constant must intersect the rest of the curve C in at least three points.) Two such maps $f : C \to \mathbb{P}^r$ and $f' : C' \to \mathbb{P}^r$ are said to be the same if there exists an isomorphism $\alpha : C \to C'$ with $f' \circ \alpha = f$. There's an analogous notion of a family of stable maps, and the Kontsevich space $\overline{M}_{g,0}(\mathbb{P}^r, d)$ is a coarse moduli space for the functor of families of stable maps. Note that we're taking the quotient by automorphisms of the domain, but not of the image, so that $\overline{M}_{g,0}(\mathbb{P}^r, d)$ shares with the Hilbert scheme $\mathcal{H}_{dm-g+1}(\mathbb{P}^r)$ a common subset parametrizing smooth curves $C \subset \mathbb{P}^r$ of degree d and genus g.

There are naturally variants of this: the space $\overline{M}_{g,n}(\mathbb{P}^r, d)$ parametrizes maps $f : C \to \mathbb{P}^r$ with C a nodal curve having n marked distinct smooth points $p_1, \ldots, p_n \in C$. (Here an automorphism of f is an automorphism of C fixing the points p_i and commuting with f; the condition of stability is thus that any smooth, rational component C_0 of C on which f is constant must have at least three distinguished points, counting both marked points and points of intersection with the rest of the curve C.) More generally, for any projective variety X and numerical equivalence class $\beta \in \mathrm{Num}_1(X)$, we have a space $\overline{M}_{g,n}(X, \beta)$ parametrizing maps $f : C \to X$ with fundamental class $f_*[C] = \beta$, again with C nodal and f having finite automorphism group.

The remarkable aspect of the Kontsevich space is simply that it is indeed proper: in other words, if $\mathcal{C} \subset \Delta \times \mathbb{P}^r$ is a flat family of subschemes of $\mathbb{P}^r$ parametrized by a smooth, one-dimensional base Δ, and the fiber C_t is a smooth curve for $t \neq 0$, then no matter what the singularities of C_0 there is a unique stable map $f : \tilde{C}_0 \to \mathbb{P}^r$ which is the limit of the inclusions $\iota_t : C_t \hookrightarrow \mathbb{P}^r$. Note that this limiting stable map $f : \tilde{C}_0 \to \mathbb{P}^r$ depends on the family, not just on the scheme C_0; the import of this in practice is that the Kontsevich space is often locally a blow-up of the Hilbert scheme along loci of curves with singularities worse than nodes. (This is not to say we have in general a regular map from the Kontsevich space to the Hilbert scheme; as we'll see in the examples below, the limiting stable map $f : \tilde{C}_0 \to \mathbb{P}^r$ doesn't determine the flat limit C_0 either.) We'll see how this plays out in four relatively simple cases below.

2.2.1. Plane conics One indication of how useful the Kontsevich space can be is that, in the case of $\overline{M}_0(\mathbb{P}^2, 2)$ (that is, plane conics), the Kontsevich space is actually equal to the space of complete conics:

To begin with, if $C \subset \mathbb{P}^2$ is a conic of rank 2 or 3—that is, anything but a double line—then the inclusion map $\iota : C \hookrightarrow \mathbb{P}^2$ is a stable map; thus the open set $W \subset \mathbb{P}^5$ of such conics is likewise an open subset of the Kontsevich space $\overline{M}_0(\mathbb{P}^2, 2)$.

But when a family $\mathcal{C} \subset \Delta \times \mathbb{P}^2$ of conics specializes to a double line $C_0 = 2L$, the limiting stable map is a finite, degree 2 map $f : C \to L$, with C either isomorphic to $\mathbb{P}^1$, or two copies of $\mathbb{P}^1$ meeting at a point. Such a map is characterized, up to automorphisms of the domain curve, by its branch divisor $B \subset L$, a divisor of degree 2. (If B consists of two distinct points, $C \cong \mathbb{P}^1$, while if $B = 2p$ for some $p \in L$, the curve C will be reducible.) Thus we have a birational morphism

$$\pi : \overline{M}_0(\mathbb{P}^2,2) \to \mathcal{H}_{2m+1}(\mathbb{P}^2) = \mathbb{P}^5$$

from the Kontsevich space to the Hilbert scheme, with 2-dimensional fibers over the locus in $\mathbb{P}^5$ corresponding to double lines.

2.2.2. Conics in space By contrast, there is not a regular map in either direction between the Hilbert scheme of conics in space and the Kontsevich space $\overline{M}_0(\mathbb{P}^3,2)$. Of course there is a common open set: its points correspond to reduced conics—that is, embedded nodal curves of degree 2. To see that this does not extend to a regular map in either direction, note first that, as before, if $\mathcal{C} \subset \Delta \times \mathbb{P}^3$ is a family of conics specializing to a double line C_0, the limiting stable map is a finite, degree 2 cover $f : C \to L$, and this cover is not determined by the flat limit C_0 of the schemes $C_t \subset \mathbb{P}^3$. But on the other hand the scheme C_0 is again a complete intersection of a plane and a quadric surface, which is to say it lies in a unique plane H; and this plane is not determined by the data of the map f.

The relationship in this case between the Hilbert scheme and the Kontsevich space is what's called in higher-dimensional birational geometry a *flip*: the Kontsevich space is obtained from the Hilbert scheme $\mathcal{H}$ by blowing up the locus of double lines, and then blowing down the exceptional divisor along another ruling. (The blow-up of $\mathcal{H}$ along the double line locus could be described as the space of pairs $(H; (C, C^*))$, where $H \subset \mathbb{P}^3$ is a plane and (C, C^*) a complete conic in $H \cong \mathbb{P}^2$.)

2.2.3. Plane cubics Here, we do have a regular map from the Kontsevich space $\overline{M}_1(\mathbb{P}^2, 3)$ to the Hilbert scheme $\mathcal{H}_{3m}(\mathbb{P}^2) \cong \mathbb{P}^9$, and it does some interesting things: it blows up the locus of triple lines, much as in the example of plane conics, and the locus of cubics consisting of a double line and a line as well. But it also blows up the locus of cubics with a cusp, and cubics consisting of a conic and a tangent line, and these are trickier: the blow-up along the locus of cuspidal cubics, for example, can be obtained either by three blow-ups with smooth centers, or one blow-up along a nonreduced scheme supported on this locus.

But what we really want to illustrate here is that the Kontsevich space $\overline{M}_1(\mathbb{P}^2, 3)$ is not irreducible—in fact, it's not even 9-dimensional! For example, maps of the form $f : C \to \mathbb{P}^2$ with C consisting of the union of an elliptic curve E and a copy of $\mathbb{P}^1$, with f mapping $\mathbb{P}^1$ to a nodal plane cubic C_0 and mapping E to a smooth point of C_0 form a 10-dimensional family of stable maps; in fact, these form an open subset of a second irreducible component of $\overline{M}_1(\mathbb{P}^2, 3)$. And there's also a third component, whose general member $f : C \to \mathbb{P}^2$ has domain C an elliptic curve, with two $\mathbb{P}^1$s attached, with the map f contracting the elliptic curve and sending the two rational tails to a line and a conic in $\mathbb{P}^2$.

2.2.4. Twisted cubics Here the shoe is on the other foot. The Hilbert scheme $\mathcal{H} = \mathcal{H}_{3m+1}$ has, as we saw, a second irreducible component besides the closure $\mathcal{H}_0$ of the locus of actual twisted cubics, and the presence of this component makes it difficult to work with. For example, it takes quite a bit of analysis to see that $\mathcal{H}_0$ is smooth, since we have no simple way of describing its tangent space; see [12] for details. By contrast, the Kontsevich space is irreducible, and has only relatively mild (finite quotient) singularities.

2.2.5. Report Card for the Kontsevich Space As with the Hilbert scheme, there are difficulties in using the Kontsevich space:

(1) **It has extraneous components**. These arise in a completely different way from the extraneous components of the Hilbert scheme, but they're there. A typical example of an extraneous component of the Kontsevich space $\overline{M}_g(\mathbb{P}^r, d)$ would consist of maps $f : C \to \mathbb{P}^r$ in which C was the union of a rational curve $C_0 \cong \mathbb{P}^1$, mapping to a rational curve of degree d in $\mathbb{P}^r$, and C_1 an arbitrary curve of genus g meeting C_0 in one point and on which f was constant; if the curve C_1 does not itself admit a nondegenerate map of degree d to $\mathbb{P}^r$, this map can't be smoothed.

So, using the Konsevitch space rather than the Hilbert scheme doesn't solve this problem, but it does provide a frequently useful alternative: there are situations where the Kontsevich space has extraneous components and the Hilbert scheme not—like the case of plane cubics described above—and also situations where the reverse is true, such as the case of twisted cubics.

(2) **No one knows what's in the closure of the locus of smooth curves**. This, unfortunately, remains an issue with the Kontsevich space. Even in the case of the space $\overline{M}_g(\mathbb{P}^2, d)$ parametrizing plane curves, where it might be hoped that the Kontsevich space would provide a better compactification of the Severi variety than simply taking its closure in the space $\mathbb{P}^N$ of all plane curves of degree d, the fact that we don't know which stable maps are smoothable represents a real obstacle to its use.

(3) **It has points corresponding to highly singular schemes, and these tend to be in turn highly singular points of $\overline{M}_g(\mathbb{P}^r, d)$**. Still true; but in this respect, at least, it might be said that the Kontsevich space represents an improvement over the Hilbert scheme: even when the image $f(C)$ of a stable map $f : C \to \mathbb{P}^r$ is highly singular, the fact that the domain of the map is at worst nodal makes the deformation theory of the map relatively tractable.

2.3. Caveat: should we restrict attention to smooth curves, or reduced ones?

Before we leave the topic of the choice of parameter space, we should mention one other choice. We have declared that we are primarily interested in the space $\mathcal{H}^\circ_{g,r,d}$ parametrizing smooth, irreducible, nondegenerate curves $C \subset \mathbb{P}^r$ of degree d and genus g. But there is at least one plausible alternative: we could replace the condition of smoothness with the weaker condition of being simply reduced.

Now, it may seem at first that. as long as we're concerned only with coarse invariants of the space $\mathcal{H}^\circ_{g,r,d}$—enumerating its irreducible components, and their dimensions—it shouldn't matter whether we work with $\mathcal{H}^\circ_{g,r,d}$ or the larger open subset $\mathcal{H}'_{g,d,r}$ of reduced, irreducible and nondegenerate curves, and for the most part this is true: the vast majority of components of $\mathcal{H}'_{g,r,d}$ have dense open subsets lying in $\mathcal{H}^\circ_{g,r,d}$. But not always: while you may have to work to locate them, there are irreducible components of $\mathcal{H}'_{g,r,d}$ whose general point corresponds to a reduced but singular curve. This can happen for local or global reasons: Hartshorne, in [8], shows that there exist reduced and irreducible curves $C \subset \mathbb{P}^3$ that are not smoothable; and in the following section we'll see how to construct examples of curves in higher-dimensional space $\mathbb{P}^r$ that have just one node as singularity and that are still not smoothable.

We have chosen here to work with the more restricted class of smooth curves. But this is largely for reasons of convenience: if we allow singularities, the question arises of whether to talk in terms of the arithmetic genus or the geometric genus; and while the former is clearly the more natural choice when we're talking about the Hilbert scheme of curves of higher genus (as we will in the following Section), the Brill-Noether-theoretic approach to curves of lower genus that we take in Section 4 is better suited to working with the geometric genus. The fact is, many of the statements we make, and the techniques we employ, can be extended to the larger class of reduced, irreducible and nondegenerate curves, but they're more complicated; and in the interests of an intelligible exposition we've opted to do everything in the simpler setting.

3. The existence problem

We start with the first question: when there exists a smooth, irreducible and nondegenerate curve of degree d and genus g in $\mathbb{P}^r$. This was answered for $r = 3$ by Halphen in the 19th century, though there was a gap in his argument; the correct proof was given in [4]. A wonderful survey of our knowledge of curves in $\mathbb{P}^3$ is given in [7] and [9]. The answer for $r = 4$ and 5 was given in [13]; and we have a conjectured answer—or at least an algorithm for determining the answer in any given case—in general; we'll describe this conjecture below.

This story has been told many times, so we'll be brief here, outlining the main ideas of the analysis. We'll assume throughout that $r \geqslant 3$.

3.1. Curves of maximal genus

The first step toward an answer to the existence problem, bounding from above the genus of an irreducible, nondegenerate curve of degree d in $\mathbb{P}^r$, was taken by Castelnuovo in the late 19th century (the case $r = 3$ was done by Halphen already). To paraphrase, Castelnuovo starts by observing that the genus of a curve $C \subset \mathbb{P}^r$ of degree d may be bounded in terms of the Hilbert function h_Γ of a general hyperplane section $\Gamma = C \cap \mathbb{P}^{r-1}$ of C. Specifically, he observes that for large m_0,

$$g = m_0 d - h_C(m_0) + 1,$$

and since

$$h_C(m) - h_C(m-1) \geqslant h_\Gamma(m)$$

for all m, this leads to the inequality

$$g \leqslant \sum_{m=1}^{\infty} (d - h_\Gamma(m)). \tag{3.1}$$

In English, *the genus* g *of* C *is bounded by the sum over* $m \geqslant 1$ *of the failure of* Γ *to impose independent conditions on polynomials of degree* m. We see moreover that we have equality in the inequality above whenever C is projectively normal, that is, the linear series cut on C by hypersurfaces of each degree m are all complete.

The problem then is to bound from below the Hilbert function of a collection Γ of d points in $\mathbb{P}^n$. Since these points will be a general hyperplane section of an irreducible, nondegenerate curve, we can assume that they're in *uniform position*, meaning that any two subsets $\Gamma', \Gamma'' \subset \Gamma$ of the same cardinality have the same Hilbert function[1]. Castelnuovo gives a remarkably naive bound: if $d \geqslant mn + 1$, we can exhibit a hypersurface of degree m containing any given mn points of Γ but not a given $(mn+1)^{\text{st}}$ simply by taking a union of m hyperplanes, each containing n of the mn points. Thus we have

$$h_\Gamma(m) \geqslant \min\{mn + 1, d\}.$$

What's remarkable is that, crude as the estimate may seem at first, it's sharp: any configuration $\Gamma \subset \mathbb{P}^n$ lying on a rational normal curve has exactly this Hilbert function.

[1] Considering only the values $h_{\Gamma'}(1)$, this amounts to saying that the points of Γ are in linear general position, which is in fact all that Castelnuovo assumes. The fact that the points of a general hyperplane section $\Gamma = H \cap C$ of an irreducible curve are in the uniform position follows from the observation that as H varies the monodromy group of the points of Γ is the full symmetric group.

Having arrived at this inequality for the Hilbert function of $\Gamma \subset \mathbb{P}^{r-1}$, Castelnuovo plugs it in to the inequality on g above to arrive at a bound

$$g \leqslant \pi_0(d, r),$$

where π_0 is defined as follows: we set

$$m = \left[\frac{d-1}{r-1}\right] \quad \text{and} \quad \epsilon = d - 1 - m(r-1)$$

(that is, ϵ is the remainder of $d-1$ under division by $r-1$; in particular, $0 \leqslant \epsilon \leqslant r-2$); we then set

$$\pi_0(d, r) = \binom{m}{2}(r-1) + m\epsilon.$$

Thus, for fixed r the bound $\pi_0(d, r)$ starts out at 0 for $d = r$ (the smallest value of d possible, given that $C \subset \mathbb{P}^r$ is nondegenerate), and increases by 1 every time d increases by 1 up to $d = 2r - 1$, where the value $\pi_0(2r-1, r) = r - 1$. (This reflects the fact that in this degree range, every curve C is nonspecial by Clifford's theorem; thus $g \leqslant d - r$.) Passing from $d = 2r - 1$ to $d = 2r$ the bound π_0 jumps by 2 to $\pi_0(2r, r) = r + 1$, as we see canonical curves—the first nonspecial curves—appear; it then continues to increase by 2 every time d increases, up to $d = 3r - 2$. In the next range $3r - 2 \leqslant d \leqslant 4r - 3$ it increases by 3, and so on; asymptotically, we see that

$$\pi_0(d, r) \sim \frac{d^2}{2(r-1)} \quad \text{as } d \to \infty.$$

Moreover, Castelnuovo's bound $g \leqslant \pi_0(d, r)$ is sharp. The observation above that the minimal possible Hilbert function of the general hyperplane section Γ of C is achieved by configurations lying on a rational normal curve suggests where to look for curves of maximal genus: at curves C lying on a surface $S \subset \mathbb{P}^r$ of degree $r-1$; that is, rational normal scrolls or (in case $r = 5$) the quadratic Veronese surface. Now, we know the Picard group of a rational normal surface scroll $S \subset \mathbb{P}^r$—it's freely generated by the classes of a hyperplane section H and a fiber F of the ruling, with intersection pairing

$$H^2 = r - 1; \quad H \cdot F = 1 \quad \text{and} \quad F^2 = 0.$$

Moreover, we know the canonical class: it's

$$K_S = -2H + (r-3)F,$$

so that we know both the degree $d = a(r-1) + b$ the genus

$$g = \binom{a}{2}(r-1) + (a-1)(b-1)$$

of a smooth curve in any class $aH + bF$ (and the dimension of the linear system in which it moves). It's straightforward then to look for classes of maximal arithmetic genus among those with a given degree d: we find that

(1) When $\epsilon \neq 0$, there's a unique class $(m+1)H-(r-2-\epsilon)F$ with arithmetic genus $g = \pi_0(d,r)$; and

(2) When $\epsilon = 0$, there are two: the above, and the class $mH+F$.

Checking that there are indeed smooth curves in these classes, Castelnuovo deduces that his bound $g \leqslant \pi_0(d,r)$ is sharp.

And now comes the truly wonderful part. Rather than stopping there, Castelnuovo goes on to prove a remarkable converse to his bound on the Hilbert function of a configuration $\Gamma \subset \mathbb{P}^n$ in uniform position: he shows that if $d \geqslant 2n+3$, any such configuration with $h_\Gamma(2) = 2n+1$ must lie on a rational normal curve! In the present application, this implies that a curve $C \subset \mathbb{P}^r$ of maximal genus $\pi_0(d,r)$ must lie on a surface $S \subset \mathbb{P}^r$ of degree $r-1$; that is, either a rational normal scroll or (if $r=5$) the quadratic Veronese surface. Since we know what the class $\mathcal{O}_S(C) \in \operatorname{Pic}(S)$ must be, we can describe the variety $\mathcal{H}^\circ_{g,r,d}$ in this case as a projective bundle over the space parametrizing scrolls, with the appropriate modification when $\epsilon = 0$ or $r = 3$ or 5. Altogether, we arrive at the following statement:

Theorem 3.2. *Let $\mathcal{H}^\circ_{g,r,d}$ be as above the closure in the Hilbert scheme of the locus of smooth, irreducible, nondegenerate curves $C \subset \mathbb{P}^r$ of degree d and genus g.*

(1) *If $g > \pi_0(d,r)$, then $\mathcal{H}^\circ_{g,r,d} = \emptyset$.*

(2) *If $g = \pi_0(d,r)$ and $d \geqslant 2r+1$, then $\mathcal{H}^\circ_{g,r,d}$ has between one and three irreducible components, as follows:*

(a) *For any r and $d \geqslant 2r+1$, $\mathcal{H}^\circ_{g,r,d}$ has an irreducible component of dimension*

$$\binom{m+2}{2}(r-1)-(m+2)(r-3-\epsilon)+r^2+2r-7.$$

(b) *If $\epsilon = 0$ and $r \neq 3$ then $\mathcal{H}^\circ_{g,r,d}$ has an additional irreducible component, of dimension*

$$\binom{m+1}{2}(r-1)+2(m+1)+r^2+2r-7;$$

and

(c) *If $r = 5$ and the degree $d = 2k$ is even, then $\mathcal{H}^\circ_{g,r,d}$ has an additional irreducible component, of dimension*

$$\binom{k+2}{2}+26.$$

In the interests of brevity, this statement does not include everything we know about the geometry of the curves in question (for example, what other linear series/alternative embeddings do such curves possess?), or for that matter about the geometry of $\mathcal{H}^\circ_{g,r,d}$ (is it connected, and if so, where do the components intersect?). Limited as it is, it will serve as a model for the sort of basic information we might wish to have about $\mathcal{H}^\circ_{g,r,d}$.

3.2. Curves of high genus

The next step in this investigation is to ask whether some of the ideas and techniques used to describe curves of maximal genus may be applied more broadly. The answer is clearly "yes," though there's still quite a gap between what we suspect to be true and what we can prove.

3.2.1. The first step Briefly, the starting point is to ask a fairly naive question:

Question 3.3. *If, as Castelnuovo tells us, the smallest possible Hilbert function of a configuration $\Gamma \subset \mathbb{P}^n$ of d points in uniform position is the function*

$$h_0(m) = \min\{mn + 1, d\},$$

then what's the second smallest?

The answer to this seemingly ill-posed question turns out to be surprisingly clear and geometric: just as the smallest possible Hilbert function h_Γ is achieved by a configuration lying on a rational normal curve $D \subset \mathbb{P}^n$, the second smallest is achieved by a configuration lying on an elliptic normal curve $E \subset \mathbb{P}^n$ of degree $n+1$. We'll talk more below about a conjectured extension of this idea and its significance; for now, let's see how it applies to the problem of describing curves of near-maximal genus.

Let's start by saying what is the Hilbert function of a configuration $\Gamma \subset E \subset \mathbb{P}^n$ on an elliptic normal curve. This is not completely determined (as it was in the case of a rational normal curve): in case d is a multiple of $n+1$, the points Γ may or may not comprise a complete intersection with E. The smallest possible Hilbert function, however, is clear: for $m < d/(n+1)$, we have $h_\Gamma(m) = h_E(m)$; for $m > d/(n+1)$ we have $h_\Gamma(m) = d$, and for $m = d/(n+1)$ we have $h_\Gamma(m) = d - 1$ if Γ is the intersection of E with a hypersurface of degree m and d otherwise. In sum, we have

$$h_\Gamma(m) \geqslant h_1(m) = \begin{cases} m(n+1), & \text{if } m < d/(n+1); \\ d-1, & \text{if } m = d/(n+1), \text{ and} \\ d, & \text{if } m > d/(n+1) \end{cases}$$

As before, if we have a curve $C \subset \mathbb{P}^r$ of degree d, and we assume that the Hilbert function h_Γ is strictly greater than the minimal h_0, we can plug this inequality into Castelnuovo's inequality 3.1 above to obtain a bound on the genus of C:

$$g \leqslant \pi_1(d, r),$$

where π_1 is defined as follows: we set

$$m_1 = \left[\frac{d-1}{r}\right] \quad \text{and} \quad \epsilon_1 = d - 1 - mr;$$

we than define

$$\pi_1(d,r) = \binom{m_1}{2} r + m_1(\epsilon_1 + 1) + \begin{cases} 1, & \text{if } \epsilon_1 = r-1 \text{ (that is, } r|d\text{); and} \\ 0, & \text{otherwise.} \end{cases}$$

Note that asymptotically

$$\pi_1(d,r) \sim \frac{d^2}{2r} \quad \text{as } d \to \infty,$$

so π_1 is in fact substantially less than π_0.

Now, suppose we have a curve $C \subset \mathbb{P}^r$, of degree $d \geqslant 2r+3$, and suppose that the genus g of C is strictly greater than $\pi_1(d,r)$. It follows that the Hilbert function of the hyperplane section $\Gamma = C \cap H$ of C must be the minimal Hilbert function h_0 and hence, by Castelnuovo's lemma, that Γ must lie on a rational normal curve in $H \cong \mathbb{P}^{r-1}$. Now, we observe that

$$\pi_1(d,r) \geqslant \pi_0(d,r+1),$$

so that the hypothesis that $g > \pi_1(d,r)$ implies that C is not the linear projection of a nondegenerate curve in $\mathbb{P}^{r+1}$ (that is, C is linearly normal). This is equivalent to saying that $h^1(\mathcal{I}_{C,\mathbb{P}^r}(1)) = 0$, and implies in particular, via the sequence

$$0 \to \mathcal{I}_{C,\mathbb{P}^r}(1) \to \mathcal{I}_{C,\mathbb{P}^r}(2) \to \mathcal{I}_{\Gamma,H}(2) \to 0$$

that *every quadric* $Q \subset H$ *containing the section* $\Gamma = C \cap H$ *of* C *is the restriction to* H *of a quadric* $Q' \subset \mathbb{P}^r$ *containing* C. The intersection of these quadrics in $\mathbb{P}^r$ is then an irreducible, nondegenerate surface $S \subset \mathbb{P}^r$ of minimal degree $r-1$, that is, either a rational normal scroll or the Veronese surface in $\mathbb{P}^5$.

The point is, we know the Picard groups of such surfaces; we know what sort of curves live on them, and the dimensions of the linear series in which they move. Since we also know the families parametrizing scrolls and the Veronese surface, we can give accordingly a complete answer to the existence problem in the range $g > \pi_1(d,r)$. We sum this up in the following Theorem, which in fact subsumes the earlier Theorem 3.2:

Theorem 3.4. *Let* $r \geqslant 3$ *and* $d \geqslant 2r+3$; *let* g *be any integer strictly greater than* $\pi_1(d,r)$.

(1) *If* $r \neq 3, 5$, *then the irreducible components of the variety* $\mathcal{H}^\circ_{g,r,d}$ *are in one-to-one correspondence with the integer solutions of the pair of equations*

$$d = a(r-1) + b$$

and

$$g = \binom{a}{2}(r-1) + (a-1)(b-1),$$

with the general point $[C] \in \mathcal{H}(a,b)$ *corresponding to a curve* $C \subset \mathbb{P}^r$ *lying on a rational normal surface scroll* S *and having class* $C \sim aH + bF \in \mathrm{Pic}(S)$. *The component* $\mathcal{H}(a,b)$ *has dimension*

$$\dim \mathcal{H}(a,b) = \binom{a+1}{2}(r-1) + (a+1)(b+1) + r^2 + 2r - 7.$$

(2) *If* $r = 5$ *and* $d = 2k$ *is even, and* $g = \pi_0(d,r)$, *there is an additional component of* $\mathcal{H}^\circ_{g,r,d}$ *corresponding to smooth plane curves of degree* k *on the Veronese surface* $S \cong \mathbb{P}^2 \subset \mathbb{P}^5$; *and*

(3) *If* $r = 3$, *the components* $\mathcal{H}(a,b)$ *and* $\mathcal{H}(a+b,-b)$ *are identified.*

3.2.2. The conjectured solution in general Having found the two smallest possible Hilbert functions of configurations of points in general position (and having characterized geometrically the configurations achieving them), it seems natural to try to extend this; if we can, the hope might be, we can classify curves in a larger range of genera than $g > \pi_1(d,r)$.

Indeed, we can do exactly that, at least conjecturally. As suggested by the argument above, the key idea is that Castelnuovo's theorem characterizing configurations of points with minimal Hilbert function as those lying on rational normal curves is not an isolated fact. Rather it's a phenomenon observed in many cases and conjectured in many others that if a configuration $\Gamma \subset \mathbb{P}^n$ of points in uniform position fails by a large amount to impose independent conditions on polynomials (that is, it has very small Hilbert function) *it does so because it lies on a curve of low degree.* Castelnuovo's Theorem is the extremal case of this, but more generally we have the

Conjecture 3.5. *Let* $\Gamma \subset \mathbb{P}^n$ *be a nondegenerate configuration of* d *points in uniform position and* α *an integer with* $0 \leqslant \alpha \leqslant n-2$. *If* $d \geqslant 2n+3+2\alpha$ *and*

$$h_\Gamma(2) \leqslant 2n+1+\alpha$$

then Γ *lies on a curve* D *of degree at most* $n+\alpha$ *in* $\mathbb{P}^n$.

Note that under the hypotheses of the conjecture, the curve D will appear as the intersection of the quadrics $Q \subset \mathbb{P}^n$ containing Γ, and indeed it's relatively easy to show the conjecture holds once we know that the intersection of the quadrics containing Γ is positive-dimensional. The case $\alpha = 0$ is Castelnuovo's theorem; the case $\alpha = 1$ (which was implicitly invoked above in the description of the "second smallest" Hilbert function of a configuration in uniform position) was done by Fano and by Eisenbud-Harris in [5], and the case $\alpha = 2$ by Petrakiev in [11][2]; the general formula was given by Chiantini, Ciliberto and Di Gennaro in [1]

[2]Petrakiev actually assumes a stronger form of uniform position: he says a reduced zero-dimensional subscheme $\Gamma \subset \mathbb{P}^n_k$ of projective space over a field k is in *symmetric position* if the Galois group of $\overline{k}$ over k acts as the full symmetric group on the points of $\Gamma(\overline{k})$. This applies in the present circumstance if we view the generic hyperplane section of a curve $C \subset \mathbb{P}^n$ as a scheme over the function field $k = K(\mathbb{P}^{n*})$ of the dual projective space.

The significance of the conjecture, in the present context, is that it tells us that curves of high genus in projective space $\mathbb{P}^r$ must lie on surfaces of low degree; and since we know a good deal about the geometry of such surfaces, this in turn allows us to answer both what degrees and genera may occur, and to say something about the number and dimensions of the corresponding components of the Hilbert scheme.

To indicate how this might go, we first write down the smallest possible Hilbert function of a collection Γ of $d \geqslant 2n+3+2\alpha$ points on a curve D of degree $n+\alpha$ in $\mathbb{P}^n$. We first set

$$m_\alpha = \left[\frac{d-1}{n+\alpha}\right] \quad \text{and} \quad \epsilon_\alpha = d-1-m_\alpha(n+\alpha),$$

and we introduce a new function

$$\mu_\alpha = \max(0, \left[\frac{\alpha - n + 1 + \epsilon_\alpha}{2}\right])$$

$$h_\alpha(\ell) = \begin{cases} \ell(n+\alpha) - \alpha + 1, & \text{if } \ell \leqslant m_\alpha; \\ d - \mu_\alpha, & \text{if } \ell = m_\alpha + 1, \text{ and} \\ d, & \text{if } \ell \geqslant m_a + 2 \end{cases}$$

and we observe that if $\Gamma \subset D \subset \mathbb{P}^n$ lies on an irreducible, nondegenerate curve D of degree $n+\alpha$, then the Hilbert function $h_\Gamma \geqslant h_\alpha$. (Here the function $h_\alpha(\ell)$ is simply the Hilbert function of D in the range $\ell \leqslant m_\alpha$, where for degree reasons a hypersurface of degree ℓ contains Γ if and only if it contains D; the value $h_\alpha(m_\alpha+1)$ reflects Clifford's theorem applied to the linear series cut on D by hypersurfaces of degree $m_\alpha + 1$ containing Γ.)

Given this, suppose now that $C \subset \mathbb{P}^r$ is a curve of degree d and genus g lying on a surface $S \subset \mathbb{P}^r$ of degree $r-1+\alpha$, with $\alpha \leqslant r-2$. Applying the inequality above to the Hilbert function of its hyperplane section $\Gamma \subset \mathbb{P}^{r-1}$, we arrive at the inequality

$$g \leqslant \pi_\alpha(d, r)$$

where

$$\pi_\alpha(d,r) = \binom{m_\alpha}{2}(r-1+\alpha) + m_\alpha(\epsilon_\alpha + \alpha) + \mu_\alpha$$

and m_α, ϵ_α and μ_α are as above, with $n = r-1$.

Beyond this range things get messy: a configuration $\Gamma \subset \mathbb{P}^n$ with $h_\Gamma(2) = 3n$ may not lie on a curve of small degree at all; it may simply lie on a rational normal surface scroll $X \subset \mathbb{P}^n$, which also has $h_X(2) = 3n$. (The rational normal scroll has the smallest Hilbert function of any irreducible, nondegenerate surface in $\mathbb{P}^n$.) Moreover, if it does lie on a small curve, that curve might have degree $2n$ rather than $2n-1$—it could be a canonical curve in $\mathbb{P}^n$, which imposes the same number of conditions on quadrics as a normal curve of degree $2n-3$. Nonetheless, we can push our conjecture one step further. We define m_{r-1} and ϵ_{r-1} as above, but now set $\nu = [\epsilon/2]$ and

$$\pi_{r-1}(d,r) = \binom{m_{r-1}}{2}(2r-2) + m_{r-1}(\epsilon_{r-1}+r) + \nu + \begin{cases} 2, & \text{if } (2r-2)|d; \text{ and} \\ 0, & \text{otherwise} \end{cases}.$$

This is in fact the maximal possible genus of a curve of degree d on a K3 surface $S \subset \mathbb{P}^r$. Note that asymptotically (for fixed r, as $d \to \infty$) we have

$$\pi_{r-1}(d,r) \sim \frac{d^2}{4r-4},$$

so this is roughly half of Castelnuovo's bound $\pi_0(d,r)$.

Finally, once we get into genera $g \leqslant \pi_{r-1}(d,r)$ below the maximum possible on a K3, it seems that *every genus occurs*: though it hasn't, to my knowledge, been verified, K3 surfaces can have large enough Picard groups that we should be able to find, for any d and $g \leqslant \pi_{r-1}(d,r)$, a polarized K3 surface (S, L) of degree $L^2 = 2r-2$ in $\mathbb{P}^r$ with a divisor class C with $C \cdot L = d$ and $C^2 = 2g-2$. (If $g > \pi_{r-1}(d,r)$, this is ruled out by the index theorem on S.)

To sum up, we have the

Conjecture 3.6. *Let π_α be as defined above, for $\alpha = 0, 1, \ldots, r-1$.*

(1) *If $C \subset \mathbb{P}^r$ is a smooth, irreducible, nondegenerate curve of degree $d \geqslant 2n+2\alpha+1$ and genus g, and*

$$g > \pi_\alpha(d,r),$$

then C must lie on a surface $S \subset \mathbb{P}^r$ of degree at most $r-2+\alpha$. In particular, if $g > \pi_{r-1}(d,r)$, then C must lie on a birationally ruled surface.

(2) *For any $g \leqslant \pi_{r-1}(d,r)$, there exist smooth, irreducible, nondegenerate curves of degree $d \geqslant 2n+2\alpha+1$ and genus g.*

Note that the second statement could very well be true in a larger range than the stated $[0, \pi_{r-1}(d,r)]$: as α gets larger, surfaces of degree $r-1+\alpha$ in $\mathbb{P}^r$ can have large Picard number, which is to say they may contain curves of a given degree having many different genera. The point is, for any given r and $\alpha \leqslant r-1$, we can classify all surfaces of degree $r-2+\alpha$ in $\mathbb{P}^r$, and say what are the possible degrees and genera of smooth curves on them. Thus (except the for the small exceptional range $2n+1 \leqslant d < 2n+2\alpha+1$), we have conjecturally a way of deciding the existence problem for any g, r and d.

3.3. Non-smoothable nodal curves

Before moving on the opposite end of the spectrum of possible genera, I want to give one additional application. As was mentioned in Section 2, one consequence of this analysis—the part that's actually been proved—is a construction of components of the Hilbert scheme whose general point corresponds to an irreducible, nondegenerate curve $C \subset \mathbb{P}^r$ having just a node as singularity. As we'll see, we can in fact generate many such examples; for concreteness, however, we'll just do one specific one.

Consider accordingly curves of degree 32 in $\mathbb{P}^{10}$, with (arithmetic) genus exactly equal to $\pi_1(32,10) = 36$. Such a curve, by the above, must lie on a surface of degree 9 or 10 in $\mathbb{P}^{10}$. We can check readily that there are no such curves on a rational normal scroll in $\mathbb{P}^{10}$—in other words, there are no integer solutions to the equations

$$9a + b = 32 \quad \text{and} \quad 9\binom{a}{2} + (a-1)(b-1) = 36.$$

Then such a curve will necessarily lie on a surface $S \subset \mathbb{P}^{10}$ of degree 10 in $\mathbb{P}^{10}$.

Now, an irreducible, nondegenerate surface of degree r in $\mathbb{P}^r$ must a priori be one of three types: a del Pezzo surface; the projection of a rational normal scroll from $\mathbb{P}^{r+1}$; or a cone over an elliptic normal curve. But in this case S can't be a projection of a scroll from $\mathbb{P}^{11}$, since $\pi(32,11) = 33 < 36$. And—this is the kicker—there are no del Pezzo surfaces in $\mathbb{P}^r$ for $r > 9$. So S must be the cone over an elliptic normal curve in $\mathbb{P}^9$. Next, we observe that *there are no smooth curves of degree 32 on such a surface*: if a curve meets a general ray of the cone at m points away from the vertex p of the cone, it must have multiplicity $d - 10m$ at p. In the present circumstances, curves of degree 32 and arithmetic genus 36 do exist on such cones—they're residual to 8 lines of the cone in the intersection of the cone with a quartic hypersurface—and the general such curve has an ordinary node at p and is smooth elsewhere.[3]

I don't know of any similar curves—reduced and irreducible curves $C \subset \mathbb{P}^r$ with just one node as singularity, that are not smoothable—when $r \leqslant 9$.

4. Curves of low and intermediate genus

In the preceding section, we discussed curves of relatively large genus—roughly half or more of Castelnuovo's bound—and found that, conjecturally at least, we could say which degree and genera occurred, and describe the irreducible components of the space parametrizing them. Not nearly as much can be said about curves of lower genus in general, but there are some things known at the opposite end of the spectrum—that is, about curves of low genus. In this section, we'll discuss what we

---

[3]I'm cheating a little bit here: C seemingly might lie on a surface $S \subset \mathbb{P}^{10}$ of degree 10 with a double line. Such a surface is indeed the projection of a rational normal scroll $\tilde{S} \subset \mathbb{P}^{11}$, and so C would be the projection of a curve $\tilde{C} \subset \tilde{S} \subset \mathbb{P}^{11}$; but because the map $\tilde{S} \to S$ is not an isomorphism the map $\tilde{C} \to C$ need not be either; thus it's a priori possible that the genus of $\tilde{C}$ could be less than 36. But we can check that the map $\tilde{S} \to S$ is an isomorphism outside a conic curve $D \subset \tilde{S}$, which maps two-to-one onto the double line of S; and C doesn't meet D enough times to induce three extra nodes. The same logic applies to the possibility that the map $\tilde{S} \to S$ fails to be an isomorphism over an isolated point of S.

can say about these and also indicate where some of the major open problems lie. As before, we assume throughout that $r \geqslant 3$.

4.1. Curves of low genus

4.1.1. The nonspecial range $g \leqslant (d-1)/2$ The simplest case to consider is the range $d \geqslant 2g+1$, or equivalently $g \leqslant (d-1)/2$. In this range every line bundle L of degree d on a curve of genus g is nonspecial—so that $r(L) = d - g$—and very ample. We thus have a tower of maps interpolating between the open subset $\mathcal{H}^\circ_{g,r,d}$ of the Hilbert scheme and the moduli space M_g of curves

$$\mathcal{H}^\circ_{g,r,d} \xrightarrow{\alpha} \mathcal{G}^r_{d,g} \xrightarrow{\beta} \mathcal{P}_{d,g} \xrightarrow{\gamma} M_g,$$

where

(1) $\mathcal{P}_{d,g}$ is the universal Picard variety of degree d over M_g, that is, the moduli space of pairs (C, L) with C a smooth curve of genus g and L a line bundle of degree d on C; and

(2) $\mathcal{G}^r_{d,g}$ is the variety of linear systems—that is, the moduli space of triples (C, L, V) with C a smooth curve of genus g, L a line bundle of degree d on C and $V \subset H^0(L)$ a vector space of dimension $r+1$.

The point is, each of the maps α, β and γ are very regular. To start on the right, the map γ is surjective, with fibers isomorphic to abelian varieties of dimension g; if we restrict ourselves to the open subset $M^\circ_g \subset M_g$ of automorphism-free curves, it's a bundle in the topological, if not the algebro-geometric sense. In particular, given that M_g is irreducible of dimension $3g-3$, we see that $\mathcal{P}_{d,g}$ is irreducible of dimension $4g-3$.

Next, the map β is similarly well-behaved. To begin with, $\mathcal{G}^r_{d,g}$ is empty when $d < g + r$, by Riemann-Roch (we are assuming here that $d \geqslant 2g-1$); otherwise, it's a Grassmannian bundle, with fibers $G(r+1, d-g+1)$. In particular, we see that $\mathcal{G}^r_{d,g}$ is irreducible, of dimension

$$\dim(\mathcal{G}^r_{d,g}) = 4g - 3 + (r+1)(g-d-r).$$

Finally, the map α is dominant—it's image is the open subset of $\mathcal{G}^r_{d,g}$ consisting of very ample linear series—and is a PGL_{r+1}-bundle over that open set. We conclude the

Proposition 4.1. *Suppose that* $d \geqslant 2g+1$. *The open subset* $\mathcal{H}^\circ_{g,r,d}$ *of the Hilbert scheme is empty if* $d < g + r$, *and otherwise irreducible of dimension*

$$\begin{aligned}\dim(\mathcal{H}^\circ_{g,r,d}) = h(g,r,d) &= 4g - 3 + (r+1)(g-d-r) + r^2 + 2r \\ &= 4g - 4 + (r+1)(d-g+1).\end{aligned}$$

In fact, we can also see from this that $\mathcal{H}^\circ_{g,r,d}$ is smooth in this range: while the moduli space M_g may be singular, in an étale or analytic neighborhood of a point $[C] \in M_g$ the map $\mathcal{H}^\circ_{g,r,d} \to M_g$ factors through the deformation space of C, which is smooth.

In the following, we'll refer to $h(g, r, d)$ as the "expected dimension" of the Hilbert scheme. One reason for this is worth pointing out. If $C \subset \mathbb{P}^r$ is a smooth curve of degree d and genus g, the degree of the normal bundle is given by

$$\begin{aligned} c_1(N_{C/\mathbb{P}^r}) &= c_1(T_{\mathbb{P}^r}) - c_1(T_C) \\ &= (r+1)d + 2g - 2. \end{aligned}$$

It follows by Riemann-Roch that the Euler characteristic

$$\begin{aligned} \chi(N_{C/\mathbb{P}^r}) &= c_1(N_{C/\mathbb{P}^r}) - (\operatorname{rank} N_{C/\mathbb{P}^r})(g-1) \\ &= (r+1)d + 2g - 2 - (r+1)(g-1) \\ &= h(g, r, d). \end{aligned}$$

In fact, we can see directly that $h^1(N_{C/\mathbb{P}^r}) = 0$—the normal bundle of C is a quotient of the restricted tangent bundle $T_{\mathbb{P}^r}|_C$, which is in turn a quotient of $\mathcal{O}_C(1)^{\oplus(r+1)}$, which is nonspecial—so that $h^0(N_{C/\mathbb{P}^r}) = h(g, r, d)$ and the Hilbert scheme is smooth of this dimension.

4.1.2. The generically nonspecial range $g \leqslant d - r$. Actually, the spaces and maps above all exist in general, and the analysis carried out in case $d \geqslant 2g + 1$ is really the paradigm of our approach throughout this section. The main difference is that as the degree drops—that is, as the genus increases relative to d—we know less and less about the geometry of the maps α, β and γ.

For example, suppose we drop the assumption that $d \geqslant 2g + 1$ and assume only that $d \geqslant g + r$. It's still true that a general line bundle of degree d on a smooth curve C of genus g is nonspecial and very ample, so that the same picture as above obtains over an open subset $U \subset \mathcal{P}_{d,g}$: the map β expresses the preimage $\beta^{-1}(U)$ as a Grassmannian bundle over U, with fibers isomorphic to $G(r+1, d-g+1)$, and the map α in turn expresses the preimage $(\beta \circ \alpha)^{-1}(U) \subset \mathcal{H}^\circ_{g,r,d}$ as a PGL_{r+1}-bundle over an open subset $V \subset \beta^{-1}(U) \subset \mathcal{G}^r_{d,g}$. It follows as before that $\mathcal{H}^\circ_{g,r,d}$ has a unique irreducible component dominating $\mathcal{P}_{d,g}$, and that this component has the expected dimension $h(g, r, d)$. We'll call this irreducible component the *principal component* of $\mathcal{H}^\circ_{g,r,d}$.

The difference here is that, when $d \leqslant 2g - 2$, there may well exist special line bundles L of degree d, with $h^0(L) > d - g + 1$. Over this (closed) locus in $\mathcal{P}_{d,g}$, the fiber dimension of the map β will jump; if it jumps enough, relative to the codimension of this locus in $\mathcal{P}_{d,g}$, it may mean that there are additional irreducible components of $\mathcal{G}^r_{d,g}$ and correspondingly of $\mathcal{H}^\circ_{g,r,d}$.

Can this actually happen? The answer is yes, and examples are not hard to come by. For example, suppose we look at trigonal curves; that is, curves C with a line bundle L of degree 3 with $h^0(L) = 2$. The first thing to observe is that, for a general trigonal curve, for $m \leqslant (g-4)/2$ the linear series $K_C - mL$ will have degree $2g-2-3m$ and dimension

$$h^0(K_C - mL) = g - 2m,$$

and will be very ample. These statements follow from the fact that the canonical model of a trigonal curve lies on a rational normal surface scroll $X \subset \mathbb{P}^{g-1}$ (with the lines of the ruling cutting out the linear series $|L|$), and that for general trigonal curves this scroll is balanced: that is, isomorphic to either $\mathbb{F}_0$ or $\mathbb{F}_1$ depending on the parity of g. The linear series $|K_C - mL|$ is then cut out by hyperplanes containing m lines of the ruling of X; for $m \leqslant g/2$ these lines are linearly independent (given that X is balanced)and the statement about $h^0(K_C - mL)$ follows. As for the fact that $|K_C - mL|$ is very ample, this follows from the fact that for $m \leqslant (g-4)/2$ and X balanced the projection of X from m lines of its ruling extends to a regular isomorphism of X with a scroll $\tilde{X} \subset \mathbb{P}^{g-1-2m}$; the curve $C \subset X$ goes along for the ride.

Now let's count dimensions. Choose integers g and r, and consider curves $C \subset \mathbb{P}^r$ that are trigonal curves embedded in $\mathbb{P}^r$ by an r-dimensional subseries of the linear system $|K_C - mL|$ for some m; naturally, we have to assume that $3 \leqslant r \leqslant g-1-2m$. To estimate the dimension of the corresponding locus $\mathcal{H}_0$ in $\mathcal{H}^{\circ}_{g,r,d}$ (with $d = 2g-2-3m$), we observe first that the locus of trigonal curves has dimension $2g+1$ in M_g. Now, for a general trigonal curve, the family of r-dimensional linear subseries of $|K_C - mL|$ will have dimension

$$\dim \mathbb{G}(r, g-1-2m) = (r+1)(g-2m-r-1),$$

and by what we said in the last paragraph the general such series will be very ample. For each such series we have a family of curves in $\mathbb{P}^r$ parametrized by PGL_{r+1}, so that the dimension of the locus $\mathcal{H}_0 \subset \mathcal{H}^{\circ}_{g,r,d}$ of such curves will be

$$\begin{aligned}\dim \mathcal{H}_0 &= 2g - 1 + (r+1)(g-2m-r-1) + (r+1)^2 - 1 \\ &= 2g + (r+1)(g-2m).\end{aligned}$$

By contrast, the expected dimension $h(g,r,d)$ of $\mathcal{H}^{\circ}_{g,r,d}$ will be

$$\begin{aligned}h(g,r,d) &= 4g - 4 + (r+1)(d-g+1) \\ &= 4g - 4 + (r+1)(2g-2-3m-g+1) \\ &= 4g - 4 + (r+1)(g-3m-1).\end{aligned}$$

Thus, the actual dimension of $\mathcal{H}_0$ will be greater than or equal to $h(g,r,d)$ whenever

$$2g + 4 \leqslant (r+1)(m+1).$$

Given that we want to be in the range

$$2g - 2 - 3m = d \geqslant g + r,$$

so that m is less than a third of g, we need to take r at least 6 in order to satisfy this inequality; but for any value of $r \geqslant 6$ there are lots of g and m to choose from. The simplest example is to take $r = 6$, $g = 47$ and $d = 53$ (that is, $m = 13$). For any such g, r and m, then, the locus of trigonal curves will lie in a component of $\mathcal{H}^\circ_{g,r,d}$ other than the principal component, and in particular we see that $\mathcal{H}^\circ_{g,r,d}$ is not irreducible.

The bottom line is that $\mathcal{H}^\circ_{g,r,d}$ may be reducible even in the range $d \geqslant g + r$. What is known in this range is that $\mathcal{H}^\circ_{g,r,d}$ is irreducible when $r = 3$ or 4 (an excellent source is the paper [10], which also deals with the next range as well). It's also shown in [5] that $\mathcal{H}^\circ_{g,r,d}$ is irreducible when d satisfies the much stronger inequality

$$d > \frac{2r-1}{r+1} g + 1.$$

4.1.3. The Brill-Noether range $g \leqslant \frac{r+1}{r}(d-r)$. The classical Brill-Noether theorem says that a general curve C of genus g will have a line bundle L of degree d with $h^0(L) \geqslant r + 1$ if and only if

$$\rho(g, r, d) = g - (r + 1)(g - d + r) \geqslant 0;$$

that is, when $g \leqslant \frac{r+1}{r}(d - r)$. Moreover, when $\rho \geqslant 0$ the locus $W^r_d(C) \subset \mathrm{Pic}^d(C)$ of such line bundles will have dimension ρ. In addition, if $\rho > 0$ the variety $W^r_d(C)$ is irreducible (again, for general C); this is due to Fulton and Lazarsfeld. And in case $\rho = 0$—so that $W^r_d(C)$ will consist of finitely many points for general C—a result of Eisenbud-Harris says that as C varies the associated monodromy acts transitively on the points of $W^r_d(C)$. Finally, it's also true that, for C general and $r \geqslant 3$, a general line bundle $L \in W^r_d(C)$ will be very ample.

Taken together, these statements imply that, whenever $\rho(g, r, d) \geqslant 0$ there will be a unique component of $\mathcal{G}^r_{d,g}$ dominating the moduli space, and correspondingly a unique component of $\mathcal{H}^\circ_{g,r,d}$ dominating M_g; and this component will have dimension

$$\begin{aligned} 3g - 3 + \rho(g, r, d) + (r + 1)^2 - 1 &= 3g - 3 + g - (r + 1)(g - d + r) + (r + 1)^2 - 1 \\ &= 4g - 4 + (r + 1)(d - g + 1) \\ &= h(g, r, d) \end{aligned}$$

We can thus sum up one aspect of the behavior of $\mathcal{H}^\circ_{g,r,d}$ in the range $\rho \geqslant 0$ in the

Proposition 4.2. *For any d, g and r such that $\rho(g, r, d) \geqslant 0$, there is a unique component $\mathcal{H}_0$ of $\mathcal{H}^\circ_{g,r,d}$ dominating M_g, and it has the expected dimension $h(g, r, d)$.*

As before, we'll call $\mathcal{H}_0$ the *principal component* of $\mathcal{H}^\circ_{g,r,d}$. The main difference between this range and the previous is simply that the principal component in this range doesn't dominate $\mathcal{P}_{d,g}$; rather, it maps onto the proper subvariety of $\mathcal{P}_{d,g}$ whose fiber over a point [C] is the variety $W^r_d(C)$.

Again, there may be additional irreducible components of $\mathcal{H}^\circ_{g,r,d}$ in this range, though work of Keem and Ein (see for example [2]) says that this does not in fact happen when $r = 3$ or 4: in other words, the locus of smooth, irreducible, nondegenerate curves in $\mathbb{P}^3$ or $\mathbb{P}^4$ with given degree, and genus satisfying $\rho(g,r,d) \geqslant 0$ is irreducible.

4.2. Curves of intermediate genus

4.2.1. Beyond the Brill-Noether range: $\rho < 0$ And here is where our knowledge of the geometry of curves in projective space falls off precipitously, and where many fascinating and mysterious questions arise.

The key fact here is that, as the genus g rises in relation to d and r, the "expected dimension" $h(g,r,d)$ of the Hilbert scheme becomes increasingly irrelevant. Indeed, from the formula for h:

$$h(g,r,d) = 4g - 4 + (r+1)(d-g+1)$$

we see that, for a given $r \geqslant 4$, given that g is bounded only by a function $\pi_0(d,r)$ that is quadratic in d, this may well be negative! (In particular, the very existence of curves of high genus, in the sense of Section 3 above, is in violation of the expected dimension of $\mathcal{H}$.) We have, in other words, no a priori expectation of what $\mathcal{H}^\circ_{g,r,d}$ should look like. Nonetheless, there are a number of observed phenomena that are worth investigating; we'll discuss two of them here.

4.2.2. Families of curves of low codimension in moduli The statement of the Brill-Noether theorem itself says absolutely nothing about linear series of degree d and dimension r with $\rho(g,r,d) < 0$ beyond the fact that a general curve doesn't have any. The proof of the theorem, however, does suggest that something more may be true: that when $\rho < 0$, the locus of curves that do have a g^r_d should have codimension $-\rho$ in M_g. Again, this is visibly false in general—if it were remotely true, curves of near-maximal genus wouldn't exist—but the interesting thing is *it does seem to hold in low codimension in* M_g.

Briefly: it's easy to exhibit components of the Hilbert scheme $\mathcal{H}^\circ_{g,r,d}$ having larger than the expected dimension. In fact, most of the families of curves you can write down are of this type: complete intersections in $\mathbb{P}^r$; curves on a fixed surface $S \subset \mathbb{P}^r$; determinantal curves; and curves obtained from any of these by liaison—in all these cases, except for some low-degree instances, the corresponding components of the Hilbert scheme will have dimension strictly greater than the expected. But all of these components of $\mathcal{H}^\circ_{g,r,d}$ are very special, in the sense that their images in M_g have codimension on the order of g or more.

This gives rise to a question, and a conjectured answer. The question first:

Question 4.3. *For each* g, *what is the smallest integer* $\beta(g)$ *such that there exists a component of* $\mathcal{H}^{\circ}_{g,r,d}$ *having dimension strictly greater than* $h(g,r,d)$, *and whose image in* M_g *has codimension* β*?*

And, based simply on lots of experimentation, an answer:

Conjecture 4.4. *If* $\mathcal{H}_0$ *is any component of* $\mathcal{H}^{\circ}_{g,r,d}$ *having dimension strictly greater than* $h(g,r,d)$, *its image in* M_g *has codimension at least* $g-4$.

To see that this is sharp (if true), we can look at curves with class $3H + bF$ on a rational normal scroll in $\mathbb{P}^r$. These are curves of degree $3(r-1)+b$ and genus

$$g = 3r + 2b - 5;$$

the expected dimension of the corresponding component $\mathcal{H}$ of the Hilbert scheme is thus

$$\begin{aligned} h(g,r,d) &= 4(g-1) + (r+1)(d-g+1) \\ &= 15r - (r-7)b - 21. \end{aligned}$$

On the other hand, the actual dimension is

$$h^0(\mathcal{O}_X(B)) - 1 + \dim PGL_{r+1} - \dim \mathrm{Aut}(X) = 4b + r^2 + 8r - 9,$$

which exceeds the expected for a range of values of b. Finally, the dimension of the image of $\mathcal{H}$ in M_g is

$$\dim \mathcal{H} - \dim PGL_{r+1} = 6r + 4b - 9;$$

it's codimension in M_g is accordingly

$$\begin{aligned} 3g - 3 - \dim \mathcal{H} &= 3(3r + 2b - 5) - 3 - (6r + 4b - 9) \\ &= 3r + 2b - 9 \\ &= g - 4. \end{aligned}$$

(This is not totally surprising: the image is just the trigonal locus.)

There is an analogous question that includes families of linear series that are not necessarily very ample; this is

Question 4.5. *For each* g, *what is the smallest integer* $\gamma(g)$ *such that there exists a component of* $\mathcal{G}^r_{d,g}$ *having dimension strictly greater than* $3g - 3 + \rho(g,r,d)$ *whose image in* M_g *has codimension* γ*?*

Note that the conjectured answer to the first question is visibly false in this case: for example, for large g the locus of curves with a g^2_8 contains the locus of tetragonal curves; since

$$\rho(g,2,8) = g - 3(g - 8 + 2) = 18 - 2g < \rho(g,1,4) = g - 2(g - 4 + 1) = 6 - g,$$

this has the wrong dimension, but codimension $g - 6$ in M_g.

4.2.3. Lower bounds on the dimension of components of $\mathcal{H}^{\circ}_{g,r,d}$ There is another mysterious phenomenon related to the fact that the expected dimension $h(g, r, d)$ of the Hilbert scheme becomes negative when g is large. When the genus gets large, so that $h(g, r, d)$ is small or negative, we have no a priori lower bound on the dimension of $\mathcal{H}^{\circ}_{g,r,d}$, or on the dimension of its image in M_g—there is nothing to prevent these from being arbitrarily small. (Well, given that the projective automorphism group of a curve of positive genus is zero-dimensional, for $g > 0$ the dimension of any component of $\mathcal{H}^{\circ}_{g,r,d}$ must be at least $\dim PGL_{r+1} = r^2 + 2r$, but that's all we know.)

But this doesn't seem to occur in reality. For example, consider curves $C \subset \mathbb{P}^r$ of degree d and near maximal genus $g > \pi_1(d, r)$, in the sense of Section 3 above. This is the range where $h(g, r, d)$ is most negative. But as we saw, such curves lie on rational normal scrolls $X \subset \mathbb{P}^r$; and a quick comparison of the adjunction formula

$$g = \frac{C \cdot C + K_X \cdot C}{2} + 1$$

and the Riemann-Roch, which tells us that the dimension of the linear system $|C|$ on X is given by

$$\dim |C| = \frac{C \cdot C - K_X \cdot C}{2},$$

together with the fact that K_X is negative on a general scroll, tells us that *the dimension of each component of $\mathcal{H}^{\circ}_{g,r,d}$ in this range is at least g.*

What goes on in between is a mystery. We can define a *Hilbert subvariety* of the moduli space M_g of curves to be the closure of the image of a component of $\mathcal{H}^{\circ}_{g,r,d}$ for some d and r; and similarly define a *Brill-Noether subvariety* to be the closure of the image of a component of $\mathcal{G}^r_{d,g}$. (Note that a Hilbert subvariety is always a Brill-Noether subvariety, but not necessarily vice versa.) We then ask:

Question 4.6. *What is the smallest possible dimension $\lambda(g)$ of a Hilbert subvariety, or the smallest possible dimension $\mu(g)$ of a Brill-Noether subvariety, of M_g?*

I don't know if these are extremal, but the smallest components I know of $\mathcal{H}^{\circ}_{g,r,d}$ relative to the dimension $3g - 3$ of M_g—at least, when d and g are allowed to grow large with respect to r—are complete intersections of $r - 1$ hypersurfaces of degree e in $\mathbb{P}^r$. The canonical bundle of such a curve C is $K_C \cong \mathcal{O}_C(-r-1+e(r-1))$, so that the genus is given by

$$\begin{aligned} g &= \frac{e^{r-1}}{2}(e(r-1) - r - 1) + 1 \\ &\sim \frac{r-1}{2} e^r. \end{aligned}$$

The dimension of the corresponding component $\mathcal{H}$ of the Hilbert scheme, on the other hand, is

$$\dim G\left(r-1, \binom{r+e}{r}\right) = (r-1)\left(\binom{r+e}{r} - r + 1\right)$$
$$\sim (r-1)\frac{e^r}{r!},$$

where the $\sim$ in both cases represents the asymptotic behavior as r is fixed and $e \to \infty$. Asymptotically, then, the dimension of the Hilbert scheme is on the order of

$$\dim \mathcal{H} \sim \frac{2}{r!} g.$$

Again, I don't know if it's possible to do better. (Bear in mind that this is purely an asymptotic statement; it doesn't give any indication of what $\lambda(g)$ may be for a given g.)

To give you an idea of the woeful state of our ignorance, we don't even know whether an isolated point of M_g can be a Hilbert subvariety! To formulate it slightly differently, call a smooth curve $C \subset \mathbb{P}^r$ a *rigid curve* if it has no deformations in $\mathbb{P}^r$ other than those given by translates of C by PGL_{r+1}. We ask

Question 4.7. *Do there exist rigid curves other than rational normal curves?*

The number of adults who believe in the existence of rigid curves other than rational normal curves is comparable to the number who believe in Santa Claus, which makes it highly embarrassing that we can't prove there aren't any.

(We can also make a more restrictive definition: we say $C \subset \mathbb{P}^r$ is *infinitesimally rigid* if it has no first-order deformations in $\mathbb{P}^r$ other than those induced by vector fields on $\mathbb{P}^r$. We *still* can't prove these don't exist.)

Finally, there is a related question concerning Weierstrass points, posed by Jonathan Wahl. We can stratify the universal curve $M_{g,1}$ over M_g by Weierstrass semigroup: for every sub-semigroup $H \subset \mathbb{N}$ of index g, we let $W_H \subset M_{g,1}$ be the locus of pairs (C, p) where the Weierstrass semigroup of $p \in C$ is H. The *weight* of a semigroup H is an upper bound on the codimension of the corresponding locus $W_H \subset M_{g,1}$; but since the weight of a Weierstrass point on a curve of genus g can be as large as $\binom{g}{2}$, this doesn't tell us much in general. The question is: do there exist rigid Weierstrass points—that is, zero-dimensional loci W_H? And if not, what is the smallest dimension of a stratum $W_H \subset M_{g,1}$?

References

[1] L. Chiantini, C. Ciliberto, and V. Di Gennaro. The genus of projective curves. *Duke Math. J.*, **70**(2), 229–245, 1993. Available at `http://dx.doi.org/10.1215/S0012-7094-93-07003-2`. ← 15

[2] L. Ein. The irreducibility of the Hilbert scheme of smooth space curves, in: Algebraic geometry, Bowdoin, 1985 (Brunswick, Maine, 1985), 83–87. *Proc. Sympos. Pure Math.*, **46**, Amer. Math. Soc., Providence, RI, 1987. ← 23

[3] W. Fulton and R. Pandharipande. Notes on stable maps and quantum cohomology, in: Algebraic geometry—Santa Cruz 1995, 45–96. *Proc. Sympos. Pure Math.*, **62**, Amer. Math. Soc., Providence, RI, 1997. ← 5

[4] L. Gruson and C. Peskine. Genre des courbes de l'espace projectif. II. *Ann. Sci. École Norm. Sup. (4)*, **15**(3), 401–418, 1982. Available at `http://www.numdam.org/item?id=ASENS_1982_4_15_3_401_0`. ← 9

[5] J. Harris and D. Eisenbud. *Curves in Projective Space*, Les Presses de l'Université de Montréal, Montreal, 1982. ← 15, 22

[6] R. Hartshorne. Algebraic geometry. *Graduate Texts in Mathematics*, **52**, Springer, New York, 1977. ← 1

[7] R. Hartshorne. On the classification of algebraic space curves, in : Vector bundles and differential equations (Proc. Conf., Nice, 1979), 83–112. *Progr. Math.*, **7**, Birkhäuser Boston, Mass., 1980. ← 9

[8] R. Hartshorne. Une courbe irréductible non lissifiable dans P^3. *C. R. Acad. Sci. Paris Sér. I Math.*, **299**(5), 133–136, 1984. ← 9

[9] R. Hartshorne. On the classification of algebraic space curves. II, in: *Algebraic geometry, Bowdoin, 1985 (Brunswick, Maine, 1985)*, 145–164. Proc. Sympos. Pure Math., **46**, Amer. Math. Soc., Providence, RI, 1987. ← 9

[10] C. Keem. Reducible Hilbert scheme of smooth curves with positive Brill-Noether number. *Proc. Amer. Math. Soc.*, **122**(2), 349–354, 1994. Available at `http://dx.doi.org/10.2307/2161023`. ← 22

[11] I. Petrakiev. Castelnuovo theory via Gröbner bases. *J. Reine Angew. Math.*, **619**, 49–73, 2008. Available at `http://dx.doi.org/10.1515/CRELLE.2008.040`. ← 15

[12] R. Piene and M. Schlessinger. On the Hilbert scheme compactification of the space of twisted cubics. *Amer. J. Math.*, **107**(4), 761–774, 1985. Available at `http://dx.doi.org/10.2307/2374355`. ← 8

[13] J. Rathmann. The genus of curves in P^4 and P^5. *Math. Z.*, **202**(4), 525–543, 1989. Available at `http://dx.doi.org/10.1007/BF01221588`. ← 9

[14] R. Vakil. Murphy's law in algebraic geometry: badly-behaved deformation spaces. *Invent. Math.*, **164**(3), 569–590, 2006. Available at `http://dx.doi.org/10.1007/s00222-005-0481-9`. ← 5

Harvard University

E-mail address: `harris@math.harvard.edu`

Global topology of the Hitchin system

Tamás Hausel

Abstract. Here we survey several results and conjectures on the cohomology of the total space of the Hitchin system: the moduli space of semi-stable rank n and degree d Higgs bundles on a complex algebraic curve C. The picture emerging is a dynamic mixture of ideas originating in theoretical physics such as gauge theory and mirror symmetry, Weil conjectures in arithmetic algebraic geometry, representation theory of finite groups of Lie type and Langlands duality in number theory.

Contents

2000 *Mathematics Subject Classification.* Primary 14D20; Secondary 14J32 14C30 20C33 .

Key words and phrases. moduli of vector bundles, Hitchin system, non-Abelian Hodge theory, mirror symmetry, Hodge theory, finite groups of Lie type.

1. Introduction

Studying the topology of moduli spaces in algebraic geometry could be considered the first approximation of understanding the moduli problem. We start with an example which is one of the original examples of moduli spaces constructed by Mumford [71] in 1962 using Geometric Invariant Theory. Let $\mathcal{N}$ denote the moduli space of semi-stable rank n degree d stable bundles on a smooth complex algebraic curve C; which turns out to be a projective variety. In 1965 Narasimhan and Seshadri [75] proved that the space $\mathcal{N}$ is canonically diffeomorphic with the manifold

$$\mathcal{N}_B := \{A_1, B_1, \ldots, A_g, B_g \in U(n) \,|[A_1, B_1] \ldots [A_g, B_g] = \zeta_n^d I\}/U(n),$$

of twisted representations of the fundamental group of C to $U(n)$, where $\zeta_n = \exp(2\pi i/n)$. Newstead [76] in 1966 used this latter description to determine the Betti numbers of $\mathcal{N}_B$ when $n = 2$ and $d = 1$. Using ideas from algebraic number theory for function fields Harder [41] in 1970 counted the rational points $\mathcal{N}(\mathbb{F}_q)$ over a finite field when $n = 2$ and $d = 1$ and compared his formulae to Newstead's results to find that the analogue of the Riemann hypothesis, the last remaining Weil conjectures, holds in this case. By the time Harder and Narasimhan [42] in 1975 managed to count $\#\mathcal{N}(\mathbb{F}_q)$ for any n and coprime d the last of the Weil conjectures had been proved by Deligne [23] in 1974, and their result in turn yielded [24] recursive formulae for the Betti numbers of $\mathcal{N}(\mathbb{C})$.

The same recursive formulae were found using a completely different method by Atiyah and Bott [4] in 1981. They studied the topology of $\mathcal{N}$ in another reincarnation with origins in theoretical physics. The Yang-Mills equations on $\mathbb{R}^4$ appeared in theoretical physics as certain non-abelian generalization to Maxwell's equation; and were used to describe aspects of the standard model of the physics of elementary particles. Atiyah and Bott considered the analogue of these equations in 2 dimensions, more precisely the solution space $\mathcal{N}_{YM}$ of Yang-Mills connections on a differentiable Hermitian vector bundle of rank n and degree d on the Riemann surface C modulo gauge transformations. Once again we have canonical diffeomorphisms $\mathcal{N}_{YM} \cong \mathcal{N}_B \cong \mathcal{N}$. Atiyah–Bott [4] used this gauge theoretical approach to study the topology of $\mathcal{N}_{YM}$ obtaining in particular recursive formulae for the Betti numbers which turned out to be essentially the same as those arising from Harder-Narasimhan's arithmetic approach. Atiyah-Bott's approach has been greatly generalized in Kirwan's work [61] to study the cohomology of Kähler quotients in differential geometry and the closely related GIT quotients in algebraic geometry. By now we have a fairly comprehensive understanding of the cohomology (besides Betti numbers: the ring structure, torsion, K-theory etc) of $\mathcal{N}$ or more generally of compact Kähler and GIT quotients.

In this survey we will be interested in a hyperkähler analogue of the above questions. It started with Hitchin's seminal paper [51] in 1987. Hitchin [51] investigated the two-dimensional reduction of the four dimensional Yang-Mills equations over a Riemann surface, what we now call Hitchin's self-duality equations on a rank n and degree d bundle on the Riemann surface C. Using gauge theoretical methods one can construct $\mathcal{M}_{\mathrm{Hit}}$ the space of solutions to Hitchin's self-duality equations modulo gauge transformations. $\mathcal{M}_{\mathrm{Hit}}$ can be considered as the hyperkähler analogue of $\mathcal{N}_{\mathrm{YM}}$. When n and d are coprime $\mathcal{M}_{\mathrm{Hit}}$ is a smooth, albeit non-compact, differentiable manifold with a natural hyperkähler metric. The latter means a metric with three Kähler forms corresponding to complex structures I, J and K, satisfying quaternionic relations.

Hitchin [51] found that the corresponding algebraic geometric moduli space is the moduli space $\mathcal{M}_{\mathrm{Dol}}$ of semi-stable Higgs bundles (E, ϕ) of rank n and degree d and again we have the natural diffeomorphism $\mathcal{M}_{\mathrm{Hit}} \cong \mathcal{M}_{\mathrm{Dol}}$ which in fact is complex analytical in the complex structure I of the hyperkähler metric on $\mathcal{M}_{\mathrm{Hit}}$. $\mathcal{M}_{\mathrm{Dol}}$ is the hyperkähler analogue of $\mathcal{N}$. Using a natural Morse function on $\mathcal{M}_{\mathrm{Hit}}$ Hitchin was able to determine the Betti numbers of $\mathcal{M}_{\mathrm{Hit}}$ when $n = 2$ and $d = 1$, and this work was extended by Gothen [37] in 1994 to the $n = 3$ case. Hitherto the Betti numbers for $n > 3$ have not been found, but see §6.

In this paper we will survey several approaches to get cohomological information on $\mathcal{M}_{\mathrm{Hit}}$. The main difficulty lies in the fact that $\mathcal{M}_{\mathrm{Hit}}$ is non-compact. However as Hitchin [51] says

"... the moduli space of all solutions turns out to be a manifold with an extremely rich geometric structure".

Due to this surprisingly rich geometrical structures on $\mathcal{M}_{\mathrm{Hit}}$ we will have several different approaches to study $H^*(\mathcal{M}_{\mathrm{Hit}})$, some motivated by ideas in theoretical physics, some by arithmetic and some by Langlands duality.

As mentioned $\mathcal{M}_{\mathrm{Hit}}$ has a natural hyperkähler metric and in complex structure J it turns out to be complex analytically isomorphic to the character variety

$$\mathcal{M}_B := \{ A_1, B_1, \ldots, A_g, B_g \in GL_n(\mathbb{C}) \mid [A_1, B_1] \ldots [A_g, B_g] = \zeta_n^d I \} // GL_n(\mathbb{C})$$

of twisted representations of the fundamental group of C to GL_n. The new phenomenon here is that this is a complex variety. Thus cohomological information could be gained by counting points of it over finite fields. Using the character table of the finite group $GL_n(\mathbb{F}_q)$ of Lie type this was accomplished in [47]. The calculation leads to some interesting conjectures on the Betti numbers of $\mathcal{M}_B$. This approach was surveyed in [44]. We will also mention this approach in §4 but here will focus on its connections to one more aspect of the geometry of $\mathcal{M}_{Hit}$. Namely, that it carries the structure of an integrable system.

This integrable system, called the Hitchin system, was defined in [52] by taking the characteristic polynomial of the Higgs field, and has several remarkable properties. First of all it is a completely integrable system with respect to the holomorphic symplectic structure on $\mathcal{M}_{Dol}$ arising from the hyperkähler metric. Second it is proper, in particular the generic fibers are Abelian varieties. In this paper we will concentrate on the topological aspects of the Hitchin system on $\mathcal{M}_{Dol}$.

The recent renewed interest in the Hitchin system could be traced back to two major advances. One is Kapustin–Witten's [58] proposal in 2006, that S-duality gives a physical framework for the geometric Langlands correspondence. In fact, [58] argues that this S-duality reduces to mirror symmetry for Hitchin systems for Langlands dual groups. This point of view has been expanded and explained in several papers such as [40], [33], [92].

One of our main motivations to study $H^*(\mathcal{M}_{Dol})$ is to try to prove the topological mirror symmetry proposal of [49], which is a certain agreement of Hodge numbers of the mirror Hitchin systems; which we now understand as a cohomological shadow of Kapustin–Witten's S-duality (see §5.5.3). This topological mirror symmetry proposal is mathematically well-defined and could be tested by using the existing techniques of studying the cohomology of $H^*(\mathcal{M}_{Dol})$. Indeed it is a theorem when $n = 2,3$ using [51] and [37]. Below we will introduce the necessary formalism to discuss these conjectures and results.

Most recently Ngô proved [78] the fundamental lemma in the Langlands program in 2008 by a detailed study of the topology of the Hitchin map over a large open subset of its image. Surprisingly, Ngô's geometric approach will be intimately related to our considerations, even though Ngô's main application is the study of the cohomology of a *singular fiber* of the Hitchin map. We will explain at the end of this survey some connections of Ngô's work to our studies of the global topology of the Hitchin map.

We should also mention that both Kapustin–Witten and Ngô study Hitchin systems for general reductive groups, for simplicity we will concentrate on the groups GL_n, SL_n and PGL_n in this paper.

Our studies will lead us into a circle of ideas relating arithmetic and the Langlands program to the physical ideas from gauge theory, S-duality and mirror symmetry in the study of the global topology of the Hitchin system. This could be considered the hyperkähler analogue of the fascinating parallels[1] between the arithmetic approach of Harder–Narasimhan and the gauge theoretical approach of Atiyah–Bott in the study of $H^*(\mathcal{N})$.

---

[1]For a recent survey on these parallels see [1].

Acknowledgement The basis of this survey are the notes taken by Gergely Bérczi, Michael Gröchenig and Geordie Williamson from a preparatory lecture course "Mirror symmetry, Langlands duality and the Hitchin system" the author gave in Oxford in Hilary term 2010. I would like to thank them and the organizers of several lecture courses where various subsets of these notes were delivered in 2010. These include the Spring school on "Geometric Langlands and gauge theory" at CRM Barcelona, a lecture series at the Geometry and Quantum Field Theory cluster at the University of Amsterdam, the "Summer School on the Hitchin fibration" at the University of Bonn and the Simons Lecture Series at Stony Brook. I would also like to thank Mark de Cataldo, Iain Gordon, Jochen Heinloth, Daniel Huybrechts, Luca Migliorini, David Nadler, Tony Pantev and especially an anonymous referee for helpful comments. The author was supported by a Royal Society University Research Fellowship.

1.1. Mirror symmetry

In order to motivate the topological mirror symmetry conjectures of [49], we give some background information on three of the main topics involved starting with mirror symmetry.

Considerations of mirror symmetry appeared in various forms in string theory at the end of the 1980's. It entered into the realm of mathematics via the work [12] of Candelas-de la Ossa-Green-Parkes in 1991 by formulating mathematically precise (and surprising) conjectures on the number of rational curves in certain Calabi-Yau 3-folds.

Mathematically, mirror symmetry relates the symplectic geometry of a Calabi-Yau manifold X to the complex geometry of its mirror Calabi-Yau Y of the same dimension. Such an unexpected duality between two previously separately studied fields in geometry has caught the interest of several mathematicians working in related fields. The literature of mirror symmetry is vast, here we only mention monographs and conference proceedings on the subject such as [20, 70, 55].

First aspect of checking a mirror symmetry proposal is the *topological mirror test*

$$h^{p,q}(X) = h^{\dim Y - p,q}(Y) \tag{1.1}$$

that the Hodge diamonds have to be mirror to each other. If we introduce the notation

$$E(X; x, y) = \sum (-1)^{i+j} h^{i,j}(X) x^i y^j$$

the symmetry translates to

$$E(X; x, y) = x^{\dim Y} E(Y; 1/x, y). \tag{1.2}$$

Here we will be interested in the case when the Calabi-Yau manifolds are hyperkähler. We note that for a *compact* hyperkähler manifold X wedging with powers of the holomorphic (2,0) symplectic form induces $h^{p,q}(X) = h^{\dim X - p,q}(X)$ and so the above topological mirror test simplifies to the agreement

$$h^{p,q}(X) = h^{p,q}(Y) \tag{1.3}$$

of Hodge numbers when both X and Y are hyperkähler.

Kontsevich [62] in 1994 suggested that mirror symmetry is underlined by a more fundamental *homological mirror symmetry*, which identifies two derived categories

$$\mathcal{D}^b(\mathrm{Fuk}(X,\omega)) \sim \mathcal{D}^b(\mathrm{Coh}(Y,I))$$

the Fukaya category $\mathcal{D}^b(\mathrm{Fuk}(X,\omega))$ of certain decorated Lagrangian subvarieties of the symplectic manifold (X,ω) and the derived category of coherent sheaves $\mathcal{D}^b(\mathrm{Coh}(Y,I))$ on the mirror Y, considered as a complex variety. This suggestion has been checked in several examples [81, 83] and has been the starting point of a large body of mathematical research.

A more geometrical proposal is contained in the work of Strominger-Yau-Zaslow [87] in 1996. They suggested a geometrical construction how to obtain the mirror Y from any given Calabi-Yau X. It suggested that Y can be constructed as the moduli space of certain special Lagrangian submanifolds of X together with a flat line bundle on it; the picture arising then can be described as

$$\begin{array}{ccc} X & & Y \\ & {\scriptstyle \chi_X}\searrow \quad \swarrow{\scriptstyle \chi_Y} & \\ & B & \end{array}$$

where χ_X and χ_Y are special Lagrangian fibrations on X and Y respectively, with generic fibers dual middle-dimensional tori. Until this [87] proposal there was no general geometrical conjectures how one might construct Y from X. Even though there have been a fair amount of work today we do not have an example where the geometrical proposal of [87] is completely implemented for the original case of Calabi-Yau 3-folds. But see [39] for a survey of recent developments.

We conclude by noting that many mathematical predictions of mirror symmetry have been confirmed but we still have no general understanding yet.

1.2. Langlands duality

Here we give some technically simplified remarks as a sketch of a basic introduction to a few ideas in the Langlands program in number theory; more details and references can be found in the surveys [32, 35].

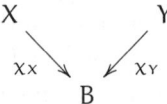

As a first approximation, the Langlands correspondence aims to describe a central object in number theory: the absolute Galois group $\mathrm{Gal}(\overline{\mathbb{Q}}/\mathbb{Q})$ via representation theory. More precisely let G be a reductive group (over the complex numbers these are just the complexifications of compact Lie groups) and ${}^L G$ its Langlands dual, which one can obtain by dualizing the root datum of G. For example for the groups of our concern in the present paper: ${}^L\mathrm{GL}_n = \mathrm{GL}_n$; ${}^L\mathrm{SL}_n = \mathrm{PGL}_n$, ${}^L\mathrm{PGL}_n = \mathrm{SL}_n$. Langlands in 1967 conjectured that one can find a correspondence between the set isomorphism classes of certain continuous homomorphisms $\mathrm{Gal}(\overline{\mathbb{Q}}/\mathbb{Q}) \to G(\overline{\mathbb{Q}}_l)$ (for $G = GL_n$ these are just n-dimensional representations) and automorphic (certain infinite dimensional) representations of ${}^L G(\mathbb{A}_{\mathbb{Q}})$ over the ring of adèles $\mathbb{A}_{\mathbb{Q}}$. The motivation for understanding the representations of $\mathrm{Gal}(\overline{\mathbb{Q}}/\mathbb{Q})$ is that it describes the absolute Galois group itself via the Tannakian formalism.

Langlands built his programme as a non-abelian generalization of the adèlic description of class field theory which can be understood as the $G = GL_1$ case of the above. Indeed representations of $\mathrm{Gal}(\overline{\mathbb{Q}}/\mathbb{Q})$ to GL_1 describe the abelianization $\mathrm{Gal}_{ab}(\overline{\mathbb{Q}}/\mathbb{Q})$ which describes all finite abelian extensions of $\mathbb{Q}$. In the case of $G = GL_2$ elliptic curves enter naturally, via the action of the absolute Galois group on their two-dimensional first étale cohomology. The corresponding objects on the automorphic side are modular forms, and the Langlands program in this case can be seen to reduce to the Shimura-Taniyama-Weil conjecture which is now a theorem due to the work of Wiles and others.

The adèlic description of class field theory was partly motivated by the deep analogy between $\mathbb{Q}$, or more generally number fields (i.e. finite extensions of the rationals), and the function field $\mathbb{F}_q(C)$, where $C/\mathbb{F}_q$ is an algebraic curve. This analogy proved powerful in attacking problems in algebraic number theory and in the Langlands program specifically. Many of the conjectured properties of number fields can be formulated for function fields, where one can use the techniques of algebraic geometry and succeed even in situations where the number field case remains open. Such an example is the Riemann hypothesis, which is still open over $\mathbb{Q}$, but was proved over function fields $\mathbb{F}_q(C)$ of curves by Weil [91] in 1941 and for any algebraic variety by Deligne [23] in 1974.

For an example of the more harmonious interaction between the number field and function field side of algebraic number theory we mention that Ngô's recent proof [78] of the fundamental lemma for $\mathbb{F}_q(C)$ (where he used the topology of the Hitchin system, for more details see §5.5.4) yielded the fundamental lemma for the number field case, due to previous work by Waldspurger [90].

If we replace $\mathbb{F}_q$ by $\mathbb{C}$, i.e. consider the complex function field $\mathbb{C}(C)$ for complex curves $C/\mathbb{C}$, i.e. Riemann surfaces, then we lose some of the analogies, but still one can reformulate a version of the Langlands correspondence mentioned above. This is the *Geometric Langlands Correspondence* as was proposed by [64, 8]. Originally this was viewed as a correspondence between isomorphism classes of G-local systems on C (analogues of the representations of the absolute Galois group) and certain Hecke eigensheaves (the analogues of automorphic representations) on the stack $\mathrm{Bun}_{{}^L G}$ of ${}^L G$-bundles on X.

In these lectures, the point of view that we will adopt will be that as a cohomological shadow of these considerations of the Geometric Langlands programme one can extract agreements of certain Hodge numbers of Hitchin systems for Langlands dual groups.

1.3. Hitchin system

Here we collect some of the basic ideas of completely integrable Hamiltonian systems and comment on the history of a large class of them: the Hitchin systems. An extensive account for the former can be found in [3] while [26] details the latter.

Recall that a Hamiltonian system is given by a symplectic manifold (X^{2d}, ω) and a Hamiltonian - or energy- function $H : X \to \mathbb{R}$. The corresponding X_H Hamiltonian vector field is defined by the property

$$dH = \omega(X_H, .).$$

The dynamics of the Hamiltonian system is given by the flow — the one parameter group of diffeomorphisms — generated by X_H.

A function $f : X \to \mathbb{R}$ is called a *first integral* if

$$X_H f = \omega(X_f, X_H) = 0$$

holds. The condition is equivalent to say that f is constant along the flow of the system. Note that $\omega(X_H, X_H) = 0$ as ω is alternating and so H is constant along the flow - which is sometimes referred to as the law of conservation of energy.

We say that the Hamiltonian system is *completely integrable* if there is a map

$$f = (H = f_1, \ldots, f_d) : X \to \mathbb{R}^d,$$

that is generic (meaning that f is generically a submersion) such that $\omega(X_{f_i}, X_{f_j}) = 0$. The generic fibre of f then has an action of $\mathbb{R}^d = \langle X_{f_1}, \ldots, X_{f_d} \rangle$ and so when f is proper the generic fibre can be identified with a torus $(S^1)^d$. In such cases one has a fairly good control of the dynamics of the system (on the generic fiber it is just an affine motion) hence the name integrable. Several examples arise in classical mechanics such as the Euler and Kovalevskaya tops and the spherical pendulum. For more details see [3].

Here we will be concerned with the complexified or algebraic version of integrable Hamiltonian systems. Thus we consider a complex 2d dimensional manifold X with a holomorphic symplectic 2-form. We now say that $f : X \to \mathbb{C}^d$ is an algebraically completely integrable system if f is generically a submersion and if $\omega(X_{f_i}, X_{f_j}) = 0$.

If f turns out to be proper, the generic fiber then will become a torus with a complex structure — in the algebraic case — an Abelian variety. This is the case for a large class of examples: the *Hitchin systems*. As we will see in more detail below the Hitchin system is attached to the cotangent bundle of the moduli space of stable G-bundles on a complex curve X. Originally it appeared in Hitchin's study [51] of the 2-dimensional reduction of the Yang-Mills equations from 4-dimensions. Here we will follow a more algebraic approach.

2. Higgs bundles and the Hitchin system

2.1. The moduli space of vector bundles on a curve

Let C be a complex projective curve of genus $g > 1$. We fix integers $n > 0$ and $d \in \mathbb{Z}$. We assume throughout that $(d, n) = 1$. Using Geometric Invariant Theory [71, 77] one can construct $\mathcal{N}^d$; the moduli space of isomorphism classes of stable rank n degree d vector bundles (equivalently GL_n-bundles) on C.

We recall that a vector bundle is called *semistable* if every subbundle F satisfies

$$\mu(F) = \frac{\deg F}{\operatorname{rk} F} \leqslant \mu(E) = \frac{\deg E}{\operatorname{rk} E}$$

A vector bundle is *stable* if one has strict inequality above for all proper subbundles.

In general one has to be careful in constructing such moduli spaces as special care has to be taken for the non-trivial automorphisms of strictly semi-stable objects. However, as we assume $(d, n) = 1$ the notions of semi-stability and stability clearly agree. In particular, we can conclude that $\mathcal{N}^d$ is smooth and projective of dimension $d_n = n^2(g-1)+1$.

Consider the determinant morphism

$$\det : \mathcal{N}^d \to \operatorname{Jac}^d(C)$$

which sends a vector bundle of rank n to its highest exterior power $\Lambda^n E$. Choose $\Lambda \in \operatorname{Jac}^d(C)$ and define $\check{\mathcal{N}}^\Lambda := \det^{-1}(\Lambda)$. When $\Lambda = \mathcal{O}_C$ is the trivial bundle, the vector bundles in $\check{\mathcal{N}}^\Lambda$ are exactly the SL_n-bundles, for general Λ we can think of points in $\check{\mathcal{N}}^\Lambda$ as "twisted SL_n-bundles". Tensoring with an nth root of $\Lambda_1\Lambda_2^{-1}$ gives an isomorphism $\check{\mathcal{N}}^{\Lambda_1} \cong \check{\mathcal{N}}^{\Lambda_2}$ thus the isomorphism class of $\check{\mathcal{N}}^\Lambda$ does not depend on the choice of $\Lambda \in \operatorname{Jac}^d(C)$. We often abuse notation and write $\check{\mathcal{N}}^d$ instead of $\check{\mathcal{N}}^\Lambda$.

The abelian variety $\operatorname{Pic}^0(C) = \operatorname{Jac}^0(C)$ acts on $\mathcal{N}^d$ via

$$(L, E) \mapsto L \otimes E.$$

As Seshadri showed in [84] the quotient of a normal variety by an Abelian variety always exist, so we can define the moduli space of degree d PGL_n bundles:

$$\hat{\mathcal{N}}^d := \mathcal{N}^d / \operatorname{Pic}^0(C).$$

The embedding $\check{\mathcal{N}}^\Lambda \subset \mathcal{N}^d$ induces

$$\hat{\mathcal{N}}^d \cong \check{\mathcal{N}}^\Lambda / \Gamma.$$

Here $\Gamma := \operatorname{Pic}^0(C)[n]$ is the group of n-torsion points of the Jacobian, isomorphic to the finite group $\mathbb{Z}_n^{2g}$. It acts on $\check{\mathcal{N}}^\Lambda$ by tensorization. Hence $\hat{\mathcal{N}}^d$ is a projective orbifold.

2.1.1. Cohomology of moduli spaces of bundles The cohomology[2] of $\mathcal{N}^d$, $\check{\mathcal{N}}^d$ and $\hat{\mathcal{N}}^d$ is well understood. The structure of the cohomology rings can be described by finding universal generators and all the relations between them [4, 88, 30].

Here we only comment on the additive structure in more detail. In 1975 Harder and Narasimhan [42] obtained recursive formulae for the number of points of these varieties over finite fields. It is then possible to use the Weil conjectures (which had been proven the year before by Deligne [23]) to obtain formulae for the Betti numbers. In 1981 Atiyah and Bott gave a different gauge-theoretic proof [4].

The main application in Harder and Narasimhan's paper is the following:

Theorem 2.1 ([42]). *The finite group* Γ *acts trivially on* $H^*(\check{\mathcal{N}}^d)$. *In particular, we have* $H^*(\check{\mathcal{N}}^d) = H^*(\hat{\mathcal{N}}^d)$.

Remark 2.2. This result is difficult to prove and relies on showing that the varieties $\check{\mathcal{N}}^d$ and $\hat{\mathcal{N}}^d$ have the same number of points over finite fields. The analogue of this result is false in the context of the moduli space of Higgs bundles as was already observed by Hitchin in [51] for $n = 2$. Interestingly for us, this will lead to the non-triviality of our topological mirror tests.

2.2. The Hitchin system

We now consider the Hitchin system, which will be an integrable system on the cotangent bundle to the moduli spaces considered in the previous section. As in the previous section, fix n and d and abbreviate $\mathcal{N} := \mathcal{N}^d$.

The cotangent bundle $T^*\mathcal{N}$ is an algebraic variety. The ring of regular functions $\mathbb{C}[T^*\mathcal{N}]$ turns out to be finitely-generated as will be proven below. The *Hitchin map* then is simply the affinization:

$$\chi : T^*\mathcal{N} \to \mathcal{A} = \mathrm{Spec}(\mathbb{C}[T^*\mathcal{N}]). \tag{2.3}$$

We now describe this map more explicitly. For a point $[E] \in \mathcal{N}$ standard deformation theory gives us an identification

$$T_{[E]}\mathcal{N} = H^1(C, \mathrm{End}(E)).$$

Applying Serre duality we obtain

$$T^*_{[E]}\mathcal{N} = H^0(C, \mathrm{End}(E) \otimes K),$$

where K denotes the canonical bundle of C. An element

$$\phi \in H^0(C, \mathrm{End}(E) \otimes K)$$

is called a *Higgs field*. Morally, it can be thought of as a matrix of one-forms on the curve. As such if we consider the characteristic polynomial of $(E, \phi) \in T^*\mathcal{N}$ then it will have the form

$$t^n + a_1 t^{n-1} + \cdots + a_n$$

[2]In this paper cohomology is with rational coefficients; unless indicated otherwise.

where $a_i \in H^0(K^i)$. For example $a_n \in H^0(K^n)$ is the determinant of the Higgs field.

As we will prove below the Hitchin map (2.3) then has the explicit description

$$\begin{aligned} \chi : T^*\mathcal{N} &\to \mathcal{A} := \bigoplus_{i=1}^n H^0(K^i) \\ (E,\phi) &\mapsto (a_1, a_2, \ldots, a_n) \end{aligned}$$

The affine space $\mathcal{A}$ is called the *Hitchin base.*

In the SL_n-case we have

$$T^*_{[E]}\check{\mathcal{N}}^d = H^0(\mathrm{End}_0(E) \otimes K)$$

that is, a covector at E is given by a *trace free* Higgs field. Thus in this case the Hitchin base is

$$\mathcal{A}^0 := \bigoplus_{i=2}^{n} H^0(C, K^i).$$

and the Hitchin map

$$\check{\chi} : T^*\check{\mathcal{N}}^d \to \mathcal{A}^0. \tag{2.4}$$

As the characteristic polynomial of the Higgs field does not change when the Higgs bundle is tensored with a line bundle, the action of Γ on $T^*\check{\mathcal{N}}$ is along the fibers of $\check{\chi}$ and so $\check{\chi}$ descends to the quotient which gives the PGL_n Hitchin map:

$$\hat{\chi} : (T^*\check{\mathcal{N}})/\Gamma \to \hat{\mathcal{A}} = \mathcal{A}^0.$$

Recall that $T^*\mathcal{N}$ is an algebraic symplectic variety with the canonical Liouville symplectic form.

Theorem 2.5 (Hitchin, 1987). *If $\chi_i, \chi_j \in \mathbb{C}[T^*\mathcal{N}]$ are two coordinate functions, then they Poisson commute, i.e. $\omega(X_{\chi_i}, X_{\chi_j}) = 0$. We have $\dim(\mathcal{A}) = \dim(\mathcal{N})$ and the generic fibres of χ are open subsets of abelian varieties. Therefore we have an algebraically completely integrable Hamiltonian system.*

As a next step we will projectivize the Hitchin map $\chi : T^*\mathcal{N} \to \mathcal{A}$. Recall that a complex point in $T^*\mathcal{N}$ is given by a pair (E,ϕ). In order to projectivize we need to allow E to become unstable.

Definition 2.6. *A* Higgs bundle *is a pair (E,ϕ) where E is a vector bundle and $\phi \in H^0(C, \mathrm{End}(E) \otimes K)$ is a Higgs field.*

The definition for semi-stability and stability for Higgs-bundles is almost the same as for vector bundles except we only consider ϕ-invariant subbundles. The moduli-space of semi-stable Higgs bundles is denoted by $\mathcal{M}^d_{\mathrm{Dol}}$ and often abbreviated as $\mathcal{M}^d$ or even $\mathcal{M}$. It is a non-singular quasi-projective variety, having $T^*\mathcal{N}$ as an open subvariety.

It is straightforward to extend $\chi : \mathcal{M}^d \to \mathcal{A}$. The following result shows that we have succeeded in projectivizing the Hitchin map:

Theorem 2.7 (Hitchin 1987 [51], Nitsure 1991 [79], Faltings 1993 [31]). *χ is a proper algebraically completely integrable Hamiltonian system. Its generic fibres are abelian varieties.*

Remark 2.8. Note that $T^*\mathcal{N} \subset \mathcal{M}$ is open dense with complement of codimension greater than 1 [52, Proposition 6.1.iv]. Thus $\mathbb{C}[T^*\mathcal{N}] \cong \mathbb{C}[\mathcal{M}]$. Therefore the affinization of $\mathcal{M}$ (which must be the Hitchin map as it is proper to an affine space) restricts to the affinization of $T^*\mathcal{N}$, which justifies our unorthodox introduction of the Hitchin map above.

2.3. Hitchin systems for SL_n and PGL_n

It is straightforward now to compactify the SL_n-Hitchin map $T^*\check{\mathcal{N}}^d \to \mathcal{A}^0$. We consider $\check{\mathcal{M}}^{\Lambda}_{Dol}$ the moduli space of isomorphism classes of semi-stable Higgs bundle (E, ϕ) of rank n, $\det E = \Lambda$ and trace-free

$$\phi \in H^0(\mathrm{End}_0(E) \otimes K)$$

Higgs field. Again the isomorphism class of $\check{\mathcal{M}}^{\Lambda}_{Dol}$ only depends on d, so we will simplify our notation to $\check{\mathcal{M}}^d_{Dol}$, $\check{\mathcal{M}}^d$ or even $\check{\mathcal{M}}$ for the SL_n Higgs moduli space.

As in the GL_n-case the Hitchin map $\check{\chi} : \check{\mathcal{M}}^d \to \mathcal{A}^0$ is given by the coefficients of the characteristic polynomial, it is proper and a completely integrable system. We also have that $T^*\check{\mathcal{N}}^d \subset \check{\mathcal{M}}^d$ open and dense.

Let us recall the two constructions of the moduli space of PGL_n-Higgs bundles. The cotangent bundle $T^*\,\mathrm{Pic}^0(C) = \mathrm{Pic}^0(C) \times H^0(C, K)$ is a group. It acts on $\mathcal{M}^d$ by

$$(L, \varphi) \cdot (E, \phi) \mapsto (L \otimes E, \varphi + \phi)$$

This induces an action of $\Gamma = \mathrm{Pic}^0[n]$ on $\check{\mathcal{M}}^d$. Then we may either define the PGL_n-moduli space as

$$\hat{\mathcal{M}}^d = \mathcal{M}^d / T^*\,\mathrm{Pic}^0(C) \cong \chi^{-1}(\mathcal{A}^0)/\,\mathrm{Pic}^0(C)$$

or equivalently as the orbifold

$$\hat{\mathcal{M}}^d = \check{\mathcal{M}}^d/\Gamma.$$

The second quotient tells us that we obtain an orbifold. Since $\check{\chi}$ is compatible with the Γ action, we obtain a well-defined proper Hitchin map

$$\hat{\chi} : \hat{\mathcal{M}}^d = \check{\mathcal{M}}^d/\Gamma \to \mathcal{A}^0 = \hat{\mathcal{A}}.$$

All the three Hitchin maps χ, $\check{\chi}$ and $\hat{\chi}$ we defined above are proper algebraically completely integrable systems, therefore as explained in §1.1.3 we should expect the generic fibers to become compact tori, and as we are in the algebraic situation: Abelian varieties. We will see below §3.3.2 that this is indeed the case.

2.4. Cohomology of Higgs moduli spaces

Compared to the moduli spaces of bundles $\mathcal{N}$ we have less information on the cohomology of $\mathcal{M}$. Only the case of $n = 2$ is understood completely. In this case [50] describes the universal generators of $H^*(\mathcal{M})$ and [48] describes the relations among the generators. Universal generators were found in [68] for all n, but we have not even a conjecture about the ring structure of $H^*(\mathcal{M})$.

The Betti numbers of $\mathcal{M}$ are known only when $n = 2$ by the work of Hitchin [51] and for $n = 3$ by the work of Gothen using a Morse theoretical technique which we will describe below §3.3.8 in more detail. There is a conjecture of the Betti numbers for all n in [47]. But for $n > 3$ the Betti numbers are not known, although see §6.

Crucially for us the analogue of Theorem 2.1 does not hold. Thus Γ acts non-trivially on $H^*(\check{\mathcal{M}})$ and so $H^*(\check{\mathcal{M}}) \not\cong H^*(\hat{\mathcal{M}})$ as was already observed by Hitchin in [51] for $n = 2$. One can conjecturally describe the non-trivial part of the Γ-module $H^*(\check{\mathcal{M}})$ by using ideas from mirror symmetry. We proceed by detailing these ideas.

3. Topological mirror symmetry for Higgs bundles

The goal of this section is to establish the global picture:

$$\begin{array}{ccc} \check{\mathcal{M}} & & \hat{\mathcal{M}} \\ & \overset{\check{\chi}}{\searrow} \quad \overset{\hat{\chi}}{\swarrow} & \\ & \mathcal{A}^0 & \end{array}$$

where $\check{\chi}$ and $\hat{\chi}$ are the Hitchin maps. The important point is that the generic fibres are (torsors for) dual abelian varieties. The reason we want to do this is the following. As we mentioned in the introduction $\mathcal{M} = \mathcal{M}_{\mathrm{Dol}}$ is isomorphic with the hyperkähler manifold $\mathcal{M}_{\mathrm{Hit}}$ in complex structure I. If we change complex structure we also have a moduli space interpretation of $\mathcal{M}_{\mathrm{Hit}}$ in complex structure J. Namely we can identify [51, §9] the complex manifold $(\mathcal{M}_{\mathrm{Hit}}, J)$ with a certain moduli space $\mathcal{M}_{\mathrm{DR}}$ of flat connections on C. In this complex structure

$$\begin{array}{ccc} \check{\mathcal{M}}_{\mathrm{DR}} & & \hat{\mathcal{M}}_{\mathrm{DR}} \\ & \overset{\check{\chi}}{\searrow} \quad \overset{\hat{\chi}}{\swarrow} & \\ & \mathcal{A}^0 & \end{array} \tag{3.1}$$

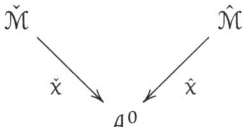

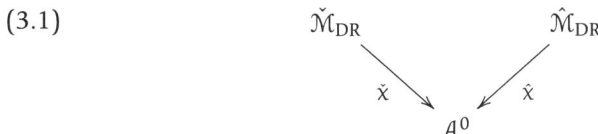

the fibres of the Hitchin map become dual special Lagrangian tori. This is the setting proposed by Strominger-Yau-Zaslow [87] for mirror symmetry as discussed in Subsection 1.1. We can thus expect $\check{\mathcal{M}}_{\mathrm{DR}}$ and $\hat{\mathcal{M}}_{\mathrm{DR}}$ to be mirror symmetric. In the physics literature such a mirror symmetry was first suggested by [10] in 1994 and more recently by Kapustin–Witten [58] in 2006. Below we will be aiming at checking the agreements of certain Hodge numbers of $\check{\mathcal{M}}_{DR}$ and $\hat{\mathcal{M}}_{DR}$ which can be called topological mirror symmetry.

3.1. Spectral curves

The simple idea of describing a polynomial by its zeroes leads to the notion of spectral curve of a Higgs bundle (E, ϕ) or more generally of its characteristic polynomial $a \in \mathcal{A}$. Recall that it has the form

$$a = t^n + a_1 t^{n-1} + \cdots + a_n,$$

where $a_i \in H^0(K^i)$. What should be the spectrum of such a polynomial? Look at one point $p \in C$, there we get $\phi_p : E_p \to E_p \otimes K_p$, we expect of an eigenvalue ν_p of ϕ_p to satisfy that there exists $0 \neq \nu \in E_p$ with the property $\phi_p(\nu) = \nu_p \nu$. Thus, we need $\nu_p \in K_p$. We do now consider all eigenvalues as a subset of the total space X of the bundle $K \to C$, and want to identify it with the complex points of a scheme. The resulting object will be called the spectral curve corresponding to $a \in \mathcal{A}$ and denoted by C_a. The picture is this:

$$\begin{array}{ccc} C_a & \subset & X \\ & \searrow^{\pi_a} & \downarrow \pi \\ & & C \end{array}$$

To construct the scheme structure on the spectral curve C_a, note that there exists a tautological section $\lambda \in H^0(X, \pi_a^* K)$ satisfying $\lambda(x) = x$. We can now pullback the sections a_i to X and obtain a section

$$s_a = \lambda^n + a_1 \lambda^{n-1} + \cdots + a_n \in H^0(X, \pi_a^* K^n)$$

Clearly C_a equals the zero set of this section, i.e. $C_a = s_a^{-1}(0)$, which comes naturally with a scheme structure.

3.2. Generic fibres of the Hitchin map

The fibers of the Hitchin map can be complicated: reducible even non-reduced. But for generic $a \in \mathcal{A}$ the spectral curve C_a is smooth and the corresponding fiber of the Hitchin map is also smooth and has a nice description. Here we review this description. For more details see [52, §8] and [7].

If (E, ϕ) is a Higgs bundle with characteristic polynomial a, then the pull-back of E to C_a will have a natural subsheaf M given generically by the eigenspace $\langle \nu \rangle \in K_p$ of $\nu_p \in C_a$. As C_a is non-singular M becomes an invertible sheaf, a line bundle. A more precise definition is to take

$$M := \ker\{\pi_a^*(\phi) - \lambda \mathrm{Id} : \pi_a^*(E) \to \pi_a^*(E \otimes K)\}.$$

One can recover E from M by the formula

$$E = \pi_{a*}(L), \tag{3.2}$$

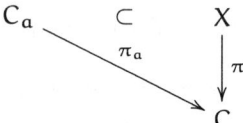

where $L = M(\Delta)$ and Δ is the ramification divisor of π_a (cf. [7, Remark 3.7]). Then equation (3.2) can be considered the *eigenspace decomposition* of ϕ on E.

More generally we want to get a correspondence between line bundles on C_a and Higgs bundles on C with characteristic polynomial a. Starting with a line bundle L on C_a we do at least know that $\pi_{a*}(L)$ is a torsion free sheaf on C, but since C is a non-singular curve this means that it is actually a vector bundle of rank n, which is the degree of the covering $\pi_a : C_a \to C$. Recall the canonical section $\lambda \in H^0(X, \pi_a^* K)$. It gives us a homomorphism

$$L \xrightarrow{\lambda} L \otimes \pi_a^* K.$$

We can now push this forward to the curve C to obtain

$$E = \pi_{a*}(L) \xrightarrow{\pi_{a*}(\lambda)} \pi_{a*}(L \otimes \pi_a^* K) = \pi_{a*}(L) \otimes K.$$

This way from a line bundle L on C we get a Higgs bundle $E \to E \otimes K$.

As C_a is integral, (E, φ) cannot have any sub Higgs bundle (as its spectral curve would be a one-dimensional subscheme of C_a), hence it is automatically stable. Thus if we set

$$d' := d + n(n-1)(g-1)$$

then we get a map:

$$\begin{gathered} \mathrm{Pic}^{d'}(C_a) \to \mathcal{M}^d \\ L \mapsto (\pi_{a*}(L), \pi_{a*}(\lambda)) \end{gathered}$$

To see that the Higgs bundle $(\pi_{a*}(L), \pi_{a*}(\lambda))$ has characteristic polynomial a recall that

$$\lambda^n + a_{n-1}\lambda^{n-1} + \cdots + a_0 = 0$$

holds on C_a. So if we push-forward this equation to C we obtain

$$\pi_{a*}(\lambda)^n + a_{n-1}\pi_{a*}(\lambda)^{n-1} + \cdots + a_0 = 0.$$

Now the Cayley-Hamilton theorem implies the assertion.

Theorem 3.3 (Hitchin 1987, Beauville-Narasimhan-Ramanan 1989). *When $a \in \mathcal{A}_{reg}$ ($\Leftrightarrow C_a$ is non-singular) we have $\chi^{-1}(a) \cong \mathrm{Pic}^{d'}(C_a)$.*

We need some modifications for SL_n and PGL_n. In the SL_n-case we have $a \in \mathcal{A}^0$, we need to find the line bundles L, such that $\pi_{a*}(L)$ has the right determinant. Define $\mathrm{Prym}^{d'}(C) \subset \mathrm{Jac}^{d'}(C_a)$ by

$$L \in \mathrm{Prym}^{d'}(C_a) \Leftrightarrow \det \pi_{a*}(L) = \Lambda$$

It is clear that for all $a \in \mathcal{A}^0_{reg}$, the Hitchin fibre satisfies $\check{\chi}^{-1}(a) \cong \mathrm{Prym}^{d'}(C_a)$.

For PGL_n we have $\hat{\chi}^{-1}(a) \cong \mathrm{Prym}^{d'}(C_a)/\Gamma$. This makes sense since for $L_\gamma \in \mathrm{Pic}(C)[n]$ we do have

$$\det(\pi_{a*}(\pi_a^*(L_\gamma) \otimes L)) = \det(L_\gamma \otimes \pi_{a*}(L)) = L_\gamma^n \otimes \det(\pi_{a*}L) = \det(\pi_{a*}L).$$

To summarize, the fibres of the Hitchin map are given:

- For GL_n: By Thm 3.3, for $a \in \mathcal{A}_{reg}$
$$\mathcal{A}_a := \chi^{-1}(a) \simeq \mathrm{Jac}^{d'}(C_a).$$
- For SL_n: following the definitions it is straightforward that for $a \in \mathcal{A}^0_{reg}$
$$\check{\mathcal{A}}_a := \check{\chi}^{-1}(a) \simeq \mathrm{Prym}^{d'}(C_a).$$
- For PGL_n: There are two ways of thinking of the Hitchin fibre:
$$\hat{\mathcal{A}}_a := \mathrm{Prym}^{d'}(C)/\Gamma \simeq \mathrm{Jac}^{d'}(C_a)/\,\mathrm{Pic}^0(C),$$
where $\mathrm{Pic}^0(C)$ acts on $\mathrm{Jac}^{d'}(C_a)$ by tensoring with the pull-back line bundle. A short computation shows that the $\Gamma \subset \mathrm{Pic}^0(C)$ action preserves $\mathrm{Prym}^{d'}(C)$.

3.3. Symmetries of the Hitchin fibration

We will see in this subsection how natural Abelian varieties act on the regular fibers of the three Hitchin map, giving them a torsor structure. Again, we study the GL_n, SL_n, PGL_n cases separately.

3.3.1. For GL_n Fixing $a \in \mathcal{A}_{reg}$, tensor product defines a simply transitive action of $\mathrm{Pic}^0(C_a)$ on $\mathrm{Jac}^d(C_a)$, and therefore $\mathcal{M}_a$ is a torsor for $P_a := \mathrm{Pic}^0(C_a)$.

3.3.2. For SL_n Fix $a \in \mathcal{A}^0_{reg}$, we have the (ramified) spectral covering map $\pi : C_a \to C$.

Definition 3.4. *The norm map*

$$\mathrm{Nm}_{C_a/C} : \mathrm{Pic}^0(C_a) \to \mathrm{Pic}^0(C) \tag{3.5}$$

is defined by any of the following three equivalent way:

(1) *Using divisors. For any divisor* D *on* C_a *we have*
$$\mathrm{Nm}_{C_a/C}(\mathcal{O}(D)) = \mathcal{O}(\pi_{a*}D),$$
where $\pi_a : C_a \to C$ *is the projection. Then one can show that the norm of a principal divisor will be a principal divisor (using the norm map between the function fields* $\mathbb{C}(C_a) \to \mathbb{C}(C)$ *- this justifies the name "norm map") - thus inducing a well-defined norm map as in* (3.5). *This definition points out why* Nm *is a group homomorphism.*

(2) *For* $L \in \mathrm{Pic}^0(C_a)$ *define*
$$\mathrm{Nm}_{C_a/C}(L) = \det(\pi_{a*}(L)) \otimes \det{}^{-1}(\pi_{a*}\mathcal{O}_{C_a}).$$

(3) *Using the fact that* $\mathrm{Pic}^0(C), \mathrm{Pic}^0(C_a)$ *are Abelian varieties, we can define the norm map as the dual of the pull-back map*
$$\pi_a^* : \mathrm{Pic}^0(C) \to \mathrm{Pic}^0(C_a),$$

that is

$$\mathrm{Nm}_{C_a/C} = \check{\pi}_a^* : \mathrm{Pic}^0(C_a) \simeq \check{\mathrm{Pic}}^0(C_a) \to \check{\mathrm{Pic}}^0(C) \simeq \mathrm{Pic}^0(C).$$

Here recall that for an Abelian variety A *the dual* $\hat{A}$ *Abelian variety denotes the moduli space of degree* 0 *line bundles on* A *and that the dual of the Jacobian of a smooth curve is itself.*

Let

$$\mathrm{Prym}^0(C_a) := \ker(\mathrm{Nm}_{C_a/C}) \tag{3.6}$$

denote the kernel of the norm map, which is an Abelian subvariety of $\mathrm{Pic}^0(C)$. Then $\mathrm{Prym}^0(C_a)$ acts on $\mathrm{Prym}^d(C_a) = \check{\mathcal{M}}_a$, and $\check{\mathcal{M}}_a$ is a torsor for $\check{P}_a := \mathrm{Prym}^0(C_a)$.

3.3.3. For PGL_n

In this case

$$\hat{\mathcal{M}}_a = \check{\mathcal{M}}_a/\Gamma = \mathcal{M}_a/\mathrm{Pic}^0(C)$$

is a torsor for $\hat{P}_a := \check{P}_a/\Gamma = P_a/\mathrm{Pic}^0(C)$.

To complete the SYZ picture (3.1) we show that $\check{P}_a$ and $\hat{P}_a$ are dual abelian varieties.

3.4. Duality of the Hitchin fibres

Take the short exact sequence

$$0 \to \mathrm{Prym}^0(C_a) \hookrightarrow \mathrm{Pic}^0(C_a) \overset{\mathrm{Nm}_{C_a/C}}{\twoheadrightarrow} \mathrm{Pic}^0(C) \to 0$$

and dualize. Since $\mathrm{Pic}^0(C)$ and $\mathrm{Pic}^0(C_a)$ are isomorphic to their duals, we get

$$0 \leftarrow \check{\mathrm{Prym}}^0(C_a) \twoheadleftarrow \mathrm{Pic}^0(C_a) \overset{\pi_a^*}{\hookleftarrow} \mathrm{Pic}^0(C) \leftarrow 0,$$

and therefore

$$\check{\check{P}}_a = \mathrm{Pic}^0(C_a)/\mathrm{Pic}^0(C) = \hat{P}_a,$$

that is $\check{P}_a$ and $\hat{P}_a$ are duals. (See [49, Lemma (2.3)] for more details.)

This is the first reflection of mirror symmetry. To summarize, we can state

Theorem 3.7 ([49]). *For a regular* $a \in \mathcal{A}^0_{reg}$ *the fibers* $\check{\mathcal{M}}_a$ *and* $\hat{\mathcal{M}}_a$ *are torsors for dual Abelian varieties* $\check{P}_a$ *and* $\hat{P}_a$*, respectively.*

We can state this theorem more precisely using the language of gerbes. To that end here is a short summary.

3.5. Gerbes on $\check{\mathcal{M}}$ and $\hat{\mathcal{M}}$

Here we sketch a quick definition of gerbes for more details see [49],[88] and [25]. Let A be a sheaf of Abelian groups on a variety X. The typical examples are $\mathcal{O}_X^\times$, and the constant sheaves $\underline{\mu_n}, \underline{U(1)}$, where μ_n is the group of mth roots of unity. Note that $\underline{\mu_n} \subset \mathcal{O}_X^\times$ and $\underline{\mu_n} \subset \underline{U(1)}$.

Definition 3.8. *An A-torsor is a sheaf F of sets on X together with an action of A, such that F is locally isomorphic with A. In particular, when nonempty, $\Gamma(U, F)$ is a torsor for $\Gamma(U, A)$ for all open $U \subset X$.*

Examples:

- $\mathcal{O}_X^\times$-torsor = line bundle
- $\underline{U(1)}$-torsor = flat unitary line bundle
- $\underline{\mu_n}$-torsor = μ_n-Galois cover

Note that the natural tensor category structure on the category of torsors $\mathrm{Tors}_A(U)$ endows it with a group-like structure. Moreover, the automorphism of an A-torsor is an element of $\Gamma(A)$.

Definition 3.9. *An A-gerbe B is a sheaf of categories (which is roughly what one would think) so that locally $B|_U$ becomes the analogue of a torsor over $\mathrm{Tors}_A(U)$.*

Let $(\mathbb{E}, \Phi)$ be a universal Higgs-bundle on $\check{\mathcal{M}} \times C$, where

$$\Phi \in H^0(\mathrm{End}_0\, \mathbb{E} \otimes \pi_a^*(K_C)),$$

and $\mathbb{E}_c = \mathbb{E}|_{\check{\mathcal{M}} \times \{c\}}$ be the fiber over $c \in C$. Such a universal bundle exists by our running assumption $(d, n) = 1$. Then

$$c_1(\mathbb{E}_c) \in H^2(\check{\mathcal{M}}, \mathbb{Z}) \simeq \mathbb{Z} \text{ is a generator modulo } n. \tag{3.10}$$

Note that $\mathbb{E}$ is not unique: it can be tensored by $L \in \mathrm{Pic}(\check{\mathcal{M}})$, but this property always holds.

Let $\mathbb{P}\mathbb{E}_c \to \check{\mathcal{M}}$ be the corresponding PGL_n-bundle. Let $\check{B}$ be the μ_n-gerbe of liftings of $\mathbb{P}\mathbb{E}_c$ as an SL_n-bundle. Because for every lifting as a GL_n-bundle (3.10) holds, there is no global lifting as an SL_n-bundle and so $\check{B}$ is not a trivial gerbe. But for $a \in \mathcal{A}_{reg}$ it turns out that $c_1(\mathbb{P}\mathbb{E}_c)|_{\check{\mathcal{M}}_a}$ is 0 mod n, and so $\mathbb{P}\mathbb{E}_c|_{\check{\mathcal{M}}_a}$ can be lifted as an SL_n-bundle, therefore $\check{B}|_{\check{\mathcal{M}}_a}$ is a trivial gerbe.

Finally we note that the action of Γ on $\check{\mathcal{M}}$ can be lifted to $\check{B}$ to get an orbifold gerbe $\hat{B}$ on $\hat{\mathcal{M}}$. As $\check{B}$ and $\hat{B}$ are $\mu_m \subset U(1)$-gerbes we can consider the induced $U(1)$-gerbes as well.

Theorem 3.11 ([49]). *One can identify the set of trivializations*

$$\mathrm{Triv}^{U(1)}(\check{B}^e|_{\check{\mathcal{M}}_a^d}) \simeq \hat{\mathcal{M}}_a^e$$

as $\check{P}_a$-torsors. Similarly,

$$\mathrm{Triv}^{U(1)}(\hat{B}^d|_{\hat{\mathcal{M}}_a^e}) \simeq \check{\mathcal{M}}_a^d.$$

Remark 3.12. The analogue of this result for arbitrary pair of Langlands dual groups was handled by Donagi–Pantev in [27]. There they also implement the fiberwise Fourier-Mukai transform over the regular locus which will be discussed in §5.5.3. The case of G_2 was considered by Hitchin in [54] in detail.

This Theorem 3.11 can be interpreted as the twisted version of the Strominger-Yau-Zaslow proposal suggested in [53]. Thus we have established the picture

$$\begin{array}{ccc} \check{\mathcal{M}}^d_{DR} & & \hat{\mathcal{M}}^e_{DR} \\ & \searrow^{\check{\chi}} \quad {}^{\hat{\chi}}\swarrow & \\ & \mathcal{A}^0 & \end{array}$$

with $\check{\chi}$ and $\hat{\chi}$ two special Lagrangian fibrations with dual tori as generic fibers according to Theorem 3.11. Thus the pair $(\check{\mathcal{M}}^d_{DR}, \check{B}^e)$ and $(\check{\mathcal{M}}^e_{DR}, \hat{B}^d)$ can considered mirror symmetric in the twisted SYZ sense of [53] Hitchin.

3.6. The stringy Serre polynomial of an orbifold

As our spaces satisfy a suitable version of the Strominger-Yau-Zaslow mirror symmetry proposal we need a definition of Hodge numbers of non-projective varieties to be able to formulate topological mirror symmetry for our mirror pairs. For more details on these Hodge numbers see [22, 6, 49, 47, 80].

Definition 3.13. *Let X be a complex algebraic variety. The Serre polynomial (virtual Hodge polynomial) is defined as*

$$E(X;u,v) = \sum_{p,q,i} (-1)^i h^{p,q}(Gr^W_{p+q} H^i_c(X)) u^p v^q$$

where $Gr^W_{p+q} H^i_c(X)$ is the $p+q$th graded piece of the weight filtration of $H^i_c(X)$. This has a Hodge structure of weight $p+q$, and $h^{p,q}$ is the corresponding Hodge number. This so-called mixed Hodge structure was constructed by Deligne [22].

Remark 3.14. We say that the mixed Hodge structure is pure if $h^{p,q} \neq 0$ implies that $p+q=i$. For example, smooth projective varieties have pure MHS on their cohomology. One can also show [44, Theorems 2.1,2.2] that the spaces $\mathcal{M}^d_{Dol}$ and $\mathcal{M}^d_{DR}$ also have pure MHS on their cohomology.

The definition of the virtual Hodge polynomial is the first step of formulating the topological mirror test. In order to obtain the correct Hodge numbers of the orbifold $\hat{\mathcal{M}}$. We need to define stringy Hodge numbers twisted by a gerbe. Let X be smooth, and assume that a finite group Γ acts on X. Then we can define the *stringy Serre polynomial* of the orbifold X/Γ as follows:

$$E_{st}(X/\Gamma), u, v) = \sum_{[\gamma]\in[\Gamma]} E(X_\gamma/C_\gamma; u, v)(uv)^{F(\gamma)}, \tag{3.15}$$

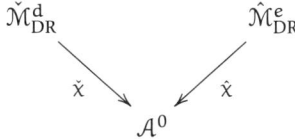

where

- $[\gamma]$ is a conjugacy class of Γ
- X_γ is the fixed point set, C_γ is the centralizer of γ in Γ, acting on X_γ.
- $F(\gamma)$ is the Fermionic shift, defined as $F(\gamma) = \sum w_i$, where γ acts on $TX|_{X_\gamma}$ with eigenvalues $e^{2\pi i w_i}$, $w_i \in [0,1)$.

Remark 3.16. These orbifold Hodge numbers can be considered the cohomological shadow of the stringy derived category of coherent sheaves on the variety X/Γ, or more topologically the stringy K-theory of X/Γ. Both can be simply defined by considering Γ-equivariant coherent sheaves or Γ-equivariant complex vector bundles on X. See [56] for more details.

Another important property of $E_{st}(X/\Gamma; u, v)$ is the following theorem:

Theorem 3.17 **([63, 6]).** *If* $\pi : Y \to X/\Gamma$ *is a crepant resolution, i.e.* $\pi^*\omega_{X/\Gamma} = \omega_Y$, *then* $E_{st}(X/\Gamma; u, v) = E(Y; u, v)$.

Remark 3.18. This is a nice way to see what the stringy Hodge numbers mean, in terms of the Hodge numbers of (any) crepant resolution. However our orbifolds never have crepant resolutions, because the generic singular points are infinitesimally modelled by quotients of symplectic representations of $\mathbb{Z}_n$ which are not generated by symplectic reflections as in [89, Theorem 1.2]. So we will be stuck with the definition in (3.15).

Finally let B be a Γ-equivariant $U(1)$-gerbe on X. Then more generally, we define

$$E_{st}^{B}(X/\Gamma; u, v) = \sum_{[\gamma]\in[\Gamma]} E(X_\gamma/C_\gamma, L_{B,\gamma}; u, v)(uv)^{F(\gamma)}, \tag{3.19}$$

where $L_{B,\gamma}$ is the local system on X_γ given by B.

These B-twisted stringy Hodge numbers can again be considered as the cohomological shadow of a certain derived category of B-twisted Γ-equivariant sheaves on X or the corresponding B-twisted equivariant K-theory.

3.7. Topological mirror test

We can now formulate our original topological mirror symmetry

Conjecture 3.20 **([49]).** *For* $(d, n) = (e, n) = 1$

$$E(\check{\mathcal{M}}^d_{DR}; u, v) = E_{st}^{\hat{B}^d}(\hat{\mathcal{M}}^e_{DR}; u, v),$$

where $\hat{B}^d$ *is the* Γ*-equivariant gerbe on* $\check{\mathcal{M}}^e$ *appearing in Theorem 3.11.*

Remark 3.21. One can object that this conjecture does not feature the change in the indices of the Hodge numbers as in the original topological mirror test (1.1) between projective Calabi-Yau manifolds. Our, rather hand-waving, argument for this in [49] was that for *compact* hyperkähler manifolds (1.1) is equivalent with (1.3) that is agreement of Hodge numbers. This was only what we could offer to explain the agreement of Hodge numbers in Conjecture 3.20, even though our examples are not compact, and their Hodge numbers do not possess any non-trivial symmetries. As will be explained after Conjecture 5.9 some recent developments lead to a solution of this problem.

As $\mathcal{M}_{\mathrm{DR}}$ can be algebraically deformed to $\mathcal{M}_{\mathrm{Dol}}$, inside nice compactifications, it is not surprising that we could prove that their mixed Hodge structure is isomorphic:

Theorem 3.22 ([49, Theorem 6.2,6.3]). *For* $(d,n)=(e,n)=1$

$$E(\check{\mathcal{M}}^d_{\mathrm{DR}};u,v)=E(\check{\mathcal{M}}^d_{\mathrm{Dol}};u,v)$$

and similarly

$$E_{\mathrm{st}}^{\hat{B}^d}(\hat{\mathcal{M}}^e_{\mathrm{DR}};u,v)=E_{\mathrm{st}}^{\hat{B}^d}(\hat{\mathcal{M}}^e_{\mathrm{Dol}};u,v).$$

Thus we have an equivalent form of Conjecture 3.20, the so-called Dolbeault version of topological mirror symmetry:

Conjecture 3.23 ([49]). *For* $(d,n)=(e,n)=1$

$$E(\check{\mathcal{M}}^d_{\mathrm{Dol}};u,v)=E_{\mathrm{st}}^{\hat{B}^d}(\hat{\mathcal{M}}^e_{\mathrm{Dol}};u,v),$$

where $\hat{B}^d$ *is the* Γ*-equivariant gerbe on* $\check{\mathcal{M}}^e$ *appearing in Theorem 3.11.*

As we will sketch in §5.5.3 this conjecture could be interpreted as a cohomological shadow of some equivalence of derived categories of sheaves on the Hitchin systems which arise in the Geometric Langlands programme.

This latter conjecture we were able to prove in the following cases:

Theorem 3.24 ([49]). *Conjecture 3.23 and so Conjecture 3.20 are valid for* $n=2$ *and* $n=3$.

In §3.3.8 below we will sketch the computation in the $n=2$ case.

Now we will unravel the meaning of Conjecture 3.23. Recall that Γ acts on $\check{\mathcal{M}}^d$ and hence Γ also acts on $H^*_c(\check{\mathcal{M}}^d)$. We get a decomposition

$$H^*_c(\check{\mathcal{M}}^d)=\bigoplus_{\kappa\in\hat{\Gamma}}H^*_c(\check{\mathcal{M}}^d)_\kappa$$

which is compatible with the mixed Hodge structure. Therefore we can write

$$E(\check{\mathcal{M}}^d;u,v)=\sum_{\kappa\in\hat{\Gamma}}E_\kappa(\check{\mathcal{M}}^d;u,v), \tag{3.25}$$

where

$$E_\kappa(\check{\mathcal{M}}^d;u,v):=\sum_{p,q,i}(-1)^i h^{p,q}(Gr^W_{p+q}H^i_c(\check{\mathcal{M}}^d)_\kappa)u^p v^q.$$

We can also expand the RHS of Conjecture 3.23. By the definition of the stringy Serre polynomial, and because Γ is commutative, we have

$$E_{st}^{\hat{B}^d}(\hat{\mathcal{M}}^e; u, v) = \sum_{\gamma \in \Gamma} E(\check{\mathcal{M}}^e_\gamma/\Gamma; L_{\hat{B}^d,\gamma}, u, v)(uv)^{F(\gamma)}$$

Thus the unraveled Conjecture 3.23 takes the form:

$$\sum_{\kappa \in \hat{\Gamma}} E_\kappa(\check{\mathcal{M}}^d; u, v) = \sum_{\gamma \in \Gamma} E(\check{\mathcal{M}}^e_\gamma/\Gamma; L_{\hat{B}^d,\gamma}, u, v)(uv)^{F(\gamma)} \tag{3.26}$$

We note that there are the same number of terms in (3.26); and in fact there is a canonical way to identify them. Note that Γ is canonically isomorphic to $H^1(C, \mathbb{Z}_n)$, where C is our underlying curve. It follows that Poincaré duality gives us a canonical pairing

$$w : \Gamma \times \Gamma \to H^2(C, \mathbb{Z}_n) = \mathbb{Z}_n,$$

the so-called Weil pairing. This allows us to identify $w : \Gamma \to \hat{\Gamma}$. This identification leads to the *refined topological mirror symmetry test*:

Conjecture 3.27. *For* $\kappa \in \hat{\Gamma}$ *we have*

$$E_\kappa(\check{\mathcal{M}}^d; u, v) = E(\check{\mathcal{M}}^e_\gamma/\Gamma, L_{\hat{B}^d,\gamma}; u, v)(uv)^{F(\gamma)}$$

where $\gamma = w(\kappa)$.

This again holds for $n = 2,3$ the case of $n = 2$ will be discussed in the next section. As we will see in §5.5.4 this refined conjecture is closely related to Ngô's geometric approach to the fundamental lemma.

Here we point out the case of the trivial character $\kappa = 1$. In that case the refined topological mirror symmetry test is:

Conjecture 3.28. *When* $(d, n) = (e, n) = 1$

$$E_1(\check{\mathcal{M}}^d; u, v) = E(\check{\mathcal{M}}^e/\Gamma; u, v)$$

or equivalently

$$E(\hat{\mathcal{M}}^d; u, v) = E(\hat{\mathcal{M}}^e; u, v).$$

Remark 3.29. This conjecture is in sharp contrast to the dependence on degree of $H^*(\hat{\mathcal{N}}^d)$. In fact another application of Harder–Narasimhan [42, Theorem 3.3.2] implies that the Betti numbers of $\hat{\mathcal{N}}^d$ and $\hat{\mathcal{N}}^e$ are different provided $0 < d < e < n/2$. The smallest such situation is when $n = 5$, $d = 1$ and $e = 2$.

Remark 3.30. As Theorem 3.11 holds for all d and e not necessarily coprime to n it is conceivable that something like Conjecture 3.28 should hold even when d (and/or e) is not coprime to n. It is however unclear what cohomology theory we should calculate on the singular moduli space $\check{\mathcal{M}}^d$, for $(d, n) \neq 1$, to produce the agreement in Conjecture 3.28. For $n = 2$ Batyrev's extension [5] of stringy cohomology of

$\check{\mathcal{M}}^0$ was calculated in [60], while stacky cohomology of $\check{\mathcal{M}}^0$ was calculated in [21]. In either case the corresponding generating functions are not polynomials; thus Conjecture 3.28 for $n = 2$, $d = 0$, $e = 1$, as it stands, cannot hold for them.

3.8. Topological mirror symmetry for $n = 2$

Consider the circle action of $\mathbb{C}^\times$ on the Higgs moduli space by rescaling the Higgs field. That is, $\lambda \cdot (E, \phi) \mapsto (E, \lambda \cdot \phi)$. We can study the corresponding Morse stratification and by Morse theory we obtain the decomposition

$$H^*(\check{\mathcal{M}}) = \bigoplus_{F_i \subset \check{\mathcal{M}}^{\mathbb{C}^\times}} H^{*-\mu_i}(F_i). \tag{3.31}$$

Here the sum is over the connected components of the fixed point set $\check{\mathcal{M}}^{\mathbb{C}^\times}$, and μ_i denotes the Morse index of F_i with respect to the $\mathbb{C}^\times$-action. Note that (3.31) is a decomposition as Γ-modules.

The components F_i have been described for $n = 2$ by Hitchin [51], and by Gothen [37] for $n = 3$. The case $n = 4$ seems quite hard; but for recent progress see §6.

One obvious fixed point locus is $F_0 \cong \check{\mathcal{N}}$, consisting of stable bundles with zero Higgs field. However this component doesn't contribute to the variant part as the Γ-action on $H^*(\check{\mathcal{N}})$ is trivial by Theorem 2.1.

From now on in this section we assume $n = 2$ and $d = 1$. Then the other components can be labelled by $i = 1, \dots, g - 1$ and consist of isomorphism classes of Higgs bundles of the form

$$F_i = \{(E, \phi) \mid E \cong L_1 \oplus L_2, \deg(L_1) = i, \phi = \begin{pmatrix} 0 & 0 \\ \varphi & 0 \end{pmatrix}, 0 \neq \varphi \in H^0(L_1^{-1} L_2 K)\}.$$

Now stability forces $\deg L_2 = 1 - i$ where $i > 0$, because L_2 is a ϕ-invariant subbundle. We can associate to $(E, \phi) \in F_i$ the divisor of φ in $S^{2g-2i-1}(C)$ yielding a $2^{2g} : 1$ covering

$$F_i \to S^{2g-2i-1}(C).$$

This covering is given by the free action of Γ on F_i.

Theorem 3.32 ([51](7.13)). *The Γ-action on $H^*(F_i)$ is only non-trivial in the middle degree $2g - 2i - 1$. We have*

$$\dim H^{2g-2i-1}_{var}(F_i) = (2^g - 1) \binom{2g - 2}{2g - 2i - 1}$$

Moreover, if $\kappa \in \hat{\Gamma}^*$ then

$$\dim H^{2g-2i-1}_{\kappa}(F_i) = \binom{2g - 2}{2g - 2i - 1}.$$

We now consider the stringy side. Recall $\hat{\mathcal{M}} = \check{\mathcal{M}}/\Gamma$ and let $\gamma \in \Gamma^* = \Gamma \setminus \{0\}$. Then γ leads to a connected covering

$$\pi_\gamma : C_\gamma \to C$$

with Galois group $\mathbb{Z}_2$. Consider the commutative diagram

$$\begin{array}{ccccc}
T^*\operatorname{Jac}^d(C_\gamma) & \xrightarrow{(\pi_\gamma)_*} & \mathcal{M}^d & & \supset \check{\mathcal{M}}^d = \det^{-1}(\Lambda, 0) \\
\downarrow \cong & & \downarrow \det & & \\
T^*\operatorname{Jac}^d(C_\gamma) & \xrightarrow{\operatorname{Nm}_{C_\gamma/C}} & T^*\operatorname{Jac}^d(C) & & \ni (\Lambda, 0)
\end{array}$$

From this diagram [49, Corollary 7.3] we have that $\check{\mathcal{M}}_\gamma$ is a torsor for $T^*\operatorname{Prym}(C_\gamma/C)^0$; where $\operatorname{Prym}(C_\gamma/C)^0 := (\operatorname{Nm}^{-1}_{C_\gamma/C}(\mathcal{O}_C))^0$ is the connected component of the Prym variety. We can then calculate that

$$\dim H^{2g-2i+1}(\check{\mathcal{M}}_\gamma/\Gamma, L_{\hat{B}^d,\gamma}) = \binom{2g-2}{2g-2i-1},$$

when $i = 1, \ldots, g-1$ and is zero otherwise. Note that the presence of the gerbe means that we see only the odd degrees. It follows that in this case we indeed have the refined topological mirror symmetry

$$E_\kappa(\check{\mathcal{M}}; u, v) = E(\check{\mathcal{M}}_\gamma/\Gamma, L_{\hat{B},\gamma}; u, v)(uv)^{F(\gamma)} \tag{3.33}$$

when $\kappa = w^{-1}(\gamma)$.

Remark 3.34. What is surprising about the agreement in (3.33) is that the left hand side comes from the fixed point components F_i for $i > 0$, because $F_0 = \mathcal{N}$ has no variant cohomology by Theorem 2.1. While the right hand side comes solely F_0, because the Γ action is free on F_i for $i > 0$. Thus in order for (3.33) to hold there is a remarkable agreement of cohomological data coming on one hand from the symmetric products $\{F_i\}_{i>0}$ of the curve and on the other hand from the moduli space of stable bundles $\check{\mathcal{N}}_\gamma \subset \check{\mathcal{N}} = F_0$.

One may give a similar proof for (3.33), when $n = 3$ using Gothen's work [37]. The proof along these lines for higher n remains incomplete.

We now introduce a more successful arithmetic technique to prove a different version of topological mirror symmetry.

4. Topological mirror symmetry for character varieties

As mentioned above in another complex structure $\mathcal{M}_{\mathrm{Dol}}$ can be identified with $\mathcal{M}_{\mathrm{DR}}$ a certain moduli space of twisted flat GL_n-connections on C. A point in $\mathcal{M}_{\mathrm{DR}}$ represents a certain twisted flat connection on a rank n bundle. One can take its monodromy yielding a twisted representation of the fundamental group of C. This leads to the character variety of the Riemann surface C.

Definition 4.1. *The* character variety for GL_n *is defined as the affine GIT quotient:*

$$\mathcal{M}_B^d := \{(A_1, B_1, \ldots, A_g, B_g) \in GL_n(\mathbb{C}) \mid [A_1, B_1] \ldots [A_g, B_g] = \zeta_n^d I_n\} // GL_n$$

The character variety for SL_n *is the space*

$$\check{\mathcal{M}}_B^d := \{(A_1, B_1, \ldots, A_g, B_g) \in SL_n(\mathbb{C}) \mid [A_1, B_1] \ldots [A_g, B_g] = \zeta_n^d I_n\} // GL_n$$

where the action is always by simultaneous conjugation on all factors, and $\zeta_n = e^{2\pi i/n}$. *The* character variety for PGL_n *is defined as*

$$\hat{\mathcal{M}}_B^d := \check{\mathcal{M}}_B^d / \Gamma = \mathcal{M}_B^d / (\mathbb{C}^\times)^{2g}.$$

Here $\Gamma := \mathbb{Z}_n^{2g} \subset (\mathbb{C}^\times)^{2g}$ *and* $\Gamma \subset (\mathbb{C}^\times)^{2g}$ *acts on* $\check{\mathcal{M}}_B^d \subset \mathcal{M}_B^d$ *by multiplying the matrices* A_i, B_i *by scalars.*

Remark 4.2. For the PGL_n-character variety we could have started to consider

$$\hat{\mathcal{M}}_B := \{(A_1, B_1, \ldots, A_g, B_g) \in PGL_n(\mathbb{C}) \mid [A_1, B_1] \ldots [A_g, B_g] = I_n\} // PGL_n.$$

This variety has n components, depending on which order n central element in GL_n will agree with the product of the commutators of the GL_n-representatives of the PGL_n elements A_i, B_i. For $\hat{\mathcal{M}}_B^d$ we picked the component corresponding to the central element $\zeta_n^d I_n$.

One can prove [47] that if d and n are coprime, $\mathcal{M}_B^d$ and $\check{\mathcal{M}}_B^d$ are non-singular, and so $\hat{\mathcal{M}}_B^d$ is an orbifold.

Taking the monodromy will give a complex analytical isomorphism $\mathcal{M}_{DR}^d \cong \mathcal{M}_B^d$ the so-called Riemann-Hilbert correspondence. We have the following more general *non-abelian Hodge theorem*

Theorem 4.3 ([86, 19, 29]). *There are canonical diffeomorphisms in all our cases* GL_n, SL_n *and* PGL_n*:*

$$\hat{\check{\mathcal{M}}}{}_{Dol}^d \cong \hat{\check{\mathcal{M}}}{}_{DR}^d \cong \hat{\check{\mathcal{M}}}{}_B^d$$

Because $\mathcal{M}_{DR}$ and $\mathcal{M}_B$ are analytically isomorphic it follows that the twisted SYZ mirror symmetry proposal is satisfied by the pair $\check{\mathcal{M}}_B^d$ and $\hat{\mathcal{M}}_B^e$ as well. We may thus formulate the Betti-version of the topological mirror symmetry conjecture [44]:

Conjecture 4.4. *For* $(d, n) = (e, n) = 1$

$$E(\check{\mathcal{M}}_B^d; u, v) = E_{st}^{\hat{B}^d}(\hat{\mathcal{M}}_B^e; u, v),$$

where $\hat{B}$ *is the* Γ*-equivariant gerbe on* $\check{\mathcal{M}}_B^e$ *analogous to the one in §3.3.5.*

Note that the mixed Hodge structures on $H^*(\mathcal{M}_B^d) \cong H^*(\mathcal{M}_{DR}^d)$ are different, in particular unlike the MHS on $H^*(\mathcal{M}_B^d)$, the MHS on $H^*(\mathcal{M}_{DR}^d)$ is pure. Thus Conjecture 4.4 is different from Conjecture 3.20 and the equivalent Conjecture 3.23.

We will also need the refined version:

Conjecture 4.5. *For* $\kappa \in \hat{\Gamma}$ *we have*

$$E_\kappa(\check{\mathcal{M}}_B^d; u, v) = E(\check{\mathcal{M}}_\gamma^e/\Gamma, L_{\hat{B}^d,\gamma}; u, v)(uv)^{F(\gamma)}$$

where $\gamma = w(\kappa)$.

Interestingly, for $(n, e) = (n, d) = 1$ the character varieties $\hat{\mathcal{M}}_B^d$ and $\hat{\mathcal{M}}_B^e$ are Galois conjugate via an automorphism of the complex numbers sending ζ_n^d to ζ_n^e. This Galois conjugation induces an isomorphism

$$H^*(\hat{\mathcal{M}}_B^d) \cong H^*(\hat{\mathcal{M}}_B^e) \tag{4.6}$$

preserving mixed Hodge structures. Therefore the $\kappa = 1$ case of refined topological mirror symmetry for character varieties follows:

Theorem 4.7. *When* $(d, n) = (e, n) = 1$

$$E_1(\check{\mathcal{M}}_B^d/; u, v) = E(\check{\mathcal{M}}_B^e/\Gamma; u, v)$$

or equivalently

$$E(\hat{\mathcal{M}}_B^d; u, v) = E(\hat{\mathcal{M}}_B^e; u, v).$$

Remark 4.8. In fact one can prove that the universal generators of $H^*(\hat{\mathcal{M}}_B^d)$ are mapped to the corresponding ones of $H^*(\hat{\mathcal{M}}_B^e)$, and consequently one can prove that the Galois conjugation (4.6) preserves even the mixed Hodge structure on $H^*(\hat{\mathcal{M}}_{\mathrm{Dol}}^d) \cong H^*(\hat{\mathcal{M}}_{\mathrm{Dol}}^e)$; which implies Conjecture 3.28. Thus in the original topological mirror symmetry Conjecture 3.23 we see that the first terms corresponding to the trivial elements in Γ and $\hat{\Gamma}$ at least agree; so we can concentrate on proving Conjecture 3.27 for non-trivial characters κ. This is the first non-trivial application of considering character varieties.

Remark 4.9. One application in [42, Theorem 3.3.2] was to show that for $(d, n) = (e, n) = 1$ the Betti numbers of $\hat{\mathcal{N}}^d$ and $\hat{\mathcal{N}}^e$ agree only when $d+e$ or $d-e$ is divisible by n. This is markedly different behaviour from Theorem 4.7 which shows that the Betti numbers of $\hat{\mathcal{M}}^d$ and $\hat{\mathcal{M}}^e$ agree as long as $(d, n) = (e, n) = 1$.

In the next section we also offer an arithmetic technique which can be used efficiently to check Conjecture 4.4 in all cases.

4.1. An arithmetic technique to calculate Serre polynomials

Recall the definition of the Serre polynomial of a complex variety X:

$$E(X; u, v) = \sum_{i,p,q} (-1)^i h^{p,q}(\mathrm{Gr}^W_{p+q} H^i_c(X)) u^p v^q$$

where $W_0 \subseteq W_1 \subseteq \ldots \subseteq W_i \subseteq \ldots \subseteq W_{2k} = H^k(X) := H^k(X; \mathbb{Q})$ is the weight filtration.

By [47, Corollary 4.1.11] $\mathcal{M}_B$ has a Hodge-Tate type MHS, that is, $h^{p,q} \neq 0$ unless $p = q$ in its MHS. In this case the Serre polynomial is a polynomial of uv, i.e

$$E(X; u, v) = E(X, uv) := \sum_{i,k} (-1)^i \dim(Gr_k^W H_c^i(X))(uv)^k, \tag{4.10}$$

but the MHS is not pure, i.e there is $k \neq i$ such that $h^{(k/2,k/2)} \neq 0$.

Roughly speaking a variety X can be defined over the integers if one can arrange the defining equations to have integer coefficients. Then one can consider those equations in finite fields $\mathbb{F}_q$, and can count these solutions to get the number $|X(\mathbb{F}_q)|$. We say that such a variety X is polynomial count if $|X(\mathbb{F}_q)|$ is polynomial in q. One can define polynomial count varieties even if they can only be defined over more general finitely generated rings than $\mathbb{Z}$. For more technical details and precise statements see [47]. Here we have

Theorem 4.11 ([47, Appendix by Katz]). *For a polynomial count variety X*

$$E(X/\mathbb{C}, q) = |X(\mathbb{F}_q)|.$$

Example. Define $\mathbb{C}^* = \mathbb{C}\backslash\{0\}$ over $\mathbb{Z}$ as the subscheme $\{xy = 1\}$ of $\mathbb{A}^2$. Then

$$E(\mathbb{C}^*; q) = |\mathbb{F}_q^*| = q - 1.$$

Since $H_c^2(\mathbb{C}^*)$ has weight 2 and $H_c^1(\mathbb{C}^*)$ has weight 0, substitution to (4.10) gives indeed $q - 1$ for $E(\mathbb{C}^*; q)$.

4.2. Arithmetic harmonic analysis on $\mathcal{M}_B$

By Fourier transform on a finite group G one gets the following Frobenius-type formula:

$$\left|\left\{a_1, b_1, \ldots, a_g, b_g \in G \left| \prod [a_i, b_i] = z\right.\right\}\right| = \sum_{\chi \in Irr(G)} \frac{|G|^{2g-1}}{\chi(1)^{2g-1}} \chi(z).$$

Therefore assuming that $\zeta_n \in \mathbb{F}_q^*$, i.e $n|q-1$, we get

$$|\mathcal{M}_B^d(\mathbb{F}_q)| = (q-1) \sum_{\chi \in Irr(GL_n(\mathbb{F}_q))} \frac{|GL_n(\mathbb{F}_q)|^{2g-2}}{\chi(1)^{2g-2}} \cdot \frac{\chi(\zeta_n^d \cdot I)}{\chi(1)}. \tag{4.12}$$

Irreducible characters of $GL_n(\mathbb{F}_q)$ have a combinatorial description by Green [38] from 1955. Consequently $|\mathcal{M}_B^d(\mathbb{F}_q)|$ can be calculated explicitly [47]. It turns out to be a polynomial, so Katz's Theorem 4.11 applies and (4.12) gives the Serre polynomial.

The same Frobenius-type formula is valid in the SL_n-case:

$$|\check{\mathcal{M}}_B^d(\mathbb{F}_q)| = \sum_{\chi \in Irr(GL_n(\mathbb{F}_q))} \frac{|SL_n(\mathbb{F}_q)|^{2g-2}}{\chi(1)^{2g-2}} \cdot \frac{\chi(\zeta_n^d \cdot I)}{\chi(1)}, \tag{4.13}$$

Here the character table of $\mathrm{SL}_n(\mathbb{F}_q)$ is trickier. After much work of Lusztig, the character table of $\mathrm{Irr}(\mathrm{SL}_n(\mathbb{F}_q))$ has only been completed by Bonnafé [11] and Shoji [85] in 2006. However for $\chi \in \mathrm{Irr}(\mathrm{GL}_n(\mathbb{F}_q))$ the splitting

$$\chi|_{\mathrm{SL}_n(\mathbb{F}_q)} = \sum \chi_i$$

into irreducible characters χ_i of $\mathrm{SL}_n(\mathbb{F}_q)$ is evenly spread out on $\zeta_n^d \cdot I$; meaning that $\chi_i(\zeta_n^d \cdot I) = \chi_j(\zeta_n^d \cdot I)$. This way the evaluation of (4.13) is possible and was done by Mereb [69] in 2010. He too obtained a polynomial for $|\check{\mathcal{M}}_B^d(\mathbb{F}_q)|$, and by Katz's theorem Theorem 4.11 this gives a formula for $E(\check{\mathcal{M}}_B^d(\mathbb{F}_q); q)$. With similar techniques one can also evaluate the κ-components $E_\kappa(\check{\mathcal{M}}_B^d; u, v)$ in the LHS of Conjecture 4.5.

In order to check the refined topological mirror symmetry Conjecture 4.5 we need also to determine $E(\hat{\mathcal{M}}_{B,\gamma}^e, L_{\hat{B}^d,\gamma}; q)$. It is simple to do when n is a prime; leading to a proof of Conjecture 4.5 and so to Conjecture 4.4 in this case. For composite n's an ongoing work [45] evaluates these by similar (twisted) arithmetic techniques. This seems to match with Mereb's result, which is expected to give the proof of Conjecture 4.4.

4.3. The case $n = 2$ for TMS-B

Here we show how the topological mirror symmetry works for $\check{\mathcal{M}}_B$ when $n = 2$. It is instructive to compare it to the arguments in §3.3.8.

The variant part of the Serre polynomial of $\check{\mathcal{M}}_B$ is the difference of the full Serre polynomial and the invariant part. This difference in turn can be evaluated using the character tables of $\mathrm{SL}_2(\mathbb{F}_q)$ and $\mathrm{GL}_2(\mathbb{F}_q)$ respectively. We get that this difference

$$\begin{aligned}
\sum_{1\neq\kappa\in\hat{\Gamma}} E_\kappa(\check{\mathcal{M}}_B) &= E(\check{\mathcal{M}}_B) - E_1(\check{\mathcal{M}}_B) = E(\check{\mathcal{M}}_B) - E(\mathcal{M}_B)/(q-1)^{2g} \\
&= \sum_{\chi\in\mathrm{Irr}(\mathrm{SL}_2(\mathbb{F}_q))} \frac{|\mathrm{SL}_2(\mathbb{F}_q)|^{2g-2}}{\chi(1)^{2g-2}} \cdot \frac{\chi(-I)}{\chi(1)} - \sum_{\chi\in\mathrm{Irr}(\mathrm{GL}_2(\mathbb{F}_q))} \frac{|\mathrm{PGL}_2(\mathbb{F}_q)|^{2g-2}}{(q-1)\chi(1)^{2g-2}} \cdot \frac{\chi(-I)}{\chi(1)} \\
&= (2^{2g}-1)q^{2g-2}\left(\frac{(q-1)^{2g-2}-(q+1)^{2g-2}}{2}\right) \\
&= \sum_{i=1}^{g-1} (2^{2g}-1)\binom{2g-2}{2i-1} q^{2g-3+2i},
\end{aligned} \tag{4.14}$$

is exactly given by those 4 irreducible characters of $\mathrm{SL}_2(\mathbb{F}_q)$ which arise from irreducible characters of $\mathrm{GL}_2(\mathbb{F}_q)$ which split into two irreducibles over $\mathrm{SL}_2(\mathbb{F}_q)$.

The mapping class group of C acts by automorphisms on $\Gamma \cong H^1(C, \mathbb{Z}_2)$ so that the induced action on the set $\hat{\Gamma}^* = \hat{\Gamma} \setminus \{1\}$ with $(2^{2g}-1)$ elements is transitive. This way we can argue that $E_{\kappa_1}(\check{\mathcal{M}}_B) = E_{\kappa_2}(\check{\mathcal{M}}_B)$ for any two $\kappa_1, \kappa_2 \in \hat{\Gamma}^*$, thus we can conclude

$$E_\kappa(\check{\mathcal{M}}_B) = q^{2g-2}\left(\frac{(q-1)^{2g-2}-(q+1)^{2g-2}}{2}\right) \tag{4.15}$$

for $\kappa \in \hat{\Gamma}^*$.

Now $\check{\mathcal{M}}_B^\gamma$ can be identified with $(\mathbb{C}^\times)^{2g-2}$, which is the Betti version of $\check{\mathcal{M}}^\gamma$ from §3.3.8, which was isomorphic to the cotangent bundle of the identity component of the Prym variety. Now the Γ-equivariant local system $L_{B,\gamma}$ kills exactly the even cohomology and so we get

$$E(\check{\mathcal{M}}_B^\gamma/\Gamma, L_{\hat{B},\gamma}) = \frac{(q-1)^{2g-2}-(q+1)^{2g-2}}{2}. \tag{4.16}$$

This proves:

Theorem 4.17. *When $n = 2$ Conjecture 4.5 and so Conjecture 4.4 hold.*

As the character tables of $SL_2(\mathbb{F}_q)$ an d $GL_2(\mathbb{F}_q)$ were already known to Schur [82] and Jordan [57] in 1907; in principle the mirror symmetry pattern (4.14) above could have been checked more than 100 years ago! Also the agreement of (4.15) and (4.16) up to a q-power, that is Conjecture 4.5, when written in terms of character sums has a somewhat similar form as the fundamental lemma. Indeed in our last section §5.5.4 we will argue that Conjecture 4.5 and the fundamental lemma have a common geometrical root.

5. Solving our problems

Although the arithmetic technique discussed in the previous session [45] is capable of proving the Betti version of the topological mirror symmetry Conjecture 4.4, it still introduces a different set of conjectures. This raises the question: which one is the "right" one, that is the true consequence of mirror symmetry, the De Rham Conjecture 3.20 and the equivalent Dolbeault Conjecture 3.23 or the Betti version Conjecture 4.4 of the topological mirror symmetries?

In the next section we will see that in fact they have a common generalization which at the same time also solves the earlier "agreement of Hodge numbers" problem discussed in Remark 3.21.

5.1. Hard Lefschetz for weight and perverse Filtrations

We start with the observation [47, Corollary 3.5.3] for GL_n and PGL_n and [69] for SL_n that

$$E(\hat{\check{\mathcal{M}}}_B; 1/q) = q^{\dim} E(\hat{\check{\mathcal{M}}}_B; q) \tag{5.1}$$

i.e. that the Serre polynomials of our character varieties are palindromic. (Here dim is the dimension of the appropriate character variety.) It is interesting to note that ultimately this is due to Alvis-Curtis duality in the character theory of finite groups of Lie-type.

We recall the weight filtration:

$$W_0 \subset \cdots \subset W_i \subset \cdots \subset W_{2k} = H^k(\mathcal{M}_B)$$

on the ordinary cohomology $H^*(\mathcal{M}_B)$. The palindromicity (5.1) then lead us to the following Curious Hard Lefschetz Conjecture in [47, Conjecture 4.2.7]:

$$(5.2) \qquad \begin{array}{rcl} L^l : Gr^W_{\dim -2l} H^{i-l}(\mathcal{M}_B) & \xrightarrow{\cong} & Gr^W_{\dim +2l} H^{i+l}(\mathcal{M}_B) \\ x & \mapsto & x \cup \alpha^l \end{array},$$

where $\dim = \dim(\mathcal{M}_B)$ and $\alpha \in W_4H^2(\mathcal{M}_B)$. For $n = 2$ this was proved in [47, §5.3]. The conjecture (5.2) is curious because because $\mathcal{M}_B$ is an affine, thus non-projective, variety and α is a weight 4 and type (2,2) class, instead of the usual weight 2 and type (1,1) class of the Hard Lefschetz theorem. However there is a situation where a similar Hard Lefschetz theorem was observed.

Namely, [14] introduce the perverse filtration:

$$P_0 \subset \cdots \subset P_i \subset \dots P_k(X) \cong H^k(X)$$

for $f : X \to Y$ proper X and Y smooth, quasi-projective. Originally they define it using the BBDG-decomposition theorem of $f_*(\underline{\mathbb{Q}})$ into perverse sheaves. But in [15] they prove a more elementary equivalent definition in the case when Y is additionally affine. Take $Y_0 \subset \cdots \subset Y_i \subset \dots Y_d = Y$ s.t. Y_i sufficiently generic with $\dim(Y_i) = i$ then the perverse filtration is given by

$$P_{k-i-1}H^k(X) = \ker(H^k(X) \to H^k(f^{-1}(Y_i))).$$

Now the Relative Hard Lefschetz Theorem [14, Theorem 2.3.3] holds:

$$\begin{array}{rcl} L^l : Gr^P_{\mathrm{rdim}-l} H^*(X) & \xrightarrow{\cong} & Gr^P_{\mathrm{rdim}+l} H^{*+2l}(X) \\ x & \mapsto & x \cup \alpha^l \end{array}$$

where rdim is the relative dimension of f and $\alpha \in W_2H^2(X)$ is a relative ample class.

Recall from Theorem 2.7 that the Hitchin map

$$\begin{array}{rcl} \chi : \mathcal{M}_{Dol} & \to & \mathcal{A} \\ (E, \phi) & \mapsto & \mathrm{charpol}(\phi) \end{array}$$

is proper, thus induces perverse filtration on $H^*(\mathcal{M}_{Dol})$. A nice explanation for the curious Hard Lefschetz conjecture (5.2) would be if the following conjecture held.

Conjecture 5.3. *We have* $P = W$, *more precisely,* $P_k(\mathcal{M}_{Dol}) \cong W_{2k}(\mathcal{M}_B)$ *under the isomorphism* $H^*(\mathcal{M}_{Dol}) \cong H^*(\mathcal{M}_B)$ *from non-Abelian Hodge theory.*

In [13] it was proved that

Theorem 5.4. $P = W$ *when* $G = GL_2, PGL_2, SL_2$.

Remark 5.5. The proof of Theorem 5.4 was accomplished by a careful study of the topology of the Hitchin map, which paralleled special cases of results of Ngô in his proof [78] of the fundamental lemma. Additionally we had to use all the previously established results [43, 50, 48, 47] on the cohomology of these $n = 2$ varieties. Interestingly for the proof for SL_2 we had to use results which were discussed here for the topological mirror symmetry presented in §3.3.8 and §4.4.3 in this paper. We will now explain why this connection to mirror symmetry is not surprising.

5.2. Perverse topological mirror symmetry

We define the perverse Serre polynomial as

$$PE(\mathcal{M}_{Dol}; u, v, q) := \sum q^k E(Gr_k^P(H_c^*(\mathcal{M}_{Dol})); u, v),$$

and the $\hat{B}^d$-twisted stringy perverse Serre polynomial as

$$PE_{st}^{\hat{B}^d}(\hat{\mathcal{M}}_{Dol}^e; u, v, q) := \sum_{\gamma \in \Gamma} PE(\check{\mathcal{M}}_{Dol,\gamma}^e/\Gamma, L_{\hat{B}^d}; u, v, q)(uvq)^{F(\gamma)}.$$

By Definition 3.13 and Theorem 3.22 we have

$$PE(\mathcal{M}_{Dol}; u, v, 1) = E(\mathcal{M}_{Dol}; u, v) = E(\mathcal{M}_{DR}; u, v). \tag{5.6}$$

Conjecture 5.3 that $P = W$ then would imply

$$PE(\mathcal{M}_{Dol}; 1, 1, q) = E(\mathcal{M}_B; q) \tag{5.7}$$

and Relative Hard Lefschetz [14, Theorem 2.3.3] shows

$$PE(\mathcal{M}_{Dol}; u, v, q) = (uvq)^{\dim} PE\left(\mathcal{M}_{Dol}; u, v; \frac{1}{quv}\right). \tag{5.8}$$

Note that although the original Hodge numbers of $\mathcal{M}_{Dol}$ did not possess any non-trivial symmetry, this refined version with the perverse filtration does. So in fact with this definition we can write down our most general form of the topological mirror symmetry conjectures.

Conjecture 5.9. $PE\left(\check{\mathcal{M}}_{Dol}^d; x, y, q\right) = (xyq)^{\dim} PE_{st}^{\hat{B}^d}\left(\hat{\mathcal{M}}_{Dol}^e; x, y, \frac{1}{qxy}\right)$

Remark 5.10. This most general form of our topological mirror symmetry conjecture solves the two problems we encountered before.

First, we see that this version of the topological mirror symmetry conjecture implies Conjecture 5.9 via (5.6) and Conjecture 4.4 if we assume Conjecture 5.4 and thus (5.7). Thus Conjecture 5.9 is a common generalization of Conjectures 3.20, 3.23 and 4.4.

Second, relative hard Lefschetz endows the perverse Hodge numbers with the symmetry (5.8), and so one can formulate topological mirror symmetry as in Conjecture 5.9. On the level of Hodge numbers this conjecture takes the form:

$$h_p^{i,j}(\check{\mathcal{M}}_{Dol}^d) = h_{st,\hat{p}}^{i+(\hat{p}-p)/2, j+(\hat{p}-p)/2}(\hat{\mathcal{M}}_{Dol}^e, \hat{B}^d), \tag{5.11}$$

where $\hat{p} = \dim(\check{\mathcal{M}}^d_{\mathrm{Dol}}) - p$ is the opposite perversity, the Hodge numbers are defined as

$$h_p^{i,j}(\check{\mathcal{M}}^d_{\mathrm{Dol}}) := \dim(H^{i,j}(\mathrm{Gr}_p^W H^{i+j}(\check{\mathcal{M}}^d_{\mathrm{Dol}})))$$

and similarly for the stringy extension on the right hand side. This form (5.11) is now more reminiscent of the original topological mirror symmetry (1.1). One can also compare the functional equation forms (1.2) and Conjecture 5.9.

The last ingredient which is missing to completely justify our topological mirror symmetry conjectures is to show that indeed these are cohomological shadows of the S-duality in the work of Kapustin–Witten [58]. Such an argument will be sketched in the next section.

5.3. Topological mirror symmetry as cohomological shadow of S-duality

Recall that Kapustin and Witten [58] suggest that the Geometrical Langlands program is S-duality reduced to 2 dimensions. This simplifies to T-duality, as first suggested by [10]. In turn by the SYZ proposal we get to mirror symmetry between $\check{\mathcal{M}}_{\mathrm{DR}}$ and $\hat{\mathcal{M}}_{\mathrm{DR}}$.

Now Kontsevich's [62] homological mirror symmetry conjecture suggests that

$$\mathcal{D}^b(\mathrm{Coh}(\check{\mathcal{M}}_{\mathrm{DR}})) \sim \mathcal{D}^b(\mathrm{Fuk}(\hat{\mathcal{M}}_{\mathrm{DR}})) \tag{5.12}$$

the derived category of coherent sheaves on $\check{\mathcal{M}}_{\mathrm{DR}}$ is equivalent with a certain Fukaya category on $\hat{\mathcal{M}}_{\mathrm{DR}}$. This latter is not straightforward to define but recent work of Nadler-Zaslow [74, 72] relates a certain Fukaya category of T^*X and a category of D-modules on X, for a compact real analytical manifold X. Thus we may imagine that this result might extend to give an equivalence of the right hand side of (5.12) with some category of D-modules on the stack $\mathrm{Bun}_{\mathrm{PGL}_n}$ of PGL_n bundles on C. The mathematical content of [58] maybe phrased that the combination of this latter Nadler-Zaslow type equivalence with the homological mirror symmetry in (5.12) leads to the proposed Geometric Langlands program of [64, 8]. As explained in [28] in a certain semi-classical limit (5.12) should become

$$\mathcal{D}^b(\mathrm{Coh}(\check{\mathcal{M}}_{\mathrm{Dol}})) \sim \mathcal{D}^b(\mathrm{Coh}(\hat{\mathcal{M}}_{\mathrm{Dol}})),$$

an equivalence of the derived categories of sheaves on Hitchin systems for Langlands dual groups. By recent work of Arinkin [2] it is expected that there is a geometrical fibrewise Fourier-Mukai transform at least for integral spectral curves. It means that there should be a Poincaré bundle $\mathcal{P}$ on the fibered product $\check{\mathcal{M}}_{\mathrm{Dol}} \times_{\mathcal{A}^0} \hat{\mathcal{M}}_{\mathrm{Dol}}$ such that the associated fiberwise Fourier-Mukai transform would identify

$$\mathcal{FM} = \hat{\pi}_*(\mathcal{P} \otimes \check{\pi}^*) : \mathcal{D}^b(\mathrm{Coh}(\check{\mathcal{M}}_{\mathrm{Dol}})) \xrightarrow{\sim} \mathcal{D}^b(\mathrm{Coh}(\hat{\mathcal{M}}_{\mathrm{Dol}})). \tag{5.13}$$

One can argue that the cohomological shadow of such a Fourier-Mukai transform for orbifolds should be defined in stringy cohomology. Also if one twists the above Fourier-Mukai transform by adding gerbes, as discussed e.g. in [9], then we should see stringy cohomology twisted with gerbes. Also we should expect the cohomological shadow of (5.13) to be compatible with the perverse filtration. All in all, the cohomological shadow of (5.13) should identify

$$(5.14) \qquad S : H_p^{r,s}(\check{\mathcal{M}}^d_{\mathrm{Dol}}) \cong H^{r+(\hat{p}-p)/2, s+(\hat{p}-p)/2}_{st,\hat{p}}(\hat{\mathcal{M}}^e_{\mathrm{Dol}}; \hat{B}^d).$$

In fact, this statement over the regular locus $\mathcal{A}_{\mathrm{reg}}$ can be proved. Moreover comparing supports of (5.14) over $\mathcal{A}^0$ or also by a Fourier transform argument on Γ one can deduce the refined version of (5.14). Namely for $\kappa \in \hat{\Gamma}$ and $\gamma = w(\kappa) \in \Gamma$ we have:

$$(5.15) \quad S : H_p^{r,s}(\check{\mathcal{M}}^d_{\mathrm{Dol}})_\kappa \cong H^{r+(\hat{p}-p)/2-F(\gamma), s+(\hat{p}-p)/2-F(\gamma)}_{\hat{p}-F(\gamma)}(\check{\mathcal{M}}^e_{\mathrm{Dol},\gamma}/\Gamma; L_{\hat{B}^d}).$$

This way we can argue that the cohomological shadow of S-duality reduced to 2-dimensions and in the semi-classical limit should yield our Topological Mirror Symmetry Conjecture 5.9:

$$PE\left(\check{\mathcal{M}}^d_{\mathrm{Dol}}; x, y, q\right) = (xyq)^{\dim} PE^{\hat{B}^d}_{st}\left(\hat{\mathcal{M}}^e_{\mathrm{Dol}}; x, y, \frac{1}{qxy}\right).$$

Remark 5.16. More delicate structures of the Kapustin–Witten reduced S-duality [58] have been mathematically implemented in recent works of Yun [93], where Ngô's techniques from [78] were havily used.

Finally, in the last section, we explain a connection between our topological mirror symmetry conjectures and Ngô's work [78] on the fundamental lemma in the Langlands program.

5.4. From topological mirror symmetry to the fundamental lemma

Ngô's celebrated[3] proof [78] of the fundamental lemma is the culmination of a series of geometrical advances in the understanding of orbital integrals including [59, 36, 65, 66, 67]. The proof proceeds by studying the Hitchin fibration over the so-called elliptic locus. In the case of SL_n, which we will be only discussing here, this means the locus $\mathcal{A}_{\mathrm{ell}} \subset \mathcal{A}^0$ containg characteristics $a \in \mathcal{A}^0$ so that the corresponding spectral curve C_a is integral, i.e. irreducible and reduced. In particular, it contains the locus we studied in this survey $\mathcal{A}_{\mathrm{reg}}$ where C_a is smooth. He considers the degree 0 Hitchin fibration $\check{\chi}_{\mathrm{ell}} : \check{\mathcal{M}}^0_{\mathrm{ell}} = \check{\chi}^{-1}(\mathcal{A}_{\mathrm{ell}}) \to \mathcal{A}_{\mathrm{ell}}$ over the elliptic locus. An important ingredient in Ngô's proof is the BBDG decomposition theorem of the derived push forward $\check{\chi}_{\mathrm{ell}*}(\underline{\mathbb{Q}})$ of the constant sheaf on $\check{\mathcal{M}}_{\mathrm{ell}}$ into perverse sheaves. He proves the so-called *support theorem*, that in certain cases, including the Hitchin fibration, the perverse components of $\check{\chi}_{\mathrm{ell}*}(\underline{\mathbb{Q}})$ are determined by a small open subset of $\mathcal{A}_{\mathrm{ell}}$. The proof then is achieved by checking the geometrical formula (5.17) below over this small open subset of $\mathcal{A}_{\mathrm{ell}}$; yielding the statement over the whole $\mathcal{A}_{\mathrm{ell}}$.

An important further geometrical insight of the paper [78] is that $\check{\chi}_{\mathrm{ell}*}(\underline{\mathbb{Q}})$ should be understood with respect to a certain symmetry of the Hitchin fibration,

[3]A detailed survey of the statement and some of the proof of the fundamental lemma could be found in [73].

which we already studied in §2.3.3 over $\mathcal{A}_{\mathrm{reg}}$. Similarly to our definition of the Prym variety $\check{P}_a$ for a smooth spectral curve C_a in (3.6), we can define the norm map $\mathrm{Nm}_{C_a/C} : \mathrm{Pic}^0(C_a) \to \mathrm{Pic}^0(C)$ and Prym variety $\check{P}_a := \ker(\mathrm{Nm}_{C_a/C})$ for an integral spectral curve $\pi_a : C_a \to C$ as well. Again similarly to the smooth case one can construct an action of $\check{P}_a$ on $\check{\mathcal{M}}_a := \check{\chi}^{-1}(a)$ when $a \in \mathcal{A}_{\mathrm{ell}}$. This way we get an action of the group scheme $\check{P}_{\mathrm{ell}}$ on $\check{\mathcal{M}}_{\mathrm{ell}}$. This symmetry of the Hitchin fibration will induce an action of $\check{P}_{\mathrm{ell}}$ on $\check{\chi}_{\mathrm{ell}*}(\underline{\mathbb{Q}})$. As the group scheme $\check{P}^0_{\mathrm{ell}}$, the connected component of the identity in $\check{P}_{\mathrm{ell}}$, acts trivially on cohomology; this action of $\check{P}_{\mathrm{ell}}$ on $\check{\chi}_{\mathrm{ell}*}(\underline{\mathbb{Q}})$ will factor through an action of the group scheme of components $\underline{\Gamma} := \check{P}_{\mathrm{ell}}/\check{P}^0_{\mathrm{ell}}$. This turns out to be a finite group scheme with stalk at a agreeing with Γ_a the group of components of $\check{P}_a$.

The finite group scheme $\underline{\Gamma}$ also connects nicely with our finite group $\Gamma = \mathrm{Pic}^0(C)[n]$. Namely one can easily show that for $\gamma \in \Gamma$ the pull back $\pi_a^*(\gamma) \in \check{P}_a$. This way we get a map

$$f : \Gamma \to \Gamma_a,$$

which is shown to be surjective in [46], where the kernel is also explicitly described. If we now consider a character $\kappa \in \hat{\Gamma}_a$, then we get a character $\kappa f \in \hat{\Gamma}$ and a corresponding $\gamma = w(\kappa f) \in \Gamma$. Then the stalk of a sheaf version of the refined S-duality in (5.15) for the $d = 0$ case over $\mathcal{A}_{\mathrm{ell}}$ followed by relative Hard Lefschetz leads to the isomorphism

$$H_p^{r,s}(\check{\mathcal{M}}_a)_\kappa \cong H_{p-F(\gamma)}^{r-F(\gamma),s-F(\gamma)}(\check{\mathcal{M}}_{a,\gamma}/\Gamma) \tag{5.17}$$

This formula, which we derived here from the cohomological shadow of Kapustin–Witten's reduced S-duality, can be identified with the stalk of Ngô's main geometric stabilization formula [78, Theorem 6.4.2] in the case of SL_n. As Ngô argues in [78], when (5.17) is proved in positive characteristic, and one takes the alternating trace of the Frobenius automorphism on both sides of (5.17), then the resulting formula can be seen to imply the fundamental lemma in the Langlands program in the function field case and in turn by Waldspurger's work [90] in the number field case.

As explained in §5.5.3 the hope is that one can push (5.17) or more precisely a sheaf version underlying (5.15) from $\mathcal{A}_{\mathrm{ell}}$ over the whole of the SL_n-Hitchin base $\mathcal{A}^0$. The SL_n-case of the work of Chaudouard and Laumon [16, 17] managed, by extending Ngô's techniques, to do this over $\mathcal{A}_{\mathrm{red}}$ that is managed to prove (5.17) for reduced, but possibly reducible, spectral curves C_a. This way [16, 17] lead to a proof of the so-called weighted fundamental lemma, which again by earlier results of Waldspurger and others completed the proof of the full endoscopic functoriality principle of Langlands. For us however it remains to extend (5.15) over the whole of $\mathcal{A}^0$, including non-reduced and reducible spectral curves, which will yield our topological mirror symmetry Conjecture 5.9.

Details of the arguments in the last two sections §5.5.3 and §5.5.4 will appear elsewhere.

6. Conclusion

In this paper we surveyed some techniques to obtain cohomological information on the topology of the total space of the Hitchin system. We painted a picture where ideas from physics and number theory were combined into a dynamic mix. Although these techniques are fairly powerful, still they have not yet lead to complete understanding. In particular, the most general conjectures are still open.

More recently there have appeared work by physicists Diaconescu et al. [18] about a new string theory framework for several conjectures relating to the topology of the total space of the GL_n-Hitchin system. Besides the links to the conjectured formulae in [47, 44] a picture is emerging which relates the main conjecture of [47, 44] with a certain version of the Gopukamar-Vafa conjecture, which ultimately can be phrased as strong support of our pivotal $P = W$ Conjecture 5.3. This way [18] uncovers close connections of our conjectures with Gromov-Witten, Donaldson-Thomas and Pandharipande-Thomas invariants of certain local Calabi-Yau 3-folds. There have been considerable progress on the latter invariants lately in the mathematics literature thus we can well hope that with this new point of view we will be able to progress our understanding of the problems surveyed in this paper.

We finish by mentioning another promising new work [34] where they manage to extend the original Morse theory method of Hitchin for $n = 2$ and Gothen for $n = 3$ to higher n using a motivic view point - originating in the number theoretic approach of Harder–Narasimhan [42]. In particular, their calculations have been done for $n = 4$ which are in agreement with the conjectures in [47, 44] and this paper.

When we add these two very recent approaches, again one originating in physics and one in number theory, to the mix of ideas surveyed in this paper, we can be sure that new exciting results and ideas will be found on questions relating to the global topology of the Hitchin system in the foreseeable future.

References

[1] A. Aravind, D. Brent, and K. Frances. Yang-Mills theory and Tamagawa numbers: the fascination of unexpected links in mathematics. *Bull. Lond. Math. Soc.*, **40**(4), 533–567, 2008. arXiv:0801.4733 ← 32

[2] D. Arinkin. Autoduality of compactified Jacobians for curves with plane singularities, (preprint arXiv:1001.3868v2) ← 60

[3] V. I. Arnold. Mathematical methods of classical mechanics, Second edition. *Graduate Texts in Mathematics*, **60**, Springer-Verlag, New York, 1989. ← 36

[4] M. F. Atiyah and R. Bott. The Yang-Mills equations over Riemann surfaces. *Philos. Trans. Roy. Soc. London Ser. A*, **308**, 1983. ← 30, 38

[5] V. Batyrev. Stringy Hodge numbers of varieties with Gorenstein canonical singularities. *Integrable systems and algebraic geometry (Kobe/Kyoto, 1997)*, 1–32, *World Sci. Publ.*, River Edge, NJ, 1998. arXiv:alg-geom/9711008 ← 50

[6] V. V. Batyrev and D. Dais. Strong McKay correspondence, string-theoretic Hodge numbers and mirror symmetry. *Topology* , **35**, 901–929, 1996. arXiv: alg-geom/9410001 ← 47, 48

[7] A. Beauville, M. S. Narasimhan, and S. Ramanan. Spectral curves and the generalised theta divisor *Jour. für die Reine und Ang. Math.*, **398**, 169–179, 1989. ← 42, 43

[8] A. Beilinson and V. Drinfeld. Quantization of Hitchin's integrable system and Hecke eigensheaves. (ca. 1995). http://www.math.uchicago.edu/~mitya/langlands.html. ← 35, 60

[9] O. Ben-Bassat. Twisting derived equivalences. *Trans. Amer. Math. Soc.*, **361**(10), 5469–5504, 2009. arXiv:math/0606631 ← 60

[10] M. Bershadsky, A. Johansen, V. Sadov, and C. Vafa. Topological reduction of 4D SYM to 2D σ-models. *Nucl. Phys. B*, **448**, 166–186, 1995. arXiv: hep-th/9501096 ← 41, 60

[11] C. Bonnafé. Sur les caractères des groupes réductifs finis à centre non connexe: applications aux groupes spéciaux linéaires et unitaires. *Astérisque*, (306):vi+165, 2006. arXiv:math/0504078 ← 56

[12] P. Candelas, X. C. de la Ossa, P. S. Green, and L. Parkes. A pair of Calabi-Yau manifolds as an exactly soluble superconformal theory. *Nuclear Phys. B*, **359**(1), 21–74, 1991. ← 33

[13] M. A. de Cataldo, T. Hausel, and L. Migliorini. Topology of Hitchin systems and Hodge theory of character varieties: the case A_1, (preprint arXiv:1004.1420) ← 58

[14] M. A. de Cataldo and L. Migliorini. The Hodge Theory of Algebraic maps. *Ann. Scient. Éc. Norm. Sup.*, 4e série, t. **38**, 693–750, 2005. arXiv:math/0306030 ← 58, 59

[15] M. A. de Cataldo and L. Migliorini. The perverse filtration and the Lefschetz hyperplane theorem. *Ann. of Math.* (2), **171**(3), 2089–2113, 2010. arXiv: 0805.4634 ← 58

[16] P.-H. Chaudouard and G. Laumon. Le lemme fondamental pondéré. I. Constructions géométriques. *Compositio Math.*, **146** (6), 1416–1506, 2010. arXiv:0902.2684 ← 62

[17] P.-H. Chaudouard and G. Laumon. Le lemme fondamental pondéré. II. Énoncés cohomologiques. (preprint arXiv:0912.4512) ← 62

[18] W. Chuang, D. E. Diaconescu, and G. Pan. Wallcrossing and Cohomology of The Moduli Space of Hitchin Pairs. (preprint arXiv:1004.4195v4) ← 63

[19] K. Corlette. Flat G-bundles with canonical metrics. *J. Differential Geom.*, **28**(3), 361–382, 1988. ← 53

[20] D. Cox and S. Katz. Mirror symmetry and algebraic geometry. *Mathematical Surveys and Monographs*, **68**, American Mathematical Society, Providence, RI, 1999. ← 33

[21] G. Daskalopoulos, J. Weitsman, R. Wentworth, and G. Wilkin. Morse Theory and Hyperkähler Kirwan Surjectivity for Higgs Bundles. (To appear in *J. Diff. Geom.*) arXiv:math/0701560 ← 51

[22] P. Deligne. Théorie de Hodge II. *Inst. Hautes Études Sci. Publ. Math.* **40**, 5–57, 1971. ← 47

[23] P. Deligne. La conjecture de Weil. I.*Inst. Hautes Études Sci. Publ. Math.*, **43**, 273–307, 1974. ← 30, 35, 38

[24] V. Desale and S. Ramanan. Poincaré polynomials of the variety of stable bundles. *Math. Ann.*, **216**(3), 233–244, 1975. ← 30

[25] R. Donagi and D. Gaitsgory. The gerbe of Higgs bundles. *Transform. Groups*, 7(2), 109–153, 2002. arXiv:math.AG/0005132. ← 46

[26] R. Donagi and E. Markman. Spectral covers, algebraically completely integrable, Hamiltonian systems, and moduli of bundles, in: *Integrable Systems and Quantum Groups*, 1-119, Springer, 1996. arXiv:alg-geom/9507017 ← 36

[27] R. Donagi and T. Pantev. Langlands duality for Hitchin systems, (preprint arXIv:math/0604617) ← 47

[28] R. Donagi and T. Pantev. Geometric Langlands and non-abelian Hodge theory, in: Geometry, analysis, and algebraic geometry: forty years of the Journal of Differential Geometry, 85–116. *Surv. Differ. Geom.*, 13, Int. Press, Somerville, MA, 2009. ← 60

[29] S. K. Donaldson. Twisted harmonic maps and the self-duality equations, *Proc. London Math. Soc.* (3), **55**(1), 127–131, 1987. ← 53

[30] R. Earl and F. Kirwan. Complete sets of relations in the cohomology rings of moduli spaces of holomorphic bundles and parabolic bundles over a Riemann surface, *Proc. London Math. Soc.* (3), **89**(3), 570–622, 2004. arXiv:math/0305345 ← 38

[31] G. Faltings. Stable G-bundles and projective connections. *J. Algebraic Geom.*, **2**, 1993. ← 40

[32] E. Frenkel. Lectures on the Langlands program and conformal field theory. *Frontiers in number theory, physics, and geometry. II*, 387–533, Springer, Berlin, 2007. arXiv:hep-th/0512172 ← 34

[33] E. Frenkel and E. Witten. Geometric endoscopy and mirror symmetry. *Commun. Number Theory Phys.*, **2** (1), 113–283, 2008. arXiv:0710.5939 ← 32

[34] O. Garcia-Prada, J. Heinloth, and A. Schmitt. On the motives of moduli of chains and Higgs bundles, arXiv:1104.5558 ← 63

[35] S. Gelbart. An elementary introduction to the Langlands program. *Bull. Amer. Math. Soc.*, **10**, 177–219, 1984. ← 34

[36] M. Goresky, R. Kottwitz, and R. Macpherson. Homology of affine Springer fibers in the unramified case. *Duke Math. J.*, **121**(3), 509–561, 2004. arXiv: math/0305144 ← 61

[37] P. B. Gothen. The Betti numbers of the moduli space of stable rank 3 Higgs bundles on a Riemann surface. *Internat. J. Math.* , **5**, 861–875, 1994. ← 31, 32, 51, 52

[38] J. A. Green. The characters of the finite general linear groups. *Trans. Amer. Math. Soc.*, **80**, 402–447, 1955. ← 55

[39] M. Gross. The Strominger-Yau-Zaslow conjecture: from torus fibrations to degenerations. *Algebraic geometry—Seattle 2005*. Part 1, 149–192, Proc. Sympos. Pure Math., 80, Part 1, Amer. Math. Soc., Providence, RI, 2009. arXiv:0802.3407 ← 34

[40] S. Gukov and E. Witten. Gauge theory, ramification, and the geometric Langlands program. *Current developments in mathematics*,35–180. Int. Press, Somerville, MA, 2008. arXiv:hep-th/0612073 ← 32

[41] G. Harder. Eine Bemerkung zu einer Arbeit von P. E. Newstead. *J. Reine Angew. Math.*, **242**, 16–25, 1970. ← 30

[42] G. Harder and M. S. Narasimhan. On the cohomology groups of moduli spaces of vector bundles on curves. *Math. Ann.*, **212**, (1974/75). ← 30, 38, 50, 54, 63

[43] T. Hausel. Vanishing of intersection numbers on the moduli space of Higgs bundles. *Adv.Theor.Math.Phys.*, **2**, 1011–1040. arXiv:math/9805071 ← 59

[44] T. Hausel. Mirror symmetry and Langlands duality in the non-Abelian Hodge theory of a curve, in : Geometric Methods in Algebra and Number Theory. *Progress in Mathematics*, Vol. **235**. Fedor Bogomolov, Yuri Tschinkel (Eds.), Birkhäuser 2005 arXiv:math.AG/0406380 ← 32, 47, 53, 63

[45] T. Hausel, M. Mereb, and F. Rodriguez-Villegas. Topological mirror symmetry in the character table of $SL_n(\mathbb{F}_q)$, (*in preparation*) ← 56, 57

[46] T. Hausel and C. Pauly. Prym varieties of spectral covers, arXiv:1012.4748 ← 62

[47] T. Hausel and F. Rodriguez-Villegas. Mixed Hodge polynomials of character varieties. *Inv. Math.*, **174**(3), 555–624, 2008. arXiv:math.AG/0612668 ← 31, 41, 47, 53, 55, 57, 58, 59, 63

[48] T. Hausel and M. Thaddeus. Relations in the cohomology ring of the moduli space of rank 2 Higgs bundles. *Journal of the American Mathematical Society*, **16**, 303–329, 2003. arXiv:math.AG/0003094 ← 41, 59

[49] T. Hausel and M. Thaddeus. Mirror symmetry, Langlands duality, and the Hitchin system. *Invent. Math.*, **153**(1), 197–229, 2003. arXiv:math/0205236 ← 32, 33, 45, 46, 47, 48, 49, 52

[50] T. Hausel and M. Thaddeus. Generators for the cohomology ring of the moduli space of rank 2 Higgs bundles . *Proc. London Math. Soc.*, **88**, 632–658, 2004. arXiv:math.AG/0003093 ← 41, 59

[51] N. Hitchin. The self-duality equations on a Riemann surface. *Proc. London Math. Soc. (3)*, **55**(1), 59–126, 1987. ← 31, 32, 36, 38, 40, 41, 51

[52] N. Hitchin. Stable bundles and integrable systems. *Duke Math. J.*, **54**, 1987. ← 32, 40, 42

[53] N. Hitchin. Lectures on special Lagrangian submanifolds, *Winter School on Mirror Symmetry, Vector Bundles and Lagrangian Submanifolds (Cambridge, MA, 1999)*, 151–182, Amer. Math. Soc., 2001. arXiv:math/9907034 ← 47

[54] N. Hitchin. Langlands duality and G2 spectral curves. *Q. J. Math.*, **58**(3), 319-344, 2007. ← 47

[55] K. Hori, S. Katz, A. Klemm, R. Pandharipande, R. Thomas, C. Vafa, R. Vakil, and E. Zaslow. Mirror symmetry, with a preface by Vafa. Clay. *Mathematics Monographs*, **1**. American Mathematical Society, Providence, RI; Clay Mathematics Institute, Cambridge, MA, 2003. ← 33

[56] T. J. Jarvis, R. Kaufmann, and T. Kimura. Stringy K-theory and the Chern character. *Invent. Math.*, **168**(1), 23–81, 2007. arXiv:math/0502280 ← 48

[57] H. Jordan. Group characters of various types of linear groups. *Amer. J. Math*, **29**, 387, 1907. ← 57

[58] K. Kapustin and E. Witten. Electric-magnetic duality and the geometric Langlands program. *Commun. Number Theory Phys.*, **1**(1), 1–236, 2007. arXiv:hep-th/0604151 ← 32, 41, 60, 61

[59] D. Kazhdan and G. Lusztig. Fixed point varieties on affine flag manifolds. Israel J. Math., **62**(2), 129–168, 1988. ← 61

[60] Y.-H. Kiem and S.-B. Yoo. The stringy E-function of the moduli space of Higgs bundles with trivial determinant. *Math. Nachr.*, **281**(6), 817–838, 2008. arXiv:math/0507007 ← 51

[61] F. C. Kirwan. Cohomology of quotients in symplectic and algebraic geometry. *Mathematical Notes*, **31**, Princeton University Press, 1984. ← 31

[62] M. Kontsevich. Homological algebra of mirror symmetry. *Proceedings of the International Congress of Mathematicians*, Vol. 1, 2 (Zürich, 1994). 120–139, (Basel), Birkhäuser, 1995. arXiv:alg-geom/9411018 ← 34, 60

[63] M. Kontsevich. Motivic Integration. *Lecture at Orsay*, 1995. ← 48

[64] G. Laumon. Correspondance de Langlands géométrique pour les corps de fonctions. *Duke Math. J.*, **54**, 309–359, 1987. ← 35, 60

[65] G. Laumon. Fibres de Springer et jacobiennes compactifiées, in: Algebraic geometry and number theory. *Progr. Math.*, **253**, 515–563. Birkhäuser Boston, Boston, MA, 2006. arXiv:math/0204109 ← 61

[66] G. Laumon. Sur le lemme fondamental pour les groupes unitaires, preprint arXiv:math/0212245. ← 61

[67] G. Laumon and B. C. Ngô. Le lemme fondamental pour les groupes unitaires. *Ann. of Math.* (2), **168**(2), 477–573, 2008. arXiv:math/0404454 ← 61

[68] E. Markman. Generators of the cohomology ring of moduli spaces of sheaves on symplectic surfaces. *J. Reine Angew. Math.*, **544**, 61–82, 2002. `arXiv:math/0009109` ← 41

[69] M. Mereb. On the E-polynomials of a family of Character Varieties, PhD thesis University of Texas at Austin, 2010. `arXiv:1006.1286v1` ← 56, 57

[70] Mirror symmetry I.-V. *AMS/IP Studies in Advanced Mathematics*, **1,9,23,33,38**, American Mathematical Society, Providence, RI; International Press, Somerville, MA. ← 33

[71] D. Mumford. Projective invariants of projective structures and applications. *Proc. Intern. Cong. Math., Stockholm*, 526–530, 1962. ← 30, 37

[72] D. Nadler. Microlocal branes are constructible sheaves. *Selecta Math.*, **15**(4), 563–619, 2009. ← 60

[73] D. Nadler. The Geometric Nature of the Fundamental Lemma, (preprint `arXiv:1009.1862v3`) ← 61

[74] D. Nadler and E. Zaslow. Constructible sheaves and the Fukaya category. *J. Amer. Math. Soc.*, **22**(1), 233–286, 2009. `arXiv:math/0604379` ← 60

[75] M. S. Narasimhan and C. S. Seshadri. Stable and unitary vector bundles on a compact Riemann surface. *Ann. of Math.*, **82**, 540–567, 1965. ← 30

[76] P. Newstead. Topological properties of some spaces of stable bundles. *Topology*, **6**, 241–262, 1967. ← 30

[77] P. Newstead. *Introduction to moduli problems and orbit spaces*, Tata Inst. Bombay, 1978. ← 37

[78] B. C. Ngô. Le lemme fondamental pour les algèbres de Lie. *Publ. Math. Inst. Hautes Études Sci.*, **111**, 1–169, 2010. `arXiv:0801.0446` ← 32, 35, 59, 61, 62

[79] N. Nitsure. Moduli space of semistable pairs on a curve. *Proc. London Math. Soc. (3)*, **62**(2), 275–300, 1991. ← 40

[80] C. Peters and J. Steenbrink. *Mixed Hodge Structures*. Ergebnisse der Mathematik, Springer, 2008. ← 47

[81] A. Polishchuk and E. Zaslow. Categorical mirror symmetry: the elliptic curve. *Adv. Theor. Math. Phys.*, **2**(2), 443–470, 1998. `arXiv:math/9801119` ← 34

[82] I. Schur. Untersuchungen über die Darstellung der endlichen Gruppen durch Gebrochene Lineare Substitutionen. *J. Reine Angew. Math.*, **132**, 85–137, 1907. ← 57

[83] P. Seidel. Homological mirror symmetry for the quartic surface, preprint `arXiv:math/0310414` ← 34

[84] C. S. Seshadri. Quotient space by an abelian variety. *Math. Ann.*, **152**, 185–194, 1963. ← 37

[85] T. Shoji. Lusztig's conjecture for finite special linear groups. *Represent. Theory*, **10**, 164–222 (electronic), 2006. `arXiv:math/0502180` ← 56

[86] C. T. Simpson. Higgs bundles and local systems. *Publ. Math. I.H.E.S.*, **75**, 5-95, 1992. ← 53

[87] A. Strominger, S.-T. Yau, and E. Zaslow. Mirror symmetry is T-duality. *Nuclear Phys. B*, **479**, 243–259, 1996. arXiv:hep-th/9606040 ← 34, 41

[88] M. Thaddeus. Floer cohomology with gerbes, in: Enumerative invariants in algebraic geometry and string theory, 105–141. *Lecture Notes in Math.*, 1947, Springer, Berlin, 2008. ← 38, 46

[89] M. Verbitsky. Holomorphic symplectic geometry and orbifold singularities. *Asian J. Math.*, **4**(3), 553–563, 2000. arXiv:math/9903175 ← 48

[90] J.-L. Waldspurger. Endoscopie et changement de caractéristique. *J. Inst. Math. Jussieu*, **5**(3), 423–525, 2006. ← 35, 62

[91] A. Weil. On the Riemann hypothesis in function fields. *Proc. Nat. Acad.Sci. U. S. A.*, **27**, 345–347, 1941. ← 35

[92] E. Witten. Mirror symmetry, Hitchin's equations, and Langlands duality. *The many facets of geometry*, 113–128. Oxford Univ. Press, Oxford, 2010. arXiv:0802.0999 ← 32

[93] Z. Yun. Towards a Global Springer Theory I,II,III, (preprints: arXiv:0810.2146, arXiv:0904.3371, arXiv:0904.3372) ← 61

24-29 St Giles', Mathematical Institute, Oxford, OX1 3LB, United Kingdom

Current address: Section de Mathématiques, École Polytechnique Fédéral de Lausanne, Section 8, CH-1015 Lausanne, Switzerland

E-mail address: tamas.hausel@epfl.ch

Differential forms on singular spaces, the minimal model program, and hyperbolicity of moduli stacks

Stefan Kebekus

In memory of Eckart Viehweg

Abstract. The Shafarevich Hyperbolicity Conjecture, proven by Arakelov and Parshin, considers a smooth, projective family of algebraic curves over a smooth quasi-projective base curve Y. It asserts that if Y is of special type, then the family is necessarily isotrivial.

This survey discusses hyperbolicity properties of moduli stacks and generalisations of the Shafarevich Hyperbolicity Conjecture to higher dimensions. It concentrates on methods and results that relate moduli theory with recent progress in higher dimensional birational geometry.

Contents

2000 *Mathematics Subject Classification.* Primary 14D22; Secondary 14D05.

Key words and phrases. hyperbolicity properties of moduli spaces, minimal model program, differential forms on singular spaces.

1. Introduction

1.1. The Shafarevich hyperbolicity conjecture

1.1.1. Statement In his contribution to the 1962 International Congress of Mathematicians, Igor Shafarevich formulated an influential conjecture, considering smooth, projective families $f^\circ : X^\circ \to Y^\circ$ of curves of genus $g > 1$, over a fixed smooth quasi-projective base curve Y°. One part of the conjecture, known as the "hyperbolicity conjecture", gives a sufficient criterion to guarantee that any such family is isotrivial. The conjecture was shown in two seminal works by Parshin and Arakelov, including the following special case.

Theorem 1.1 (Shafarevich Hyperbolicity Conjecture, [55], [52, 1]). *Let $f^\circ : X^\circ \to Y^\circ$ be a smooth, complex, projective family of curves of genus $g > 1$, over a smooth quasi-projective base curve Y°. If Y° is isomorphic to one of the following varieties,*

- *the projective line $\mathbb{P}^1$,*
- *the affine line $\mathbb{A}^1$,*
- *the affine line minus one point $\mathbb{C}^*$, or*
- *an elliptic curve,*

then any two fibres of f° are necessarily isomorphic.

Notation-Assumption 1.2. *Throughout this paper, a* family *is a flat morphism of algebraic varieties with connected fibres. We always work over the complex number field.*

Remark 1.3. Following standard convention, we refer to Theorem 1.1 as "Shafarevich hyperbolicity conjecture" rather than "Arakelov-Parshin theorem". The reader interested in a complete picture is referred to [60, p. 253ff], where all parts of the Shafarevich conjecture are discussed in more detail.

Formulated in modern terms, Theorem 1.1 asserts that any morphism from a smooth, quasi-projective curve Y° to the moduli stack of algebraic curves is necessarily constant if Y° is one of the special curves mentioned in the theorem. If Y° is a quasi-projective variety of arbitrary dimension, then any morphism from Y° to the moduli stack contracts all rational and elliptic curves, as well as all affine lines and $\mathbb{C}^*$s that are contained in Y°.

We refer to [16, Sect. 16.E.1] for a discussion of the relation between the Shafarevich hyperbolicity conjecture and the notions of Brody– and Kobayashi hyperbolicity.

1.1.2. Aim and scope This survey is concerned with generalisations of the Shafarevich hyperbolicity conjecture to higher dimensions, concentrating on methods and results that relate moduli– and minimal model theory. We hope that the methods presented here will be applicable to a much wider ranges of problems, in moduli theory and elsewhere. The list of problems that we would like to address include the following.

Questions 1.4. Apart from the quasi-projective curves mentioned above, what other varieties admit only constant maps to the moduli stack of curves? What about moduli stacks of higher dimensional varieties? Given a variety Y°, is there a good geometric description of the subvarieties that will always be contracted by any morphism to any reasonable moduli stack?

Much progress has been achieved in the last years and several of the questions can be answered today. It turns out that there is a close connection between the minimal model program of a given quasi-projective variety Y°, and its possible morphisms to moduli stacks. Some of the answers obtained are in fact consequences of this connection.

In the limited number of pages available, we say almost nothing about the history of higher dimensional moduli, or about the large body of important works that approach the problem from other points of view. Hardly any serious attempt is made to give a comprehensive list of references, and the author apologises to all those whose works are not adequately represented here, be it because of the author's ignorance or simply because of lack of space.

The reader who is interested in a broader overview of higher dimensional moduli, its history, complete references, and perhaps also in rigidity questions for morphisms to moduli stacks is strongly encouraged to consult the excellent surveys found in this handbook and elsewhere, including [16, 40, 60]. A gentle and very readable introduction to moduli of higher dimensional varieties is also found in [41], while Viehweg's book [59] serves as a standard technical reference for the construction of moduli spaces.

Most relevant notions and facts from minimal model theory can either be found in the introductory text [46], or in the extremely clear and well-written reference book [35]. Recent progress in minimal model theory is surveyed in [16].

1.2. Outline of this paper

Section 2 introduces a number of conjectural generalisations of the Shafarevich hyperbolicity conjecture and gives an overview of the results that have been obtained in this direction. In particular, we mention results relating the moduli map and the minimal model program of the base of a family.

Sections 3 and 4 introduce the reader to methods that have been developed to attack the conjectures mentioned in Section 2. While Section 3 concentrates on positivity results on moduli spaces and on Viehweg and Zuo's construction of (pluri-)differential forms on base manifolds of families, Section 4 summarises results concerning differential forms on singular spaces. Both sections contain sketches of proofs which aim to give an idea of the methods that go into the proofs, and which might serve as a guideline to the original literature. The introduction to Section 4 motivates the results on differential forms by explaining a first strategy of proof for a special case of a (conjectural) generalisation of the Shafarevich hyperbolicity conjecture. Following this plan of attack, a more general case is treated in the concluding Section 5, illustrating the use of the methods introduced before.

1.3. Acknowledgements

This paper is dedicated to the memory of Eckart Viehweg. Like so many other mathematicians of his age group, the author benefited immensely from Eckart's presence in the field, his enthusiasm, guidance and support. Eckart will be remembered as an outstanding mathematician, and as a fine human being.

The work on this paper was partially supported by the DFG Forschergruppe 790 "Classification of algebraic surfaces and compact complex manifolds". Patrick Graf kindly read earlier versions of this paper and helped to remove several errors and misprints. Many of the results presented here have been obtained in joint work of Sándor Kovács and the author. The author would like to thank Sándor for innumerable discussions, and for a long lasting collaboration. He would also like to thank the anonymous referee for careful reading and for numerous suggestions that helped to improve the quality of this survey.

Not all the material presented here is new, and some parts of this survey have appeared in similar form elsewhere. The author would like to thank his coauthors for allowing him to use material from their joint research papers. The first subsection of every chapter lists the sources that the author was aware of.

2. Generalisations of the Shafarevich hyperbolicity conjecture

2.1. Families of higher dimensional varieties

Given its importance in algebraic and arithmetic geometry, much work has been invested to generalise the Shafarevich hyperbolicity conjecture, Theorem 1.1. Historically, the first generalisations have been concerned with families $f^\circ : X^\circ \to Y^\circ$ where Y° is still a quasi-projective curve, but where the fibres of f° are allowed to have higher dimension. The following elementary example shows, however, that Theorem 1.1 cannot be generalised naïvely, and that additional conditions must be posed.

Example 2.1 (Counterexample to the Shafarevich hyperbolicity conjecture for higher dimensional fibers). Consider a smooth projective surface Y of general type which contains a rational or elliptic curve $C \subset Y$. Assume that the automorphism group of Y fixes the curve C pointwise. Examples can be obtained by choosing any surface of general type and then blowing up sufficiently many points in sufficiently general position —each blow-up will create a rational curve and lower the number of automorphisms. Thus, if c_1 and $c_2 \in C$ are any two distinct points, then the surfaces Y_{c_i} obtained by blowing up the points c_i are non-isomorphic.

In order to construct a proper family, consider the product $Y \times C$ with its projection $\pi : Y \times C \to C$ and with the natural section $\Delta \subset Y \times C$. If X is the blow-up of $Y \times C$ in Δ, then we obtain a smooth, projective family $f : X \to C$ of surfaces of general type, with the property that no two fibres are isomorphic.

It can well be argued that Counterexample 2.1 is not very natural, and that the fibres of the family f would trivially be isomorphic if they had not been blown up artificially. This might suggest to consider only families that are "not the blow-up of something else". One way to make this condition is precise is to consider only *families of minimal surfaces*, i.e., surfaces F whose canonical bundle K_F is semiample. In higher dimensions, it is often advantageous to impose a stronger condition and consider only *families of canonically polarised manifolds*, i.e., manifolds F whose canonical bundle K_F is ample.

Hyperbolicity properties of families of minimal surfaces and families of minimal varieties have been studied by a large number of people, including Migliorini [47], Kovács [37, 38] and Oguiso-Viehweg [50]. For families of canonically polarised manifolds, the analogue of Theorem 1.1 has been shown by Kovács in the algebraic setup [39]. Combining algebraic arguments with deep analytic methods, Viehweg and Zuo prove a more general Brody hyperbolicity theorem for moduli spaces of canonically polarised manifolds which also implies an analogue of Theorem 1.1, [62].

Theorem 2.2 (Hyperbolicity for families of canononically polarized varieties, [39, 62]). *Let $f^\circ : X^\circ \to Y^\circ$ be a smooth, complex, projective family of canonically polarised varieties of arbitrary dimension, over a smooth quasi-projective base curve Y°. Then the conclusion of the Shafarevich hyperbolicity conjecture, Theorem 1.1, holds.* □

2.2. Families over higher dimensional base manifolds

This paper discusses generalisations of the Shafarevich hyperbolicity conjecture to families over higher dimensional base manifolds. To formulate any generalisation, two points need to be clarified.

(1) We need to define a higher dimensional analogue for the list of quasi-projective curves given in Theorem 1.1.
(2) Given any family $f^\circ : X^\circ \to Y^\circ$ over a higher dimensional base, call two points $y_1, y_2 \in Y^\circ$ equivalent if the fibres $(f^\circ)^{-1}(y_1)$ and $(f^\circ)^{-1}(y_2)$ are isomorphic. If Y° is a curve, then either there is only one equivalence class, or all equivalence classes are finite. For families over higher dimensional base manifolds, the equivalence classes will generally be subvarieties of

arbitrary dimension. We will need to have a quantitative measure for the number of equivalence classes and their dimensions.

The problems outlined above justify the definition of the *logarithmic Kodaira dimension* and of the *variation of a family*, respectively. Before coming to the generalisations of the Shafarevich hyperbolicity conjecture in Section 2.2.3 below, we recall the definitions for the reader's convenience.

2.2.1. The logarithmic Kodaira dimension The logarithmic Kodaira dimension generalises the classical notion of Kodaira dimension to the category of quasi-projective varieties.

Definition 2.3 (Logarithmic Kodaira dimension). *Let Y° be a smooth quasi-projective variety and Y a smooth projective compactification of Y° such that $D := Y \setminus Y^\circ$ is a divisor with simple normal crossings. The logarithmic Kodaira dimension of Y°, denoted by $\kappa(Y^\circ)$, is defined to be the Kodaira-Iitaka dimension of the line bundle $\mathcal{O}_Y(K_Y + D) \in \operatorname{Pic}(Y)$. A quasi-projective variety Y° is called of* log general type *if $\kappa(Y^\circ) = \dim Y^\circ$, i.e., the divisor $K_Y + D$ is big.*

It is a standard fact of logarithmic geometry that a compactification Y with the described properties exists, and that the logarithmic Kodaira dimension $\kappa(Y^\circ)$ does not depend on the choice of the compactification.

Observation 2.4. The quasi-projective curves listed in Theorem 1.1 are precisely those curves Y° with logarithmic Kodaira dimension $\kappa(Y^\circ) \leqslant 0$.

2.2.2. The variation of a family The following definition provides a quantitative measure of the *birational* variation of a family. Note that the definition is meaningful even in cases where no moduli space exists.

Definition 2.5 (Variation of a family, cf. [58, Introduction]). *Let $f^\circ : X^\circ \to Y^\circ$ be a projective family over an irreducible base Y°, and let $\overline{\mathbb{C}(Y^\circ)}$ denote the algebraic closure of the function field of Y°. The variation of f°, denoted by $\operatorname{Var}(f^\circ)$, is defined as the smallest integer ν for which there exists a subfield K of $\overline{\mathbb{C}(Y^\circ)}$, finitely generated of transcendence degree ν over $\mathbb{C}$ and a K-variety F such that $X \times_{Y^\circ} \operatorname{Spec} \overline{\mathbb{C}(Y^\circ)}$ is birationally equivalent to $F \times_{\operatorname{Spec} K} \operatorname{Spec} \overline{\mathbb{C}(Y^\circ)}$.*

Remark 2.6. In the setup of Definition 2.5, assume that all fibres if Y° are canonically polarised complex manifolds. Then coarse moduli schemes are known to exist, [59, Thm. 1.11], and the variation is the same as either the dimension of the image of Y° in moduli, or the rank of the Kodaira-Spencer map at the general point of Y°. Further, one obtains that $\operatorname{Var}(f^\circ) = 0$ if and only if all fibres of f° are isomorphic. In this case, the family f° is called "isotrivial".

2.2.3. Viehweg's conjecture Using the notion of "logarithmic Kodaira dimension" and "variation", the Shafarevich hyperbolicity conjecture can be reformulated as follows.

Theorem 2.7 (Reformulation of Theorem 1.1). *If* $f^\circ : X^\circ \to Y^\circ$ *is any smooth, complex, projective family of curves of genus* $g > 1$, *over a smooth quasi-projective base curve* Y°, *and if* $\mathrm{Var}(f^\circ) = \dim Y^\circ$, *then* $\kappa(Y^\circ) = \dim Y^\circ$.

Aiming to generalise the Shafarevich hyperbolicity conjecture to families over higher dimensional base manifolds, Viehweg has conjectured that this reformulation holds true in arbitrary dimension.

Conjecture 2.8 (Viehweg's conjecture, [60, 6.3]). *Let* $f^\circ : X^\circ \to Y^\circ$ *be a smooth projective family of varieties with semiample canonical bundle, over a quasi-projective manifold* Y°. *If* f° *has maximal variation, then* Y° *is of log general type. In other words,*

$$\mathrm{Var}(f^\circ) = \dim Y^\circ \;\Rightarrow\; \kappa(Y^\circ) = \dim Y^\circ.$$

Viehweg's conjecture has been proven by Sándor Kovács and the author in case where Y° is a surface, [27, 26], or a threefold, [29]. The methods developed in these papers will be discussed, and an idea of the proof will be given later in this paper, cf. the outline of this paper given in Section 1.2 on page 73.

Theorem 2.9 (Viehweg's conjecture for families over threefolds, [29, Thm. 1.1]). *Viehweg's conjecture holds in case where* $\dim Y^\circ \leqslant 3$. □

For families of *canonically polarised* varieties, much stronger results have been obtained, giving an explicit geometric explanation of Theorem 2.9.

Theorem 2.10 (Relationship between the moduli map and the MMP, [29, Thm. 1.1]). *Let* $f^\circ : X^\circ \to Y^\circ$ *be a smooth projective family of canonically polarised varieties, over a quasi-projective manifold* Y° *of dimension* $\dim Y^\circ \leqslant 3$. *Let* Y *be a smooth compactification of* Y° *such that* $D := Y \setminus Y^\circ$ *is a divisor with simple normal crossings.*

Then any run of the minimal model program of the pair (Y, D) *will terminate in a Kodaira or Mori fibre space whose fibration factors the moduli map birationally.*

Remark 2.11. Neither the compactification Y nor the minimal model program discussed in Theorem 2.10 is unique. When running the minimal model program, one often needs to choose the extremal ray that is to be contracted.

In order to explain the statement of Theorem 2.10, let $\mathfrak{M}$ be the appropriate coarse moduli space whose existence is shown, e.g. in [59, Thm. 1.11]. Further, let $\mu^\circ : Y^\circ \to \mathfrak{M}$ be the moduli map associated with the family f°, and let $\mu : Y \dashrightarrow \mathfrak{M}$ be the associated rational map from the compactification Y. If $\lambda : Y \dashrightarrow Y_\lambda$ is a rational map obtained by running the minimal model program, and if $Y_\lambda \to Z_\lambda$ is the associated Kodaira or Mori fibre space, then Theorem 2.10 asserts the existence of a map $Z_\lambda \dashrightarrow \mathfrak{M}$ that makes the following diagram commutative,

$$\begin{array}{ccc} Y & \overset{\lambda}{\underset{\text{MMP of the pair }(Y,D)}{\dashrightarrow}} & Y_\lambda \\ {\scriptstyle\text{moduli map induced by } f^\circ}\ \Big\downarrow & & \Big\downarrow\ {\scriptstyle\text{Kodaira or Mori fibre space}} \\ \mathfrak{M} & \underset{\exists !}{\dashleftarrow} & Z_\lambda . \end{array}$$

Now, if we assume in addition that $\kappa(Y^\circ) \geqslant 0$, then the minimal model program terminates in a Kodaira fibre space whose base Z_λ has dimension $\dim Z_\lambda = \kappa(Y^\circ)$, so that $\operatorname{Var}(f^\circ) \leqslant \kappa(Y^\circ)$. If we assume that $\kappa(Y^\circ) = -\infty$, then the minimal model program terminates in proper Mori fibre space and we obtain that $\dim Z_\lambda < \dim Y$ and $\operatorname{Var}(f^\circ) < \dim Y^\circ$. The following refined answer to Viehweg's conjecture is therefore an immediate corollary of Theorem 2.10.

Corollary 2.12 (Refined answer to Viehweg's conjecture, [29, Cor. 1.3]). *Let* $f^\circ : X^\circ \to Y^\circ$ *be a smooth projective family of canonically polarised varieties, over a quasi-projective manifold* Y° *of dimension* $\dim Y^\circ \leqslant 3$. *Then either*

(1) $\kappa(Y^\circ) = -\infty$ *and* $\operatorname{Var}(f^\circ) < \dim Y^\circ$, *or*

(2) $\kappa(Y^\circ) \geqslant 0$ *and* $\operatorname{Var}(f^\circ) \leqslant \kappa(Y^\circ)$. □

Remark 2.13. Corollary 2.12 asserts that any family of canonically polarised varieties over a base manifold Y° with $\kappa(Y^\circ) = 0$ is necessarily isotrivial.

Remark 2.14. Corollary 2.12 has also been shown in case where Y° is a *projective* manifold of arbitrary dimension, conditional to the standard conjectures of minimal model theory[1], cf. [28, Thm. 1.4]. A very short proof that does not rely on minimal model theory has been announced by Patakfalvi as this paper goes to print, [53].

Example 2.15 (Optimality of Corollary 2.12 in case $\kappa(Y^\circ) = -\infty$). To see that the result of Corollary 2.12 is optimal in case $\kappa(Y^\circ) = -\infty$, let $f_1^\circ : X_1^\circ \to Y_1^\circ$ be any family of canonically polarised varieties with $\operatorname{Var}(f_1^\circ) = 2$, over a smooth surface Y_1° (which may or may not be compact). Setting $X^\circ := X_1^\circ \times \mathbb{P}^1$ and $Y^\circ := Y_1^\circ \times \mathbb{P}^1$, we obtain a family $f^\circ = f_1^\circ \times \mathrm{Id}_{\mathbb{P}^1} : X^\circ \to Y^\circ$ with variation $\operatorname{Var}(f^\circ) = 2$, and with a base manifold Y° of Kodaira dimension $\kappa(Y^\circ) = -\infty$.

Example 2.16 (Related and complementary results in case $\kappa(Y^\circ) = -\infty$). In the setup of Corollary 2.12, if Y° is a projective Fano manifold, then a fundamental result of Campana and Kollár-Miyaoka-Mori asserts that Y° is rationally connected, [33, V. Thm. 2.13]. In other words, given any two points x, y in Y°, there exists a rational curve $C \subset Y^\circ$ which contains both x and y. Recalling from Theorem 2.2 that families over rational curves are isotrivial, it follows immediately that the family f° is necessarily isotrivial itself.

---

[1] i.e., existence and termination of the minimal model program and abundance

A much stronger version of this result has been shown by Lohmann, [45]. Given a projective variety Y and a $\mathbb{Q}$-divisor D such that (Y, D) is a divisorially log terminal (=dlt)[2] pair, consider the smooth quasi-projective variety

$$Y^\circ := (Y \setminus \operatorname{supp}\lfloor D \rfloor)_{\mathrm{reg}}.$$

Lohmann shows that if (Y, D) is log-Fano, that is, if the $\mathbb{Q}$-divisor $-(K_Y+D)$ is ample, then any family of canonically polarized varieties over Y° is necessarily isotrivial. The proof relies on a generalization of Araujo's result [2] which relates extremal rays in the moving cone of a variety with fiber spaces that appear at the end of the minimal model program. Lohmann shows that the moduli map factorizes through any of the fibrations obtained in this way.

2.2.4. Campana's conjecture In a series of papers, including [5, 6], Campana introduced the notion of "geometric orbifolds" and "special varieties". Campana's language helps to formulate a very natural generalisation of Theorem 1.1, which includes the cases covered by the Viehweg Conjecture 2.8, and gives (at least conjecturally) a satisfactory geometric explanation of isotriviality observed in some families over spaces that are not covered by Conjecture 2.8.

Before formulating the conjecture, we briefly recall the precise definition of a special logarithmic pair for the reader's convenience. We take the classical Bogomolov-Sommese Vanishing Theorem as our starting point. We refer to [22, 11] or to the original reference [8] for an explanation of the sheaf $\Omega^p_Y(\log D)$ of logarithmic differentials.

Theorem 2.17 (Bogomolov-Sommese Vanishing Theorem, cf. [11, Sect. 6]). *Let Y be a smooth projective variety and $D \subset Y$ a reduced (possibly empty) divisor with simple normal crossings. If $p \leqslant \dim Y$ is any number and $\mathcal{A} \subseteq \Omega^p_Y(\log D)$ any invertible subsheaf, then the Kodaira-Iitaka dimension of $\mathcal{A}$ is at most p, i.e., $\kappa(\mathcal{A}) \leqslant p$.* □

In a nutshell, we say that a pair (Y, D) is special if the inequality in the Bogomolov-Sommese Vanishing Theorem is always strict.

Definition 2.18 (Special logarithmic pair). *In the setup of Theorem 2.17, a pair (Y, D) is called* special *if the strict inequality $\kappa(\mathcal{A}) < p$ holds for all p and all invertible sheaves $\mathcal{A} \subseteq \Omega^p_Y(\log D)$. A smooth, quasi-projective variety Y° is called special if there exists a smooth compactification Y such that $D := Y \setminus Y^\circ$ is a divisor with simple normal crossings and such that the pair (Y, D) is special.*

Remark 2.19 (Special quasi-projective variety). It is an elementary fact that if Y° is a smooth, quasi-projective variety and Y_1, Y_2 two smooth compactifications such that $D_i := Y_i \setminus Y^\circ$ are divisors with simple normal crossings, then (Y_1, D_1) is special if and only if (Y_2, D_2) is. The notion of special should thus be seen as a property of the quasi-projective variety Y°.

---

[2] We refer to [35, Sect. 2.3] for the definition of a *dlt* pair, and for related notions concerning singularities of pairs that are relevant in minimal model theory.

Fact 2.20 (Examples of special manifolds, cf. [5, Thms. 3.22 and 5.1]). Rationally connected manifolds and manifolds X with $\kappa(X) = 0$ are special. □

With this notation in place, Campana's conjecture can be formulated as follows.

Conjecture 2.21 (Campana's conjecture, [6, Conj. 12.19]). *Let* $f : X° \to Y°$ *be a smooth family of canonically polarised varieties over a smooth quasi-projective base. If* $Y°$ *is special, then the family* f *is isotrivial.*

In analogy with the construction of the maximally rationally connected quotient map of uniruled varieties, Campana constructs in [5, Sect. 3] an almost-holomorphic "core map" whose fibres are special in the sense of Definition 2.18. Like the MRC quotient, the core map is uniquely characterised by certain maximality properties, [5, Thm. 3.3], which essentially say that the core map of X contracts almost all special subvarieties contained in X. If Campana's Conjecture 2.21 holds, this would imply that the core map always factors the moduli map, similar to what we have seen in Section 2.2.3 above,

$$\begin{array}{ccc} Y & \xrightarrow{\text{core map, almost holomorphic}} & Z \\ {\scriptstyle \text{moduli map induced by } f°} \downarrow & \nearrow & \\ \mathfrak{M}. & \xleftarrow{\exists !} & \end{array}$$

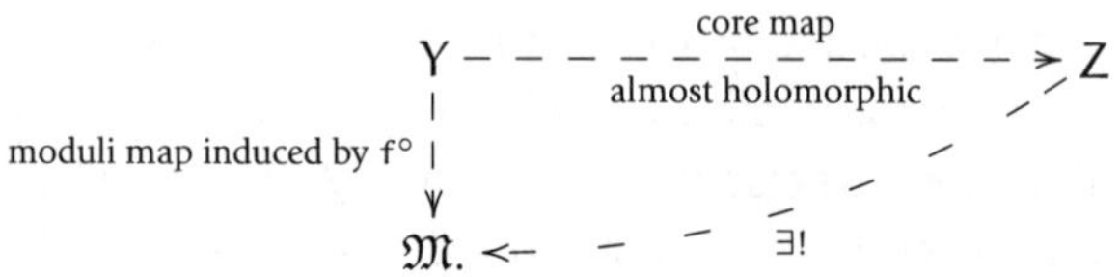

As with Viehweg's Conjecture 2.8, Campana's Conjecture 2.21 has been shown for surfaces [24] and threefolds [23].

Theorem 2.22 (Campana's conjecture in dimension three, [23, Thm. 1.5]). *Campana's Conjecture 2.21 holds if* $\dim Y° \leqslant 3$. □

2.3. Conjectures and open problems

Viehweg's Conjecture 2.8 and Campana's Conjecture 2.21 have been shown for families over base manifolds of dimension three or less. As we will see in Section 5, the restriction to three-dimensional base manifolds comes from the fact that minimal model theory is particularly well-developed for threefolds, and from our limited ability to handle differential forms on singular spaces of higher dimension. We do not believe that there is a fundamental reason that restricts us to dimension three, and we do believe that the relationship between the moduli map and the MMP found in Theorem 2.10 will hold in arbitrary dimension.

Conjecture 2.23 (Relationship between the moduli map and the MMP). *Let* $f° : X° \to Y°$ *be a smooth projective family of canonically polarised varieties, over a quasi-projective manifold* $Y°$. *Let* Y *be a smooth compactification of* $Y°$ *such that* $D := Y \setminus Y°$ *is a divisor with simple normal crossings. Then any run of the minimal model program of the pair* (Y, D) *will terminate in a Kodaira or Mori fibre space whose fibration factors the moduli map birationally.*

Conjecture 2.24 (Refined Viehweg conjecture, cf. [27, Conj. 1.6]). *Corollary 2.12 holds without the assumption that* $\dim Y^\circ \leqslant 3$.

Given the current progress in minimal model theory, a proof of Conjectures 2.23 and 2.24 does no longer seem out of reach.

3. Techniques I: Existence of Pluri-differentials on the base of a family

3.1. The existence result

Throughout the present Section 3, we consider a smooth projective family $f^\circ : X^\circ \to Y^\circ$ of projective, canonically polarised complex manifolds, over a smooth complex quasi-projective base. We assume that the family is not isotrivial, and fix a smooth projective compactification Y of Y° such that $D := Y \setminus Y^\circ$ is a divisor with simple normal crossings. In this setup, Viehweg and Zuo have shown the following fundamental result asserting the existence of many logarithmic pluri-differentials on Y.

Theorem 3.1 (Existence of pluri-differentials on Y, [61, Thm. 1.4(i)]). *Let* $f^\circ : X^\circ \to Y^\circ$ *be a smooth projective family of canonically polarised complex manifolds, over a smooth complex quasi-projective base. Assume that the family is not isotrivial and fix a smooth projective compactification* Y *of* Y° *such that* $D := Y \setminus Y^\circ$ *is a divisor with simple normal crossings.*

Then there exists a number $m > 0$ *and an invertible sheaf* $\mathcal{A} \subseteq \operatorname{Sym}^m \Omega^1_Y(\log D)$ *whose Kodaira-Iitaka dimension is at least the variation of the family,* $\kappa(\mathcal{A}) \geqslant \operatorname{Var}(f^\circ)$. □

Remark 3.2. Observe that the Shafarevich hyperbolicity conjecture, Theorem 1.1, follows as an immediate corollary of Theorem 3.1.

Remark 3.3. A somewhat weaker version of Theorem 3.1 holds for families of projective manifolds with only semiample canonical bundle if one assumes additionally that the family is of maximal variation, i.e., that $\operatorname{Var}(f^\circ) = \dim Y^\circ$, cf. [61, Thm. 1.4(iv)].

As we will see in Section 5, the "Viehweg-Zuo" sheaf $\mathcal{A}$ is one of the crucial ingredients in the proofs of Viehweg's and Campana's conjectures for families over threefolds, Theorems 2.9, 2.10 and 2.22. A careful review of Viehweg and Zuo's construction reveals that the "Viehweg-Zuo sheaf" $\mathcal{A}$ comes from the coarse moduli space $\mathfrak{M}$, at least generically. The precise statement, given in Theorem 3.6, uses the following notation.

Notation 3.4 (Differentials coming from moduli space generically). *Let* $\mu : Y^\circ \to \mathfrak{M}$ *be the moduli map associated with the family* f°, *and consider the subsheaf* $\mathcal{B} \subseteq \Omega^1_Y(\log D)$, *defined on presheaf level as follows: if* $U \subseteq Y$ *is any open set and* $\sigma \in H^0(U, \Omega^1_Y(\log D))$ *any section, then* $\sigma \in H^0(U, \mathcal{B})$ *if and only if the restriction* $\sigma|_{U'}$ *is in the image of the* *differential map*

$$d\mu|_{U'} : \mu^*\big(\Omega^1_{\mathfrak{M}}\big)|_{U'} \longrightarrow \Omega^1_{U'},$$

where $U' \subseteq U \cap Y^\circ$ *is the open subset where the moduli map* μ *has maximal rank.*

Remark 3.5. By construction, it is clear that the sheaf $\mathcal{B}$ is a saturated subsheaf of $\Omega^1_Y(\log D)$, i.e., that the quotient sheaf $\Omega^1_Y(\log D)/\mathcal{B}$ is torsion free. We say that $\mathcal{B}$ is the saturation of Image$(d\mu)$ in $\Omega^1_Y(\log D)$.

Theorem 3.6 (Refinement of the Viehweg-Zuo Theorem 3.1, [24, Thm. 1.4]). *In the setup of Theorem 3.1, there exists a number* $m > 0$ *and an invertible subsheaf* $\mathcal{A} \subseteq \operatorname{Sym}^m \mathcal{B}$ *whose Kodaira-Iitaka dimension is at least the variation of the family,* $\kappa(\mathcal{A}) \geqslant \operatorname{Var}(f^\circ)$.

Theorem 3.6 follows without too much work from Viehweg's and Zuo's original arguments and constructions, which are reviewed in Section 3.2 below. Compared with Theorem 3.1, the refined Viehweg-Zuo theorem relates more directly to Campana's Conjecture 2.21 and other generalizations of the Shafarevich conjecture. To illustrate its use, we show in the surface case how Theorem 3.6 reduces Campana's Conjecture 2.21 to the Viehweg Conjecture 2.8, for which a positive answer is known.

Corollary 3.7 (Campana's conjecture in dimension two). *Conjecture 2.21 holds if* $\dim Y^\circ = 2$.

Proof. We maintain the notation of Conjecture 2.21 and let $f : X^\circ \to Y^\circ$ be a smooth family of canonically polarised varieties over a smooth quasi-projective base, with Y° a special surface. Since Y° is special, it is not of log general type, and the solution to Viehweg's conjecture in dimension two, [29, Thm. 1.1], gives that $\operatorname{Var}(f^\circ) < 2$.

We argue by contradiction, suppose that $\operatorname{Var}(f^\circ) = 1$ and choose a compactification (Y, D) as in Definition 2.18. By Theorem 3.6 there exists a number $m > 0$ and an invertible subsheaf $\mathcal{A} \subseteq \operatorname{Sym}^m \mathcal{B}$ such that $\kappa(\mathcal{A}) \geqslant 1$. However, since $\mathcal{B}$ is saturated in the locally free sheaf $\Omega^1_Y(\log D)$, it is reflexive, [51, Claim on p. 158], and since $\operatorname{Var}(f^\circ) = 1$, the sheaf $\mathcal{B}$ is of rank 1. Thus $\mathcal{B} \subseteq \Omega^1_Y(\log D)$ is an invertible subsheaf, [51, Lem. 1.1.15, on p. 154], and Definition 2.18 of a special pair gives that $\kappa(\mathcal{B}) < 1$, contradicting the fact that $\kappa(\mathcal{A}) \geqslant 1$. It follows that $\operatorname{Var}(f^\circ) = 0$ and that the family is hence isotrivial. □

Outline of this section Given its importance in the theory, we give a very brief synopsis of Viehweg-Zuo's proof of Theorem 3.1, showing how the theorem follows from deep positivity results[3] for push-forward sheaves of relative dualizing sheaves, and for kernels of Kodaira-Spencer maps, respectively. Even though no proof of the refined Theorem 3.6, is given, it is hoped that the reader who chooses to read Section 3.2 will believe that Theorem 3.6 follows with some extra work by essentially the same methods.

---

[3]The positivity results in question are formulated in Theorem 3.10 and Fact 3.22, respectively.

The reader who is interested in a detailed understanding, including is referred to the papers [32], [61], and to the survey [60]. The overview contained in this section and the facts outlined in Section 3.2.5 can perhaps serve as a guideline to [61].

Many of the technical difficulties encountered in the full proof of Theorem 3.1 vanish if f° is a family of curves. The proof becomes very transparent in this case. In particular, it is very easy to see how the Kodaira-Spencer map associated with the family f° transports the positivity found in push-forward sheaves into the sheaf of differentials $\Omega^1_Y(\log Y)$. After setting up notation in Section 3.2.1, we have therefore included a Section 3.2.2 which discusses the curve case in detail.

Most of the material presented in the current Section 3, including the synopsis of Viehweg-Zuo's construction, is taken without much modification from the paper [24]. The presentation is inspired in part by [60].

3.2. A synopsis of Viehweg-Zuo's construction

3.2.1. Setup of notation Throughout the present Section 3.2, we choose a smooth projective compactification X of X° such that the following holds:

(1) The difference $\Delta := X \setminus X^\circ$ is a divisor with simple normal crossings.
(2) The morphism f° extends to a projective morphism $f : X \to Y$.

It is then clear that $\Delta = f^{-1}(D)$ set-theoretically. Removing a suitable subset $S \subset Y$ of codimension $\operatorname{codim}_Y S \geqslant 2$, the following will then hold automatically on $Y' := Y \setminus S$ and $X' := X \setminus f^{-1}(S)$, respectively.

(3) The restricted morphism $f' := f|_{X'}$ is flat.
(4) The divisor $D' := D \cap Y'$ is smooth.
(5) The divisor $\Delta' := \Delta \cap X'$ is a relative normal crossing divisor, i.e. a normal crossing divisor whose components and all their intersections are smooth over the components of D'.

In the language of Viehweg-Zuo, [61, Def 2.1(c)], the restricted morphism $f' : X' \to Y'$ is a "good partial compactification of f°".

Remark 3.8 (Restriction to a partial compactification). Let $\mathcal{G}$ be a locally free sheaf on Y, and let $\mathcal{F}' \subseteq \mathcal{G}|_{Y'}$ be an invertible subsheaf. Since $\operatorname{codim}_Y S \geqslant 2$, there exists a unique extension of the sheaf $\mathcal{F}'$ to an invertible subsheaf $\mathcal{F} \subseteq \mathcal{G}$ on Y. Furthermore, the restriction map $H^0(Y, \mathcal{F}) \to H^0(Y', \mathcal{F}')$ is an isomorphism. In particular, the notion of Kodaira-Iitaka dimension makes sense for the sheaf $\mathcal{F}'$, and $\kappa(\mathcal{F}') = \kappa(\mathcal{F})$.

We denote the relative dimension of X over Y by $n := \dim X - \dim Y$.

3.2.2. Idea of proof of Theorem 3.1 for families of curves Before sketching the proof of Theorem 3.1 in full generality, we illustrate the main idea in a particularly simple setting.

Simplifying Assumptions 3.9. Throughout the present introductory Section 3.2.2, we maintain the following simplifying assumptions in addition to the assumptions made in Theorem 3.1.

(1) The quasi-projective variety Y° is in fact projective. In particular, we assume that $X = X^\circ$, $f = f^\circ$, that $D = \emptyset$ and that $\Delta = \emptyset$.
(2) The family $f : X \to Y$ is a family of curves of genus $g > 1$. In particular, we have that $(T_{X/Y})^* = \Omega^1_{X/Y} = \omega_{X/Y}$, where $T_{X/Y}$ is the kernel of the derivative $Tf : T_X \to f^*(T_Y)$.
(3) The variation of f° is maximal, that is, $\operatorname{Var}(f^\circ) = \dim Y^\circ$.

The proof of Theorem 3.1 sketched here uses positivity of the push-forward of relative dualizing sheaves as its main input. The positivity result required is discussed in Viehweg's survey [60, Sect. 1–3], where positivity is obtained as a consequence of generalised Kodaira vanishing theorems. The reader interested in a broader overview might also want to look at the remarks and references in [43, Sect. 6.3.E], as well as the papers [32, 64]

Theorem 3.10 (Positivity of push-forward sheaves, cf. [61, Prop. 3.4.(i)]). *Under the Simplifying Assumptions 3.9, the push-forward sheaf* $f_*\big(\omega_{X/Y}^{\otimes 2}\big)$ *is locally free of positive rank. If* $\mathcal{A} \in \operatorname{Pic}(Y)$ *is any ample line bundle, then there exist numbers* $N, M \gg 0$ *and a sheaf morphism*

$$\phi : \mathcal{A}^{\oplus M} \to \operatorname{Sym}^N f_*\big(\omega_{X/Y}^{\otimes 2}\big)$$

which is surjective at the general point of Y. □

For the reader's convenience, we recall two other facts used in the proof, namely the existence of a Kodaira-Spencer map, and Serre duality in the relative setting.

Theorem 3.11 (Kodaira-Spencer map, cf. [63, Sect. 9.1.2] or [20, Sect. 6.2]). *Under the Simplifying Assumptions 3.9, since* $\operatorname{Var}(f) > 0$, *there exists a non-zero sheaf morphism* $\kappa : T_Y \to R^1 f_*(T_{X/Y})$ *which measures the variation of the isomorphism classes of fibres in moduli.* □

Theorem 3.12 (Serre duality in the relative setting, cf. [44, Sect. 6.4]). *Under the simplifying Assumptions 3.9, if* $\mathcal{F}$ *is any coherent sheaf on* X, *then there exists a natural isomorphism* $f_*(\mathcal{F}^* \otimes \omega_{X/Y}) \cong \big(R^1 f_*(\mathcal{F})\big)^*$. □

Proof of Theorem 3.1 under the Simplifying Assumptions 3.9. Consider the dual, say $\kappa^* : \big(R^1 f_*(T_{X/Y})\big)^* \to (T_Y)^*$, of the (non-trivial) Kodaira-Spencer map discussed in Theorem 3.11. Recalling that $T^*_{X/Y}$ equals the relative dualizing sheaf $\omega_{X/Y}$, and using the relative Serre duality Theorem 3.12, the sheaf morphism κ^* is naturally identified with a non-zero morphism

$$\kappa^* : f_*\big(\omega_{X/Y}^{\otimes 2}\big) \to \Omega^1_Y. \tag{3.13}$$

Choosing an ample line bundle $\mathcal{A} \in \operatorname{Pic}(Y)$ and sufficiently large and divisible numbers $N, M \gg 0$, Theorem 3.10 yields a sequence of sheaf morphisms

$$\mathcal{A}^{\oplus M} \xrightarrow[\text{gen. surjective}]{\phi} \operatorname{Sym}^N f_*(\omega_{X/Y}^{\otimes 2}) \xrightarrow[\text{non-trivial}]{\operatorname{Sym}^N(\kappa^*)} \operatorname{Sym}^N \Omega_Y^1,$$

whose composition $\mathcal{A}^{\oplus M} \to \operatorname{Sym}^N \Omega_Y^1$ is clearly not the zero map. Consequently, we obtain a non-trivial map $\mathcal{A} \to \operatorname{Sym}^N \Omega_Y^1$, finishing the proof of Theorem 3.1 under the Simplifying Assumptions 3.9. □

The proof outlined above uses the dual of the Kodaira-Spencer map as a vehicle to transport the positivity which exists in $f_*(\omega_{X/Y}^{\otimes 2})$ into the sheaf Ω_Y^1 of differential forms on Y. If f was a family of surfaces rather than a family of curves, then Serre duality cannot easily be used to identify the dual of $R^1 f_*(T_{X/Y})$ with a push-forward sheaf of type $f_*(\omega_{X/Y}^{\otimes \bullet})$, or any with other sheaf whose positivity is well-known. To overcome this problem, Viehweg suggested to replace the Kodaira-Spencer map κ by sequences of more complicated sheaf morphisms $\tau_{p,q}^0$ and τ^k, constructed in Sections 3.2.3 and 3.2.4 below. To motivate the slightly involved construction of these maps, we recall without proof a description of the classical Kodaira-Spencer map.

Construction 3.14. Under the Simplifying Assumptions 3.9, consider the standard sequence of relative differential forms on X,

$$0 \to f^*\Omega_Y^1 \to \Omega_X^1 \to \Omega_{X/Y}^1 \to 0, \tag{3.15}$$

and its twist with the invertible sheaf $\omega_{X/Y}^*$,

$$0 \to f^*\Omega_Y^1 \otimes \omega_{X/Y}^* \to \Omega_X^1 \otimes \omega_{X/Y}^* \to \underbrace{\Omega_{X/Y}^1 \otimes \omega_{X/Y}^*}_{\cong \mathcal{O}_X} \to 0.$$

Using that $f_*(\mathcal{O}_X) = \mathcal{O}_Y$, the first connecting morphism associated with this sequence then reads

$$\mathcal{O}_Y \to \Omega_Y^1 \otimes R^1 f_*(\omega_{X/Y}^*) =: \mathcal{F}. \tag{3.16}$$

The sheaf $\mathcal{F}$ is naturally isomorphic to the sheaf $\operatorname{Hom}\big(T_Y, R^1 f_*(T_{X/Y})\big)$. To give a morphism $\mathcal{O}_Y \to \mathcal{F}$ is thus the same as to give a map $T_Y \to R^1 f_*(T_{X/Y})$, and the morphism obtained in (3.16) is the same as the Kodaira-Spencer map discussed in Theorem 3.11.

Observe also that Serre duality yields a natural identification of $\mathcal{F}$ with the sheaf $\operatorname{Hom}\big(f_*(\omega_{X/Y}^{\otimes 2}), \Omega_Y^1\big)$. To give a morphism $\mathcal{O}_Y \to \mathcal{F}$ it is thus the same as to give a map $f_*(\omega_{X/Y}^{\otimes 2}) \to \Omega_Y^1$. The morphism obtained in this way from (3.16) is of course the morphism κ^* of Equation (3.13).

3.2.3. Proof of Theorem 3.1, construction of the $\tau^0_{p,q}$ In the general setting of Theorem 3.1 where the simplifying Assumptions 3.9 do not generally hold, the starting point of the Viehweg-Zuo construction is the standard sequence of relative logarithmic differentials associated to the flat morphism f' which generalises Sequence (3.15) from above,

$$0 \to (f')^*\Omega^1_{Y'}(\log D') \to \Omega^1_{X'}(\log \Delta') \to \Omega^1_{X'/Y'}(\log \Delta') \to 0. \tag{3.17}$$

We refer to [10, Sect. 4] for a discussion of Sequence (3.17), and for a proof of the fact that the cokernel $\Omega^1_{X'/Y'}(\log \Delta')$ is locally free. By [19, II, Ex. 5.16], Sequence (3.17) defines a filtration of the pth exterior power,

$$\Omega^p_{X'}(\log \Delta') = F^0 \supseteq F^1 \supseteq \cdots \supseteq F^p \supseteq F^{p+1} = \{0\},$$

with $F^r/F^{r+1} \cong (f')^*\big(\Omega^r_{Y'}(\log D')\big) \otimes \Omega^{p-r}_{X'/Y'}(\log \Delta')$. Take the first sequence induced by the filtration,

$$0 \longrightarrow F^1 \longrightarrow F^0 \longrightarrow F^0/F^1 \longrightarrow 0,$$

modulo F^2, and obtain

$$\begin{aligned} 0 \longrightarrow (f')^*\big(\Omega^1_{Y'}(\log D')\big) \otimes \Omega^{p-1}_{X'/Y'}(\log \Delta') \longrightarrow F^0/F^2 \longrightarrow \\ \longrightarrow \Omega^p_{X'/Y'}(\log \Delta') \longrightarrow 0. \end{aligned} \tag{3.18}$$

Setting $\mathcal{L} := \Omega^n_{X'/Y'}(\log \Delta')$, twisting Sequence (3.18) with $\mathcal{L}^{-1}$ and pushing down, the connecting morphisms of the associated long exact sequence give maps

$$\tau^0_{p,q} : F^{p,q} \longrightarrow F^{p-1,q+1} \otimes \Omega^1_{Y'}(\log D'),$$

where $F^{p,q} := R^q f'_*\big(\Omega^p_{X'/Y'}(\log \Delta') \otimes \mathcal{L}^{-1}\big)/\text{torsion}$. Set $\mathcal{N}^{p,q}_0 := \ker(\tau^0_{p,q})$.

3.2.4. Alignment of the $\tau^0_{p,q}$ The morphisms $\tau^0_{p,q}$ and $\tau^0_{p-1,q+1}$ can be composed if we tensor the latter with the identity morphism on $\Omega^1_{Y'}(\log D')$. More specifically, we consider the following morphisms,

$$\underbrace{\tau^0_{p,q} \otimes \mathrm{Id}_{\Omega^1_{Y'}(\log D')^{\otimes q}}}_{=:\tau_{p,q}} : F^{p,q} \otimes \big(\Omega^1_{Y'}(\log D')\big)^{\otimes q} \to F^{p-1,q+1} \otimes \big(\Omega^1_{Y'}(\log D')\big)^{\otimes q+1},$$

and their compositions

$$\underbrace{\tau_{n-k+1,k-1} \circ \cdots \circ \tau_{n-1,1} \circ \tau_{n,0}}_{=:\tau^k} : F^{n,0} \to F^{n-k,k} \otimes \big(\Omega^1_{Y'}(\log D')\big)^{\otimes k}. \tag{3.19}$$

3.2.5. Fundamental facts about τ^k and $\mathcal{N}^{p,q}_0$ Theorem 3.1 is shown by relating the morphism $\tau^0_{p,q}$ with the structure morphism of a Higgs-bundle coming from the variation of Hodge structures associated with the family f°. Viehweg's positivity results of push-forward sheaves of relative differentials, as well as Zuo's results on the curvature of kernels of generalised Kodaira-Spencer maps are the main input here. Rather than recalling the complicated line of argumentation, we simply state two central results from the argumentation of [61].

Fact 3.20 (Factorization via symmetric differentials, [61, Lem. 4.6]). For any k, the morphism τ^k factors via the symmetric differentials $\mathrm{Sym}^k\,\Omega^1_{Y'}(\log D') \subseteq (\Omega^1_{Y'}(\log D'))^{\otimes k}$. More precisely, the morphism τ^k takes its image in $F^{n-k,k} \otimes \mathrm{Sym}^k\,\Omega^1_{Y'}(\log D')$. □

Consequence 3.21. Using Fact 3.20 and the observation that $F^{n,0} \cong \mathcal{O}_{Y'}$, we can therefore view τ^k as a morphism

$$\tau^k : \mathcal{O}_{Y'} \longrightarrow F^{n-k,k} \otimes \mathrm{Sym}^k\,\Omega^1_{Y'}(\log D').$$

While the proof of Fact 3.20 is rather elementary, the following deep result is at the core of Viehweg-Zuo's argument. Its role in the proof of Theorem 3.1 is comparable to that of the Positivity Theorem 3.10 discussed in Section 3.2.2.

Fact 3.22 (Negativity of $\mathcal{N}_0^{p,q}$, [61, Claim 4.8]). Given any numbers p and q, there exists a number k and an invertible sheaf $\mathcal{A}' \in \mathrm{Pic}(Y')$ of Kodaira-Iitaka dimension $\kappa(\mathcal{A}') \geqslant \mathrm{Var}(f^0)$ such that $(\mathcal{A}')^* \otimes \mathrm{Sym}^k\big((\mathcal{N}_0^{p,q})^*\big)$ is generically generated. □

3.2.6. End of proof To end the sketch of proof, we follow [61, p. 315] almost verbatim. By Fact 3.22, the trivial sheaf $F^{n,0} \cong \mathcal{O}_{Y'}$ cannot lie in the kernel $\mathcal{N}_0^{n,0}$ of $\tau^1 = \tau^0_{n,0}$. We can therefore set $1 \leqslant m$ to be the largest number with $\tau^m(F^{n,0}) \neq \{0\}$. Since m is maximal, $\tau^{m+1} = \tau_{n-m,m} \circ \tau^m \equiv 0$ and

$$\mathrm{Image}(\tau^m) \subseteq \ker(\tau_{n-m,m}) = \mathcal{N}_0^{n-m,m} \otimes \mathrm{Sym}^m\,\Omega^1_{Y'}(\log D').$$

In other words, τ^m gives a non-trivial map

$$\tau^m : \mathcal{O}_{Y'} \cong F^{n,0} \longrightarrow \mathcal{N}_0^{n-m,m} \otimes \mathrm{Sym}^m\,\Omega^1_{Y'}(\log D').$$

Equivalently, we can view τ^m as a non-trivial map

$$\tau^m : (\mathcal{N}_0^{n-m,m})^* \longrightarrow \mathrm{Sym}^m\,\Omega^1_{Y'}(\log D'). \tag{3.23}$$

By Fact 3.22, there are many morphisms $\mathcal{A}' \to \mathrm{Sym}^k\big((\mathcal{N}_0^{n-m,m})^*\big)$, for k large enough. Together with (3.23), this gives a non-zero morphism $\mathcal{A}' \to \mathrm{Sym}^{k\cdot m}\,\Omega^1_{Y'}(\log D')$.

We have seen in Remark 3.8 that the sheaf $\mathcal{A}' \subseteq \mathrm{Sym}^{k\cdot m}\,\Omega^1_{Y'}(\log D')$ extends to a sheaf $\mathcal{A} \subseteq \mathrm{Sym}^{k\cdot m}\,\Omega^1_{Y}(\log D)$ with $\kappa(\mathcal{A}) = \kappa(\mathcal{A}') \geqslant \mathrm{Var}(f^\circ)$. This ends the proof of Theorem 3.1. □

3.3. Open problems

In spite of its importance, little is known about further properties that the Viehweg-Zuo sheaves $\mathcal{A}$ might have.

Question 3.24. For families of higher-dimensional manifolds, how does the Viehweg-Zuo construction behave under base change? Does it satisfy any universal properties at all? If not, is there a "natural" positivity result for base spaces of families that does satisfy good functorial properties?

In the setup of Theorem 3.1, if $Z^\circ \subset Y^\circ$ is any closed submanifold, then the associated Viehweg-Zuo sheaves $\mathcal{A}$, constructed for the family $f^\circ : X^\circ \to Y^\circ$, and $\mathcal{A}_Z$, constructed for the restricted family $f^\circ_Z : X^\circ \times_{Y^\circ} Z^\circ \to Z^\circ$, may differ. In particular, it is not clear that $\mathcal{A}_Z$ is the restriction of $\mathcal{A}$, and the sheaves $\mathcal{A}$ and $\mathcal{A}_Z$ may live in different symmetric products of their respective Ω^1's.

One likely source of non-compatibility with base change is the choice of the number m in Section 3.2.6 ("largest number with $\tau^m(F^{n,0}) \neq \{0\}$"). It seems unlikely that this definition behaves well under base change.

Question 3.25. For families of higher-dimensional manifolds, are there distinguished subvarieties in moduli space that have special Viehweg-Zuo sheaves, perhaps contained in particularly high/low symmetric powers of Ω^1? Does the lack of a restriction morphism induce a geometric structure on the moduli space?

The refinement of the Viehweg-Zuo Theorem, presented in Theorem 3.6 above, turns out to be important for the applications that we have in mind. It is, however, not clear to us if the sheaf $\mathcal{B}$ which appears in Theorem 3.6 is really optimal.

Question 3.26. Prove that the sheaf $\mathcal{B} \subseteq \Omega^1_Y(\log D)$ is the smallest sheaf possible for which Theorem 3.6 holds, or else find the smallest possible sheaf. For instance, does Theorem 3.6 admit a natural improvement if we replace $\Omega^1_Y(\log D)$ by a suitable sheaf of orbifold differentials, using Campana's language of geometric orbifolds?

4. Techniques II: Reflexive differentials on singular spaces

4.1. Motivation

4.1.1. A special case of the Viehweg conjecture To motivate the results presented in this section, let $f^\circ : X^\circ \to Y^\circ$ be a smooth, projective family of canonically polarised varieties over a smooth, quasi-projective base manifold, and assume that the family f° is of maximal variation, i.e., that $\mathrm{Var}(f^\circ) = \dim Y^\circ$. As before, choose a smooth compactification $Y \supseteq Y^\circ$ such that $D := Y \setminus Y^\circ$ is a divisor with only simple normal crossings.

To prove Viehweg's conjecture, we need to show that the logarithmic Kodaira dimension of Y° is maximal, i.e., that $\kappa(Y^\circ) = \dim Y^\circ$. In particular, we need to rule out the possibility that $\kappa(Y^\circ) = 0$. As we will see in the proof of Proposition 4.1 below, a relatively elementary argument exists in cases where the Picard number of Y is one, $\rho(Y) = 1$. We refer the reader to [21, Sect. I.1] for the notion of semistability and for a discussion of the Harder-Narasimhan filtration used in the proof.

Proposition 4.1 (Partial answer to Viehweg's conjecture in case $\rho(Y) = 1$). *In the setup described above, if we additionally assume that* $\rho(Y) = 1$, *then* $\kappa(Y^\circ) \neq 0$.

Proof. We argue by contradiction and assume that both $\kappa(Y^\circ) = 0$ and that $\rho(Y) = 1$. Let $\mathcal{A} \subseteq \mathrm{Sym}^m \Omega^1_Y(\log D)$ be the big invertible sheaf whose existence is guaranteed by the Viehweg-Zuo construction, Theorem 3.1. Since $\rho(Y) = 1$, the sheaf $\mathcal{A}$ is actually ample.

As a first step, observe that the log canonical bundle $K_Y + D$ must be torsion, i.e., that there exists a number $m' \in \mathbb{N}^+$ such that $\mathcal{O}_Y\big(m' \cdot (K_Y + D)\big) \cong \mathcal{O}_Y$. This follows from the assumption that $\kappa(K_Y + D) = 0$ and from the observation that on a projective manifold with $\rho = 1$, any invertible sheaf which admits a non-zero section is either trivial or ample. In particular, we obtain that the divisor $K_Y + D$ is numerically trivial.

Next, let $C \subseteq Y$ be a general complete intersection curve in the sense of Mehta-Ramanathan, cf. [21, Sect. II.7]. The numerical triviality of $K_Y + D$ will then imply that

$$(K_Y + D).C = c_1\big(\Omega^1_Y(\log D)\big).C = c_1\big(\mathrm{Sym}^m\, \Omega^1_Y(\log D)\big).C = 0.$$

On the other hand, since $\mathcal{A}$ is ample, we have that $c_1(\mathcal{A}).C > 0$. In summary, we obtain that the symmetric product sheaf $\mathrm{Sym}^m\, \Omega^1_Y(\log D)$ is not semistable. Since we are working in characteristic zero, this implies that the sheaf of Kähler differentials $\Omega^1_Y(\log D)$ will likewise not be semistable, and contains a destabilising subsheaf $\mathcal{B} \subseteq \Omega^1_Y(\log D)$ with $c_1(\mathcal{B}).C > 0$, cf. [21, Cor. 3.2.10]. Since the intersection number $c_1(\mathcal{B}).C$ is positive, the rank r of the sheaf $\mathcal{B}$ must be strictly less than $\dim Y$, and its determinant is an ample invertible subsheaf of the sheaf of logarithmic r-forms,

$$\det \mathcal{B} \subseteq \Omega^r_Y(\log D).$$

This, however, contradicts the Bogomolov-Sommese Vanishing Theorem 2.17 and therefore ends the proof. □

4.1.2. Application of minimal model theory The assumption that $\rho(Y) = 1$ is not realistic. In the general situation, where $\rho(Y)$ can be arbitrarily large, we will apply the minimal model program to the pair (Y, D). As we will sketch in Section 5, assuming that the standard conjectures of minimal model theory hold true, a run of the minimal model program for a suitable choice of a boundary divisor will yield a diagram,

$$\begin{array}{ccc} Y & \overset{\lambda}{\underset{\text{minimal model program}}{\dashrightarrow}} & Y_\lambda \\ & & \pi \big\downarrow \text{fibre space} \\ & & Z_\lambda, \end{array}$$

where $\lambda : Y \dashrightarrow Y_\lambda$ is a birational map whose inverse does not contract any divisors, and where either $\rho(Y_\lambda) = 1$ and Z_λ is a point, or where Y_λ has the structure of a proper Mori– or Kodaira fibre space. In the first case, we can try to copy the proof of Proposition 4.1 above. In the second case, we can use the fibre structure and try to argue inductively.

The main problem that arises when adopting the proof of Proposition 4.1 is the presence of singularities. Both the space Y_λ and the cycle-theoretic image divisor

$D_\lambda \subset Y_\lambda$ will generally be singular, and the pair (Y_λ, D_λ) will generally be dlt. This leads to two difficulties.

(1) The sheaf $\Omega^1_{Y_\lambda}(\log D_\lambda)$ of logarithmic Kähler differentials is generally not pure in the sense of [21, Sect. 1.1]. Accordingly, there is no good notion of stability that would be suitable to construct a Harder-Narasimhan filtration.
(2) The Viehweg-Zuo construction does not work for singular varieties. The author is not aware of any method suitable to prove positivity results for Kähler differentials, or prove the existence of sections in any symmetric product of $\Omega^1_{Y_\lambda}(\log D_\lambda)$.

The aim of the present Section 4 is to show that that both problems can be overcome if we replace the sheaf $\Omega^1_{Y_\lambda}(\log D_\lambda)$ of Kähler differentials by its double dual $\Omega^{[1]}_{Y_\lambda}(\log D_\lambda) := \left(\Omega^1_{Y_\lambda}(\log D_\lambda)\right)^{**}$. We refer to [54, Sect. 1.6] for a discussion of the double dual in this context, and to [51, II. Sect. 1.1] for a thorough discussion of reflexive sheaves. The following notation will be useful in the discussion.

Notation 4.2 (Reflexive tensor operations). *Let X be a normal variety and $\mathcal{A}$ a coherent sheaf of $\mathcal{O}_X$-modules. Given any number $n \in \mathbb{N}$, set $\mathcal{A}^{[n]} := (\mathcal{A}^{\otimes n})^{**}$, $\mathrm{Sym}^{[n]}\mathcal{A} := (\mathrm{Sym}^n \mathcal{A})^{**}$. If $\pi : X' \to X$ is a morphism of normal varieties, set $\pi^{[*]}(\mathcal{A}) := (\pi^*\mathcal{A})^{**}$. In a similar vein, set $\Omega^{[p]}_X := (\Omega^p_X)^{**}$ and $\Omega^{[p]}_X(\log D) := (\Omega^p_X(\log D))^{**}$.*

Notation 4.3 (Reflexive differential forms). *A section in $\Omega^{[p]}_X$ or $\Omega^{[p]}_X(\log D)$ will be called a* reflexive form *or a* reflexive logarithmic form, *respectively.*

Fact 4.4 (Torsion freeness and Harder-Narasimhan filtration). Reflexive sheaves are torsion free and therefore pure. In particular, a Harder-Narasimhan filtration exists for $\Omega^{[p]}_X(\log D)$ and for the symmetric products $\mathrm{Sym}^{[n]}\,\Omega^1_X(\log D)$.

Fact 4.5 (Extension over small sets). If X is a normal space, if $\mathcal{A}$ is any reflexive sheaf on X and if $Z \subset X$ any set of $\mathrm{codim}_X Z \geqslant 2$, then the restriction map

$$H^0(X, \mathcal{A}) \to H^0(X \setminus Z, \mathcal{A})$$

is in fact isomorphic. We say that "sections in $\mathcal{A}$ extend over the small set Z".

If $U := X \setminus Z$ is the complement of Z, with inclusion map $\iota : U \to X$, it follows immediately that $\mathcal{A} = \iota_*(\mathcal{A}|_U)$. In a similar vein, if $\mathcal{B}_U$ is any locally free sheaf on U, its push-forward sheaf $\iota_*(\mathcal{B}_U)$ will always be reflexive.

4.1.3. Outline of this section It follows almost by definition that sheaves of reflexive differentials have very good push-forward properties. In Section 4.2 we will use these properties to overcome one of the problems mentioned above and to produce Viehweg-Zuo sheaves of reflexive differentials on singular spaces. Perhaps more importantly, we will in Section 4.3 recall extension results for log canonical varieties. These results show that reflexive differentials often admit a pull-back map, similar to the standard pull-back of Kähler differentials. A generalisation of the

Bogomolov-Sommese vanishing theorem to log canonical varieties follows as a corollary.

In Section 4.4, we recall that some of the most important constructions known for logarithmic differentials on snc pairs also work for reflexive differentials on dlt pairs. This includes the existence of a residue sequence. For our purposes, this makes reflexive differentials almost as useful as regular differentials in the theory of smooth spaces. As we will roughly sketch in Section 5, these results will allow to adapt the proof of Proposition 4.1 to the singular setup, and will give a proof of Viehweg's Conjecture 2.8, at least for families over base manifolds of dimension $\leqslant 3$. Section 4.5 gives a brief sketch of the proof of the pull-back result of Section 4.3. We end by mentioning a few open problems and conjectures.

Some of the material presented in the current Section 4, including Section 4.4 and all the illustrations, is taken without much modification from the paper [15]. Section 4.2 follows the paper [29].

4.2. Existence of a push-forward map

Fact 4.5 implies that any Viehweg-Zuo sheaf which exists on a pair (Z, Δ) of a smooth variety and a reduced divisor with simple normal crossing support immediately implies the existence of a Viehweg-Zuo sheaf of reflexive differentials on any minimal model of (Z, Δ), and that the Kodaira-Iitaka dimension only increases in the process. To formulate the result precisely, we briefly recall the definition of the Kodaira-Iitaka dimension for reflexive sheaves.

Definition 4.6 (Kodaira-Iitaka dimension of a sheaf, [29, Not. 2.2]). *Let Z be a normal projective variety and $\mathcal{A}$ a reflexive sheaf of rank one on Z. If $h^0(Z, \mathcal{A}^{[n]}) = 0$ for all $n \in \mathbb{N}$, then we say that $\mathcal{A}$ has Kodaira-Iitaka dimension $\kappa(\mathcal{A}) := -\infty$. Otherwise, set*

$$M := \{n \in \mathbb{N} \mid h^0(Z, \mathcal{A}^{[n]}) > 0\},$$

recall that the restriction of $\mathcal{A}$ to the smooth locus of Z is locally free and consider the natural rational mapping

$$\phi_n : Z \dashrightarrow \mathbb{P}(H^0(Z, \mathcal{A}^{[n]})^*) \text{ for each } n \in M.$$

The Kodaira-Iitaka dimension of $\mathcal{A}$ is then defined as

$$\kappa(\mathcal{A}) := \max_{n \in M}(\dim \overline{\phi_n(Z)}).$$

With this notation, the main result concerning the push-forward is then formulated as follows.

Proposition 4.7 (Push forward of Viehweg-Zuo sheaves, [29, Lem. 5.2]). *Let (Z, Δ) be a pair of a smooth variety and a reduced divisor with simple normal crossing support. Assume that there exists a reflexive sheaf $\mathcal{A} \subseteq \operatorname{Sym}^{[n]} \Omega^1_Z(\log \Delta)$ of rank one. If $\lambda : Z \dashrightarrow Z'$ is a birational map whose inverse does not contract any divisor, if Z' is normal and Δ'*

is the (necessarily reduced) cycle-theoretic image of Δ*, then there exists a reflexive sheaf* $\mathcal{A}' \subseteq \mathrm{Sym}^{[n]}\,\Omega^1_{Z'}(\log\Delta')$ *of rank one, and of Kodaira-Iitaka dimension* $\kappa(\mathcal{A}') \geqslant \kappa(\mathcal{A})$.

Proof. The assumption that λ^{-1} does not contract any divisors and the normality of Z' guarantee that $\lambda^{-1} : Z' \dashrightarrow Z$ is a well-defined embedding over an open subset $U \subseteq Z'$ whose complement has codimension $\mathrm{codim}_{Z'}(Z' \setminus U) \geqslant 2$, cf. Zariski's main theorem [19, V 5.2]. In particular, $\Delta'|_U = (\lambda^{-1}|_U)^{-1}(\Delta)$. Let $\iota : U \to Z'$ denote the inclusion and set $\mathcal{A}' := \iota_*\big((\lambda^{-1}|_U)^*\mathcal{A}\big)$ —this sheaf is reflexive by Fact 4.5. We obtain an inclusion of reflexive sheaves, $\mathcal{A}' \subseteq \mathrm{Sym}^{[n]}\,\Omega^1_{Z'}(\log\Delta')$. By construction, we have that $h^0(Z', \mathcal{A}'^{[m]}) \geqslant h^0(Z, \mathcal{A}^{[m]})$ for all $m > 0$, hence $\kappa(\mathcal{A}') \geqslant \kappa(\mathcal{A})$. □

Given the importance of the Viehweg-Zuo construction, Theorem 3.1, we will call the sheaves $\mathcal{A}$ which appear in Proposition 4.7 "Viehweg-Zuo sheaves".

Notation 4.8 (Viehweg-Zuo sheaves). *Let* (Z, Δ) *be a pair of a smooth variety and a reduced divisor with simple normal crossing support, and let* $n \in \mathbb{N}$ *be any number. A reflexive sheaf* $\mathcal{A} \subseteq \mathrm{Sym}^{[n]}\,\Omega^1_Z(\log\Delta)$ *of rank one will be called a "Viehweg-Zuo sheaf".*

4.3. Existence of a pull-back morphism, statement and applications

Kähler differentials are characterised by a number of universal properties, one of the most important being the existence of a pull-back map: if $\gamma : Z \to X$ is any morphism of algebraic varieties and if $p \in \mathbb{N}$, then there exists a canonically defined sheaf morphism

$$d\gamma : \gamma^*\Omega^p_X \to \Omega^p_Z. \tag{4.9}$$

The following example illustrates that for sheaves of reflexive differentials on normal spaces, a pull-back map does not exist in general.

Example 4.10 (Pull-back morphism for dualizing sheaves, cf. [15, Ex. 4.2]). Let X be a normal Gorenstein variety of dimension n, and let $\gamma : Z \to X$ be any resolution of singularities. Observing that the sheaf of reflexive n-forms is precisely the dualizing sheaf, $\Omega^{[n]}_X \simeq \omega_X$, it follows directly from the definition of canonical singularities that X has canonical singularities if and only if a pull-back morphism $d\gamma : \gamma^*\Omega^{[n]}_X \to \Omega^n_Z$ exists.

Together with Daniel Greb, Sándor Kovács and Thomas Peternell, the author has shown that a pull-back map for reflexive differentials always exists if the target is log canonical.

Theorem 4.11 (Pull-back map for differentials on lc pairs, [15, Thm. 4.3]). *Let* (X, D) *be an log canonical pair, and let* $\gamma : Z \to X$ *be a morphism from a normal variety* Z *such that the image of* Z *is not contained in the reduced boundary or in the singular locus, i.e.,*

$$\gamma(Z) \not\subseteq (X, D)_{\mathrm{sing}} \cup \mathrm{supp}\lfloor D \rfloor.$$

If $1 \leqslant p \leqslant \dim X$ *is any index and*

$$\Delta := \textit{largest reduced Weil divisor contained in } \gamma^{-1}\big(\textit{non-klt locus}\big),$$

then there exists a sheaf morphism,

$$d\gamma : \gamma^* \Omega_X^{[p]}(\log\lfloor D \rfloor) \to \Omega_Z^{[p]}(\log \Delta),$$

that agrees with the usual pull-back morphism (4.9) *of Kähler differentials at all points* $p \in Z$ *where* $\gamma(p) \notin (X, D)_{\text{sing}} \cup \operatorname{supp}\lfloor D \rfloor$.

Remark 4.12. If follows from the definition of klt, [35, Def. 2.34], that the components of D which appear with coefficient one are always contained in the non-klt locus of (X, D). In particular, the divisor Δ defined in Theorem 4.11 always contains the codimension-one part of $\gamma^{-1}\big(\operatorname{supp}\lfloor D \rfloor\big)$.

The assertion of Theorem 4.11 is rather general and perhaps a bit involved. For klt spaces, the statement reduces to the following simpler result.

Theorem 4.13 (Pull-back map for differentials on klt spaces). *Let X be a normal klt variety*[4]*, and let* $\gamma : Z \to X$ *be a morphism from a normal variety Z such that the image* $\gamma(Z)$ *is not entirely contained in the singular locus of X. If* $1 \leqslant p \leqslant \dim X$ *is any index then there exists a sheaf morphism,*

$$d\gamma : \gamma^* \Omega_X^{[p]} \to \Omega_Z^{[p]},$$

that agrees on an open set with the usual pull-back morphism of Kähler differentials. □

Extension properties of differential forms that are closely related to the existence of pull-back maps have been studied in the literature, mostly considering only special values of p. Using Steenbrink's generalization of the Grauert-Riemenschneider vanishing theorem as their main input, similar results were shown by Steenbrink and van Straten for varieties X with only isolated singularities and for $p \leqslant \dim X - 2$, without any further assumption on the nature of the singularities, [57, Thm. 1.3]. Flenner extended these results to normal varieties, subject to the condition that $p \leqslant \operatorname{codim} X_{\text{sing}} - 2$, [12]. Namikawa proved the extension properties for $p \in \{1, 2\}$, in case X has canonical Gorenstein singularities, [49, Thm. 4]. In the case of finite quotient singularities similar results were obtained in [25]. For a log canonical pair with reduced boundary divisor, the cases $p \in \{1, \dim X - 1, \dim X\}$ were settled in [14, Thm. 1.1].

A related setup where the pair (X, D) is snc, and where $\pi : \widetilde{X} \to X$ is the composition of a finite Galois covering and a subsequent resolution of singularities has been studied by Esnault and Viehweg. In [9] they obtain in their special setting similar results and additionally prove vanishing of higher direct image sheaves.

A brief sketch of the proof of Theorem 4.11 is given in Section 4.5 below. The proof uses a strengthening of the Steenbrink vanishing theorem, which follows from

[4]More precisely, we should say "Let X be a normal variety such that the pair $(X, \emptyset)$ is klt..."

local Hodge-theoretic properties of log canonical singularities, in particular from the fact that log canonical spaces are Du Bois. These methods are combined with results available only for special classes of singularities, such as the recent progress in minimal model theory and partial classification of singularities that appear in minimal models.

4.3.1. Applications Theorem 4.11 has many applications useful for moduli theory. We mention two applications which will be important in our context. The first application generalises the Bogomolov-Sommese vanishing theorem to singular spaces.

Corollary 4.14 (Bogomolov-Sommese vanishing for lc pairs, [15, Thm. 7.2]). *Let* (X, D) *be a log canonical pair, where* X *is projective. If* $\mathcal{A} \subseteq \Omega_X^{[p]}(\log\lfloor D \rfloor)$ *is a* $\mathbb{Q}$*-Cartier reflexive subsheaf of rank one, then* $\kappa(\mathcal{A}) \leqslant p$.

Remark 4.15 (Notation used in Corollary 4.14). The number $\kappa(\mathcal{A})$ appearing in the statement of Corollary 4.14 is the generalised Kodaira-Iitaka dimension introduced in Definition 4.6. A reflexive sheaf $\mathcal{A}$ is rank one is called $\mathbb{Q}$-Cartier if there exists a number $n \in \mathbb{N}^+$ such that the nth reflexive tensor product $\mathcal{A}^{[n]}$ is invertible.

Proof of Corollary 4.14 in a special case. We prove Corollary 4.14 only in the special case where the sheaf $\mathcal{A} \subseteq \Omega_X^{[p]}(\log\lfloor D \rfloor)$ is invertible. The reader interested in a full proof is referred to the original reference [15].

Let $\gamma : Z \to X$ be any resolution of singularities, and let $\Delta \subset Z$ be the reduced divisor defined in Theorem 4.11 above. Theorem 4.11 will then assert the existence of an inclusion

$$\gamma^*(\mathcal{A}) \to \Omega_Z^p(\log \Delta),$$

and the standard Bogomolov-Sommese vanishing result, Theorem 2.17, applies to give that $\kappa\big(\gamma^*(\mathcal{A})\big) \leqslant p$. Since $\mathcal{A}$ is invertible, and since γ is birational, it is clear that $\kappa\big(\gamma^*(\mathcal{A})\big) = \kappa(\mathcal{A})$, finishing the proof. □

Warning 4.16. Taking the double dual of a sheaf does generally *not* commute with pulling back. Since reflexive tensor products were used in Definition 4.6 to define the Kodaira-Iitaka dimension of a sheaf, it is generally false that the Kodaira-Iitaka dimension stays invariant when pulling a sheaf $\mathcal{A}$ back to a resolution of singularities. The proof of Corollary 4.14 which is given in the simple case where $\mathcal{A}$ is invertible does therefore not work without substantial modification in the general setup where $\mathcal{A}$ is only $\mathbb{Q}$-Cartier.

The second application of Theorem 4.11 concerns rationally chain connected singular spaces. Rationally chain connected manifolds are rationally connected, and do not carry differential forms. Building on work of Hacon and M^c^Kernan, [17], we show that the same holds for reflexive forms on klt pairs.

Corollary 4.17 (Reflexive differentials on rationally chain connected spaces, [15, Thm. 5.1]). *Let (X, D) be a klt pair. If X is rationally chain connected, then X is rationally connected, and $H^0(X, \Omega_X^{[p]}) = 0$ for all $p \in \mathbb{N}, 1 \leqslant p \leqslant \dim X$.*

Proof. Choose a log resolution of singularities, $\pi : \widetilde{X} \to X$ of the pair (X, D). Since klt pairs are also dlt, a theorem of Hacon-McKernan, [17, Cor. 1.5(2)], applies to show that X and $\widetilde{X}$ are both rationally connected. In particular, it follows that $H^0(\widetilde{X}, \Omega_{\widetilde{X}}^p) = 0$ for all $p > 0$ by [33, IV. Cor. 3.8].

Since (X, D) is klt, Theorem 4.11 asserts that there exists a pull-back morphism $d\pi : \pi^*\Omega_X^{[p]} \to \Omega_{\widetilde{X}}^p$. As π is birational, $d\pi$ is generically injective and since $\Omega_X^{[p]}$ is torsion-free, this means that the induced morphism on the level of sections is injective:

$$\pi^* : H^0(X, \Omega_X^{[p]}) \to H^0(\widetilde{X}, \Omega_{\widetilde{X}}^p) = 0.$$

The claim then follows. □

4.4. Residue theory and restrictions for differentials on dlt pairs

Logarithmic Kähler differentials on snc pairs are canonically defined. They are characterised by strong universal properties and appear accordingly in a number of important sequences, filtered complexes and other constructions. First examples include the following:

(1) the pull-back property of differentials under arbitrary morphisms,
(2) relative differential sequences for smooth morphisms,
(3) residue sequences associated with snc pairs, and
(4) the description of Chern classes as the extension classes of the first residue sequence.

Reflexive differentials do in general not enjoy the same universal properties as Kähler differentials. However, we have seen in Theorem 4.11 that reflexive differentials do have good pull-back properties if we are working with log canonical pairs. In the present Section 4.4, we would like to make the point that each of the other properties listed above also has a very good analogue for reflexive differentials, as long as we are working with dlt pairs. This makes reflexive differential extremely useful in practise. In a sense, it seems fair to say that "reflexive differentials and dlt pairs are made for one another".

4.4.1. The relative differential sequence for snc pairs Here we recall the generalisation of the standard sequence for relative differentials, [19, Prop. II.8.11], to the logarithmic setup. For this, we introduce the notion of an *snc morphism* as a logarithmic analogue of a smooth morphism. Although *relatively snc divisors* have long been used in the literature, cf. [8, Sect. 3], we are not aware of a good reference that discusses them in detail. Recall that a pair (X, D) is called an "snc pair" if X is smooth, and if the divisor D is reduced and has only simple normal crossing support.

Notation 4.18 (Intersection of boundary components). *Let (X, D) be a pair of a normal space X and a divisor D, where D is written as a sum of its irreducible components $D = \alpha_1 D_1 + \ldots + \alpha_n D_n$. If $I \subseteq \{1, \ldots, n\}$ is any non-empty subset, we consider the scheme-theoretic intersection $D_I := \cap_{i \in I} D_i$. If I is empty, set $D_I := X$.*

Remark 4.19 (Description of snc pairs). In the setup of Notation 4.18, it is clear that the pair (X, D) is snc if and only if all D_I are smooth and of codimension equal to the number of defining equations: $\operatorname{codim}_X D_I = |I|$ for all I where $D_I \neq \emptyset$.

Definition 4.20 (Snc morphism, relatively snc divisor, [61, Def. 2.1]). *If (X, D) is an snc pair and $\phi : X \to T$ a surjective morphism to a smooth variety, we say that D is* relatively snc, *or that ϕ is* an snc morphism of the pair (X, D) *if for any set I with $D_I \neq \emptyset$ all restricted morphisms $\phi|_{D_I} : D_I \to T$ are smooth of relative dimension* $\dim X - \dim T - |I|$.

Remark 4.21 (Fibers of an snc morphisms). If (X, D) is an snc pair and $\phi : X \to T$ is any surjective snc morphism of (X, D), it is clear from Remark 4.19 that if $t \in T$ is any point, with preimages $X_t := \phi^{-1}(t)$ and $D_t := D \cap X_t$ then the pair (X_t, D_t) is again snc.

Remark 4.22 (All morphisms are generically snc). If (X, D) is an snc pair and $\phi : X \to T$ is any surjective morphism, it is clear from generic smoothness that there exists a dense open set $T^\circ \subseteq T$, such that $D \cap \phi^{-1}(T^\circ)$ is relatively snc over T°.

Let (X, D) be a reduced snc pair, and $\phi : X \to T$ an snc morphism of (X, D), as introduced in Definition 4.20. In this setting, the standard pull-back morphism of 1-forms extends to yield the following exact sequence of locally free sheaves on X,

$$0 \to \phi^*\Omega^1_T \to \Omega^1_X(\log D) \to \Omega^1_{X/T}(\log D) \to 0, \tag{4.23}$$

called the "relative differential sequence for logarithmic differentials". We refer to [10, Sect. 4.1] [8, Sect. 3.3] or [4, p. 137ff] for a more detailed explanation. For forms of higher degrees, the sequence (4.23) induces filtration

$$\Omega^p_X(\log D) = \mathcal{F}^0(\log) \supseteq \mathcal{F}^1(\log) \supseteq \cdots \supseteq \mathcal{F}^p(\log) \supseteq \{0\} \tag{4.24}$$

with quotients

$$0 \to \mathcal{F}^{r+1}(\log) \to \mathcal{F}^r(\log) \to \phi^*\Omega^r_T \otimes \Omega^{p-r}_{X/T}(\log D) \to 0 \tag{4.25}$$

for all r. We refer to [19, Ex. II.5.16] for the construction of (4.24).

The main result of this section, Theorem 4.26, gives analogues of (4.23)–(4.25) in case where (X, D) is dlt. In essence, Theorem 4.26 says that all properties of the relative differential sequence still hold on dlt pairs if one removes from X a set Z of codimension $\operatorname{codim}_X Z \geqslant 3$.

Theorem 4.26 (Relative differential sequence on dlt pairs, [15, Thm. 10.6]). *Let (X, D) be a dlt pair with X connected. Let $\phi : X \to T$ be a surjective morphism to a normal* *variety* T. *Then, there exists a non-empty smooth open subset* $T^\circ \subseteq T$ *with preimages* $X^\circ = \phi^{-1}(T^\circ)$, $D^\circ = D \cap X^\circ$, *and a filtration*

$$\Omega^{[p]}_{X^\circ}(\log\lfloor D^\circ \rfloor) = \mathcal{F}^{[0]}(\log) \supseteq \cdots \supseteq \mathcal{F}^{[p]}(\log) \supseteq \{0\} \tag{4.27}$$

on X° *with the following properties.*

(1) *The filtrations* (4.24) *and* (4.27) *agree wherever the pair* $(X^\circ, \lfloor D^\circ \rfloor)$ *is snc, and* ϕ *is an snc morphism of* $(X^\circ, \lfloor D^\circ \rfloor)$.
(2) *For any* r, *the sheaf* $\mathcal{F}^{[r]}(\log)$ *is reflexive, and* $\mathcal{F}^{[r+1]}(\log)$ *is a saturated subsheaf of* $\mathcal{F}^{[r]}(\log)$.
(3) *For any* r, *there exists a sequence of sheaves of* $\mathcal{O}_{X^\circ}$*-modules,*

$$0 \to \mathcal{F}^{[r+1]}(\log) \to \mathcal{F}^{[r]}(\log) \to \phi^*\Omega^r_{T^\circ} \otimes \Omega^{[p-r]}_{X^\circ/T^\circ}(\log\lfloor D^\circ \rfloor) \to 0,$$

which is exact and analytically locally split in codimension 2.
(4) *There exists an isomorphism* $\mathcal{F}^{[p]}(\log) \simeq \phi^*\Omega^p_{T^\circ}$.

Remark 4.28 (Notation used in Theorem 4.26). If S is any complex variety, we call a sequence of sheaf morphisms,

$$0 \to \mathcal{A} \to \mathcal{B} \to \mathcal{C} \to 0, \tag{4.29}$$

"exact and analytically locally split in codimension 2" if there exists a closed subvariety $C \subset S$ of codimension $\operatorname{codim}_S C \geqslant 3$ and a covering of $S \setminus C$ by subsets $(U_i)_{i \in I}$ which are open in the analytic topology, such that the restriction of (4.29) to $S \setminus C$ is exact, and such that the restriction of (4.29) to any of the open sets U_i splits. We refer to Footnote 2 on Page 79 for references concerning the notion of a "dlt pair".

Idea of proof of Theorem 4.26. We give only a very rough and incomplete idea of the proof of Theorem 4.26. To construct the filtration in (4.27), one takes the filtration (4.24) which exists on the open set $X \setminus X_{\text{sing}}$ wherever the morphism ϕ is snc, and extends the sheaves to reflexive sheaves that are defined on all of X. It is then not very difficult to show that the sequences (4.26.3) are exact and locally split away from a subset $Z \subset X$ of codimension $\operatorname{codim}_X Z \geqslant 2$. The main point of Theorem 4.26 is, however, that it suffices to remove from X a set of codimension $\operatorname{codim}_X Z \geqslant 3$. For this, a careful analysis of the codimension-two structure of dlt pairs, cf. [15, Sect. 9], proves to be key. □

4.4.2. Residue sequences for reflexive differential forms A very important feature of logarithmic differentials is the existence of a residue map. In its simplest form consider a smooth hypersurface $D \subset X$ in a manifold X. The residue map is then the cokernel map in the exact sequence

$$0 \to \Omega^1_X \to \Omega^1_X(\log D) \to \mathcal{O}_D \to 0.$$

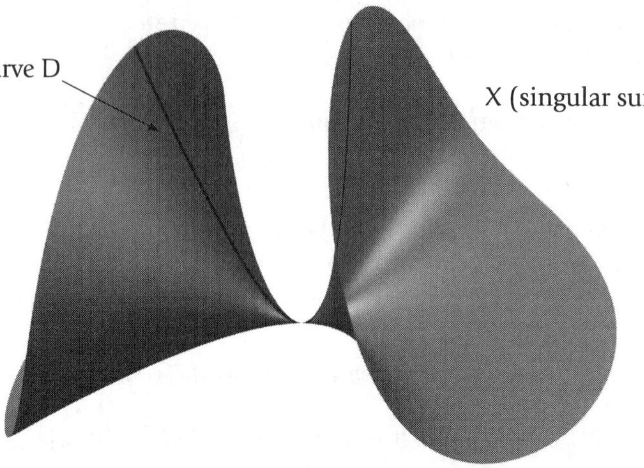

Figure 1. A setup for the residue map on singular spaces.

More generally, consider a reduced snc pair (X, D). Let $D_0 \subseteq D$ be any irreducible component and recall from [11, 2.3(b)] that there exists a residue sequence,

$$0 \to \Omega_X^p(\log(D - D_0)) \longrightarrow \Omega_X^p(\log D) \xrightarrow{\rho^p} \Omega_{D_0}^{p-1}(\log D_0^c) \to 0,$$

where $D_0^c := (D - D_0)|_{D_0}$ denotes the "restricted complement" of D_0. More generally, if $\phi : X \to T$ is an snc morphism of (X, D) we have a relative residue sequence

$$(4.30) \quad 0 \to \Omega_{X/T}^p(\log(D - D_0)) \longrightarrow \Omega_{X/T}^p(\log D) \xrightarrow{\rho^p} \Omega_{D_0/T}^{p-1}(\log D_0^c) \to 0.$$

The sequence (4.30) is not a sequence of locally free sheaves on X, and its restriction to D_0 will never be exact on the left. However, an elementary argument, cf. [27, Lem. 2.13.2], shows that restriction of (4.30) to D_0 induces the following exact sequence

$$(4.31) \qquad 0 \to \Omega_{D_0/T}^p(\log D_0^c) \xrightarrow{i^p} \Omega_{X/T}^p(\log D)|_{D_0} \xrightarrow{\rho_D^p} \Omega_{D_0/T}^{p-1}(\log D_0^c) \to 0,$$

which is very useful for inductive purposes. We recall without proof the following elementary fact about the residue sequence.

Fact 4.32 (Residue map as a test for logarithmic poles). If $\sigma \in H^0\big(X, \Omega_{X/T}^p(\log D)\big)$ is any reflexive form, then $\sigma \in H^0\big(X, \Omega_{X/T}^p(\log(D - D_0))\big)$ if and only if $\rho^p(\sigma) = 0$.

If the pair (X, D) is not snc, no residue map exists in general. However, if (X, D) is dlt, then [35, Cor. 5.52] applies to show that D_0 is normal, and an analogue of the residue map ρ^p exists for sheaves of reflexive differentials. To illustrate the problem we are dealing with, consider a normal space X that contains a smooth Weil divisor $D = D_0$, similar to the one sketched in Figure 1. One can easily construct examples where the singular set $Z := X_{sing}$ is contained in D and has codimension 2 in X, but codimension one in D. In this setting, a reflexive form $\sigma \in H^0(D_0, \Omega_X^{[p]}(\log D_0)|_{D_0})$ is simply the restriction of a logarithmic form defined outside of Z, and the form $\rho^{[p]}(\sigma)$ is the extension of the well-defined form $\rho^p(\sigma|_{D_0\setminus Z})$ over Z, as a rational form with poles along $Z \subset D_0$. If the singularities of X are bad, it will generally happen that the extension $\rho^{[p]}(\sigma)$ has poles of arbitrarily high order. Theorem 4.33 asserts that this does not happen when (X, D) is dlt.

Theorem 4.33 (Residue sequences for dlt pairs, [15, Thm. 11.7]). *Let* (X, D) *be a dlt pair with* $\lfloor D \rfloor \neq \emptyset$ *and let* $D_0 \subseteq \lfloor D \rfloor$ *be an irreducible component. Let* $\phi : X \to T$ *be a surjective morphism to a normal variety* T *such that the restricted map* $\phi|_{D_0} : D_0 \to T$ *is still surjective. Then, there exists a non-empty open subset* $T^\circ \subseteq T$, *such that the following holds if we denote the preimages as* $X^\circ = \phi^{-1}(T^\circ)$, $D^\circ = D \cap X^\circ$, *and the "complement" of* D_0° *as* $D_0^{\circ,c} := (\lfloor D^\circ \rfloor - D_0^\circ)|_{D_0^\circ}$.

(1) *There exists a sequence*

$$0 \to \Omega^{[r]}_{X^\circ/T^\circ}(\log(\lfloor D^\circ \rfloor - D_0^\circ)) \to \Omega^{[r]}_{X^\circ/T^\circ}(\log\lfloor D^\circ \rfloor) \xrightarrow{\rho^{[r]}} \Omega^{[r-1]}_{D_0^\circ/T^\circ}(\log D_0^{\circ,c}) \to 0$$

which is exact in X° *outside a set of codimension at least* 3. *This sequence coincides with the usual residue sequence* (4.30) *wherever the pair* (X°, D°) *is snc and the map* $\phi^\circ : X^\circ \to T^\circ$ *is an snc morphism of* (X°, D°).

(2) *The restriction of Sequence (4.33.1) to* D_0 *induces a sequence*

$$0 \to \Omega^{[r]}_{D_0^\circ/T^\circ}(\log D_0^{\circ,c}) \to \Omega^{[r]}_{X^\circ/T^\circ}(\log\lfloor D^\circ \rfloor)|^{**}_{D_0^\circ} \xrightarrow{\rho^{[r]}_{D_0^\circ}} \Omega^{[r-1]}_{D_0^\circ/T^\circ}(\log D_0^{\circ,c}) \to 0$$

which is exact on D_0° *outside a set of codimension at least* 2 *and coincides with the usual restricted residue sequence* (4.31) *wherever the pair* (X°, D°) *is snc and the map* $\phi^\circ : X^\circ \to T^\circ$ *is an snc morphism of* (X°, D°). □

As before, the proof of Theorem 4.33 relies on our knowledge of the codimension-two structure of dlt pairs. Fact 4.32 and Theorem 4.33 together immediately imply that the residue map for reflexive differentials can be used to check if a reflexive form has logarithmic poles along a given boundary divisor.

Remark 4.34 (Residue map as a test for logarithmic poles). In the setting of Theorem 4.33, if $\sigma \in H^0(X, \Omega_X^{[p]}(\log\lfloor D \rfloor))$ is any reflexive form, then $\sigma \in H^0(X, \Omega_X^{[p]}(\log\lfloor D \rfloor - D_0))$ if and only if $\rho^{[p]}(\sigma) = 0$.

4.4.3. The residue map for 1-forms Let X be a smooth variety and $D \subset X$ a smooth, irreducible divisor. The first residue sequence (4.30) of the pair (X, D) then reads

$$0 \to \Omega_D^1 \to \Omega_X^1(\log D)|_D \xrightarrow{\rho^1} \mathcal{O}_D \to 0,$$

and we obtain a connecting morphism of the long exact cohomology sequence,

$$\delta : H^0(D, \mathcal{O}_D) \to H^1(D, \Omega^1_D).$$

In this setting, the standard description of the first Chern class in terms of the connecting morphism, [19, III. Ex. 7.4], asserts that

$$c_1(\mathcal{O}_X(D)|_D) = \delta(\mathbf{1}_D) \in H^1(D, \Omega^1_D), \tag{4.35}$$

where $\mathbf{1}_D$ is the constant function on D with value one. Theorem 4.36 generalises Identity (4.35) to the case where (X, D) is a reduced dlt pair with irreducible boundary divisor.

Theorem 4.36 (Description of Chern classes, [15, Thm. 12.2]). *Let (X, D) be a dlt pair, $D = \lfloor D \rfloor$ irreducible. Then, there exists a closed subset $Z \subset X$ with $\operatorname{codim}_X Z \geqslant 3$ and a number $m \in \mathbb{N}$ such that mD is Cartier on $X^\circ := X \setminus Z$, such that $D^\circ := D \cap X^\circ$ is smooth, and such that the restricted residue sequence*

$$0 \to \Omega^1_D \to \Omega^{[1]}_X(\log D)|^{**}_D \xrightarrow{\rho_D} \mathcal{O}_D \to 0 \tag{4.37}$$

defined in Theorem 4.33 is exact on D°. Moreover, for the connecting homomorphism δ in the associated long exact cohomology sequence

$$\delta : H^0(D^\circ, \mathcal{O}_{D^\circ}) \to H^1(D^\circ, \Omega^1_{D^\circ})$$

we have

$$\delta(m \cdot \mathbf{1}_{D^\circ}) = c_1\left(\mathcal{O}_{X^\circ}(mD^\circ)|_{D^\circ}\right). \tag{4.38}$$

4.5. Existence of a pull-back morphism, idea of proof

The proof of Theorem 4.11 is rather involved. To illustrate the idea of the proof, we concentrate on a very special case, and give only indications what needs to be done to handle the general setup.

4.5.1. Simplifying assumptions and setup of notation The following simplifying assumptions will be maintained throughout the present Section 4.5.

Simplifying Assumptions 4.39. The space X has dimension $n := \dim X \geqslant 3$. It is klt, has only one single isolated singularity $x \in X$, and the divisor D is empty. The morphism $\gamma : Z \to X$ is a resolution of singularities, whose exceptional set $E \subset Z$ is a divisor with simple normal crossing support.

To prove Theorem 4.11, we need to show in essence that reflexive differential forms $\sigma \in H^0(X, \Omega^{[p]}_X)$ pull back to give differential forms $\tilde{\sigma} \in H^0(Z, \Omega^p_Z)$. The following observation, an immediate consequence of Fact 4.5, turns out to be key.

Observation 4.40. To give a reflexive differential $\sigma \in H^0(X, \Omega^{[p]}_X)$, it is equivalent to give a differential form $\sigma^\circ \in H^0(X \setminus X_{sing}, \Omega^p_X)$, defined on the smooth locus of X. Since the resolution map identifies the open subvarieties $Z \setminus E$ and $X \setminus X_{sing}$, we see that to give a reflexive differential $\sigma \in H^0(X, \Omega_X^{[p]})$, it is in fact equivalent to give a differential form $\widetilde{\sigma}^\circ \in H^0(Z \setminus E, \Omega_Z^p)$.

In essence, Observation 4.40 says that to show Theorem 4.11, we need to prove that the natural restriction map

$$H^0(Z, \Omega_Z^p) \to H^0(Z \setminus E, \Omega_Z^p) \tag{4.41}$$

is in fact surjective. In other words, we need to show that any differential form on Z, which is defined outside of the γ-exceptional set E, automatically extends across E, to give a differential form defined on all of Z. This is done in two steps. We first show that the restriction map

$$H^0(Z, \Omega_Z^p(\log E)) \to H^0(Z \setminus E, \Omega_Z^p(\log E)) = H^0(Z \setminus E, \Omega_Z^p) \tag{4.42}$$

is surjective. In other words, we show that any differential form on Z, defined outside of E, extends as a form with logarithmic poles along E. Secondly, we show that the natural inclusion map

$$H^0(Z, \Omega_Z^p) \to H^0(Z, \Omega_Z^p(\log E)) \tag{4.43}$$

is likewise surjective. In other words, we show that globally defined differentials forms on Z, which are allowed to have logarithmic poles along E, really do not have any poles. Surjectivity of the morphisms (4.42) and (4.43) together will then imply surjectivity of (4.41), finishing the proof of Theorem 4.11.

The arguments used to prove surjectivity of (4.42) and (4.43), respectively, are of rather different nature. We will sketch the arguments in Sections 4.5.2 and 4.5.3 below.

4.5.2. Surjectivity of the restriction map (4.42) Under the Simplifying Assumptions 4.39, surjectivity of the map (4.42) has essentially been shown by Steenbrink and van Straten, [57]. We give a brief synopsis of their line of argumentation and indicate additional steps of argumentation required to handle the general setting. To start, recall from [19, III ex. 2.3e] that the map (4.42) is part of the standard sequence that defines cohomology with supports,

$$\begin{aligned} \cdots \to H^0(Z, \Omega_Z^p(\log E)) \to H^0(Z \setminus E, \Omega_Z^p(\log E)) \to \\ \to H_E^1(Z, \Omega_Z^p(\log E)) \to \cdots \end{aligned} \tag{4.44}$$

We aim to show that the last term in (4.44) vanishes. There are two main ingredients to the proof: formal duality and Steenbrink's vanishing theorem.

Theorem 4.45 (Formal duality theorem for cohomology with support, [18, Chapt. 3, Thm. 3.3]). *Under the Assumptions 4.39, if $\mathcal{F}$ is any locally free sheaf on Z and $0 \leqslant j \leqslant \dim Z$ any number, then there exists an isomorphism*

$$\big((R^j\gamma_*\mathcal{F})_x\big)^{\widehat{}} \cong H_E^{n-j}(Z, \mathcal{F}^* \otimes \omega_Z)^*,$$

where $\widehat{\ }$ *denotes completion with respect to the maximal ideal* $\mathfrak{m}_x$ *of the point* $x \in X$, *and where* $n = \dim X = \dim Z$. □

A brief introduction to formal duality, together with a readable, self-contained proof of Theorem 4.45 is found in [14, Appendix A] while Hartshorne's lecture notes [18] are the standard reference for these matters.

Theorem 4.46 (Steenbrink vanishing, [56, Thm. 2.b]). *If* p, q *are any two numbers with* $p + q > \dim Z$, *then* $R^q\gamma_*(\mathcal{J}_E \otimes \Omega_Z^p(\log E)) = 0$. □

Remark 4.47. Steenbrink's vanishing theorem is proven using local Hodge theory of isolated singularities. For $p = n$, the sheaves Ω_Z^n and $\mathcal{J}_E \otimes \Omega_Z^n(\log E)$ are isomorphic. In this case, the Steenbrink vanishing theorem reduces to Grauert-Riemenschneider vanishing, [13].

Setting $\mathcal{F} := \mathcal{J}_E \otimes \Omega_Z^{n-p}(\log E)$ and using that $\mathcal{F}^* \otimes \omega_Z \cong \Omega_Z^p(\log E)$, formal duality and Steenbrink vanishing together show that $H_E^1(Z, \Omega_Z^p(\log E)) = 0$, for $p < \dim Z - 1$, proving surjectivity of (4.42) for these values of p. The other cases need to be treated separately.

case $\mathbf{p = n}$: After passing to an index-one cover, surjectivity of (4.42) in case $p = n$ follows almost directly from the definition of klt, cf. [14, Sect. 5].

case $\mathbf{p = n - 1}$: In this case one uses the duality between Ω_Z^{n-1} and the tangent sheaf T_Z, and the fact that any section in the tangent sheaf of X always lifts to the canonical resolution of singularities, cf. [14, Sect. 6].

General case The argument outlined above, using formal duality and Steenbrink vanishing, works only because we were assuming that the singularities of X are isolated. In the general case, where the Simplifying Assumptions 4.39 do not necessarily hold, this is not necessarily the case. In order to deal with non-isolated singularities, one applies a somewhat involved cutting-down procedure, as indicated in Figure 2. This way, it is often possible to view non-isolated log canonical singularities a family of isolated singularities, where surjectivity of (4.42) can be shown on each member of the family. To conclude that it holds on all of Z, the following strengthening of Steenbrink vanishing is required.

Theorem 4.48 (Steenbrink-type vanishing for log canonical pairs, [15, Thm. 14.1]). *Let* (X, D) *be a log canonical pair of dimension* $n \geqslant 2$. *If* $\gamma : Z \to X$ *is a log resolution of singularities with exceptional set* E *and*

$$\Delta := \operatorname{supp}(E + \gamma^{-1}\lfloor D \rfloor),$$

then $R^{n-1}\gamma_*(\Omega_Z^p(\log \Delta) \otimes \mathcal{O}_Z(-\Delta)) = 0$ *for all* $0 \leqslant p \leqslant n$. □

The proof of Theorem 4.48 essentially relies on the fact that log canonical pairs are Du Bois, [34]. The Du Bois property generalises the notion of rational singularities. For an overview, see [42].

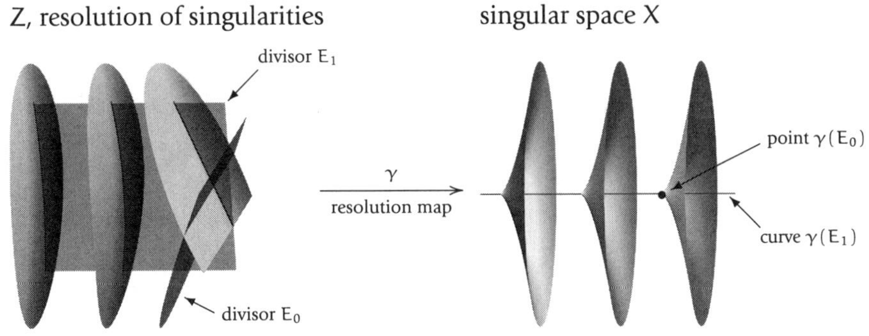

The figure sketches a situation where X is a threefold whose singular locus is a curve. Near the general point of the singular locus, the variety X looks like a family of isolated surfaces singularities. The exceptional set E of the resolution map γ contains two irreducible divisors E_0 and E_1.

Figure 2. Non-isolated singularities

4.5.3. Surjectivity of the inclusion map (4.43) Let $\sigma \in H^0(Z, \Omega_Z^p(\log E))$ be any differential form on Z that is allowed to have logarithmic poles along E. To show surjectivity of the inclusion map (4.43), we need to show that σ really does not have any poles along E. To give an idea of the methods used to prove this, we consider only the case where $p > 1$. We discuss two particularly simple cases first.

The case where E is irreducible Assume that E is irreducible. To show that σ does not have any logarithmic poles along E, recall from Fact 4.32 that it suffices to show that σ is in the kernel of the residue map

$$\rho^p : H^0(Z, \Omega_Z^p(\log E)) \to H^0(E, \Omega_E^{p-1}).$$

On the other hand, we know from a result of Hacon-McKernan, [17, Cor. 1.5(2)], that E is rationally connected, so that $H^0(E, \Omega_E^{p-1}) = 0$. This clearly shows that σ is in the kernel of ρ^p and completes the proof when E is irreducible.

The case where (Z, E) admits a simple minimal model program In general, the divisor E need not be irreducible. Let us therefore consider the next difficult case where E is reducible with two components, say $E = E_1 \cup E_2$. The resolution map γ will then factor via a γ-relative minimal model program of the pair (Z, E), which we assume for simplicity to have the following particularly special form, sketched[5] also in Figure 3.

$$Z = Z_0 \xrightarrow[\text{contracts } E_1 \text{ to a point}]{\lambda_1} Z_1 \xrightarrow[\text{contracts } E_{2,1} := (\lambda_1)_*(E_2) \text{ to a point}]{\lambda_2} X.$$

[5]The computer code used to generate the images in Figure 3 is partially taken from [3].

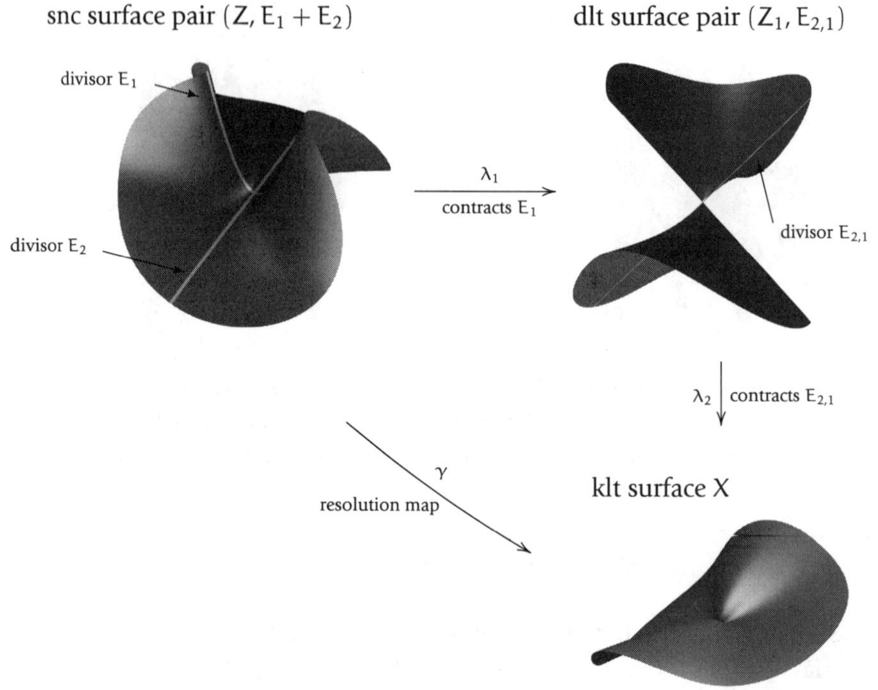

This sketch shows a resolution of an isolated klt surface singularity, and the decomposition of the resolution map given by the minimal model program of the snc pair $(Z, E_1 + E_2)$.

Figure 3. Resolution of an isolated klt surface singularity

In this setting, the arguments outlined above apply verbatim to show that σ has no poles along the divisor E_1. To show that σ does not have any poles along the remaining component E_2, observe that it suffices to consider the induced reflexive form on the possibly singular space Z_1, say $\sigma_1 \in H^0\big(Z_1, \Omega^{[p]}_{Z_1}(\log E_{2,1})\big)$, where $E_{2,1} := (\lambda_1)_*(E_2)$, and to show that σ_1 does not have any poles along $E_{2,1}$. For that, we follow the same line of argument once more, accounting for the singularities of the pair $(X_1, E_{2,1})$.

The pair $(X_1, E_{2,1})$ is dlt, and it follows from adjunction that the divisor $E_{2,1}$ is necessarily normal, [35, Cor. 5.52]. Using the residue map for reflexive differentials on dlt pairs that was constructed in Theorem 4.33,

$$\rho^{[p]} : H^0\big(X_1, \Omega^{[p]}_{Z_1}(\log E_{2,1})\big) \to H^0\big(E_{2,1}, \Omega^{[p-1]}_{E_{2,1}}\big),$$

we have seen in Remark 4.34 that it suffices to show that $\rho^{[p]}(\sigma_1) = 0$. Because the morphism λ_2 contracts the divisor $E_{2,1}$ to a point, the result of Hacon-McKernan will again apply to show that $E_{2,1}$ is rationally connected. Even though there are numerous examples of rationally connected spaces that carry non-trivial reflexive forms, we claim that in our special setup we do have the vanishing

$$H^0\big(E_{2,1}, \Omega^{[p-1]}_{E_{2,1}}\big) = 0. \tag{4.49}$$

For this, recall from the adjunction theory for Weil divisors on normal spaces, [36, Chapt. 16 and Prop. 16.5] and [7, Sect. 3.9 and Glossary], that there exists a Weil divisor D_E on the normal variety $E_{2,1}$ which makes the pair $(E_{2,1}, D_E)$ klt. Now, if we knew that the extension theorem would hold for the pair $(E_{2,1}, D_E)$, we can prove the vanishing statement (4.49), arguing exactly as in the proof of Corollary 4.17, where we show the non-existence of reflexive forms on rationally connected klt spaces as a corollary of the Pull-Back Theorem 4.11. Since $\dim E_{2,1} < \dim X$, this suggests an inductive proof, beginning with easy-to-prove extension theorems for reflexive forms on surfaces, and working our way up to higher-dimensional varieties. The proof in [15] follows this inductive pattern.

The general case To handle the general case, where the Simplifying Assumptions 4.39 do not necessarily hold true, we need to work with pairs (X, D) where D is not necessarily empty, the γ-relative minimal model program might involve flips, and the singularities of X need not be isolated. All this leads to a slightly protracted inductive argument, heavily relying on cutting-down methods and outlined in detail in [15, Sect. 19].

4.6. Open problems

In view of the Viehweg-Zuo construction, it would be very interesting to know if a variant of the Pull-Back Theorem 4.11 holds for symmetric powers of $\Omega^{[1]}_X(\log D)$, or for other tensor powers. As shown by examples, cf. [14, Ex. 3.1.3], the naïve generalisation of Theorem 4.11 is wrong. Still, it seems conceivable that a suitable generalisation, perhaps formulated in terms of Campana's orbifold differentials, might hold. However, note that several of the key ingredients used in the proof of Theorem 4.11, including Steenbrink's vanishing theorem, rely on (local) Hodge theory, for which no version is known for tensor powers of differential forms.

Question 4.50. Is there a formulation of the Pull-Back Theorem 4.11 that holds for symmetric and other tensor powers of differential forms?

Examples suggest that the Pull-Back Theorem 4.11 is optimal, and that the class of log canonical pairs is the natural class of spaces where a pull-back theorem can hold.

Question 4.51. To what extend is the Pull-Back Theorem 4.11 optimal? Is there a version of the pull-back theorem that does not require the log canonical divisor $K_X + D$ to be $\mathbb{Q}$-Cartier? If we are interested only in special values of p, is the divisor Δ the smallest possible?

The last question concerns the generalisation of the Bogomolov-Sommese vanishing theorem. One of the main difficulties with its current formulation is the requirement that the sheaf $\mathcal{A}$ be $\mathbb{Q}$-Cartier. We have seen in Section 4.1 how interesting reflexive subsheaves $\mathcal{A} \subseteq \Omega_X^{[p]}$ can often be constructed using the Harder-Narasimhan filtration. Unless the space X is $\mathbb{Q}$-factorial, there is, however, no way to guarantee that a sheaf constructed this way will actually be $\mathbb{Q}$-Cartier. The property to be $\mathbb{Q}$-factorial, however, is not stable under taking hyperplane sections and difficult to guarantee in practise.

Question 4.52. Is there a version of the generalised Bogomolov-Sommese vanishing theorem, Corollary 4.14, that does not require the sheaf $\mathcal{A}$ to be $\mathbb{Q}$-Cartier?

5. Viehweg's conjecture for families over threefolds, sketch of proof

5.1. A special case of the Viehweg conjecture

We conclude this paper by sketching a proof of the Viehweg Conjecture 2.8 in one special case, illustrating the use of the methods introduced in Sections 3 and 4. As in Section 4.1 we consider a family $f^\circ : X^\circ \to Y^\circ$ of canonically polarised varieties over a quasi-projective threefold. Assuming that f° is of maximal variation, we would like to show that the logarithmic Kodaira dimension $\kappa(Y^\circ)$ cannot be zero.

Proposition 5.1 (Partial answer to Viehweg's conjecture). *Let* $f^\circ : X^\circ \to Y^\circ$ *be a smooth, projective family of canonically polarised varieties over a smooth, quasi-projective base manifold of dimension* $\dim Y^\circ = 3$. *Assume that the family* f° *is of maximal variation, i.e., that* $\mathrm{Var}(f^\circ) = \dim Y^\circ$. *Then* $\kappa(Y^\circ) \neq 0$.

The proof of Proposition 5.1 follows the line of argumentation outlined in Section 4.1. We prove that the Picard number of a suitable minimal model cannot be one, thereby exhibiting a fibre space structure to which induction can be applied. The presentation follows [29, Sect. 9].

5.2. Sketch of proof

In essence, we follow the line of argument sketched in Section 4.1. We argue by contradiction, i.e., we maintain the assumptions of Proposition 5.1 and assume in addition that $\kappa(Y^\circ) = 0$.

5.2.1. Setup of notation As before, choose a smooth compactification $Y \supseteq Y^\circ$ such that $D := Y \setminus Y^\circ$ is a divisor with only simple normal crossings. Let $\lambda : Y \dashrightarrow Y_\lambda$ be the rational map obtain by a run of the minimal model program for the pair (Y, D) and set $D_\lambda := \lambda_*(D)$. The following is then known to hold.

(1) The variety Y_λ is normal and $\mathbb{Q}$-factorial.
(2) The variety Y_λ is log terminal. The pair (Y_λ, D_λ) is dlt.
(3) There exists a number m' such that $m'(K_{Y_\lambda} + D_\lambda) \equiv 0$. In particular, the divisor $K_{Y_\lambda} + D_\lambda$ is numerically trivial.

By Viehweg-Zuo's Theorem 3.1, there exists a number $m > 0$ and a big invertible sheaf $\mathcal{A} \subseteq \operatorname{Sym}^m \Omega^1_Y(\log D)$. As we have seen in Proposition 4.7, this induces a reflexive sheaf $\mathcal{A}_\lambda \subseteq \operatorname{Sym}^{[m]} \Omega^1_{Y_\lambda}(\log D_\lambda)$ of rank one and Kodaira-Iitaka dimension $\kappa(\mathcal{A}_\lambda) = \dim Y_\lambda$.

5.2.2. The Harder-Narasimhan filtration of $\Omega^{[1]}_{Y_\lambda}(\log D_\lambda)$ As in Section 4.1 above, we employ the Harder-Narasimhan filtration to obtain additional information about the space Y_λ.

Claim 5.2. *The divisor D_λ is not empty.*

Proof. For simplicity, we prove Claim 5.2 only in case where the canonical divisor K_{Y_λ} is Cartier, and where the space Y_λ therefore has only canonical singularities. For a proof in the general setup, the same line of argumentation applies after passing to a global index-one cover. We argue by contradiction and assume that $D_\lambda = 0$.

As before, let $C \subseteq Y_\lambda$ be a general complete intersection curve in the sense of Mehta-Ramanathan, cf. [21, Sect. II.7]. Since the general complete intersection curve C avoids the singular locus of Y_λ, we obtain that the restricted sheaf of Kähler differentials $\Omega^1_{Y_\lambda}|_C$ as well as its dual $\mathcal{T}_{Y_\lambda}|_C$, the restriction of the tangent sheaf, are locally free. Further, the numerical triviality of $K_{Y_\lambda} \equiv K_{Y_\lambda} + D_\lambda$ implies that

$$K_{Y_\lambda}.C = c_1\big(\Omega^{[1]}_{Y_\lambda}(\log D_\lambda)\big).C = c_1\big(\operatorname{Sym}^{[m]} \Omega^1_{Y_\lambda}(\log D_\lambda)\big).C = 0.$$

On the other hand, since $\mathcal{A}_\lambda$ is big, we have that $c_1(\mathcal{A}_\lambda).C > 0$. As in the proof of Proposition 4.1, this implies that the restricted sheaves $\Omega^1_{Y_\lambda}|_C$ as well as its dual $\mathcal{T}_{Y_\lambda}|_C$, are not semistable. The maximal destabilising subsheaf of $\mathcal{T}_{Y_\lambda}|_C$ is semistable and of positive degree, hence ample. In this setup, a variant [31, Cor. 5] of Miyaoka's uniruledness criterion [48, Cor. 8.6] applies to give the uniruledness of Y_λ. For more details on this criterion, see the survey [30].

To finish the argument, let $r : W \to Y_\lambda$ be a resolution of singularities. Since uniruledness is a birational property, the space W is uniruled and therefore has Kodaira-dimension $\kappa(W) = -\infty$. On the other hand, since Y_λ has only canonical singularities, the $\mathbb{Q}$-linear equivalence class of the canonical bundle K_W is given as

$$K_W \equiv r^*(K_{Y_\lambda}) + (\text{effective, } r\text{-exceptional divisor}).$$

But because K_{Y_λ} is $\mathbb{Q}$-linearly equivalent to the trivial divisor, we obtain that $\kappa(W) \geqslant 0$, a contradiction. □

5.2.3. Further contractions Claim 5.2 implies that $K_{Y_\lambda} \equiv -D_\lambda$ and it follows that for any rational number $0 < \varepsilon < 1$,

$$\kappa\big(K_{Y_\lambda} + (1-\varepsilon)D_\lambda\big) = \kappa\big(\varepsilon K_{Y_\lambda}\big) = \kappa\big(Y_\lambda\big) = -\infty. \tag{5.3}$$

Now choose one ε and run the log minimal model program for the dlt pair $(Y_\lambda, (1-\varepsilon)D_\lambda)$. This way one obtains morphisms and birational maps as follows

$$Y_\lambda \overset{\mu}{\underset{\text{minimal model program}}{\dashrightarrow}} Y_\mu \overset{\pi}{\underset{\text{Mori fibre space}}{\longrightarrow}} Z.$$

Again, let $D_\mu := \mu_*(D_\lambda)$ be the cycle-theoretic image of D_λ. The main properties of Y_μ and D_μ are summarised as follows.

(1) The variety Y_μ is normal and $\mathbb{Q}$-factorial.
(2) The variety Y_μ is log terminal. The pair $(Y_\mu, (1-\varepsilon)D_\mu)$ is dlt.
(3) The divisor $K_{Y_\mu} + D_\mu$ is numerically trivial.
(4) There exists a reflexive sheaf $\mathcal{A}_\mu \subseteq \mathrm{Sym}^{[m]}\,\Omega^1_{Y_\mu}(\log D_\mu)$ of rank one and Kodaira-Iitaka dimension $\kappa(\mathcal{A}_\mu) = \dim Y_\mu$.

In fact, more is true.

Claim 5.4. *The pair (Y_μ, D_μ) is log canonical.*

Proof. Since $K_{Y_\lambda} + D_\lambda \equiv 0$, some positive multiples of K_{Y_λ} and $-D_\lambda$ are numerically equivalent. For any two rational numbers $0 < \varepsilon', \varepsilon'' < 1$, the divisors $K_{Y_\lambda} + (1-\varepsilon')D_\lambda$ and $K_{Y_\lambda} + (1-\varepsilon'')D_\lambda$ are thus again numerically equivalent up to a positive rational multiple.

The birational map μ is therefore a minimal model program for the pair $(Y_\lambda, (1-\varepsilon)D_\lambda)$, independently of the number ε chosen in its construction. It follows that (Y_μ, D_μ) is a limit of dlt pairs and therefore log canonical. □

5.2.4. The fibre space structure of Y_μ Another application of the "Harder-Narasimhan-trick" exhibits a fibre structure of Y_μ.

Claim 5.5. *The Picard-number $\rho(Y_\mu)$ is larger than one. In particular, the map $Y_\mu \to Z$ is a proper fibre space whose fibres are proper subvarieties of Y_μ.*

Proof. As before, let $C \subseteq Y_\mu$ be a general complete intersection curve. Again, the existence of the Viehweg-Zuo sheaf $\mathcal{A}_\mu$ implies that the sheaf of reflexive differentials $\Omega^{[1]}_{Y_\mu}(\log D_\mu)$ is not semistable, and contains a destabilising subsheaf $\mathcal{B}_\mu \subseteq \Omega^{[1]}_{Y_\mu}(\log D_\mu)$ with $c_1(\mathcal{B}_\mu).C > 0$. Since the intersection number $c_1(\mathcal{B}_\mu).C$ is positive, the rank r of the sheaf $\mathcal{B}_\mu$ must be strictly less than $\dim Y_\mu$, and its determinant is a subsheaf of the sheaf of logarithmic r-forms,

$$\det \mathcal{B}_\mu \subseteq \Omega^{[r]}_{Y_\mu}(\log D_\mu) \quad \text{with} \quad c_1(\det \mathcal{B}_\mu).C > 0 \quad \text{and} \quad r < \dim Y_\mu.$$

If $\rho(Y_\mu) = 1$, then the sheaf $\det \mathcal{B}_\mu$ would necessarily be $\mathbb{Q}$-ample, violating the Bogomolov-Sommese vanishing theorem for log canonical pairs, Corollary 4.14. This finishes the proof of Claim 5.5. □

Now, if $F \subset Y_\mu$ is a general fibre of π and $D_F := D_\mu \cap F$, then F is a normal curve or surface, and the pair (F, D_F) is log canonical and has Kodaira dimension $\kappa(K_F + D_F) = 0$. By [35, Prop. 4.11], the variety F is even $\mathbb{Q}$-factorial. It is then possible to argue by induction: assuming that Viehweg's conjecture holds for families over surfaces, one obtains that the restriction of the family f° to the strict transform $(\mu \circ \lambda)_*^{-1}(F)$ cannot be of maximal variation. Since the fibres dominate the variety, this contradicts the assumption that the family f° is of maximal variation, and therefore finishes the sketch of proof of Proposition 5.1.

The reader interested in more details is referred to [29, Sect. 9], where a stronger statement is shown, proving that any family over a base manifold with $\kappa(Y^\circ) = 0$ is actually isotrivial.

References

[1] S. J. Arakelov. Families of algebraic curves with fixed degeneracies. *Izv. Akad. Nauk SSSR Ser. Mat.*, **35**, 1269–1293, 1971. MR 0321933 (48 #298) ← 72

[2] C. Araujo. The cone of pseudo-effective divisors of log varieties after Batyrev. *Math. Z.*, **264**(1), 179–193, 2010. DOI:10.1112/S0010437X09004321. ← 79

[3] J. Baum. *Aufblasungen und Desingularisierungen*. Staatsexamensarbeit, Universität zu Köln, 2007. ← 103

[4] J. Bertin, J.-P. Demailly, L. Illusie, and C. Peters. Introduction to Hodge theory, *SMF/AMS Texts and Monographs*, **8**, American Mathematical Society, Providence, RI, 2002, Translated from the 1996 French original by James Lewis and Peters. MR 1924513 (2003g:14009) ← 96

[5] F. Campana. Orbifolds, special varieties and classification theory. *Ann. Inst. Fourier (Grenoble)*, **54**(3), 499–630, 2004. MR 2097416 (2006c:14013) ← 79, 80

[6] F. Campana. *Orbifoldes spéciales et classification biméromorphe des variétés Kählériennes compactes*, preprint arXiv:0705.0737v5, October 2008. ← 79, 80

[7] A. Corti et al.. Flips for 3-folds and 4-folds. *Oxford Lecture Series in Mathematics and its Applications*, **35**, Oxford University Press, Oxford, 2007. MR 2352762 (2008j:14031) ← 105

[8] P. Deligne. Équations différentielles à points singuliers réguliers. *Lecture Notes in Mathematics*, **163**, Springer-Verlag, Berlin, 1970. MR 54 #5232 ← 79, 95, 96

[9] H. Esnault and E. Viehweg. Revêtements cycliques, Algebraic threefolds (Varenna, 1981). *Lecture Notes in Math.*, **947**, 241–250, Springer, Berlin, 1982. MR 672621 (84m:14015) ← 93

[10] H. Esnault and E. Viehweg. Effective bounds for semipositive sheaves and for the height of points on curves over complex function fields. *Compositio Math.*, **76**(1–2), 69–85, 1990. Algebraic geometry (Berlin, 1988). MR 1078858 (91m:14038) ← 86, 96

[11] H. Esnault and E. Viehweg. Lectures on vanishing theorems. *DMV Seminar*, **20**, Birkhäuser Verlag, Basel, 1992. Available at the author's web site http://www.uni-due.de/~mat903/books.html. MR 1193913 (94a:14017) ← 79, 98

[12] H. Flenner. Extendability of differential forms on nonisolated singularities *Invent. Math.*, **94**(2), 317–326, 1988. MR 958835 (89j:14001) ← 93

[13] H. Grauert and O. Riemenschneider. Verschwindungssätze für analytische Kohomologiegruppen auf komplexen Räumen. *Invent. Math.*, **11**, 263–292, 1970. MR 0302938 (46 #2081) ← 102

[14] D. Greb, S. Kebekus, and S. J. Kovács. Extension theorems for differential forms, and Bogomolov-Sommese vanishing on log canonical varieties. *Compos. Math.*, **146**, 193–219, 2010. DOI:10.1112/S0010437X09004321. ← 93, 102, 105

[15] D. Greb, S. Kebekus, S. J. Kovács, and T. Peternell. Differential forms on log canonical spaces. *Publ. Math. IHES.*, **114**(1), 87–169, 2011. DOI:10.1007/s10240-011-0036-0. An extended version with additional graphics is available as arXiv:1003.2913. ← 91, 92, 94, 95, 96, 97, 99, 100, 102, 105

[16] C. D. Hacon and S. J. Kovács. *Classification of Higher Dimensional Algebraic Varieties*, Birkhäuser, May 2010. ← 72, 73

[17] C. D. Hacon and J. M[c]Kernan. On Shokurov's rational connectedness conjecture. *Duke Math. J.*, **138**(1), 119–136, 2007. MR 2309156 (2008f:14030) ← 94, 95, 103

[18] R. Hartshorne. Ample subvarieties of algebraic varieties, Notes written in collaboration with C. Musili. *Lecture Notes in Mathematics*, **156**, Springer-Verlag, Berlin, 1970. MR 0282977 (44 #211) ← 101, 102

[19] R. Hartshorne. Algebraic geometry. *Graduate Texts in Mathematics*, **52**, Springer-Verlag, New York, 1977. MR 0463157 (57 #3116) ← 86, 92, 95, 96, 100, 101

[20] D. Huybrechts. Complex geometry: An introduction. *Universitext*, Springer-Verlag, Berlin, 2005. MR 2093043 (2005h:32052) ← 84

[21] D. Huybrechts and M. Lehn. The geometry of moduli spaces of sheaves. *Aspects of Mathematics*, **E31**, Friedr. Vieweg & Sohn, Braunschweig, 1997. MR 1450870 (98g:14012) ← 88, 89, 90, 107

[22] S. Iitaka. Algebraic geometry. *Graduate Texts in Mathematics*, **76**, Springer-Verlag, New York, 1982; An introduction to birational geometry of algebraic varieties. *North-Holland Mathematical Library*, **24**. MR 637060 (84j:14001) ← 79

[23] K. Jabbusch and S. Kebekus. Families over special base manifolds and a conjecture of Campana. *Math. Z.*, **269**(3), 847–878, 2011. DOI:10.1007/ s00209-010-0758-6. ← 80

[24] K. Jabbusch and S. Kebekus. Positive sheaves of differentials coming from coarse moduli spaces, *Ann. Inst. Fourier (Grenoble)*, **61**(6), 2277-2290, 2011. DOI:10.5802/aif.2673. arXiv:0904.2445. ← 80, 82, 83

[25] A. J. de Jong and J. Starr. Cubic fourfolds and spaces of rational curves. *Illinois J. Math.*, **48**(2), 415–450, 2004. MR 2085418 (2006e:14007) ← 93

[26] S. Kebekus and S. J. Kovács. *The structure of surfaces mapping to the moduli stack of canonically polarized varieties*, preprint arXiv:0707.2054, July 2007. ← 77

[27] S. Kebekus and S. J. Kovács. Families of canonically polarized varieties over surfaces. *Invent. Math.*, **172**(3), 657–682, 2008. DOI:10.1007/s00222-008-0128-8. MR 2393082 ← 77, 81, 98

[28] S. Kebekus and S. J. Kovács. Families of varieties of general type over compact bases. *Adv. Math.*, **218**(3), 649–652, 2008. DOI:10.1016/j.aim.2008.01.005. MR 2414316 (2009d:14042) ← 78

[29] S. Kebekus and S. J. Kovács. The structure of surfaces and threefolds mapping to the moduli stack of canonically polarized varieties. *Duke Math. J.*, **155**(1), 1–33, 2010. ← 77, 78, 82, 91, 106, 109

[30] S. Kebekus and L. Solá Conde. Existence of rational curves on algebraic varieties, minimal rational tangents, and applications, *Global aspects of complex geometry*, 359–416, Springer, Berlin, 2006. MR 2264116 ← 107

[31] S. Kebekus, L. Solá Conde and M. Toma. Rationally connected foliations after Bogomolov and McQuillan. *J. Algebraic Geom.*, **16**(1), 65–81, 2007. ← 107

[32] J. Kollár. Higher direct images of dualizing sheaves. I. *Ann. of Math. (2)*, **123**(1), 11–42, 1986. MR 825838 (87c:14038) ← 83, 84

[33] J. Kollár. Rational curves on algebraic varieties, in: Ergebnisse der Mathematik und ihrer Grenzgebiete. 3. Folge. *A Series of Modern Surveys in Mathematics*, **32**, Springer-Verlag, Berlin, 1996. MR 1440180 (98c:14001) ← 78, 95

[34] J. Kollár and S. J. Kovács. Log canonical singularities are Du Bois. *J. Amer. Math. Soc.*, **23**, 791–813, 2010. DOI:10.1090/S0894-0347-10-00663-6 ← 102

[35] J. Kollár and S. Mori. Birational geometry of algebraic varieties. *Cambridge Tracts in Mathematics*, **134**, Cambridge University Press, Cambridge, 1998, With the collaboration of C. H. Clemens and A. Corti, Translated from the 1998 Japanese original. MR 2000b:14018 ← 73, 79, 93, 98, 104, 108

[36] J. Kollár et al.. *Flips and Abundance for Algebraic Threefolds*. Société Mathématique de France, Paris, 1992, Papers from the Second Summer Seminar on Algebraic Geometry held at the University of Utah, Salt Lake City, Utah, August 1991, Astérisque No. 211 (1992). MR 1225842 (94f:14013) ← 105

[37] S. J. Kovács. Smooth families over rational and elliptic curves. *J. Algebraic Geom.*, **5**(2), 369–385, 1996. Erratum: *J. Algebraic Geom.*, **6**(2), 391, 1997. MR 1374712 (97c:14035) ← 75

[38] S. J. Kovács. On the minimal number of singular fibres in a family of surfaces of general type. *J. Reine Angew. Math.*, **487**, 171–177, 1997. MR 1454264 (98h:14038) ← 75

[39] S. J. Kovács. Algebraic hyperbolicity of fine moduli spaces. *J. Algebraic Geom.*, **9**(1), 165–174, 2000. MR 1713524 (2000i:14017) ← 75

[40] S. J. Kovács. Subvarieties of moduli stacks of canonically polarized varieties. *Proceedings of Symposia in Pure Mathematics*, American Mathematical Society, Providence, RI, 2006, to appear. ← 73

[41] S. J. Kovács. Young person's guide to moduli of higher dimensional varieties, in: Algebraic geometry—Seattle 2005. Part 2, 711–743. *Proc. Sympos. Pure Math.*, **80**, Amer. Math. Soc., Providence, RI, 2009. MR 2483953 (2009m:14051) ← 73

[42] S. J. Kovács and K. Schwede. *Hodge theory meets the minimal model program: a survey of log canonical and Du Bois singularities*, preprint, 2009. arXiv:0909.0993v1 ← 102

[43] R. Lazarsfeld. Positivity in algebraic geometry. II, in: Ergebnisse der Mathematik und ihrer Grenzgebiete. 3. Folge. *A Series of Modern Surveys in Mathematics [Results in Mathematics and Related Areas. 3rd Series. A Series of Modern Surveys in Mathematics]*, **49**, Springer-Verlag, Berlin, 2004. Positivity for vector bundles, and multiplier ideals. MR 2095472 ← 84

[44] Q. Liu. Algebraic geometry and arithmetic curves. *Oxford Graduate Texts in Mathematics*, **6**, Oxford University Press, Oxford, 2002. Translated from the French by Reinie Erné, Oxford Science Publications. MR 1917232 (2003g:14001) ← 84

[45] D. Lohmann. *Families of canonically polarized manifolds over log Fano varieties*, preprint arXiv:1107.4545, July 2011. ← 79

[46] K. Matsuki. Introduction to the Mori program. *Universitext*, Springer-Verlag, New York, 2002. MR 2002m:14011 ← 73

[47] L. Migliorini. A smooth family of minimal surfaces of general type over a curve of genus at most one is trivial. *J. Algebraic Geom.*, **4**(2), 353–361, 1995. MR 1311355 (95m:14023) ← 75

[48] Y. Miyaoka. Deformations of a morphism along a foliation and applications. Algebraic geometry, Bowdoin, 1985 (Brunswick, Maine, 1985), *Proc. Sympos. Pure Math.*, **46**, 245–268. Amer. Math. Soc., Providence, RI, 1987. MR 927960 (89e:14011) ← 107

[49] Y. Namikawa. Extension of 2-forms and symplectic varieties. *J. Reine Angew. Math.*, **539**, 123–147, 2001. MR 1863856 (2002i:32011) ← 93

[50] K. Oguiso and E. Viehweg. On the isotriviality of families of elliptic surfaces. *J. Algebraic Geom.*, **103**, 569–598, 2001. MR 1832333 (2002d:14054) ← 75

[51] C. Okonek, M. Schneider, and H. Spindler. Vector bundles on complex projective spaces. *Progress in Mathematics*, **3**, Birkhäuser Boston, Mass., 1980. MR 561910 (81b:14001) ← 82, 90

[52] A. N. Parshin. Algebraic curves over function fields. I. *Izv. Akad. Nauk SSSR Ser. Mat.*, **32**, 1191–1219, 1968. MR 0257086 (41 #1740) ← 72

[53] Z. Patakfalvi. *Viehweg's hyperbolicity conjecture is true over compact bases*, preprint arXiv:1109.2835, September 2011. ← 78

[54] M. Reid. Young person's guide to canonical singularities, Algebraic geometry, Bowdoin, 1985 (Brunswick, Maine, 1985). *Proc. Sympos. Pure Math.*, **46**, 345–414, Amer. Math. Soc., Providence, RI, 1987. MR 927963 (89b:14016) ← 90

[55] I. R. Shafarevich. Algebraic number fields. *Proc. Internat. Congr. Mathematicians (Stockholm, 1962)*, 163–176, Inst. Mittag-Leffler, Djursholm, 1963, English translation: *Amer. Math. Soc. Transl. (2)*, **31**, 25–39, 1963. MR 0202709 (34 #2569) ← 72

[56] J. Steenbrink. Vanishing theorems on singular spaces. *Astérisque*, **130**, 330–341, 1985. Differential systems and singularities (Luminy, 1983). MR 804061 (87j:14026) ← 102

[57] J. Steenbrink and D. van Straten. Extendability of holomorphic differential forms near isolated hypersurface singularities. *Abh. Math. Sem. Univ. Hamburg*, **55**, 97–110, 1985. MR 831521 (87j:32025) ← 93, 101

[58] E. Viehweg. Weak positivity and the additivity of the Kodaira dimension for certain fibre spaces, in: Algebraic varieties and analytic varieties (Tokyo, 1981). *Adv. Stud. Pure Math.*, **1**, 329–353, North-Holland, Amsterdam, 1983. MR 715656 (85b:14041) ← 76

[59] E. Viehweg. Quasi-projective moduli for polarized manifolds. *Ergebnisse der Mathematik und ihrer Grenzgebiete (3) [Results in Mathematics and Related Areas (3)]*, **30**, Springer-Verlag, Berlin, 1995. MR 1368632 (97j:14001) ← 73, 76, 77

[60] E. Viehweg. Positivity of direct image sheaves and applications to families of higher dimensional manifolds, in: School on Vanishing Theorems and Effective Results in Algebraic Geometry (Trieste, 2000). *ICTP Lect. Notes*, **6**, 249–284, Abdus Salam Int. Cent. Theoret. Phys., Trieste, 2001. Available on the ICTP web site. MR 1919460 (2003f:14024) ← 72, 73, 77, 83, 84

[61] E. Viehweg and K. Zuo. Base spaces of non-isotrivial families of smooth minimal models, in: Complex geometry (Göttingen, 2000), 279–328, Springer, Berlin, 2002. MR 1922109 (2003h:14019) ← 81, 83, 84, 86, 87, 96

[62] E. Viehweg and K. Zuo.On the Brody hyperbolicity of moduli spaces for canonically polarized manifolds. *Duke Math. J.*, **118**(1), 103–150, 2003. MR 1978884 (2004h:14042) ← 75

[63] C. Voisin. Hodge theory and complex algebraic geometry. I, english ed. *Cambridge Studies in Advanced Mathematics*, **76**, Cambridge University Press, Cambridge, 2007. Translated from the French by Leila Schneps. MR 2451566 (2009j:32014) ← 84

[64] K. Zuo. On the negativity of kernels of Kodaira-Spencer maps on Hodge bundles and applications. *Asian J. Math.*, **4**(1), 279–301, 2000). Kodaira's issue. MR 1803724 (2002a:32011) ← 84

Stefan Kebekus, Albert-Ludwigs Universität Freiburg, Mathematisches Institut, Eckerstraße 1, 79104 Freiburg im Breisgau, Germany

E-mail address: stefan.kebekus@math.uni-freiburg.de

Contractible extremal rays on $\overline{M}_{0,n}$

Seán Keel and James McKernan

Contents

1. Introduction and statement of results

One of the richest objects of study in higher dimensional algebraic geometry is the Mori-Kleiman (closed) cone of curves, $\overline{NE_1}(M)$, defined as the closed convex cone in $H_2(M, \mathbb{R})$ generated by classes of irreducible curves on M. A lot of geometric information about M is encoded in the cone of curves. For example the possibilities for maps with connected fibres are determined by the cone's faces. Not surprisingly, $\overline{NE_1}(M)$ is difficult to compute. Even when M is well understood, it can be difficult to find generators for the cone, that is, to find all of the "edges", or to use the technical term, "extremal rays" ("edge" is potentially misleading as portions of the cone may be curved—see the last paragraph of this Introduction for the distinction). Indeed, it is not even generally obvious whether or not a given curve spans an extremal ray. One expects better luck understanding rays on which $-c_1(M) = K_M$ (or more generally log terminal $K_M + \Delta$) are negative, as these are described by the powerful cone and contraction theorems of Mori-Kawamata-Shokurov: each is generated by a smooth rational curve, and can be "contracted", i.e. there is a map (with domain M) whose fibral curves are precisely the curves whose homology class lies on the extremal ray. Thus as a first step in computing the cone, it is natural to consider such "negative" extremal rays, or more generally, rays which can be contracted.

Here we consider $\overline{M}_{0,n}$, the moduli space of stable n-pointed rational curves, as well as $\widetilde{M}_{0,n}$ the quotient of $\overline{M}_{0,n}$ by the natural symmetric group action, which is (an irreducible component of) the moduli space of log pairs (see [1]).

The locus of points in $\overline{M}_{0,n}$ corresponding to a curve with at least $k+1$ components has pure codimension k; we call its irreducible components the **vital codimension k-cycles**. Vital divisors, curves, k-cycles etc. are analogously defined. By a vital cycle in $\widetilde{M}_{0,n}$ we mean the image of a vital cycle in $\overline{M}_{0,n}$. It is relatively easy to check that the vital cycles generate the Chow group. It is natural to wonder if much more is true:

Question 1.1. *Is every effective cycle linearly equivalent to an effective sum of vital cycles?*

This was first posed to us by William Fulton. In the interest of drama, we will refer to (1.1) as Fulton's conjecture. Here we consider only the cases of curves and divisors. As homological and linear equivalence are the same on $\overline{M}_{0,n}$, the conjecture in these cases is equivalent to the statement that vital cycles generate all extremal rays of $\overline{NE}_1$ and $\overline{NE}^1$, the cones of curves and divisors. We prove this for $\overline{NE}^1(\widetilde{M}_{0,n})$ and for contractible extremal rays of $\overline{NE}_1(\overline{M}_{0,n})$.

Let $D \subset \overline{M}_{0,n}$ be the boundary, i.e. the sum of the vital divisors. Let $D = \sum B_i$ be its decomposition into S_n orbits (there are $[n/2]$ such orbits). For a subvariety $Z \subset \overline{M}_{0,n}$, let $\tilde{Z}$ be its image (with reduced structure) in $\widetilde{M}_{0,n}$.

Here are precise statements of our results:

Theorem 1.2. *Let R be an extremal ray of the cone of curves $\overline{NE}_1(\overline{M}_{0,n})$. Then R is spanned by a vital curve under any of the following conditions*

(1) *There is a morphism $f: \overline{M}_{0,n} \longrightarrow Y$, contracting R, with $\rho(Y) = \rho(\overline{M}_{0,n}) - 1$, and such that the exceptional locus of f is not a curve.*
(2) $(K_{\overline{M}_{0,n}} + G) \cdot R < 0$, *where G is an effective boundary whose support is contained in D.*
(3) $n \leqslant 7$.

Of course (1.2.3) says Fulton's conjecture holds for curves, provided $n \leqslant 7$. We were able to prove much stronger results for $\widetilde{M}_{0,n}$ (especially (1.3.1-2)):

Theorem 1.3.

(1) *The cone of effective divisors $NE^1(\widetilde{M}_{0,n})$ is simplicial, generated by the $\tilde{B}_i$.*
(2) *An effective divisor on $\widetilde{M}_{0,n}$ fails to be big iff its support is a proper subset of $\tilde{D}$. Any non-trivial nef divisor is big.*
(3) *The cone of curves of $\overline{NE}_1(\widetilde{M}_{0,n})$ is generated by curves in $\tilde{D}$.*

Now suppose $n \leqslant 11$.

(4) $NE_1(\widetilde{M}_{0,n})$ *is a finite rational polyhedron, with edges spanned by images of vital curves,*
(5) *Every proper face is contractible by a log Mori fibre space. In particular every nef divisor is eventually free.*
(6) *The divisor $\sum_{i=2}^{[n/2]} r_i \tilde{B}_i$ is nef (resp. ample) iff*

$$r_{a+b} + r_{a+c} + r_{b+c} - r_a - r_b - r_c - r_d$$

is non-negative (resp. strictly positive), for all positive integers a, b, c *and* d, *with* $n = a + b + c + d$ *(where we define* $r_1 = 0$ *and* $r_i = r_{n-i}$ *for* $i > [n/2]$*).*

(By a simplicial cone we mean a cone over a simplex, i.e. a polyhedral cone whose edges are linearly independent)

The spaces $\overline{M}_{0,n}$ and $\widetilde{M}_{0,n}$ are interesting from a number of viewpoints. They are closely related to the moduli space of curves, $\overline{\mathcal{M}}_g$. A finite quotient of $\overline{M}_{0,n}$ occurs as a locus of degenerate curves in the boundary of $\overline{\mathcal{M}}_g$, while $\overline{M}_{0,n}$ is the base of the complete Hurwitz scheme (see [2]) which can be used, for example, to prove that $\overline{\mathcal{M}}_g$ is irreducible. By [3], $\overline{M}_{0,n}$ parametrizes degenerations of rational normal curves. Generalisations of $\overline{M}_{0,n}$ are important for Quantum Cohomology calculations, see [11]. $\overline{M}_{0,n}$ is useful for studying fibrations with general fibre $\mathbb{P}^1$, as in particular it can sometimes be used in lieu of a minimal model program. Kawamata exploits this in [5] to prove additivity of log Kodaira dimension for one dimensional fibres, and in [6] to prove a codimension two subadjunction formula.

We note that there is an explicit construction of $\overline{M}_{0,n}$, as a blow up of $\mathbb{P}^{n-3}$ along a sequence of simple centres (see (3.1)). In particular $\overline{M}_{0,5}$ is a del Pezzo of degree five, $\overline{M}_{0,6}$ is log Fano, and $\overline{M}_{0,7}$ is nearly log Fano, in the sense that $-K_{\overline{M}_{0,7}}$ is effective. We do not know of such an explicit construction of $\widetilde{M}_{0,n}$, and we have in general a much weaker grasp on its geometry (though a much stronger grasp on its cones). Note by (1.3.3), $\widetilde{M}_{0,n}$ admits no nontrivial fibrations. See also (3.7).

Despite the fact that the blowup construction gives an easy computation of some invariants of $\overline{M}_{0,n}$, one cannot expect the same for the cone of curves. For example the blow up of $\mathbb{P}^2$ in eight points has a finite polyhedral cone of curves, but one can choose a ninth point in such a way that the blow up has a cone with infinitely many edges. We do not use the blow up description in any significant way in our proof of (1.2-3).

As we note in (3.5) $-K_{\widetilde{M}_{0,n}}$ and $-K_{\overline{M}_{0,n}}$ are not effective for $n \geqslant 8$. In view of this, the cases of (1.3) for $8 \leqslant n \leqslant 11$ are interesting in that they give examples of non log Fano varieties, for which every face of the cone of curves is none the less contractible. From this perspective, (1.1), if true, would really be rather surprising. Each vital curve (indeed every vital cycle) is smooth and rational. The cone of curves of a log Fano is generated by rational curves, but one does not expect this in general, even for a rational variety. For example, let S be the blow up of $\mathbb{P}^2$ in a large number of general points. As observed by Kollár, and independently by Caporaso and Harris, K_S is strictly negative on rational curves, but of course $K_S^2 < 0$, so K_S must be positive on some curves (but see the remark after (2.4)).

If (1.1) holds for curves, then one can describe the ample cones of $\overline{M}_{0,n}$ or $\widetilde{M}_{0,n}$ by a series of inequalities analogous to those in (1.3), using (4.3). One can then describe, at least in theory, the cone of curves, since it is dual to the ample cone. As an example of the complexity of these cones, $NE_1(\overline{M}_{0,7})$ is a polyhedral cone of dimension 42 with 350 edges (see (4.3) and (4.6)).

Fulton's conjecture implies every vital curve spans an extremal ray and each is $K_{\overline{M}_{0,n}} + G$ negative for some G as in (1.2.2) (see (4.6)). So by the contraction theorem [7] each vital curve is contracted by a map of relative Picard number one. For $n \geqslant 9$ every vital curve deforms. So if (1.1) holds, then (1.2) contains all the possibilities for extremal rays, and (1.2.1) has all the possibilities for $n \geqslant 9$.

Here is a brief outline of our proofs of (1.2-3). It turns out each component of D is a product $\overline{M}_{0,i} \times \overline{M}_{0,j}$, for $i, j < n$ (see §3). For (1.2) we proceed by induction, the main work is to show the extremal ray R is in the subcone generated by curves in D. For this our main tool is (2.2). (1.3) follows from (1.2) and some simple intersection calculations: one set to show that $NE^1(\widetilde{M}_{0,n})$ is simplicial, and a second to show that for $n \leqslant 11$ every face of the cone contracts to a log Mori fibre space.

§2 contains some results about the cone spanned by curves lying in a divisor. Most of the results of §2 are of independent interest (in particular (2.3-2.6)), and hopefully have broader applications. For this reason we work in greater generality. §3 contains the necessary ingredients to apply some of the results of §2 to $D \subset \overline{M}_{0,n}$. Intersection products of various vital cycles are easy to compute, and the pairing between divisors and curves is described in §4. §5 finishes the proof of (1.3).

We would like to say a few words about other seemingly natural approaches to (1.1). For curves, it is enough (in fact equivalent) to show that if a divisor intersects all vital curves non-negatively, then it is nef. By induction it is sufficient to show that such a divisor is linearly equivalent to an effective sum of vital divisors. As the vital divisors generate the Picard group, the intersection conditions give a finite collection of simple inequalities on the coefficients. Unfortunately the combinatorics are intimidating, and we were not able to make any progress in this direction, even for $n = 6$.

Throughout we will use the main results of the minimal model program, the contraction theorem, the cone theorem etc., as well as the established notation as set out in [7]. We also use elementary properties and notions of cones from Chapter II.4 of [10]. In particular, by an extremal ray R of a closed convex cone W we mean a one dimensional subcone with the property that if $x + y \in R$ for $x, y \in W$ then $x, y \in R$. We note that every closed convex cone is the convex hull of its extremal rays. By an *edge* of W, we mean an extremal ray R which is also the intersection of W with a hyperplane. All spaces are assumed to be of finite type over $\mathbb{C}$. Unless otherwise stated, by a divisor we mean an $\mathbb{R}$-divisor. In [13] the main results of the MMP are extended to $\mathbb{R}$-divisors. However we only need one such result (see (2.2)).

2. The cone spanned by curves inside a divisor

We first introduce some notation and definitions. Let D be a reduced Weil divisor inside the projective $\mathbb{Q}$-factorial klt variety M of dimension n. Let W be the closed subcone of $\overline{NE}_1(M)$ generated by curves lying in D.

We are interested in extremal rays that lie outside of W and moreover under what conditions $W = \overline{NE}_1(M)$.

Definitions 2.1. *We say that* D *has* **anti-nef normal bundle** *if for every curve* $C \subset D$, $C \cdot D \leqslant 0$.

We will say an extremal ray R *of* $\overline{NE}_1(M)$ *is* **log extremal** *if there exists a klt divisor* $K_M + \Delta$ *such that* $(K_M + \Delta) \cdot R < 0$.

Log extremal rays are very special: By the cone and contraction Theorems they are spanned by rational curves C, and there is a morphism $f : M \longrightarrow Y$ contracting C such that $f_*(\mathcal{O}_M) = \mathcal{O}_Y$ and $\rho(Y) = \rho(M) - 1$. In particular, every log extremal ray is an edge of $\overline{NE}_1(M)$.

The following result of Shokurov will prove useful:

Lemma 2.2 (Shokurov). *Let* X *be a projective variety, and let* $L \in N^1(X)$ *be a nef class (not necessarily rational) with* $L^{\dim(X)} > 0$. *Then* L *is in the interior of* $NE^1(X)$.

Proof. This is implied by the proof of (6.17) of [13]. □

(2.2) has some interesting corollaries:

Corollary 2.3. *If the components of* D *span* $\overline{NE}^1(M)$ *then* $W = \overline{NE}_1(M)$.

Proof. Let $D = \sum D_i$ be the decomposition of D into irreducible components.

Let A be an ample divisor with support in D, and let $R \subset \overline{NE}_1(M)$ be an edge. Assume $R \not\subset W$. Let L be a nef class supporting R. $L|D$ is ample. Since L is an effective sum of D_i, $L^{\dim M} > 0$ thus by (2.2), R cannot be numerically effective. Since the D_i generate $\overline{NE}^1(M)$, $R \cdot D_i < 0$ for some i. But this implies $R \in W$, a contradiction. □

Proposition 2.4. *Let* G *be an effective* $\mathbb{Q}$*-divisor, with non-empty support* D.

Let R *be an edge of* $\overline{NE}_1(M)$, *which does not lie in* W. *If* $(K_M + G) \cdot R \leqslant 0$ *then* R *is log extremal and* $K_M \cdot R \leqslant 0$.

In particular, if $-(K_M + G)$ *is nef then* $\overline{NE}_1(M)$ *is spanned by* W *and log extremal rays* R, *such that* $K_M \cdot R \leqslant 0$.

Proof. Let R be an edge of $\overline{NE}_1(M)$, not lying in W. In particular $R \cdot D_i \geqslant 0$ and so $K_M \cdot R \leqslant 0$.

On the other hand we are done if $K_M \cdot R < 0$. Thus we may assume $K_M \cdot R = 0$. Let $L \in N^1(M)$ be a nef class supporting R. Then L is strictly positive on $W \setminus 0$ and so by compactness of a slice of W, $L + \epsilon D$ is nef and supports R for $0 < \epsilon \ll 1$. As L is ample on D $L^{n-1} \cdot D > 0$. In particular we can replace L by $L + \epsilon D$ and assume $L^n > 0$. Then by (2.2) $R \cdot V < 0$ for some effective Weil divisor V. But $(K_M + \epsilon V) \cdot R < 0$ and $(K_M + \epsilon V)$ is klt for $0 < \epsilon \ll 1$. □

Remark. (2.4) is interesting even in the case of a surface. For example pick a cubic in $\mathbb{P}^2$ and blow up as many points as you like along the cubic. Let M be the resulting surface and D the strict transform of the cubic. (2.4) then says that D union all the -2-curves and -1-curves generate the cone of curves of M.

Proposition 2.5. *Let* $f : M \longrightarrow Y$ *be a proper surjection from a smooth projective variety* M *to a normal variety* Y *with* $f_*(\mathcal{O}_M) = \mathcal{O}_Y$, *and* $\rho(Y) = \rho(M) - 1$. *Suppose* D *has ample support and each irreducible component of* D *has anti-nef normal bundle.*

If $f|D$ *is finite then* f *is birational, and its exceptional locus is a curve.*

Proof. Suppose on the contrary that there is an irreducible surface E whose image has dimension at most one.

Let $D = \sum_i D_i$ be the decomposition of D into irreducible components. Note the assumptions on Picard number imply that any class in $N^1(M)$ which is zero on some fibral curve, is pulled back from $N^1(Y)$.

Since D has ample support $I = D \cap E$ is non-empty. As $f|D$ is finite, I and each $D_i \cap E = D_i \cap I$, is an effective $\mathbb{Q}$-Cartier divisor of E, and in particular, is purely one dimensional. Thus if I meets D_i, it has an irreducible component contained in D_i. Since D has ample support, and $f|D$ is finite, E contracts to an irreducible curve $C \subset f(D) = f(I)$ and $f|I$ is finite.

Claim. We can find two irreducible components B_1, B_2 of I and (after renaming) two divisors D_1, D_2 with $B_i \subset D_i$ such that $B_i \cdot D_j \geqslant 0$ (for $i \neq j$) and at least one inequality is strict:

Choose an irreducible component B_1 of I contained in a maximal number of D_i. Suppose (after reordering) $D_1, D_2, \ldots, D_k$ are the components of D containing B_1. Since the D_i have anti-nef normal bundles, and D has ample support, for some $j > k$ we have $D_j \cdot B_1 > 0$. Let B_2 be an irreducible component of $D_j \cap I$. By the choice of B_1 we can assume (after reordering) that $B_2 \not\subset D_1$. Now set $D_2 = D_j$.

This establishes the claim.

Since D_1, D_2 each meet a fibre, we can choose $\lambda > 0$ such that $D_1 - \lambda D_2$ is pulled back from Y. Let $J = (D_1 - \lambda D_2)|E$. Then $J \cdot B_1 \leqslant 0$ and $J \cdot B_2 \geqslant 0$, and one inequality is strict. Since J is pulled back from C, and the B_i are multi-sections, this is a contradiction. □

Remark.

(1) The assumption on the relative Picard number in (2.5) is necessary; it cannot be replaced by the weaker assumption that f is the contraction of an extremal ray. For example consider $M = E \times E$ for an elliptic curve E, $D = F_1 + F_2$ the sum of the two fibres and $f : M \longrightarrow E$ the addition map.

(2) The assumption on Picard number holds when f is the contraction of a log extremal ray.

(3) One can not rule out the final possibility. For example: Let M be a del Pezzo surface whose cone of curves is not a simplex (e.g. blow up $\mathbb{P}^2$ at three non-collinear points). Let D be a sum of $\rho(M)$ -1-curves with ample

support (any effective class is a sum of at most $\rho(M)$ extremal rays, and all the extremal rays are -1-curves). Let f blow down some other -1-curves.

Lemma 2.6. *Suppose* M *is smooth of dimension at least three, every component of* D *has anti-nef normal bundle, and* D *has ample support. Let* G *be a nonempty effective* $\mathbb{Q}$*-divisor whose support lies in* D.

(1) *Let* R *be an edge of* $\overline{NE}_1(M)$. *If either* $(K_M + G) \cdot R < 0$, *or the dimension of* M *is at least four and* $(K_M + G) \cdot R \leqslant 0$ *then* $R \in W$.
(2) *If* $-(K_M + G)$ *is nef, and either the support of* G *is exactly* D *or the dimension of* M *is at least four, then* $W = \overline{NE}_1(M)$.

Proof. Let R be an edge of $\overline{NE}_1(M)$, and suppose $R \notin W$ but $(K_M + G) \cdot R \leqslant 0$. Then by (2.4) we know that R is spanned by a contractible rational curve C. (1) and (2) now follow easily from (2.5) and the observation that if $K_M \cdot C < 0$ and M is a threefold (resp. $K_M \cdot C \leqslant 0$ and M has dimension at least four) then C deforms inside M (see II.1.13 of [10]). □

We will use the following technical result in the next section.

Lemma 2.7. *Let* $N \subset M$ *be a normal divisor and suppose that* $N^1(M) \longrightarrow N^1(N)$ *is surjective. Let* $f : M \longrightarrow Y$ *be a map to a normal projective variety with* $f_*(\mathcal{O}_M) = \mathcal{O}_Y$, *and* $\rho(Y) = \rho(M) - 1$. *Let* $g : N \longrightarrow Z$ *be the Stein factorisation of* $f|_N$. *If* f|N *is not finite, then* $\rho(Z) = \rho(N) - 1$.

Proof. f contracts an extremal ray R. Suppose f|N is not finite. Then $R \in N_1(N)$. Suppose that $L \in N^1$ and $L \cdot R = 0$. Since every class in $N^1(D)$ extends to M, $L|_N$ is pulled back from Z and the result follows. □

3. Geometry of $\overline{M}_{0,n}$ and $\widetilde{M}_{0,n}$.

We will use (a slight modification of) the notation of, as well as several simple facts from pgs. 551–554 of [8]. For the readers convenience we will recall the most important ideas:

A vital divisor is determined by a partition of $\{1, 2, \ldots, n\}$ into disjoint subsets T, T^c, each containing at least two elements. The generic point of the corresponding vital divisor D_{T,T^c} is a curve with two irreducible components, with the labels of T on one component, and the labels of T^c on the other. There is a canonical isomorphism

$$D_{T,T^c} = M_{T \cup \{b\}} \times M_{T^c \cup \{b\}}$$

where e.g. by $M_{T\cup\{b\}}$ we mean a copy of $\overline{M}_{0,|T|+1}$ with the indices labeled by the elements of T, with b an extra index, corresponding to the singular point. We indicate the two projections by π_T and π_{T^c}.

The vital divisors have normal crossings, and each vital codimension k-cycle is uniquely expressible as a complete intersection of vital divisors. Each vital k-cycle has an expression as a product of $\overline{M}_{0,i}$ analogous to that for the vital divisors. In particular, under the above decomposition, any vital curve of D_{T,T^c} is a product of a vital curve on one factor, with a vital point on the second.

Proposition 3.1 (Kapranov). *For each index* $i \in \{1,2,\ldots,n\}$ *there is a birational map* $q_i : \overline{M}_{0,n} \longrightarrow \mathbb{P}^{n-3}$ *with the following properties:*

(1) q_i *is a composition of blow ups along smooth centres, constructed as follows. Fix* $n-1$ *general points, and blow up successively (from lowest to highest dimensional) the (strict transforms) of every linear subspace spanned by any subset of these points.*
(2) q_i *takes vital cycles to linear spaces spanned by the chosen points.*
(3) *If* $i \in T$ *then* $q_i|_{D_{T,T^c}} = q_i \circ \pi_T$ *for* $i \in T$.
(4) *If* F *is the general fibre of the map* $\overline{M}_{0,n} \longrightarrow \overline{M}_{0,n-1}$ *given by dropping the* i^{th} *point, then* $q_i(F)$ *is a rational normal curve.*
(5) q_i *is a composition of smooth blow downs, blowing down iteratively the (images of) the divisors* D_{T,T^c} *with* $i \notin T$, *and* $|T| = 3,4,\ldots,n-2$.

Proof. See [4]. □

Lemma 3.2. *Let* ϕ *be an element of* Aut $(\mathbb{P}^1)$ *of finite order* p *and let* Z *be the closure of the locus of* n*-tuples of distinct elements of* $\mathbb{P}^1$ *whose coordinates are permuted by* ϕ. *Let* q *be a general element of* Z. *If* ϕ *fixes* j *coordinates of* q *then the dimension of* Z *at* q *is* $(n-j)/p$.

Proof. Let $G \subset \text{Aut}(\mathbb{P}^1)$ be the subgroup generated by ϕ. Then G has a non-trivial finite orbit, from which it follows that G has exactly two fixed points, and after changing coordinates (so the fixed points are 0 and ∞) $\phi : \mathbb{A}^1 \to \mathbb{A}^1$ is multiplication by a pth root of unity. The coordinates divide into orbits, each of which is either a fixed point, or has exactly p elements. The result follows. □

Lemma 3.3. S_n *acts freely in codimension one on* $M_n \setminus D$ *for* $n \geqslant 7$, *and faithfully for* $n \geqslant 5$.

The action of S_4 *on* M_4 *factors through the action on the set of partitions of* $\{1,2,3,4\}$ *into disjoint subsets of two elements. Nontrivial elements of the kernel are of form* $(ij)(kl)$ *for* i, j, k, l *distinct.*

Proof. The claims about the S_4 action are easily checked, and left to the reader.

So assume $n \geqslant 5$. We bound the dimension of the locus of points in $M_n \setminus D$ which are fixed by some element of S_n. Equivalently, we bound the dimension of the locus $Z \subset M_n \setminus D$ of n-tuples (modulo automorphisms) whose coordinates are permuted by some automorphism of $\mathbb{P}^1$.

Let $\mathcal{U} \subset \mathbb{P}^{1 \times n}$ be the locus of distinct points, and let

$$\mathcal{T} = \{(q, \phi, a, b) | q \in \mathcal{U}, \phi \in \text{Aut}(\mathbb{P}^1), a \neq b \in \mathbb{P}^1 \text{s.t. } \phi \text{ permutes } q \text{ and fixes}(a, b)\}.$$

Replace $\mathcal{T}$ by any one of its irreducible components.

Let ϕ be a general point of the image of $\mathrm{pr} : \mathcal{T} \to \mathrm{Aut}(\mathbb{P}^1)$, and let $q \in \mathrm{pr}^{-1}(\phi)$. By (3.2) $\mathrm{pr}^{-1}(\phi)$ has dimension $(n-j)/p$ at q, while the fibre of $\mathcal{T} \to U$ has dimension three. Thus at the image of q, U has dimension $(n-j)/p - 1$. The result follows. □

Let $B_i = \sum_{|T|=i} D_{T,T^c}$ for $2 \leqslant i \leqslant k = [n/2]$. B_i is the orbit under S_n of any D_{T,T^c} with $|T| = i$.

Lemma 3.4. *For* $n \geqslant 7$ *the quotient map* $q : \overline{M}_{0,n} \longrightarrow \widetilde{M}_{0,n}$ *is unramified in codimension one outside of* B_2, *and has ramification index two along* B_2.

Proof. Suppose $\sigma \in S_n$ fixes each point of the irreducible divisor $G \subset \overline{M}_{0,n}$. By (3.3), $G = D_{T,T^c}$ for some T preserved by σ. Since the action of σ on $M_{T\cup\{b\}}$ factors through the subgroup of $S_{|T|+1}$ which fixes b, it follows from (3.3) that $T = \{i,j\}$ and $\sigma = (i,j)$. □

We will use the following formulae, essentially from [12]:

Lemma 3.5.

$$K_{\overline{M}_{0,n}} + \sum_{j=2}^{k}\left(2 - \frac{j(n-j)}{n-1}\right) B_j = 0 = K_{\widetilde{M}_{0,n}} + \left(\frac{1}{2} + \frac{1}{(n-1)}\right)\widetilde{B}_2 + \sum_{j=3}^{k}\left(2 - \frac{j(n-j)}{n-1}\right)\widetilde{B}_j$$

In particular $-K_{\widetilde{M}_{0,n}}$ *(resp.* $-K_{\overline{M}_{0,n}}$*) is pseudo-effective iff* $n \leqslant 7$.

Proof. The first formula is Proposition 1 of [12], the second follows easily from the first and (3.4) and the last statement then follows from (4.8). In fact we may use (4.3) to prove the first formula in a similar way to the way it is derived in [12].

However it is possible to prove the first formula in an entirely elementary way, using (3.1). Indeed the image D' of D is the union of $\binom{n-1}{2}$ hyperplanes, and the coefficients of B_i are easily identified as the discrepancies of the divisor $K_{\mathbb{P}^{n-3}} + (2/(n-1))D'$. □

Lemma 3.6. $K_{\overline{M}_{0,n}} + D$ *is ample and is linearly equivalent to an effective divisor with the same support as* D.

Proof. We proceed by induction on n. The result is easy for $n = 4$.

By (3.5), $K_{\overline{M}_{0,n}} + D$ is linearly equivalent to an effective divisor Γ with support D.

Note that $(K_{\overline{M}_{0,n}} + D)|D_T$ is the tensor product of the "same expressions" pulled back from the two components in the product description of D_T. Thus by induction $(K_{\overline{M}_{0,n}} + D)|D$ is ample.

It is easy to see that D meets (set theoretically) every curve. Use induction and consider the map $f : \overline{M}_{0,n} \longrightarrow \overline{M}_{0,n-1}$, observe that D meets every fibral curve, and note that $D \supset f^{-1}(D(\overline{M}_{0,n-1}))$.

Thus $(K_{\overline{M}_{0,n}} + D) \cdot C > 0$ for all curves C.

It follows that $K_{\overline{M}_{0,n}} + D$ is nef, and nef and big by induction. Thus by the base point free theorem (applied to the big and nef klt divisor $K_{\overline{M}_{0,n}} + D - \epsilon\Gamma$) $m(K_{\overline{M}_{0,n}} + D)$ is basepoint free for $m \gg 0$. Since it intersects every curve positively, it is thus ample. □

The results above have some interesting geometric consequences:

Remarks 3.7.

(1) By (3.1.1) $\overline{M}_{0,5}$ is isomorphic to $\mathbb{P}^2$ blown up at four points. Thus it is a del Pezzo surface of degree five. It is interesting to note that $K_{\overline{M}_{0,5}} + D = -K_{\overline{M}_{0,5}}$ is very ample and defines the anticanonical embedding of $\overline{M}_{0,5}$ inside $\mathbb{P}^5$. For any n, if C is a vital curve, then $(K_{\overline{M}_{0,n}} + D) \cdot C = 1$. Thus it would be nice to know if $K_{\overline{M}_{0,n}} + D$ is very ample, for if it were, then under the corresponding map, every vital curve would be embedded as a line.

(2) Note that $\widetilde{M}_{0,5}$ is a log del Pezzo of rank one. It is easy to compute, using (3.2) that $\widetilde{M}_{0,5}$ has two quotient singularities, one of index two and the other of index five. It is then easy, from the classification of log del Pezzos, (see [9]) to conclude that $\widetilde{M}_{0,5}$ has one A_1-singularity and one singularity of type (2,3).

(3) Note that the map $D_{T,T^c} \longrightarrow \widetilde{M}_{0,n}$ factors through $M_{|T|+1,|T^c|+1}/S_{|T|} \times S_{|T^c|}$, but not through the quotient by $S_{|T|+1} \times S_{|T^c|+1}$. Thus an inductive study of $\overline{NE}_1(\widetilde{M}_{0,n})$ is problematic. In particular one cannot obtain the analog of (3.9) as below.

(4) By (3.1), $q_i^*(\mathcal{O}(1))$ is numerically equivalent to an effective divisor with support exactly D. It follows by (3.6) that for any curve $C \subset D$ there is some vital divisor which is negative on C.

Lemma 3.8. *For any projective variety* T, $N_1(\overline{M}_{0,n} \times T) = N_1(\overline{M}_{0,n}) \times N_1(T)$ *under the map induced by the two projections. The same map induces an isomorphism*

$$NE_1(\overline{M}_{0,n} \times T) = NE_1(\overline{M}_{0,n}) \times NE_1(T)$$

Proof. This follows from Theorem 2 of [8]. □

Corollary 3.9. *Fulton's conjecture for* $NE^1(\overline{M}_{0,n})$ *implies the conjecture for* $NE_1(\overline{M}_{0,n})$.

Proof. Immediate from (2.2) and (3.8). □

Proof of (1.2). As we are going to use induction it is actually more convenient to prove a slightly stronger result. Let M be any product of $\overline{M}_{0,i}$. We will prove (1.2) for M. By a vital cycle on M we mean a product of vital cycles on each component. We will continue to use the same notation, so for example by D_T we mean the inverse image of this divisor from a projection onto one of the components of M.

Let m be the dimension of M. When $m \leqslant 2$ it is easy to check that vital curves generate the cone (see (3.7.1)).

D has ample support by (3.6), and each D_{T,T^c} has anti-nef normal bundle by (4.5).

When $m \leqslant 4$ then we can apply (3.5), and (2.6) inductively, to prove (1.2.3).

Let R be an extremal ray satisfying either (1.2.1) or (1.2.2). We show R is spanned by a vital curve by induction on m which we may assume is at least 5.

Note M retracts onto any vital cycle, thus if Z is any vital cycle, the restriction $N^1(M) \to N^1(Z)$ is surjective. In particular if R is in the image of (the injection) $i : \overline{NE_1}(Z) \to \overline{NE_1}(M)$, then R spans an extremal ray on Z.

Assume R satisfies (1.2.2). By (2.6), R spans an extremal ray on some D_T. Since D_T has anti-nef normal bundle, we can increase its coefficient in G to one, restrict to D_T, and apply adjunction and induction.

Assume R satisfies (1.2.1). By (2.5), R is spanned by a curve $C \subset D_T$. By (3.7.4) we may assume $C \cdot D_T < 0$, so C does not deform away from D_T. By (2.7), $C \subset D_T$ is contracted by a map of relative Picard number one, and so we can apply induction. □

4. Intersecting vital curves and divisors.

By a *marked point* of an n-pointed curve, we either mean one of the singular points of the curve, or one of the labeled points $p_1, p_2, \dots, p_n$.

Notation 4.1. *Let C be a vital curve. Let $G = G(C)$ be the n-pointed stable curve corresponding to the generic point of C. G has $n-3$ components, all but one of which contain 3 marked points, and exactly one of which contains 4 marked points. We call this last component $Q = Q(C)$, the distinguished component of G. Let $s(C)$ be the number of singular points on Q, $l(C)$ be the number of labeled points. C determines a decomposition of $\{1,2,\dots,n\}$ into 4 disjoint subsets: $G \setminus Q$ has exactly $s(C)$ connected components. We decompose $\{1,2,\dots,n\}$ into those labeled points on each of the components. Additionally we take the singleton sets for each of the $l(C)$ labeled points on G. We call this decomposition P_C.*

There are $n-4$ singular points on G (intersection points of two components). Each singular $p \in G$ defines a decomposition, by letting T_p and T_p^c be the labels on the two connected components of $G \setminus \{p\}$. C is the complete intersection $\bigcap_{p \in \mathrm{Sing}(G)} D_{T_p,T_p^c}$. Let A_{T_p} and $A_{T_p^c}$ be the connected components of $G \setminus \{p\}$.

Lemma 4.2. *Let C be a vital curve.*

(1) *For $p \in \mathrm{Sing}(G)$*

$$\pi_{T_p} : D_{T_p,T_p^c} \longrightarrow M_{T_p \cup \{b\}}.$$

contracts the vital curve C iff $A_{T_p^c}$ contains the generic point of $Q(C)$.

(2) *q_i contracts C iff i is not one of the labeled points of $Q(C)$ (in particular any C with $l(C) = 0$ is contracted).*

Proof. (1) is immediate and (2) follows from (1) and (3.1.3). □

Lemma 4.3. P_C *uniquely determines the numerical class of* C. $K_{\overline{M}_{0,n}} \cdot C = 2 - l(C)$. *For any vital divisor* D_{T,T^c} *we have:*

(1) $D_{T,T^c} \cdot C = -1$ *iff* T *or* T^c *is one of the equivalence classes of* P_C. *Equivalently, iff* T *or* T^c *is* T_p *for some singular point* $p \in Q$.
(2) $D_{T,T^c} \cdot C = 1$ *iff* T *or* T^c *is the union of two equivalence classes.*
(3) *Otherwise* $D_{T,T^c} \cdot C = 0$.

Proof. Since the vital divisors generate Pic $(\overline{M}_{0,n})$ the description of $D_{T,T^c} \cdot C$ implies the first statement. The expression for $K_{\overline{M}_{0,n}} \cdot C$ follows from the expression for $D_{T,T^c} \cdot C$ using the adjunction formula, since C is a complete intersection of vital divisors.

Fix $p \in \mathrm{Sing}(G)$ and let S be the intersection of the D_{T_q,T_q^c} for $q \neq p$. Then $C \cdot D_{T_p,T_p^c}$ is the self intersection of C in S. S is a vital surface, and so it is either $\overline{M}_{0,5}$ (which is $\mathbb{P}^2$ blown up in 4 points) or $\overline{M}_{0,4} \times \overline{M}_{0,4}$ (which is $\mathbb{P}^1 \times \mathbb{P}^1$), and C is a vital curve in S. In the first case C is a -1-curve, and in the second a fibre of one of the two projections. Let γ be the pointed stable curve corresponding to a generic point of S. In the first case γ has one component with 5 marked points, and in the second case, two components each with 4 marked points. G is obtained as the limit as two of the marked points (on the same component) come together at p. It's clear that the first case occurs iff $p \in Q$, whence (1). Note the argument shows that if $C \subset D_{T,T^c}$ then $C \cdot D_{T,T^c}$ is either 0 or 1.

If $D_{T,T^c} \cdot C > 0$ then $D_{T,T^c} \cap C$ is a vital point of $C = \overline{M}_{0,4}$, i.e. a reduced point, thus $D_{T,T^c} \cdot C = 1$. This occurs if T or T^c is a union of two equivalence classes of P_C and every vital divisor of C can be obtained in this way. Since each vital cycle is uniquely a complete intersection of vital divisors, this gives (2).

Since the possibilities with $C \cdot D_{T,T^c}$ nonzero are classified by (1) and (2), (3) follows. □

Corollary 4.4. *The numerical class of* $\widetilde{C}$ *is determined by the cardinalities of the subsets in* P_C. *If these cardinalities are* a, b, c, d *then*

$$C \cdot \sum r_i B_i = -r_a - r_b - r_c - r_d + r_{a+b} + r_{a+c} + r_{a+d}$$

where we define $r_1 = 0$ *and* $r_i = r_{n-i}$ *for* $i > [n/2]$.

Lemma 4.5.

$$N_{D_{T,T^c}} \overline{M}_{0,n} = (q_b \circ \pi_T)^*(\mathcal{O}(-1)) \otimes (q_b \circ \pi_{T^c})^*(\mathcal{O}(-1)).$$

Proof. Since the vital curves generate N_1 we only need to check how both sides intersect a vital curve $C \subset D_{T,T^c}$. By (3.1.2), and (4.3) the possible values of these intersections are 0 and -1, and it is enough to show show $D_{T,T^c} \cdot C = -1$ iff one of the two maps $q_b \circ \pi_T$ or $q_b \circ \pi_{T^c}$ fails to contract C.

By (4.2.1) we may assume that π_T is finite on C (otherwise switch T and T^c). By (4.2.1) and (4.3.1), $D_{T,T^c} \cdot C = -1$ iff b is a labeled point of $Q(\pi_T(C))$, thus by (4.2.2), iff q_b is finite on $\pi_T(C)$. □

Lemma 4.6. *Let C be a vital curve, and let*

$$D_C = \sum_{p \in \mathrm{Sing}(G) \cap Q} D_{T_p, T_p^c}.$$

Then $(K_{\overline{M}_{0,n}} + D_C) \cdot C = -2$. *Further,* $K_{\overline{M}_{0,n}} + D + 1/sD_C$ *intersects vital curves non-negatively, and vanishes on exactly those vital curves numerically equivalent to* C.

Proof. Immediate from (4.3). □

4.6.1 Remark. (1.1) and the basepoint free theorem imply $K+D+1/sD_C$ is eventually free, and thus C spans an extremal ray. Presumably this could be checked directly.

The following is immediate:

Lemma 4.7. *Let* $T \subset \{1,2,\dots,n\}$ *with* $|T| \geqslant 3, |T^c| \geqslant 2$. *For* $i \in T$,

$$D_{T\setminus\{i\}, T^c \cup \{i\}}|_{D_{T,T^c}} = D_{ib, T\setminus\{i\}} \times M_{T^c \cup \{b\}}$$

under the canonical product decomposition. There is no other vital divisor with the same restriction.

Lemma 4.8. *Suppose there is a numerical equality*

$$\sum_{i=2}^{k} m_i B_i \sim F$$

and either F *is nef, or both sides are effective and have no divisor common to their supports. Then*

$$\begin{aligned} rm_{r-1} &\geqslant (r-2)m_r && \text{for } 3 \leqslant r \leqslant k \\ (n-r)m_{r+1} &\geqslant (n-r-2)m_r && \text{for } 2 \leqslant r \leqslant k-1. \end{aligned}$$

When the left hand side is effective, it is either trivial, or has support exactly D. (1.3.1-4) *hold.*

Proof. We prove the first inequality, the argument for the second is analogous. The final remarks follow from the inequalities.

Choose T with $|T| = r$. Let Z_r be the general fibre of

$$M_{T \cup \{b\}} \longrightarrow M_T.$$

Let $p \in M_{T^c \cup \{b\}}$ be a general point, and let $D_{T,T^c} \supset C_r = Z_r \times \{p\}$. By (3.1), (4.5) and (4.7) we have

$$C_r \cdot B_i = \begin{cases} r \text{ if } i = r-1 \\ -(r-2) \text{ if } i = r \\ 0 \text{ otherwise} \end{cases}$$

The inequality is obtained by intersecting both sides with C_r.

Note that (1.3.1-3) follow immediately and that (1.3.4) then follows from (2.3). □

5. $NE_1(\widetilde{M}_{0,n})$ for small n

Given (4.8), it is natural to hope that every nef divisor on $\widetilde{M}_{0,n}$ is eventually free. The obvious approach is to try to use the basepoint free theorem, and thus to realise some positive multiple of a big nef class E (pulled back from $\widetilde{M}_{0,n}$) as a klt divisor $K_{\overline{M}_{0,n}} + \Delta$.

Lemma 5.1. *If E is a big nef class on a normal $\mathbb{Q}$-factorial variety M, and there is a divisor Δ with $K_M + \Delta$ klt and numerically equivalent to a positive multiple of E, then the extremal subcone of $\overline{NE}_1(M)$ supported by E is rational polyhedral, and is contracted by a log Mori fibre space. If $M = \overline{M}_{0,n}$, the subcone supported by E is spanned by vital curves.*

Proof. By (2.2) we have $E = A+Z$ where A is ample and Z is effective. If $V \subset \overline{NE}_1(M)$ is the extremal subcone supported by E, then $K_M + \Delta + \epsilon Z$ is negative on $V \setminus 0$. Thus the result follows from the cone and contraction theorems, together with (1.2). □

Let E be a nef divisor on $\overline{M}_{0,n}$, pulled back from $\widetilde{M}_{0,n}$.

By (3.5), for $n \leqslant 7$ the conditions of (5.1) are satisfied (if $K+\Gamma$ is klt and trivial, let $\Delta = \Gamma + \epsilon E$).

In general, by (3.5) and (4.8), replacing E by a large multiple one has $E = K_{\overline{M}_{0,n}} + \Delta$ for some Δ supported on D. We can try to make Δ a boundary by subtracting off part of E, thus we are lead to consider:

Definition-Lemma 5.2. *Let E be a non-trivial nef class on $\overline{M}_{0,n}$, pulled back from $\widetilde{M}_{0,n}$ with $n \geqslant 8$. Then there is a unique effective class Δ_E with the following properties*

(1) *Δ_E has support a proper subset of D*
(2) *$K_{\overline{M}_{0,n}} + \Delta_E = \lambda E$ for some $\lambda > 0$.*

Proof. For any λ, $-K_{\overline{M}_{0,n}} + \lambda E$ is pulled back from $\widetilde{M}_{0,n}$, thus by (4.8), (1) is the requirement that Δ_E be on the boundary of NE^1. Since E is in the interior of NE^1, and by (3.5), $-K_{\overline{M}_{0,n}} \notin NE^1$, the result is clear. □

Notation: *For the next corollary, define the integer function $f(a, b, c, d)$ to be 2 minus the number of variables equal to one.*

We will say that P_n holds if for a given integer n the following implication holds:

Let $r_1, r_2, \ldots, r_{n-1}$ be a collection of non-negative real numbers, with $r_1 = 0$, $r_i = r_{n-i}$, and $r_j = 0$ for some $2 \leqslant j \leqslant k$. If

$$f(a, b, c, d) + r_{a+b} + r_{a+c} + r_{a+d} \geqslant r_a + r_b + r_c + r_d$$

for every set of positive integers a, b, c, d with $n = a + b + c + d$, then $r_i < 1$ for all i.

Corollary 5.3. *Δ_E is a pure boundary for every non-trivial nef class pulled back from $\widetilde{M}_{0,n}$ iff P_n holds.*

Proof. By (4.8) P_n is equivalent to the statement: If $\sum r_i B_i$ has support a proper subset of D and $(K_{\overline{M}_{0,n}} + \sum r_i B_i) \cdot C \geqslant 0$ for all vital curves C, then $\sum r_i B_i$ is a pure boundary. Thus the only thing to show is that if Δ_E is a pure boundary for every non-trivial nef class, then the images of vital curves generate $NE_1(\widetilde{M}_{0,n})$. This follows from (5.1). □

For a given n it is straightforward to check whether or not P_n holds:

Lemma 5.4. P_n *holds for* $8 \leqslant n \leqslant 11$.

Proof. We will check P_9. The cases $n = 8$, 10 and 11 are similarly checked.

Let $r_1, r_2, \ldots, r_8$ be a collection of non negative numbers, as in the definition of P_9. From the sums

$$1 + 2 + 3 + 3 = 9$$

$$1 + 1 + 1 + 6 = 9$$

$$1 + 1 + 2 + 5 = 9$$

we obtain the inequalities

$$1 + 2r_4 \geqslant r_3 + r_2$$

$$2r_3 \geqslant r_4$$

$$3r_2 \geqslant r_3 + 1$$

The result follows easily by considering in turn the possibilities $r_4 = 0$, $r_3 = 0$ and $r_2 = 0$. □

Observe that (1.3) follows from (5.1), (5.3) and (5.4).

There are examples to show that P_n fails for all $n \geqslant 12$.

References

[1] V. Alexeev. Moduli spaces $M_{g,n}(W)$ for surfaces, in: Higher-dimensional complex varieties (Trento, 1994), 1–22, de Gruyter, Berlin, 1996. ← 115

[2] J. Harris and D. Mumford. On the Kodaira Dimension of the Moduli Space of Curves. *Invent. Math.*, **67**, 23–88, 1982. ← 117

[3] M. M. Kapranov. Veronese Curves and Grothendieck-Knudson Moduli Space $\overline{M}_{0,n}$. *J. of Algebraic Geometry*, **2**, 239–262, 1993. ← 117

[4] M. M. Kapranov. Chow Quotients of Grassmannians. I., in: I.M. Gelfand Seminar, S. Gelfand, S. Gindikin eds.. *Advances in Soviet Mathematics*, **16**, part 2., A.M.S., 1993, 29–110. ← 122

[5] Y. Kawamata. Addition formula of logarithmic Kodaira dimensions for morphisms of relative dimension one, in: Algebraic Geometry Kyoto 1977, 207–217, Kinokuniya, 1978. ← 117

[6] Y. Kawamata. Subadjunction of Log Canonical Divisors for a Subvariety of Codimension 2, in: Birational algebraic geometry (Baltimore, MD, 1996), 79–88, *Contemp. Math.*, **207**, Amer. Math. Soc., Providence, RI, 1997. ← 117

[7] Y. Kawamata, K. Matsuda, and K. Matsuki. Introduction to the minimal model program. *Adv. Stud. Pure Math*, **10**, 283–360, 1987. ← 118

[8] S. Keel. Intersection Theory of Moduli Space of Stable N-pointed Curves of Genus Zero. *Transactions of the AMS*, **330**, 545–574, 1992. ← 121, 124

[9] S. Keel and J. McKernan. Rational Curves on Quasiprojective Varieties. *Mem. Amer. Math. Soc.*, **140**, no.669, 1999. ← 124

[10] J. Kollár. Rational curves on projective varieties. *Springer Ergebnisse der Mathematik*, **32**, 1996. ← 118, 121

[11] M. Kontsevich and Y. Manin. Gromov-Witten Classes, Quantum Cohomology, and Enumerative Geometry. *Comm. Math. Phys.* **164**(3), 525–562, 1994. ← 117

[12] R. Pandharipande. The Canonical Class of $\overline{M}_{0,n}(\mathbb{P}^d)$ and Enumerative Geometry. *Internat. Math. Res. Notices*, **4**, 173–186, 1997. ← 123

[13] V. V. Shokurov. 3-Fold Log Models, in Algebraic geometry, 4. *J. Math. Sci.*, **81**(3), 2667–2699, 1996. ← 118, 119

Department of Mathematics, University of Texas, at Austin, Austin TX 78712
E-mail address: keel@math.utexas.edu

Department of Mathematics, MIT, 77 Massachusetts Avenue, Cambridge, MA 02139, USA
E-mail address: mckernan@math.mit.edu

Moduli of varieties of general type

János Kollár

Contents

1. Introduction

1.1 (Moduli theories). The development of a good moduli theory consists of four basic steps.

(1.1.1) *Identify a class of objects whose moduli theory is nice.*

In some cases the answer is obvious, for instance, we should study the moduli theory of smooth projective curves. In other cases, it took some time to understand what the correct objects should be: Abelian varieties should be replaced by *polarized* Abelian varieties, K3 surfaces by *marked* K3 surfaces and vector bundles by *stable* vector bundles. We see later that smooth, projective varieties of general type should be replaced by their canonical models. We discuss this in Section 2.

In all these cases, we get non-proper moduli spaces. This is inconvenient for many reasons, for instance it is hard to count objects with various properties. To remedy this, one should look for a compactification whose points have clear geometric meaning.

(1.1.2) *Choose a larger class of objects to form a proper moduli space.*

The choice of these objects is usually neither obvious nor unique. It was not until the 1960's that the importance of this step was understood and stable curves and semi-stable sheaves were identified and studied in detail. For surfaces of general type the right class was described in [35] and for polarized Abelian varieties in [4]. The solution for varieties of general type is treated in Section 3.

(1.1.3) *Establish the correct moduli functor.*

Once a class of objects **V** is established, the corresponding moduli functor is usually declared to consist of all flat families whose fibers are in **V**. However, for varieties of general type, allowing all flat families gives the *wrong* moduli functor. The problem and the solution are analyzed in Section 4.

(1.1.4) *Study the resulting moduli spaces and their applications.*

The moduli of curves and the moduli of semi-stable vector bundles on curves appear in many contexts and by now they established themselves as two of the richest applications of algebraic geometry. We are only at the beginning of the development of the moduli of higher dimensional varieties; the basic results are outlined in Section 5. I hope to see many more applications in the future.

These notes are intended to give a survey of the subject, stressing key examples and results. The forthcoming books [31, 32] aim to give a complete treatment.

Definition 1.2 (Moduli functors). Let **V** be a "reasonable" class of projective varieties (or schemes, or sheaves, or ...). For the next definition we only need to assume that if $K \supset k$ is a field extension then a k-variety X_k is in **V** iff $X_K := X_k \times_{\operatorname{Spec} k} \operatorname{Spec} K$ is in **V**. Define the corresponding moduli functor as

$$\mathcal{V}arieties_{\mathbf{V}}(T) := \left\{ \begin{array}{c} \text{Flat families } X \to T \text{ such that} \\ \text{every fiber is in } \mathbf{V}, \\ \text{modulo isomorphisms over } T. \end{array} \right\} \tag{1.2.1}$$

(As noted in (1.1.3), we will need to impose additional restrictions eventually, but for now let us ignore these.)

1.3 (Moduli spaces). We say that a scheme $\operatorname{Moduli}_{\mathbf{V}}$, or, more precisely, a flat morphism

$$u : \operatorname{Univ}_{\mathbf{V}} \to \operatorname{Moduli}_{\mathbf{V}}$$

is a *fine moduli space* for the functor $\mathcal{V}arieties_{\mathbf{V}}$ if, for every scheme T, pulling back gives an equality

$$\mathcal{V}arieties_{\mathbf{V}}(T) = \operatorname{Mor}(T, \operatorname{Moduli}_{\mathbf{V}}).$$

Our aim is to understand all families whose fibers are in **V** and a fine moduli space presents the answer in the most succinct way.

Applying the definition to $T = \operatorname{Spec} K$, where K is a field, we see that every fiber of $u : \operatorname{Univ}_{\mathbf{V}} \to \operatorname{Moduli}_{\mathbf{V}}$ is in **V** and the K-points of $\operatorname{Moduli}_{\mathbf{V}}$ are in one-to-one correspondence with the K-isomorphism classes of objects in **V**.

We consider the existence of a fine moduli space as the ideal possibility. Unfortunately, it is rarely achieved, mainly due to the presence of automorphisms. When there is no fine moduli space, we still can ask for a scheme that best approximates its properties. Therefore, we look for schemes M for which there is a natural transformation of functors

$$T_M : \mathcal{V}arieties(*) \longrightarrow \operatorname{Mor}(*, M).$$

Such schemes certainly exist, for instance, if we work over a field k then we can always take $M = \operatorname{Spec} k$. All schemes M for which T_M exists form an inverse system which is closed under fiber products. Thus, as long as we are not unlucky, there is a universal (or largest) scheme with this property. Though it is not usually done, it should be called the *categorical moduli space*.

This object can be rather useless in general. For instance, fix n, d and let $H_{n,d}$ be the class of all hypersurfaces of degree d in $\mathbb{P}^{n+1}_k$ up to isomorphisms. The categorical moduli space $\mathrm{HypSurf}_{n,d}$ exists and it is $\operatorname{Spec} k$. Indeed, there is a family of hypersurfaces over the linear system $|\mathcal{O}_{\mathbb{P}^n}(d)| \cong \mathbb{P}^N$ where $N = \binom{n+d}{n} - 1$. Thus we get a surjective morphism $\mathbb{P}^N \to \mathrm{HypSurf}_{n,d}$ which is constant on the $GL(n+1)$-orbits. Any morphism $\mathbb{P}^N \to X$ that maps a curve to a point is constant, hence $\mathrm{HypSurf}_{n,d} = \operatorname{Spec} k$.

In order to get something reasonable, we impose extra conditions. A scheme Moduli_V is a *coarse moduli space* for $\mathbf{V}$ if the following hold.

(1) There is a natural transformation of functors

$$\mathrm{ModMap} : \mathcal{V}arieties_V(*) \longrightarrow \mathrm{Mor}(*, \mathrm{Moduli}_V),$$

(2) Moduli_V is universal satisfying (1), and

(3) for any algebraically closed field $K \supset k$,

$$\mathrm{ModMap} : \mathcal{V}arieties_V(\operatorname{Spec} K) \xrightarrow{\cong} \mathrm{Mor}(\operatorname{Spec} K, \mathrm{Moduli}_V) = \mathrm{Moduli}_V(K)$$

is an isomorphism (of sets).

In many cases, the naturally occurring moduli spaces have a further very useful property.

(4) There is a family $V_U \to U$ in $\mathbf{V}$ such that the corresponding moduli map $\mathrm{ModMap}(V_U) : U \to \mathrm{Moduli}_V$ is surjective, open and quasi finite.

Following woodworking terminology, I propose to call a moduli space satisfying conditions (1–4) a *bastard* moduli space.

1.4 (Problems with the moduli of smooth varieties).

In contrast with curves, the moduli theory for higher dimensional smooth varieties can be very badly behaved, as shown by the following examples.

(1.4.1) (Ruled surfaces) Let C be a smooth curve and L a line bundle on C that is generated by 2 sections f, g. On $S := C \times \mathbb{A}^1_t$, with first projection π_1, consider the exact sequence

$$0 \to \pi_1^* L^{-1} \stackrel{(t,f,g)}{\longrightarrow} \pi_1^* L^{-1} + \mathcal{O}_S + \mathcal{O}_S \to Q \to 0.$$

Q is a rank 2 vector bundle on $C \times \mathbb{A}^1_t$. We can view $\mathbb{P}_{C\times\mathbb{A}^1_t} Q$ is a $\mathbb{P}^1$-bundle over S, or as a flat family of ruled surfaces over $\mathbb{A}^1$.

If $t \neq 0$ then $t : \pi_1^* L^{-1} \to \pi_1^* L^{-1}$ is an isomorphism, thus $Q_t \cong \mathcal{O}_C + \mathcal{O}_C$. If $t = 0$ then we get

$$Q_0 \cong L^{-1} + \operatorname{coker}\left[L^{-1} \stackrel{(f,g)}{\to} \mathcal{O}_C + \mathcal{O}_C\right] \cong L^{-1} + L.$$

Thus we get a flat family of smooth ruled surfaces whose general member is $\mathbb{P}^1 \times C$ and whose special member is $\mathbb{P}_C(L^{-1} + L)$. In a coarse moduli space over $\mathbb{C}$ both of these should correspond to $\mathbb{C}$-points, but the above family shows that the moduli point $[\mathbb{P}_C(L^{-1} + L)]$ is in the closure of the moduli point $[\mathbb{P}^1 \times C]$. This is impossible, hence the coarse moduli space can not be a scheme. (This is a quite good starting point for introducing stacks, where such things are possible. However, the main problems in the moduli theory of varieties of general type do not seem to involve such sticky stacky issues, so I have stuck to schemes and algebraic spaces as moduli.)

One can be even more specific for $C = \mathbb{P}^1$. Minimal ruled surfaces over $\mathbb{P}^1$ are $\mathbb{F}_m := \mathbb{P}_{\mathbb{P}^1}(\mathcal{O}_{\mathbb{P}^1} + \mathcal{O}_{\mathbb{P}^1}(m))$ for $m \geqslant 0$. The "moduli space" has 2 connected components, corresponding to even and odd values of m. There are no closed points in this "moduli space;" the closure of $[\mathbb{F}_m]$ consists of all the points $\{[\mathbb{F}_m], [\mathbb{F}_{m+2}], [\mathbb{F}_{m+4}], \dots\}$. This is worked out in detail in [10].

(1.4.2) (Abelian, elliptic and K3 surfaces)

A general problem in all these cases is that a typical deformation of such an algebraic surface over $\mathbb{C}$ is a non-algebraic complex analytic surface. Thus any algebraic theory captures only a small part of the full moduli theory.

(1.4.3) For Abelian varieties and for K3 surfaces, the moduli spaces look very strange topologically. For instance, the 3-dimensional space of Kummer surfaces is dense in the 20-dimensional space of all K3 surfaces [38].

This can be corrected by fixing a basis in $H^2(*, \mathbb{Z})$, but it is not clear how similar tricks work in general. Also, as it happens already for stable curves, we would like to consider families where not all fibers are homeomorphic to each other. Then it is no clear what one means by " fixing a basis in $H^2(*, \mathbb{Z})$."

(1.4.4) (Repeated blow-ups lead to non-separatedness)

Let $f : X \to B$ be a smooth family of projective surfaces over a smooth (affine) pointed curve $b \in B$. Let $C_1, C_2, C_3 \subset X$ be three sections of f, all passing through a point $x_b \in X_b$ that intersect at x_b with independent tangent directions and are disjoint elsewhere.

Set $X^1 := B_{C_1} B_{C_2} B_{C_3} X$, where we first blow-up $C_3 \subset X$, then the birational transform of C_2 in $B_{C_3} X$ and finally the birational transform of C_1 in $B_{C_2} B_{C_3} X$. Similarly, set $X^2 := B_{C_1} B_{C_3} B_{C_2} X$. Since the C_i are sections, all these blow-ups are smooth families of projective surfaces over B.

Write the tangent vector of C_i at x_b as $(v_i, 1)$ where v_i is tangent to the fiber X_b. One can see that X_b^1 (resp. X_b^2) is obtained from X_b by first blowing up $x_b \in X_b$ and then on the exceptional curve blowing up the points corresponding to $v_2 - v_3$

and $v_1 - v_3$ (resp. $v_1 - v_2$ and $v_3 - v_2$). Thus, if $v_1 - v_3$ and $v_1 - v_2$ are linearly independent, then

(1) all the fibers are smooth, projective surfaces,

(2) $X^1 \to B$ and $X^2 \to B$ are isomorphic over $B \setminus \{b\}$ but

(3) the natural birational map $X^1_b \to X_b \leftarrow X^2_b$ is *not* an isomorphism.

So if X_b has no birational automorphisms then the fibers X^1_b and X^2_b are not isomorphic. (Note that if we stop at $B_{C_2}B_{C_3}X$ and $B_{C_3}B_{C_2}X$, then we have isomorphic fibers but the two families are not isomorphic, as in the next example.)

This type of behavior happens every time we look at deformations of a surface with at least 3 points blown-up.

(1.4.5) (Non-separatedness for minimal resolutions.)

Let $X_0 := (f(x_1, \ldots, x_4) = 0) \subset \mathbb{P}^3$ be a surface of degree n that has an ordinary double point at $p = (0{:}0{:}0{:}1)$ as its sole singularity and contains the pair of lines $(x_1x_2 = x_3 = 0)$. Let g be homogeneous of degree $n-1$ such that x_4^{n-1} appears in it with nonzero coefficient. Consider the family of surfaces

$$X := \big(f(x_1, \ldots, x_4) + t x_3 g(x_1, \ldots, x_4) = 0\big) \subset \mathbb{P}^3_x \times \mathbb{A}^1_t.$$

Note that X_t is smooth for general $t \neq 0$ and X contains the pair of smooth surfaces $(x_1x_2 = x_3 = 0)$.

For $i = 1,2$, let $X^i := B_{(x_i,x_3)}X$ denote the blow-up of $(x_i = x_3 = 0)$ with induced morphisms $\pi_i : X^i \to X$ and $f_i : X^i \to \mathbb{A}^1$. There is a natural birational map $\phi := \pi_2^{-1} \circ \pi_1 : X^1 \dashrightarrow X^2$. Let B_pX denote the blow-up of $p = \big((0{:}0{:}0{:}1),0\big)$ with exceptional divisor $E \subset B_pX$. One checks that

(1) all the fibers are smooth, projective minimal models,

(2) $X^1 \to \mathbb{A}^1$ and $X^2 \to \mathbb{A}^1$ are isomorphic over $\mathbb{A}^1 \setminus \{0\}$,

(3) the fibers X^1_0 and X^2_0 are isomorphic, but

(4) $X^1 \to \mathbb{A}^1$ and $X^2 \to \mathbb{A}^1$ are *not* isomorphic.

While it is not clear from our construction, similar problems happen for any smooth family of surfaces where the general fiber has ample canonical class and a special fiber has nef (but not ample) canonical class, see [7, 9, 39].

(1.4.6) (Non-projective families.)

In the above example, assume that $n \geqslant 4$ and let $h(x_1, \ldots, x_4)$ be general of degree n. Consider the family of surfaces

$$X := \big(f(x_1, \ldots, x_4) + t^2 h(x_1, \ldots, x_4) = 0\big) \subset \mathbb{P}^3_x \times \mathbb{A}^1_t.$$

There is no smooth algebraic surface contained in X passing through p, but there is such a local analytic smooth surface. Blowing it up we get a complex analytic family $X' \to \mathbb{A}^1$ where every fiber is a smooth projective surface yet the family itself is not projective over $\mathbb{A}^1$.

1.5 (Answers to these problems). The problems (1.4.1–3) come from the global geometry of the varieties that we work with.

The current assumption is that the moduli problem of uniruled varieties is usually pathological. Furthermore, any general attempt to create a good moduli functor ends up with a theory that is not compatible with birational equivalence. (There are examples, like the moduli of smooth hypersurfaces of degree n in $\mathbb{P}^n$ for $n \geqslant 4$, where the biregular moduli theory ends up being birationally invariant. However, even in these cases, it is not clear that a sensible compactification exists.)

For varieties with trivial canonical class one should get a nice moduli theory only after a suitable "rigidification." This can consist of choosing a basis in $H^2(*, \mathbb{Z})$ or fixing an ample divisor. The compactification question is mostly still unsolved. For instance, a geometrically meaningful compactification of the moduli of K3 surfaces is yet to be found.

The problems (1.4.4–6) are more local. The aim of these notes is to explain how to deal with them for varieties of general type. The solution is to work with *canonical models* instead of smooth varieties of general type.

Following [17] (see [37, Sec.2.1.C] for details) a smooth projective variety X of dimension n is of *general type* if the following equivalent conditions hold:

(1) $h^0(X, \mathcal{O}_X(mK_X)) \geqslant \epsilon \cdot m^n$ for some $\epsilon > 0$ and $m \gg 1$.

(2) $\operatorname{Proj} R(X, K_X)$ has dimension n.

(3) The natural map $X \dashrightarrow \operatorname{Proj} R(X, K_X)$ is birational.

The main reason, however, why we do not study the moduli functor of smooth varieties up to isomorphism is that, in dimensions two and up, smooth projective varieties do not form the *smallest* basic class. Given any smooth projective variety X, one can blow up any set of points or subvarieties $Z \subset X$ to get another smooth projective variety $B_Z X$ which is very similar to X. Therefore, the basic object should be not a single smooth projective variety but a whole *birational equivalence class* of smooth projective varieties. Thus it would be better to work with smooth, proper families $X \to S$ modulo birational equivalence over S. That is, with the moduli functor

$$\mathrm{GenType}_{bir}(S) := \left\{ \begin{array}{c} \text{Smooth, proper families } X \to S, \\ \text{every fiber is of general type,} \\ \text{modulo birational equivalence over } S. \end{array} \right\} \tag{1.5.4}$$

In essence this is what we end up doing, but it is very cumbersome to deal with birational equivalence over a base scheme. (Even the definition of "birational equivalence over a base" needs clarification. Do we want the total spaces to be birational, the corresponding fibers to be birational or something else?)

The following result, or rather, its proof, shows the way to a good moduli theory for varieties of general type.

Proposition 1.6. *Let* $f_i : X^i \to B$ *be two smooth families of projective varieties over a smooth curve* B *such that the canonical classes* K_{X^i} *are* f_i*-ample. Then every isomorphism between the generic fibers* $\phi : X^1_{k(B)} \cong X^2_{k(B)}$ *extends to an isomorphism* $\Phi : X^1 \cong X^2$.

Proof. Let $\Gamma \subset X^1 \times_B X^2$ be the closure of the graph of ϕ. Let $Y \to \Gamma$ be the normalization, with projections $p_i : Y \to X^i$ and $f : Y \to B$. We use the canonical class to compare the X^i. Since the X^i are smooth,

$$K_Y \sim p_i^* K_{X^i} + E_i \quad \text{where } E_i \text{ is effective and } p_i\text{-exceptional.} \tag{1.6.1}$$

Since $(p_i)_* \mathcal{O}_Y(mE_i) = \mathcal{O}_{X^i}$ for every $m \geqslant 0$, we get that

$$\begin{aligned}(f_i)_* \mathcal{O}_{X^i}(mK_{X^i}) &= (f_i)_*(p_i)_* \mathcal{O}_{X^i}(mp_i^* K_{X^i}) &&= \\ &= (f_i)_*(p_i)_* \mathcal{O}_{X^i}(mp_i^* K_{X^i} + mE_i) &&= \\ &= (f_i)_*(p_i)_* \mathcal{O}_Y(mK_Y) &&= f_* \mathcal{O}_Y(mK_Y).\end{aligned}$$

Since the K_{X^i} are f_i-ample, $X^i = \operatorname{Proj}_B \sum_{m \geqslant 0} (f_i)_* \mathcal{O}_{X^i}(mK_{X^i})$. Putting these together, we get the isomorphism

$$\begin{aligned}\Phi : X^1 &\cong \operatorname{Proj}_B \textstyle\sum_{m \geqslant 0} (f_1)_* \mathcal{O}_{X^1}(mK_{X^1}) \cong \\ &\cong \operatorname{Proj}_B \textstyle\sum_{m \geqslant 0} f_* \mathcal{O}_Y(mK_Y) \cong \\ &\cong \operatorname{Proj}_B \textstyle\sum_{m \geqslant 0} (f_2)_* \mathcal{O}_{X^2}(mK_{X^2}) \cong X^2. \qquad \square\end{aligned}$$

Note that the smoothness of the X^i is used only through the pull-back formula (1.6.1). This leads to the first major definition:

Preliminary Definition 1.7. Let B be a smooth curve, X a normal variety and $f : X \to B$ a non-constant projective morphism. Assume that

(1) $m_0 K_X$ is Cartier for some $m_0 > 0$. (This is needed to make sense of (2) and also to define the pull-back in (3).)
(2) $m_0 K_X$ is f-ample.
(3) If $p : Y \to X$ is a resolution of singularities then

$$m_0 K_Y \sim p^*(m_0 K_X) + E(m_0)$$

where $E(m_0)$ is effective and p-exceptional.

We do not want to keep carrying the m_0 along, thus we switch to $\mathbb{Q}$-divisors and $\mathbb{Q}$-linear equivalence and write (3) as

(3′) If $p : Y \to X$ is a resolution of singularities then

$$K_Y \sim_{\mathbb{Q}} p^* K_X + E$$

where E is an effective, p-exceptional $\mathbb{Q}$-divisor.

With these assumptions we can make the following informal definitions (to be made precise in (2.1) and (3.1)). If a family $f : X \to B$ as above satisfies the assumptions (1–3′) then

(4) a "general" fiber of f is a *canonical model* and
(5) a "special" fiber of f is a *semi-log-canonical model* if (1–3′) continue to hold after base change $X' := X \times_B B' \to B'$, for every smooth curve B'.

In this area, "semi" refers to allowing non-normal schemes and "log" refers to allowing exceptional divisors with coefficients $\geqslant -1$, see (3.1).

(Note that (1–2) are inherited by every base change, thus the only question is (3′). Already for curves, this condition is necessary. Indeed, let $(f(x, y, z) = 0)$ be any plane curve with isolated singularities and $(g(x, y, z) = 0)$ a general smooth plane curve. Then

$$\big(f(x, y, z) + tg(x, y, z) = 0\big) \subset \mathbb{P}^2 \times \mathbb{A}^1_t$$

is smooth near the $t = 0$ fiber. After the base change $t = s^m$ we get

$$\big(f(x, y, z) + s^m g(x, y, z) = 0\big) \subset \mathbb{P}^2 \times \mathbb{A}^1_s$$

and now (3′) is satisfied iff $(f(x, y, z) = 0)$ has only ordinary nodes. At points of multiplicity $\geqslant 3$, we can take $m = 3$ and check this by one blow up, but at a cusp one needs to take $m \geqslant 6$ and several blow-ups. This is essentially the classification of Du Val singularities of surfaces; see [34, Sec.4.2] for detailed computations.)

Basic principles

The moduli theory of higher dimensional varieties of general type is governed by the following four basic principles.

Principle 1.8. *Canonical models are the correct higher dimensional analogs of smooth projective curves of genus* $\geqslant 2$.

Principle 1.9. *Semi-log-canonical models are the correct higher dimensional analogs of stable projective curves of genus* $\geqslant 2$.

Principle 1.10. *Flat families of canonical models form the correct higher dimensional open moduli problem; see (4.5).*

Principle 1.11. *Flat families of semi-log-canonical models do* **not** *form the correct higher dimensional compactified moduli problem.*

The correct answer will be given in Section 4.

2. Canonical models

As noted in (1.8), canonical models are the basic objects of our moduli theory.

Definition 2.1. A normal projective variety X is a *canonical model* if

(1) m_0K_X is Cartier for some $m_0 > 0$,

(2) for every (equivalently, for one) resolution of singularities $f : X' \to X$ there is an effective, f-exceptional $\mathbb{Q}$-divisor E such that

$$K_{X'} \sim_{\mathbb{Q}} f^*K_X + E,$$

(3) K_X is ample.

Singularities satisfying (1–2) are called *canonical.*

This assumption implies that the *canonical ring* of X

$$R(X, K_X) := \sum_{m \geqslant 0} H^0(X, \mathcal{O}_X(mK_X))$$

is isomorphic to the canonical ring of X'

$$R(X', K_{X'}) := \sum_{m \geqslant 0} H^0(X', \mathcal{O}_{X'}(mK_{X'})).$$

In particular,

$$X = \operatorname{Proj}_k R(X, K_X) = \operatorname{Proj}_k R(X', K_{X'})$$

is the unique canonical model in the birational equivalence class of X.

Now we know [8, 41] that the canonical ring $R(X, K_X)$ of a smooth projective variety of general type is always finitely generated, thus we can define the *canonical model of* X as

$$X^{can} := \operatorname{Proj}_k \sum_{m \geqslant 0} H^0(X, \mathcal{O}_X(mK_X)),$$

which is a projective variety. It is not obvious, but true, that X^{can} satisfies the properties (1–3) [39].

Definition 2.2 (Moduli functor of canonical models). The moduli functor of canonical models is

$$\mathrm{CanMod}(S) := \left\{ \begin{array}{c} \text{Flat, proper families } X \to S, \\ \text{every fiber is a canonical model,} \\ \text{modulo isomorphisms over } S. \end{array} \right\} \tag{2.2.1}$$

This is an improved version of the birational moduli functor $\mathrm{GenType}_{bir}(*)$ (1.5.1).

(Traditionally this was considered to be the obviously correct definition, but, in view of Principle 1.11, it needs an explanation. For details, see (4.5).)

By a theorem of [40], in a smooth, proper family of varieties of general type the canonical rings form a flat family and so do the canonical models. Thus there is a natural transformation

$$T_{\text{CanMod}} : \mathrm{GenType}_{bir}(*) \to \mathrm{CanMod}(*).$$

If k is a field then $T_{\text{CanMod}} : \mathrm{GenType}_{bir}(\operatorname{Spec} k) \to \mathrm{CanMod}(\operatorname{Spec} k)$ is an isomorphism of sets since any canonical model is birational to a smooth projective variety. However, not every flat family of canonical surfaces over a smooth curve is obtained as the canonical model of a smooth family of projective surfaces.

Canonical curves are exactly the smooth, projective curves of genus $\geqslant 2$. Canonical surfaces have at worst Du Val singularities (also called rational double points). Starting with dimension 3, we get more complicated singularities. For instance,

$$(x_0^{a_0} + \cdots + x_n^{a_n} = 0) \subset \mathbb{A}^{n+1}$$

is canonical iff $\sum \frac{1}{a_i} > 1$ and a quotient of $\mathbb{A}^n$ be a finite subgroup G without quasi-reflections is canonical iff for every $g \in G$ $(g \neq 1)$ with eigenvalues $e^{2\pi i c_j}$ (where $0 \leqslant c_j < 1$) we have $\sum_j c_j \geqslant 1$.

The most important general property of canonical singularities is the following. For short proofs see [34, Sec.5.1] or [27].

Theorem 2.3. [13] *Let X be a canonical model over a field of characteristic 0. Then X has rational singularities. That is, $R^i f_* \mathcal{O}_Y = 0$ for $i > 0$ for every resolution of singularities $f : Y \to X$.*

Using the covering trick of [39] this implies that the reflexive hulls $\omega_X^{[m]}$ are CM (=Cohen-Macaulay) for every m. (See [34, Sec.5.5] for a quick introduction to CM sheaves.)

3. Semi-log-canonical models

First we translate (1.7.5) into a proper definition.

Let B be a smooth curve, X a normal variety and $f : X \to B$ a non-constant projective morphism. When is the fiber X_b a semi-log-canonical model?

By [21] there is a smooth pointed curve $b' \in B'$ and a finite morphism $(b' \in B') \to (b \in B)$ such that the base change $f' : X' := X \times_B B' \to B'$ has a resolution $\pi : Y' \to X'$ such that $(f' \circ \pi)^{-1}(b') \subset Y'$ is a reduced simple normal crossing divisor. We can also assume that the birational transform $Y'_{b'} := \pi_*^{-1} X'_{b'}$ is smooth. Thus $(f' \circ \pi)^{-1}(b') = Y'_{b'} + F$ where F is a reduced simple normal crossing divisor. By (1.7.3),

$$K_{Y'} \sim_{\mathbb{Q}} \pi^* K_{X'} + E'$$

where E' is effective. Since $Y'_{b'} + F' = \pi^*(X'_{b'})$, we can rewrite the above as

$$K_{Y'} + Y'_{b'} \sim_{\mathbb{Q}} \pi^*(K_{X'} + X'_{b'}) + E' - F.$$

Restricting to $Y'_{b'}$ we get

$$K_{Y'_{b'}} \sim_{\mathbb{Q}} \pi^* K_{X'_{b'}} + (E' - F)|_{Y'_{b'}}.$$

Since E' is effective, its restriction is again effective, but the restriction of $-F$ brings in negative coefficients. However, none of these is smaller than -1 since F and $Y'_{b'}$ intersect transversally.

Definition 3.1. A reduced, projective variety X is a *semi-log-canonical model* or *slc model* iff the following hold.

(1) $m_0 K_X$ is Cartier for some $m_0 > 0$.
(2) X satisfies Serre's condition S_2 and has only nodes in codimension 1.
(3) For every resolution of singularities $f : X' \to X$ there is a $\mathbb{Q}$-divisor $E = \sum_i a_i E_i$ such that

$$K_{X'} \sim_{\mathbb{Q}} f^* K_X + \textstyle\sum_i a_i E_i \quad \text{and } a_i \geqslant -1 \text{ for every } i.$$

(4) K_X is ample.

Clarifications: In order to define K_X, note that by (2), ω_X is locally free outside a codimension 2 subset of X, hence it corresponds to a linear equivalence class K_X of Weil divisors which are Cartier outside a codimension 2 subset of X.

One has to be a little careful with E because of the nodes on X. Denote the nodal divisor by $D \subset X$. If we normalize a node, its preimage is 2 points. Corresponding to D, on X' there is a unique divisor D' that is a double cover of D. In (3), E has an f-exceptional part but it also has to contain this divisor D' with coefficient -1.

This definition combines a global condition (4) with purely local conditions (1–3). Singularities satisfying (1–3) are called *semi-log-canonical* or *slc*.

For slc models it is usually better to use *semi-resolutions*, that is, a proper birational morphism $g : X^s \to X$ such that X^s has only double normal crossing points $(xy = 0) \subset \mathbb{C}^{n+1}$ and pinch points $(x^2 = y^2 z) \subset \mathbb{C}^{n+1}$ and g maps the double locus of X^s birationally on the double locus of X; see [26] for details. Let E denote the (reduced) exceptional divisor of a semi-resolution g. Then the canonical ring of X

$$R(X, K_X) := \sum_{m \geqslant 0} H^0(X, \mathcal{O}_X(mK_X))$$

is isomorphic to the *semi-log-canonical ring* of X^s

$$R(X^s, K_{X^s} + E) := \sum_{m \geqslant 0} H^0(X^s, \mathcal{O}_{X^s}(mK_{X^s} + mE)).$$

This actually creates a lot of problems since semi-log-canonical rings are not always finitely generated [30].

It is a quite subtle theorem that semi-log-canonical models actually satisfy the preliminary definition (1.7.5). This is proved in [36, 17.4] and [19].

To get a feeling for semi-log-canonical, let us review the classification of slc surface singularities.

Singularities of semi-log-canonical surfaces

It is convenient to describe the singularities of log canonical surfaces by the dual graph of their minimal resolution. That is, given a singularity $(s \in S)$ with minimal resolution $g : X \to S$ we draw a graph Γ whose vertices are the g-exceptional curves and two vertices are connected by an edge iff the corresponding curves intersect. We use the number $-(E_i \cdot E_i)$ to represent a vertex. In our examples, save in (3.2.4.a), all the exceptional curves are isomorphic to $\mathbb{P}^1$.

Let $\det(\Gamma)$ denote the determinant of the negative of the intersection matrix of the dual graph. This matrix is positive definite for exceptional curves. For instance, if $\Gamma = \{2 \;-\; 2 \;-\; 2\}$ then

$$\det(\Gamma) = \det \begin{pmatrix} 2 & -1 & 0 \\ -1 & 2 & -1 \\ 0 & -1 & 2 \end{pmatrix} = 4.$$

For more details concerning the lists below, see [36, Sec.3] or [35].

3.2 (List of log canonical surface singularities).

Each case includes all previous ones.

(3.2.1) Terminal = smooth.

(3.2.2) Canonical = Du Val (= rational double point).

(3.2.3) Log terminal = quotient of $\mathbb{C}^2$ by a finite subgroup of $GL(2,\mathbb{C})$ that acts freely outside the origin. The order of the group is $\det(\Gamma)$. A more detailed list is the following:

(a) (Cyclic quotient)

$$c_1 - \cdots - c_n$$

(b) (Dihedral quotient) Here $n \geqslant 2$ with dual graph

$$\begin{array}{ccc} & & 2 \\ & & / \\ c_1 - \cdots - c_n & & \\ & & \backslash \\ & & 2 \end{array}$$

(c) (Other quotients) The dual graph has 1 fork (with Γ_i as in (a))

$$\begin{array}{c} \Gamma_1 - c_0 - \Gamma_2 \\ | \\ \Gamma_3 \end{array}$$

with 3 cases for $(\det(\Gamma_1), \det(\Gamma_2), \det(\Gamma_3))$:

(Tetrahedral) (2,3,3)
(Octahedral) (2,3,4)
(Icosahedral) (2,3,5).

(3.2.4) Log canonical

(a) (Simple elliptic) $\Gamma = \{E\}$ has a single vertex which is a smooth elliptic curve with self intersection $\leqslant -1$.

(b) (Cusp) Γ is a circle of smooth rational curves, at least one of them with with $c_i \geqslant 3$. (The cases $n = 1, 2$ are somewhat special.)

$$\begin{array}{ccccc} & c_n & - \cdots - & c_{r+1} & \\ / & & & & \backslash \\ c_1 & & & & c_r \\ \backslash & & & & / \\ & c_2 & - \cdots - & c_{r-1} & \end{array}$$

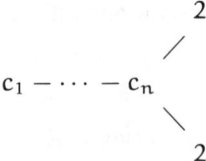

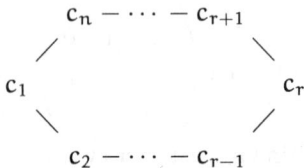

(c) ($\mathbb{Z}/2$-quotient of a cusp or simple elliptic) Γ has 2 forks.

$$\begin{array}{ccccc} 2 & & & & 2\\ & \diagdown & & \diagup & \\ & & c_1 - \cdots - c_n & & \\ & \diagup & & \diagdown & \\ 2 & & & & 2 \end{array}$$

(d) (Other quotients of a simple elliptic) The dual graph is as in (3.2.3.c) with 3 possibilities for $\big(\det(\Gamma_1), \det(\Gamma_2), \det(\Gamma_3)\big)$:

($\mathbb{Z}/3$-quotient) (3,3,3)
($\mathbb{Z}/4$-quotient) (2,4,4)
($\mathbb{Z}/6$-quotient) (2,3,6).

If X is a non-normal semi-log-canonical surface singularity, then we describe its normalization $\bar{X}$ together with the preimage of the double curve $\bar{B} \subset \bar{X}$.

The *extended dual graph* $(\Gamma, \bar{B})$ has an additional vertex (represented by $\bullet$) for each local branch of $\bar{B}$ connected to C_i if $(\bar{B} \cdot C_i) \neq 0$.

3.3 (List of semi-log-canonical surface singularities). There are 3 irreducible cases. (The number on some edges is the different, which we do not define here [36, Sec.16]. Their role is explained in (3.3.4).

(3.3.1) (Cyclic quotient, one branch of $\bar{B}$)

$$\bullet \overset{1-\frac{1}{\det\Gamma}}{\text{———}} c_1 - \cdots - c_n$$

(3.3.2) (Cyclic quotient, two branches of $\bar{B}$)

$$\bullet \overset{1}{-} c_1 - \cdots - c_n \overset{1}{-} \bullet$$

(3.3.3) (Dihedral quotient) Here $n \geqslant 2$ with dual graph

$$\begin{array}{ccc} & & 2\\ & & \diagup \\ \bullet \overset{1}{-} c_1 - \cdots - c_n & & \\ & & \diagdown \\ & & 2 \end{array}$$

(3.3.4) (Reducible cases) We can take several components as above and glue them together along two local branches of $\bar{B}$. The gluing is allowed only if we see the same numbers on the edges.

Thus we can glue 2 copies as in (3.3.1) as long as both have the same $\det(\Gamma)$ or we can take any number of germs as in (3.3.2), make a chain out of them and then either turn the chain into a circle or end it with copies of (3.3.3). To end a chain, we are also allowed to glue a local branch of $\bar{B}$ to itself by an involution. For instance, $\bullet \; - \; 1$ glued to itself gives the pinch point $(x^2 = y^2 z) \subset \mathbb{A}^3$.

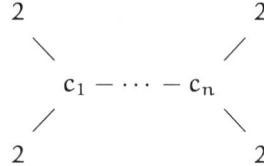

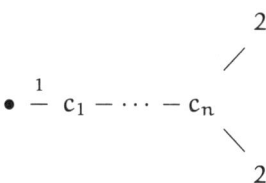

Note: The above dual graphs are correct in any characteristic, the descriptions as quotients are correct as long as the characteristic does not divide the order of the group mentioned.

Du Bois singularities

Semi-log-canonical singularities need not be rational, not even CM (=Cohen-Macaulay) and their most important property is that they are Du Bois. After some examples, we discuss Du Bois singularities and their useful properties.

Example 3.4. It is easy to see that a cone over a smooth variety $X \subset \mathbb{P}^N$ is log canonical iff $K_X \sim_{\mathbb{Q}} r \cdot H$ for some $r \leqslant 0$ where H is the hyperplane class. (See [28, Sec.4] for more details and examples.) For us the interesting case is when $K_X \sim_{\mathbb{Q}} 0$ (hence $r = 0$). For these, the cone is CM (resp. rational) iff $H^i(X, \mathcal{O}_X) = 0$ for $0 < i < \dim X$ (resp. for $0 < i \leqslant \dim X$). Thus we see the following.

(1) A cone over an Abelian variety A is CM iff $\dim A = 1$.
(2) A cone over a K3 surface is CM but not rational.
(3) A cone over an Enriques surface is CM and rational.
(4) A cone over a smooth Calabi-Yau complete intersection is CM but not rational.

The concept of Du Bois singularities was introduced by Steenbrink in [42] as a weakening of rationality. The precise definition is rather involved, but our main applications rely only on the following consequence.

Theorem 3.5. [33], [31, Chap.6] *Let X be a proper slc scheme over $\mathbb{C}$. Then the natural map*

$$H^i(X^{an}, \mathbb{C}) \to H^i(X^{an}, \mathcal{O}_{X^{an}}) \cong H^i(X, \mathcal{O}_X)$$

is surjective for all i. (In fact, with a functorial splitting.)

In studying moduli questions, it is very useful to know that certain numerical invariants are locally constant. Many of these follow from the Du Bois property, via the following base-change theorem [12, 11].

Proposition 3.6. *Let $f : X \to S$ be a flat, proper morphism. Assume that the fiber X_s is Du Bois for some $s \in S$. Then there is an open neighborhood $s \in S^0 \subset S$ such that, for all i,*

(1) *$R^i f_* \mathcal{O}_X$ is locally free and compatible with base change over S^0 and*
(2) *$s \mapsto H^i(X_s, \mathcal{O}_{X_s})$ is a locally constant function on S^0.*

Proof. By Cohomology and Base Change [16, III.12.11], the theorem is equivalent to proving that the restriction maps

$$\phi^i_s : R^i f_* \mathcal{O}_X \to H^i(X_s, \mathcal{O}_{X_s}) \tag{3.6.3}$$

are surjective for every i. By the Theorem on Formal Functions [16, III.11.1], it is enough to prove this when S is replaced by any 0-dimensional scheme S_n whose closed point is s.

Thus assume from now on that $f_n : X_n \to S_n$ is a flat proper morphism, that $s \in S_n$ is the only closed point and that X_s is Du Bois. Then $H^0(S_n, R^i f_* \mathcal{O}_X) = H^i(X_n, \mathcal{O}_{X_n})$, hence we can identify the ϕ_s^i with the maps

$$\psi^i : H^i(X_n, \mathcal{O}_{X_n}) \to H^i(X_s, \mathcal{O}_{X_s}) \tag{3.6.4}$$

By the Lefschetz principle we may assume that everything is defined over $\mathbb{C}$. By GAGA (cf. [16, App.B]), both sides of (3.6.4) are unchanged if we replace X_n by the corresponding analytic space X_n^{an}. Let $\mathbb{C}_{X_n}$ (resp. $\mathbb{C}_{X_s}$) denote the sheaf of locally constant functions on X_n (resp. X_s) and $j_n : \mathbb{C}_{X_n} \to \mathcal{O}_{X_n}$ (resp. $j_s : \mathbb{C}_{X_s} \to \mathcal{O}_{X_s}$) the natural inclusions. We have a commutative diagram

$$\begin{array}{ccc} H^i(X_n, \mathbb{C}_{X_n}) & \stackrel{\alpha^i}{\to} & H^i(X_s, \mathbb{C}_{X_s}) \\ j_n^i \downarrow & & \downarrow j_s^i \\ H^i(X_n, \mathcal{O}_{X_n}) & \stackrel{\psi^i}{\to} & H^i(X_s, \mathcal{O}_{X_s}) \end{array}$$

Note that α^i is an isomorphism since the inclusion $X_s \hookrightarrow X_n$ is a homeomorphism and j_s^i is surjective since X_s is Du Bois. Thus ψ^i is also surjective. □

One can greatly extend the scope of (3.6) as follows. Let L be an f-semi-ample line bundle on X (that is, L^m is f-generated by global sections for some $m > 0$). By taking the mth root of a general section of L^m, one obtains a cyclic ramified covering $\pi : Y \to X$ such that $\pi_* \mathcal{O}_Y = \sum_{r=0}^{m-1} L^{-r}$ and $f \circ \pi : Y \to S$ also has slc fiber over s. (See [34, Sec.2.4] for details.) Applying (3.6) to $f \circ \pi : Y \to S$ implies the following.

Corollary 3.7. *Let $f : X \to S$ be a proper and flat morphism with slc fibers over closed points; S connected. Let L be an f-semi-ample line bundle on X. Then, for all i,*

(1) $R^i f_*(L^{-1})$ *is locally free and compatible with base change and*
(2) $H^i(X_s, L_s^{-1})$ *is independent of* $s \in S$. □

Choose L to be f-ample above. By [34, 5.72], X_s is CM iff $H^i(X_s, L_s^{-m}) = 0$ for all $m \gg 1$ and $i < \dim X$. The latter properties are deformation invariant for slc fibers by (3.7). Thus we conclude:

Corollary 3.8. *Let $f : X \to S$ be a projective and flat morphism with slc fibers over closed points; S connected. Then, if one fiber of f is CM then all fibers of f are CM.* □

(Note that for arbitrary flat, projective morphisms $f : X \to S$, the set of points $s \in S$ such that the fiber X_s is CM is open, but usually not closed.)

The next example shows that non-CM varieties occur among the irreducible components of smoothable, CM and slc varieties.

Example 3.9. Here is an example of a stable family of projective varieties $\{Y_t : t \in T\}$ such that

(1) Y_t is smooth, projective for $t \neq 0$,
(2) K_{Y_t} is ample and Cartier for every t,

(3) Y_0 is slc and CM,
(4) the irreducible components of Y_0 are normal, but
(5) one of the irreducible components of Y_0 is not CM.

Let Z be a smooth Fano variety of dimension $n \geqslant 2$ such that $-K_Z$ is very ample, for instance $Z = \mathbb{P}^2$. Set $X := \mathbb{P}^1 \times Z$ and view it as embedded by $|-K_X|$ into $\mathbb{P}^N$ for suitable N. Let $C(X) \subset \mathbb{P}^{N+1}$ be the cone over X.

Let $M \in |-K_Z|$ be a smooth member and consider the following divisors in X:

$$D_0 := \{(0:1)\} \times Z, \ D_1 := \{(1:0)\} \times Z \quad \text{and} \quad D_2 := \mathbb{P}^1 \times M.$$

Note that $D_0 + D_1 + D_2 \sim -K_X$. Let $E_i \subset C(X)$ denote the cone over D_i. Then $E_0 + E_1 + E_2$ is a hyperplane section of $C(X)$ and $\big(C(X), E_0 + E_1 + E_2\big)$ is lc.

For some $m > 0$, let $H_m \subset C(X)$ be a general intersection with a degree m hypersurface. Then

$$\big(C(X), E_0 + E_1 + E_2 + H_m\big)$$

is snc outside the vertex and is lc at the vertex. Set $Y_0 := E_0 + E_1 + E_2 + H_m$. Since $Y_0 \sim \mathcal{O}_{C(X)}(m+1)$, we can view it as a slc limit of a family of smooth hypersurface sections $Y_t \subset C(X)$.

The cone over X is CM, hence its hyperplane section $E_0 + E_1 + E_2 + H_m$ is also CM. However, E_2 is not CM. To see this, note that E_2 is the cone over $\mathbb{P}^1 \times M$ and, by the Küneth formula,

$$H^i(\mathbb{P}^1 \times M, \mathcal{O}_{\mathbb{P}^1 \times M}) = H^i(M, \mathcal{O}_M) = \begin{cases} k & \text{if } i = 0, n-1, \\ 0 & \text{otherwise.} \end{cases}$$

Thus E_2 is not CM by (3.4).

Using (3.7) for $i = \dim X_s$ and duality, we conclude that $H^0\big(X_s, \omega_{X_s} \otimes L_s^m\big)$ is independent of $s \in S$ for every $m \geqslant 1$. As in [16, III.9.9], we get from this the following.

Proposition 3.10. *Let $f : X \to S$ be a projective, flat morphism with slc fibers over closed points. Then $\omega_{X/S}$ exists and is compatible with base change. That is, for any $g : T \to S$ the natural map*

$$g_X^* \omega_{X/S} \to \omega_{X_T/T} \quad \textit{is an isomorphism}$$

where $g_X : X_T := X \times_S T \to X$ is the first projection. □

(This is interesting already when S is a smooth curve, in which case $\omega_{X/S} = \omega_X \otimes f^* \omega_S^{-1}$ always exists but usually it is not compatible with base change, not even for flat morphisms with normal fibers.

The method of (3.10) seems like a very complicated way to prove that $\omega_{X/S}$ behaves as expected, but, as far as I can tell, this was not known before. A proof for non-projective algebraic maps is given in [27]. I do not know how to prove (3.10) for analytic morphisms $f : X \to S$.)

If the fibers X_s are CM, then $H^i(X_s, \omega_{X_s} \otimes L_s)$ is dual to $H^{n-i}(X_s, L_s^{-1})$, and the following is clear. In general, a more detailed inductive argument is needed [32].

Corollary 3.11. *Let $f : X \to S$ be a projective, flat morphism with slc fibers over closed points; S connected. Then, for all i,*

(1) $R^i f_* \omega_{X/S}$ *is locally free and compatible with base change and*
(2) $H^i(X_s, \omega_{X_s})$ *is independent of* $s \in S$. □

4. Moduli of semi-log-canonical models

Let us illustrate (1.11) with the an example of a flat, projective family of surfaces with log canonical singularities over the pair of lines $(xy = 0) \subset \mathbb{C}^2$ such that over one line we have smooth surfaces with ample canonical class and over the other line we have smooth elliptic surfaces.

Example 4.1 (Jump of Kodaira dimension). There are 2 families of non-degenerate degree 4 smooth surfaces in $\mathbb{P}^5$.

One family consists of Veronese surfaces $\mathbb{P}^2 \subset \mathbb{P}^5$ embedded by $\mathcal{O}(2)$. The general member of the other family is $\mathbb{P}^1 \times \mathbb{P}^1 \subset \mathbb{P}^5$ embedded by $\mathcal{O}(2,1)$, special members are embeddings of the ruled surface $\mathbb{F}_2$. The two families are distinct since

$$K^2_{\mathbb{P}^2} = 9 \quad \text{and} \quad K^2_{\mathbb{P}^1 \times \mathbb{P}^1} = 8.$$

For both of these surface, a smooth hyperplane section is a degree 4 rational normal curve in $\mathbb{P}^4$.

Let $T_0 \subset \mathbb{P}^5$ be the cone over the degree 4 rational normal curve in $\mathbb{P}^4$. T_0 has a log canonical (even log terminal) singularity and $K^2_{T_0} = 9$.

For us the interesting feature is that one can write T_0 as a limit of smooth surfaces in two distinct ways, corresponding to the two possibilities of writing the degree 4 rational normal curve in $\mathbb{P}^4$ as a hyperplane section of a surface.

From the first family, we get T_0 as the special fiber of a flat family whose general fiber is $\mathbb{P}^2$. This family is denoted by $\{T_t : t \in \mathbb{C}\}$. From the second family, we get T_0 as the special fiber of a flat family whose general fiber is $\mathbb{P}^1 \times \mathbb{P}^1$. This family is denoted by $\{T'_t : t \in \mathbb{C}\}$. (In general, one needs to worry about the possibility of getting embedded points at the vertex, but in both cases the special fiber is indeed T_0. Degenerating a variety $X \subset \mathbb{P}^n$ to the cone over a hyperplane section results in embedded points at the vertex iff $H^1(X, \mathcal{O}_X(m)) \neq 0$ for some $m \geqslant 0$.)

Note that K^2 is constant in the family $\{T_t : t \in \mathbb{C}\}$ but jumps at $t = 0$ in the family $\{T'_t : t \in \mathbb{C}\}$.

Next we take a suitable cyclic cover of the two families to get similar examples with ample canonical class.

Let $\pi_0 : S_0 \to T_0$ be a double cover, ramified along a smooth quartic hypersurface section. Note that $K_{T_0} \sim_{\mathbb{Q}} -\frac{3}{2}H$ where H is the hyperplane class. Thus, by the

Hurwitz formula,

$$K_{S_0} \sim_{\mathbb{Q}} \pi_0^*(K_{T_0} + 2H) \sim_{\mathbb{Q}} \tfrac{1}{2}\pi_0^* H.$$

So S_0 has ample canonical class and $K_{S_0}^2 = 2$. Since π_0 is étale over the vertex of T_0, S_0 has 2 singular points, locally (in the analytic or étale topology) isomorphic to the singularity on T_0. Thus S_0 is a log canonical surface.

Both of the smoothings lift to smoothings of S_0.

From the family $\{T_t : t \in \mathbb{C}\}$ we get a smoothing $\{S_t : t \in \mathbb{C}\}$ where $\pi_t : S_t \to \mathbb{P}^2$ is a double cover, ramified along a smooth octic. Thus S_t is smooth, $K_{S_t} \sim_{\mathbb{Q}} \pi_t^* \mathcal{O}_{\mathbb{P}^2}(1)$ is ample and $K_{S_t}^2 = 2$. From the family $\{T_t' : t \in \mathbb{C}\}$ we get a smoothing $\{S_t' : t \in \mathbb{C}\}$ where $\pi_t' : S_t' \to \mathbb{P}^1 \times \mathbb{P}^1$ is a double cover, ramified along a smooth curve of bidegree $(8,4)$. One of the families of lines on $\mathbb{P}^1 \times \mathbb{P}^1$ pulls back to an elliptic pencil on S_t' and $K_{S_t'}^2 = 0$.

In order to exclude such examples, we concentrate on the Hilbert function of a slc model.

Definition 4.2 (Hilbert function of slc models). Let X be an slc model. Note that ω_X is locally free outside a subscheme $Z \subset X$ such that Z has codimension $\geqslant 2$. Hence the reflexive hull $\omega_X^{[m]} := (\omega_X^{\otimes m})^{**}$ is isomorphic to $\omega_X^{\otimes m}$ over $X \setminus Z$. The *Hilbert function* of X is

$$H_X(m) := \chi(X, \omega_X^{[m]}).$$

If $\omega_X^{[N]}$ is locally free, then

$$\omega_X^{[m_0+mN]} \cong \omega_X^{[m_0]} \otimes \omega_X^{[mN]},$$

thus $H_X(m_0 + mN)$ is a polynomial in m and so $H_X(m)$ is a polynomial in m whose coefficients are periodic functions (with period N).

We view $\chi(X, \omega_X^{[m]})$ as the basic numerical invariant of X. It is then natural to insist that they stay constant in "good" families of slc models. Over a reduced base, this is enough to get the correct definition.

Definition 4.3 (Moduli of slc models over reduced bases). Let $H(m)$ be an integer valued function. On *reduced* schemes, the moduli functor of semi-log-canonical models with Hilbert function H is

$$\mathrm{SlcMod}_H(S) := \left\{ \begin{array}{c} \text{Flat, proper families } X \to S\text{, fibers are slc models with} \\ \text{ample canonical class and Hilbert function } H(m), \\ \text{modulo isomorphisms over } S. \end{array} \right\}$$

Over an arbitrary base, let $f : X \to S$ be a flat, proper family of slc models. Note that $\omega_{X/S}$ is locally free outside a subscheme $Z \subset X$ such that $Z \cap X_s$ has codimension $\geqslant 2$ in each fiber. Each $\omega_{X/S}^{\otimes m}$ is also locally free on $X \setminus Z$, hence it has a reflexive hull $\omega_{X/S}^{[m]}$.

If $s \in S$ is a general point, then $\omega_{X/S}^{[m]}|_{X_s} \cong \omega_{X_s}^{[m]}$ but for an arbitrary $s \in S$ we only have a restriction map

$$r_s : \omega_{X/S}^{[m]}|_{X_s} \to \omega_{X_s}^{[m]}$$

which is, in general, neither injective nor surjective. The best way to ensure that every fiber of X_s has the same Hilbert function is to require these restriction maps to be isomorphisms for every $s \in S$. (It turns out that this is the only way, that is, the kernel and the cokernel of r_s can not cancel each other for every s, unless they are both zero.) This leads to our final definition.

Definition 4.4 (Moduli of slc models). Let $H(m)$ be an integer valued function. The moduli functor of semi-log-canonical models with Hilbert function H is

$$\mathrm{SlcMod}_H(S) := \left\{ \begin{array}{c} \text{Flat, proper families } X \to S \text{, fibers are slc models with} \\ \text{ample canonical class and Hilbert function } H(m), \\ \omega_{X/S}^{[m]} \text{ is flat over } S \text{ and commutes with base change} \\ \text{for every } m \text{, modulo isomorphisms over } S. \end{array} \right\}$$

Aside 4.5. We can now explain Principle 1.10. The reason is that for flat families of canonical models, $\omega_{X/S}^{[m]}$ is automatically flat over S and commutes with base change. This follows from two special properties of canonical singularities. For simplicity, consider a flat family $X \to \operatorname{Spec} k[\epsilon]$ whose special fiber X_0 is affine.

First we use that, as a result of the classification of canonical surface singularities (3.2.2), there is an open subset $j : U \hookrightarrow X$ whose complement Z has codimension $\geqslant 3$ such that ω_{U_0} is locally free. Thus we have an exact sequence

$$0 \to \epsilon \cdot \omega_{U_0}^m \to \omega_U^m \to \omega_{U_0}^m \to 0.$$

By pushing it forward, we get

$$0 \to \epsilon \cdot j_*\omega_{U_0}^m \to j_*\omega_U^m \to j_*\omega_{U_0}^m \to \epsilon \cdot R^1 j_*\omega_{U_0}^m$$

As noted after (2.3), $\omega_X^{[m]}$ is a CM sheaf, hence has depth $\geqslant 3$ at every point of Z. Therefore,

$$R^1 j_*\omega_{U_0}^m = H^1(U_0, \omega_{U_0}^m) = H^2_{Z_0}(X_0, \omega_{X_0}^{[m]}) = 0.$$

This implies that $\omega_X^{[m]}$ equals $j_*\omega_U^m$ and it is flat over $k[\epsilon]$.

Now that we have the correct definition, we need to prove that the corresponding deformation theory is reasonable. The key result is the following.

Theorem 4.6. *Let $f : X \to S$ be flat, projective morphism whose fibers are slc models. Let H be an integer valued function.*

Then there is a locally closed embedding $S_H \hookrightarrow S$ such that a morphism $g : T \to S$ factors through S_H iff $X \times_S T \to T$ is in $\mathrm{SlcMod}_H(T)$.

For surfaces, a proof of this is outlined in [14], a general solution is in [1]. Another approach relies on hulls, introduced in [25].

Definition 4.7. Let X be a scheme over a field k and F a coherent sheaf on X. Set $n := \dim \operatorname{Supp} F$. The *hull* of F is the unique $q : F \to F^{[**]}$ such that

(1) q is an isomorphism at all n-dimensional points of $\operatorname{Supp} F$,
(2) q is surjective at all $(n-1)$-dimensional points of $\operatorname{Supp} F$,
(3) $F^{[**]}$ is S_2.

One can construct $F^{[**]}$ as follows. First replace F by $F/\operatorname{tors}_{n-1}(F)$ where $\operatorname{tors}_{n-1}(F)$ is the largest subsheaf whose support has dimension $\leqslant n-1$. Then there is a closed subscheme $Z \subset \operatorname{Supp} F$ of codimension $\geqslant 2$ such that $F/\operatorname{tors}_{n-1}(F)$ is S_2 on $X \setminus Z$. Let $j : X \setminus Z \to X$ be the open embedding and take

$$F^{[**]} = j_*\Big(\big(F/\operatorname{tors}_{n-1}(F)\big)|_{X\setminus Z}\Big).$$

Thus the hull of a nonzero sheaf is also nonzero, in contrast with the reflexive hull which kills all torsion sheaves. If X itself is normal, F is coherent and $\operatorname{Supp} F = X$, then $F^{[**]}$ is the usual reflexive hull F^{**} of F.

Definition 4.8. Let $f : X \to S$ be a morphism and F a coherent sheaf. A *hull* of F over S is a coherent sheaf G together with a map $q : F \to G$ such that,

(1) G is flat over S and
(2) for every $s \in S$, the induced map $q_s : F_s \to G_s$ is a hull (4.7).

It is easy to see that a hull is unique if it exists.

It is clear from the definition that hulls are preserved by base change. That is, if $g : T \to S$ is a morphism, $X_T := X \times_S T$ and $g_X : X_T \to X$ the first projection then $g_X^* q : g_X^* F \to g_X^* G$ is also a hull.

Definition 4.9. Let $f : X \to S$ be a projective morphism and F a coherent sheaf on X. For a scheme $g : T \to S$ set $\mathit{Hull}(F)(T) = 1$ if $g_X^* F$ has a hull over S and $\mathit{Hull}(F)(T) = \emptyset$ if $g_X^* F$ does not have a hull over S, where $g_X : T \times_S X \to X$ is the projection.

The main existence theorem is the following.

Theorem 4.10 (Flattening decomposition for hulls). [25] *Let $f : X \to S$ be a projective morphism and F a coherent sheaf on X. Then*

(1) *$\mathit{Hull}(F)$ has a fine moduli space* $\operatorname{Hull}(F)$.
(2) *The structure map $\eta : \operatorname{Hull}(F) \to S$ is a locally closed decomposition, that is, η is one-to-one and onto on geometric points and a locally closed embedding on every connected component.*

Applying (4.10) to the relative dualizing sheaf gives the following result.

Corollary 4.11. *Let $f : X \to S$ be projective and equidimensional. Assume that there is a closed subscheme $Z \subset X$ such that* $\operatorname{codim}(X_s, Z \cap X_s) \geqslant 2$ *for every $s \in S$, $(X \setminus Z) \to S$ is flat and $\omega_{X/S}$ is locally free on $X \setminus Z$. Then, for any m there is a locally closed decomposition $S_m \to S$ such that for any $g : T \to S$ the following are equivalent*

(1) $\omega^{[m]}_{X\times_S T/T}$ *is flat over* T *and commutes with base change.*
(2) g *factors through* $S_m \to S$.

Proof. We claim that $S_m = \operatorname{Hull}(\omega^{\otimes m}_{X/S})$. Given $g : T \to S$, let

$$j_T : X \times_S T \setminus Z \times_S T \to X \times_S T$$

be the inclusion. Then

$$\omega^{[m]}_{X\times_S T/T} = (j_T)_* g^*_X \omega^{\otimes m}_{X\setminus Z/S}.$$

If $T \mapsto \omega^{[m]}_{X\times_S T/T}$ commutes with restrictions to the fibers of $X \times_S T \to T$, then $\omega^{[m]}_{X\times_S T/T}$ has S_2 fibers, hence $\omega^{[m]}_{X\times_S T/T}$ is the hull of $\omega^{\otimes m}_{X\times_S T/T}$.

Conversely, if $\omega^{\otimes m}_{X\times_S T/T}$ has a hull then it is $\omega^{[m]}_{X\times_S T/T}$ and it commutes with further base changes by (4.8). □

In order to prove (4.6), choose N such that $\omega^{[N]}_{X_s}$ is locally free for every $s \in S$. For $1 \leqslant i \leqslant N$, let $S_i \to S$ be as in (4.11). Take T to be the fiber product $S_1 \times_S \cdots \times_S S_N \to S$. Then $\omega^{[m]}_{X\times_S T/T}$ is flat over T and commutes with base change for every m. Thus S_H is the disjoint union of those connected components of T where the Hilbert function is H. □

5. Coarse moduli spaces

Having defined the correct moduli functor for slc models, we can now get down to studying its properties and the corresponding moduli spaces.

5.1 (Valuative criterion of separatedness). The functor SlcMod satisfies the valuative criterion of separatedness; this is essentially (1.6).

5.2 (Valuative criterion of properness). The short answer is that SlcMod satisfies the valuative criterion of properness, but some warnings are in order.

We proceed very much as for curves. We start with a family of canonical models over an open curve $X^0 \to B^0$. By the semi-stable reduction theorem of [21], after a base change $C^0 \to B^0$ and extending the family over a proper curve $C \supset C^0$, there is a resolution $g : Y \to C$ all of whose fibers are reduced simple normal crossing divisors. Finally we replace Y by its relative canonical model

$$Y^c := \operatorname{Proj}_C \sum_{m \geqslant 0} g_*(\omega^m_{Y/C}).$$

It is not hard to see that $Y^c \to C$ is in SlcMod(C) extending $X \times_{B^0} C^0 \to C^0$.

This establishes the valuative criterion of properness if canonical models are dense in the moduli of slc models. We probably mostly care about the irreducible components where canonical models are dense, so we could take this as the final answer.

However, not all irreducible components are such, and it would be better to understand all of them.

So let us start with a family of slc models $X^0 \to B^0$. We can proceed as above, but instead of the relative canonical model of Y we need to take the relative semi-log-canonical model. As we noted, semi-log-canonical rings are not always finitely generated [30].

A possible solution is to normalize the family, construct the models of the normalization over C and then try to reconstruct the desired extension of the original family. This is actually quite subtle, see [32].

5.3 (Existence of coarse moduli spaces). Fix a function H and an integer m. Let $\mathcal{S}lc\mathcal{M}od_{H,m}$ be the moduli functor of slc models with Hilbert function H for which $\omega^{[m]}$ is locally free, very ample and has no higher cohomologies. All of these thus embed into $\mathbb{P}^N$ for $N = H(m) - 1$. We use a variant of (4.6) to show that there is a locally closed subscheme $S_{H,m}$ of the Hilbert scheme $\mathrm{Hilb}(\mathbb{P}^N)$ that parametrizes families of m-canonically embedded slc models with Hilbert function H.

The general quotient theorems of [24], [20] apply and we obtain the coarse moduli space $\mathrm{SlcMod}_{H,m}$ of $\mathcal{S}lc\mathcal{M}od_{H,m}$ as the geometric quotient $S_{H,m}/\mathrm{Aut}(\mathbb{P}^N)$.

Finally we let m run through the sequence $2!, 3!, 4!, \ldots$ to get an increasing sequence of coarse moduli spaces whose union is the coarse moduli space SlcMod_H. For now we know only that it is a separated algebraic space which is locally of finite type.

5.4 (Properness). We saw that SlcMod_H satisfies the valuative criterion of properness, hence it is proper iff it is of finite type.

The irreducible components where the canonical models are dense were studied by [18]. He proves that one can control the procedure outlined in (5.2) uniformly. Thus every such irreducible component is of finite type, hence proper.

With some modifications, this implies that every irreducible component of SlcMod_H is proper.

Thus the only remaining question is: can there by infinitely many irreducible components?

To illustrate some of the difficulties, let us consider a much simpler question: can we bound the number of irreducible components of a slc surface S with Hilbert function H?

For curves the answer is easy. If $C = \cup_i C_i$ then

$$2g(C) - 2 = \deg \omega_C = \textstyle\sum_i \deg(\omega_C|_{C_i}).$$

Each C_i on the right hand side contributes at least 1, hence there are at most $2g-2$ irreducible components. In the surface case, we have something very similar. If $S = \cup_i S_i$ then we can compute the self intersection of the canonical class as

$$(K_S \cdot K_S) = \textstyle\sum_i (K_S|_{S_i} \cdot K_S|_{S_i}).$$

The unexpected problem is that K_S is only a $\mathbb{Q}$-Cartier divisor, hence each summand on the right hand side is a positive rational number, not an integer.

We are, however, saved if the contributions on the right are bounded away from 0. This, and much more that is needed for boundedness was proved in [2] and improved in [6]. The lower bound $\frac{1}{1764}$ was established in [23]. (I do not know the optimal bound, but $\big(\mathbb{P}(3,4,5), (x^3y + y^2z + z^2x = 0)\big)$ has $(K_S + D)^2 = \frac{1}{60}$.)

In higher dimensions, recent work of [15] establishes a lower bound for $\big(K_X|_{X_i}\big)^n$; the methods are likely to give boundedness as well.

5.5 (Projectivity). The method of [22] shows that every proper subscheme of SlcMod is projective. For m sufficiently divisible, the 1-dimensional vector spaces $\det H^0(X, \omega_X^{[m]})$ naturally glue together to an ample line bundle.

The proof uses the Nakai-Moishezon ampleness criterion, thus it works only for proper schemes.

It seems very hard to give quasi-projectivity criteria. For instance, [30] gives an example of a normal crossing surface S with a line bundle L and normalization $\pi : \bar{S} \to S$ such that π^*L is ample yet L is not ample, in fact no power of L is generated by global sections.

6. Moduli of slc pairs

In dimension 1, it is useful to study not just the moduli of curves but also the moduli of pointed curves. Similarly, in higher dimensions, one should consider the moduli of pairs (X, Δ) where $\Delta = \sum_i a_i D_i$ is a linear combination of divisors with coefficients $0 \leqslant a_i \leqslant 1$. These were first investigated in [3].

The first task is to define slc singularities of pairs. This is actually quite natural, see [36].

By contrast, finding the correct analog of (4.4) turns out to be a quite thorny problem. Instead of going into details, let me just present a key example, due to Hassett, which shows that in general we can not view a deformation of a pair as first a deformation of X and then a deformation of the D_i. One must view (X, Δ) as an indivisible unit.

Example 6.1. Let $S \subset \mathbb{P}^5$ be the cone over the degree 4 rational normal curve. Fix $r \geqslant 1$ and let D_S be the sum of $2r$ lines. Then $(S, \frac{1}{r}D_S)$ is lc and $\big(K_S + \frac{1}{r}D_S\big)^2 = 4$.

There are two different deformations of the pair (S, D_S).

(6.1.1) First, set $P := \mathbb{P}^2$ and let D_P be the sum of r general lines. Then $(P, \frac{1}{r}D_P)$ is lc (even canonical if $r \geqslant 2$) and $\big(K_P + \frac{1}{r}D_P\big)^2 = 4$. The usual smoothing of $S \subset \mathbb{P}^5$ to the Veronese surface gives a family $f : (X, D_X) \to \mathbb{P}^1$ with general fiber (P, D_P) and special fiber (S, D_S). We can concretely realize this as deforming $(P, D_P) \subset \mathbb{P}^5$ to the cone over a general hyperplane section. Note that for any general D_S there is a choice of lines D_P such that the above limit is exactly D_S.

The total space (X, D_X) is the cone over (P, D_P) (blown up along curve) and X is $\mathbb{Q}$-factorial. The structure sheaf of an effective divisor on X is CM.

In particular, D_S is a flat limit of D_P. Since the D_P is a plane curve of degree r, we conclude that

$$\chi(\mathcal{O}_{D_S}) = \chi(\mathcal{O}_{D_P}) = -\frac{r(r-3)}{2}.$$

(6.1.2) Second, set $Q := \mathbb{P}^1 \times \mathbb{P}^1$ and let A, B denote the classes of the 2 rulings. Let D_Q be the sum of r lines from the A-family. Then $(Q, \frac{1}{r}D_Q)$ is canonical and $(K_Q + \frac{1}{r}D_Q)^2 = 4$. The usual smoothing of $S \subset \mathbb{P}^5$ to $\mathbb{P}^1 \times \mathbb{P}^1$ embedded by $H := A + 2B$ gives a family $g : (Y, D_Y) \to \mathbb{P}^1$ with general fiber (Q, D_Q) and special fiber (S, D_S). We can concretely realize this as deforming $(Q, D_Q) \subset \mathbb{P}^5$ to the cone over a general hyperplane section.

The total space (Y, D_Y) is the cone over (Q, D_Q) (blown up along curve) and Y is not $\mathbb{Q}$-factorial. However, $K_Q + \frac{1}{r}D_Q \sim_{\mathbb{Q}} -H$, thus $K_Y + \frac{1}{r}D_Y$ is $\mathbb{Q}$-Cartier and $(Y, \frac{1}{r}D_Y)$ is lc.

In this case, however, D_Q is not a flat limit of D_P for $r > 1$; one also gets embedded points at the vertex. This follows, for instance, from comparing their Euler characteristic:

$$\chi(\mathcal{O}_{D_S}) = -\frac{r(r-3)}{2} \quad \text{and} \quad \chi(\mathcal{O}_{D_Q}) = r.$$

(6.1.3) Because of their role in the canonical algebra, we are also interested in the sheaves $\mathcal{O}(mK + \lfloor \frac{m}{r}D \rfloor)$.

Let H_P be the hyperplane class of $P \subset \mathbb{P}^5$ (that is, 2 times a line $L \subset P$) and write $m = br + a$ where $0 \leqslant a < r$. One computes that

$$\begin{aligned}
\chi(P, \mathcal{O}_P(mK_P + \lfloor \tfrac{m}{r}D_P \rfloor + nH_P)) &= \tbinom{2n-2m+2}{2} - a(2n-2m+1) + \tbinom{a}{2},\\
\chi(S, \mathcal{O}_S(mK_S + \lfloor \tfrac{m}{r}D_S \rfloor + nH_S)) &= \tbinom{2n-2m+2}{2} - a(2n-2m+1) + \tbinom{a}{2},\\
\chi(Q, \mathcal{O}_Q(mK_Q + \lfloor \tfrac{m}{r}D_Q \rfloor + nH_Q)) &= \tbinom{2n-2m+2}{2} - a(2n-2m+1).
\end{aligned}$$

From this we conclude that the restriction of $\mathcal{O}_Y(mK_Y + \lfloor mD_Y \rfloor)$ to the central fiber S agrees with $\mathcal{O}_S(mK_S + \lfloor mD_S \rfloor)$ only if $a \in \{0,1\}$, that is when $m \equiv 0,1 \mod r$. The if part was clear from the beginning. Indeed, if $a = 0$ then $\mathcal{O}_Y(mK_Y + \lfloor mD_Y \rfloor)$ is locally free and if $a = 1$ then $\mathcal{O}_Y(mK_Y + \lfloor mD_Y \rfloor)$ is $\mathcal{O}_Y(K_Y)$ tensored with a locally free sheaf. Both of these commute with restrictions.

In the other cases we only get an injection

$$\mathcal{O}_Y(mK_Y + \lfloor mD_Y \rfloor)|_S \hookrightarrow \mathcal{O}_S(mK_S + \lfloor mD_S \rfloor)$$

whose quotient is a torsion sheaf of length $\binom{a}{2}$ supported at the vertex.

There are several ways to overcome these problems; several of them will be discussed in [32].

(1) Embedded points do not appear if all the coefficients a_i are $> \frac{1}{2}$ [29].
(2) By wiggling the a_i suitably, one again avoids embedded points.

(3) Fix m such that $\mathcal{O}_X(mK_X + m\Delta)$ is locally free. One can identify a pair (X, Δ) with the corresponding map $\omega_X^{\otimes m} \to \mathcal{O}_X(mK_X + m\Delta)$. It turns out to be easier to deal with the moduli of triples $\left(X, \omega_X^{\otimes m} \to L\right)$ for some line bundle L.

(4) The branch varieties of [5] give another approach.

Acknowledgments. Parts of this paper were written in connection with a lecture series on moduli at IHP, Paris. I thank my audience and especially C. Voisin for the invitation, useful comments and corrections. Partial financial support was provided by the NSF under grant number DMS-0758275.

References

[1] Dan Abramovich and Brendan Hassett. Stable varieties with a twist, in: Classification of algebraic varieties, 1–38. *EMS Ser. Congr. Rep.*, Eur. Math. Soc., Zürich, 2011. MR 2779465 ← 149

[2] Valery Alexeev. Boundedness and K^2 for log surfaces. *Internat. J. Math.*, **5**(6), 779–810, 1994. MR 1298994 (95k:14048) ← 153

[3] Valery Alexeev. Moduli spaces $M_{g,n}(W)$ for surfaces, in: Higher-dimensional complex varieties (Trento, 1994),1–22, de Gruyter, Berlin, 1996. MR 1463171 (99b:14010) ← 153

[4] Valery Alexeev. Complete moduli in the presence of semiabelian group action. *Ann. of Math. (2)*, **155**(3), 611–708, 2002. MR 1923963 (2003g:14059) ← 131

[5] Valery Alexeev and Allen Knutson. *Jour. Reine Angew. Math.*, **639**, 29-71, 2010. ← 155

[6] Valery Alexeev and Shigefumi Mori. Bounding singular surfaces of general type, in: Algebra, arithmetic and geometry with applications (West Lafayette, IN, 2000), 143–174, Springer, Berlin, 2004. MR 2037085 (2005f:14077) ← 153

[7] Michael Artin. Algebraic construction of Brieskorn's resolutions. *J. Algebra*, **29**, 330–348, 1974. MR 0354665 (50 #7143) ← 135

[8] Caucher Birkar, Paolo Cascini, Christopher D. Hacon, and James McKernan. Existence of minimal models for varieties of log general type. *J. Amer. Math. Soc.*, **23**(2), 405–468, 2010. MR 2601039 ← 139

[9] Egbert Brieskorn. Die Auflösung der rationalen Singularitäten holomorpher Abbildungen, *Math. Ann.*, **178**, 255–270, 1968. MR 0233819 (38 #2140) ← 135

[10] F. Catanese. Moduli of algebraic surfaces, in: Theory of moduli (Montecatini Terme, 1985). *Lecture Notes in Math.*, **1337**, 1–83, Springer, Berlin, 1988. MR 963062 (89i:14031) ← 134

[11] Philippe Du Bois. Complexe de de Rham filtré d'une variété singulière. *Bull. Soc. Math. France*, **109**(1), 41–81, 1981. MR 613848 (82j:14006) ← 144

[12] Philippe Du Bois and Pierre Jarraud. Une propriété de commutation au changement de base des images directes supérieures du faisceau structural. *C. R. Acad. Sci. Paris Sér. A*, **279**, 745–747, 1974. MR 0376678 (51 #12853) ← 144

[13] R. Elkik. Rationalité des singularités canoniques. *Inv. Math.*, **64**, 1–6, 1981. ← 140

[14] Paul Hacking. Compact moduli of plane curves. *Duke Math. J.*, **124**(2), 213–257, 2004. MR 2078368 (2005f:14056) ← 149

[15] C. Hacon, J. McKernan, and C. Xu. On the birational automorphisms of varieties of general type, to appear in Annals of Math. arXiv:1011.1464, 2010. ← 153

[16] Robin Hartshorne. Algebraic geometry. *Graduate Texts in Mathematics*, **52**, Springer-Verlag, New York, 1977. MR 0463157 (57 #3116) ← 144, 145, 146

[17] Shigeru Iitaka. On D-dimensions of algebraic varieties. *J. Math. Soc. Japan*, **23**, 356–373, 1971. MR 0285531 (44 #2749) ← 136

[18] Kalle Karu. Minimal models and boundedness of stable varieties. *J. Algebraic Geom.*, **9**(1), 93–109, 2000. MR 1713521 (2001g:14059) ← 152

[19] Masayuki Kawakita. Inversion of adjunction on log canonicity. *Invent. Math.*, **167**(1), 129–133, 2007. MR 2264806 (2008a:14025) ← 141

[20] Seán Keel and Shigefumi Mori. Quotients by groupoids. *Ann. of Math. (2)*, **145**(1), 193–213, 1997. MR 1432041 (97m:14014) ← 152

[21] G. Kempf, F. F. Knudsen, D. Mumford, and B. Saint-Donat. Toroidal embeddings. I. *Lecture Notes in Mathematics*, **339**, Springer-Verlag, Berlin, 1973. MR 0335518 (49 #299) ← 140, 151

[22] János Kollár. Projectivity of complete moduli. *J. Differential Geom.*, **32**(1), 235–268, 1990. MR 1064874 (92e:14008) ← 153

[23] János Kollár. Log surfaces of general type; some conjectures, in: Classification of algebraic varieties (L'Aquila, 1992), 261–275. *Contemp. Math.*, **162**, Amer. Math. Soc., Providence, RI, 1994. MR 1272703 (95c:14042) ← 153

[24] János Kollár. Quotient spaces modulo algebraic groups. *Ann. of Math. (2)*, **145**(1), 33–79, 1997. MR 1432036 (97m:14013) ← 152

[25] János Kollár. Hulls and husks, 2008. MR arXiv:0805.0576 ← 149, 150

[26] János Kollár. Semi log resolutions, arXiv:0812.3592, 2008. ← 141

[27] János Kollár. A local version of the Kawamata-Viehweg vanishing theorem, *Pure and Applied Mathematics Quarterly*, **7** 1477-1494 arXiv:1005.4843, 2012. ← 140, 146

[28] János Kollár. Exercises in the birational geometry of algebraic varieties, in: Analytic and algebraic geometry (J. McNeal and M. Mustaţă, eds.), 495–524. *IAS/Park City Math. Ser.*, **17**, AMS, 2010. ← 144

[29] János Kollár. Seminormal log centers and deformations of pairs, to appear in Proceedings of the Edinburgh Mathematical Society arXiv: 1103.0528, 2011. ← 154

[30] János Kollár. Two examples of surfaces with normal crossing singularities. *Sci. China Math.*, **54**(8), 1707–1712, 2011. ← 141, 152, 153

[31] János Kollár. *Singularities of the minimal model program*, Cambridge University Press, Cambridge, 2012, With the collaboration of S. Kovács (to appear). ← 132, 144

[32] János Kollár. *Moduli of varieties of general type*, (book in preparation), 2013. ← 132, 147, 152, 154

[33] János Kollár and Sándor J. Kovács. Log canonical singularities are Du Bois. *J. Amer. Math. Soc.*, **23**(3), 791–813, 2010. MR 2629988 ← 144

[34] János Kollár and Shigefumi Mori. Birational geometry of algebraic varieties. *Cambridge Tracts in Mathematics*, **134**, Cambridge University Press, Cambridge, 1998. With the collaboration of C. H. Clemens and A. Corti, Translated from the 1998 Japanese original. MR 1658959 (2000b:14018) ← 138, 140, 145

[35] János Kollár and N. I. Shepherd-Barron. Threefolds and deformations of surface singularities. *Invent. Math.*, **91**(2), 299–338, 1988. MR 922803 (88m:14022) ← 131, 141

[36] J. Kollár et al. Flips and abundance for algebraic threefolds. *Soc. Math. France, Astérisque*, **211**, 1992. ← 141, 143, 153

[37] Robert Lazarsfeld. Positivity in algebraic geometry. I-II. *Ergebnisse der Mathematik und ihrer Grenzgebiete. 3. Folge.*, **48–49**, Springer-Verlag, Berlin, 2004. MR 2095471 (2005k:14001a) ← 136

[38] I. I. Pjateckiĭ-Šapiro and I. R. Šafarevič. Torelli's theorem for algebraic surfaces of type K3. *Izv. Akad. Nauk SSSR Ser. Mat.*, **35**,530–572, 1971. MR 0284440 (44 #1666) ← 134

[39] Miles Reid. Canonical 3-folds. Journées de Géometrie Algébrique d'Angers, Juillet 1979/Algebraic Geometry, Angers, 1979, Sijthoff & Noordhoff, Alphen aan den Rijn, 1980, 273–310. MR 605348 (82i:14025) ← 135, 139, 140

[40] Yum-Tong Siu. Invariance of plurigenera. *Invent. Math.*, **134**(3), 661–673, 1998. MR 1660941 (99i:32035) ← 139

[41] Yum-Tong Siu. Finite generation of canonical ring by analytic method. *Sci. China Ser. A*, **51**(4), 481–502, 2008. MR 2395400 ← 139

[42] J. H. M. Steenbrink. Mixed Hodge structures associated with isolated singularities, in: Singularities, Part 2 (Arcata, Calif., 1981), 513–536. *Proc. Sympos. Pure Math.*, **40**, Amer. Math. Soc., Providence, RI, 1983. MR 713277 (85d:32044) ← 144

Princeton University, Princeton NJ 08544-1000
E-mail address: kollar@math.princeton.edu

Singularities of stable varieties

Sándor J Kovács

Contents

1. Introduction

The theory of moduli of curves has been extremely successful and part of this success is due to the compactification of the moduli space of smooth projective curves by the moduli space of stable curves. A similar construction is desirable in higher dimensions but unfortunately the methods used for curves do not produce the same results in higher dimensions. In fact, even the definition of what *stable* should mean is not entirely clear a priori. In order to construct modular compactifications of moduli spaces of higher dimensional canonically polarized varieties one must understand

Supported in part by NSF Grant DMS-0856185, and the Craig McKibben and Sarah Merner Endowed Professorship in Mathematics at the University of Washington.

the possible degenerations that would produce this desired compactification that itself is a moduli space of an enlarged class of canonically polarized varieties.

The main purpose of the present article is to discuss the relevant issues that arise in higher dimensions and how these lead us to the definition of stable varieties and stable families. Particular emphasis is placed on understanding the singularities of stable varieties including some recent results.

The structure of the article is the following: In §2 and §3 I review the relevant properties of stable curves and their families, including the admissible singularities of the total spaces of stable families. §4 shows how generalizing the properties of these total spaces leads to the right generalization of stable singularities in higher dimensions. In §5 I review the construction and main properties of canonical sheaves and divisors. §6 is devoted to the singularities of the minimal model program, mainly from a moduli theoretic point of view. In §7 I define some important basic notions and recall some fundamental theorems such as Grothendieck duality and Kodaira vanishing. §8 and §9 are concerned with the definition and basic properties of rational and Du Bois singularities respectively. In §10 I review the most important criteria for rational and Du Bois singularities organized around the principle that a natural morphism in the derived category should admit a left inverse essentially only if it is a quasi-isomorphism combined by a push-forward map admitting a section by a trace map. In §11 I review the applications of the results in §10 to stable families and in §12 what is known about the deformation theory of stable singularities.

Without trying to be comprehensive, here is a list of relevant references on background. In order to study higher dimensional varieties one should be familiar with the main techniques of birational geometry. The standard reference for this is [37] and for some more recent results the reader may consult [15]. For moduli spaces of higher dimensional smooth varieties a good reference is [62]. For moduli spaces of stable varieties one may refer to [27, 38, 29]. A light introduction to the ideas involved is contained in [46].

Definitions and notation 1.1. Let k be an algebraically closed field of characteristic 0. Unless otherwise stated, all objects will be assumed to be defined over k. A *scheme* will refer to a scheme of finite type over k and unless stated otherwise, a *point* refers to a closed point.

For a morphism $f : Y \to S$ and another morphism $T \to S$, the symbol Y_T will denote $Y \times_S T$. In particular, for $t \in S$ I will write $Y_t = f^{-1}(t)$. In addition, if $T = \operatorname{Spec} F$, then Y_T will also be denoted by Y_F.

Let X be a scheme and $\mathscr{F}$ an $\mathscr{O}_X$-module. The mth *reflexive power* of $\mathscr{F}$ is the double dual (or reflexive hull) of the mth tensor power of $\mathscr{F}$:

$$\mathscr{F}^{[m]} := (\mathscr{F}^{\otimes m})^{**}.$$

A *line bundle* on X is an invertible $\mathscr{O}_X$-module. A $\mathbb{Q}$-*line bundle* $\mathscr{L}$ on X is a reflexive $\mathscr{O}_X$-module of rank 1 one of whose reflexive power is a line bundle, i.e., there exists an $m \in \mathbb{N}_+$ such that $\mathscr{L}^{[m]}$ is a line bundle. The smallest such m is called the *index* of $\mathscr{L}$.

For the advanced reader: whenever I mention Weil divisors, assume that X is S_2 and think of a *Weil divisorial sheaf*, that is, a rank 1 reflexive $\mathscr{O}_X$-module which is locally free in codimension 1. For flatness issues consult [33, Theorem 2].

For the novice: whenever I mention Weil divisors, assume that X is normal and adopt the definition [18, p.130]. For the adventurous novice: This is mainly interesting for canonical divisors. Read §5.

For a Weil divisor D on X, its associated *Weil divisorial sheaf* is the $\mathscr{O}_X$-module $\mathscr{O}_X(D)$ defined on the open set $U \subseteq X$ by the formula

$$\Gamma(U, \mathscr{O}_X(D)) = \left\{ \frac{a}{b} \,\middle|\, \begin{array}{l} a, b \in \Gamma(U, \mathscr{O}_X),\ b \text{ is not a zero divisor} \\ \qquad \text{anywhere on } U, \text{ and } \ D + \operatorname{div}(a) - \operatorname{div}(b) \geqslant 0 \end{array} \right\}$$

and made into a sheaf by the natural restriction maps.

A Weil divisor D on X is a *Cartier divisor*, if its associated Weil divisorial sheaf, $\mathscr{O}_X(D)$ is a line bundle. If the associated Weil divisorial sheaf, $\mathscr{O}_X(D)$ is a $\mathbb{Q}$-line bundle, then D is a *$\mathbb{Q}$-Cartier divisor*. The latter is equivalent to the property that there exists an $m \in \mathbb{N}_+$ such that mD is a Cartier divisor.

The symbol $\sim$ stands for *linear* and $\equiv$ for *numerical equivalence* of divisors.

Let $\mathscr{L}$ be a line bundle on a scheme X. It is said to be *generated by global sections* if for every point $x \in X$ there exists a global section $\sigma_x \in H^0(X, \mathscr{L})$ such that the germ σ_x generates the stalk $\mathscr{L}_x$ as an $\mathscr{O}_X$-module. If $\mathscr{L}$ is generated by global sections, then the global sections define a morphism

$$\phi_{\mathscr{L}} \colon X \to \mathbb{P}^N = \mathbb{P}\left(H^0(X, \mathscr{L})\right).$$

$\mathscr{L}$ is called *semi-ample* if $\mathscr{L}^m$ is generated by global sections for $m \gg 0$. $\mathscr{L}$ is called *ample* if it is semi-ample and $\phi_{\mathscr{L}^m}$ is an embedding for $m \gg 0$. A line bundle $\mathscr{L}$ on X is called *big* if the global sections of $\mathscr{L}^m$ define a rational map $\phi_{\mathscr{L}^m} \colon X \dashrightarrow \mathbb{P}^N$ such that X is birational to $\phi_{\mathscr{L}^m}(X)$ for $m \gg 0$. Note that in this case $\mathscr{L}^m$ is not necessarily generated by global sections, so $\phi_{\mathscr{L}^m}$ is not necessarily defined everywhere. I will leave it to the reader the make the obvious adaptation of these notions for the case of $\mathbb{Q}$-line bundles.

If it exists, then a *canonical divisor* of a scheme X is denoted by K_X and the *canonical sheaf* of X is denoted by ω_X. See §5 for more.

A smooth projective variety X is of *general type* if ω_X is big. It is easy to see that this condition is invariant under birational equivalence between smooth projective varieties. An arbitrary projective variety is of *general type* if so is a desingularization of it.

A projective variety is *canonically polarized* if ω_X is ample. Notice that if a smooth projective variety is canonically polarized, then it is of general type.

Further definitions will be given in later sections. In particular, for the definition of *Cohen-Macaulay* and *Gorenstein* see §5.

2. Stable curves

First I will recall the definition and main properties of families of stable curves and then subsequently investigate how these may be generalized to higher dimensions.

Definition 2.1. [16, 2.12] A *stable curve* is a connected projective curve that

(2.1.1) has only nodes as singularities; and,
(2.1.2) has only finitely many automorphisms.

The finiteness condition on the automorphism group is equivalent to either one the following:

(2.1.2a) Every smooth rational component of the curve meets the other components in at least 3 points.
(2.1.2b) The dualizing sheaf of the curve is ample.

With respect to (2.1.2b) note that nodes are local complete intersections and hence a stable curve is Gorenstein by definition. In particular its dualizing sheaf exists and it is a line bundle, and hence it makes sense to ask whether it is ample.

The fact that the moduli functor of stable curves gives a good compactification of the moduli functor of smooth curves hinges on the stable reduction theorem:

Theorem 2.2. [16, 3.47],[25] *Let* B *be a smooth curve,* $0 \in \mathrm{B}$ *a point, and* $\mathrm{B}^\circ = \mathrm{B} \setminus \{0\}$. *Let* $\mathrm{X}^\circ \to \mathrm{B}^\circ$ *be a flat family of stable curves of genus* $\geqslant 2$. *Then there exists a branched cover* $\mathrm{B}' \to \mathrm{B}$ *totally ramified over* 0 *and a family* $\mathrm{X}' \to \mathrm{B}'$ *of stable curves extending the fiber product* $\mathrm{X}^\circ \times_{\mathrm{B}^\circ} \mathrm{B}'$. *Moreover, any two such extensions are dominated by a third. In particular, their special fibers, that is, the preimage of* 0 *in* B', *are isomorphic.*

Note that being a family of stable curves implies that $\mathrm{X}' \to \mathrm{B}'$ does not have any multiple fibers. On the other hand, one cannot expect to have a smooth total space, X', for this family, although its singularities are the mildest possible: In general X' will have Du Val singularities (of type A). This follows from an explicit computation of the versal deformation space of a node. These singularities may be resolved by successive blowing ups resulting in an exceptional divisor consisting of a chain of rational curves, each appearing with multiplicity 1 in the fiber of the blown up surface over the point $0 \in \mathrm{B}$. This leads to semi-stable reduction where one only requires the curves in the family to be semi-stable, that is, instead of (2.1.2a) one only requires that every smooth rational component of the curve meets the other components in at least 2 points, but in exchange one obtains that one may require the total space of the family be smooth.

In the next statement I collect the ideas from these observations that will be important in our quest to understand stable varieties in higher dimensions.

Observation 2.3. For a stable family of curves, $X \to B$, let $\widetilde{X} \to X$ be a resolution of singularities and $0 \in B$ a point. Then

(2.3.1) $\omega_{X/B}$ is relatively ample;
(2.3.2) the special fiber X_0 is uniquely determined by the rest of the family;
(2.3.3) X has Du Val singularities; and
(2.3.4) $\widetilde{X} \to B$ has reduced fibers;

3. Canonical models

Next we will investigate how stability may be generalized to higher dimensions. For a more detailed study and many other results see [38].

First let us consider our goals. One wants to find a class of singularities that allows us to define a moduli functor that would compactify the moduli functor of smooth canonically polarized varieties. In other words, one wants to define *stable* varieties as canonically polarized varieties with singularities only from this particular class and one would like that any family of smooth canonically polarized varieties over a punctured curve have a unique stable limit, possibly over a branched covering which is totally ramified over the punctured point.

Taking into account previous observations in the case of families of curves, this means that one would like to achieve a notion of stable families such that (2.3.1) and (2.3.2) remain true. The first of these conditions is simply saying that stable varieties should be canonically polarized. This is both reasonable and expected and if one is familiar with the construction of moduli spaces via the Hilbert scheme (see for instance [62]) then one can see that this is also necessary for other reasons as well. The second condition, that is, uniqueness of specialization is important with regard to the moduli space one hopes to construct eventually: this condition is essentially saying that this moduli space would be separated, surely a condition one would like to have.

The other two conditions in (2.3), namely (2.3.3) and (2.3.4) are actually the ones that will help us figure out the right class of singularities having the desired properties mentioned above.

It turns out that (2.3.1) and (2.3.3) combined implies the uniqueness of specialization, that is, once one has (2.3.1), then (2.3.3) actually implies (2.3.2). The last condition, (2.3.4) will be useful in determining what class of singularities would the fibers need to have in order for the total space to have the kind of singularities that are the appropriate generalization of Du Val singularities in the case of families of curves. We will investigate this further in §4.

Du Val singularities, also known as *rational double points*, or *canonical Gorenstein surface singularities* may be defined a number of ways, see [6] for fifteen of these. The original definition of them is actually the one that generalizes well to higher dimensions.

In the following I will need to use the canonical sheaf on singular varieties. If X is Cohen-Macaulay, then a dualizing sheaf exists and the canonical sheaf may be defined as that. For the definition in more general settings please see §5.

Definition 3.1. Let X be a normal variety and assume that it admits a canonical sheaf ω_X which is a line bundle. (This holds for example if X is Gorenstein). Then X has *canonical* singularities if for a resolution of singularities $\phi : \widetilde{X} \to X$ one has the following:

$$\phi^*\omega_X \subseteq \omega_{\widetilde{X}}.$$

If $\dim X = 2$, these are also called *Du Val* singularities.

Remark 3.2. The assumption that ω_X is a line bundle is in fact not necessary to define canonical singularities, but it makes the definition simpler. We will later extend the definition to a larger class.

Notice that the (injective) morphism $\phi^*\omega_X \to \omega_{\widetilde{X}}$ does not always exist. However, if a non-zero morphism like that exists, then it is necessarily injective cf. (3.3)

Even though such a morphism does not always exist, it is easy to see that it does if X is smooth. Indeed, in that case there exists a natural morphism induced by the pull-back of differential forms $\phi^*\Omega_X \to \Omega_{\widetilde{X}}$ and taking determinants implies the existence of a non-zero morphism $\phi^*\omega_X \to \omega_{\widetilde{X}}$.

Lemma 3.3. *Let Y be an irreducible variety, $\mathscr{L}$ and $\mathscr{F}$ torsion-free sheaves on Y, and $\alpha : \mathscr{L} \to \mathscr{F}$ a non-zero morphism. If $\mathscr{L}$ has rank 1, then α must be injective.*

Proof. Let $\mathscr{K} = \ker \alpha$ and $\mathscr{I} = \operatorname{im} \alpha$. If α is non-zero, then $\mathscr{I}$ is a non-zero subsheaf of the torsion-free $\mathscr{F}$. Since $\mathscr{L}$ is rank 1 (at the general point of Y) it follows that so is $\mathscr{I}$. Therefore α is generically injective which implies that $\mathscr{K}$ is a torsion sheaf. However, $\mathscr{L}$ is also torsion-free and hence $\mathscr{K} = 0$. □

Corollary 3.4. *Let X be a normal variety and assume that it admits a canonical sheaf ω_X which is a line bundle. If for a resolution of singularities $\phi : \widetilde{X} \to X$ there exists a non-zero morphism $\phi^*\omega_X \to \omega_{\widetilde{X}}$, then X has canonical singularities. In particular, if X is smooth, then it has canonical singularities and in the definition of canonical singularities if the required condition holds for a single resolution of singularities, then it holds for all of them.*

Proof. Left to the reader. □

This leads us to another interesting condition that characterizes canonical singularities of Gorenstein varieties.

Lemma 3.5. *Let X be a normal variety and assume that it admits a canonical sheaf ω_X which is a line bundle and $\phi : \widetilde{X} \to X$ a resolution of singularities. Then the following are equivalent:*

(3.5.1) *X has canonical singularities;*

(3.5.2) $\phi_*\omega_{\widetilde{X}} \simeq \omega_X$; *and*

(3.5.3) $\phi_*\omega_{\widetilde{X}}^{\otimes m} \simeq \omega_X^{\otimes m}$ *for all* $m \geqslant 0$.

Proof. First assume that $\phi_*\omega_{\widetilde{X}} \simeq \omega_X$. Notice that there always exists a natural morphism $\phi_*\omega_{\widetilde{X}} \to \omega_X$, which is injective by (3.3), so this condition could be phrased by saying that "the natural morphism $\phi_*\omega_{\widetilde{X}} \to \omega_X$ is surjective". In fact, the point of the condition is that this isomorphism implies that there exists a non-zero morphism $\omega_X \to \phi_*\omega_{\widetilde{X}}$ and via adjointness of ϕ^* and ϕ_* that implies the existence of a non-zero morphism $\phi^*\omega_X \to \omega_{\widetilde{X}}$, which in turn implies that X has canonical singularities.

Now assume that X has canonical singularities, that is, there exists an injective morphism $\phi^*\omega_X \to \omega_{\widetilde{X}}$. It follows that the line bundle $\omega_{\widetilde{X}} \otimes \phi^*\omega_X^{-1}$ corresponds to an effective Cartier divisor E on $\widetilde{X}$, so one obtains the expression:

$$\omega_{\widetilde{X}} \simeq \phi^*\omega_X \otimes \mathscr{O}_{\widetilde{X}}(E).$$

Since X is normal it also follows that E is ϕ-exceptional and hence

$$\omega_{\widetilde{X}}\big|_E \simeq \mathscr{O}_E(E).$$

Therefore for any $m \geqslant 0$ one has the following short exact sequence:

$$0 \longrightarrow \phi^*\omega_X^{\otimes m} \longrightarrow \omega_{\widetilde{X}}^{\otimes m} \longrightarrow \mathscr{O}_{mE}(mE) \longrightarrow 0\,.$$

In order to finish the proof one needs to prove that $\phi_*\mathscr{O}_{mE}(mE) = 0$. This is easy to prove for surfaces, since the fact that E is exceptional implies that its self-intersection is negative, hence the sheaf $\mathscr{O}_{mE}(mE)$ has no global sections. The statement in arbitrary dimension follows by a simple induction on the dimension considering general hyperplane sections. For details see [24, 1-3-2]. □

Combining canonical singularities with canonical polarization leads to the notion of *canonical models*:

Theorem 3.6. *Let X be a variety with canonical singularities and $\phi : \widetilde{X} \to X$ a resolution of singularities. Assume that ω_X is ample. Then X is isomorphic to the* canonical model *of $\widetilde{X}$. In particular one has that*

$$X \simeq \operatorname{Proj} \bigoplus_{m \geqslant 0} H^0(\widetilde{X}, \omega_{\widetilde{X}}^{\otimes m}).$$

Proof. Since ω_X is ample, it follows easily that

$$X \simeq \operatorname{Proj} \bigoplus_{m \geqslant 0} H^0(X, \omega_X^{\otimes m}),$$

and $H^0(X, \omega_X^{\otimes m}) \simeq H^0(\widetilde{X}, \omega_{\widetilde{X}}^{\otimes m})$ for any $m \geqslant 0$ by (3.5.3). □

The same proof provides a relative version of this statement:

Theorem 3.7. *Let $f : X \to B$ be a proper flat morphism and $\phi : \widetilde{X} \to X$ a resolution of singularities. Let $\widetilde{f} = f \circ \phi$ and assume that X has canonical singularities, B is a smooth* *curve and* $\omega_{X/B}$ *is relatively ample with respect to* f. *Then one has a natural* B*-isomorphism*

$$X/B \simeq \big(\mathrm{Proj}_B \bigoplus_{m\geqslant 0} \widetilde{f}_*\omega_{\widetilde{X}/B}^{\otimes m}\big)/B.$$

Proof. Since $\omega_{X/B}$ is relatively ample, it follows that

$$X/B \simeq \big(\mathrm{Proj}_B \bigoplus_{m\geqslant 0} f_*\omega_{X/B}^{\otimes m}\big)/B,$$

and $f_*\omega_{X/B}^{\otimes m} \simeq f_*\omega_X^{\otimes m} \otimes \omega_B^{-m} \simeq \widetilde{f}_*\omega_{\widetilde{X}}^{\otimes m} \otimes \omega_B^{-m} \simeq \widetilde{f}_*\omega_{\widetilde{X}/B}^{\otimes m}$ for any $m \geqslant 0$ by (3.5.3). □

Corollary 3.8. *Let* B *be a smooth curve,* $0 \in B$ *a point, and* $B^\circ = B \setminus \{0\}$. *Let* $f : X \to B$ *and* $f' : X' \to B$ *be two proper flat morphisms such that restricting* f *and* f′ *over* B° *gives isomorphic families, i.e.,* $(X \times_B B^\circ)/B^\circ \simeq (X' \times_B B^\circ)/B^\circ$ *as* B°*-schemes. If both* X *and* X′ *have canonical singularities and both* $\omega_{X/B}$ *and* $\omega_{X'/B}$ *are relatively ample, then* $X/B \simeq X'/B$ *as* B*-schemes. In particular, the special fibers of* f *and* f′ *are isomorphic:* $X_0 \simeq X_0'$.

Proof. Let $\widetilde{X}$ be a common resolution of singularities of X and X′ with resolution morphisms be $\phi : \widetilde{X} \to X$ and $\phi' : \widetilde{X} \to X'$. It follows that then $f \circ \phi = f' \circ \phi'$ so one may denote this morphism by $\widetilde{f}$ and so

$$X/B \simeq \big(\mathrm{Proj}_B \bigoplus_{m\geqslant 0} \widetilde{f}_*\omega_{\widetilde{X}/B}^{\otimes m}\big)/B \simeq X'/B$$

by (3.7). □

The important conclusion to draw from this is that in order to guarantee uniqueness of specialization one should require that a stable family has a relatively ample canonical sheaf and its total space has canonical singularities.

Observation 3.9. For a stable family $X \to B$ over a smooth curve B let $\widetilde{X} \to X$ be a resolution of singularities and $0 \in B$ a point. Then one expects the following conditions to hold:

(3.9.1) $\omega_{X/B}$ is relatively ample;
(3.9.2) X has canonical singularities; and
(3.9.3) $\widetilde{X} \to B$ has reduced fibers;

Notice that I dropped the condition that "the special fiber X_0 is uniquely determined by the rest of the family" from (2.3) not because we no longer need it but because (3.9.1) and (3.9.2) imply it.

In the next section we will investigate what the third condition (3.9.3) gives us with regard to the singularities of the fibers.

4. Stable singularities

Let $f : X \to B$ be a flat morphism over a smooth curve B, $\phi : \widetilde{X} \to X$ a resolution of singularities and $0 \in B$ a point. Assume that X has canonical singularities and $\widetilde{f} : \widetilde{X} \to B$ has reduced fibers.

One would like to understand the condition this places on the singularities of X_0, the special fiber of f. To this end let us assume that ϕ is an embedded resolution of $X_0 \subset X$ and such that $\phi^* X_0 = \widetilde{X}_0 \subset \widetilde{X}$ is an snc divisor. Notice that by assumption B is a smooth curve, so X_0 is a Cartier divisor and hence pulling it back makes sense. Furthermore, assume that $\widetilde{f}$ has reduced fibers so $\phi^* X_0 = \widetilde{X}_0$ itself is an snc divisor not just that its support is one.

We saw in the proof of (3.5) that X having canonical singularities implies, and in fact is equivalent to, that

$$\omega_{\widetilde{X}} \simeq \phi^* \omega_X(E) \tag{4.1}$$

for some effective ϕ-exceptional divisor $E \subset \widetilde{X}$.

Since ϕ is an embedded resolution of $X_0 \subset X$, $\widetilde{X}_0$ contains a union of components $\widehat{X}_0$ that gives a resolution of singularities $\widehat{\phi}_0 = \phi|_{\widehat{X}_0} : \widehat{X}_0 \to X_0$. One cannot, however, expect $\widetilde{X}_0$ be equal to $\widehat{X}_0$, so one obtains that

$$\phi^* X_0 = \widetilde{X}_0 = \widehat{X}_0 + F, \tag{4.2}$$

where F is the effective ϕ-exceptional divisor formed by the unions of the components of $\widetilde{X}_0$ not contained in $\widehat{X}_0$. Since $\widetilde{X}$ is smooth, all of these are Cartier divisors.

By adjunction one has that $\omega_{\widehat{X}_0} \simeq \omega_{\widetilde{X}}(\widehat{X}_0)|_{\widehat{X}_0}$ and $\omega_{X_0} \simeq \omega_X(X_0)|_{X_0}$. Combining this with (4.1) and (4.2) leads to the isomorphism

$$\begin{aligned} \omega_{\widehat{X}_0} &\simeq \omega_{\widetilde{X}}(\widehat{X}_0)|_{\widehat{X}_0} \simeq \phi^* \omega_X(E + \phi^* X_0 - F)|_{\widehat{X}_0} \simeq \\ &\simeq \widehat{\phi}_0^* \left(\omega_X(X_0)|_{X_0} \right) \otimes \mathscr{O}_{\widehat{X}_0} \left((E - F)|_{\widehat{X}_0} \right) \simeq \widehat{\phi}_0^* \omega_{X_0} \otimes \mathscr{O}_{\widehat{X}_0} \left((E - F)|_{\widehat{X}_0} \right) \end{aligned}$$

Now let $\widehat{E}_0 = E|_{\widehat{X}_0}$ and $\widehat{F}_0 = F|_{\widehat{X}_0}$. Then one obtains that

$$\widehat{\phi}_0^* \omega_{X_0} \subseteq \omega_{\widehat{X}_0}(\widehat{F}_0) \tag{4.3}$$

This is not quite the definition of canonical singularities, but a somewhat weaker condition. Notice however that while we did not know much about the multiplicities of the components of E other than that they are non-negative, we do know that $\widetilde{X}_0 = \widehat{X}_0 + F$ is an snc divisor and hence so is $\widehat{F}_0 = F \cap \widehat{X}_0 \subset \widehat{X}_0$. This is an important detail. This means that although $\widehat{\phi}_0^* \omega_{X_0}$ does not necessarily admit a non-zero morphism to $\omega_{\widehat{X}_0}$, it does admit an embedding to a slightly larger sheaf. This leads to the definition of *(semi) log canonical singularities* see §6 for more details.

Observe that the above computation works backwards as well, so we actually found what we were looking for: a condition on the singularities of the fibers instead of a condition on the singularities of the total space.

5. The dualizing sheaf versus the canonical divisor

In order to construct moduli spaces one needs a polarization of our objects. The (essentially only) natural choice of a line bundle on an abstract smooth projective variety is the canonical bundle. This is the main reason we are studying *canonically* polarized varieties. When one extends our moduli problem in order to have compact moduli spaces one still needs a canonical polarization. However, the dualizing sheaf, even if it exists, is not necessarily a line bundle. Therefore, a discussion of how one produces canonical polarizations on stable varieties is in order. Below we will use many of the notions and notation from (1.1) but we also need a few more.

Definition 5.1. A finitely generated non-zero module M over a noetherian local ring R is called *Cohen-Macaulay* if its depth over R is equal to its dimension. For the definition of depth and dimension I refer the reader to [1]. The ring R is called *Cohen-Macaulay* if it is a Cohen-Macaulay module over itself.

Let X be a scheme and $x \in X$ a point. One says that X has *Cohen-Macaulay* singularities at x (or simply X is *CM* at x), if the local ring $\mathcal{O}_{X,x}$ is Cohen-Macaulay.

If in addition, X admits a dualizing sheaf ω_X which is a line bundle in a neighbourhood of x, then X is *Gorenstein* at x.

The scheme X is *Cohen-Macaulay* (resp. *Gorenstein*) if it is *Cohen-Macaulay* (resp. *Gorenstein*) at x for all $x \in X$.

If X is Cohen-Macaulay, then it admits a dualizing sheaf. However, stable varieties are not necessarily Cohen-Macaulay, so one needs a more sophisticated approach.

Stable varieties are projective and projective varieties admit *dualizing complexes*: If $X \subseteq \mathbb{P}^N$ and $d = \dim X$, then

$$\omega_X^{\bullet} \simeq_{\text{qis}} \mathcal{R}\mathcal{H}om_{\mathbb{P}^N}(\mathcal{O}_X, \omega_{\mathbb{P}^N}[N]).$$

Using this dualizing complex one can always define the *canonical sheaf*:

$$\omega_X := h^{-d}(\omega_X^{\bullet})$$

In fact, this allows us to define the canonical sheaf of any quasi-projective variety, or more generally any locally closed subset of a variety that admits a dualizing complex. For more on this the reader is referred to [17, 3].

Suppose $U \subseteq X$ is an open subset of the projective variety X. Then let

$$\omega_U^{\bullet} := \omega_X^{\bullet}|_U.$$

Remark 5.2. Note that X is Cohen-Macaulay if and only if

$$\omega_X^{\bullet} \simeq_{\text{qis}} \omega_X[d],$$

that is, if the only non-zero cohomology sheaf of $\omega_X^{\bullet}$ is the $-d^{\text{th}}$ (and d still denotes $\dim X$). In this case the canonical sheaf is isomorphic to the *dualizing sheaf*.

X is Gorenstein if and only if it is Cohen-Macaulay and ω_X is a line bundle.

For a normal variety X the usual way to define the canonical sheaf is different but produces the same sheaf. Being normal is equivalent to being R_1 and S_2, that is, X is normal if and only if it is non-singular in codimension 1 and satisfies Serre's S_2 condition.

Let $U = X \setminus \operatorname{Sing} X$ be the locus where X is non-singular and $\iota : U \hookrightarrow X$ its natural embedding to X. Then one may define the canonical=dualizing sheaf of U as the determinant of the cotangent bundle, i.e., the sheaf of top differential forms, $\omega_U = \det \Omega_U$. Then the usual definition of the canonical sheaf of X is $\omega'_X := \iota_* \omega_U$. It is relatively easy to see that both ω_X and ω'_X are reflexive and agree in codimension 1, so they are actually isomorphic (cf. [19, §1]).

$$\begin{array}{ccc} \omega_X & \xrightarrow{\simeq} & \omega'_X \\ \| & & \| \\ h^{-d}(\omega^\bullet_X) & & \iota_*\omega_U. \end{array}$$

Indeed, since $\iota : U \hookrightarrow X$ is an open embedding, the restriction of the dualizing complex of X to U is the dualizing complex of U:

$$\omega^\bullet_X\big|_U \simeq_{\mathrm{qis}} \omega^\bullet_U.$$

In particular, since restriction to U is an exact functor, ome also has

$$\omega_X\big|_U \simeq \omega_U.$$

Recall that X is assumed to be normal. In that case the R_1 condition implies that $\operatorname{codim}_X(X \setminus U) \geqslant 2$ and the S_2 condition combined with the fact that ω_X is reflexive implies that then

$$\omega_X \simeq \iota_*\left(\omega_X\big|_U\right) \simeq \iota_*\omega_U. \tag{5.3}$$

Possibly some readers are more familiar with this isomorphism in the divisor setting.

Let X be an irreducible normal variety and $\iota : U \hookrightarrow X$ the non-singular locus as above. A *canonical divisor* K_X of X is a Weil divisor whose associated *Weil divisorial sheaf*,

$$\mathscr{O}_X(K_X) := \{f \in K(X) | K_X + \operatorname{div}(f) \geqslant 0\},$$

is isomorphic to the canonical sheaf ω_X. This is usually defined the following way: Define ω_U as above. As U is non-singular, ω_U is a line bundle and hence corresponds to a Cartier divisor. Let $K_U = \sum \lambda_i K_i$ denote a Weil divisor associated to this Cartier divisor. Let $\overline{K}_i$ denote the closure of K_i in X and let

$$K_X := \sum \lambda_i \overline{K}_i.$$

Since $\operatorname{codim}_X(X \setminus U) \geqslant 2$, this is the unique Weil divisor on X for which $K_X\big|_U = K_U$. By the same argument as in the paragraph preceding (5.3) it follows that

$$\omega_X \simeq \mathscr{O}_X(K_X).$$

As already clear from the case of curves, when working with objects on the boundary of the moduli space one is forced to work with non-normal schemes. We will need one more important detail to make this work. Notice that U being non-singular is not essential in the above constructions. Since one knows how to define the canonical sheaf of a quasi-projective variety, one does not need U to be non-singular for that. The only place where we used the non-singularity of U was to establish that ω_U is a line bundle. In other words, we may replace the condition of U being non-singular with assuming that its canonical sheaf is a line bundle. In particular, assuming that U is Gorenstein will do the trick and then we still have that

$$\omega_X \simeq \iota_* \left(\omega_X|_U\right) \simeq \iota_* \omega_U. \tag{5.4}$$

The precise condition we need in order to be able to define stabile varieties is the following.

Definition 5.5. A variety is called G_1 if it is Gorenstein in codimension 1.

If X is G_1 and S_2 then everything said about the canonical sheaf of normal varieties above works the same way. In particular, one may talk about a *canonical divisor* K_X which is a Weil divisor that is Cartier in codimension 1. In fact, if X is G_1 and S_2, then one does not need to assume that X admits a dualizing complex and one does not need to define the canonical sheaf that way:

Definition 5.6. Let X be a scheme that is G_1 and S_2 and $\iota : U \hookrightarrow X$ be an open set such that $\operatorname{codim}_X(X \setminus U) \geqslant 2$ and U is Gorenstein. Then

$$\omega_X := \iota_* \omega_U$$

is called the *canonical sheaf* of X.

Lemma 5.7. *If X admits a dualizing complex, using the above definition for ω_X, one still has that*

$$\omega_X \simeq h^{-d}(\omega_X^{\bullet}),$$

where $d = \dim X$. *In particular, the two definitions of the dualizing sheaf agree.*

Proof. This follows from (5.4). □

Remark 5.8. We are now in a perfect position to take a deep breath, make a few observations, and lose any inhibition we might have against working with non-normal varieties. Being normal is the same as being R_1 and S_2 and we are replacing that with being G_1 and S_2. In other words, we are not going wild with all kinds of weird schemes. As far as our canonical divisors are concerned we are not much worse off than working with normal varieties. The main thing to keep in mind is that our varieties may be singular along a divisor. This means that for example one has to be careful when working with Weil divisors. However, the extent of this is essentially that by the G_1 assumption ω_X is a line bundle near the general points of the 1-codimensional part of the singular locus of X and hence we may choose canonical divisors whose support does not contain any components of that 1-codimensional singular locus. This implies that X is non-singular at the general points of these canonical divisors, so we may work with them as we are used to work with Weil divisors. In addition, we will put even more restrictions on our singularities. In particular, our stable varieties will only have double normal crossings in codimension 1. These are arguably the simplest non-normal singularities and they are also Gorenstein.

As indicated at the beginning of this section, in order to construct our moduli spaces one needs a canonical polarization on our stable varieties. The obvious assumption would be to require that stable varieties are Gorenstein. This works in dimension 1, but not in higher dimension. Consider a cone over a quartic rational scroll in $\mathbb{P}^5$. Then a general pencil of hyperplanes defines a family of smooth varieties degenerating to one that is not Gorenstein; a cone over a quartic rational curve in $\mathbb{P}^4$. For a more detailed explanation of this example see A. Taking a branched cover over a general high degree hypersurface section of the cone one obtains a family of smooth canonically polarized varieties degenerating to one with the same kind of singularities as above. This example shows that if one sticks to Gorenstein singularities, or even just to those for which ω_X is a line bundle, one will not get a compact moduli space.

So, if ω_X is not a line bundle, how does one get a "canonical polarization"? The point is that even though one cannot assume that ω_X is a line bundle, may assume that some power of it is. Of course, since ω_X is not a line bundle, one has to be careful what "power" means. Tensor powers of non-locally free sheaves tend to get even worse. For instance, tensor powers of torsion-free, or even reflexive sheaves may have torsion or co-torsion. Also, we want the power to be still associated to a Weil divisor. In other words, we want it to be a reflexive sheaf, i.e., we need to take reflexive powers:

Definition 5.9. Let X be a scheme that admits a canonical sheaf ω_X. (For instance it admits a dualizing complex or it is G_1 and S_2). Then one defines the *pluricanonical sheaves* of X as the reflexive powers of the canonical sheaf of X:

$$\omega_X^{[m]} := \left(\omega_X^{\otimes m}\right)^{**}.$$

Lemma 5.10. *Let X be a scheme that is G_1 and S_2. Then for any $m \in \mathbb{Z}$,*

$$\omega_X^{[m]} \simeq \mathscr{O}_X(mK_X).$$

Proof. Let $\iota : U \hookrightarrow X$ be an open dense subset of X such that $\operatorname{codim}_X(X \setminus U) \geqslant 2$ and $\omega_X\big|_U \simeq \omega_U$ is a line bundle. It follows that $\omega_X^{\otimes m}\big|_U \simeq \omega_U^{\otimes m}$ is a line bundle, and hence

$$\omega_X^{[m]}\big|_U \simeq \mathscr{O}_X(mK_X)\big|_U.$$

Since both $\omega_X^{[m]}$ and $\mathscr{O}_X(mK_X)$ are reflexive, this means that they are isomorphic cf. [20, 1.11]. □

This means that if X is G_1 and S_2, then one may work with pluricanonical divisors the same way as if X was normal.

Remark 5.11. Talking about Weil divisors on non-normal schemes is tricky, because in order to define the multiplicity of a function along a prime divisor and hence define the notion of linear equivalence of Weil divisors, one needs the local rings of general points of these prime divisors to be DVRs. Therefore, one only considers prime divisors that are not contained in the singular locus of the ambient scheme. The condition G_1 ensures that the canonical sheaf may be represented by a Weil divisor that satisfies this requirement.

We are now ready to introduce the notion that allows us to have canonical polarizations even if ω_X is not a line bundle.

Definition 5.12. Let X be a scheme that admits a canonical sheaf ω_X. Then, as in (1.1), ω_X is called a *$\mathbb{Q}$-line bundle* if some pluricanonical sheaf $\omega_X^{[m]}$ is a line bundle.

As a direct consequence of (5.10) one obtains:

Lemma 5.13. *Let X be a scheme that is G_1 and S_2. Then K_X is $\mathbb{Q}$-Cartier if and only ω_X is a $\mathbb{Q}$-line bundle.*

6. Singularities of the minimal model program

It is time to take a more detailed look at the singularities we have encountered and give precise definitions. For an excellent introduction to this topic the reader is urged to take a thorough look at Miles Reid's *Young person's guide to canonical singularities* [56]. For the precise theory the standard reference is [37] and for recent results one may consult [15].

6.1. Log canonical singularities

As we have already seen in the case of stable curves, in order to construct compact moduli spaces one must deal with non-normal singularities as that is the nature of degenerations: normalization does not work in families. However, as a warm-up, let us first define the normal and more traditional singularities that are relevant in the minimal model program. This will help understanding the somewhat more technical definitions required to deal with the non-normal case.

Definition 6.1. Let X be a normal variety such that K_X is $\mathbb{Q}$-Cartier and $\phi : \widetilde{X} \to X$ a resolution of singularities with a normal crossing exceptional divisor $E = \cup E_i$. One would like to compare the canonical divisors of $\widetilde{X}$ and X. Since ϕ is an isomorphism on an open set this means that the relative canonical divisor, that is, the difference between $K_{\widetilde{X}}$ and the pull-back of K_X is a divisor supported entirely on the exceptional locus. However, as K_X is not necessarily Cartier one may not be able to pull it back. One may pull back a multiple of it, so one compares that to the same multiple of $K_{\widetilde{X}}$. Then one divides the difference by the appropriate power. Notice that this way one may actually define the pull-back of K_X as a $\mathbb{Q}$-divisor:

$$\phi^* K_X := \frac{1}{m}\phi^*(mK_X),$$

where m is such that mK_X is Cartier. Then one may indeed compare the canonical divisors of $\widetilde{X}$ and X:

$$K_{\widetilde{X}} \sim_{\mathbb{Q}} \phi^* K_X + \sum a_i E_i.$$

where $a_i \in \mathbb{Q}$. Then

$$X \text{ has } \begin{array}{c} \textit{terminal} \\ \textit{canonical} \\ \textit{log terminal} \\ \textit{log canonical} \end{array} \text{ singularities if } \begin{array}{c} a_i > 0. \\ a_i \geqslant 0. \\ a_i > -1. \\ a_i \geqslant -1. \end{array}$$

for all i and any resolution ϕ as above.

Remark 6.2. We saw in §4 that the "right" class of singularities for the total space of a stable family is that of *canonical singularities* and that this leads to the fibers having *log canonical singularities*. Here we extended the definition of canonical singularities from varieties whose canonical sheaf is a line bundle to those whose canonical sheaf is a $\mathbb{Q}$-line bundle. We will generalize these definitions to include the non-normal relatives of these singularities in §§6.4 which will be the right class for "stable singularities".

Next we will see further evidence supporting this claim.

Example 6.3. This is an auxiliary example that I will use later.
Let $\Xi = (x^d + y^d + z^d + tw^d = 0) \subseteq \mathbb{P}^3_{x:y:z:w} \times \mathbb{A}^1_t$. The special fiber Ξ_0 is a cone over a smooth plane curve of degree d and the general fiber Ξ_t, for $t \neq 0$, is a smooth surface of degree d in $\mathbb{P}^3$.

Fact 6.4. Let W be a smooth variety and $X = X_1 \cup X_2 \subseteq W$ such that X_1 and X_2 are Cartier divisors in W. Then by adjunction

$$\begin{aligned} K_X &\sim (K_W + X_1 + X_2)\big|_X \\ K_{X_1} &\sim (K_W + X_1)\big|_{X_1} \\ K_{X_2} &\sim (K_W + X_2)\big|_{X_2}, \end{aligned}$$

and hence

$$\begin{aligned} K_X\big|_{X_1} &\sim K_{X_1} + X_2\big|_{X_1} \\ K_X\big|_{X_2} &\sim K_{X_2} + X_1\big|_{X_2}. \end{aligned}$$

It turns out that these latter equalities are true under more general conditions and hence they allow one to check when the canonical divisor of a reducible variety is ample by working with the canonical divisor of its irreducible components.

Example 6.5. As before, let $f : X \to B$ be a flat morphism, B a smooth curve, and $\phi : \widetilde{X} \to X$ a resolution of singularities. In this example assume that X_0 is a normal projective surface with K_{X_0} ample and an isolated singular point $P \in \operatorname{Sing} X_0$ such that X_0 is isomorphic to a cone $\Xi_0 \subseteq \mathbb{P}^3$ as in Example 6.3 locally analytically near P. Assume further that X is smooth. One would like to see whether one may resolve the singular point $P \in X_0$ and still stay within our moduli problem, i.e., that K would remain ample. For this purpose one may assume that P is the only singular point of X_0.

Because of the assumption on the singularities one may assume that ϕ is the blowing up of $P \in X$ and let $\widehat{X}_0$ denote the strict transform of X_0 on $\widetilde{X}$. Then $\widetilde{X}_0 = \widehat{X}_0 \cup E$ where $E \simeq \mathbb{P}^2$ is the exceptional divisor of the blow up. Clearly, $\phi : \widehat{X}_0 \to X_0$ is the blow up of P on X_0, so it is a smooth surface and $\widehat{X}_0 \cap E$ is isomorphic to the degree d curve over which X is locally analytically a cone.

One would like to determine the condition on d that ensures that the canonical divisor of $\widetilde{X}_0$ is still ample. According to (6.4) this means that one needs that $K_E + \widehat{X}_0|_E$ and $K_{\widehat{X}_0} + E|_{\widehat{X}_0}$ be ample. As $E \simeq \mathbb{P}^2$, $\omega_E \simeq \mathscr{O}_{\mathbb{P}^2}(-3)$, so $\mathscr{O}_E(K_E + \widehat{X}_0|_E) \simeq \mathscr{O}_{\mathbb{P}^2}(d-3)$. This is ample if and only if $d > 3$.

As this computation is local near P the only relevant issue about the ampleness of $K_{\widehat{X}_0} + E|_{\widehat{X}_0}$ is whether it is ample in a neighbourhood of $E_0 := E|_{\widehat{X}_0}$. By (6.6) this is equivalent to asking when $(K_{\widehat{X}_0} + E_0) \cdot E_0$ is positive.

Claim 6.6. Let Z be a smooth projective surface with non-negative Kodaira dimension and $\Gamma \subset Z$ an effective divisor. If $(K_Z + \Gamma) \cdot C > 0$ for every proper curve $C \subset Z$, then $K_Z + \Gamma$ is ample.

Proof. By the assumption on the Kodaira dimension there exists an $m > 0$ such that mK_Z is effective, hence so is $m(K_Z + \Gamma)$. Then by the assumption on the intersection number, $(K_Z+\Gamma)^2 > 0$, so the statement follows by the Nakai-Moishezon criterium. □

Now, observe that by the adjunction formula $(K_{\widehat{X}_0}+E_0)\cdot E_0 = \deg K_{E_0} = d(d-3)$ as E_0 is isomorphic to a plane curve of degree d. Again, one obtains the same condition as above and thus conclude that $K_{\widetilde{X}_0}$ is ample if and only if $d > 3$.

Since the objects that one considers in the current moduli problem must have an ample canonical class, one may only replace X_0 by $\widetilde{X}_0$ if $d > 3$. For our moduli problem this means that one has to allow cone singularities over curves of degree $d \leqslant 3$. The singularity one obtains for $d = 2$ is a rational double point, but the singularity for $d = 3$ is not, it is not even rational.

In fact, the above calculation tells us more. One has that $K_{\widehat{X}_0} = \phi^* K_{X_0} + aE_0$ for some $a \in \mathbb{Z}$. To compute a, first recall that $\deg K_{E_0} = d(d-3)$ and $E_0^2 = -d$. Then

$$\deg K_{E_0} = (K_{\widehat{X}_0} + E_0) \cdot E_0 = (\phi^* K_{X_0} + (a+1)E_0) \cdot E_0 = (a+1)E_0^2 = -(a+1)d.$$

Therefore $a = 2-d$. In other words, the condition obtained above, that one needs to allow cone singularities over plane curves of degree $d \leqslant 3$ is equivalent to allowing log canonical singularities cf. (6.1).

I have mentioned that stable singularities are not necessarily Cohen-Macaulay. Until we identified the actual class we want to call stable this was more or less an empty statement. By now, it is rather clear that log canonical singularities will belong to the class we are looking for, so we might as well point to an example of non-CM log canonical singularities.

Example 6.7. Let X be a cone over an abelian variety of dimension at least 2. Then X is log canonical, but not Cohen-Macaulay.

As mentioned several times, one also has to deal with some non-normal singularities and in fact in the example in (6.5) one does not really need that X be normal. In the next few subsections we will see examples of non-normal singularities that one has to handle. In particular, we will see that one has to allow the non-normal cousins of log canonical singularities. These are called *semi-log canonical* singularities and the reader can find their definition in (6.4).

6.2. Normal crossings

A *normal crossing* singularity is one that is locally analytically (or formally) isomorphic to the intersection of coordinate hyperplanes in a linear space. In other words, it is a singularity locally analytically defined as $(x_1x_2\cdots x_r = 0) \subseteq \mathbb{A}^n$ for some $r \leqslant n$. In particular, as opposed to the curve case, for surfaces it allows for triple intersections. However, triple (or higher) intersections may be "semi-resolved": Let $X = (xyz = 0) \subseteq \mathbb{A}^3$. Blow up the origin $O \in \mathbb{A}^3$, $\sigma : \mathrm{Bl}_O\mathbb{A}^3 \to \mathbb{A}^3$ and consider the strict transform of X, $\sigma : \widetilde{X} \to X$. Observe that $\widetilde{X}$ has only double normal crossings and the morphism σ is an isomorphism over $X \setminus \{O\}$. Therefore, this is a semi-resolution as defined in (6.11.4). Double normal crossings cannot be resolved the same way, because the double locus is of codimension 1, so any morphism from any space with any kind of singularities that are not double normal crossings would fail to be an isomorphism in codimension 1.

Since normal crossings are (analytically) locally defined by a single equation, they are Gorenstein and hence the canonical sheaf ω_X is still a line bundle and so it makes sense to require it to be ample.

These singularities already appear for stable curves, so it is not surprising that they are still here. As one wants to understand degenerations of one's preferred families, one has to allow (at least) normal crossings.

Another important point to remember about normal crossings is that they are *not* normal. For some interesting and perhaps surprising examples of surfaces with normal crossings see [32].

6.3. Pinch points

Another non-normal singularity that can occur as the limit of smooth varieties is the *pinch point*. It is locally analytically defined as $(x_1^2 = x_2x_3^2) \subseteq \mathbb{A}^n$ $(n \geqslant 3)$. This singularity is a double normal crossing away from the pinch point. Its normalization is smooth, but blowing up the pinch point does not make it any better as shown by the example that follows.

Example 6.8. Let $X = (x_1^2 = x_2^2 x_3) \subseteq \mathbb{A}^3$, where x_1, x_2, x_3 are linear coordinates on $\mathbb{A}^3$, $O = (0,0,0)$ and compute $\mathrm{Bl}_O X$. First, recall that

$$\mathrm{Bl}_O \mathbb{A}^3 = \{(x_1, x_2, x_3) \times [y_1 : y_2 : y_3] | x_i y_j = x_j y_i \text{ for } i, j = 1,2,3\} \subset \mathbb{A}^3 \times \mathbb{P}^2,$$

where y_1, y_2, y_3 are homogenous coordinates on $\mathbb{P}^2$.

(6.8.1) Assume that $y_1 = 1$. Then $x_2 = x_1 y_2$ and $x_3 = x_1 y_3$ and the equation of the preimage of X becomes $x_1^2 = x_1^3 y_2^2 y_3$. This breaks up into $x_1^2 = 0$ and $1 = x_1 y_2^2 y_3$. The former equation defines the exceptional divisor and the latter defines the strict transform of X, i.e., $\mathrm{Bl}_O X$. This does not have any points over $O \in X$, so on this chart, the blow up morphism $\mathrm{Bl}_O X \to X$ is an isomorphism and $\mathrm{Bl}_O X$ is smooth.

(6.8.2) Assume that $y_2 = 1$. Then $x_1 = x_2 y_1$ and $x_3 = x_2 y_3$ and the equation of the preimage of X becomes $x_2^2 y_1^2 = x_2^3 y_3$. This breaks up into $x_2^2 = 0$ and $y_1^2 = x_2 y_3$. Again, the former equation defines the exceptional divisor and the latter the strict transform of X, $\mathrm{Bl}_O X$. Notice that on this chart a coordinate system is given by x_2, y_1, y_3 and the equation defines a quadric cone. Then blowing up the vertex of the cone gives a resolution on this chart.

(6.8.3) Assume that $y_3 = 1$. Then $x_1 = x_3 y_1$ and $x_2 = x_3 y_2$ and the equation of the preimage of X becomes $x_3^2 y_1^2 = x_3^3 y_2^2$. This breaks up as $x_3^2 = 0$ and $y_1^2 = y_2^2 x_3$. Again, the former equation defines the exceptional divisor and the latter the strict transform of X, $\mathrm{Bl}_O X$. Notice that on this chart a coordinate system is given by x_3, y_1, y_3 and the latter equation is the same as the one we started with. So, $\mathrm{Bl}_O X$ again has a pinch point.

This computation shows that the blow-up of a pinch point will be, if anything, more singular, than the original and at best it can be resolved to be a pinch point again.

From this example one concludes that a pinch point cannot be resolved or even just made somewhat "better" by only trying to change it over the pinch point. It may only be resolved by taking the normalization. As in the case of double normal crossings, this is not an isomorphism in codimension 1.

Observation 6.9. Double normal crossings and pinch points share the following interesting properties:

(6.9.1) Their normalization is smooth.

(6.9.2) The normalization morphism is *not* an isomorphism in codimension 1.

(6.9.3) It is not possible to find a partial resolution that is an isomorphism in codimension 1 that would make them better in any reasonable sense.

Remark 6.10. Notice that all normal crossings share the first two properties, but, in dimension at least 2, not the third one as they may be partially resolved to double normal crossings.

One concludes that double normal crossing and pinch point singularities are unavoidable. However, at the same time, they should be viewed as the simplest non-normal singularities. In fact, in some sense they are much simpler than most normal singularities.

Furthermore, all other singularities can be resolved to these: Any reduced scheme admits a partial resolution to a scheme with only double normal crossings and pinch points such that the resolution morphism is an isomorphism wherever the original scheme is smooth, or has only double normal crossings or pinch points [34]. Of course, this only gives a partial resolution that is an isomorphism in codimension 1 if the scheme one starts with has double normal crossings in codimension 1 already. However, this turns out to be a condition one can achieve.

We will discuss relevant partial resolutions in more detail in (6.11).

6.4. Semi-log canonical singularities

Next, I will make the definition of the non-normal version of log canonical singularities precise.

Definition 6.11. Let X be a scheme of dimension n and $x \in X$ a closed point.

(6.11.1) $x \in X$ is a *double normal crossing* if it is locally analytically (or formally) isomorphic to the singularity

$$\{0 \in (x_0x_1 = 0)\} \subseteq \{0 \in \mathbb{A}^{n+1}\},$$

where $n \geqslant 1$.

(6.11.2) $x \in X$ is a *pinch point* if it is locally analytically (or formally) isomorphic to the singularity

$$\{0 \in (x_0^2 = x_1^2x_2)\} \subseteq \{0 \in \mathbb{A}^{n+1}\},$$

where $n \geqslant 2$.

(6.11.3) X is *semi-smooth* if all closed points of X are either smooth, or a double normal crossing, or a pinch point. In this case, unless X is smooth, $D_X := \operatorname{Sing} X \subseteq X$ is a smooth $(n-1)$-fold. If $\nu : \widetilde{X} \to X$ is the normalization, then $\widetilde{X}$ is smooth and $\widetilde{D}_X := \nu^{-1}(D_X) \to D_X$ is a double cover ramified along the pinch locus. Furthermore, the definition implies that if X is semi-smooth, then it is Gorenstein. In particular, it admits a canonical sheaf ω_X which is a line bundle.

(6.11.4) A morphism, $\phi : Y \to X$ is a *semi-resolution* if

- ϕ is proper,
- Y is semi-smooth,
- no component of D_Y is ϕ-exceptional, and

- there exists a closed subset $Z \subseteq X$, with $\operatorname{codim}(Z, X) \geqslant 2$ such that

$$\phi|_{\phi^{-1}(X \setminus Z)} : \phi^{-1}(X \setminus Z) \xrightarrow{\simeq} X \setminus Z$$

is an isomorphism.

Let E denote the exceptional divisor (i.e., the codimension 1 part of the exceptional set, not necessarily the whole exceptional set) of ϕ. Then ϕ is a *good semi-resolution* if $E \cup D_Y$ is a divisor with global normal crossings on Y.

(6.11.5) X has *semi-log canonical* (*slc*) (resp. *semi-log terminal* (*slt*)) singularities if
- (a) X is reduced,
- (b) X is S_2,
- (c) X admits a canonical sheaf ω_X, which is a $\mathbb{Q}$-line bundle of index m, and
- (d) there exists a good semi-resolution of singularities $\phi : \widetilde{X} \to X$ with exceptional divisor $E = \cup E_i$ such that $\omega_{\widetilde{X}}^m \simeq \phi^* \omega_X^{[m]} \otimes \mathscr{O}_{\widetilde{X}}(m \cdot \sum a_i E_i)$ with $a_i \in \mathbb{Q}$ and $a_i \geqslant -1$ (resp. $a_i > -1$) for all i.

Remark 6.12. A semi-smooth scheme has at worst hypersurface singularities, so in particular it is Gorenstein. This means that condition (6.11.5d) implies that X is G_1. In other words it follows that X admits a canonical sheaf. However, (6.11.5d) cannot be stated without assuming this first. On the other hand, it means that one may assume that X is G_1 instead. In other words, without loss of generality one may define slc (resp. slt) singularities as those satisfying that

(6.12.1) X is reduced,
(6.12.2) X is G_1 and S_2,
(6.12.3) ω_X is a $\mathbb{Q}$-line bundle of index m, and
(6.12.4) there exists a good semi-resolution of singularities $\phi : \widetilde{X} \to X$ with exceptional divisor $E = \cup E_i$ such that $\omega_{\widetilde{X}}^m \simeq \phi^* \omega_X^{[m]} \otimes \mathscr{O}_{\widetilde{X}}(m \cdot \sum a_i E_i)$ with $a_i \in \mathbb{Q}$ and $a_i \geqslant -1$ (resp. $a_i > -1$) for all i.

(6.13) Furthermore, once one assumes that X is G_1 and S_2, one may work with canonical divisors instead of canonical sheaves. In other words, we may also define slc (resp. slt) singularities as those satisfying that

(6.13.1) X is reduced,
(6.13.2) X is G_1 and S_2,
(6.13.3) K_X is $\mathbb{Q}$-Cartier, and
(6.13.4) there exists a good semi-resolution of singularities $\phi : \widetilde{X} \to X$ with exceptional divisor $E = \cup E_i$ such that $K_{\widetilde{X}} \equiv \phi^* K_X + \sum a_i E_i$ with $a_i \in \mathbb{Q}$ and $a_i \geqslant -1$ (resp. $a_i > -1$) for all i.

Again, (6.11.4) implies that X is G_1, but one needs that assumption even to work with K_X. Of course, instead of G_1, one may start by assuming that X admits a semi-resolution, then conclude that canonical divisors may be defined and then go on with the definition.

Remark 6.14. It is relatively easy to prove that if X has semi-log canonical (resp. semi-log terminal) singularities, then the condition in (6.11.5d) follows for *all* good semi-resolutions.

Remark 6.15. One may further generalize the notion of semi log canonical and define *weakly semi log canonical* singularities as those that are seminormal, S_2 and with an appropriately chosen divisor on the normalization, that pair is log canonical. In this context semi-log canonical singularities are exactly those weakly semi-log canonical divisors that are G_1. For the precise definition and more details on these singularities and their relationships see [53].

Remark 6.16. In the definition of a semi-resolution, one could choose to require that the exceptional set be a divisor. This leads to slightly different notions. It is still to be seen whether this variation leads to anything interesting (that is, anything interesting that is different from all the interesting things the definition above leads to). For more on singularities related to semi-resolutions see [38], [39], and [34].

Now we are ready to define stable varieties in arbitrary dimensions.

Definition 6.17. A variety X is called *stable* if

(6.17.1) X is projective,
(6.17.2) X has semi log canonical singularities, and
(6.17.3) ω_X is an ample $\mathbb{Q}$-line bundle.

Remark 6.18. Notice that if $\dim X = 1$, then this is equivalent with the previous definition of a stable curve (2.1).

We should also revisit the definition of *stable families*. As opposed to the case of curves, our stable varieties are canonically polarized by a $\mathbb{Q}$-line bundle and not a line bundle. As far as embedding into a projective space, computing intersection numbers, and pulling back pluricanonical sheaves are concerned this does not cause a big difference. However it introduces an additional element to which one has to pay attention when dealing with families.

We do not simply want a family of canonically polarized varieties but a family where these canonical polarizations are compatible. In other words, we want a relative canonical polarization of the family that restricts to the canonical polarization of the members of the family. In particular, we want that for a stable family $X \to B$,

$$\omega_{X/B}\big|_{X_b} \simeq \omega_{X_b} \quad \text{for all } b \in B. \tag{6.19}$$

It turns out that for curves this follows from the other assumptions and as a matter of fact we have also (secretly) assumed it during our quest for stable varieties cf. (3.9).

The only point to keep in mind is that now if one wants to define stable families only using properties of the fibers, as in the case of curves, then one might lose this condition accidentally. For an example that this can actually happen, that is, that there exists families of stable varieties that are not stable families in the sense of our earlier requirements see A.

Definition 6.20. A morphism $f : X \to B$ is called a *weakly stable family* if it satisfies the following conditions:

(6.20.1) f is flat and projective

(6.20.2) $\omega_{X/B}$ is a relatively ample $\mathbb{Q}$-line bundle

(6.20.3) X_b has semi log canonical singularities for all $b \in B$.

This definition actually still hides one very important detail. The fact that $\omega_{X/B}$ is a $\mathbb{Q}$-line bundle means that it has an index, that is, an integer $N \in \mathbb{N}$ such that $\omega_{X/B}^{[N]}$ is a line bundle and this is the smallest positive reflexive power of $\omega_{X/B}$ which is a line bundle. It follows that then (cf. [21, 2.6]),

$$\omega_{X/B}^{[N]}|_{X_b} \simeq \omega_{X_b}^{[N]}. \tag{6.21}$$

In particular, ω_{X_b} is a $\mathbb{Q}$-line bundle of index m for some m that divides N. This means that X_b may appear in weakly stable families whose relative canonical sheaf is a $\mathbb{Q}$-line bundle of index N for any multiple of m. This actually leads to a problem with respect to the moduli spaces of these families. There may be weakly stable families all of whose members have canonical sheaves of index m, but the relative canonical sheaf of the family has index $N > m$. In other words one might encounter families that are admissible as families of varieties of index N but not as families of varieties of index m, even though all members have index m. A reasonable resolution of this problem is to ask that besides (6.21) a similar restriction should hold for all reflexive powers of the relative canonical sheaf.

Definition 6.22. A weakly stable family $f : X \to B$ is called a *stable family* if it satisfies *Kollár's condition*, that is, for any $m \in \mathbb{N}$

$$\omega_{X/B}^{[m]}|_{X_b} \simeq \omega_{X_b}^{[m]}.$$

Remark 6.23. Notice that it is always true that the double dual of the restriction of the relative pluricanonical sheaf is the corresponding pluricanonical sheaf of the fiber:

$$\left(\omega_{X/B}^{[m]}|_{X_b}\right)^{**} \simeq \omega_{X_b}^{[m]},$$

so the main content of Kollár's condition is that the restriction of all pluricanonical sheaves have to be reflexive. For more on the definition of stable families and the corresponding moduli functors see [46, §7].

Remark 6.24. Notice further that Kollár's condition includes condition (6.19). Interestingly, it is not obvious that even this simple condition holds for weakly stable families. It holds for families of curves since stable curves are Gorenstein. It also holds for families of surfaces since stable surfaces are Cohen-Macaulay on account of being S_2 and this condition holds for families of Cohen-Macaulay varieties cf. [3, 3.5.1].

However, stable varieties of dimension $\geqslant 3$ are not necessarily Cohen-Macaulay (6.7), so it is absolutely not obvious weather the relative canonical sheaf is invariant under base change. It turns out that this is actually true by (11.3) cf. [36]. To see that this invariance under base change for weakly stable families is highly non-trivial the reader is referred to the examples in [54] that show that this statement is sharp in some reasonable sense.

7. Duality and vanishing

In this section I will first state two fundamental theorems that will be used later and then list a few vanishing theorems that are important in both the minimal model program and higher dimensional moduli theory.

Before anything else, we need a few definitions.

Definition 7.1. Let X be a complex scheme (i.e., a scheme of finite type over $\mathbb{C}$) of dimension n. Let $D_{\mathrm{filt}}(X)$ denote the derived category of filtered complexes of $\mathscr{O}_X$-modules with differentials of order $\leqslant 1$ and $D_{\mathrm{filt,coh}}(X)$ the subcategory of $D_{\mathrm{filt}}(X)$ of complexes K, such that for all i, the cohomology sheaves of $\mathrm{Gr}^i_{\mathrm{filt}}K$ are coherent cf. [4], [14]. Let $D(X)$ and $D_{\mathrm{coh}}(X)$ denote the derived categories with the same definition except that the complexes are assumed to have the trivial filtration. The superscripts $+, -, b$ carry the usual meaning (bounded below, bounded above, bounded). Isomorphism in these categories is denoted by $\simeq_{\mathrm{qis}}$. A sheaf $\mathscr{F}$ is also considered as a complex $\mathscr{F}^\bullet$ with $\mathscr{F}^0 = \mathscr{F}$ and $\mathscr{F}^i = 0$ for $i \neq 0$. If K is a complex in any of the above categories, then $h^i(K)$ denotes the i-th cohomology sheaf of K.

The right derived functor of an additive functor F, if it exists, is denoted by $\mathcal{R}F$ and $\mathcal{R}^iF$ is short for $h^i \circ \mathcal{R}F$. Furthermore, $\mathbb{H}^i$ will denote $\mathcal{R}^i\Gamma$, where Γ is the functor of global sections. Note that according to this terminology, if $\phi\colon Y \to X$ is a morphism and $\mathscr{F}$ is a coherent sheaf on Y, then $\mathcal{R}\phi_*\mathscr{F}$ is the complex whose cohomology sheaves give rise to the usual higher direct images of $\mathscr{F}$.

Similarly, the left derived functor of an additive functor F, if it exists, is denoted by $\mathcal{L}F$ and $\mathcal{L}^iF$ is short for $h^i \circ \mathcal{L}F$.

The next two theorems are very important in studying cohomological properties of singular varieties.

Theorem 7.2 (Grothendieck Duality) [17, VII]. *Let $\phi : Y \to X$ be a proper morphism between finite dimensional noetherian schemes that admit dualizing complexes. Then for any bounded complex $G \in D^b(Y)$,*

$$\mathcal{R}\phi_*\mathcal{R}\mathcal{H}om_Y(G, \omega_Y^\bullet) \simeq_{qis} \mathcal{R}\mathcal{H}om_X(\mathcal{R}\phi_*G, \omega_X^\bullet).$$

Theorem 7.3 (Adjointness of ϕ_* and ϕ^*) [17, II.5.10]. *Let $\phi : Y \to X$ be a proper morphism. Then for any bounded complexes $F \in D^b(X)$ and $G \in D^b(Y)$,*

$$\mathcal{R}\phi_*\mathcal{R}\mathcal{H}om_Y(\mathcal{L}\phi^*F, G) \simeq_{qis} \mathcal{R}\mathcal{H}om_X(F, \mathcal{R}\phi_*G).$$

Vanishing theorems have played a central role in algebraic geometry for the last couple of decades, especially in classification theory. Kollár [28] gives an introduction to the basic use of vanishing theorems as well as a survey of results and applications available at the time. For more recent results one should consult [10, 11, 7, 31, 58, 42, 43, 44, 45]. Because of the availability of those surveys, I will only recall statements that are important for the present article. Nonetheless, any discussion of vanishing theorems should start with the fundamental vanishing theorem of Kodaira.

Theorem 7.4 [26]. *Let Y be a smooth complex projective variety and $\mathscr{L}$ an ample line bundle on Y. Then*

$$H^i(Y, \omega_Y \otimes \mathscr{L}) = 0 \textit{ for } i \neq 0.$$

This has been generalized in several ways, but as noted above I will only state what I use in this article. For the many other generalizations the reader is invited to peruse the above references.

The original statement of Kodaira was generalized to allow semi-ample and big line bundles in place of ample ones by Grauert and Riemenschneider.

Theorem 7.5 [13]. *Let Y be a smooth complex projective variety and $\mathscr{L}$ a semi-ample and big line bundle on Y. Then*

$$H^i(Y, \omega_Y \otimes \mathscr{L}) = 0 \textit{ for } i \neq 0.$$

This also has a relative version:

Theorem 7.6 [13]. *Let Y be a smooth complex variety, $\phi : Y \to X$ a projective birational morphism, and $\mathscr{L}$ a semi-ample line bundle on Y. Then*

$$\mathcal{R}^i\phi_* (\omega_Y \otimes \mathscr{L}) = 0 \textit{ for } i \neq 0.$$

By Serre duality both (7.4) and (7.5) has a dual version:

Theorem 7.7. *Let Y be a smooth complex projective variety and $\mathscr{L}$ a semi-ample and big line bundle on Y. Then*

$$H^j(Y, \mathscr{L}^{-1}) = 0 \textit{ for } j \neq \dim Y.$$

What would be the dual version of (7.6) in the same spirit? Instead of Serre duality one would have to use Grothendieck duality:

Let Y be a smooth complex variety of dimension d, $\phi : Y \to X$ a projective morphism, and $\mathscr{L}$ a semi-ample line bundle on Y. Then

$$(7.8) \quad \mathcal{R}\mathcal{H}om_X(\mathcal{R}\phi_*(\omega_Y \otimes \mathscr{L}), \omega_X^{\bullet}) \simeq_{\text{qis}}$$
$$\simeq_{\text{qis}} \mathcal{R}\phi_* \mathcal{R}\mathcal{H}om_Y(\omega_Y \otimes \mathscr{L}, \omega_Y[d]) \simeq_{\text{qis}} \mathcal{R}\phi_* \mathscr{L}^{-1}[d]$$

In the case of (7.4) and (7.5) $X = \operatorname{Spec} \mathbb{C}$, so $\omega_X^{\bullet} \simeq_{\text{qis}} \mathbb{C}$. Then the the left hand side is quasi-isomorphic to the dual of $\mathcal{R}\phi_*(\omega_X \otimes \mathscr{L}) \simeq_{\text{qis}} \phi_*(\omega_X \otimes \mathscr{L}) \simeq H^0(Y, \omega_X \otimes \mathscr{L})$. Therefore $h^i(\mathcal{R}\phi_* \mathscr{L}^{-1}[d]) = 0$ for $i \neq 0$. This is how (7.7) follows: $\mathcal{R}^j\phi_* \mathscr{L}^{-1} = H^j(Y, \mathscr{L}^{-1}) = 0$ for $j \neq d$.

In the case ϕ is birational there is a shift by d on both side so the expected dual form of this vanishing would be

$$\mathcal{R}^j\phi_*\mathcal{L}^{-1} = 0 \text{ for } j \neq 0. \tag{7.9}$$

However, this does not always hold. To see this let us consider the simplest semi-ample line bundle, $\mathcal{O}_Y$. Then $\mathcal{R}^i\phi_*\omega_Y = 0$ for $i \neq 0$ by (7.6), so (7.8) reduces to the following:

$$\mathcal{R}\mathcal{H}om_X(\phi_*\omega_Y, \omega_X^{\bullet}) \simeq_{\text{qis}} \mathcal{R}\phi_*\mathcal{O}_Y[d] \tag{7.10}$$

Now suppose that X is normal and $\omega_Y \simeq \mathcal{O}_Y$. Then it follows that if (7.9) holds for $\mathcal{L} = \mathcal{O}_Y$, then $\omega_X^{\bullet}$ has only one non-zero cohomology sheaf and hence X is Cohen-Macaulay. In other words, if X is normal, but not CM and Y has a trivial canonical bundle, then (7.9) does not hold with $\mathcal{L} = \mathcal{O}_Y$ or more generally with $\mathcal{L} = \phi^*\mathcal{M}$ for any line bundle $\mathcal{M}$ on X.

The point is that the dual form of the relative Grauert-Riemenschneider vanishing theorem is a singularity condition on the target of the morphism in question. Notice that (7.9) follows from (7.10) for $\mathcal{L} = \mathcal{O}_X$ if X is Cohen-Macaulay and $\phi_*\omega_Y \simeq \omega_X$. It turns out that this defines a very important class of singularities which is the topic of the next section.

8. Rational singularities

Rational singularities are among the most important classes of singularities. The essence of rational singularities is that their cohomological behavior is very similar to that of smooth points. For instance, vanishing theorems can be easily extended to varieties with rational singularities. Establishing that a certain class of singularities is rational opens the door to using very powerful tools on varieties with those singularities.

Definition 8.1. Let X be a normal variety and $\phi : Y \to X$ a resolution of singularities. X is said to have *rational* singularities if $\mathcal{R}^i\phi_*\mathcal{O}_Y = 0$ for all $i > 0$, or equivalently if the natural map $\mathcal{O}_X \to \mathcal{R}\phi_*\mathcal{O}_Y$ is a quasi-isomorphism.

The notion of *irrational centers* is very closely related. For the definition and basic properties see [49].

A very useful property of rational singularities is that they are Cohen-Macaulay. In fact, this is part of Kempf's characterization of rational singularities:

Theorem 8.2. [25, p.50] *Let X be a normal variety and $\phi : Y \to X$ a resolution of singularities. Then X has rational singularities if and only if X is Cohen-Macaulay and $\phi_*\omega_Y \simeq \omega_X$.*

Proof. Let $d = \dim X$. If X has rational singularities, then

$$\omega_X^{\bullet} \simeq_{\text{qis}} \mathcal{R}\mathcal{H}om_X(\mathcal{O}_X, \omega_X^{\bullet}) \simeq_{\text{qis}} \mathcal{R}\mathcal{H}om_X(\mathcal{R}\phi_*\mathcal{O}_Y, \omega_X^{\bullet}) \simeq_{\text{qis}} \\ \simeq_{\text{qis}} \mathcal{R}\phi_*\mathcal{R}\mathcal{H}om_Y(\mathcal{O}_Y, \omega_Y[d]) \simeq_{\text{qis}} \mathcal{R}\phi_*\omega_Y[d] \simeq_{\text{qis}} \phi_*\omega_Y[d],$$

which implies that X has to be Cohen-Macaulay and $\omega_X \simeq \phi_*\omega_Y$.

Similarly, if X is Cohen-Macaulay and $\omega_X \simeq \phi_*\omega_Y$, then $\omega_X^{\bullet} \simeq_{\text{qis}} \phi_*\omega_Y[d]$ and so

$$\mathcal{R}\phi_*\mathcal{O}_Y \simeq_{\text{qis}} \mathcal{R}\phi_*\mathcal{R}\mathcal{H}om_Y(\omega_Y[d], \omega_Y^{\bullet}) \simeq_{\text{qis}} \mathcal{R}\mathcal{H}om_X(\mathcal{R}\phi_*\omega_Y[d], \omega_X^{\bullet}) \simeq_{\text{qis}} \\ \simeq_{\text{qis}} \mathcal{R}\mathcal{H}om_X(\phi_*\omega_Y[d], \omega_X^{\bullet}) \simeq_{\text{qis}} \mathcal{R}\mathcal{H}om_X(\omega_X^{\bullet}, \omega_X^{\bullet}) \simeq_{\text{qis}} \mathcal{O}_X$$

shows that X has rational singularities. □

A very important fact is that log terminal singularities are rational:

Theorem 8.3 [9]. *Let X be a variety with log terminal singularities. Then X has rational singularities.*

This is actually an easy consequence of a characterization theorem that will be stated later in (10.2). The proof will be given in (10.6) after the necessary notation is introduced in §10.

In particular, canonical singularities are rational and as a corollary one obtains that the total space of a stable family should have rational singularities.

Now we may repeat the investigation that helped us figure out what kind of singularities stable varieties should have. Previously we figured that if the total space has canonical singularities then the fibers should have semi log canonical singularities. Next we would like to see what it means for the fibers that the total space of the family has rational singularities.

So, let $f : X \to B$ be a family of reduced varieties such that B is a smooth curve and X has rational singularities. Let $b \in B$ a fixed point and let $\phi : Y \to X$ be a resolution of singularities such that $\operatorname{supp}(\operatorname{Exc}(\phi) \cup \phi^{-1}X_b)$ is a simple normal crossing divisor. Observe that by assumption and construction $X_b = f^*b$ is a Cartier divisor and $Y_b = \phi^*X_b$. Following the spirit of our assumption on stable families (3.9) assume that Y_b is reduced, that is, $Y_b = \phi^{-1}X_b$. One also has the following commutative diagram of distinguished triangles:

$$\begin{array}{ccccccc}
\mathcal{O}_X(-X_b) & \longrightarrow & \mathcal{O}_X & \longrightarrow & \mathcal{O}_{X_b} & \xrightarrow{+1} & \\
\downarrow{\scriptstyle \alpha_1} & & \downarrow{\scriptstyle \alpha_2} & & \downarrow{\scriptstyle \alpha_3} & & \\
\mathcal{R}\phi_*\mathcal{O}_Y(-Y_b) & \longrightarrow & \mathcal{R}\phi_*\mathcal{O}_Y & \longrightarrow & \mathcal{R}\phi_*\mathcal{O}_{Y_b} & \xrightarrow{+1} &
\end{array}$$

Notice that if the horizontal morphisms in this diagram are the usual natural morphisms, then α_3 is uniquely determined by α_1 and α_2 by (B.2).

As X has rational singularities,

$$\alpha_2 : \mathcal{O}_X \to \mathcal{R}\phi_*\mathcal{O}_Y$$

is a quasi-isomorphism. Since $\mathcal{O}_Y(-Y_b) \simeq \phi^*\mathcal{O}_X(-X_b)$ it follows by the projection formula that

$$\alpha_1 = \alpha_2 \otimes \mathrm{id}_{\mathcal{O}_X(-X_b)} : \mathcal{O}_X(-X_b) \to \mathcal{R}\phi_*\mathcal{O}_Y(-Y_b) \simeq_{\text{qis}} \mathcal{R}\phi_*\mathcal{O}_Y \otimes \mathcal{O}_X(-X_b)$$

is also a quasi-isomorphism. Therefore the triangulated category version of the 9-lemma (see B) implies that

$$\alpha_3 : \mathcal{O}_{X_b} \to \mathcal{R}\phi_*\mathcal{O}_{Y_b}$$

is also a quasi-isomorphism. Note that this does not mean that X_b has rational singularities as Y_b is not a resolution of singularities, it is in general not even birational to X_b. However, it definitely means that these singularities are not too far from rational singularities.

We found in §4 that one cannot expect the fibers to have the same type of singularities as the total space, just as one cannot expect all hyperplane sections of varieties in general to have the same type of singularities as the original varieties. Similarly here one cannot expect to have the members of the family have rational singularities. However, just as in §4, one finds that the singularities of the fibers are not too much worse. These are called Du Bois singularities and we will get acquainted with them in the next few sections. Notice that the condition we obtained here is almost identical to the one given by Schwede's criterion in (9.8).

9. DB singularities

Du Bois singularities are probably harder to appreciate than rational singularities at first, but they are equally important. Their main importance comes from two facts: They are not too far from rational singularities, that is, they share many of their properties, but the class of Du Bois singularities is more inclusive than that of rational singularities. For instance, log canonical singularities are Du Bois, but not necessarily rational.

Du Bois singularities are defined via Deligne's Hodge theory and so their strong connection to the singularities of the minimal model program might seem unexpected. Nevertheless, they play a very important role. We will need a little preparation before we can define these singularities, but first I would like to mention a few facts to underline their importance.

The concept of Du Bois singularities, abbreviated as *DB*, was introduced by Steenbrink in [59] as a weakening of rationality. The following statement is a direct consequence of the definition and this is the most important property of a DB singularity:

Theorem 9.1. *Let X be a proper scheme of finite type over $\mathbb{C}$. If X has only DB singularities, then the natural map*

$$H^i(X^{an}, \mathbb{C}) \to H^i(X^{an}, \mathscr{O}_{X^{an}}) \cong H^i(X, \mathscr{O}_X)$$

is surjective for all i.

In fact, this essentially characterizes Du Bois singularities shown by the next theorem. For details see [48].

Theorem 9.2. [48] *Let X be a projective variety over $\mathbb{C}$. Then X has only Du Bois singularities if and only if for any $L \subseteq X$ general (global) complete intersection subvariety*

$$\dim_{\mathbb{C}} H^i(L, \mathscr{O}_L) \leqslant \dim_{\mathbb{C}} \mathrm{Gr}_F^0 H^i(L, \mathbb{C}),$$

where $\mathrm{Gr}_F^0 H^i(L, \mathbb{C})$ is the graded quotient associated to Deligne's Hodge filtration on $H^i(L, \mathbb{C})$.

Using [5, Lemme 1], (9.1) implies the following:

Corollary 9.3. *Let $f : X \to B$ be a proper, flat morphism of complex varieties with B connected. Assume that X_b has only DB singularities for all $b \in B$. Then $h^i(X_b, \mathscr{O}_{X_b})$ is independent of $b \in B$ for all* i.

This will be important later.

The starting point of the precise definition is Du Bois's construction, following Deligne's ideas, of the generalized de Rham complex, which is called the *Deligne-Du Bois complex*. Recall, that if X is a smooth complex algebraic variety of dimension n, then the sheaves of differential p-forms with the usual exterior differentiation give a resolution of the constant sheaf $\mathbb{C}_X$. I.e., one has a complex of sheaves,

$$\mathscr{O}_X \xrightarrow{d} \Omega_X^1 \xrightarrow{d} \Omega_X^2 \xrightarrow{d} \Omega_X^3 \xrightarrow{d} \cdots \xrightarrow{d} \Omega_X^n \simeq \omega_X,$$

which is quasi-isomorphic to the constant sheaf $\mathbb{C}_X$ via the natural map $\mathbb{C}_X \to \mathscr{O}_X$ given by considering constants as holomorphic functions on X. Recall that this complex *is not* a complex of quasi-coherent sheaves. The sheaves in the complex are quasi-coherent, but the maps between them are not $\mathscr{O}_X$-module morphisms. Notice however that this is actually not a shortcoming; as $\mathbb{C}_X$ is not a quasi-coherent sheaf, one cannot expect a resolution of it in the category of quasi-coherent sheaves.

The Deligne-Du Bois complex is a generalization of the de Rham complex to singular varieties. It is a complex of sheaves on X that is quasi-isomorphic to the constant sheaf, $\mathbb{C}_X$. The terms of this complex are harder to describe but its properties, especially cohomological properties are very similar to the de Rham complex of smooth varieties. In fact, for a smooth variety the Deligne-Du Bois complex is quasi-isomorphic to the de Rham complex, so it is indeed a direct generalization.

The construction of this complex, $\underline{\Omega}_X^{\bullet}$, is based on simplicial resolutions. The reader interested in the details is referred to the original article [4]. Note also that a simplified construction was later obtained in [2] and [14] via the general theory of polyhedral and cubic resolutions. An easily accessible introduction can be found in [60]. Other useful references are the recent book [55] and the survey [52]. I will actually not use these resolutions here. They are needed for the construction, but if one is willing to believe the listed properties (which follow in a rather straightforward way from the construction) then one should be able to follow the material presented here.

Recently Schwede found a simpler alternative construction of (part of) the Deligne-Du Bois complex that does not need a simplicial resolution (9.8). This allows one to define Du Bois singularities (9.5) without needing simplicial resolutions and it is quite useful in applications. For applications of the Deligne-Du Bois complex and Du Bois singularities other than the ones listed here see [61], [30, Chapter 12], [40, 42].

The word "hyperresolution" will refer to either a simplicial, polyhedral, or cubic resolution. Formally, the construction of $\underline{\Omega}^{\bullet}_X$ is essentially the same regardless the type of resolution used and no specific aspects of either types will be used.

The following definition is included to make sense of the statements of some of the forthcoming theorems. It can be safely ignored if the reader is not interested in the detailed properties of the Deligne-Du Bois complex and is willing to accept that it is a very close analog of the de Rham complex of smooth varieties.

Theorem 9.4 [4, 6.3, 6.5]. *Let X be a complex scheme of finite type. Then there exists a unique object $\underline{\Omega}^{\bullet}_X \in \operatorname{Ob} D_{\mathrm{filt}}(X)$ such that using the notation*

$$\underline{\Omega}^p_X := \operatorname{Gr}^p_{\mathrm{filt}} \underline{\Omega}^{\bullet}_X[p],$$

it satisfies the following properties

(9.4.1)

$$\underline{\Omega}^{\bullet}_X \simeq_{qis} \mathbb{C}_X.$$

(9.4.2) *$\underline{\Omega}$ is functorial, i.e., if $\phi\colon Y \to X$ is a morphism of complex schemes of finite type, then there exists a natural map ϕ^* of filtered complexes*

$$\phi^*\colon \underline{\Omega}^{\bullet}_X \to \mathcal{R}\phi_*\underline{\Omega}^{\bullet}_Y.$$

Furthermore, $\underline{\Omega}^{\bullet}_X \in \operatorname{Ob}\left(D^b_{\mathrm{filt,coh}}(X)\right)$ and if ϕ is proper, then ϕ^* is a morphism in $D^b_{\mathrm{filt,coh}}(X)$.

(9.4.3) *Let $U \subseteq X$ be an open subscheme of X. Then*

$$\underline{\Omega}^{\bullet}_X\big|_U \simeq_{qis} \underline{\Omega}^{\bullet}_U.$$

(9.4.4) *If X is proper, then there exists a spectral sequence degenerating at E_1 and abutting to the singular cohomology of X:*

$$E_1^{pq} = \mathbb{H}^q\left(X, \underline{\Omega}^p_X\right) \Rightarrow H^{p+q}(X, \mathbb{C}).$$

(9.4.5) *If* $\varepsilon_\bullet : X_\bullet \to X$ *is a hyperresolution, then*

$$\underline{\Omega}^\bullet_X \simeq_{qis} R\varepsilon_{\bullet *}\Omega^\bullet_{X_\bullet}.$$

In particular, $h^i\left(\underline{\Omega}^p_X\right) = 0$ *for* $i < 0$.

(9.4.6) *There exists a natural map,* $\mathcal{O}_X \to \underline{\Omega}^0_X$*, compatible with* (9.4.2).

(9.4.7) *If* X *is smooth, then*

$$\underline{\Omega}^\bullet_X \simeq_{qis} \Omega^\bullet_X.$$

In particular,

$$\underline{\Omega}^p_X \simeq_{qis} \Omega^p_X.$$

(9.4.8) *If* $\phi : Y \to X$ *is a resolution of singularities, then*

$$\underline{\Omega}^{\dim X}_X \simeq_{qis} \mathcal{R}\phi_*\omega_Y.$$

It turns out that the Deligne-Du Bois complex behaves very much like the de Rham complex for smooth varieties. Observe that (9.4.4) says that the Hodge-to-de Rham spectral sequence works for singular varieties if one uses the Deligne-Du Bois complex in place of the de Rham complex. This has far reaching consequences and if the associated graded pieces $\underline{\Omega}^p_X$ turn out to be computable, then this single property leads to many applications.

The natural map $\mathcal{O}_X \to \underline{\Omega}^0_X$ given by (9.4.6) may be considered as an invariant of the singularities of X. Clearly, if X is smooth, then it is a quasi-isomorphism, but it may be a quasi-isomorphism even if X is not smooth. In fact, we are interested in situations when this map is a quasi-isomorphism. When X is proper over $\mathbb{C}$, such a quasi-isomorphism implies that the natural map

$$H^i(X^{an}, \mathbb{C}) \to H^i(X, \mathcal{O}_X) = \mathbb{H}^i(X, \underline{\Omega}^0_X)$$

is surjective because of the degeneration at E_1 of the spectral sequence in (9.4.4) (cf. (9.1)). Notice that this condition is crucial for proving Kodaira-type vanishing theorems cf. [30, §9], [15, 3.H].

Following Du Bois, Steenbrink was the first to study this condition and he christened this property after Du Bois. It should be noted that many of the ideas that play important roles in this theory originated from Deligne. Unfortunately the now standard terminology does not reflect this.

Definition 9.5. A scheme X is said to have *Du Bois* singularities (or *DB* singularities for short) if the natural map $\mathcal{O}_X \to \underline{\Omega}^0_X$ from (9.4.6) is a quasi-isomorphism.

Remark 9.6. If $\varepsilon : X_\bullet \to X$ is a hyperresolution of X then X has Du Bois singularities if and only if the natural map $\mathcal{O}_X \to \mathcal{R}\varepsilon_{\bullet *}\mathcal{O}_{X_\bullet}$ is a quasi-isomorphism.

A relative version of this notion for pairs was defined in [47].

Example 9.7. It is easy to see that smooth points are Du Bois and Deligne proved that normal crossing singularities are Du Bois as well cf. [5, Lemme 2(b)].

I will finish this section with Schwede's characterization of DB singularities. This condition makes it possible to define DB singularities without hyperresolutions, derived categories, etc. It makes it easier to get acquainted with these singularities, but it is still useful to know the original definition for many applications.

Theorem 9.8 [57]. *Let X be a reduced separated scheme of finite type over a field of characteristic zero. Assume that* $X \subseteq Z$ *where Z is smooth and let* $\phi : W \to Z$ *be a proper birational map with W smooth and where* $Y = \phi^{-1}(X)_{\mathrm{red}}$*, the reduced pre-image of X, is a simple normal crossings divisor (or in fact any scheme with DB singularities). Then X has DB singularities if and only if the natural map* $\mathcal{O}_X \to \mathcal{R}\phi_*\mathcal{O}_Y$ *is a quasi-isomorphism.*

In fact, one can say more. There is a quasi-isomorphism $\mathcal{R}\phi_*\mathcal{O}_Y \xrightarrow{\simeq_{qis}} \underline{\Omega}^0_X$ *such that the natural map* $\mathcal{O}_X \to \underline{\Omega}^0_X$ *can be identified with the natural map* $\mathcal{O}_X \to \mathcal{R}\phi_*\mathcal{O}_Y$.

Notice that this condition is the one obtained at the end of the previous section. Given that and our earlier findings on (semi-) log canonical singularities it may not come as a surprise that Kollár had conjectured a strong connection between these singularities. As canonical singularities are rational one should expect a similar implication between log canonical and Du Bois:

Conjecture 9.9 [39, 1.13] (Kollár's Conjecture). *Log canonical singularities are Du Bois.*

This conjecture has been recently confirmed in [36]. For more see §10 and in particular (10.15).

10. The splitting principle

The moral of this section can be summarized by the following principle:

The Splitting Principle. *Morphisms do not split accidentally.*

Remark 10.1. It is customary to casually use the word "splitting" to explain the statements of the theorems that follow. However, the reader should be warned that one has to be careful with the meaning of this, because these "splittings" take place in the derived category, which is not abelian. For this reason, in the statements of the theorems below I use the terminology that a morphism admits a *left inverse.* In an abelian category this condition is equivalent to "splitting" and being a direct component (of a direct sum). With a slight abuse of language I labeled these as "Splitting theorems" cf. (10.2), (10.7) and (10.14).

The first theorem I will recall is a criterion for a singularity to be rational.

Theorem 10.2 [41] **(Splitting theorem I).** *Let* $\phi : Y \to X$ *be a proper morphism of varieties over* $\mathbb{C}$ *and* $\rho : \mathcal{O}_X \to \mathcal{R}\phi_*\mathcal{O}_Y$ *the associated natural morphism. Assume that Y has rational singularities and* ρ *has a left inverse, i.e., there exists a morphism (in the derived category of* $\mathcal{O}_X$*-modules)* $\rho' : \mathcal{R}\phi_*\mathcal{O}_Y \to \mathcal{O}_X$ *such that* $\rho' \circ \rho$ *is a quasi-isomorphism of* $\mathcal{O}_X$ *with itself. Then X has only rational singularities.*

Remark 10.3. Note that ϕ in the theorem does not have to be birational or even generically finite. It follows from the conditions that it is surjective.

Corollary 10.4. *Let X be a complex variety and $\phi : Y \to X$ a resolution of singularities. If $\mathcal{O}_X \to \mathcal{R}\phi_*\mathcal{O}_Y$ has a left inverse, then X has rational singularities.*

Corollary 10.5. *Let X be a complex variety and $\phi : Y \to X$ a finite morphism. If Y has rational singularities, then so does X.*

Using this criterion it is quite easy to prove that log terminal singularities are rational (8.3). For related statements see [37, 5.22] and the references therein.

Theorem 10.6 (= Theorem 8.3). *Let X be a variety with log terminal singularities. Then X has rational singularities.*

Proof. [41] The question is local, so one may restrict to a neighbourhood of a point. Then the index 1 cover $\pi : \widetilde{X} \to X$ is a finite morphism onto X. In particular, π_* is exact and the natural morphism $\mathcal{O}_X \to \pi_*\mathcal{O}_{\widetilde{X}}$ has a left inverse by the construction of the index 1 cover.

Therefore, by (10.5) it is enough to prove that $\widetilde{X}$ has rational singularities and so one may assume that X has canonical singularities and ω_X is a line bundle.

Let $\phi : Y \to X$ be a resolution of singularities of X. Since X has canonical singularities and ω_X is a line bundle, there exists a non-trivial morphism

$$\iota : \mathcal{L}\phi^*\omega_X \simeq_{\mathrm{qis}} \phi^*\omega_X \to \omega_Y.$$

Its adjoint morphism on X, $\omega_X \to \mathcal{R}\phi_*\omega_Y$, is a quasi-isomorphism by (3.5) and (7.6)

Applying $\mathcal{R}\mathcal{H}om_Y(\_\_, \omega_Y)$ to ι and using (7.3), one obtains the following diagram which defines ρ':

$$\begin{array}{ccccc}
\mathcal{R}\phi_*\mathcal{R}\mathcal{H}om_Y(\omega_Y, \omega_Y) & \xrightarrow{\simeq_{\mathrm{qis}}} & \mathcal{R}\phi_*\mathcal{R}\mathcal{H}om_Y(\mathcal{L}\phi^*\omega_X, \omega_Y) & \xrightarrow{\simeq_{\mathrm{qis}}} & \mathcal{R}\mathcal{H}om_X(\omega_X, \mathcal{R}\phi_*\omega_Y) \\
\uparrow \simeq_{\mathrm{qis}} & & & & \downarrow \simeq_{\mathrm{qis}} \\
\mathcal{R}\phi_*\mathcal{O}_Y & & \xrightarrow{\quad\rho'\quad} & & \mathcal{O}_X.
\end{array}$$

The last quasi-isomorphism uses the fact that $\mathcal{R}\phi_*\omega_Y \simeq_{\mathrm{qis}} \omega_X$. It is easy to see that $\rho' \circ \rho$ acts trivially on $\mathcal{O}_X$ and hence the statement follows by 10.2 (or (10.4)). □

There is a criterion for DB singularities that is similar to the one in (10.2):

Theorem 10.7 [40, 2.3] **(Splitting theorem II)**. *Let X be a complex variety. If $\mathcal{O}_X \to \underline{\Omega}^0_X$ has a left inverse, then X has DB singularities.*

This criterion has several important consequences. Here is one of them:

Corollary 10.8 [40, 2.6]. *Let X be a complex variety with rational singularities. Then X has DB singularities.*

Proof. Let $\phi : Y \to X$ be a resolution of singularities. Then since Y is smooth the natural map $\rho : \mathcal{O}_X \to \mathcal{R}\phi_*\mathcal{O}_Y$ factors through $\underline{\Omega}^0_X$ by (9.4.6). Then, since X has rational singularities, ρ is a quasi-isomorphism, so one obtains that the natural map $\mathcal{O}_X \to \underline{\Omega}^0_X$ has a left inverse. Therefore, X has DB singularities by (10.7). □

Recently a few more criterions have been found for DB singularities. The next one resembles Kempf's criterion for rational singularities (8.2) and shows that indeed DB singularities may be considered a close generalization of rational singularities.

Theorem 10.9 [53, 3.1]. *Let X be a normal Cohen-Macaulay scheme of finite type over $\mathbb{C}$. Let $\phi : Y \to X$ be a resolution of singularities such that the (reduced) exceptional set G is a simple normal crossing divisor. Then X has DB singularities if and only if $\phi_*\omega_Y(G) \simeq \omega_X$.*

Related results have been obtained in the non-normal Cohen-Macaulay case, see [53] for details.

Remark 10.10. The submodule $\phi_*\omega_Y(G) \subseteq \omega_X$ is independent of the choice of the log resolution. Thus this submodule may be viewed as an invariant that partially measures how far a scheme is from being DB (compare with [12]).

As an easy corollary, one obtains another proof that rational singularities are DB (this time via the Kempf-criterion for rational singularities).

Corollary 10.11 [41]. *Let X be a complex variety with rational singularities. Then X has DB singularities.*

Proof. Since X has rational singularities, it is Cohen-Macaulay and normal. Then $\phi_*\omega_Y = \omega_X$ but one also has $\phi_*\omega_Y \subseteq \phi_*\omega_Y(G) \subseteq \omega_X$, and thus $\phi_*\omega_Y(G) = \omega_X$ as well. The statement now follows from Theorem 10.9. □

One also sees immediately that log canonical singularities coincide with DB singularities in the Gorenstein case.

Corollary 10.12 [40, 3.6][53, 3.16]. *Suppose that X is Gorenstein and normal. Then X is DB if and only if X is log canonical.*

Proof. X is easily seen to be log canonical if and only if $\phi_*\omega_{Y/X}(G) \simeq \mathcal{O}_X$. The projection formula then completes the proof. □

In fact, a slightly jazzed up version of this argument can be used to show that every Cohen-Macaulay log canonical pair is DB:

Corollary 10.13 [53, 3.16]. *CM log canonical singularities are DB.*

We will see below that it is actually not necessary to assume CM in the previous theorem. However, the characterization of DB singularities in (10.9) is still useful on its own.

Theorem 10.14 [36, 1.6] **(Splitting theorem III)**. *Let $\phi : Y \to X$ be a proper morphism between reduced schemes of finite type over $\mathbb{C}$. Let $W \subseteq X$ be a closed reduced subscheme* *with ideal sheaf* $\mathscr{I}_{W\subseteq X}$ *and* $F := \phi^{-1}(W) \subset Y$ *with ideal sheaf* $\mathscr{I}_{F\subseteq Y}$*. Assume that the natural map* ρ

$$\mathscr{I}_{W\subseteq X} \underset{\rho}{\overset{\rho'}{\rightleftarrows}} \mathcal{R}\phi_* \mathscr{I}_{F\subseteq Y}$$

admits a left inverse ρ'*, that is,* $\rho' \circ \rho = \mathrm{id}_{\mathscr{I}_{W\subseteq X}}$*. Then if* Y, F*, and* W *all have DB singularities, then so does* X*.*

A somewhat more general version of this was proved in [50].

This criterion forms the cornerstone of the proof of the following theorem:

Theorem 10.15 [36, 1.5]. *Let* $\phi : Y \to X$ *be a proper surjective morphism with connected fibers between normal varieties. Assume that* Y *has log canonical singularities and* $K_Y \sim_{\mathbb{Q},\phi} 0$*. Then* X *is DB.*

Corollary 10.16 [36, 1.4]. *Log canonical singularities are DB.*

For the proofs and more general statements, please see [36]. Also, note that this statement holds in a more general situation, namely it is in fact true that already semi-log canonical singularities are DB. This is proved in [35, §6.2]

Remark 10.17. Notice that in (10.14) it is not required that ϕ be birational. On the other hand the assumptions of the theorem and [41, Thm 1] imply that if $Y \setminus F$ has rational singularities, e.g., if Y is smooth, then $X \setminus W$ has rational singularities as well.

This theorem is used in [36] to derive various consequences, some of which are formally unrelated to DB singularities. I will mention some of these in the sequel, but the interested reader should look at the original article to obtain the full picture.

11. Stable families

The connection between log canonical and DB singularities has many useful applications in moduli theory. I list a few below without proof.

Theorem 11.1 [36, 7.8,7.9,7.13]. *Let* $f : X \to B$ *be a flat projective morphism of complex varieties with* B *connected and such that* X_b *has log canonical singularities for all* $b \in B$*. Then*

(11.1.1) $h^i(X_b, \mathscr{O}_{X_b})$ *is independent of* $b \in B$ *for all* i*.*
(11.1.2) *If one fiber of* f *is Cohen-Macaulay, then all fibers are Cohen-Macaulay.*
(11.1.3) *The cohomology sheaves* $h^i(\omega_f^{\bullet})$ *are flat over* B*, where* $\omega_f^{\bullet}$ *denotes the relative dualizing complex of* f*.*

For arbitrary flat, proper morphisms, the set of fibers that are Cohen-Macaulay is open, but not necessarily closed. Thus the key point of (11.1.2) is to show that this set is also closed.

The generalization of these results to the semi log canonical case turns out to be straightforward, but it needs some foundational work to extend some of the results used here to the semi log canonical case. This is done in [35, §6.2]. The general case then implies that each connected component of the moduli space of stable log varieties either parameterizes only Cohen-Macaulay or parameterizes only non-Cohen-Macaulay objects.

Notice that this still does not mean that one should abandon the non-Cohen-Macaulay objects. There exists smooth projective varieties of general type whose log canonical model is not Cohen-Macaulay and one should naturally prefer to have a moduli space that includes these. Nevertheless, it is very useful to know that if the general fiber is Cohen-Macaulay, then so is the special fiber.

(11.1) is proved using (10.15), (9.3) and the following theorem. Before I can state that theorem I need a simple definition. Let $f : X \to B$ be a flat morphism. One says that f is a *DB family* if X_b is DB for all $b \in B$.

Theorem 11.2 [36, 7.9]. *Let $f : X \to B$ be a projective DB family and $\mathscr{L}$ a relatively ample line bundle on X. Then*

(11.2.1) *the sheaves $h^{-i}(\omega_f^{\bullet})$ are flat over B for all i,*

(11.2.2) *the sheaves $f_*(h^{-i}(\omega_f^{\bullet}) \otimes \mathscr{L}^{\otimes q})$ are locally free and compatible with arbitrary base change for all i and for all $q \gg 0$, and*

(11.2.3) *for any base change morphism $\vartheta : T \to B$ and for all i,*

$$\left(h^{-i}(\omega_f^{\bullet})\right)_T \simeq h^{-i}(\omega_{f_T}^{\bullet}).$$

Let me emphasize a special case of this theorem. This had been known for families of CM varieties, so for instance for stable families of relative dimension at most 2.

Corollary 11.3. *Let $f : X \to B$ be a weakly stable family. Then $\omega_{X/B}$ commutes with arbitrary base change.*

For related results that show that this statement is sharp in a certain sense see [54].

12. Deformations of DB singularities

Given the importance of DB singularities in moduli theory it is a natural question whether they are invariant under small deformation.

It is relatively easy to see from the construction of the Deligne-Du Bois complex that a general hyperplane section (or more generally, the general member of a base point free linear system) on a variety with DB singularities again has DB singularities. Therefore the question of deformation follows from the following.

Conjecture 12.1 [61]. *Let $D \subset X$ be a reduced Cartier divisor and assume that D has only DB singularities in a neighborhood of a point $x \in D$. Then X has only DB singularities in a neighborhood of the point x.*

This conjecture was confirmed for isolated Gorenstein singularities by Ishii [22]. Also note that rational singularities satisfy this property, see [8].

One also has the following easy corollary of the results presented earlier:

Theorem 12.2. *Assume that* X *is Gorenstein and* D *is normal. Then the statement of* (12.1) *is true.*

Proof. The question is local so one may restrict to a neighborhood of x. If X is Gorenstein, then so is D as it is a Cartier divisor. Then D is log canonical by (10.12), and then X is also log canonical by inversion of adjunction [23]. (Recall that if D is normal, then so is X along D). Therefore X is also DB. □

Remark 12.3. It is claimed in [41, 3.2] that the conjecture holds in full generality. Unfortunately, the proof published there is not complete. It works as long as one assumes that the non-DB locus of X is contained in D. For instance, one may assume that this is the case if the non-DB locus is isolated.

The problem with the proof is the following: it is stated that by taking sufficiently general hyperplane sections one may assume that the non-DB locus is isolated. However, this is incorrect. One may only assume that the *intersection* of the non-DB locus of X with D is isolated. If one takes a further general section then it will miss the intersection point and then it is not possible to make any conclusions about that case.

Until very recently the best known result with regard to this conjecture had been the following:

Theorem 12.4 [41, 3.2]. *Let* $D \subset X$ *be a reduced Cartier divisor and assume that* D *has DB singularities in a neighborhood of a point* $x \in D$ *and that* $X \setminus D$ *has DB singularities. Then* X *has only DB singularities in a neighborhood of* x.

After submitting this article, but fortunately before it went to press the above conjecture has been settled by the author of this paper and Karl Schwede:

Theorem 12.5 [51, 4.1,4.2]. *Conjecture 12.1 holds, that is: If* $D \subset X$ *is a reduced Cartier divisor such that* D *has only DB singularities in a neighborhood of a point* $x \in D$, *then* X *has only DB singularities in a neighborhood of the point* x.

Experience shows that divisors not in general position tend to have worse singularities than the ambient space in which they reside. Therefore one would in fact expect that if $X \setminus D$ and D are nice (e.g., they have DB singularities), then perhaps X is even better behaved.

We have also seen that rational singularities are DB and at least Cohen-Macaulay DB singularities are not so far from being rational cf. (10.9). The following result of Schwede supports this philosophical point.

Theorem 12.6 [57, 5.1]. *Let* X *be a reduced scheme of finite type over a field of characteristic zero,* D *a Cartier divisor that has DB singularities and assume that* $X \setminus D$ *is smooth. Then* X *has rational singularities (in particular, it is Cohen-Macaulay).*

Let me conclude with a conjectural generalization of this statement:

Conjecture 12.7. *Let X be a reduced scheme of finite type over a field of characteristic zero, D a Cartier divisor that has DB singularities and assume that $X \setminus D$ has rational singularities. Then X has rational singularities (in particular, it is Cohen-Macaulay).*

Essentially the same proof as in (12.2) shows that this is also true under the same additional hypotheses.

Theorem 12.8. *Assume that X is Gorenstein and D is normal. Then the statement of (12.7) is true.*

Proof. If X is Gorenstein, then so is D as it is a Cartier divisor. Then by (10.12) D is log canonical. Then X is also log canonical near D by inversion of adjunction [23].

As X is Gorenstein and $X \setminus D$ has rational singularities, it follows that $X \setminus D$ has canonical singularities. Then X has only canonical singularities everywhere. This can be seen by observing that D is a Cartier divisor and examining the discrepancies that lie over D for (X, D) as well as for X. Therefore, by (8.3) [9] X has only rational singularities along D. □

Appendix A. The $\mathbb{Q}$-Cartier condition in families

Let $R \subseteq \mathbb{P}^4$ be a quartic rational normal curve, i.e., the image of the embedding of $\mathbb{P}^1$ into $\mathbb{P}^4$ by the global sections of $\mathcal{O}_{\mathbb{P}^1}(4)$.

Let $T \subseteq \mathbb{P}^5$ be a quartic rational scroll, i.e., the image of the embedding of $\mathbb{P}^1 \times \mathbb{P}^1$ into $\mathbb{P}^5$ by the global sections of $\mathcal{O}_{\mathbb{P}^1 \times \mathbb{P}^1}(1,2)$. Then R is a hyperplane section of T. Indeed, let f_1 and f_2 denote the divisor classes of the two rulings on T and let $H \subseteq \mathbb{P}^5$ be a general hyperplane. Then $C := H \cap T$ is a smooth curve such that $C \sim_T f_1 + 2f_2$. Then by the adjunction formula $2g(C) - 2 = (-2f_1 - 2f_2 + C) \cdot C = -2$, hence $C \simeq \mathbb{P}^1$. Furthermore, then $C^2 = 4$, so $\mathcal{O}_T(1,2)|_C \simeq \mathcal{O}_C(4)$. Therefore C is a quartic rational curve in $H \simeq \mathbb{P}^4$, and thus it may be identified with R.

Let $C_R \subseteq \mathbb{P}^5$ be the projectivized cone over R in $\mathbb{P}^5$ and $C_T \subseteq \mathbb{P}^6$ the projectivized cone over T in $\mathbb{P}^6$. Then as R is a hyperplane section of T, it follows that both T and C_R are hyperplane sections of C_T, so T is a smoothing of C_R.

Let $V \subseteq \mathbb{P}^5$ be a Veronese surface, i.e., the image of the Veronese embedding; the embedding of $\mathbb{P}^2$ into $\mathbb{P}^5$ by the global sections of $\mathcal{O}_{\mathbb{P}^2}(2)$. Let $D \subset V$ be the image of a smooth conic of $\mathbb{P}^2$. Then D is a hyperplane section of V and it is also a rational normal quartic curve in $\mathbb{P}^4$ so it can also be identified with R. Therefore, the same way as above, using C_V, the cone over V, one sees that V is also a smoothing of C_R.

It is relatively easy, and thus left to the reader, to compute that C_R has log terminal singularities. In particular, this type of singularity is among those that appear on stable varieties. In fact, considering a cyclic covering [37, 2.50] branched over a highly divisible relatively very ample divisor gives a family of stable varieties with the same kind of singularities as the ones that appear here. This can be applied for both of the families coming from C_T and C_V.

The problem this example points to is that if one allows arbitrary families, then one may get unwanted results. For example, using the families derived from C_T and C_V would mean that $T \simeq \mathbb{P}^1 \times \mathbb{P}^1$ and $V \simeq \mathbb{P}^2$ should be considered to have the same deformation type (or the same statement for the surfaces of general type on the cyclic cover mapping to these fibers). However, there are obviously no smooth families that they both belong to, they are topologically very different. For instance, $K_T^2 = 8$ while $K_V^2 = 9$.

The crux of the matter is that K_{C_T} is *not* $\mathbb{Q}$-Cartier and consequently the family obtained from it is not a *(weakly) stable family* as defined in (6.20) and (6.22). This is actually an important point: the canonical classes of the members of the family are $\mathbb{Q}$-Cartier, but the relative canonical class of the family is not $\mathbb{Q}$-Cartier. In particular, the canonical divisors of the members of the family are not consistent.

The family obtained from C_V *has* a $\mathbb{Q}$-Cartier canonical class and consequently ensures that the canonical divisors of the members of the family are similar to some extent. Among other things this implies that $K_{C_R}^2 = 9$. One may also use an actual parametrization of C_R to verify this fact independently. It is interesting to note that K_{C_R} is $\mathbb{Q}$-Cartier, but not Cartier even though its self-intersection number is an integer.

Appendix B. The nine lemma in triangulated categories

For lack of an appropriate reference the following pseudo-trivial theorem is proved here for the reader's convenience.

Theorem B.1. *Let $A, B, C, A', B', C', A'', B''$ be objects in a triangulated category $\mathfrak{T}$ and assume that there exists a commutative diagram in which the first two rows and the first two columns form distinguished triangles:*

(B.1.1)
$$\begin{array}{ccccccc}
A & \xrightarrow{\phi} & B & \xrightarrow{\psi} & C & \xrightarrow{+1} & \\
{\scriptstyle\alpha}\downarrow & & {\scriptstyle\beta}\downarrow & & & & \\
A' & \xrightarrow{\phi'} & B' & \xrightarrow{\psi'} & C' & \xrightarrow{+1} & \\
{\scriptstyle\alpha'}\downarrow & & {\scriptstyle\beta'}\downarrow & & & & \\
A'' & & B'' & & & & \\
{\scriptstyle +1}\downarrow & & {\scriptstyle +1}\downarrow & & & &
\end{array}$$

Then there exist a morphism $\gamma : C \to C'$ with mapping cone C'', i.e., such that

$$C \longrightarrow C' \longrightarrow C'' \xrightarrow{+1}$$

is a distinguished triangle, and morphisms $A'' \to B''$, $B'' \to C''$, $C'' \to A''[1]$ such that

$$A'' \longrightarrow B'' \longrightarrow C'' \xrightarrow{+1}$$

is a distinguished triangle, and the diagram

(B.1.2)
$$
\begin{array}{ccccccc}
A & \xrightarrow{\phi} & B & \xrightarrow{\psi} & C & \xrightarrow{+1} & \\
\alpha\downarrow & & \beta\downarrow & & \gamma\downarrow & & \\
A' & \xrightarrow{\phi'} & B' & \xrightarrow{\psi'} & C' & \xrightarrow{+1} & \\
\alpha'\downarrow & & \beta'\downarrow & & \gamma'\downarrow & & \\
A'' & \longrightarrow & B'' & \longrightarrow & C'' & \xrightarrow{+1} & \\
+1\downarrow & & +1\downarrow & & +1\downarrow & &
\end{array}
$$

is commutative. Furthermore, if the triangulated category $\mathfrak{T}$ is a derived category and C and C' are such that $h^i(\mathsf{C}) = 0$ for $i \neq 0$ and $h^j(\mathsf{C}') = 0$ for $j < 0$, then γ is uniquely determined by the original diagram (B.1.1).

Proof. The proof consists of repeated applications of the octahedral axiom.

First consider the composition $\mathsf{A} \to \mathsf{B} \to \mathsf{B}'$ and let D be an object that completes this morphism to a distinguished triangle.

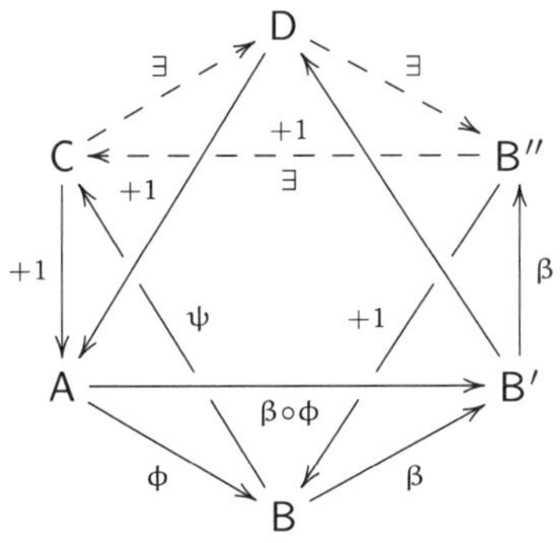

Then by the octahedral axiom there exist morphisms as indicated on the above diagram such that

$$\mathsf{C} \longrightarrow \mathsf{D} \longrightarrow \mathsf{B} \xrightarrow{+1}$$

is a distinguished triangle.

Next, consider the composition $\mathsf{A} \to \mathsf{A}' \to \mathsf{B}'$. Since $\phi' \circ \alpha = \beta \circ \phi$, D is still an object that completes this composition to a distinguished triangle.

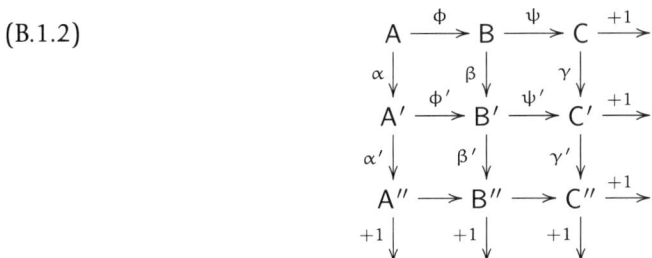

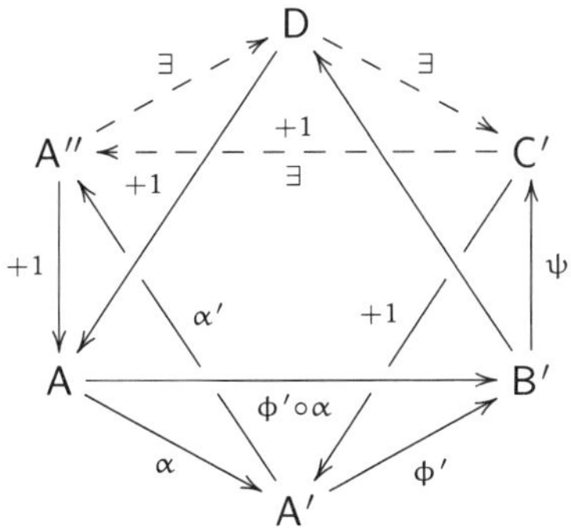

Then by the octahedral axiom there exist morphisms as indicated on the above diagram such that

$$A'' \longrightarrow D \longrightarrow C' \xrightarrow{+1}$$

is a distinguished triangle.

Finally, consider the composition $C \to D \to C'$ using the morphisms obtained by the above two applications of the octahedral axiom. Let

$$\gamma : C \to C'$$

be defined as this composition and C'' its mapping cone.

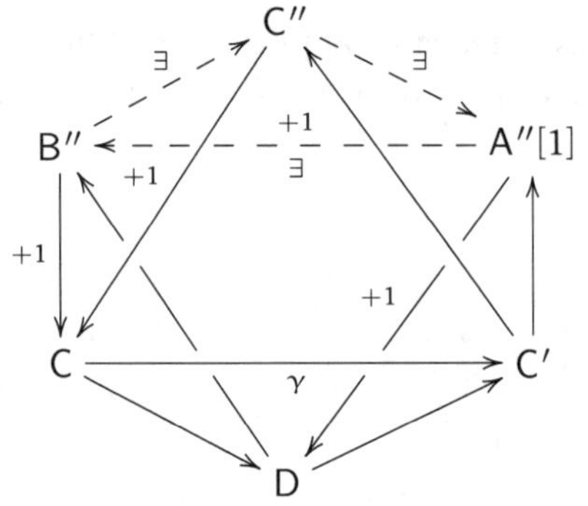

Then by the octahedral axiom there exist morphisms as indicated on the above diagram such that

$$B'' \longrightarrow C'' \longrightarrow A''[1] \xrightarrow{+1} ,$$

and hence

$$A'' \longrightarrow B'' \longrightarrow C'' \xrightarrow{+1}$$

are distinguished triangles. The fact that the diagram (B.1.2) is commutative follows from the construction and the uniqueness of γ in the indicated case follows from Lemma B.2. □

Lemma B.2. [36, 2.2.4] *Let* C, C' *objects in a derived category such that* $h^i(C) = 0$ *for* $i \neq 0$ *and* $h^j(C') = 0$ *for* $j < 0$. *Then any morphism* $\gamma : C \to C'$ *is uniquely determined by* $h^0(\gamma)$.

Proof. By the assumption, the morphism $\gamma : C \to C'$ may be represented by a morphism of complexes $\widetilde{\gamma} : \widetilde{C} \to \widehat{C}$, where $C \simeq \widetilde{C}$ such that $\widetilde{C}^0 = h^0(C)$ and $\widetilde{C}^i = 0$ for all $i \neq 0$, and $C' \simeq \widehat{C}$ such that $h^0(\widehat{C}) \subseteq \widehat{C}^0$. However $\widetilde{\gamma}$ has only one non-zero term, $h^0(\gamma)$. □

References

[1] W. Bruns and J. Herzog. Cohen-Macaulay rings. *Cambridge Studies in Advanced Mathematics*, vol. **39**, Cambridge University Press, Cambridge, 1993. MR 1251956 (95h:13020) ← 168

[2] J. A. Carlson. Polyhedral resolutions of algebraic varieties. *Trans. Amer. Math. Soc.*, **292**(2), 595–612, 1985. MR 808740 (87i:14008) ← 186

[3] B. Conrad. Grothendieck duality and base change. *Lecture Notes in Mathematics*, vol. **1750**, Springer-Verlag, Berlin, 2000. MR 1804902 (2002d:14025) ← 168, 180

[4] P. Du Bois. Complexe de de Rham filtré d'une variété singulière. *Bull. Soc. Math. France*, **109**(1), 41–81, 1981. MR 613848 (82j:14006) ← 181, 186, 187

[5] P. Du Bois and P. Jarraud. Une propriété de commutation au changement de base des images directes supérieures du faisceau structural. *C. R. Acad. Sci. Paris Sér. A*, **279**, 745–747, 1974. MR 0376678 (51 #12853) ← 186, 188

[6] A. H. Durfee. Fifteen characterizations of rational double points and simple critical points. *Enseign. Math.* (2), **25**(1–2), 131–163, 1979. MR 543555 (80m:14003) ← 163

[7] L. Ein. Multiplier ideals, vanishing theorems and applications, in: Algebraic geometry—Santa Cruz 1995, 203–219. *Proc. Sympos. Pure Math.*, vol. **62**, Amer. Math. Soc., Providence, RI, 1997. MR 1492524 (98m:14006) ← 182

[8] R. Elkik. Singularités rationnelles et déformations. *Invent. Math.*, **47**(2), 139–147, 1978. MR 501926 (80c:14004) ← 193

[9] R. Elkik. Rationalité des singularités canoniques. *Invent. Math.*, **64**(1), 1–6, 1981. MR 621766 (83a:14003) ← 184, 195

[10] H. Esnault and E. Viehweg. Logarithmic de Rham complexes and vanishing theorems. *Invent. Math.*, **86**(1), 161–194, 1986. MR 853449 (87j:32088) ← 182

[11] H. Esnault and E. Viehweg. Lectures on vanishing theorems. *DMV Seminar*, vol. **20**, Birkhäuser Verlag, Basel, 1992. MR 1193913 (94a:14017) ← 182

[12] O. Fujino. Theory of non-lc ideal sheaves-basic properties. *Kyoto J. Math.*, **50**(2), 225–245, 2011. ← 191

[13] H. Grauert and O. Riemenschneider. Verschwindungssätze für analytische Kohomologiegruppen auf komplexen Räumen. *Invent. Math.*, **11**, 263–292, 1970. MR 0302938 (46 #2081) ← 182

[14] F. Guillén, V. Navarro Aznar, P. Pascual Gainza, and F. Puerta. Hyperrésolutions cubiques et descente cohomologique, *Lecture Notes in Mathematics*, vol. **1335**, Springer-Verlag, Berlin, 1988. Papers from the Seminar on Hodge-Deligne Theory held in Barcelona, 1982. MR 972983 (90a:14024) ← 181, 186

[15] C. D. Hacon and S. J. Kovács. Classification of higher dimensional algebraic varieties. *Oberwolfach Seminars*, Vol. **41**, Birkhäuser Verlag, Basel, 2010. MR 2675555 ← 160, 172, 188

[16] J. Harris and I. Morrison. Moduli of curves. *Graduate Texts in Mathematics*, vol. **187**, Springer-Verlag, New York, 1998. MR 1631825 (99g:14031) ← 162

[17] R. Hartshorne. Residues and duality. Lecture notes of a seminar on the work of A. Grothendieck, given at Harvard 1963/64. With an appendix by P. Deligne. *Lecture Notes in Mathematics*, No. **20**, Springer-Verlag, Berlin, 1966. MR 0222093 (36 #5145) ← 168, 181

[18] R. Hartshorne. Algebraic geometry. *Graduate Texts in Mathematics*,**52**, Springer-Verlag, New York, 1977. MR 0463157 (57 #3116) ← 161

[19] R. Hartshorne. Stable reflexive sheaves. *Math. Ann.*, **254**(2), 121–176, 1980. MR 597077 (82b:14011) ← 169

[20] R. Hartshorne. Generalized divisors on Gorenstein schemes. *Proceedings of Conference on Algebraic Geometry and Ring Theory in honor of Michael Artin, Part III (Antwerp, 1992)*, Vol. **8**, 287–339, 1994. MR 1291023 (95k:14008) ← 171

[21] B. Hassett and S. J. Kovács. Reflexive pull-backs and base extension. *J. Algebraic Geom.*, **13**(2), 233–247, 2004. MR 2047697 (2005b:14028) ← 180

[22] S. Ishii. Small deformations of normal singularities. *Math. Ann.*, **275** (1), 139–148, 1986. MR 849059 (87i:14003) ← 193

[23] M. Kawakita. Inversion of adjunction on log canonicity. *Invent. Math.*, **167**(1), 129–133, 2007. MR 2264806 (2008a:14025) ← 194, 195

[24] Y. Kawamata, K. Matsuda, and K. Matsuki. Introduction to the minimal model problem, in: Algebraic geometry, Sendai, 1985. *Adv. Stud. Pure Math.*, Vol. **10**, 283–360, North-Holland, Amsterdam, 1987. MR 946243 (89e:14015) ← 165

[25] G. Kempf, F. F. Knudsen, D. Mumford, and B. Saint-Donat. Toroidal embeddings. I. *Lecture Notes in Mathematics*, Vol. **339**, Springer-Verlag, Berlin, 1973. MR 0335518 (49 #299) ← 162, 183

[26] K. Kodaira. On a differential-geometric method in the theory of analytic stacks. *Proc. Nat. Acad. Sci. U.S.A.*, **39**, 1268–1273, 1953. MR 0066693 (16,618b) ← 182

[27] J. Kollár. Toward moduli of singular varieties. *Compositio Math.*, **56**(3), 369–398, 1985. MR 814554 (87e:14009) ← 160

[28] J. Kollár. Vanishing theorems for cohomology groups, in: Algebraic geometry, Bowdoin, 1985 (Brunswick, Maine, 1985). *Proc. Sympos. Pure Math.*, Vol. **46**,233–243, Amer. Math. Soc., Providence, RI, 1987. MR 927959 (89j:32039) ← 182

[29] J. Kollár. Projectivity of complete moduli. *J. Differential Geom.*, **32**(1), 235–268, 1990. MR 1064874 (92e:14008) ← 160

[30] J. Kollár. Shafarevich maps and automorphic forms. *M. B. Porter Lectures*, Princeton University Press, Princeton, NJ, 1995. MR 1341589 (96i:14016) ← 187, 188

[31] J. Kollár. Singularities of pairs, in: Algebraic geometry—Santa Cruz 1995. *Proc. Sympos. Pure Math.*, Vol. **62**, 221–287, Amer. Math. Soc., Providence, RI, 1997. MR 1492525 (99m:14033) ← 182

[32] J. Kollár. Two examples of surfaces with normal crossing singularities, 2007. arXiv:0705.0926v2 [math.AG] ← 175

[33] J. Kollár. Hulls and husks, 2008. arXiv:0805.0576v2 [math.AG] ← 161

[34] J. Kollár. Semi log resolutions, 2008. arXiv:0812.3592v1 [math.AG] ← 177, 179

[35] J. Kollár. Singularities of the minimal model program. To be published by Cambridge University Press, Cambridge, 2013. With the collaboration of Sándor J. Kovács. ← 192, 193

[36] J. Kollár and S. J. Kovács. Log canonical singularities are Du Bois. *J. Amer. Math. Soc.*, **23**(3), 791–813, 2010. doi:10.1090/S0894-0347-10-00663-6 ← 181, 189, 191, 192, 193, 198

[37] J. Kollár and S. Mori. Birational geometry of algebraic varieties. *Cambridge Tracts in Mathematics*, Vol. **134**, Cambridge University Press, Cambridge, 1998. With the collaboration of C. H. Clemens and A. Corti, Translated from the 1998 Japanese original. MR 1658959 (2000b:14018) ← 160, 172, 190, 195

[38] J. Kollár and N. I. Shepherd-Barron. Threefolds and deformations of surface singularities. *Invent. Math.*, **91**(2), 299–338, 1988. MR 922803 (88m:14022) ← 160, 163, 179

[39] J. Kollár et al.. Flips and abundance for algebraic threefolds. *Société Mathématique de France*, Paris, 1992. Papers from the Second Summer Seminar on Algebraic Geometry held at the University of Utah, Salt Lake City, Utah, August 1991, Astérisque No. 211 (1992). MR 1225842 (94f:14013) ← 179, 189

[40] S. J. Kovács. Rational, log canonical, Du Bois singularities: on the conjectures of Kollár and Steenbrink. *Compositio Math.*, **118**(2), 123–133, 1999. MR 1713307 (2001g:14022) ← 187, 190, 191

[41] S. J. Kovács. A characterization of rational singularities, *Duke Math. J.*, **102**(2), 187–191, 2000. MR 1749436 (2002b:14005) ← 189, 190, 191, 192, 194

[42] S. J. Kovács. Rational, log canonical, Du Bois singularities. II. Kodaira vanishing and small deformations. *Compositio Math.*, **121**(3), 297–304, 2000. MR 1761628 (2001m:14028) ← 182, 187

[43] S. J. Kovács. Logarithmic vanishing theorems and Arakelov-Parshin boundedness for singular varieties. *Compositio Math.*, **131**(3), 291–317, 2002. MR 1905025 (2003a:14025) ← 182

[44] S. J. Kovács. Families of varieties of general type: the Shafarevich conjecture and related problems, in: Higher dimensional varieties and rational points (Budapest, 2001), 133–167. *Bolyai Soc. Math. Stud.*, Vol. **12**, Springer, Berlin, 2003. MR 2011746 (2004j:14041) ← 182

[45] S. J. Kovács. Vanishing theorems, boundedness and hyperbolicity over higher-dimensional bases. *Proc. Amer. Math. Soc.*, **131**(11), 3353–3364, 2003 (electronic). MR 1990623 (2004f:14047) ← 182

[46] S. J. Kovács. Young person's guide to moduli of higher dimensional varieties, in: Algebraic geometry—Seattle 2005. Part 2, 711–743. *Proc. Sympos. Pure Math.*, Vol. **80**, Amer. Math. Soc., Providence, RI, 2009. MR 2483953 ← 160, 180

[47] S. J. Kovács. DB pairs and vanishing theorems, *Kyoto J. Math.*, Nagata Memorial Issue, **51**(1), 47–69, 2011. ← 188

[48] S. J. Kovács. The intuitive definition of du bois singularities, to appear in Geometry and Arithmetics (C. Faber, G. Farkas, and R. de Jong, eds.), *EMS Series on Congress Reports*, European Mathematical Society, 2012. ← 186

[49] S. J. Kovács. Irrational centers, *Pure and Applied Mathematics Quarterly*, Special Issue: In memory of Eckart Viehweg, **7**(4), 1495–1516, 2011. ← 183

[50] S. J. Kovács. The splitting principle and singularities, Compact moduli spaces and vector bundles, 195–204, *Contemp. Math.*, Vol. **564**, Amer. Math. Soc., Providence, RI, 2012. ← 192

[51] S. J. Kovács and K. Schwede. Du Bois singularities deform, preprint, 2011. `arXiv:1107.2349v1 [math.AG]` ← 194

[52] S. J. Kovács and K. Schwede. Hodge theory meets the minimal model program: a survey of log canonical and Du Bois singularities, in: Topology of Stratified Spaces, 51–94 *Math. Sci. Res. Inst. Publ.*, vol. **58**, Cambridge Univ. Press, Cambridge, 2011. ← 187

[53] S. J. Kovács, K. Schwede, and K. E. Smith. The canonical sheaf of Du Bois singularities. *Adv. Math.*, **224**(4), 1618–1640, 2010. MR 2646306 ← 179, 191

[54] Zs. Patakfalvi. Base change for the relative canonical sheaf in families of normal varieties, 2010. `arXiv:1005.5207v1 [math.AG]` ← 181, 193

[55] C. A. M. Peters and J. H. M. Steenbrink. Mixed Hodge structures, in: Ergebnisse der Mathematik und ihrer Grenzgebiete. 3. Folge. *A Series of Modern Surveys in Mathematics [Results in Mathematics and Related Areas. 3rd Series. A Series of Modern Surveys in Mathematics]*, Vol. **52**, Springer-Verlag, Berlin, 2008. MR 2393625 ← 187

[56] M. Reid. Young person's guide to canonical singularities, in: Algebraic geometry, Bowdoin, 1985 (Brunswick, Maine, 1985), 345–414. *Proc. Sympos. Pure Math.*, Vol. **46**, Amer. Math. Soc., Providence, RI, 1987. MR 927963 (89b:14016) ← 172

[57] K. Schwede. A simple characterization of Du Bois singularities, *Compos. Math.*, **143**(4), 813–828, 2007. MR 2339829 (2008k:14034) ← 189, 194

[58] K. E. Smith. Vanishing, singularities and effective bounds via prime characteristic local algebra, in: Algebraic geometry—Santa Cruz 1995, 289–325. *Proc. Sympos. Pure Math.*, Vol. **62**, Amer. Math. Soc., Providence, RI, 1997. MR 1492526 (99a:14026) ← 182

[59] J. H. M. Steenbrink. Mixed Hodge structures associated with isolated singularities, in: Singularities, Part 2 (Arcata, Calif., 1981), 513–536. *Proc. Sympos.*

Pure Math., Vol. **40**, Amer. Math. Soc., Providence, RI, 1983. MR 713277 (85d:32044) ← 185

[60] J. H. M. Steenbrink. Vanishing theorems on singular spaces. Astérisque (1985), no. 130, 330–341, Differential systems and singularities (Luminy, 1983). MR 804061 (87j:14026) ← 187

[61] J. H. M. Steenbrink. Mixed Hodge structures associated with isolated singularities, in :Singularities, Part 2 (Arcata, Calif., 1981), 513–536. *Proc. Sympos. Pure Math.*, Vol. **40**, Amer. Math. Soc., Providence, RI, 1983. MR 713277 (85d:32044) ← 187, 193

[62] E. Viehweg. Quasi-projective moduli for polarized manifolds. *Ergebnisse der Mathematik und ihrer Grenzgebiete (3) [Results in Mathematics and Related Areas (3)]*, Vol. **30**, Springer-Verlag, Berlin, 1995. MR 1368632 (97j:14001) ← 160, 163

University of Washington, Department of Mathematics, 354350, Seattle, WA 98195-4350, USA
E-mail address: skovacs@uw.edu
URL: http://www.math.washington.edu/~kovacs

Soliton equations and the Riemann-Schottky problem

I. Krichever and T. Shiota

Contents

1. Introduction

Novikov's conjecture on the Riemann-Schottky problem: *the Jacobians of smooth algebraic curves are precisely those indecomposable principally polarized abelian varieties (ppavs) whose theta-functions provide solutions to the Kadomtsev-Petviashvili (KP) equation,* was the first evidence of nowadays well-established fact: connections between the algebraic geometry and the modern theory of integrable systems is beneficial for both sides.

The purpose of this paper is twofold. Our first goal is to present a proof of the strongest known characterization of a Jacobian variety in this direction: *an indecomposable ppav* X *is the Jacobian of a curve if and only if its Kummer variety* K(X) *has a trisecant line* [36, 37]. We call this characterization *Welters' (trisecant) conjecture* after the work of Welters [58]. It was motivated by Novikov's conjecture and Gunning's celebrated theorem [26]. The approach to its solution, proposed in [36], is general enough to be applicable to a variety of Riemann-Schottky-type problems. In [25, 35] it was used for a characterization of principally polarized Prym varieties. The latter problem is almost as old and famous as the Riemann-Schottky problem but is

Research is supported in part by National Science Foundation under the grant DMS-04-05519 and by The Ministry of Education and Science of the Russian Federation (contract 02.740.11.5194).

much harder. In some sense the Prym varieties may be geometrically the easiest-to-understand ppavs beyond Jacobians, and studying them may be a first step towards understanding the geometry of more general abelian varieties as well.

Our second and primary objective is to take this opportunity to elaborate on motivations underlining the proposed solution of the Riemann-Schottky problem, to introduce a certain circle of ideas and methods, developed in the theory of soliton equations, and to convince the reader that they are algebro-geometric in nature, simple and universal enough to be included in the Handbook of moduli. The results appeared in this article have already been published elsewhere.

Riemann-Schottky problem

Let $\mathbb{H}_g := \{B \in M_g(\mathbb{C}) \mid {}^tB = B,\ \mathrm{Im}(B) > 0\}$ be the Siegel upper half space. For $B \in \mathbb{H}_g$ let $\Lambda := \Lambda_B := \mathbb{Z}^g + B\mathbb{Z}^g$ and $X := X_B := \mathbb{C}^g/\Lambda_B$. Riemann's theta function

$$(1.1)\quad \theta(z) := \theta(z, B) := \sum_{m\in\mathbb{Z}^g} e^{2\pi i(m,z)+\pi i(m,Bm)}, \quad (m, z) = m_1z_1 + \cdots + m_gz_g,$$

is holomorphic and Λ-quasiperiodic in $z \in \mathbb{C}^g$, so $\Theta := \Theta_B := \theta^{-1}(0)$ defines a divisor on X. Moreover, $(X, [\Theta])$ becomes a ppav, where $[\Theta]$ denotes the algebraic equivalence class of Θ. Thus $\mathbb{H}_g/\mathrm{Sp}(2g, \mathbb{Z}) \simeq \mathcal{A}_g$, the moduli space of g-dimensional ppavs. In what follows we may denote $(X, [\Theta])$ by X for simplicity. A ppav $(X, [\Theta]) \in \mathcal{A}_g$ is said to be *indecomposable* if Θ is irreducible, or equivalently[1] if there *do not* exist $(X_i, [\Theta_i]) \in \mathcal{A}_{g_i}$ with $g_i > 0$, $i = 1, 2$, such that $X = X_1 \times X_2$ and $\Theta = \Theta_1 \times X_2 + X_1 \times \Theta_2$.

Let $\mathcal{M}_g$ be the moduli space of nonsingular curves of genus g, and let $J\colon \mathcal{M}_g \to \mathcal{A}_g$ be the Jacobi map, i.e., for $\Gamma \in \mathcal{M}_g$, $J(\Gamma)$ is $\mathrm{Pic}^0(\Gamma)$ with canonical polarization given by $W_{g-1} = \{\mathcal{L} \in \mathrm{Pic}^{g-1}(\Gamma) \mid h^0(\mathcal{L}) = h^1(\mathcal{L}) > 0\}$ regarded as a divisor on $\mathrm{Pic}^0(\Gamma)$, or more explicitly: taking a symplectic basis a_i, b_i $(i = 1, \ldots, g)$ of $H_1(\Gamma, \mathbb{Z})$ and a basis $\omega_1, \ldots, \omega_g$ of the space of holomorphic 1-forms on Γ such that $\int_{a_i} \omega_j = \delta_{ij}$, we define the *period matrix* and the *Jacobian variety* of Γ by

$$B := \left(\int_{b_i} \omega_j\right) \in \mathbb{H}_g \quad \text{and} \quad J(\Gamma) := (X_B, [\Theta_B]) \in \mathcal{A}_g,$$

respectively. The latter is independent of the choice of (a_i, b_i).

$J(\Gamma)$ is indecomposable and the Jacobi map J is injective (Torelli's theorem). The *(Riemann-)Schottky problem* is the problem of characterizing the Jacobi locus $\mathcal{J}_g := J(\mathcal{M}_g)$ or its closure $\overline{\mathcal{J}_g}$ in $\mathcal{A}_g$. For $g = 2, 3$ the dimensions of $\mathcal{M}_g$ and $\mathcal{A}_g$ coincide, and hence $\overline{\mathcal{J}_g} = \mathcal{A}_g$ by Torelli's theorem. Since $\mathcal{J}_4$ is of codimension 1 in $\mathcal{A}_4$, the case $g = 4$ is the first nontrivial case of the Riemann-Schottky problem.

A nontrivial relation for the Thetanullwerte of a curve of genus 4 was obtained by F. Schottky [50] in 1888, giving a modular form which vanishes on $\mathcal{J}_4$, and hence

[1] since principal polarization means parallel translation is the only way to deform Θ, translating each component of Θ has the same effect as translating Θ as a whole.

at least a *local* solution of the Riemann-Schottky problem in $g = 4$, i.e., $\overline{\mathcal{J}_4}$ is *an irreducible component of* the zero locus $\mathcal{S}_4$ of the Schottky relation. The irreducibility of $\mathcal{S}_4$ was proved by Igusa [27] in 1981, establishing $\overline{\mathcal{J}_4} = \mathcal{S}_4$, an effective answer to the Riemann-Schottky problem in genus 4.

Generalization of the Schottky relation to a curve of higher genus, the so-called Schottky-Jung relations, formulated as a conjecture by Schottky and Jung [51], were proved by Farkas-Rauch [20]. Later, van Geemen [23] proved that the Schottky-Jung relations give a local solution of the Riemann-Schottky problem. They do not give a global solution when $g > 4$, since the variety they define has extra components already for $g = 5$ (Donagi [18]).

More recent development on the Riemann-Schottky problem, as reviewed in [1, 6, 13], includes a completely new approach of Buser and Sarnak [9] which provides an effective way to characterize *non*-Jacobians.

Fay's trisecant formula and the KP equation

Over more than 120 year-long history of the Riemann-Schottky problem, quite a few geometric characterizations of the Jacobians have been obtained. Following Mumford's review with a remark on Fay's trisecant formula [49], and the advent of soliton theory and Novikov's conjecture [29, 30, 48], much progress was made in the 1980s to characterizing Jacobians and Pryms using Fay-like formulas and KP-like equations. They are closely related to each other since Fay's formula, written as a bilinear equation for the Riemann theta function, follows from a difference analogue of the *bilinear identity*[2]

$$\oint_{k=\infty} \tau(t - [k^{-1}])\tau(t' + [k^{-1}])e^{\sum (t_i - t_i')k^i} dk = 0, \tag{1.2}$$

which itself is equivalent to the KP hierarchy [10, 11]. Equation (1.2) can also be regarded as a generating function for the Plücker relations for an infinite dimensional Grassmannian.

Compared with Igusa's work which studies the geometry of $\mathcal{S}_4$ and characterize the Jacobian *locus* $\mathcal{J}_4$, in this approach Fay-like formulas or KP-like equations are used to (in a sense) construct the curve Γ and thus characterize the Jacobian *varieties*. Therefore this approach to the Riemann-Schottky problem is also related to the Torelli theorem; however, the relation is only remote since the conditions like Fay's formula and the KP equation contain extra parameters like vector U (and the lack of

---

[2] Here $t = (t_1, t_2, \dots)$ and $t' = (t_1', t_2', \dots)$ are two sequences of formal independent variables near zero, k is a formal independent variable near infinity, $[k^{-1}] = (1/k, 1/(2k^2), \dots, 1/(nk^n), \dots)$, and τ, the so-called tau-function, is a scalar-valued unknown function of the KP hierarchy. For a quasiperiodic solution obtained from smooth curve Γ we have $\tau(t) = e^{Q(t)}\theta(\sum t_i U_i + z, B(\Gamma))$ for some quadratic form $Q(t)$, vectors $U_i \in \mathbb{C}^g$ and arbitrary $z \in \mathbb{C}^g$. Also, Fay's formula itself can in a sense be obtained from (1.2) by specializing the time variables using the so-called Miwa variables.

Prym-Torelli does not stop us from studying the Prym-Schottky problem using the analogue of this approach).

Let us first describe the trisecant formula in geometric terms. The Kummer variety $K(X)$ of $X \in \mathcal{A}_g$ is the image of the Kummer map

$$K = K_X \colon X \ni z \longmapsto \big(\Theta[\varepsilon,0](z) \mid \varepsilon \in ((1/2)\mathbb{Z}/\mathbb{Z})^g\big) \in \mathbb{CP}^{2^g-1}, \tag{1.3}$$

where $\Theta[\varepsilon,0](z) = \theta[\varepsilon,0](2z,2B)$ are the level two theta-functions with half-integer characteristics $\varepsilon \in ((1/2)\mathbb{Z}/\mathbb{Z})^g$, i.e., they equal $\theta(2(z+B\varepsilon),2B)$ up to some exponential factor so that we have

$$\theta(z+w)\theta(z-w) = \sum_{\varepsilon\in((1/2)\mathbb{Z}/\mathbb{Z})^g} \Theta[\varepsilon,0](z)\Theta[\varepsilon,0](w). \tag{1.4}$$

We have $K(-z) = K(z)$ and $K(X) \simeq X/\{\pm 1\}$.

A *trisecant* of the Kummer variety is a projective line which meets $K(X)$ at three points. *Fay's trisecant formula* states that if $X = J(\Gamma)$, then $K(X)$ has a family of trisecants parametrized by 4 points A_i, $1 \leqslant i \leqslant 4$, on Γ. Namely, identifying a point on Γ with its image under the Abel-Jacobi map $\Gamma \to \mathrm{Pic}^1(\Gamma)$ and taking $r \in \mathrm{Pic}^{-1}(\Gamma)$ such that $2r = A_4 - A_1 - A_2 - A_3$, we have:

$$K(r+A_1),\quad K(r+A_2) \quad \text{and} \quad K(r+A_3) \quad \text{are collinear,} \tag{1.5}$$

i.e.,

$$K\left(\frac{A_4+A_1-A_2-A_3}{2}\right),\quad K\left(\frac{A_4-A_1+A_2-A_3}{2}\right) \quad \text{and}$$
$$K\left(\frac{A_4-A_1-A_2+A_3}{2}\right) \quad \text{are collinear}$$

if we take the three occurrences of "division by 2" consistent with each other. In what follows, the same remark applies if division by 2 in X appears more than once in one formula, as in Theorems 1.25, 7.1.

Since we have $K(-z) = K(z)$, condition (1.5) is symmetric in all the A_i's. However, in its proof as well as its applications the four points tend to play different roles. E.g., fixing the 3 points A_1, A_2, A_3 we may regard it as a one-parameter family of trisecants parametrized by A_4 or r. Now drop the assumptions that $X = J(\Gamma)$ and $A_i \in \Gamma \subset X$: suppose X is a ppav such that (1.5) holds for some $A_1, A_2, A_3 \in X$ and infinitely many (hence a one-parameter family of) $r \in X$. Gunning proved in [26] that, under certain nondegeneracy conditions, X is then a Jacobian.

Gunning's work was extended by Welters who proved that a Jacobian variety can be characterized by the existence of a formal one-parameter family of flexes of the Kummer variety [57]. A flex of the Kummer variety is a projective line which is tangent to $K(X)$ at some point up to order 2. It is a limiting case of trisecants when the three intersection points come together.

In [2] Arbarello and De Concini showed that the assumption in Welters' characterization is equivalent to a singly infinite sequence of partial differential equations contained in the KP hierarchy, and proved that only a first finite number of equations in the sequence are sufficient, by giving an explicit bound for the number of equations, $N = [(3/2)^g g!]$, based on the degree of $K(X)$.

Novikov's conjecture

The second author's answer to Novikov's conjecture [54] illustrated how the soliton theory itself can provide natural, useful algebraic tools as well as powerful analytic tools to study the Riemann-Schottky problem, as immediately noticed by van der Geer [24], when only an early version of [54] was available:

An algebraic argument based on earlier results of Burchnall, Chaundy and the first author [8, 29, 30] characterizes the Jacobians using a commutative ring R of ordinary differential operators associated to a solution of the KP hierarchy. A simple counting argument then shows that only the first $2g+1$ time evolutions in the hierarchy are needed to obtain R. Indeed, suppose $X = \mathbb{C}^g/\Lambda$ appears as an orbit of the first $2g+1$ KP flows represented by a "linear motion" $\phi\colon \mathbb{C}^{2g+1} \to \mathbb{C}^g$ followed by the projection $\mathbb{C}^g \to X$. Then $K := \ker\phi$ is $(g+1)$-dimensional, and if $(c_i) \in K$ then $\sum_i c_i \partial\mathcal{L}/\partial t_i = 0$, hence by the definition of the KP hierarchy $Q = \sum_i c_i P_i$ commutes with $\mathcal{L}$. Any two such Q's commute with each other [52], so the $\mathbb{C}$-algebra R' generated by all such Q's is commutative. A simple counting shows that R contains an ordinary differential operator of every order $n \geqslant 2g+2$, which implies that R' is maximally commutative and hence $R' = R$, from the way of constructing it. Applying Burchnall et al's theory to R to recover the spectral curve Γ etc., we observe that $X \simeq J(\Gamma)$. The $2g+1$ KP flows yield a finite number of differential equations for the Riemann theta function θ of X, to characterize a Jacobian. As for the number of equations, an easy estimate shows that $4g^2$ is enough, although more careful argument should yield a better bound. Note that this is much smaller than Arbarello et al's estimate.

The analytic tools comes into play when one studies Novikov's conjecture, that just the first equation ($N = 1!$) of the hierarchy, i.e., the KP equation (1.12), suffices to characterize the Jacobians: in [54] various tools obtained from analytic considerations on the KP equation and family of its solutions were combined with the algebraic arguments explained above to prove the conjecture. Even Arbarello and De Concini's geometric re-proof of Novikov's conjecture [3] used the hardest analytic ingredient of [54] as it is, since it had no geometric alternative until Marini's work [45] in 1998. Analytic tools are also essential in the proofs of Welters' conjecture and its Prym analogue presented in this paper, as condition (C) in each of Theorems 1.6, 1.19, 1.25, 7.1. Note that (1.10), from which condition (C) in Theorem 1.6 follow, comes from a generalization of Calogero-Moser system.

Novikov's conjecture does not give an effective solution of the Riemann-Schottky problem by itself: since it states that X is a Jacobian if and only if

$$u = -2(\partial_x^2 \ln \theta(Ux + Vy + Wt + Z) + c)$$

satisfies (1.12) for *some* U, V, W and c, obtaining an effective solution from it involves elimination of those constants from (1.12); this is hard to do explicitly.

Welters' conjecture

Novikov's conjecture is equivalent to the statement that the Jacobians are characterized by the existence of length 3 formal jet of flexes. In [58] Welters formulated the question: *if the Kummer variety* $K(X)$ *has one trisecant, does it follow that* X *is a Jacobian ?* In fact, there are three particular cases of the Welters conjecture, corresponding to three possible configurations of the intersection points (a, b, c) of $K(X)$ and the trisecant:

(i) all three points coincide $(a = b = c)$;
(ii) two of them coincide $(a = b \neq c)$;
(iii) all three intersection points are distinct $(a \neq b \neq c \neq a)$.

Of course the first two cases can be regarded as degenerations of the general case (iii). However, when the presence of only one trisecant is assumed, all three cases are independent and require separate treatment. The proof of case (i) of Welters' conjecture was obtained by the first author in [36]:

Theorem 1.6. *An indecomposable principally polarized abelian variety* (X, θ) *is the Jacobian variety of a smooth algebraic curve of genus g if and only if there exist* g*-dimensional vectors* $U \neq 0, V, A$ *, and constants* p *and* E *such that one of the following three equivalent conditions are satisfied:*

(A) *the equality*

$$(\partial_y - \partial_x^2 + u)\, \psi = 0, \tag{1.7}$$

where

$$u = -2\partial_x^2 \ln \theta(Ux + Vy + Z), \quad \psi = \frac{\theta(A + Ux + Vy + Z)}{\theta(Ux + Vy + Z)}\, e^{p\,x + E\,y}, \tag{1.8}$$

holds, for an arbitrary vector Z*;*

(B) *for all theta characteristics* $\varepsilon \in (\frac{1}{2}\mathbb{Z}/\mathbb{Z})^g$

$$\left(\partial_V - \partial_U^2 - 2p\, \partial_U + (E - p^2)\right)\, \Theta[\varepsilon, 0](A/2) = 0$$

(here and below ∂_U, ∂_V *are the derivatives along the vectors* U *and* V*, respectively);*

(C) *on the theta-divisor* $\Theta = \{Z \in X \mid \theta(Z) = 0\}$

$$\begin{aligned}[(\partial_V\theta)^2 - (\partial_U^2\theta)^2]\partial_U^2\theta + 2[\partial_U^2\theta\partial_U^3\theta - \partial_V\theta\partial_U\partial_V\theta]\partial_U\theta \\ + [\partial_V^2\theta - \partial_U^4\theta](\partial_U\theta)^2 = 0 \pmod{\theta}\end{aligned} \tag{1.9}$$

The direct substitution of the expression (1.8) in equation (1.7) and the use of the addition formula for the Riemann theta-functions shows the equivalence of conditions (A) and (B) in the theorem. Condition (B) means that the image of the point $A/2$ under the Kummer map is an inflection point (case (i) of Welters' conjecture).

Condition (C) is the relation that is *really used* in the proof of the theorem. Formally it is weaker than the other two conditions because its derivation does not use an explicit form (1.8) of the solution ψ of equation (1.7), but requires only an existence of a meromorphic solution: consider a holomorphic function $\tau(x, y)$ of a complex variable x depending smoothly on a parameter y, and assume that in a neighborhood of a simple zero $\eta(y)$ of function τ (that is, $\tau(\eta(y), y) = 0$ and $\partial_x \tau(\eta(y), y) \neq 0$) equation (1.7) with potential $u = -2\partial_x^2 \ln \tau$ has a meromorphic solution ψ. Then the equation

$$\ddot{\eta} = 2w \tag{1.10}$$

holds, where the "dots" denote derivatives in y, and w is the third coefficient of the Laurent expansion of the function u at the point η, i.e.,

$$u(x, y) = \frac{2}{(x - \eta(y))^2} + v(y) + w(y)(x - \eta(y)) + \cdots .$$

Equations (1.10) was first derived in [4] where the assertion of the theorem was proved under the assumption[3] that the closure of the group in X generated by A coincides with X. Expanding the function θ in a neighborhood of a point $z \in \Theta := \{z \mid \theta(z) = 0\}$ such that $\partial_U \theta(z) \neq 0$, and noting that the latter condition holds on a dense subset of Θ since B is indecomposable, it is easy to see that equation (1.10) is equivalent to (1.9).

Equation (1.7) is one of the two auxiliary linear problems for the KP equation. Namely, the compatibility condition of (1.7) and the second auxiliary linear equation

$$\left(\partial_t - \partial_x^3 + \frac{3}{2} u \partial_x + w\right) \psi = 0 \tag{1.11}$$

is equivalent to the KP equation [19, 59]:

$$\frac{3}{4} u_{yy} = \frac{\partial}{\partial x}\left(u_t - \frac{1}{4} u_{xxx} - \frac{3}{2} u u_x\right). \tag{1.12}$$

For the first author, the motivation to consider not the whole KP equation but just one of its auxiliary linear problem was his earlier work [33] on the elliptic Calogero-Moser (CM) system, where it was observed for the first time that equation (1.7) is all what one needs to construct the elliptic solutions of the KP equation. Moreover, the construction of the Lax representation with a *spectral parameter* and the corresponding spectral curves of the elliptic CM system proposed in [33] can be regarded as an effective solution of the inverse problem: how to reconstruct the algebraic curve from the matrix B if its Kummer variety admits one flex with the vector U (in the

---

[3]under different additional assumptions the corresponding statement was proved in the earlier works [33, 45]

assumption of the Theorem) which *spans* an elliptic curve in the abelian variety X. Briefly, that solution of the reconstruction problem can be presented as follows:

If the vector U *spans an elliptic curve* $E \subset X$, *then the equation*

$$\theta(Ux + Vy + Z) = 0 \tag{1.13}$$

for a generic Z *has* g *simple roots* $x_i(y)$ *depending on* y *(they are just intersection points of the shifted elliptic curve* $E + Vy + Z \subset X$ *with the theta-divisor* $\Theta \subset X$*). These roots define* $g \times g$ *matrix* $L(y, z)$ *with entries given by*

$$L_{ii}(t, z) = \frac{1}{2}\dot{x}_i, \quad L_{ij} = \Phi(x_i - x_j, z), \quad i \neq j, \tag{1.14}$$

where

$$\Phi(x, z) := \frac{\sigma(z - x)}{\sigma(z)\sigma(x)} e^{\zeta(z)x}, \tag{1.15}$$

with ζ *and* σ *the standard Weierstrass functions.*

The spectral curve Γ_{cm} *of the CM system is the normalization at the point* $k = \infty, z = 0$ *of the closure in* $\mathbb{P}^1 \times E$ *of the affine curve given in* $\mathbb{C} \times (E \setminus 0)$ *by the characteristic equation*

$$R(k, z) = \det(kI + L(y, z)) = 0\,. \tag{1.16}$$

Under the assumptions of the theorem, the CM curve Γ_{cm} *does not depend on* y *and is the solution of the inverse problem.*

Without an assumption on U the proof of Theorem 1.6 is much more complex and less effective. The ultimate goal is to construct, under the assumption that the condition (C) is satisfied, a ring of commuting ordinary differential operators, because, as shown in [8], a pair of commuting differential operators L_1, L_2 satisfies an algebraic relation $R(L_1, L_2) = 0$. This is the key moment, when an algebraic curve emerges in the proof. It then remains only to show that the corresponding curve is the solution of the inverse problem.

The first step in the proof is to introduce in the problem a formal spectral parameter. It is analogous to the introduction of the spectral parameter in the Lax matrix for the elliptic CM system. This parameter k appears in the notion of a *formal wave solution* of equation (1.7).

The wave solution of (1.7) is a solution of the form

$$\psi(x, y, k) = e^{kx + (k^2 + b)y}\left(1 + \sum_{s=1}^{\infty} \xi_s(x, y)\, k^{-s}\right). \tag{1.17}$$

The aim is to show that under the assumptions of the theorem there exists a unique, up to multiplication by a constant factor $c(k)$, formal wave solution such that

$$\xi_s = \frac{\tau_s(Ux + Vy + Z, y)}{\theta(Ux + Vy + Z)}, \tag{1.18}$$

where $\tau_s(Z, y)$, is an entire function of Z.

As it was stressed above, strictly speaking the KP equation and the KP hierarchy are not present in the assumptions of the theorem, but the analytical difficulties in the construction of the formal wave solutions of (1.7) can be traced back to those in the second author's proof [54] of Novikov's conjecture.

The main idea of proof in [54] is to show that if $\tau_0 = e^{cx^2/2}\theta(Ux+Vy+Wt+Z)$ satisfies the KP equation in Hirota's form[4]

$$(D_x^4 + 3D_y^2 - 4D_xD_t)\tau_0 \cdot \tau_0 = 0\,,$$

so that $u = -2\partial_x^2\tau_0$ satisfies the KP equation (1.12), then it can be extended to a τ-function of the KP *hierarchy*, as a *global* holomorphic function of the infinite number of variables $t = (t_i) = (t_1, t_2, t_3, \dots)$, with $t_1 = x$, $t_2 = y$, $t_3 = t$. Local existence of τ directly follows from the KP equation. The global existence of the τ-function is crucial. The rest is a corollary of the KP theory and the theory of commuting ordinary differential operators developed by Burchnall-Chaundy [8] and the first author [29, 30].

The core of the problem is that there is a homological obstruction for the global existence of τ. It is controlled by the cohomology group $H^1(\mathbb{C}^g \setminus \Sigma, \mathcal{V})$, where *singular locus* Σ is defined as ∂_U-invariant subset of the theta-divisor Θ and $\mathcal{V}$ is the sheaf of ∂_U-invariant meromorphic functions on $\mathbb{C}^g \setminus \Sigma$ with poles along Θ. The hardest part of [54], as clarified in [3], is the proof that the locus Σ is empty [5].

The coefficients ξ_s of the wave function are defined recurrently by the equation $2\partial_U \xi_{s+1} = \partial_y \xi_s - \partial_U^2 \xi_s + u\xi_s$. It turned out that equation (1.9) in the condition (C) of the theorem are necessary and sufficient for the local existence of meromorphic solutions. The global existence of ξ_s is controlled by the same cohomology group $H^1(\mathbb{C}^g \setminus \Sigma, \mathcal{V})$ as above. Fortunately, in the framework of our approach there is no need to prove directly that the bad locus is empty. The first step is to construct certain wave solutions outside the bad locus. We call them λ-periodic wave solutions. They are defined uniquely up to ∂_U-invariant factor. The next step is to show that for each $Z \notin \Sigma$ the λ-periodic wave solution is a common eigenfunction of a commutative ring $\mathcal{A}^Z$ of ordinary difference operators. The coefficients of these operators are independent of ambiguities in the construction of ψ. For the generic Z the ring $\mathcal{A}^Z$ is maximal and the corresponding spectral curve Γ is Z-independent. The correspondence $j\colon Z \longmapsto \mathcal{A}^Z$ and the results of the works [8, 29, 30, 48], where a theory of rank 1 commutative rings of differential operators was developed, allows us to make the next crucial step and prove the global existence of the wave function. Namely, on $(X \setminus \Sigma)$ the wave function can be globally defined as the preimage $j^*\psi_{BA}$ under j of the Baker-Akhiezer function on Γ and then can be extended on X by usual Hartogs'

---

[4] We define $P(D_x, \dots)f \cdot f := P(\partial_{x'}, \dots)(f(x+x', \dots)f(x-x', \dots))|_{x'=\dots=0}$ for a polynomial or a power series P; a Hirota equation is an equation of the form $P(D_x, \dots)f \cdot f = 0$; see [11, 54].

[5] The first author is grateful to Enrico Arbarello for an explanation of these deep ideas and a crucial role of the singular locus Σ, which helped him to focus on the heart of the problem.

arguments. The global existence of the wave function implies that X contains an orbit of the KP hierarchy, as an abelian subvariety. The orbit is isomorphic to the generalized Jacobian $J(\Gamma) = \mathrm{Pic}^0(\Gamma)$ of the spectral curve ([54]). Therefore, the generalized Jacobian is compact. The compactness of $J(\Gamma)$ implies that the spectral curve is smooth and the correspondence j extends by linearity and defines the isomorphism $j: X \to J(\Gamma)$.

The proof of Welters' conjecture was completed in [37]. First, here is the theorem which treats case (ii) of the conjecture:

Theorem 1.19. *An indecomposable, principally polarized abelian variety* (X, θ) *is the Jacobian of a smooth curve of genus g if and only if there exist non-zero g-dimensional vectors* $U \neq A \pmod{\Lambda}$, V, *such that one of the following equivalent conditions holds:*

(A) *The differential-difference equation*

$$(1.20)\qquad (\partial_t - T + u(x,t))\,\psi(x,t) = 0, \quad T = e^{\partial_x},$$

is satisfied for

$$(1.21)\qquad u = (T-1)v(x,t), \quad v = -\partial_t \ln \theta(xU + tV + Z)$$

and

$$(1.22)\qquad \psi = \frac{\theta(A + xU + tV + Z)}{\theta(xU + tV + Z)}\, e^{xp+tE},$$

where p, E *are constants and* Z *is arbitrary.*

(B) *The equations*

$$\partial_V \Theta[\varepsilon,0]\,((A-U)/2) - e^p \Theta[\varepsilon,0]\,((A+U)/2) + E\Theta[\varepsilon,0]\,((A-U)/2) = 0$$

are satisfied for all $\varepsilon \in (\frac{1}{2}\mathbb{Z}/\mathbb{Z})^g$. *Here and below* ∂_V *is the constant vector field on* $\mathbb{C}^g$ *corresponding to the vector* V.

(C) *The equation*

$$(1.23)\qquad \partial_V\,[\theta(Z+U)\,\theta(Z-U)]\,\partial_V\theta(Z) = [\theta(Z+U)\,\theta(Z-U)]\,\partial^2_{VV}\theta(Z) \pmod{\theta}$$

is valid on the theta-divisor $\Theta = \{Z \in X \mid \theta(Z) = 0\}$.

Equation (1.20) is one of the two auxiliary linear problems for the 2D Toda lattice equation

$$(1.24)\qquad \partial_\xi \partial_\eta \varphi_n = e^{\varphi_{n-1}-\varphi_n} - e^{\varphi_n - \varphi_{n+1}},$$

which can be regarded as a partial discretization of the KP equation. The idea to use it for the characterization of the Jacobians was motivated by [36] and the first author's earlier work with Zabrodin [44], where a connection of the theory of elliptic solutions of the 2D Toda lattice equations and the theory of the elliptic Ruijsenaars-Schneider system was established. In fact, Theorem 1.19 in a slightly different form was proved in [44] under the additional assumption that the vector U *spans an elliptic curve* in X.

The equivalence of (A) and (B) is a direct corollary of the addition formula for the theta-function. The statement (B) is the second particular case of the trisecant conjecture: the line in $\mathbb{CP}^{2^g-1}$ passing through the two points $K((A-U)/2)$ and $K((A+U)/2)$ of the Kummer variety is tangent to $K(X)$ at the point $K((A-U)/2)$.

The affirmative answer to the third particular case, (iii), of Welters' conjecture is given by the following statement.

Theorem 1.25. *An indecomposable, principally polarized abelian variety (X,θ) is the Jacobian of a smooth curve of genus g if and only if there exist non-zero g-dimensional vectors $U \neq V \neq A \neq U \,(\mathrm{mod}\Lambda)$ such that one of the following equivalent conditions holds:*

(A) *The difference equation*

$$\psi(m,n+1)=\psi(m+1,n)+u(m,n)\psi(m,n) \tag{1.26}$$

is satisfied for

$$u(m,n)=\frac{\theta((m+1)U+(n+1)V+Z)\,\theta(mU+nV+Z)}{\theta(mU+(n+1)V+Z)\,\theta((m+1)U+nV+Z)} \tag{1.27}$$

and

$$\psi(m,n)=\frac{\theta(A+mU+nV+Z)}{\theta(mU+nV+Z)}\,e^{mp+nE}, \tag{1.28}$$

where p, E are constants and Z is arbitrary.

(B) *The equations*

$$\Theta[\varepsilon,0]\left(\frac{A-U-V}{2}\right)+e^{p}\Theta[\varepsilon,0]\left(\frac{A+U-V}{2}\right)=e^{E}\Theta[\varepsilon,0]\left(\frac{A+V-U}{2}\right),$$

are satisfied for all $\varepsilon\in(\frac{1}{2}\mathbb{Z}/\mathbb{Z})^g$.

(C) *The equation*

$$\theta(Z+U)\,\theta(Z-V)\,\theta(Z-U+V)+\theta(Z-U)\,\theta(Z+V)\,\theta(Z+U-V)=0 \pmod{\theta} \tag{1.29}$$

is valid on the theta-divisor $\Theta=\{Z\in X \mid \theta(Z)=0\}$.

Under the assumption that the vector U *spans an elliptic curve* in X, Theorem 1.25 was proved in [38], where the connection of the elliptic solutions of BDHE and, the so-called, elliptic nested Bethe Ansatz equations was established.

Equation (1.26) is one of the two auxiliary linear problems for the so-called bilinear discrete Hirota equation (BDHE):

$$\tau_n(l+1,m)\tau_n(l,m+1)-\tau_n(l,m)\tau_n(l+1,m+1)+\tau_{n+1}(l+1,m)\tau_{n-1}(l,m+1)=0\,. \tag{1.30}$$

At the first glance all three nonlinear equation: the KP equation, the 2D Toda equation, and the BDHE equation, look quite unlikely. But in the theory of integrable systems it is well-known that these fundamental soliton equations are in intimate relation, similar to that between all three cases of the trisecant conjecture. Namely, the KP equation is as a continuous limit of the BDHE, and the 2D Toda equation can be obtained in an intermediate step.

The structure of the statements of the last two theorems, and the structure of their proofs look almost literally identical to that in Theorem 1.6. To some extend that is correct: in all cases the first step is to construct the corresponding wave solution. The conditions (C) in all three cases play the same role. They ensure the local existence of the wave function. The key distinction between the differential and the difference cases arises at the next step. As it was mentioned above, in the case of differential equations a cohomological argument [54, Lemma 12] can be applied to glue local solutions into a global one. In the difference case there is no analog of the cohomological argument and we use a different approach. Instead of *proving* the global existence of solutions we, to some extend, *construct* them by defining first their residue on the theta-divisor. It turns out that the residue is regular on Θ outside the *singular locus* Σ. Surprisingly, it turns out that in the fully discrete case the proof of the statement that the singular locus is in fact empty can be obtained at much earlier stage than in the continuous or semi-continuous case. In part, it is due the drastic simplification in the fully discrete case of the corresponding equation on the theta-divisor (compare (1.29) with (1.9)).

Structure of the article

In the next section we introduce the basic concept of the algebro-geometric integration theory of soliton equation, that is the concept of the Baker-Akhiezer function, which is defined by its analytic properties on an algebraic curve with fixed local coordinates at marked points. The uniqueness of the Baker-Akhiezer function implies that it is a solution of certain linear differential equations. The existence of the Baker-Akhiezer function is proved by explicit theta-functional formula, which then leads to explicit theta-functional formulae for the coefficients of the corresponding equations. That proves "the only if" part in all the theorems above.

In section 3, we introduce the KP hierarchy in Sato's form as a system of commuting flows on the space of formal pseudodifferential operators. Because the flows commute, the hierarchy can be reduced to the stationary points of one of the flows (or their linear combination). That is a reduction from a spatially two-dimensional system to a spatially one-dimensional system.[6] Under this reduction,

[6] Here the term "spatially two-dimensional (resp. one-dimensional) system," also known as "sub-subholonomic (resp. subholonomic) system" or "$(2+1)$-d (resp. $(1+1)$-d) system," means the one whose "general solution" depends on functions of two variables (resp. one variable), or equivalently, on doubly infinite (resp. singly infinite) sequences of parameters in the formal power series set-up. (The word "space" is associated to the notion of free parameters because in an initial value problem of a partial differential equation the free parameters for a solution are given by its initial data, which are given on a "space-like" hypersurface.) E.g., since initial data for the KP hierarchy, i.e., $\mathcal{L}|_{t=0}$ for $\mathcal{L}$ in (4.4), are given by a singly infinite sequence of one-variable functions $\{v_s(x)\}_{s=1,2,\ldots}$ or, by expanding each $v_s(x)$ in a power series $v_s(x) = \sum_i v_{si}x^i$, a doubly infinite sequence of parameters $\{v_{si}\}_{s=1,2,\ldots;i=0,1,\ldots}$, the KP

the KP hierarchy defined first on "a space" of infinite number of functions of one variable (the coefficients of a pseudodifferential operator) is equivalent to a system of commuting flows on the space of finite number of functions of one variable. For the case of stationary points of a linear combination of the first n flows of the KP hierarchy these functions are coefficients of a differential operator L_n of order n. One may take one step further and consider stationary points of two commuting flows. It turns out that if the corresponding integers n and m are co-prime, then the corresponding orbits of the whole hierarchy are finite-dimensional and can be identified with certain subspaces of the finite-dimensional linear space of solutions to the system of ordinary differential equations:

$$(1.31) \qquad [L_n, L_m] = 0, \quad L_n = \partial_x^n + \sum_{i=0}^{n-1} u_i(x)\partial_x^i, \quad L_m = \partial_x^n + \sum_{j=0}^{m-1} v_j(x)\partial_x^j.$$

This is a setup explaining the role of commuting operators in the modern theory of integrable systems.

As a purely algebraic problem it was considered and partly solved in the remarkable works of Burchnall and Chaundy [8] in the 1920s. They proved that for any pair of such operators there exists a polynomial in two variables such that $R(L_n, L_m) = 0$. Moreover, they proved that if the orders n and m of these operators are co-prime, $(n, m) = 1$, and the algebraic curve Γ defined in $\mathbb{C}^2$ by equation $R(\lambda, \mu) = 0$ is smooth, then the commuting operators are uniquely defined by the curve and a set of g points on Γ, where g is the genus of Γ. In such a form, the solution of the problem is one of pure classification: one set is equivalent to the other. Even the attempt to obtain exact formulae for the coefficients of commuting operators had not been made. Baker proposed making the programme effective by looking at analytic properties of the eigenfunction ψ. The Baker program was rejected by the authors of [8] consciously (see the postscript of Baker's paper [5]) and all these results were forgotten for a long time.

The theory of commuting differential operators and its extension to the difference case is presented in Section 4. The outline of the proof of the trisecant conjecture is in Section 5. In Section 6 we present a solution of the characterization problem for Prym varieties which was obtained by Grushevsky and the first author ([35, 25]). The last Section 7 is devoted to a theory of abelian solutions of the soliton equation. The notion of such solutions was introduced by the authors in [41, 42], where it was shown that all of them are algebro-geometric. The theory of abelian solutions

hierarchy is a "spatially two-dimensional system." For $2 \leqslant n \in \mathbb{Z}$ the n-reduction of the KP hierarchy (KdV if $n = 2$, Boussinesq if $n = 3$, etc.) is defined by imposing the condition that $\mathcal{L}^n$ is a differential operator. Since, as an ordinary differential operator, $\mathcal{L}^n|_{t=0}$ depends on finite number of one-variable functions and hence on finite number of singly-infinite sequences, it is a "spatially one-dimensional system."

can be regarded as an extension of the results above to the case of non-principally polarized abelian varieties.

2. The Baker-Akhiezer functions – General scheme

Let Γ be a nonsingular algebraic curve of genus g with N marked points P_α and fixed local parameters $k_\alpha^{-1}(Q)$ in neighborhoods of the marked points. The basic scalar *multi-point* and *multi-variable* Baker-Akhiezer function $\psi(t, Q)$ is a function of external parameters

$$t = (t_{\alpha,i}), \quad \alpha = 1, \ldots, N; \ i = 0, \ldots; \ \sum_\alpha t_{\alpha,0} = 0, \tag{2.1}$$

only finite number of which is non-zero, and a point $Q \in \Gamma$. For each set of the external parameters t it is defined by its analytic properties on Γ.

Remark. For the simplicity we will begin with the assumption that the variables $t_{\alpha,0}$ are integers, i.e., $t_{\alpha,0} \in \mathbb{Z}$.

Lemma 2.2. *For any set of g points $\gamma_1, \ldots, \gamma_g$ in a general position there exists a unique (up to constant factor $c(t)$) function $\psi(t, Q)$, such that:*

(i) the function ψ (as a function of the variable $Q \in \Gamma$) is meromorphic everywhere except for the points P_α and has at most simple poles at the points $\gamma_1, \ldots, \gamma_g$ (if all of them are distinct);

(ii) in a neighborhood of the point P_α the function ψ has the form

$$\psi(t, Q) = k_\alpha^{t_{\alpha,0}} \exp\left(\sum_{i=1}^{\infty} t_{\alpha,i} k_\alpha^i\right)\left(\sum_{s=0}^{\infty} \xi_{\alpha,s}(t) k_\alpha^{-s}\right), \tag{2.3}$$

where $k_\alpha = k_\alpha(Q)$ is the reciprocal of a local parameter at P_α, i.e., $k_\alpha^{-1} \in \mathfrak{m}_{P_\alpha} \setminus \mathfrak{m}_{P_\alpha}^2$.

From the uniqueness of the Baker-Akhiezer function it follows that:

Theorem 2.4. *For each pair $(\alpha, n > 0)$ there exists a unique operator $L_{\alpha,n}$ of the form*

$$L_{\alpha,n} = \partial_{\alpha,1}^n + \sum_{j=0}^{n-1} u_j^{(\alpha,n)}(t)\partial_{\alpha,1}^j, \tag{2.5}$$

(where $\partial_{\alpha,n} = \partial/\partial t_{\alpha,n}$) such that

$$(\partial_{\alpha,n} - L_{\alpha,n})\, \psi(t, Q) = 0\,. \tag{2.6}$$

The idea of the proof of the theorems of this type proposed in [29], [30] is universal.

For any formal series of the form (2.3) their exists a unique operator $L_{\alpha,n}$ of the form (2.5) such that

$$(\partial_{\alpha,n} - L_{\alpha,n})\, \psi(t, Q) = O(k_\alpha^{-1}) \exp\left(\sum_{i=1}^{\infty} t_{\alpha,i} k_\alpha^i\right). \tag{2.7}$$

The coefficients of $L_{\alpha,n}$ are universal differential polynomials with respect to $\xi_{s,\alpha}$. They can be found after substitution of the series (2.3) into (2.7).

It turns out that if the series (2.3) is not formal but is an expansion of the Baker-Akhiezer function in the neighborhood of P_α the congruence (2.7) becomes an equality. Indeed, let us consider the function ψ_1

$$\psi_1 = (\partial_{\alpha,n} - L_{\alpha,n})\psi(t,Q). \tag{2.8}$$

It has the same analytic properties as ψ except for the only one. The expansion of this function in the neighborhood of P_α starts from $O(k_\alpha^{-1})$. From the uniqueness of the Baker-Akhiezer function it follows that $\psi_1 = 0$ and the equality (2.6) is proved.

Corollary 2.9. *The operators $L_{\alpha,n}$ satisfy the compatibility conditions*

$$\left[\partial_{\alpha,n} - L_{\alpha,n}, \partial_{\alpha,m} - L_{\alpha,m}\right] = 0. \tag{2.10}$$

Remark. The equations (2.10) are gauge invariant. For any function $c(t)$ operators

$$\widetilde{L}_{\alpha,n} = cL_{\alpha,n}c^{-1} + (\partial_{\alpha,n}c)c^{-1} \tag{2.11}$$

have the same form (2.5) and satisfy the same operator equations (2.10). The gauge transformation (2.11) corresponds to the gauge transformation of the Baker-Akhiezer function

$$\widetilde{\psi}(t,Q) = c(t)\psi(t,Q). \tag{2.12}$$

In addition to differential equations (2.6) the Baker-Akhiezer function satisfies an infinite system of differential-difference equations. Recall that the discrete variables $t_{\alpha,0}$ are subject to the constraint $\sum_\alpha t_{\alpha,0} = 0$. Therefore, only the first $(N-1)$ of them are independent and $t_{N,0} = -\sum_{\alpha=1}^{N-1} t_{\alpha,0}$. Let us denote by T_α, $\alpha = 1,\ldots,N-1$, the operator that shifts the arguments $t_{\alpha,0} \to t_{\alpha,0}+1$ and $t_{N,0} \to t_{N,0}-1$, respectively. For the sake of brevity in the formulation of the next theorem we introduce the operator $T_N = T_1^{-1}$.

Theorem 2.13. *For each pair $(\alpha, n>0)$ there exists a unique operator $\widehat{L}_{\alpha,n}$ of the form*

$$\widehat{L}_{\alpha,n} = T_\alpha^n + \sum_{j=0}^{n-1} v_j^{(\alpha,n)}(t)\, T_\alpha^j, \quad v_0^{(N,n)}(t) = 0, \tag{2.14}$$

such that

$$\left(\partial_{\alpha,n} - \widehat{L}_{\alpha,n}\right)\psi(t,Q) = 0. \tag{2.15}$$

The proof if identical to that in the differential case. The operators $\widehat{L}_{\alpha,n}$ are defined by congruence insuring that the resulting function satisfies all the condition of the Baker-Akhiezer function plus vanishing of one of the leading coefficients. After that the uniqueness of the Baker-Akhiezer function implies that the congruence is in fact the equality.

Corollary 2.16. *The operators $\widehat{L}_{\alpha,n}$ satisfy the compatibility conditions*

$$[\partial_{\alpha,n} - \widehat{L}_{\alpha,n}, \partial_{\alpha,m} - \widehat{L}_{\alpha,m}] = 0 . \tag{2.17}$$

It should be emphasized that the algebro-geometric construction is not a sort of abstract "existence" and "uniqueness" theorems. It provides the explicit formulae for solutions in terms of the Riemann theta-functions. They are the corollary of the explicit formula for the Baker-Akhiezer function:

Theorem 2.18. *The Baker-Akhiezer function is given by the formula*

$$\psi(t,P) = c(t)\exp\left(\sum t_{\alpha,i}\Omega_{\alpha,i}(P)\right)\frac{\theta(A(P)+\sum U_{\alpha,i}t_{\alpha,i}+Z)}{\theta(A(P)+Z)}, \tag{2.19}$$

where the sum is taken over all the indices $(\alpha, i > 0)$ and over the indices $(\alpha,0)$ with $\alpha = 1,\dots,N-1$, and:

a) $\Omega_{\alpha,i}(P)$ is the abelian integral, $\Omega_{\alpha,i}(P) = \int^P d\Omega_{\alpha,i}$, corresponding to the normalized (i.e., $\oint_{a_k} d\Omega_{\alpha,i} = 0$) meromorphic differential on Γ, which for $i > 0$ has the only pole of the form $d\Omega_{\alpha,i} = d\left(k_\alpha^i + O(1)\right)$ at the marked point P_α and for $i = 0$ has simple poles at the marked point P_α and P_N with residues ± 1, respectively;

b) $2\pi i U_{\alpha,j}$ is the vector of b-periods of the differential $d\Omega_{\alpha,j}$, i.e.,

$$U^k_{\alpha,j} = \frac{1}{2\pi i}\oint_{b_k} d\Omega_{\alpha,j};$$

c) $A(P)$ is the Abel transform, i.e., a vector with the coordinates $\int^P d\omega_k$;

d) Z is an arbitrary vector (it corresponds to the divisor of poles of Baker-Akhiezer function).

Notice that from the bilinear Riemann relations it follows that the expansion of the Abel transform near the marked point has the form

$$A(P) = A(P_\alpha) - \sum_{i=1}^{\infty}\frac{1}{i}U_{\alpha,i}k_\alpha^{-i}. \tag{2.20}$$

Example 1. One-point Baker-Akhiezer function. KP hierarchy

In the one-point case the Baker-Akhiezer function has an exponential singularity at a single point P_1 and depends on a single set of variables $t_i = t_{1,i}$. Note that in this case there is no discrete variable, $t_{1,0} \equiv 0$. Let us choose the normalization of the Baker-Akhiezer function with the help of the condition $\xi_{1,0} = 1$, i.e., an expansion of ψ in the neighborhood of P_1 equals

$$\psi(t_1, t_2, \dots, Q) = \exp\left(\sum_{i=1}^{\infty} t_i k^i\right)\left(1 + \sum_{s=1}^{\infty}\xi_s(t)k^{-s}\right). \tag{2.21}$$

Under this normalization (gauge) the corresponding operator L_n has the form

$$L_n = \partial_1^n + \sum_{i=0}^{n-2} u_i^{(n)} \partial_1^i . \tag{2.22}$$

For example, for $n = 2,3$ after redefinition $x = t_1$ we have $L_2 = \partial_x^2 - u$, $L_3 = \partial_x^3 - \frac{3}{2} u \partial_x - w$ with

$$u(x, t_2, \ldots) = 2\partial_x \xi_1(x, t_2, \ldots), \tag{2.23}$$

Therefore, if we define $y = t_2$, $t = t_3$, then $u(x, y, t, t_4, \ldots)$ satisfies the KP equation (1.12).

The normalization of the leading coefficient in (2.21) defines the the function $c(t)$ in (2.19). That gives the following formula for the normalized one-point Baker-Akhiezer function:

$$\psi(t, Q) = \exp\left(\sum t_i \Omega_i(P)\right) \frac{\theta(A(P) + \sum U_i t_i + Z)\, \theta(Z)}{\theta(\sum U_i t_i + Z)\, \theta(A(P) + Z)}, \tag{2.24}$$

(shifting Z if needed we may assumed that $A(P_1) = 0$). In order to get the explicit theta-functional form of the solution of the KP equation it is enough to take the derivative of the first coefficient of the expansion at the marked point of the ratio of theta-functions in the formula (2.24).

Using (2.20) we get the final formula for the algebro-geometric solutions of the KP hierarchy [30]

$$u(t_1, t_2, \ldots) = -2\partial_1^2 \ln \theta \left(\sum_{i=1}^{\infty} U_i t_i + Z \right) + \text{const}. \tag{2.25}$$

Example 2. Two-point Baker-Akhiezer function. 2D **Toda hierarchy**

In the two-point case the Baker-Akhiezer function has exponential singularities at two points P_α, $\alpha = 1,2$, and depends on two sets of continuous variables $t_{\alpha,i>0}$. In addition it depends on one discrete variable $n = t_{1,0} = -t_{2,0}$. Let us choose the normalization of the Baker-Akhiezer function with the help of the condition $\xi_{1,0} = 1$, i.e., in the neighborhood of P_1 the Baker-Akhiezer function has the form:

$$\psi(n, t_{\alpha,i>0}, Q) = k_1^n \exp\left(\sum_{i=1}^{\infty} t_{1,i} k_1^i \right) \left(1 + \sum_{s=1}^{\infty} \xi_{1,s}(n,t) k_1^{-s} \right), \tag{2.26}$$

and in the neighborhood of P_2

$$\psi(n, t_{\alpha,i>0}, Q) = k_2^{-n} \exp\left(\sum_{i=1}^{\infty} t_{2,i} k_2^i \right) \left(\sum_{s=0}^{\infty} \xi_{2,s}(n,t) k_1^{-s} \right). \tag{2.27}$$

According to Theorem 2.4, the function ψ satisfies two sets of differential equations. The compatibility conditions (2.10) within the each set can be regarded as two copies of the KP hierarchies. In addition the two-point Baker-Akhiezer function satisfies differential difference equation (2.14). The first two of them have the form

$$(2.28)\qquad (\partial_{1,1} - T + u)\psi = 0, \quad (\partial_{2,1} - wT^{-1})\psi = 0,$$

where

$$(2.29)\qquad u = (T-1)\xi_{1,1}(n,t), \quad w = e^{\phi_n - \phi_{n-1}}, \quad e^{\phi_n(t)} = \xi_{2,0}(n,t).$$

The compatibility condition of these equations is equivalent to the 2D Toda equation (1.24) with $\xi = t_{1,1}$ and $\eta = t_{2,1}$. The explicit formula for ϕ_n is a direct corollary of the explicit formula for the Baker-Akhiezer function. The normalization of ψ as in (2.26) defines the coefficient c in (2.19)

(2.30)

$$\psi = \exp\left(n\Omega_{1,0} + \sum t_{\alpha,i}\Omega_{\alpha,i}(P)\right) \frac{\theta(A(P) + nU + \sum U_{\alpha,i}t_{\alpha,i} + Z)\,\theta(Z)}{\theta(nU + \sum U_{\alpha,i}t_{\alpha,i} + Z)\,(\theta(A(P) + Z)}.$$

If we denote $x = 0,\ t = t_{1,1}$ and set $t_{1,i>1} = t_{2,i>0} = 0$, then up to a constant in (x,t) factor the formula (2.30) coincides with (1.22). Expanding ψ at P_1 we get the formula for the coefficient u in in the first linear equation (2.29), which coincides with (1.21). That proved "the only if" part of Theorem 1.19.

Example 3. Three-point Baker-Akhiezer function

Starting with three-point case, in which the number of discrete variables is 2, the Baker-Akhiezer function satisfies certain linear difference equations (in addition to the differential and the differential-difference equations (2.6), (2.15)). The origin of these equations is easy to explain. Indeed, if all the continuous variables vanish, $t_{\alpha,i>0} = 0$, then the Baker-Akhiezer function $\psi_{n,m}(P)$, where $n = -t_{1,0}$, $m = -t_{2,0}$, is a meromorphic function having pole of order $n+m$ at P_3 and zeros of order n and m at P_1 and P_2 respectively, i.e.,

$$(2.31)\qquad \psi_{n,m} \in H^0(D + n(P_3 - P_1) + m(P_3 - P_2)), \quad D = \gamma_1 + \cdots + \gamma_g.$$

The functions $\psi_{n+1,m}, \psi_{n,m+1}, \psi_{n,m}$ all lie in $H^0(D + (n+m+1)P_3 - nP_1 - mP_2)$. By Riemann-Roch theorem for a generic D the latter space is 2-dimensional. Hence, these functions are linear dependent, and they can be normalized such the the linear dependence takes the form (1.26). The theta-functional formula for the Baker-Akhiezer function directly implies formulae (1.27), (1.28) and proves "the only if" part of Theorem 1.25.

For the first glance it seems that everything here is within the framework of classical algebraic-geometry. What might be new brought to this subject by the soliton theory is understanding that *the discrete variables* $t_{\alpha,0}$ *can be replaced by continuous ones.* Of course, if in the formula (2.19) the variable $t_{\alpha,0}$ is not an integer, then ψ is not a single valued function on Γ. Nevertheless, because the monodromy properties of ψ do not change if the shift of the argument is integer, it satisfied the same type of linear

equations with coefficients given by the same type of formulae. It is necessary to emphasize that in such a form the difference equation becomes functional equation.
Remark. In the four-point case there is three discrete variables n, m, l. In each two of them the Baker-Akhiezer function satisfies a difference equation. Compatibility of these equations is the BDHE equation (1.30).

3. Dual Baker-Akhiezer function

The concept of the dual Baker-Akhiezer function $\psi^+(t, P)$ is universal and is at the heart of Hirota's bilinear form of soliton equations, and plays an essential role in our proof of Welters' conjecture. It is necessary to emphasize that, although the concept is universal, the definition of the dual Baker-Akhiezer function depends on a choice of *dual* divisor $D^+ = \gamma_1^+ + \cdots + \gamma_g^+$. As it will be shown later the notion of duality between divisors of ψ and ψ^+ reflects a choice of one of the variables $t_{\alpha,0}$ or $t_{\alpha,1}$. In all the cases the pole divisor D^+ of the dual Baker-Akhiezer function is defined by the equation

$$D + D^+ = K + \kappa \in J(\Gamma), \tag{3.1}$$

where K is a canonical class and κ is a certain degree 2 divisor, that encodes the type of duality. Depending on its choice, the dual Baker-Akhiezer function is then defined by the following analytic properties:

i) the function ψ^+ (as a function of the variable $P \in \Gamma$) is meromorphic everywhere except for the points P_α and has at most simple poles at the points $\gamma_1^+, \ldots, \gamma_g^+$ (if all of them are distinct);

(ii) in a neighborhood of the point P_α the function ψ has the form

$$\psi(t, Q) = k^{-t_{\alpha,0}} \exp\left(\sum_{i=1}^{\infty} -t_{\alpha,i} k_\alpha^i\right) \left(\sum_{s=0}^{\infty} \xi_{\alpha,s}^+(t) k_\alpha^{-s}\right), \quad k_\alpha = k_\alpha(Q). \tag{3.2}$$

In fact it is the same Baker-Akhiezer type function and, therefore, admits the same type of explicit theta-function formula:

$$\psi^+(t, P) = c^+(t) \exp\left(-\sum t_{\alpha,i}\Omega_{\alpha,i}(P)\right) \frac{\theta(A(P) - \sum U_{\alpha,i}t_{\alpha,i} - Z + \widehat{\kappa})}{\theta(A(P) - Z + \widehat{\kappa})}. \tag{3.3}$$

The basic type of duality and their meaning are explained below in two examples.

Example 1. One-point case. Duality for a continuous variable

The notion of dual Baker-Akhiezer function in the one point case was first introduced in [10]. In this dual divisor is defined by (3.1) where $\kappa = 2P_1$. In other words, for a generic effective degree g divisor D there exists a unique meromorphic differential $d\Omega$ with pole of degree 2 at P_1, $d\Omega = d(k_1 + O(1))$ having zeros at the points γ_s; in addition it has g more zeros that are denoted by $\gamma_1^+, \ldots, \gamma_g^+$.

The functions ψ and ψ^+s have essential singularities, their product or products of their derivatives are meromorphic functions on Γ. Moreover, from the definition of the duality it follows that after multiplication by corresponding differential $d\Omega$ one gets a meromorphic differential on Γ with the only pole at P_1. That proves the following statement.

Lemma 3.4. *Let ψ and ψ^+ be the Baker-Akhiezer function and its dual. Then the following equations hold:*

$$\text{res}_{P_1}\left(\psi^+(\partial_x^j\psi)\right) d\Omega = 0, \quad j = 0,1,\ldots. \tag{3.5}$$

Equations (3.5) allows to express the coefficients ξ_s^+ of the expansion of the dual function ψ^+ at P_1 as universal differential polynomials in terms of the coefficients ξ_s, of the Baker-Akhiezer function. The first such equation is $\xi_1 + \xi_1^+ = 0$. Another corollary of (3.5) is infinite number of bilinear identities for the theta-function, that one obtains after substitution of (2.19), (3.3) into (3.5). These identities are usually called *Hirota's bilinear equations.*

Corollary 3.6. *Let ψ be the Baker-Akhiezer function and L_i be the linear operator of the form (2.22) such that $(\partial_n - L_n)\psi = 0$. Then the dual Baker-Akhiezer function is a solution of the formal adjoint equation*

$$\psi^+(\partial_n - L_n) = 0\,. \tag{3.7}$$

Recall that the *right action* of a differential operator is defined as a formal adjoint action, i.e., $f^+\partial_i = -\partial_i f^+$ (and the left-hand side of this formula should not be confused with the more common differentiation-followed-by-multiplication construction for a differential operator). The proof of the corollary will be given in the next section.

Example 2. Two-point case. Duality for a discrete variable

In the two-point case, in which there is one discrete variable n, the dual divisor D^+ is defined by (3.1) with $\kappa = P_1 + P_2$, i.e., γ_s and $\gamma_{s'}$ are zeros of a differential $d\Omega$ having simple poles at the marked points P_1 and P_2. Without loss of generality we may assume that at these points it has residues ∓ 1.

Lemma 3.8. *Let ψ and ψ^+ be the Baker-Akhiezer function and its dual. Then the following equations hold:*

$$\text{res}_{P_1}\left(\psi^+(T^i\psi)\right) d\Omega = 0, \quad i = 1, 2, \ldots. \tag{3.9}$$

By definition of the duality, the differential on the left-hand side of (3.9) has pole only at P_1. Hence its residue vanishes. Note also that the differential $\psi^+\psi d\Omega$ has poles at P_1 and P_2. The constant c^+ in the normalization of the dual Baker-Akhiezer function is chosen such that

$$\text{res}_{P_1}\left(\psi^+\psi\right) d\Omega = 1. \tag{3.10}$$

Corollary 3.11. *Let ψ be the Baker-Akhiezer function and let $\widehat{L}_n$ be the linear operator of the form*

$$\widehat{L}_i = T^i + \sum_{j=0}^{n-1} v_j^{(n)} T^j \tag{3.12}$$

such that $(\partial_{1,i} - \widehat{L}_n)\psi = 0$. Then the dual Baker-Akhiezer function is a solution of the formal adjoint equation

$$\psi^+(\partial_{1,i} - \widehat{L}_i) = 0 . \tag{3.13}$$

As in the case of differential operators, here and below the right action of a difference operator is defined as formal adjoint action, i.e., $f^+T = T^{-1}f^+$.

4. Integrable hierarchies

In its original form equations (2.10), (2.17) is just an infinite system of partial differential equation for an infinite number of coefficients of all the operators, depending on infinite number of independent variables called "times". Of course, restricting to a finite number of variables one gets an equation or a finite number of equations for a finite number of variables. Some of them are fundamental equations of mathematical physics, and as such deserve special interest. That is true for all three basic equations mentioned above, that is KP, 2D Toda and BDHE. Our next goal is to present the hierarchies of these equations in the form of commuting flows on a certain "phase spaces" that are spaces of pseudodifferential or pseudo*difference* operators. This form is due to Sato and his coauthors [11].

KP hierarchy

Let $\mathcal{O}$ be a linear space of a formal pseudodifferential operators in the variable x, i.e., formal series

$$\mathcal{D} = \sum_{s=-N}^{\infty} v_s(x)\partial_x^{-s} . \tag{4.1}$$

By definition the coefficient v_1 at ∂_x^{-1} in (4.1) is called the residue of $\mathcal{D}$

$$v_1 := \operatorname{res}_\partial \mathcal{D}. \tag{4.2}$$

The commutator relations $\partial_x^{-1}\cdot v(x) = v(x)\partial_x^{-1} - v_x(x)\partial_x^{-2} + v_{xx}(x)\partial_x^{-2} - \ldots$ define on $\mathcal{O}$ a structure of associative ring. For any pseudodifferential operator $\mathcal{D}$ its differential part is defined as the unique differential operator such that $\mathcal{D} - \mathcal{D}_+ = \mathcal{D}_- = O(\partial_x^{-1})$, i.e., for $\mathcal{D}$ as in (4.1) its differential part is equal to

$$\mathcal{D}_+ = \sum_{s=-N}^{0} v_s(x)\partial_x^{-s} . \tag{4.3}$$

The KP hierarchy is defined on the space $\mathcal{P}$ of monic pseudodifferential operators of order 1, i.e., of the operators of the form

$$(4.4)\qquad \mathcal{L} = \partial_x + \sum_{s=1}^{\infty} v_s(x)\partial_x^{-s}.$$

Proposition 4.5. *The equations*

$$(4.6)\qquad \partial_i \mathcal{L} = [\mathcal{L}_+^i, \mathcal{L}]$$

define commuting flows on the space $\mathcal{P}$.

Proof. The left-hand side of equation (4.6) is a pseudodifferential operator $\partial_i \mathcal{L} = \sum_{s\geqslant 1}(\partial_i v_s)\partial^{-s}$ of order at most -1. Therefore, (4.6) is well-defined if and only if the right-hand side is a pseudodifferential operator of order at most -1. To show this, notice, that the identity $[\mathcal{L}^i, \mathcal{L}] = 0$ implies $[\mathcal{L}_+^i, \mathcal{L}] = -[\mathcal{L}_-^i, \mathcal{L}]$. Be definition $\mathcal{L}_-^i$ is an operator of order at most -1. Hence, $[\mathcal{L}_-^i, \mathcal{L}]$ is also of order at most -1.

For the proof of the second statement of the proposition it is necessary to show that equations (4.6) imply the equation

$$(4.7)\qquad [\partial_i - \mathcal{L}_+^i, \partial_j - \mathcal{L}_+^j] = \partial_i \mathcal{L}_+^j - \partial_j \mathcal{L}_+^i + [\mathcal{L}_+^j, \mathcal{L}_+^i] = 0.$$

The left-hand side of (4.7) is a differential operator. Therefore, in order to show that it vanish, it is enough to show that it is a pseudodifferential operator of order at most -1. From (4.6) it follows that $\partial_i \mathcal{L}^j = [\mathcal{L}_+^i, \mathcal{L}^j]$ Then using the the identity $[\mathcal{L}^i, \mathcal{L}^j] = 0$ we have

$$(4.8)\qquad \partial_i \mathcal{L}_+^j = [\mathcal{L}_+^i, \mathcal{L}^j] - \partial_i \mathcal{L}_-^j = [\mathcal{L}^j, \mathcal{L}_-^i] + O(\partial_x^{-1}) = [\mathcal{L}_+^j, \mathcal{L}_-^i] + O(\partial_x^{-1}).$$

Similarly,

$$(4.9)\qquad [\mathcal{L}_+^i, \mathcal{L}_+^j] = [\mathcal{L}_+^j, \mathcal{L}_-^i] - [\mathcal{L}_+^j, \mathcal{L}_-^i] + O(\partial_x^{-1}).$$

Substituting (4.8), (4.9) into (4.7) completes the proof of the proposition.

The operator $\mathcal{L}_+^2$ has the form $\partial_x^2 - u(x,y)$, with $u = -2v_1$ where v_1 is the coefficient at ∂_x^{-1} of $\mathcal{L}$, i.e., $v_1 = \operatorname{res}_\partial \mathcal{L}$. Equations (4.7) with $j = 2$ have the form

$$(4.10)\qquad \partial_{t_m} u = [\partial_y - \partial_x^2 + u, \mathcal{L}_+^m] = -[\partial_y - \partial_x^2 + u, \mathcal{L}_-^m] = 2\partial_x F_m,$$

where

$$F_m := \operatorname{res}_\partial \mathcal{L}^m.$$

Important remark At first glance the system (4.10) looks like a system of commuting evolution equations, but it is not. The right-hand side of (4.10) are universal differential polynomials in v_i. In general there is no way to reconstruct from one function $u(x,y)$ an infinite set of functions $v_i(x)$ of one variable. It can be done only under ceratin assumptions. In [40] that was done in the case when $u(x,y)$ is a periodic function of the variables x and y. To some extend the main part in the proof of the first case of Welter's conjecture can be seen as the proof of the equivalence of (4.6) and (4.10) in the case when u is as in the statement of Theorem 1.6.

For further use let us present some other basic notations and construction. The first one is the notion of wave function.

Lemma 4.11. *Let $\mathcal{L}$ be a monic pseudodifferential operator of the form (4.4). Then the equation $\mathcal{L}\psi = k\psi$ has a unique solution of the form*

$$\psi = e^{kx}\left(1 + \sum_{s=1}^{\infty} \xi_s(x)k^{-s}\right) \tag{4.12}$$

normalized by the condition $\xi_s(0) = 0$.

The proof is elementary. Substituting (4.12) into the equation gives a system of equations having the form $p_x \xi_s = R_s(v_k, \xi_{s'})$ with $k, s' < s$. Therefore, they uniquely define ξ_s,if the initial conditions are fixed.

The wave function is then define the wave operator

$$\Phi = 1 + \sum_{s=1}^{\infty} \varphi_s(x)\partial_x^{-s} \tag{4.13}$$

by the equation $\psi = \Phi e^{kx}$. Notice, that the last equation implies

$$\mathcal{L} = \Phi \cdot \partial_x \cdot \Phi^{-1}. \tag{4.14}$$

The formal dual wave function is given by the formula

$$\psi^+ = e^{-kx}\left(1 + \sum_{s=1}^{\infty} \xi_s^+(x)k^{-s}\right) := e^{-kx}\Phi^{-1} \tag{4.15}$$

is a solution of the formal adjoint equation $\psi^+\mathcal{L} = k\psi^+$.

The defining property of the dual wave function are equations that we proved for the dual Baker-Akhiezer function in the previous section. Namely,

Lemma 4.16. *Let ψ be a wave function and ψ^+ its dual. Then the equations*

$$\operatorname{res}_k(\psi^+(\partial_x^n\psi))\, dk = 0, \quad n = 0,1,\ldots \tag{4.17}$$

hold.

The proof is a direct corollary of the identity

$$\operatorname{res}_k\left(e^{-kx}\mathcal{D}_1\right)\left(\mathcal{D}_2 e^{kx}\right)\, dk = \operatorname{res}_\partial\left(\mathcal{D}_2\mathcal{D}_1\right), \tag{4.18}$$

which holds for any pair of pseudodifferential operators (for details see [11, 15]).

In the same way one can show that the product of the wave function and its dual is a generating series for the right-hand sides of the hierarchy (4.10).

Lemma 4.19. *The coefficients of the expansion*

$$\psi^+\psi = 1 + \sum_{s=2}^{\infty} J_s k^{-s} \tag{4.20}$$

are given by $J_{n+1} = F_n = \operatorname{res}_\partial \mathcal{L}^n$.

Proof. From the definition of $\mathcal{L}$ it follows that

$$\operatorname{res}_k \left(\psi^+(\mathcal{L}^n\psi)\right) dk = \operatorname{res}_k \left(\psi^+ k^n \psi\right) dk = J_{n+1}. \tag{4.21}$$

On the other hand, using the identity (4.18) we get

$$\operatorname{res}_k(\psi^+\mathcal{L}^n\psi)\, dk = \operatorname{res}_k \left(e^{-kx}\Phi^{-1}\right)\left(\mathcal{L}^n\Phi e^{kx}\right) dk = \operatorname{res}_\partial \mathcal{L}^n = F_n. \tag{4.22}$$

The lemma is proved.

2D Toda hierarchy

In the two-point case there are two sets of continuous variables and one discrete variable which we denote by x. It is instructive enough to consider the hierarchy of equations corresponding to one set of continuous times associated with one marked point. In this subsection we present the definition of the hierarchy of the differential-difference equations (2.17) in the form of the commuting flows on the space $\mathcal{P}$ of the pseudodifference operators of the form

$$\mathcal{L} = T + \sum_{s=0}^{\infty} w_s(x)T^{-s}, \quad T = e^{\partial_x}. \tag{4.23}$$

In the ring of the pseudodifference operators

$$\mathcal{D} = \sum_{s=-N}^{\infty} v_s(x)T^{-s} \tag{4.24}$$

the notion of the residue as follows:

$$\operatorname{res}_T \mathcal{D} := v_0 . \tag{4.25}$$

For any pseudodifferential operator $\mathcal{D}$ its positive part is defined as the difference operator such that $\mathcal{D}_- := \mathcal{D} - \mathcal{D}_+ = O(T^{-1})$, i.e., if $\mathcal{D}$ is as in (4.24), then

$$\mathcal{D}_+ := \sum_{s=-N}^{0} v_s(x)T^{-s} . \tag{4.26}$$

Proposition 4.27. *The equations*

$$\partial_i \mathcal{L} = [\mathcal{L}^i_+, \mathcal{L}] \tag{4.28}$$

define commuting flows on the space $\mathcal{P}$.

The proof of the first statement goes along the same lines as in the case of KP hierarchy. The proof of the second statement that (4.28) implies

$$[\partial_i - \mathcal{L}^i_+, \partial_j - \mathcal{L}^j_+] = 0 \tag{4.29}$$

is also identical. The first operator $\mathcal{L}_+$ is of the form $\mathcal{L}_+ = T - u$ with $u = -w_0$. The equation (4.28) for $i = 1$ gives $\partial_t u = (1-T)w_1$, where $w_1 = \operatorname{res}_T \mathcal{L}T$. Here and below $t = t_1$. For further use, let us present the equation

$$\partial_t F_m = (T-1)F^1_m, \tag{4.30}$$

where

$$F_m = \text{res}_T\, \mathcal{L}^m, \quad F_m^1 = \text{res}_T\, \mathcal{L}^m T,$$

which directly follows from the comparison of residues of two side of the equality $\partial_t \mathcal{L}^m = [\mathcal{L}_+, \mathcal{L}^m]$. The commutativity equations (4.29) imply that the evolution of u with respect to all the other times

$$\partial_{t_m} u = -(T-1)F_m^1 = -\partial_t F_m\,. \tag{4.31}$$

As in the KP case, in general the last equations can not be regarded as well-defined hierarchy on the space of one function $u(x,t)$ because the definition of F_m involves other coefficients of $\mathcal{L}$. The main part of the proof of the second case of Welters' conjecture can be seen as a reconstruction of $\mathcal{L}$ in terms of u under the assumption of Theorem 1.19.

We conclude this section by providing a necessary definitions and identities, which are just discrete analog of that above. Namely, the wave function is a solution of the equation $\mathcal{L}\psi = k\psi$ of the form

$$\psi = k^x \left(1 + \sum_s \xi_s(x) k^{-s}\right). \tag{4.32}$$

It defines a unique wave operator by the equation

$$\psi = \Phi k^x, \quad \Phi = 1 + \sum_{s=1}^{\infty} \varphi_s(x) T^{-s}. \tag{4.33}$$

Then, the dual wave function is defined by the left action of the operator Φ^{-1}: $\psi^+ = k^{-x}\Phi^{-1}$. Recall that the left action of a pseudodifference operator is the formal adjoint action under which the left action of T on a function f is $(fT) = T^{-1}f = e^{-\partial_x} f$, consistently with the left action of ∂_x on a function, $(f\partial_x) = -\partial_x f$.

Lemma 4.34. *The coefficient of the product*

$$\psi^+\psi = 1 + \sum_{s=1}^{\infty} J_s(Z,t)\, k^{-s} \tag{4.35}$$

are equal to $J_n = F_n = \text{res}_T\, \mathcal{L}^n$.

Proof. From the definition of $\mathcal{L}$ it follows that

$$\text{res}_k \left(\psi^+(\mathcal{L}^n\psi)\right) k^{-1} dk = \text{res}_k \left(\psi^+ k^n \psi\right) k^{-1} dk = J_n. \tag{4.36}$$

On the other hand, using the identity

$$\text{res}_k \left(k^{-x}\mathcal{D}_1\right)\left(\mathcal{D}_2 k^x\right) k^{-1} dk = \text{res}_T \left(\mathcal{D}_2\mathcal{D}_1\right), \tag{4.37}$$

which is the 2D Toda analogue of (4.18), we get

$$\text{res}_k(\psi^+\mathcal{L}^n\psi)k^{-1}dk = \text{res}_k \left(k^{-x}\Phi^{-1}\right)\left(\mathcal{L}^n\Phi k^x\right) k^{-1} dk = \text{res}_T\, \mathcal{L}^n = F_n\,. \tag{4.38}$$

Therefore $F_n = J_n$ and the lemma is proved.

5. Commuting differential and difference operators.

In the previous section hierarchies of the KP and 2D Toda equations were defined as systems of commuting flows on the spaces of pseudodifferential or pseudodifference operators, respectively. Consider now the subspace $\mathcal{O}_n \subset \mathcal{O}$ of operators whose n-th power is a differential (difference) operator L_n, i.e., $\mathcal{L}^n = L_n$ or equivalently $\mathcal{L}^n_- = 0$. The latter directly implies that $\partial_{t_n}\mathcal{L} = 0$. In other words the subspace $\mathcal{O}_n$ is the subspace of stationary points of the n-th flow of the hierarchy. It has finite functional dimension and can be simply identified with the space of all monic differential (difference) operators because any such operator L_n uniquely defines the corresponding pseudodifferential $\mathcal{L} = L_n^{1/n}$. The subspace $\mathcal{O}_n$ is invariant with respect to all the other flows. Their restriction on $\mathcal{O}_n$ is a closed system of evolution equations on a space of finite-number of unknown functions and can be represented in the form $\partial_i L_n = [L^{i/n}_{n,+}, L_n]$. For $n = 2$ the corresponding reduction of the KP hierarchy is equivalent to the hierarchy of the KdV equation $4u_t = 6uu_x + u_{xxx}$. An attempt to find explicit periodic solutions of the KdV equation had led Novikov in to the idea to consider further reduction to stationary points of one of the "higher" KdV flows. In terms of the original KP hierarchy that is a subspace stationary for two flows of the hierarchy (or two linear combinations of basic flows). The corresponding subspace is the space of differential order n monic ordinary differential operator L_n such that there exists operator L_m commuting with L_n of order m (not multiple of n), i.e., the space of solutions of a system (1.31). As it was mentioned in the introduction, the problem of classification of commuting ordinary differential operators as pure algebraic problem was consider in remarkable works by Burchnall and Chaundy [8].

Briefly the key points of their proof of the statement that a pair of such operators always satisfies algebraic relation

$$R(L_n, L_m) = 0, \tag{5.1}$$

are the following. The commutativity of L_n and L_m implies that the space $V(\lambda)$ of solutions of the ordinary linear equation $L_n y(x) = \lambda y(x)$ is invariant with respect to the operator L_m. The matrix elements L_m^{ij} of the corresponding finite dimensional linear operator $L_m(\lambda)$

$$L_m|_{V(\lambda)} = L_m(\lambda) \colon V(\lambda) \longmapsto V(\lambda) \tag{5.2}$$

in the canonical basis $c_i(x, \lambda, x_0) \in \mathcal{L}(\lambda)$, $\quad c_i(x, \lambda, x_0)|_{x=x_0} = \delta_{ij}$, are polynomial functions in the variable λ. They depend on the choice of the normalization point $x = x_0$, i.e., $L_m^{ij} = L_m^{ij}(\lambda, x_0)$. The characteristic polynomial

$$R(\lambda, \mu) = \det(\mu - L_m^{ij}(\lambda, x_0)) \tag{5.3}$$

is a polynomial in both variables λ and μ and does not depend on x_0.

According to the property of characteristic polynomials we have

$$R(L_n, L_m)y(x, \lambda) = 0.$$

Notice, that $R(L_n, L_m)$ is an ordinary differential operator. Therefore, if it is not equal to zero then its kernel is finite dimensional. Hence, the last equation valid for all λ implies (5.1), and the first statement of [8] is proved.

The equation $R(\lambda, \mu) = 0$ defines affine part of an algebraic curve. Let us show that it is always compactified by one *smooth* point P_0. Indeed the equation $L_n\psi = k^n\psi$ has always a unique formal wave solution, i.e., a solution of the form (4.12) normalized by the conditions $\xi_s(0)=0$. Moreover, any solution of the latter equation of the form $e^{kx} \cdot$ (Laurent series in k^{-1}) is equal to $\psi(x, k)c(k)$, where $c(k)$ is a constant Laurent series. The operator L_m commutes with L_n, therefore $L_m\psi$ is also a solution to the same equation. Hence, there exists a Laurent series

$$a_m(k) = k^m + \sum_{s=-m+1}^{\infty} a_{m,s}k^{-s} \tag{5.4}$$

such that $\mathcal{L}_m\psi = a(k)\psi(x, k)$, i.e., ψ is a formal common eigenfunction of the operators L_n, L_m. That implies the following expansion of the characteristic equation at infinity $\lambda \to \infty$:

$$R(\lambda, \mu) = \prod_{i=0}^{n-1}(\mu - a(k_i)), \quad k_i^n = \lambda. \tag{5.5}$$

Now we are ready to explain a role of the condition under which Burchnall and Chaundy where able to make the next step. Namely, the condition that orders of operators are co-prime. The leading coefficient of $a(k)$ is k^m. Hence, if $(n, m) = 1$ then in the neighborhood of the infinite (and, therefore, almost everywhere else) the operator $L_n(\lambda)$ has n-distinct eigenvalues, and is diagonalizable, i.e., for each generic point $P = (\lambda, \mu) \in \Gamma$ there is a unique eigenfunction $\psi(x, P; x_0)$ of the operators L_n, L_m normalized by the condition $\psi(x_0, P; x_0) = 1$. It can be written as

$$\psi(x, P; x_0) = \sum_{i=0}^{n-1} h_i(P, x_0)c_i(x, \lambda; x_0), \quad h_0(P, x_0) = 1, \tag{5.6}$$

where c_i are canonical basis of solution to the equation $L_n y = \lambda y$ defined above and h_i are coordinates of the eigenvector of the matrix $\mathcal{L}_m(\lambda)$. They are rational expressions in λ and μ, and, therefore are meromorphic functions of $P \in \Gamma$ (if Γ is smooth, otherwise they become meromorphic on an normalization of Γ). The functions c_i, as solutions of the initial value problem, are entire function of the variable λ. Hence, ψ in an affine part of Γ is a meromorphic function with poles that are independent of x (but depend on the normalization point $x = x_0$). If Γ is smooth than their number is equal to the genus g of Γ. By definition of the canonical basis we have that $\psi_x(x, P)\psi^{-1}(x, P)|_{x=x_0} = h_1(P, x_0)$. The asymptotic of h_1 can be easy found using the formal wave solution. It equals $h_1 = k + (O(k^{-1}))$. Therefore

$\psi = \exp\left(\int_{x_0} h_1(x,P)dx\right)$ has at P_0 exponential singularity and is a Baker-Akhiezer function (with the shift of x by x_0).

Theorem 5.7. [8, 29, 30, 48] *There is a natural correspondence*

$$\mathcal{A} \longleftrightarrow \{\Gamma, P_0, [k^{-1}]_1, \mathcal{F}\} \tag{5.8}$$

between regular at $x = 0$ commutative rings $\mathcal{A}$ of ordinary linear differential operators containing a pair of monic operators of co-prime orders, and sets of algebraic-geometrical data $\{\Gamma, P_0, [k^{-1}]_1, \mathcal{F}\}$, where Γ is an algebraic curve with a fixed first jet $[k^{-1}]_1$ of a local coordinate k^{-1} in the neighborhood of a smooth point $P_0 \in \Gamma$ and $\mathcal{F}$ is a torsion-free rank 1 sheaf on Γ such that

$$H^0(\Gamma, \mathcal{F}) = H^1(\Gamma, \mathcal{F}) = 0. \tag{5.9}$$

The correspondence becomes one-to-one if the rings $\mathcal{A}$ are considered modulo conjugation $\mathcal{A}' = g(x)\mathcal{A}g^{-1}(x)$.

Note that in [29, 30, 8] the main attention was paid to the generic case of the commutative rings corresponding to smooth algebraic curves. The invariant formulation of the correspondence given above is due to Mumford [48].

The algebraic curve Γ is called the spectral curve of $\mathcal{A}$. The ring $\mathcal{A}$ is isomorphic to the ring $A(\Gamma, P_0)$ of meromorphic functions on Γ with the only pole at the point P_0. The isomorphism is defined by the equation

$$L_a\psi_0 = a\psi_0, \quad L_a \in \mathcal{A}, \ a \in A(\Gamma, P_0). \tag{5.10}$$

Here ψ_0 is a common eigenfunction of the commuting operators. At $x = 0$ it is a section of the sheaf $\mathcal{F} \otimes \mathcal{O}(-P_0)$.

Remark. As we have seen above, the construction of the correspondence (5.8) depends on a choice of initial point $x_0 = 0$. The spectral curve and the sheaf $\mathcal{F}$ are defined by the evaluations of the coefficients of generators of $\mathcal{A}$ and a finite number of their derivatives at the initial point. In fact, the spectral curve is independent on the choice of x_0, but the sheaf does depend on it, i.e., $\mathcal{F} = \mathcal{F}_{x_0}$.

Using the shift of the initial point it is easy to show that the correspondence (5.8) extends to the commutative rings of operators whose coefficients are *meromorphic* functions of x at $x = 0$. The rings of operators having poles at $x = 0$ correspond to sheaves for which the condition (5.9) is violated.

Remark. In their original paper Burchnall and Chaundy stressed that there is no approach to a classification of commutative differential operators whose ordered are not co-prime. The classification of commutative rings of ordinary differential operators was completed in [32], where it was shown that a maximal ring $\mathcal{A}$ of commuting differential operators is uniquely defined by an algebraic curve with marked point, the first jet of local coordinate at the marked point, and if the curve is smooth by the rank k and degree rg vector bundle. In addition it depends on $r - 1$

arbitrary functions of one variable. Here k is the rank of $\mathcal{A}$ defined as the greatest common divisor of the orders of commuting operators.

Commuting difference operators

A theory of commuting difference operators containing a pair of operators of co-prime orders was developed in [48, 31]. It is analogous to the theory of rank 1 commuting (Relatively recently this theory was generalized to the case of commuting difference operators of arbitrary rank in [39].) For further use we present here the classification of commutative differential operators of the form

$$L_n = T^n + \sum_{s=1}^{n-1} u_i(x)T^i . \tag{5.11}$$

Theorem 5.12. ([48, 31]) *Let $\mathcal{A}$ be a maximum commutative ring of ordinary difference operators of the form (5.11) containing a pair of operators of co-prime orders. Then there is an irreducible algebraic curve Γ, such that the ring $\mathcal{A}^{Z}$ is isomorphic to the ring $A(\Gamma, P_+, P_-)$ of the meromorphic functions on Γ with the only pole at a smooth point P_+, vanishing at another smooth point P_-. The ring is uniquely defined by a torsion-free rank 1 sheaves $\mathcal{F}$ on Γ such that*

$$h^0(\Gamma, \mathcal{F}(nP_+ - nP_-)) = h^1(\Gamma, \mathcal{F}(nP_+ - nP_-)) = 0. \tag{5.13}$$

The correspondence becomes one-to-one if the rings $\mathcal{A}$ are considered modulo conjugation $\mathcal{A}' = g(x)\mathcal{A}g^{-1}(x)$.

Remark. As in the continuous case the construction of the correspondence depends on a choice of initial point $x_0 = 0$. The spectral curve and the sheaf $\mathcal{F}$ are defined by the evaluations of the coefficients of generators of $\mathcal{A}$ at a finite number of points of the form $x_0 + n$. In fact, the spectral curve is independent on the choice of x_0, but the sheaf does depend on it, i.e., $\mathcal{F} = \mathcal{F}_{x_0}$.

Using the shift of the initial point it is easy to show that the correspondence (5.8) extends to the commutative rings of operators whose coefficients are *meromorphic* functions of x. The rings of operators having poles at $x = 0$ correspond to sheaves for which the condition (5.13) for $n = 0$ is violated.

6. Proof of Welters' conjecture

As it was mentioned in the introduction the proof of all the particular cases of Welters' trisecant conjecture uses different hierarchies: the KP, the 2D Toda, and BDHE. In each case there are some specific difficulties but the main ideas and structures of the proof are the same. In all the cases the first step is to construct the wave solution. It is necessary to emphasize that it is not a wave solution to the ordinary pseudodifferential or pseudodifference operators discussed in Section 4. The corresponding wave solutions are defined as formal solutions to a *partial differential equation*. In this case there is no way to define such a solution in a unique way without additional assumption on a global structure of the coefficients of the equation. As an instructive example we present in this section the proof of the first particular case of Welters' conjecture, namely, the proof of Theorem 1.6.

First, we prove the implication (A) → (C). Let $\tau(x,y)$ be a holomorphic function of the variable x in some open domain $D \in \mathbb{C}$ smoothly depending on a parameter y. Suppose that for each y the zeros of τ are simple,

$$\tau(x_i(y),y)=0, \quad \tau_x(x_i(y),y)\neq 0. \tag{6.1}$$

Lemma 6.2. *([4]) If equation (1.7) with the potential* $u=-2\partial_x^2\ln\tau(x,y)$ *has a meromorphic in* D *solution* $\psi_0(x,y)$*, then equations (1.10) hold.*

Proof. Consider the Laurent expansions of ψ_0 and u in the neighborhood of one of the zeros x_i of τ:

$$\begin{aligned} u &= \frac{2}{(x-x_i)^2}+v_i+w_i(x-x_i)+\cdots; \\ \psi_0 &= \frac{\alpha_i}{x-x_i}+\beta_i+\gamma_i(x-x_i)+\delta_i(x-x_i)^2+\cdots. \end{aligned} \tag{6.3}$$

(All coefficients in these expansions are smooth functions of the variable y). Substitution of (6.3) in (1.7) gives a system of equations. The first three of them are

$$\alpha_i\dot{x}_i+2\beta_i=0;\ \dot{\alpha}_i+\alpha_i v_i+2\gamma_i=0;\ \dot{\beta}_i+v_i\beta_i-\gamma_i\dot{x}_i+\alpha_i w_i=0. \tag{6.4}$$

Taking the y-derivative of the first equation and using two others we get (1.10).

Let us show that equations (1.10) are sufficient for the existence of meromorphic wave solutions, i.e., solutions of the form (1.17).

Lemma 6.5. *Suppose that equations (1.10) for the zeros of* $\tau(x,y)$ *hold. Then there exist meromorphic wave solutions of equation (1.7) that have simple poles at* x_i *and are holomorphic everywhere else.*

Proof. Substitution of (1.17) into (1.7) gives a recurrent system of equations

$$2\xi_{s+1}'=\partial_y\xi_s+u\xi_s-\xi_s''. \tag{6.6}$$

We are going to prove by induction that this system has meromorphic solutions with simple poles at all the zeros x_i of τ.

Let us expand ξ_s at $x=x_i$:

$$\xi_s=\frac{r_s}{x-x_i}+r_{s0}+r_{s1}(x-x_i), \tag{6.7}$$

where for brevity we omit the index i in the notations for the coefficients of this expansion. Suppose that ξ_s are defined and equation (6.6) has a meromorphic solution. Then the right-hand side of (6.6) has the zero residue at $x=x_i$, i.e.,

$$\operatorname{res}_{x_i}(\partial_y\xi_s+u\xi_s-\xi_s'')=\dot{r}_s+v_i r_s+2r_{s1}=0. \tag{6.8}$$

We need to show that the residue of the next equation vanishes also. From (6.6) it follows that the coefficients of the Laurent expansion for ξ_{s+1} are equal to

$$r_{s+1} = -\dot{x}_i r_s - 2r_{s0}, \tag{6.9}$$

$$2r_{s+1,1} = \dot{r}_{s0} - r_{s1} + w_i r_s + v_i r_{s0}. \tag{6.10}$$

These equations imply

$$\dot{r}_{s+1} + v_i r_{s+1} + 2r_{s+1,1} = -r_s(\ddot{x}_i - 2w_i) - \dot{x}_i(\dot{r}_s - v_i r_s s + 2r_{s1}) = 0, \tag{6.11}$$

and the lemma is proved.

λ-periodic wave solutions

Our next goal is to fix a *translation-invariant* normalization of ξ_s which defines wave functions uniquely up to a x-independent factor. It is instructive to consider first the case of the periodic potentials $u(x+1,y) = u(x,y)$ (see details in [40]).

Equations (6.6) are solved recursively by the formulae

$$\xi_{s+1}(x,y) = c_{s+1}(y) + \xi^0_{s+1}(x,y), \tag{6.12}$$

$$\xi^0_{s+1}(x,y) = \frac{1}{2}\int_{x_0}^{x} (\partial_y \xi_s - \xi_s'' + u\xi_s)\, dx, \tag{6.13}$$

where $c_s(y)$ are *arbitrary* functions of the variable y. Let us show that the periodicity condition $\xi_s(x+1,y) = \xi_s(x,y)$ defines the functions $c_s(y)$ uniquely up to an additive constant. Assume that ξ_{s-1} is known and satisfies the condition that the corresponding function ξ^0_s is periodic. The choice of the function $c_s(y)$ does not affect the periodicity property of ξ_s, but it does affect the periodicity in x of the function $\xi^0_{s+1}(x,y)$. In order to make $\xi^0_{s+1}(x,y)$ periodic, the function $c_s(y)$ should satisfy the linear differential equation

$$\partial_y c_s(y) + B(y)\, c_s(y) + \int_{x_0}^{x_0+1} \left(\partial_y \xi^0_s(x,y) + u(x,y)\, \xi^0_s(x,y)\right) dx, \tag{6.14}$$

where $B(y) = \int_{x_0}^{x_0+1} u\, dx$. This defines c_s uniquely up to a constant.

In the general case, when u is quasi-periodic, the normalization of the wave functions is defined along the same lines.

Let $Y_U = \langle \mathbb{C}U \rangle$ be the Zariski closure of the group $\mathbb{C}U = \{Ux \mid x \in \mathbb{C}\}$ in X. Shifting Y_U if needed, we may assume, without loss of generality, that Y_U is not in the singular locus, $Y_U \not\subset \Sigma$. Then, for a sufficiently small y, we have $Y_U + Vy \notin \Sigma$ as well. Consider the restriction of the theta-function onto the affine subspace $\mathbb{C}^d + Vy$, where $\mathbb{C}^d :=$ (the identity component of $\pi^{-1}(Y_U)$), and $\pi\colon \mathbb{C}^g \to X = \mathbb{C}^g/\Lambda$ is the universal covering map of X:

$$\tau(z,y) = \theta(z + Vy), \quad z \in \mathbb{C}^d. \tag{6.15}$$

The function $u(z,y) = -2\partial_1^2 \ln \tau$ is periodic with respect to the lattice $\Lambda_U = \Lambda \cap \mathbb{C}^d$ and, for fixed y, has a double pole along the divisor $\Theta^U(y) = (\Theta - Vy) \cap \mathbb{C}^d$.

Lemma 6.16. *Let equations (1.10) for zeros of $\tau(Ux+z,y)$ hold and let λ be a vector of the sublattice $\Lambda_U = \Lambda \cap \mathbb{C}^d \subset \mathbb{C}^g$. Then:*

(i) equation (1.7) with the potential $u(Ux+z,y)$ has a wave solution of the form $\psi = e^{kx+k^2y}\phi(Ux+z,y,k)$ such that the coefficients $\xi_s(z,y)$ of the formal series

$$(6.17) \qquad \phi(z,y,k) = e^{by}\left(1+\sum_{s=1}^{\infty}\xi_s(z,y)\,k^{-s}\right)$$

are λ-periodic meromorphic functions of the variable $z \in \mathbb{C}^d$ with a simple pole at the divisor $\Theta^U(y)$,

$$(6.18) \qquad \xi_s(z+\lambda,y) = \xi_s(z,y) = \frac{\tau_s(z,y)}{\tau(z,y)};$$

(ii) $\phi(z,y,k)$ is unique up to a factor $\rho(z,k)$ that is ∂_U-invariant and holomorphic in z,

$$(6.19) \qquad \phi_1(z,y,k) = \phi(z,y,k)\rho(z,k),\ \ \partial_U\rho = 0\,.$$

Proof. The functions $\xi_s(z)$ are defined recursively by the equations

$$(6.20) \qquad 2\partial_U\xi_{s+1} = \partial_y\xi_s + (u+b)\xi_s - \partial_U^2\xi_s\,.$$

A particular solution of the first equation $2\partial_U\xi_1 = u+b$ is given by the formula

$$(6.21) \qquad 2\xi_1^0 = -2\partial_U \ln\tau + (l,z)\,b\,,$$

where (l,z) is a linear form on $\mathbb{C}^d$ given by the scalar product of z with a vector $l \in \mathbb{C}^d$ such that $(l,U) = 1$. By definition, the vector λ is in Y_U. Therefore, $(l,\lambda) \neq 0$. The periodicity condition for ξ_1^0 defines the constant b

$$(6.22) \qquad (l,\lambda)b = (2\partial_U \ln\tau(z+\lambda,y) - 2\partial_U\ln\tau(z,y))\,,$$

which depends only on a choice of the lattice vector λ. A change of the potential by an additive constant does not affect the results of the previous lemma. Therefore, equations (1.10) are sufficient for the local solvability of (6.20) in any domain, where $\tau(z+Ux,y)$ has simple zeros, i.e., outside of the set $\Theta_1^U(y) = (\Theta_1 - Vy)\cap\mathbb{C}^d$, where $\Theta_1 = \Theta\cap\partial_U\Theta$. This set does not contain a ∂_U-invariant line because any such line is dense in Y_U. Therefore, the sheaf $\mathcal{V}_0$ of ∂_U-invariant meromorphic functions on $\mathbb{C}^d \setminus \Theta_1^U(y)$ with poles along the divisor $\Theta^U(y)$ coincides with the sheaf of holomorphic ∂_U-invariant functions. That implies the vanishing of $H^1(\mathbb{C}^d \setminus \Theta_1^U(y), \mathcal{V}_0)$ and the existence of global meromorphic solutions ξ_s^0 of (6.20) which have a simple pole at the divisor $\Theta^U(y)$ (see details in [3, 54]). If ξ_s^0 are fixed, then the general global meromorphic solutions are given by the formula $\xi_s = \xi_s^0 + c_s$, where the constant of integration $c_s(z,y)$ is a holomorphic ∂_U-invariant function of the variable z.

Let us assume, as in the example above, that a λ-periodic solution ξ_{s-1} is known and that it satisfies the condition that there exists a periodic solution ξ_s^0 of the next equation. Let ξ^*_{s+1} be a solution of (6.20) for fixed ξ^0_s. Then it is easy to see that the function

$$\xi^0_{s+1}(z,y) = \xi^*_{s+1}(z,y) + c_s(z,y)\,\xi^0_1(z,y) + \frac{(l,z)}{2}\partial_y c_s(z,y) \tag{6.23}$$

is a solution of (6.20) for $\xi_s = \xi^0_s + c_s$. A choice of a λ-periodic ∂_U-invariant function $c_s(z,y)$ does not affect the periodicity property of ξ_s, but it does affect the periodicity of the function ξ^0_{s+1}. In order to make ξ^0_{s+1} periodic, the function $c_s(z,y)$ should satisfy the linear differential equation

$$(l,\lambda)\partial_y c_s(z,y) = 2\xi^*_{s+1}(z+\lambda,y) - 2\xi^*_{s+1}(z,y)\,. \tag{6.24}$$

This equation, together with an initial condition $c_s(z) = c_s(z,0)$ uniquely defines $c_s(x,y)$. The induction step is then completed. We have shown that the ratio of two periodic formal series ϕ_1 and ϕ is y-independent. Therefore, equation (6.19), where $\rho(z,k)$ is defined by the evaluation of the both sides at $y=0$, holds. The lemma is thus proven.

Corollary 6.25. *Let $\lambda_1,\dots,\lambda_d$ be a set of linear independent vectors of the lattice Λ_U and let z_0 be a point of $\mathbb{C}^d$. Then, under the assumptions of the previous lemma, there is a unique wave solution of equation (1.7) such that the corresponding formal series $\phi(z,y,k;z_0)$ is quasi-periodic with respect to Λ_U, i.e., for $\lambda \in \Lambda_U$*

$$\phi(z+\lambda,y,k;z_0) = \phi(z,y,k;z_0)\,\mu_\lambda(k) \tag{6.26}$$

and satisfies the normalization conditions

$$\mu_{\lambda_i}(k) = 1, \quad \phi(z_0,0,k;z_0) = 1\,. \tag{6.27}$$

The proof is identical to that of the part (b) of the Lemma 12 in [54]. Let us briefly present its main steps. As shown above, there exist wave solutions corresponding to ϕ which are λ_1-periodic. Moreover, from the statement (ii) above it follows that for any $\lambda' \in \Lambda_U$

$$\phi(z+\lambda,y,k) = \phi(z,y,k)\,\rho_\lambda(z,k)\,, \tag{6.28}$$

where the coefficients of ρ_λ are ∂_U-invariant holomorphic functions. Then the same arguments as in [54] show that there exists a ∂_U-invariant series $f(z,k)$ with holomorphic in z coefficients and formal series $\mu^0_\lambda(k)$ with constant coefficients such that the equation

$$f(z+\lambda,k)\rho_\lambda(z,k) = f(z,k)\,\mu_\lambda(k) \tag{6.29}$$

holds. The ambiguity in the choice of f and μ corresponds to the multiplication by the exponent of a linear form in z vanishing on U, i.e.,

$$f'(z,k) = f(z,k)\,e^{(b(k),z)}, \quad \mu'_\lambda(k) = \mu_\lambda(k)\,e^{(b(k),\lambda)}, \quad (b(k),U) = 0\,, \tag{6.30}$$

where $b(k) = \sum_s b_s k^{-s}$ is a formal series with vector-coefficients that are orthogonal to U. The vector U is in general position with respect to the lattice. Therefore, the ambiguity can be uniquely fixed by imposing $(d-1)$ normalizing conditions $\mu_{\lambda_i}(k)=1,\ i>1$ (recall that $\mu_{\lambda_1}(k)=1$ by construction).

The formal series $f\phi$ is quasi-periodic and its multipliers satisfy (6.27). Then, by that properties it is defined uniquely up to a factor which is constant in z and y. Therefore, for the unique definition of ϕ_0 it is enough to fix its evaluation at z_0 and $y=0$. The corollary is proved.

The spectral curve

The next goal is to show that λ-periodic wave solutions of equation (1.7), with u as in (1.8), are common eigenfunctions of rings of commuting operators.

Note that a simple shift $z\to z+Z$, where $Z\notin\Sigma$, gives λ-periodic wave solutions with meromorphic coefficients along the affine subspaces $Z+\mathbb{C}^d$. Theses λ-periodic wave solutions are related to each other by ∂_U-invariant factor. Therefore choosing, in the neighborhood of any $Z\notin\Sigma$, a hyperplane orthogonal to the vector U and fixing initial data on this hyperplane at $y=0$, we define the corresponding series $\phi(z+Z,y,k)$ as a *local* meromorphic function of Z and the *global* meromorphic function of z.

Lemma 6.31. *Let the assumptions of Theorem 1.6 hold. Then there is a unique pseudo-differential operator*

(6.32)
$$\mathcal{L}(Z,\partial_x)=\partial_x+\sum_{s=1}^{\infty}w_s(Z)\partial_x^{-s}$$

such that

(6.33)
$$\mathcal{L}(Ux+Vy+Z,\partial_x)\,\psi=k\psi,$$

where $\psi=e^{kx+k^2y}\phi(Ux+Z,y,k)$ is a λ-periodic solution of (1.7). The coefficients $w_s(Z)$ of $\mathcal{L}$ are meromorphic functions on the abelian variety X with poles along the divisor Θ.

Proof. Let ψ be a λ-periodic wave solution. The substitution of (6.17) in (6.33) gives a system of equations that recursively define $w_s(Z,y)$ as differential polynomials in $\xi_s(Z,y)$. The coefficients of ψ are local meromorphic functions of Z, but the coefficients of $\mathcal{L}$ are well-defined *global meromorphic functions* of on $\mathbb{C}^g\setminus\Sigma$, because different λ-periodic wave solutions are related to each other by ∂_U-invariant factor, which does not affect $\mathcal{L}$. The singular locus is of codimension $\geqslant 2$. Then Hartogs' holomorphic extension theorem implies that $w_s(Z,y)$ can be extended to a global meromorphic function on $\mathbb{C}^g$.

The translational invariance of u implies the translational invariance of the λ-periodic wave solutions. Indeed, for any constant s the series $\phi(Vs+Z,y-s,k)$ and $\phi(Z,y,k)$ correspond to λ-periodic solutions of the same equation. Therefore, they coincide up to a ∂_U-invariant factor. This factor does not affect $\mathcal{L}$. Hence, $w_s(Z,y)=w_s(Vy+Z)$.

The λ-periodic wave functions corresponding to Z and $Z+\lambda'$ for any $\lambda' \in \Lambda$ are also related to each other by a ∂_U-invariant factor:

$$\partial_U\left(\phi_1(Z+\lambda', y, k)\phi^{-1}(Z, y, k)\right) = 0\,. \tag{6.34}$$

Hence, w_s are periodic with respect to Λ and therefore are meromorphic functions on the abelian variety X. The lemma is proved.

Consider now the differential parts of the pseudodifferential operators $\mathcal{L}^m$. Let $\mathcal{L}^m_+$ be the differential operator such that $\mathcal{L}^m_- = \mathcal{L}^m - \mathcal{L}^m_+ = F_m\partial^{-1} + O(\partial^{-2})$. The leading coefficient F_m of $\mathcal{L}^m_-$ is the residue of $\mathcal{L}^m$:

$$F_m = \operatorname{res}_\partial \mathcal{L}^m. \tag{6.35}$$

From the construction of $\mathcal{L}$ it follows that $[\partial_y - \partial_x^2 + u, \mathcal{L}^n] = 0$. Hence,

$$[\partial_y - \partial_x^2 + u, \mathcal{L}^m_+] = -[\partial_y - \partial_x^2 + u, \mathcal{L}^m_-] = 2\partial_x F_m \tag{6.36}$$

(compare with (4.10)). The functions F_m are differential polynomials in the coefficients w_s of $\mathcal{L}$. Hence, $F_m(Z)$ are meromorphic functions on X. Next statement is crucial for the proof of the existence of commuting differential operators associated with u.

Lemma 6.37. *The abelian functions F_m have at most the second order pole on the divisor Θ.*

Proof. We need a few more standard constructions from the KP theory. If ψ is as in Lemma 3.8, then there exists a unique pseudodifferential operator Φ such that

$$\psi = \Phi e^{kx+k^2y}, \quad \Phi = 1 + \sum_{s=1}^{\infty} \varphi_s(Ux+Z, y)\partial_x^{-s}. \tag{6.38}$$

The coefficients of Φ are universal differential polynomials on ξ_s. Therefore, $\varphi_s(z + Z, y)$ is a global meromorphic function of $z \in C^d$ and a local meromorphic function of $Z \notin \Sigma$. Note that $\mathcal{L} = \Phi(\partial_x)\,\Phi^{-1}$.

Consider the dual wave function defined by the left action of the operator Φ^{-1}: $\psi^+ = \left(e^{-kx-k^2y}\right)\Phi^{-1}$. Recall that the left action of a pseudodifferential operator is the formal adjoint action under which the left action of ∂_x on a function f is $(f\partial_x) = -\partial_x f$. If ψ is a formal wave solution of (1.7), then ψ^+ is a solution of the adjoint equation

$$(-\partial_y - \partial_x^2 + u)\psi^+ = 0\,. \tag{6.39}$$

The same arguments, as before, prove that if equations (1.10) for poles of u hold then ξ_s^+ have simple poles at the poles of u. Therefore, if ψ is as in Lemma 6.16, then the dual wave solution is of the form $\psi^+ = e^{-kx-k^2y}\phi^+(Ux+Z, y, k)$, where the coefficients $\xi_s^+(z+Z, y)$ of the formal series

$$\phi^+(z+Z, y, k) = e^{-by}\left(1 + \sum_{s=1}^{\infty} \xi_s^+(z+Z, y)\,k^{-s}\right) \tag{6.40}$$

are λ-periodic meromorphic functions of the variable $z \in \mathbb{C}^d$ with a simple pole at the divisor $\Theta^u(y)$.

The ambiguity in the definition of ψ does not affect the product

$$\psi^+\psi = \left(e^{-kx-k^2y}\Phi^{-1}\right)\left(\Phi e^{kx+k^2y}\right). \tag{6.41}$$

Therefore, although each factor is only a local meromorphic function on $\mathbb{C}^g \setminus \Sigma$, the coefficients J_s of the product

$$\psi^+\psi = \phi^+(Z,y,k)\phi(Z,y,k) = 1 + \sum_{s=2}^{\infty} J_s(Z,y)k^{-s}. \tag{6.42}$$

are *global meromorphic functions* of Z. Moreover, the translational invariance of u implies that they have the form $J_s(Z,y) = J_s(Z+Vy)$. Each of the factors in the left-hand side of (6.42) has a simple pole on $\Theta - Vy$. Hence, $J_s(Z)$ is a meromorphic function on X with a second order pole at Θ. According to Lemma 4.19, we have $F_n = J_{n+1}$. That completes the proof of the lemma.

Let $\hat{F}$ be a linear space generated by $\{F_m,\ m = 0,1,\ldots\}$, where we set $F_0 = 1$. It is a subspace of the 2^g-dimensional space of the abelian functions that have at most second order pole at Θ. Therefore, for all but $\hat{g} = \dim \hat{F}$ positive integers n, there exist constants $c_{i,n}$ such that

$$F_n(Z) + \sum_{i=0}^{n-1} c_{i,n}F_i(Z) = 0. \tag{6.43}$$

Let I denote the subset of integers n for which there are no such constants. We call this subset the gap sequence.

Lemma 6.44. *Let $\mathcal{L}$ be the pseudodifferential operator corresponding to a λ-periodic wave function ψ constructed above. Then, for the differential operators*

$$L_n = \mathcal{L}_+^n + \sum_{i=0}^{n-1} c_{i,n}\mathcal{L}_+^{n-i} = 0,\ n \notin I, \tag{6.45}$$

the equations

$$L_n\psi = a_n(k)\psi, \quad a_n(k) = k^n + \sum_{s=1}^{\infty} a_{s,n}k^{n-s}, \tag{6.46}$$

where $a_{s,n}$ are constants, hold.

Proof. First note that from (6.36) it follows that

$$[\partial_y - \partial_x^2 + u, L_n] = 0. \tag{6.47}$$

Hence, if ψ is a λ-periodic wave solution of (1.7) corresponding to $Z \notin \Sigma$, then $L_n\psi$ is also a formal solution of the same equation. That implies the equation $L_n\psi = a_n(Z,k)\psi$, where a is ∂_u-invariant. The ambiguity in the definition of ψ does not affect a_n. Therefore, the coefficients of a_n are well-defined *global* meromorphic functions on $\mathbb{C}^g \setminus \Sigma$. The ∂_U- invariance of a_n implies that a_n, as a function of Z, is holomorphic outside of the locus. Hence it has an extension to a holomorphic function on $\mathbb{C}^g$. Equations (6.34) imply that a_n is periodic with respect to the lattice Λ. Hence a_n is Z-independent. Note that $a_{s,n} = c_{s,n},\ s \leqslant n$. The lemma is proved.

The operator L_m can be regarded as a $Z \notin \Sigma$-parametric family of ordinary differential operators L_m^Z whose coefficients have the form

$$L_m^Z = \partial_x^n + \sum_{i=1}^{m} u_{i,m}(Ux+Z)\,\partial_x^{m-i}, \quad m \notin I. \tag{6.48}$$

Corollary 6.49. *The operators L_m^Z commute with each other,*

$$[L_n^Z, L_m^Z] = 0, \quad Z \notin \Sigma. \tag{6.50}$$

From (6.46) it follows that $[L_n^Z, L_m^Z]\psi = 0$. The commutator is an ordinary differential operator. Hence, the last equation implies (6.50).

Lemma 6.51. *Let $\mathcal{A}^Z$, $Z \notin \Sigma$, be a commutative ring of ordinary differential operators spanned by the operators L_n^Z. Then there is an irreducible algebraic curve Γ of arithmetic genus $\hat{g} = \dim \hat{F}$ such that $\mathcal{A}^Z$ is isomorphic to the ring $A(\Gamma, P_0)$ of the meromorphic functions on Γ with the only pole at a smooth point P_0. The correspondence $Z \to \mathcal{A}^Z$ defines a holomorphic imbedding of $X \setminus \Sigma$ into the space of torsion-free rank 1 sheaves $\mathcal{F}$ on Γ*

$$j\colon X \setminus \Sigma \longmapsto \overline{\mathrm{Pic}}(\Gamma). \tag{6.52}$$

Proof. In order to get the statement of the theorem as a direct corollary of Theorem 5.1, it remains only to show that the ring $\mathcal{A}^Z$ is maximal. Recall, that a commutative ring $\mathcal{A}$ of linear ordinary differential operators is called maximal if it is not contained in any bigger commutative ring. Let us show that for a generic Z the ring $\mathcal{A}^Z$ is maximal. Suppose that it is not. Then there exits $\alpha \in I$, where I is the gap sequence defined above, such that for each $Z \notin \Sigma$ there exists an operator L_α^Z of order α which commutes with $L_n^Z, n \notin I$. Therefore, it commutes with $\mathcal{L}$. A differential operator commuting with $\mathcal{L}$ up to the order $O(1)$ can be represented in the form $L_\alpha = \sum_{m<\alpha} c_{i,\alpha}(Z)\mathcal{L}_+^i$, where $c_{i,\alpha}(Z)$ are ∂_1-invariant functions of Z. It commutes with $\mathcal{L}$ if and only if

$$F_\alpha(Z) + \sum_{i=0}^{n-1} c_{i,\alpha}(Z)F_i(Z) = 0, \quad \partial_U c_{i,\alpha} = 0. \tag{6.53}$$

Note the difference between (6.43) and (6.53). In the first equation the coefficients $c_{i,n}$ are constants. The λ-periodic wave solution of equation (1.7) is a common eigenfunction of all commuting operators, i.e., $L_\alpha\psi = a_\alpha(Z,k)\psi$, where $a_\alpha = k^\alpha + \sum_{s=1}^{\infty} a_{s,\alpha}(Z)k^{\alpha-s}$ is ∂_1-invariant. The same arguments as those used in the proof of equation (6.46) show that the eigenvalue a_α is Z-independent. We have $a_{s,\alpha} = c_{s,\alpha},\ s \leqslant \alpha$. Therefore, the coefficients in (6.53) are Z-independent. That contradicts the assumption that $\alpha \notin I$. The lemma is proved.

Our next goal is to prove finally the global existence of the wave function.

Lemma 6.54. *Let the assumptions of the Theorem 1.19 hold. Then there exists a common eigenfunction of the corresponding commuting operators L_n^Z of the form $\psi = e^{kx}\phi(Ux + Z, k)$ such that the coefficients of the formal series*

$$\phi(Z,k) = 1 + \sum_{s=1}^{\infty} \xi_s(Z)\, k^{-s} \tag{6.55}$$

are global meromorphic functions with a simple pole at Θ.

Proof. It is instructive to consider first the case when the spectral curve Γ of the rings $\mathcal{A}^Z$ is smooth. Then, as shown in ([29, 30]), the corresponding common eigenfunction of the commuting differential operators (the Baker-Akhiezer function), normalized by the condition $\psi_0|_{x=0} = 1$, is of the form ([29, 30])

$$\hat{\psi}_0 = \frac{\hat{\theta}(\hat{A}(P) + \hat{U}x + \hat{Z})\,\hat{\theta}(\hat{Z})}{\hat{\theta}(\hat{U}x + \hat{Z})\,\hat{\theta}(\hat{A}(P) + \hat{Z})}\, e^{x\,\Omega(P)}. \tag{6.56}$$

(compare with (2.24). Here $\hat{\theta}(\hat{Z})$ is the Riemann theta-function constructed with the help of the matrix of b-periods of normalized holomorphic differentials on Γ; $\hat{A}\colon \Gamma \to J(\Gamma)$ is the Abel-Jacobi map; Ω is the abelian integral corresponding to the second kind meromorphic differential $d\Omega$ with the only pole of the form dk at the marked point P_0 and $2\pi i\hat{U}$ is the vector of its b-periods.

Remark. Let us emphasize, that the formula (6.56) is not the result of solution of some differential equations. It is a direct corollary of analytic properties of the Baker-Akhiezer function $\hat{\psi}_0(x, P)$ on the spectral curve.

The last factors in the numerator and the denominator of (6.56) are x-independent. Therefore, the function

$$\hat{\psi}_{BA} = \frac{\hat{\theta}(\hat{A}(P) + \hat{U}x + \hat{Z})}{\hat{\theta}(\hat{U}x + \hat{Z})}\, e^{x\,\Omega(P)} \tag{6.57}$$

is also a common eigenfunction of the commuting operators.

In the neighborhood of P_0 the function $\hat{\psi}_{BA}$ has the form

$$\hat{\psi}_{BA} = e^{kx}\left(1 + \sum_{s=1}^{\infty} \frac{\tau_s(\hat{Z} + \hat{U}x)}{\hat{\theta}(\hat{U}x + \hat{Z})}\, k^{-s}\right), \quad k = \Omega, \tag{6.58}$$

where $\tau_s(\hat{Z})$ are global holomorphic functions.

According to Lemma 6.51, we have a holomorphic imbedding $\hat{Z} = j(Z)$ of $X \setminus \Sigma$ into $J(\Gamma)$. Consider the formal series $\psi = j^*\hat{\psi}_{BA}$. It is globally well-defined out of Σ. If $Z \notin \Theta$, then $j(Z) \notin \hat{\Theta}$ (which is the divisor on which the condition (5.9) is violated). Hence, the coefficients of ψ are regular out of Θ. The singular locus is at least of codimension 2. Hence, using once again Hartogs' arguments we can extend ψ on X.

If the spectral curve is singular, we can proceed along the same lines using the generalization of (6.57) given by the theory of Sato τ-function ([53]). Namely, a set of algebraic-geometrical data (5.8) defines the point of the Sato Grassmannian, and therefore, the corresponding τ-function: $\tau(t;\mathcal{F})$. It is a holomorphic function of the variables $t=(t_1,t_2,\ldots)$, and is a section of a holomorphic line bundle on $\overline{\mathrm{Pic}(\Gamma)}$.

The variable x is identified with the first time of the KP-hierarchy, $x=t_1$. Therefore, the formula for the Baker-Akhiezer function corresponding to a point of the Grassmannian ([53]) implies that the function $\hat{\psi}_{BA}$ given by the formula

$$\hat{\psi}_{BA}=\frac{\tau(x-k,-\frac{1}{2}k^2,-\frac{1}{3}k^3,\ldots;\mathcal{F})}{\tau(x,0,0,\ldots;\mathcal{F})}e^{kx} \tag{6.59}$$

is a common eigenfunction of the commuting operators defined by $\mathcal{F}$. The rest of the arguments proving the lemma are the same, as in the smooth case.

Lemma 6.60. *The linear space $\hat{F}$ generated by the abelian functions $\{F_0=1, F_m=\mathrm{res}_\partial \mathcal{L}^m\}$, is a subspace of the space H generated by F_0 and by the abelian functions $H_i=\partial_U\partial_{z_i}\ln\theta(Z)$.*

Proof. Recall that the functions F_n are abelian functions with at most second order pole on Θ. Hence, a priori $\hat{g}=\dim\hat{F}\leqslant 2^g$. In order to prove the statement of the lemma it is enough to show that $F_n=\partial_U Q_n$, where Q_n is a meromorphic function with a pole along Θ. Indeed, if Q_n exists, then, for any vector λ in the period lattice, we have $Q_n(Z+\lambda)=Q_n(Z)+c_{n,\lambda}$. There is no abelian function with a simple pole on Θ. Hence, there exists a constant q_n and two g-dimensional vectors l_n, l'_n, such that $Q_n=q_n+(l_n,Z)+(l'_n,h(Z))$, where $h(Z)$ is a vector with the coordinates $h_i=\partial_{z_i}\ln\theta$. Therefore, $F_n=(l_n,U)+(l'_n,H(Z))$.

Let $\psi(x,Z,k)$ be the formal Baker-Akhiezer function defined in the previous lemma. Then the coefficients $\varphi_s(Z)$ of the corresponding wave operator Φ (6.38) are global meromorphic functions with poles on Θ.

The left and right action of pseudodifferential operators are formally adjoint, i.e., for any two operators the equality $(e^{-kx}\mathcal{D}_1)(\mathcal{D}_2e^{kx})=e^{-kx}(\mathcal{D}_1\mathcal{D}_2e^{kx})+\partial_x(e^{-kx}(\mathcal{D}_3e^{kx}))$ holds. Here $\mathcal{D}_3$ is a pseudodifferential operator whose coefficients are differential polynomials in the coefficients of $\mathcal{D}_1$ and $\mathcal{D}_2$. Therefore, from (6.41) it follows that

$$\psi^+\psi=1+\sum_{s=2}^{\infty}F_{s-1}k^{-s}=1+\partial_x\left(\sum_{s=2}^{\infty}Q_sk^{-s}\right). \tag{6.61}$$

The coefficients of the series Q are differential polynomials in the coefficients φ_s of the wave operator. Therefore, they are global meromorphic functions of Z with poles on Θ. Lemma is proved.

The construction of multivariable Baker-Akhiezer functions presented in Section 2 for smooth curves is a manifestation of general statement valid for singular spectral curves: flows of the KP hierarchy define deformations of the commutative rings $\mathcal{A}$ of ordinary linear differential operators. The spectral curve is invariant under these flows. For a given spectral curve Γ the orbits of the KP hierarchy are isomorphic to the generalized Jacobian $J(\Gamma) = \mathrm{Pic}^0(\Gamma)$, which is the equivalence classes of zero degree divisors on the spectral curve (see details in [54, 29, 30, 53]).

As shown in Section 4, the evolution of the potential u is described by equation (4.6) The first two times of the hierarchy are identified with the variables $t_1 = x, t_2 = y$. Equations (4.6) identify the space $\hat{F}_1$ generated by the functions $\partial_u F_n$ with the tangent space of the KP orbit at $\mathcal{A}^Z$. Then, from Lemma 6.9 it follows that this tangent space is a subspace of the tangent space of the abelian variety X. Hence, for any $Z \notin \Sigma$, the orbit of the KP flows of the ring $\mathcal{A}^Z$ is in X, i.e., it defines an holomorphic imbedding:

$$i_Z : J(\Gamma) \longmapsto X. \tag{6.62}$$

From (6.62) it follows that $J(\Gamma)$ is *compact*.

The generalized Jacobian of an algebraic curve is compact if and only if the curve is *smooth* ([14]). On a smooth algebraic curve a torsion-free rank 1 sheaf is a line bundle, i.e., $\overline{\mathrm{Pic}}(\Gamma) = J(\Gamma)$. Then (6.52) implies that i_Z is an isomorphism. Note that for the Jacobians of smooth algebraic curves the bad locus Σ is empty ([54]), i.e., the imbedding j in (6.52) is defined everywhere on X and is inverse to i_Z. Theorem 1.6 is proved.

7. Characterization of the Prym varieties

To begin with let us recall the definition of Prym varieties. An involution $\sigma : \Gamma \longrightarrow \Gamma$ of a smooth algebraic curve Γ induces an involution $\sigma^* : J(\Gamma) \longrightarrow J(\Gamma)$ of the Jacobian. The kernel of the map $1+\sigma^*$ on $J(\Gamma)$ is the sum of a lower-dimensional abelian variety, called the Prym variety (the connected component of zero in the kernel), and a finite group. The Prym variety naturally has a polarization induced by the principal polarization on $J(\Gamma)$. However, this polarization is not principal, and the Prym variety admits a natural principal polarization if and only if σ has at most two fixed points on Γ — this is the case we will concentrate on.

From the point of view of integrable systems, attempts to prove the analog of Novikov's conjecture for the case of Prym varieties of algebraic curves with two smooth fixed points of involution were made in [56, 55, 7]. In [56] it was shown that Novikov-Veselov (NV) equation provides solution of the characterization problem up to possible existence of additional irreducible components. In [55, 7] the characterizations of the Prym varieties in terms of BKP and NV equations were proved only under certain additional assumptions. Moreover, in [7] an example of a ppav that is not a Prym but for which the theta function gives a solution to the BKP equation was constructed. Thus for more than 15 years it was widely accepted that Prym varieties can not be characterized with the help if integrable systems.

In [35] the first author proved that Prym varieties of algebraic curves with two smooth fixed points of involution are characterized among all ppavs by the property of their theta functions providing explicit formulas *for solutions of the integrable* 2D

Schrödinger equation, which is one of the auxiliary linear problems for the Novikov-Veselov equation.

Prym varieties possess generalizations of some properties of Jacobians. In [7] Beauville and Debarre, and in [22] Fay showed that the Kummer images of Prym varieties admit a 4-dimensional family of quadrisecant planes (as opposed to a 4-dimensional family of trisecant lines for Jacobians). Similarly to the case of Jacobians, it was then shown by Debarre in [12] that the existence of a one-dimensional family of quadrisecants characterizes Prym varieties among all ppavs. However, Beauville and Debarre in [7] constructed a ppav that is *not* a Prym but such that its Kummer image *has* a quadrisecant plane. Thus no analog of the trisecant conjecture for Prym varieties was conjectured, and the question of characterizing Prym varieties by a finite amount of geometric data (i.e., by polynomial equations for theta functions at a finite number of points) remained completely open.

In [25] S. Grushevsky and the first author proved that Prym varieties of unramified covers are characterized among all ppavs by the property of their Kummer images admitting a *symmetric pair of quadrisecant 2-planes*. That there exists such a symmetric pair of quadrisecant planes for the Kummer image of a Prym variety can be deduced from the description of the 4-dimensional family of quadrisecants, using the natural involution on the Abel-Prym curve. However, the statement that a symmetric pair of quadrisecants in fact characterizes Pryms seems completely unexpected.

The geometric characterization of Prym varieties follows from a characterization of Prym varieties among all ppavs by some theta-functional equations, which by using Riemann's bilinear addition theorem can be shown to be equivalent to the existence of a symmetric pair of quadrisecant planes. In order to obtain such a characterization of Prym varieties in [25] a *new* hierarchy of difference equations, starting from a discrete version of the Schrödinger equation was introduced, developed, and studied . The hierarchy constructed can be thought of as a discrete analog of the Novikov-Veselov hierarchy.

Theorem 7.1 (Main theorem). *An indecomposable principally polarized abelian variety* $(X, \theta) \in \mathcal{A}_g$ *lies in the closure of the locus* $\mathcal{P}_g$ *of Prym varieties of unramified double covers if and only if there exist vectors* $A, U, V, W \in \mathbb{C}^g$ *representing distinct points in* X*, none of them points of order two, and constants* $c_1, c_2, c_3, w_1, w_2, w_3 \in \mathbb{C}$ *such that one of the following equivalent conditions holds:*

(A) *The difference* 2D *Schrödinger equation*

$$\psi_{n+1,m+1} - u_{n,m}(\psi_{n+1,m} - \psi_{n,m+1}) - \psi_{n,m} = 0, \tag{7.2}$$

with

$$u_{n,m} := C_{nm} \frac{\theta((n+1)U + mV + \nu W + Z)\,\theta(nU + (m+1)V + \nu W + Z)}{\theta((n+1)U + (m+1)V + \bar{\nu} W + Z)\,\theta(nU + mV + \bar{\nu} W + Z)} \tag{7.3}$$

and

$$(7.4)\qquad \psi_{n,m} := \frac{\theta(A + nU + mV + \nu_{nm}W + Z)}{\theta(nU + mV + \overline{\nu}_{nm}W + Z)}\, w_1^n w_2^m w_3^{\nu_{nm}} \left(c_1^m c_2^n\right)^{1-2\nu_{nm}},$$

is satisfied for all $Z \in X$, *where*

(7.5)

$$\nu := \nu_{nm} := \frac{1 + (-1)^{n+m+1}}{2}, \quad \overline{\nu} := 1 - \nu, \quad C_{nm} := c_3 \left(c_2^{2n+1} c_1^{2m+1}\right)^{1-2\nu_{nm}}.$$

(B) *The following identity holds:*

$$w_1 w_2 (c_1 c_2)^{\pm 1} \widetilde{K}\left(\frac{A + U + V \mp W}{2}\right) - w_1 c_3 (w_3 c_1)^{\pm 1} \widetilde{K}\left(\frac{A + U - V \pm W}{2}\right)$$
$$+ w_2 c_3 (w_3 c_2)^{\pm 1} \widetilde{K}\left(\frac{A + V - U \pm W}{2}\right) - \widetilde{K}\left(\frac{A - U - V \mp W}{2}\right) = 0,$$

where $\widetilde{K}\colon \mathbb{C}^g \ni z \mapsto \left(\Theta[\varepsilon, 0](z)\right) \in \mathbb{C}^{2^g}$ *is a lifting of the Kummer map (1.3) to the universal covering of* X.

(C) *The two equations (one for the top choice of signs everywhere, and one for the bottom)*

$$(7.6)\qquad c_1^{\mp 2} c_3^2\, \theta(Z + U - V)\, \theta(Z - U \pm W)\, \theta(Z + V \pm W)$$
$$+ c_2^{\mp 2} c_3^2\, \theta(Z - U + V)\, \theta(Z + U \pm W)\, \theta(Z - V \pm W)$$
$$= c_1^{\mp 2} c_2^{\mp 2}\, \theta(Z - U - V)\, \theta(Z + U \pm W)\, \theta(Z + V \pm W)$$
$$+ \theta(Z + U + V)\, \theta(Z - U \pm W)\, \theta(Z - V \pm W)$$

are valid on the theta divisor $\{Z \in X : \theta(Z) = 0\}$.

A purely geometric restatement of part (B) of this result is as follows.

Corollary 7.7 (Geometric characterization of Pryms). *A ppav* $(X, \theta) \in \mathcal{A}_g$ *lies in the closure of the locus of Prym varieties of unramified (étale) double covers if and only there exist four distinct points* $p_1, p_2, p_3, p_4 \in X$, *none of them points of order two, such that the following two quadruples of points on the Kummer variety of* X:

$$\{K(p_1 + \varepsilon_2 p_2 + \varepsilon_3 p_3 + \varepsilon_4 p_4) \mid \varepsilon_i \in \{\pm 1\},\ \varepsilon_2 \varepsilon_3 \varepsilon_4 = +1\}$$

and

$$\{K(p_1 + \varepsilon_2 p_2 + \varepsilon_3 p_3 + \varepsilon_4 p_4) \mid \varepsilon_i \in \{\pm 1\},\ \varepsilon_2 \varepsilon_3 \varepsilon_4 = -1\}$$

are linearly dependent.

Equivalently, this can be stated as saying that (X, θ) *lies in the closure of the Prym if and only if there exists a pair of symmetric (under the* $z \mapsto 2p_1 - z$ *involution) quadrisecants of* $K(X)$.

At first glance the structure of the proof is the same as above. It begins with a construction of a wave solution of the discrete analog of 2D Schrödinger equation (7.2). But in fact, the hierarchy considered involves essentially a pair of functions and is thus essentially a matrix hierarchy, unlike the scalar hierarchy arising for the trisecant case. The argument is very delicate, and involves using the pair of quadrisecant conditions to recursively construct a pair of auxiliary solutions (essentially corresponding to the two components of the kernel, only one of which is the Prym). We refer the reader to [25]) for details.

Our goal for this section is to elaborate on the "only if" part of the statement of the theorem, because as a byproduct it gives new identities for theta-function which are poorly understood an seems require additional attention.

Four point Baker-Akhiezer function

Four-point Baker-Akhiezer function depends on three discrete parameters and, as was mentioned in Section 2 gives solution to the BDHE equation. For various choice of two linear combination of these variables one obtain various linear equation. In [34] (see details in [43]) it was shown that the following choice of the "discrete times" gives a a construction of algebraic-geometric 2D difference Schrödinger operators.

Let Γ be a smooth algebraic curve of genus $\hat{g}$. Fix four points $P_1^{\pm}, P_2^{\pm} \in \Gamma$, and let $\widehat{D} = \gamma_1 + \cdots + \gamma_{\hat{g}}$ be a generic effective divisor on Γ of degree $\hat{g}$. By the Riemann-Roch theorem one computes $h^0(\widehat{D} + n(P_1^+ - P_1^-) + m(P_2^+ - P_2^-)) = 1$, for any $n, m \in \mathbb{Z}$, and for $\widehat{D}$ generic. We denote by $\widehat{\psi}_{n,m}(P)$, $P \in \Gamma$ the unique section of this bundle. This means that $\widehat{\psi}_{n,m}$ is the unique up to a constant factor meromorphic function such that (away from the marked points $P_i^{\pm}$) it has poles only at γ_s, of multiplicity not greater than the multiplicity of γ_s in $\widehat{D}$, while at the points P_1^+, P_2^+ (resp. P_1^-, P_2^-) the function $\widehat{\psi}_{n,m}$ has poles (resp. zeros) of orders n and m.

If we fix local coordinates k^{-1} in the neighborhoods of marked points (it is customary in the subject to think of marked points as punctures, and thus it is common to use coordinates such that k at the marked point is infinite rather than zero), then the Laurent series for $\psi_{n,m}(P)$, for $P \in \Gamma$ near a marked point, has the form

$$\widehat{\psi}_{n,m} = k^{\pm n}\left(\sum_{s=0}^{\infty} \xi_s^{\pm}(n,m)k^{-s}\right), \quad k = k(P),\ P \to P_1^{\pm}, \tag{7.8}$$

$$\widehat{\psi}_{n,m} = k^{\pm m}\left(\sum_{s=0}^{\infty} \chi_s^{\pm}(n,m)k^{-s}\right), \quad k = k(P),\ P \to P_2^{\pm}. \tag{7.9}$$

As it was shown in Section 2 the function $\psi_{n,m}$ can be expressed as follows:

$$\widehat{\psi}_{n,m}(P) = r_{nm}\frac{\hat{\theta}(\widehat{A}(P) + n\widehat{U} + m\widehat{V} + \widehat{Z})}{\hat{\theta}(\widehat{A}(P) + \widehat{Z})}\, e^{n\widehat{\Omega}_1(P) + m\widehat{\Omega}_2(P)}, \tag{7.10}$$

where for $i = 1,2$ the differential $d\widehat{\Omega}^i \in H^0(K_\Gamma + P_i^+ + P_i^-)$ is of the third kind, normalized to have residues ∓ 1 at $P_i^\pm$ and with zero integrals over all the a-cycles, and $\widehat{\Omega}^i$ is the corresponding abelian integral; we have the following expression r_{nm} is some constant, $\widehat{U} = \widehat{A}(P_1^-) - \widehat{A}(P_1^+)$, $\widehat{V} = \widehat{A}(P_2^-) - \widehat{A}(P_2^+)$, and

$$\widehat{Z} = -\sum_s \widehat{A}(\gamma_s) + \widehat{\kappa}, \tag{7.11}$$

where $\widehat{\kappa}$ is the vector of Riemann constants.

Change of notation We use here notation $\widehat{\theta}$ for the Riemann theta-function of Γ, for later use of θ for the Prym theta function.

Theorem 7.12 ([34]). *The Baker-Akhiezer function $\widehat{\psi}_{n,m}$ given by formula (7.10) satisfies the following difference equation*

$$\widehat{\psi}_{n+1,m+1} - a_{n,m}\widehat{\psi}_{n+1,m} - b_{n,m}\widehat{\psi}_{n,m+1} + c_{n,m}\widehat{\psi}_{n,m} = 0\,. \tag{7.13}$$

Setup for the Prym construction

We now assume that the curve Γ is an algebraic curve endowed with an involution σ without fixed points; then Γ is a unramified double cover $\Gamma \longrightarrow \Gamma_0$, where $\Gamma_0 = \Gamma/\sigma$. If Γ is of genus $\widehat{g} = 2g+1$, then by Riemann-Hurwitz the genus of Γ_0 is $g+1$. From now on we assume that $g > 0$ and thus $\widehat{g} > 1$. On Γ one can choose a basis of cycles a_i, b_i with the canonical matrix of intersections $a_i \cdot a_j = b_i \cdot b_j = 0$, $a_i \cdot b_j = \delta_{ij}$, $0 \leqslant i,j \leqslant 2g$, such that under the involution σ we have $\sigma(a_0) = a_0$, $\sigma(b_0) = b_0$, $\sigma(a_j) = a_{g+j}$, $\sigma(b_j) = b_{g+j}$, $1 \leqslant j \leqslant g$. If $d\omega_i$ are normalized holomorphic differentials on Γ dual to this choice of a-cycles, then the differentials $du_j = d\omega_j - d\omega_{g+j}$, for $j = 1, \dots, g$ are odd, i.e., satisfy $\sigma^*(du_k) = -du_k$, and we call them the normalized holomorphic Prym differentials. The matrix of their b-periods

$$\Pi_{kj} = \oint_{b_k} du_j, \quad 1 \leqslant k,j \leqslant g, \tag{7.14}$$

is symmetric, has positive definite imaginary part, and defines the Prym variety

$$\mathcal{P}(\Gamma) := \mathbb{C}^g/(\mathbb{Z}^g + \Pi\mathbb{Z}^g)$$

and the corresponding Prym theta function

$$\theta(z) := \theta(z, \Pi),$$

for $z \in \mathbb{C}^g$. We assume that the marked points $P_1^\pm, P_2^\pm$ on Γ are permuted by the involution, i.e., $P_i^+ = \sigma(P_i^-)$. For further use let us fix in addition a third pair of points $P_3^\pm$, such that also $P_3^- = \sigma(P_3^+)$.

The Abel-Jacobi map $\Gamma \hookrightarrow J(\Gamma)$ induces the Abel-Prym map $A\colon \Gamma \longrightarrow \mathcal{P}(\Gamma)$ (this is the composition of the Abel-Jacobi map $\widehat{A}\colon \gamma \hookrightarrow J(\Gamma)$ with the projection $J(\Gamma) \to \mathcal{P}(\Gamma)$). There is a choice of the base point involved in defining the Abel-Jacobi map, and thus in the Abel-Prym map; let us choose this base point (such a choice is unique up to a point of order two in $\mathcal{P}(\Gamma)$) in such a way that

$$A(P) = -A(\sigma(P)). \tag{7.15}$$

Admissible divisors

An effective divisor on Γ of degree $\hat{g} - 1 = 2g$, $D = \gamma_1 + \cdots + \gamma_{2g}$, is called *admissible* if it satisfies

$$[D] + [\sigma(D)] = K_\Gamma \in J(\Gamma) \tag{7.16}$$

(where K_Γ is the canonical class of Γ), and if moreover $H^0(D + \sigma(D))$ is generated by an even holomorphic differential $d\Omega$, i.e., that

$$d\Omega(\gamma_s) = d\Omega(\sigma(\gamma_s)) = 0, \quad d\Omega = \sigma^*(d\Omega). \tag{7.17}$$

Algebraically, what we are saying is the following. The divisors D satisfying (7.16) are the preimage of the point K_Γ under the map $1 + \sigma$, and thus are a translate of the subgroup $\mathrm{Ker}(1 + \sigma) \subset J(\Gamma)$ by some vector. As shown by Mumford [46], this kernel has two components — one of them being the Prym, and the other being the translate of the Prym variety by the point of order two corresponding to the cover $\Gamma \to \Gamma_0$ as an element in $\pi_1(\Gamma_0)$. The existence of an even differential as above picks out one of the two components, and the other one is obtained by adding $A - \sigma(A)$ to the divisor of such a differential, for some A. statement.

Proposition 7.18. *For a generic vector Z the zero-divisor D of the function $\theta(A(P) + Z)$ on Γ is of degree $2g$ and satisfies the constraints (7.16) and (7.17), i.e., is admissible.*

Remark. S. Grushevsky and the first author had been unable to find a complete proof of precisely this statement in the literature. However, both Elham Izadi and Roy Smith have independently supplied them with simple proofs of this result, based on Mumford's description and results on Prym varieties. As pointed out by a referee, this result can also be easily obtained by applying Fay's proposition 4.1 in [21]. In [25] independent analytic proof was proposed which also can be seen analytic proof of some of Mumford's results.

Note that the function $\theta(A(P) + Z)$ is multi-valued on Γ, but its zero-divisor is well-defined. The arguments identical to that in the standard proof of the inversion formula (7.11) show that the zero divisor $D(Z) := \theta(A(P)+Z)$ is of degree $\hat{g}-1 = 2g$.

Lemma 7.19. *For any pair of points $P_j^\pm$ conjugate under the involution σ there exists a unique differential $d\Omega_j$ of the third kind (i.e., a dipole differential with simple poles at these points and holomorphic elsewhere), such that it has residues ∓ 1 at these points, is odd under σ, i.e., satisfies $d\Omega_j = -\sigma^*(d\Omega_j)$, and such that all of its a-periods are integral multiples of πi, i.e., such a differential $d\Omega_i$ exists for a unique set of numbers $l_0, \ldots, l_g \in \mathbb{Z}$ satisfying*

$$\oint_{a_k} d\Omega_j = \pi i\, l_k, \quad k = 0, \ldots, g. \tag{7.20}$$

Indeed, by Riemann's bilinear relations there exists a unique differential $d\Omega$ of the third kind with residues as required, and satisfying $\oint_{a_k} d\Omega = 0$ for all k. Note, however, that then $\oint_{a_k} \sigma^*(d\Omega)$ is not necessarily zero, as the image $\sigma(a_k)$ of the loop a_k, while homologous to a_{g+k} on $\tilde{\Gamma}$, is not necessarily homologic to a_{g+k} (resp. to a_0 for $\sigma(a_0)$) on $\tilde{\Gamma}\setminus\{P_j^{\pm}\}$. Thus each integral $\oint_{a_k} \sigma^*(d\Omega)$, being equal to $2\pi i$ times the winding number of $\sigma(a_k)$ around P_j^+ minus that around P_j^-, is equal to $2\pi i l_k$ for some $l_k \in \mathbb{Z}$. We now subtract from $d\Omega$ the linear combination $\pi i\,(l_0 d\omega_0 + \sum_{k=1}^g l_k(d\omega_k + d\omega_{g+k}))$ of even abelian differentials to get the desired $d\Omega_j$.

Theorem 7.21. [25] *For a generic* $D = D(Z)$ *and for each set of integers* (n, m, r) *such that*

$$n + m + r = 0 \bmod 2\,, \tag{7.22}$$

the space

$$H^0(D + n(P_1^+ - P_1^-) + m(P_2^+ - P_2^-) + r(P_3^+ - P_3^-))$$

is one-dimensional. A basis element of this space is given by

(7.23)
$$\psi_{n,m,r}(P) := h_{n,m,r}\frac{\theta(A(P) + nU + mV + rW + Z)}{\theta(A(P) + Z)}\,e^{n\Omega_1(P)+m\Omega_2(P)+r\Omega_3(P)},$$

where Ω_j *is the abelian integral corresponding to the differential* $d\Omega_j$ *defined by lemma 7.19, and* U, V, W *are the vectors of* b*-periods of these differentials, i.e.,*

$$2\pi i U_k = \oint_{b_k} d\Omega_1,\ 2\pi i V_k = \oint_{b_k} d\Omega_2,\ 2\pi i W_k = \oint_{b_k} d\Omega_3. \tag{7.24}$$

The proof is identically the same as the proof of (2.19). It is easy to check that the right-hand side of (7.23) is a single valued function on Γ having all the desired properties, and thus it gives a section of the desired bundle. Note that the constraint (7.22) is required due to (7.20), and the uniqueness of ψ up to a constant factor, i.e., the one-dimensionality of the H^0 above, is a direct corollary of the Riemann-Roch theorem.

Note that bilinear Riemann identities imply

$$2U = A(P_1^-) - A(P_1^+),\ \ 2V = A(P_2^-) - A(P_2^+),\ \ 2W = A(P_3^-) - A(P_3^+). \tag{7.25}$$

Let us compare the definition of $\widehat{\psi}_{n,m}$ defined for any curve Γ, with that of $\psi_{n,m,r}$, which is only defined for a curve with an involution satisfying a number of conditions. To make such a comparison, consider the divisor $\widehat{D} = D + P_3^+$ of degree $\hat{g} = 2g + 1$, and let $\widehat{\psi}_{n,m}$ be the corresponding Baker-Akhiezer function.

Corollary 7.26. *For the Baker-Akhiezer function* $\widehat{\psi}_{nm}$ *corresponding to the divisor* $\widehat{D} = D + P_3^+$ *we have*

$$\widehat{\psi}_{nm} = \psi_{n,m,\nu}\,, \tag{7.27}$$

where $\nu = \nu_{nm}$ *is defined in (7.5), i.e., is 0 or 1 so that* $n + m + \nu$ *is even.*

Corollary 7.28. *If $n+m$ is even, then by formulae (7.10), (7.23)*

$$\frac{\widehat{\theta}(\widehat{A}(P)+n\widehat{U}+m\widehat{V}+\widehat{Z})\,\widehat{\theta}(\widehat{A}(P_0)+\widehat{Z})}{\widehat{\theta}(\widehat{A}(P)+\widehat{Z})\,\widehat{\theta}(\widehat{A}(P_0)+n\widehat{U}+m\widehat{V}+\widehat{Z})} = \frac{\theta(A(P)+nU+mV+Z)\,\theta(A(P_0)+Z)}{\theta(A(P)+Z)\,\theta(A(P_0)+nU+mV+Z)}e^{nr_1+mr_2}, \tag{7.29}$$

where $r_i = \int_{P_0}^{P}(d\widehat{\Omega}_i - d\Omega_i)$, and we recall that $\widehat{Z} = \widehat{A}(\widehat{D}) + \widehat{\kappa}$, and Z is its image.

Remark. This equality, valid for any pair of points P, P_0 is a nontrivial identity between theta functions. The first author's attempts to derive it directly from the Schottky-Jung relations have failed so far.

Notation For brevity throughout the rest of the paper we use the notation: $\psi_{n,m} := \psi_{n,m,\nu_{nm}}$.

Lemma 7.30. [25] *The Baker-Akhiezer function $\psi_{n,m}$ given by*

$$\psi_{n,m} = \frac{\theta(A(P)+Un+Vm+\nu_{nm}W+Z)}{\theta(Un+Vm+\overline{\nu}_{nm}W+Z)\,\theta(A(P)+Z)}\cdot\frac{e^{n\Omega_1(P)+m\Omega_2(P)+\nu_{nm}\Omega_3(P)}}{e^{(2\nu_{nm}-1)(n\Omega_1(P_3^+)+m\Omega_2(P_3^+))}}, \tag{7.31}$$

where $\overline{\nu}_{nm} = 1-\nu_{nm}$ as in (7.5), satisfies the equation (7.2), i.e.,

$$\psi_{n+1,m+1} - u_{n,m}(\psi_{n+1,m} - \psi_{n,m+1}) - \psi_{n,m} = 0,$$

with $u_{n,m}$ as in (7.3), (7.5), where

$$c_1 = e^{\Omega_2(P_3^+)},\quad c_2 = e^{\Omega_1(P_3^+)},\quad c_3 = e^{\Omega_1(P_2^+)}. \tag{7.32}$$

Note that the first and the last factors in the denominator of (7.31) correspond to a special choice of the normalization constants $h_{n,m,\nu}$ in (7.23):

$$\begin{aligned}\psi_{nm}(P_3^-) &= (\theta(Z+W))^{-1}, & \nu_{nm} &= 0,\\ \psi_{nm}e^{-\Omega_3}|_{P=P_3^+} &= (\theta(Z-W))^{-1}, & \nu_{nm} &= 1.\end{aligned} \tag{7.33}$$

This normalization implies that for even $n+m$ the difference $(\psi_{n+1,m+1} - \psi_{n,m})$ equals zero at P_3^-. At the same time as a corollary of the normalization we get that $(\psi_{n+1,m} - \psi_{n,m+1})$ has no pole at P_3^+. Hence, these two differences have the same analytic properties on Γ and thus are proportional to each other (the relevant H^0 is one-dimensional by Riemann-Roch). The coefficient of proportionality u_{nm} can be found by comparing the singularities of the two functions at P_1^+.

The second factor in the denominator of the formula (7.31) does not affect equation (7.2). Hence, the lemma proves the "only if" part of the statement (A) of the main theorem for the case of smooth curves. It remains valid under degenerations to singular curves which are smooth outside of fixed points Q_k which are simple double points, i.e., to the curves of type $\{\Gamma, \sigma, Q_k\}$.

Remark. Equation (7.2) as a special reduction of (7.13) was introduced in [17]. It was shown that equation (7.13) implies a five-term equation

$$(7.34)\quad \psi_{n+1,m+1} - \tilde{a}_{nm}\psi_{n+1,m-1} - \tilde{b}_{n,m}\psi_{n-1,m+1} + \tilde{c}_{nm}\psi_{n-1,m-1} = \tilde{d}_{n,m}\psi_{n,m}$$

if and only if it is of the form (7.2). A reduction of the algebro-geometric construction proposed in [34] in the case of algebraic curves with involution having two fixed points was found. It was shown that the corresponding Baker-Akhiezer functions do satisfy an equation of the form (7.2). Explicit formulae for the coefficients of the equations in terms of Riemann theta-functions were obtained. The fact that the Baker-Akhiezer functions and the coefficients of the equations can be expressed in terms of Prym theta-functions was then obtained in [16] (only for the case of curves with involution having two fixed points) and [25], and was used in the latter to characterize Prym varieties.

The statement that $\psi_{n,m}$ satisfy (7.34) can be proved directly. Indeed all the functions involved in the equation are in

$$H^0(D + (n+1)P_1^+ - (n-1)P_1^- + (m+1)P_2^+ - (m-1)P_2^- + \nu(P_3^+ - P_3^-)).$$

By the Riemann-Roch theorem the dimension of the latter space is 4. Hence, any five elements of this space are linearly dependent, and it remains to find the coefficients of (7.34) by a comparison of singular terms at the points $P_1^\pm, P_2^\pm$.

Theorem 7.35. [25] *For any four points* A, U, V, W *on the image* $\Gamma \hookrightarrow \mathcal{P}(\Gamma)$, *and any* $Z \in \mathcal{P}(\Gamma)$ *the following equation holds:*

$$(7.36)\quad \begin{aligned} \theta(Z+W) \times [&\theta(A+U+V+Z)\,\theta(Z-U)\,\theta(Z-V) \\ &- c_1^2c_3^2\,\theta(A+U-V+Z)\,\theta(Z-U)\,\theta(Z+V) \\ &- c_2^2c_3^2\,\theta(A-U+V+Z)\,\theta(Z+U)\,\theta(Z-V) \\ &+ c_1^2c_2^2\,\theta(A-U-V+Z)\,\theta(Z+U)\,\theta(Z+V)] = \\ = \theta(A+Z) \times [&\theta(W+U+V+Z)\,\theta(Z-U)\,\theta(Z-V) \\ &- c_1^2c_3^2\,\theta(W+U-V+Z)\,\theta(Z-U)\,\theta(Z+V) \\ &- c_2^2c_3^2\,\theta(W-U+V+Z)\,\theta(Z+U)\,\theta(Z-V) \\ &+ c_1^2c_2^2\,\theta(W-U-V+Z)\,\theta(Z+U)\,\theta(Z+V)]. \end{aligned}$$

To the best of the authors' knowledge equation (7.36) is a new identity for Prym theta-functions. For Z such that $\theta(W+Z) = 0$ it is equivalent to equation (7.6) with the minus sign chosen. The second equation of the pair (7.6) can be obtained from (7.34) considered for the odd case, i.e., for $n + m = 1 \bmod 2$. Using theta functional formulas, it can be shown using (7.34) that equation (7.36) is equivalent to (7.2).

8. Abelian solutions of the soliton equations

In [41, 42] the authors introduced a notion of *abelian solutions* of soliton equations which provides a unifying framework the elliptic solutions of these equations and and algebraic-geometrical solutions of rank 1 expressible in terms of Riemann (or Prym) theta-function. A solution $u(x, y, t)$ of the KP equation is called *abelian* if it is of the form

$$u = -2\partial_x^2 \ln \tau(Ux + z, y, t), \tag{8.1}$$

where $x, y, t \in \mathbb{C}$ and $z \in \mathbb{C}^n$ are independent variables, $0 \neq U \in \mathbb{C}^n$, and for all y, t the function $\tau(\cdot, y, t)$ is a holomorphic section of a line bundle $\mathcal{L} = \mathcal{L}(y, t)$ on an abelian variety $X = \mathbb{C}^n/\Lambda$, i.e., for all $\lambda \in \Lambda$ it satisfies the monodromy relations

$$\tau(z + \lambda, y, t) = e^{a_\lambda \cdot z + b_\lambda} \tau(z, y, t), \quad \text{for some } a_\lambda \in \mathbb{C}^n,\ b_\lambda = b_\lambda(y, t) \in \mathbb{C}. \tag{8.2}$$

There are two particular cases in which a complete characterization of the abelian solutions has been known for years. The first one is the case $n = 1$ of elliptic solutions of the KP equations. The second case in which a complete characterization of abelian solutions is known is the case of indecomposable principally polarized abelian variety (ppav). The corresponding θ-function is unique up to normalization, so that Ansatz (8.1) takes the form $u = -2\partial_x^2 \ln \theta(Ux + Z(y, t) + z)$. Since the flows commute, $Z(y, t)$ must be linear in y and t: $u = -2\partial_x^2 \ln \theta(Ux + Vy + Wt + z)$. Besides these two cases of abelian solutions with known characterization, another may be worth mentioning. Let Γ be a curve, $P \in \Gamma$ a smooth point, and $\pi\colon \Gamma \to \Gamma_0$ a ramified covering map such that the curve Γ_0 has arithmetic genus $g_0 > 0$ and P is a branch point of the covering. Let $J(\Gamma) = \mathrm{Pic}^0(\Gamma)$ be the (generalized) Jacobian of Γ, let $\mathrm{Nm}\colon J(\Gamma) \to J(\Gamma_0)$ be the reduced norm map as in [47], and let

$$X = \ker(\mathrm{Nm})^0 \subset J(\Gamma)$$

be the identity component of the kernel of Nm. Suppose X is compact. By assumption we have

$$\dim J(\Gamma) - \dim X = \dim J(\Gamma_0) = g_0 > 0,$$

so that X is a proper subvariety of $J(\Gamma)$, and the polarization on X induced by that on $J(\Gamma)$ is *not* principal. and define the KP flows on $\overline{\mathrm{Pic}^{g-1}}(\Gamma)$ using the data (Γ, P, ζ).

In general, since for any $r_0 \in \mathbb{Z}_{>0}$ the space $\sum_{r \leqslant r_0} \mathbb{C}\partial/\partial t_r$ is independent of the choice of ζ, for any $\zeta \in \mathfrak{m}_P \setminus \mathfrak{m}_P^2$ and $0 < r < m$ (so in particular for $r = 1$), the r-th KP orbit of $\mathcal{F}$ is contained in $\mathcal{F} \otimes X$, and so it gives an abelian solution. Let us call this the *Prym-like* case. An important subcase of it is the quasiperiodic solutions of Novikov-Veselov (NV) or BKP hierarchies.

In the Prym-like case, just as in the NV/BKP case we can put singularities to Γ and Γ_0 in such a way that X remains compact, so it is more general than the KP quasiperiodic solutions. Recall that NV or BKP quasiperiodic solutions can be obtained from Prym varieties $\mathrm{Prym}(\Gamma, \iota)$ of curves Γ with involution ι having two fixed points. The Riemann theta function of $J(\Gamma)$ restricted to a suitable translate of $\mathrm{Prym}(\Gamma,\iota)$ becomes the square of another holomorphic function, which defines the principal polarization on $\mathrm{Prym}(\Gamma,\iota)$. The Prym theta function becomes NV or BKP tau function, whose square is a special KP tau function with all *even* times set to zero, so any KP time-translate of it

- gives an abelian solution of the KP hierarchy with $n = \dim X$ being one-half the genus $g(\Gamma)$ of Γ, and
- defines twice the principal polarization on X.

A natural question is whether these conditions characterize the (time-translates of) NV or BKP quasiperiodic solutions.

Hurwitz' formula tells us that in the Prym-like case $n = \dim(X) \geqslant g(\Gamma)/2$, where the equality holds only in the NV/BKP case. At the moment we have no examples of abelian solutions with $1 < n < g(\Gamma)/2$.

For simplicity we present here a solution to the classification problem of abelian solutions of the KP equation obtained in [41] under an additional assumption on the density of the orbit $\mathbb{C}U \bmod \Lambda$ in X.

Theorem 8.3. *Let $u(x,y,t)$ be an abelian solution of the KP such that the group $\mathbb{C}U \bmod \Lambda$ is dense in X. Then there exists a unique algebraic curve Γ with smooth marked point $P \in \Gamma$, holomorphic imbedding $j_0\colon X \to J(\Gamma)$ and a torsion-free rank 1 sheaf $\mathcal{F} \in \overline{\mathrm{Pic}^{g-1}}(\Gamma)$ where $g = g(\Gamma)$ is the arithmetic genus of Γ, such that setting with the notation $j(z) = j_0(z) \otimes \mathcal{F}$*

$$\tau(Ux + z, y, t) = \rho(z,y,t)\,\widehat{\tau}(x,y,t,0,\ldots \mid \Gamma, P, j(z)) \tag{8.4}$$

where $\widehat{\tau}(t_1,t_2,t_3,\ldots \mid \Gamma,P,\mathcal{F})$ is the KP τ-function corresponding to the data $(\Gamma,P,\mathcal{F})$, and $\rho(z,y,t) \not\equiv 0$ satisfies the condition $\partial_U \rho = 0$.

Note that if Γ is smooth then:

$$\widehat{\tau}(x,t_2,t_3,\cdots \mid \Gamma,P,j(z)) = \theta\Big(Ux + \sum V_i t_i + j(z) \,\Big|\, B(\Gamma)\Big)\, e^{Q(x,t_2,t_3,\ldots)}, \tag{8.5}$$

where $V_i \in \mathbb{C}^n$, Q is a quadratic form, and $B(\Gamma)$ is the period matrix of Γ. A linearization on $J(\Gamma)$ of the nonlinear (y,t)-dynamics for $\tau(z,y,t)$ indicates the possibility of the existence of integrable systems on spaces of theta-functions of higher level. A CM system is an example of such a system for $n = 1$.

Without the density assumption there are examples in which the KP hierarchy has basically no control beyond the closure of the orbit, showing the importance of the principal polarization in a Novikov-like conjecture in which a minimal number of equation is used to study the nature of X. Having this in mind, we may regard principally polarized Prym-Tjurin varieties [28] as a way to study analogues of Novikov's conjecture.

References

[1] E. Arbarello. Survey of Work on the Schottky Problem up to 1996. Section added to the 2nd edition of Mumford's Red Book, pp. 287–291, 301–304, *Lecture Notes in Math.*, **1358**, Springer, 1999. ← 207

[2] E. Arbarello and C. De Concini. On a set of equations characterizing Riemann matrices. *Ann. of Math. (2)*, **120**(1), 119–140, 1984. ← 208

[3] E. Arbarello and C. De Concini. Another proof of a conjecture of S.P. Novikov on periods of abelian integrals on Riemann surfaces. *Duke Math. Journal*, **54**, 163–178, 1987. ← 209, 213, 236

[4] E. Arbarello, I. Krichever, and G. Marini. Characterizing Jacobians via flexes of the Kummer Variety. *Math. Res. Lett.*, **13**(1), 109–123, 2006. ← 211, 234

[5] H. F. Baker. Note on the foregoing paper "Commutative ordinary differential operators". *Proc. Royal Soc. London*, **118**, 584–593, 1928. ← 217

[6] A. Beauville. *Le problème de Schottky et la conjecture de Novikov.* Séminaire Bourbaki, année 1986–87, Exposé 675. Astérisque 152–153 (1987), 101–112. ← 207

[7] A. Beauville and O. Debarre. Sur le problème de Schottky pour les variétés de Prym. *Ann. Scuola Norm. Sup. Pisa – Cl. Sci.*, Sér. 4, **14**(4), 613–623, 1987. ← 244, 245

[8] J. L. Burchnall and T. W. Chaundy. Commutative ordinary differential operators. I, II. *Proc. London Math. Soc.*, **21**, 420–440, 1922. and *Proc. Royal Soc. London*, **118**, 557–583, 1928. ← 209, 212, 213, 217, 230, 231, 232

[9] P. Buser and P. Sarnak. On the period matrix of a Riemann surface of large genus (with an appendix by J.H. Conway and N.J.A. Sloane). *Invent. Math.*, **117**, 27–56, 1994. ← 207

[10] I. V. Cherednik. Differential equations for the Baker-Akhiezer functions of algebraic curves. *Funct. Anal. Appl.*, **12**, 195–203, 1978. ← 207, 223

[11] E. Date, M. Jimbo, M. Kashiwara, and T. Miwa. Transformation groups for soliton equations, in: Proc. RIMS Symp. Non-linear integrable systems – classical theory and quantum theory, Kyoto, Japan, 13–16 May 1981. M. Jimbo and T. Miwa, eds. World Scientific, 1983, pp. 39–119. ← 207, 213, 225, 227

[12] O. Debarre. Vers une stratification de l'espace des modules des variétés abéliennes principalement polarisées, in: Complex algebraic varieties (Bayreuth, 1990), 71–86. *Lecture Notes in Math.*, **1507**, Springer, Berlin, 1992. ← 245

[13] O. Debarre. The Schottky problem: an update, in: Current topics in complex algebraic geometry (Berkeley, CA, 1992/93); pp. 57–64. (H. Clemens and J. Kollár, eds.) MSRI Publ. 28, Cambridge Univ. Press, Cambridge, 1995. ← 207

[14] P. Deligne and D. Mumford. The irreducibility of the space of curves of given genus.*Inst. Hautes Etudes Sci. Publ. Math.*, No.**36**, 75–109, 1969. ← 244

[15] L. A. Dickey. Soliton equations and Hamiltonian systems. *Advanced Series in Mathematical Physics*, Vol. **12**, World Scientific, Singapore, 1991. ← 227

[16] A. Doliwa. The B-quadrilateral lattice, its transformations and the algebro-geometric construction. *J. Geom. Phys.*, **57**, 1171–1192, 2007. ← 252

[17] A. Doliwa, P. Grinevich, M. Nieszporski and P.M. Santini. Integrable lattices and their sub-lattices: From the discrete Moutard (discrete Cauchy-Riemann) 4-point equation to the self-adjoint 5-point scheme. *J. Math. Phys.*, **48**, 013513, 2007. ← 252

[18] R. Donagi. Non-Jacobians in the Schottky loci. *Annals of Math.*, **126**, 193–217, 1987. ← 207

[19] V. Driuma. *JETP Letters*, **19**, 387–388, 1974. ← 211

[20] H. M. Farkas and H. E. Rauch. Period relations of Schottky type on Riemann surfaces. *Ann. of Math. (2)*,**92**, 434–461, 1970. ← 207

[21] J. D. Fay. Theta functions on Riemann surfaces. *Lecture Notes in Math.*, **352**, Springer-Verlag, Berlin-New York, 1973. ← 249

[22] J. D. Fay. On the even-order vanishing of Jacobian theta functions. *Duke Math. J.*, **51**(1), 109–132, 1984. ← 245

[23] B. van Geemen. Siegel modular forms vanishing on the moduli space of curves.*Invent. Math.*, **78**(2), 329–349, 1984. ← 207

[24] G. van der Geer. The Schottky problem, in: Arbeitstagung Bonn 1984; pp. 385–406. F. Hirzebruch et al., eds. *Lecture Notes in Math.*, **1111**, Springer, Berlin, 1985. ← 209

[25] S. Grushevsky and I. Krichever. Integrable discrete Schrödinger equations and a characterization of Prym varieties by a pair of quadrisecants. *Duke Mathematical Journal*, **152**(2), 318–371, 2010. ← 205, 217, 245, 247, 249, 250, 251, 252

[26] R. C. Gunning. Some curves in abelian varieties. *Invent. Math.*, **66**(2), 377–389, 1982. ← 205, 208

[27] J. Igusa. On the irreducibility of Schottky's divisor. *J. Fac. Sci. Univ. Tokyo Sect. IA Math.*, **28**(3), (1981), 531–545, 1982. ← 207

[28] V. Kanev. Principal polarizations of Prym-Tjurin varieties. *Compositio Math.*, **64**, 243–270, 1987. ← 254

[29] I. Krichever. Integration of non-linear equations by methods of algebraic geometry. *Funct. Anal. Appl.*, **11** (1), 12–26, 1977. ← 207, 209, 213, 218, 232, 242, 244

[30] I. Krichever. Methods of algebraic geometry in the theory of non-linear equations. *Russian Math. Surveys*, **32**(6), 185–213, 1977. ← 207, 209, 213, 218, 221, 232, 242, 244

[31] I. Krichever. Algebraic curves and non-linear difference equation. *Uspekhi Mat. Nauk*, **33**(4), 215–216, 1978. ← 233

[32] I. Krichever. Commutative rings of ordinary linear differential operators. *Funkts. Analiz i Ego Pril.*, **12** (3), 20–31 1978; *Funct. Anal. Appl.*, **12** (3), 175–185, 1978.← 232

[33] I. Krichever. Elliptic solutions of the Kadomtsev-Petviashvili equation and integrable systems of particles. *Funct. Anal. Appl.*, **14**(4), 282–290, 1980. ← 211

[34] I. Krichever. Two-dimensional periodic difference operators and algebraic geometry. *Doklady Akad. Nauk USSR*, **285**(1), 31–36, 1985. ← 247, 248, 252

[35] I. Krichever. A characterization of Prym var.ieties. *Int. Math. Res. Not.*, Art. ID 81476, 36 pp, 2006. ← 205, 217, 244

[36] I. Krichever. Integrable linear equations and the Riemann-Schottky problem, in: Algebraic Geometry and Number Theory, Birkhäuser, Boston, 2006. ← 205, 210, 214

[37] I. Krichever. Characterizing Jacobians via trisecants of the Kummer Variety. *Ann. of Math.*, **172**, 485–516, 2010. ← 205, 214

[38] I. Krichever, O. Lipan , P. Wiegmann, and A. Zabrodin. Quantum Integrable Systems and Discrete Classical Hirota Equations. *Commun. Math. Phys.*, **188**, 267–304, 1997. ← 215

[39] I. Krichever and S. Novikov. Two-dimensional Toda lattice, commuting difference operators and holomorphic vector bundles. *Uspekhi Mat. Nauk*, **58**(3), 51–88, 2003. ← 233

[40] I. Krichever and D. H. Phong. Symplectic forms in the theory of solitons. *Surveys in Differential Geometry*, **IV**, C.L. Terng and K. Uhlenbeck, eds. pp. 239–313, International Press, 1998. ← 226, 235

[41] I. Krichever and T. Shiota. Abelian solutions of the KP equation, in: Geometry, Topology and Mathematical Physics. V.M. Buchstaber and I.M. Krichever, eds. *Amer. Math. Soc. Transl.(2)*, **224**, 173–191, 2008. ← 217, 253, 254

[42] I. Krichever and T. Shiota. Abelian solutions of the soliton equations and geometry of abelian varieties, in: Liaison, Schottky Problem and Invariant Theory. M.E. Alonso, E. Arrondo, R. Mallavibarrena, I. Sols, eds. *Progress in Math.*, Vol.**280**, 197–222, Birkhäuser, 2010. ← 217, 253

[43] I. Krichever, P. Wiegmann, and A. Zabrodin. Elliptic solutions to difference non-linear equations and related many-body problems. *Comm. Math. Phys.*, **193**(2), 373–396, 1998. ← 247

[44] I. Krichever and A. V. Zabrodin,. Spin generalization of the Ruijsenaars-Schneider model, non-abelian 2D Toda chain and representations of Sklyanin algebra. *Uspekhi Mat. Nauk*, **50**(6), 3–56, 1995. ← 214

[45] G. Marini. A geometrical proof of Shiota's theorem on a conjecture of S.P. Novikov. *Compositio Math.*, **111**, 305–322, 1998. ← 209, 211

[46] D. Mumford. Theta characteristics of an algebraic curve. *Ann. Sci. École Norm. Sup. (4)*, **4**, 181–192, 1971. ← 249

[47] D. Mumford. Prym varieties I, in: *Contributions to analysis*, L. Ahlfors, I. Kra, B. Maskit and L. Nirenberg, eds., 325–350, Academic Press, 1974. ← 253

[48] D. Mumford. An algebro-geometric construction of commuting operators and of solutions to the Toda lattice equation, Korteweg-de Vries equation and related

non-linear equations, in: *Proceedings Int. Symp. Algebraic Geometry*, Kyoto, 1977. M. Nagata, ed., 115–153, Kinokuniya Book Store, Tokyo, 1978. ← 207, 213, 232, 233

[49] D. Mumford. Curves and their Jacobians. University of Michigan Press, Ann Arbor, 1975; also included in: The Red Book of Varieties and Schemes, 2nd Edition. *Lecture Notes in Math.*, **1358**, Springer, 1999. ← 207

[50] F. Schottky. Zur Theorie der Abelschen Functionen von vier Variabeln. *J. reine angew. Math.*, **102**, 304–352, 1888. ← 206

[51] F. Schottky and H. Jung. Neue Sätze über Symmetrralfunktionen und die Abel'schen Funktionen der Riemann'schen Theorie. S.-B. Preuss. Akad. Wiss. Berlin; *Phys. Math. Kl.*, **1**, 282–297, 1909. ← 207

[52] I. Schur. Über vertauschbare lineare Differentialausdrücke. *Sitzungsberichte der Berliner Mathematischen Gesellschaft*, **4**, 2–8, 1905. [I. Schur, Gesammelte Abhandlungen, Bd I, Springer, 1973]. ← 209

[53] G. Segal and G. Wilson. Loop groups and equations of KdV type. *IHES Publ. Math.*, **61**, 5–65, 1985. ← 242, 243, 244

[54] T. Shiota. Characterization of Jacobian varieties in terms of soliton equations. *Invent. Math.*, **83**(2), 333–382, 1986. ← 209, 213, 214, 216, 236, 237, 244

[55] T. Shiota. Prym varieties and soliton equations, in: Infinite-dimensional Lie algebras and groups (Luminy-Marseille, 1988), *Adv. Ser. Math. Phys.*, **7**, 407–448, Teaneck: World Sci. Publishing, 1989. ← 244

[56] I. Taimanov. Prym varieties of branch covers and nonlinear equations. *Matem. Sbornik*, **181**(7), 934–950, 1990. ← 244

[57] G. E. Welters. On flexes of the Kummer variety (note on a theorem of R. C. Gunning). *Nederl. Akad. Wetensch. Indag. Math.*, **45**(3), 501–520, 1983. ← 208

[58] G. E. Welters. A criterion for Jacobi varieties.*Ann. of Math.*, **120**(3), 497–504, 1984. ← 205, 210

[59] V. Zakharov and A. Shabat. A scheme for integrating the nonlinear equations of mathematical physics by the method of the inverse scattering problem. I. *Funkts. Analiz i Ego Pril.*, **8**(3), 45–53, 1974; [Funct. Anal. Appl., 8, no. 3 (1974) 226–235]. ← 211

Columbia University, New York, USA,

Landau Institute for Theoretical Physics, Moscow, Russia

Kharkevich Institute for Problems of Information Transmission, Moscow, Russia
E-mail address: `krichev@math.columbia.edu`

Kyoto University, Kyoto, Japan
E-mail address: `shiota@math.kyoto-u.ac.jp`

GIT and moduli with a twist

Radu Laza

Abstract. We survey the role played by GIT in the study of moduli spaces, with an emphasis on the birational geometry of GIT quotients.

Contents

1. Moduli and GIT – a brief history

Geometric Invariant Theory (GIT) is an important tool in the study of moduli spaces in algebraic geometry. In fact, Mumford said in the preface of the first edition

The author is partially supported by NSF grant DMS-0968968 and a Sloan fellowship.

of the foundational GIT book [94]: *"to construct moduli schemes for various types of algebraic objects ... appears to be, in essence, a special and highly non-trivial case"* of GIT. The modern point of view is more nuanced and in many instances moved away from GIT. For instance, stacks, already used in [30], have become the language of modern moduli theory. Furthermore, their use is essential in many situations (e.g. [3] from which we inspired our title). In another direction, the ideas pioneered by Kollár, Shepherd-Barron, and Alexeev ([73], [4]) and based on the minimal model program give an approach to the compactification of moduli spaces of higher dimensional varieties without using GIT (see the survey [71] in this volume). Finally, the log structures enlarge the notion of smoothness so that some moduli spaces are naturally compact (see the survey [1] in this volume). Yet, arguably GIT still plays an important role in moduli theory today. The twist in the title alludes to the developments since the appearance of variation of GIT quotient theory (abbreviated as VGIT in what follows), due to Dolgachev–Hu [32] and Thaddeus [105], in which the birational geometry of GIT quotients (and moduli spaces) plays an increasingly central role. This paper is a modest attempt to survey these developments.

The first instance of GIT and moduli spaces is probably the study of the moduli space of elliptic curves. Since an elliptic curve can be embedded as a cubic curve in $\mathbb{P}^2$ uniquely up to projective equivalence, an algebraic description of the moduli space is the (GIT) quotient $\mathbb{P}\operatorname{Sym}^3 V^* /\!/ \mathrm{PGL}(3)$ (with V the standard $\mathrm{PGL}(3)$ representation). It turns out that $\mathbb{P}\operatorname{Sym}^3 V^* /\!/ \mathrm{PGL}(3) \cong \mathbb{P}^1$ corresponding to the fact that the ring of $\mathrm{PGL}(3)$-invariant polynomials is isomorphic to the polynomial algebra $k[S,T]$ (with the standard j-invariant being the rational function $\frac{16S^3}{T^2+64S^3}$). The investigation of this type of problems, i.e. finding the invariants for various group actions coming from geometry, was one of the most active areas in mathematics in the nineteenth century. The highlights of that period include the computations of invariants for $n \leqslant 6$ points on $\mathbb{P}^1$, quartic curves, and cubic surfaces. Finding explicit invariants is a difficult task, and our knowledge today is not much better (see however [52] for a discussion of the ring of invariants for n ordered points on $\mathbb{P}^1$). This classical period of invariant theory essentially ended with the arrival (around 1900) of the fundamental result of Hilbert that says that the ring $\operatorname{Sym}(V^*)^{\mathrm{PGL}(n)}$ of invariant polynomials for a linear representation V is finitely generated.

The subject of invariant theory was reinvented by Mumford in the sixties. The work of Mumford [94] put the subject on firm theoretical footing and showed how to use GIT without explicitly knowing the invariants. In particular, Mumford used GIT to show that the moduli space of curves M_g is quasi-projective. Later, Mumford and Gieseker ([93]) proved that (the coarse moduli space associated to) the Deligne–Mumford compactification $\overline{M}_g$ of the moduli space of genus g curves is a projective compactification of M_g that can be constructed via GIT. Some other major results around the same time include the proof of quasi-projectivity for the moduli of surfaces of general type (Gieseker [41]) and compactifications for the moduli spaces of vector bundles over curves (Mumford, Narasimhan, Seshadri, e.g. [101]) and surfaces (Gieseker [42]), and then torsion free sheaves (Maruyama [86]).

After a very active period in the sixties and seventies, the focus in moduli theory and GIT shifted somewhat. Among the more important later results we mention the work of Kirwan on the cohomology of certain moduli spaces via GIT ([66]) and on partial desingularizations of GIT quotients ([67]). Also, Viehweg [106] proved that the moduli space of varieties of general type exists as a quasi-projective variety. It is important to note that the work of Viehweg does not continue the approach of Mumford and Gieseker for curves and surfaces. Namely, in order to prove stability for generic smooth varieties, Viehweg used a non-standard polarization on the Hilbert scheme, in effect prescribing the set of stable points. The changing of the linearization in a GIT problem is nowadays a common occurrence, but it has become so only in the next phase of the GIT story: the theory of variation of GIT quotients, discussed in the following paragraph.

A renewal of GIT occurred in the early nineties starting from a well-known, but essentially ignored, issue in the construction of GIT: the construction of a quotient $X/\!/G$ depends on the choice of a G-linearized ample line bundle $\mathcal{L}$ on X. Thaddeus [105] and independently Dolgachev–Hu [32] (and previously Brion–Procesi [21] in the toric case) analyzed this dependence and showed that it is surprisingly well behaved. Roughly speaking, there are only finitely many distinct possibilities for the GIT quotients, which are related by quite explicit birational operations. A little later, Hu–Keel [55] showed that these results are connected in a fundamental way with the most well behaved spaces in the minimal model program, the so-called *Mori dream spaces*.

The influence of the ideas introduced by the arrival of VGIT extends beyond the direct applications of the theory to moduli spaces. Namely, the surprisingly nice results of VGIT together with substantial progress in birational geometry (see [19]), suggest that the various birational models of a moduli space are related in a meaningful and useful way. More specifically, different constructions (or different choices in a construction) for a moduli space typically give different compactifications. Initially, this was thought of as a pathological feature of moduli theory. VGIT contributed to a significant shift in this view: one realized that the existence of multiple compactifications can be used to one's advantage. Typically, using a compactification with relatively simple structure, one can extract important information (e.g. cohomology) about other compactifications of more geometric interest. In other words, each compactification gives a facet of the moduli problem and taken together one gets a fuller picture. There are numerous instances of this principle in the current literature on moduli spaces, but we choose to mention here only two examples familiar to the author: (1) the search for a canonical model for the moduli space of curves $\overline{M}_g$ (e.g. [50, 49], and the survey [35] in this volume), and (2) the comparison between certain period domains and GIT quotients (e.g. [80, 81]).

Content. After a brief discussion (Sect. 2) of the constructions and results in the standard GIT situation, we survey (Sect. 3) the main results of the theory of variation of GIT quotients. We then (Sect. 4) discuss some tools (such as the numerical criterion) that can be used to understand GIT quotients in general, but focus on the VGIT situation. In Section 5, we review the relationship between VGIT and birational geometry. We close with a survey of some applications of the birational geometry of GIT quotients and moduli spaces (Sect. 6).

Acknowledgement. The author is grateful to M. Thaddeus and I. Dolgachev from whom he learned about the subject. The comments of M. Fedorchuk, N. Giansiracusa, K. Schwede, and D. Swinarski have helped improve the manuscript.

Disclaimer. The omissions and inaccuracies are solely the responsibility of the author. Also, the topics included are not intended to exhaust the subject, but rather reflect the interests and expertise of the author. In particular, one important topic not discussed here is the connection between GIT/VGIT and the moment map (see [94, Ch. 8] and [32]).

Conventions. We work over an algebraically closed field k of characteristic 0. Throughout the paper G is a reductive group acting on a quasi-projective variety X. The variety is always assumed to be normal, and quite frequently smooth. We will denote orbits by $G \cdot x$ and stabilizers by G_x. G^0 (and G_x^0) will denote the connected component of the corresponding group.

2. The GIT construction and main results

This section aims to give a brief overview of the standard GIT as developed by Mumford [94]. Other standard textbook references are [96], [97], [31], and [91].

2.1. Affine Quotients.

The simplest instance of a GIT quotient is that of a reductive group G acting linearly on an affine variety $X = \operatorname{Spec} R$. In this situation, a foundational result of Hilbert says that the ring of invariants $R^G \subset R$ is a finitely generated k-algebra ([94, Thm. A.1.0], [31, Thm. 3.2, 3.3]). Thus, it is natural to define the quotient to be the affine variety:

$$X/G := \operatorname{Spec} R^G. \tag{2.1}$$

Remark 2.2. The assumption of reductive group is essential for the finite generation of R^G (cf. Nagata [95]; see [92] for some geometric counter-examples, and [68] for a survey of results on quotients by non-reductive groups). Two special cases of reductive groups are of particular interest: G is a semi-simple group (e.g. $PGL(n)$) or a torus $(\mathbb{G}_m)^n$. In general a reductive group is built out of these two cases (see [94, App. A] for more on reductive groups).

2.1.1. Categorical vs. Geometric Quotient. In the absence of additional choices (such as characters for G, as discussed elsewhere in this survey), the definition (2.1) is essentially the only possibility in the realm of algebraic geometry. More precisely, the natural G-invariant projection $\pi : X \to X/G$ makes X/G a *universal categorical quotient* (see [94, Def. 0.7] and [94, Thm. 1.1]), i.e. any G-invariant morphism $f : X \to Y$ factors through X/G:

$$\begin{array}{ccc} X & \xrightarrow{\pi} & X/G \\ & \searrow^{f} & \downarrow{\scriptstyle \exists ! \bar{f}} \\ & & Y \end{array} .$$

However, X/G is typically not a *geometric quotient* (see [94, Def. 0.6]). In general, there is no one-to-one correspondence between the points of X/G and the orbits of G. For a simple example, consider the action of $G = \mathbb{C}^*$ on $X = \mathbb{A}^1_{\mathbb{C}} \cong \mathbb{C}$ given by the natural multiplication $(t, x) \in \mathbb{C}^* \times \mathbb{C} \to t \cdot x \in \mathbb{C}$. Since $\mathbb{C}[x]^G \cong \mathbb{C}$ (the only invariants are the constants), the quotient X/G is a point and the quotient map π is the trivial projection $\mathbb{C} \to \{\text{pt.}\}$. On the other hand, the action of G on X has two orbits: $\{0\} = G \cdot \{0\}$ and $\mathbb{C} \setminus \{0\} = G \cdot \{1\}$. The issue here is that the fibers of π are always closed in X. In our example, the orbit $\{0\}$ is closed, while the closure of the orbit $\mathbb{C} \setminus \{0\}$ is the affine line, which contains the closed orbit $\{0\}$. Thus, the two orbits map to the same point via π, showing that the orbits are not always separated by the quotient map.

2.1.2. Closed Orbits. In some sense, when G is reductive, the failure of X/G to be a geometric quotient is always of the type exemplified above. Namely, we recall the following fact about the orbits of algebraic group actions (e.g. [56, §8.3]): *each G-orbit is smooth, locally closed in X, and its boundary is a union of orbits of strictly lower dimension.* This easily gives that each fiber of π contains a unique closed orbit, namely, the orbit of minimal dimension in that fiber. Furthermore, if $G \cdot x_0$ is a closed orbit in X, then for all $x \in \pi^{-1}(\pi(x_0))$, the orbit of $G \cdot x$ contains $G \cdot x_0$ in its closure.

2.2. Projective Quotients.

One is typically interested in the action of a reductive group G on a projective variety X. Mumford constructed a GIT quotient in this situation by considering an ample line bundle $\mathcal{L}$ together with a G-linearization (i.e. essentially a lift of the G-action from X to $\mathcal{L}$; see [94, §3]). This choice gives an embedding:

$$i : X \xrightarrow{|\mathcal{L}^{\otimes k}|} \mathbb{P}^N \quad (\text{for some } k \gg 0)$$

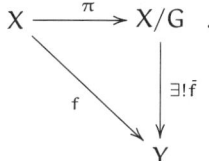

such that G acts linearly on $\mathbb{P}^N$ and the embedding i is G-equivariant. By considering affine cones, one can reduce to the affine situation. Concretely, the definition of a GIT quotient[1] in the projective case is as follows.

Definition 2.3. Let G be a reductive group acting on a projective variety X. For $\mathcal{L}$ an ample G-linearized line bundle, the associated *GIT quotient* is the projective variety:

$$X/\!/_{\mathcal{L}}G := \operatorname{Proj} \oplus_{n\geqslant 0} H^0(X, \mathcal{L}^{\otimes n})^G. \tag{2.4}$$

As in the affine case, one shows that $X/\!/G$ has good categorical properties and thus the definition (2.4) is in some sense canonical (see [94, Thm. 1.10]).

Remark 2.5. A more general situation is the relative situation: X projective over an affine variety Y, G acts equivariantly on $X \to Y$, and $\mathcal{L}$ is a relative ample line bundle on X. The above definitions can be easily adapted to this relative situation. In particular, we note that there exists a structural morphism $X/\!/_{\mathcal{L}}G \to Y/G$, where $X/\!/_{\mathcal{L}}G$ is a projective quotient (the relative version of (2.4)) and Y/G is an affine quotient (as in (2.1)). A particularly interesting case is the toric case discussed in §5.2: $X = Y = \mathbb{A}^n$, $G = T = (\mathbb{G}_m)^r$, $\mathcal{L} = \mathcal{O}_X$.

2.2.1. Semistable points. A new type of behavior appears in the case of projective quotients. Namely, the natural quotient map $\pi : X \to X/\!/_{\mathcal{L}}G$ is only a rational map: the domain of definition of π is precisely the set of points $x \in X$ such that there exists a G-invariant section $\sigma \in H^0(X, \mathcal{L}^{\otimes n})^G$ (for some n) with $\sigma(x) \neq 0$ (N.B. here X_σ is automatically affine). We call such points *semistable* and denote by $X^{ss}(\mathcal{L}) \subset X$ the corresponding open set. The points in $X^{us}(\mathcal{L}) := X \setminus X^{ss}(\mathcal{L})$ are called *unstable* and are excluded from the GIT analysis.

2.2.2. Stable points. In moduli theory, one is particularly interested in geometric quotients. The discussion of §2.1.2 applies also here. It follows that for points $x \in X^{ss}(\mathcal{L})$ such the orbit $G \cdot x$ is closed in X^{ss} and of maximal dimension (i.e. $\dim G{\cdot}x = \dim G$, or equivalently the stabilizer G_x is finite) one has $\pi^{-1}(\pi(x)) = G{\cdot}x$. One calls such points *stable points*. The set of stable points $X^s(\mathcal{L}) \subset X^{ss}(\mathcal{L})$ is an open G-invariant subset such that the induced quotient $X^s(\mathcal{L})/G$ is both a geometric and categorical quotient. If $X^s(\mathcal{L}) \neq \emptyset$, then $X^s(\mathcal{L})/G$ is an open dense subset of the quotient $X/\!/_{\mathcal{L}}G$.

Remark 2.6. To emphasize that the quotient map $X \dashrightarrow X/\!/G$ is only defined on X^{ss} and that $X/\!/G$ is only a categorical quotient, one uses the notation $/\!/$. In contrast, $X^s \to X^s/G$ satisfies all expected properties and thus the notation $/$.

---

[1] Alternatively, the general GIT quotient can be defined by gluing affine quotients (see the proof of [94, Thm. 1.10]). In the situation discussed here (esp. $\mathcal{L}$ is ample), the two approaches are equivalent.

2.3. GIT and moduli

GIT and moduli theory are closely related since in many situations it is possible to construct good parameter spaces X (e.g. Hilbert and Quot schemes) for algebraic objects (with extra rigidifying structure) on which a reductive group G acts naturally. For instance, one might be interested in (smooth) algebraic varieties of a certain type T. A typical first step in constructing a moduli space for them is to prove that all (smooth) varieties of type T have a uniform embedding $V \hookrightarrow \mathbb{P}^N$ in a large projective space (e.g. smooth curves embedded by some large power $\omega_C^{\otimes \nu}$ of the canonical bundle). If this is true, then the smooth varieties of type T are parameterized by a locally closed subset of some irreducible component X of $\mathrm{Hilb}_p(\mathbb{P}^N)$, where p is the corresponding Hilbert polynomial of V. The group $G = \mathrm{PGL}(N+1)$ acts naturally on X by change of coordinates on $\mathbb{P}^N$ (which amounts to changing the embedding $V \hookrightarrow \mathbb{P}^N$). Thus, a moduli space for varieties of type T would be roughly the GIT quotient $X/\!\!/G$, which is a projective variety. Unfortunately, it is very hard to prove that even the generic variety of type T is GIT stable. Successful applications of GIT to constructions of moduli spaces with quasi-projective coarse scheme include the important cases of abelian varieties ([94, Thm. 7.10]) and curves ([94, Cor. 7.14]). For curves, it is possible to control the stability enough to obtain that the Deligne–Mumford compactification has a projective coarse moduli space ([93, Thm. 5.1]; see [89] for a survey).

The most desirable case is when $X^s(\mathcal{L}) = X^{ss}(\mathcal{L})$. Namely, one gets that the quotient $X/\!\!/_{\mathcal{L}} G = X^s(\mathcal{L})/G$ is both a projective variety and a geometric quotient. In good situations (e.g. $\overline{M}_g$), this quotient has also a modular interpretation. If X is smooth, the quotient $X/\!\!/_{\mathcal{L}} G$, considered as a stack, is a smooth proper Deligne–Mumford stack with a projective coarse scheme (arguably, this is an ideal outcome for a moduli problem). As we will see later, quite often, $X^s(\mathcal{L}) = X^{ss}(\mathcal{L})$ happens for all generic choices of $\mathcal{L}$. Unfortunately, the natural GIT set-up for many moduli problems gives situations with $X^s(\mathcal{L}) \subsetneq X^{ss}(\mathcal{L})$. Even in those situations the fact that $X/\!\!/_{\mathcal{L}} G$ is projective can be used to one's advantage. Specifically, the properness of the quotient gives the following useful *semistable replacement property* ([93, Lem. 5.3], [102, Prop. 2.1]).

Lemma 2.7. *Let* $S = \mathrm{Spec}\, R$ *and* $S^* = \mathrm{Spec}(K)$, *where* R *is a DVR with field of fractions* K *and closed point* o. *Assume that* $S^* \to X^s/G$ *for some GIT quotient. Then, after a finite base change* $S' \to S$ *(ramified only at the special point* o*), there exists a lift* $\tilde{f} : S' \to X^{ss}$ *of* f *as in the diagram:*

$$\begin{array}{ccccccc} & & S' & \xrightarrow{\tilde{f}} & X^{ss} & & \\ & \swarrow^{f} & & & & \searrow & \\ S & \hookleftarrow & S^* & \longrightarrow & X^s/G & \hookrightarrow & X/\!\!/G \end{array} .$$

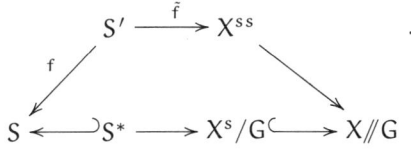

Furthermore, one can assume that $\tilde{f}(o)$ belongs to a closed orbit.

In other words, while a GIT quotient typically fails to have a modular meaning at the boundary, one can use this lemma to understand the degenerations of smooth objects and then construct or understand a good compactification of the moduli space (see [102] and [22] for some concrete applications of this principle). For a formalization, from the perspective of stacks, of the properties satisfied by the GIT quotients occurring in constructions of moduli spaces see [10].

2.4. Some concluding remarks on standard GIT

For a reductive group G acting on a projective variety X, the discussion above can be summarized as:

(1) The construction of a quotient is based on the fact that the ring of invariants R^G is finitely generated. The quotient has good functorial properties.
(2) $X/\!\!/G$ is a projective variety.
(3) The quotient map π is defined only on the semistable locus X^{ss}.
(4) Each fiber of $\pi : X^{ss} \to X/\!\!/G$ contains a unique closed orbit in X^{ss}. Furthermore, $\pi(x) = \pi(y)$ iff $\overline{G \cdot x} \cap \overline{G \cdot y} \cap X^{ss} \neq \emptyset$.
(4) X^s/G is a geometric quotient; it is an open and, if non-empty, dense subset of $X/\!\!/G$. In particular, G_x is finite and $\pi^{-1}(\pi(x)) = G \cdot x$ for $x \in X^s$.
(5) In the cases coming from moduli problems: if $X^{ss} = X^s$, then the corresponding moduli space (e.g. $\overline{M}_g$) is a Deligne–Mumford stack (and smooth if X is smooth) with the associated coarse scheme being a projective normal variety . Even when $X^{ss} \subsetneq X^s$, the projectivity of the GIT quotient is useful to analyze the degenerations of smooth objects and to compactify the moduli space.

While the definition (2.4) of the GIT quotient depends on the choice of linearization $\mathcal{L}$, we choose to ignore $\mathcal{L}$ from notation in this summary to emphasize that for a long time the dependence on $\mathcal{L}$ was essentially ignored (see however [43]). One reason for this might be that in many GIT situations coming from moduli problems there is a preferred choice for $\mathcal{L}$ (e.g. asymptotic linearizations on Hilbert schemes). A systematic investigation of the dependence on the linearization started with the toric case ([43]) and culminated with the results of Dolgachev–Hu [32] and Thaddeus [105] discussed below.

3. The main results of VGIT

In this section we review the main results of the theory of variation of GIT quotients following Thaddeus [105] and Dolgachev-Hu [32].

3.1. The space of linearizations and the partition into chambers

The first step in understanding the dependence of the GIT quotient $X/\!\!/_{\mathcal{L}}G$ on the linearization $\mathcal{L}$ is to define a GIT equivalence relation for linearizations. While the space of linearizations is essentially already understood in [94, §1.3], the fact that the natural GIT equivalence is well behaved (cf. [105, Thm. 2.4], [32, Thm. 0.2.3]) is a surprising result and one of the cornerstones of VGIT.

3.1.1. The parameter space for linearizations. A linearization consists of an underlying line bundle together with the extra data of a G-linearization ([94, Def. 1.6]). Thus, denoting by $\mathrm{Pic}^G(X)$ the space of G-linearized line bundles, there is a forgetful map $\mathrm{Pic}^G(X) \to \mathrm{Pic}(X)$, whose kernel is $\chi(G)$, the group of characters of G. However, the GIT quotient $X/\!\!/_{\mathcal{L}}G$ and the sets of stable and semistable points $X^{s(s)}(\mathcal{L})$ only depend on the algebraic equivalence class of the linearization $\mathcal{L}$ (see [105, §2], [32, Def. 2.3.4] for definitions). More precisely, one has (cf. [105, Prop. 2.1], [32, Prop. 2.3.6]):

Proposition 3.1. *If $\mathcal{L}$ is an ample linearization, then $X^{ss}(L)$, and the quotient $X/\!\!/_{\mathcal{L}}G$ regarded as a polarized variety, depend only on the G-algebraic equivalence class of $\mathcal{L}$.*

It follows that the relevant parameter space for linearizations is a finitely generated abelian group $\mathrm{NS}^G(X)$, called *the G-linearized Neron–Severi group*, which has a natural forgetful map $\mathrm{NS}^G(X) \to \mathrm{NS}(X)$ to the usual Neron–Severi group. Note however that not every line bundle $\mathcal{L}$ can be linearized, but, if X is normal[2], some power $\mathcal{L}^{\otimes n}$ can be linearized (cf. [94, Cor. 1.6]). Also, changing a G-linearized bundle $\mathcal{L}$ by a multiple does not change $X^{ss}(\mathcal{L})$ or the quotient $X/\!\!/_{\mathcal{L}}G$; it only changes the polarization of $X/\!\!/_{\mathcal{L}}G$ by the same multiple. Thus, it is preferable to work with numerical equivalences and with $\mathbb{Q}$ coefficients (*fractional linearizations* in [105]), i.e. to consider $\mathrm{NS}^G_{\mathbb{Q}}(X) := \mathrm{NS}^G(X) \otimes_{\mathbb{Z}} \mathbb{Q}$ as the parameter space for linearizations. One gets (see [32, §2.3.9] for $\mathbb{Z}$ coefficients):

Proposition 3.2. *Let G be a reductive group acting on a normal projective variety X. Then, the space of linearizations $\mathrm{NS}^G_{\mathbb{Q}}(X)$ sits in an exact sequence:*

$$0 \to \chi(G) \otimes_{\mathbb{Z}} \mathbb{Q} \xrightarrow{i} \mathrm{NS}^G_{\mathbb{Q}}(X) \xrightarrow{f} \mathrm{NS}(X) \otimes_{\mathbb{Z}} \mathbb{Q} \to 0,$$

where the inclusion i corresponds to the choices of linearization on the trivial bundle $\mathcal{O}_X$ and f is the map that forgets the data of G-linearization. In particular, $\mathrm{NS}^G_{\mathbb{Q}}(X)$ is a finite dimensional vector space of dimension $\rho(X) + \tau(G)$, where $\rho(X)$ is the Picard number of X and $\tau(G)$ is the dimension of the radical of G (a torus).

Remark 3.3. For $G = \mathrm{SL}(n)$, every line bundle carries a unique G-linearization (N.B. $\mathrm{PGL}(n)$ can be understood via the isogeny to $\mathrm{SL}(n)$, see [94, p. 33]). Thus, the choice of linearization is equivalent to the choice of an ample line bundle on X. On the other hand, for a torus $G = (\mathbb{G}_m)^n$, one has a lattice of characters $M := \chi(G) \cong \mathbb{Z}^n$. A particular case of interest for the toric case is when $X = \mathbb{A}^m$ is an affine variety. Thus, the only bundle is $\mathcal{O}_X$, and $\mathrm{NS}^G(X) \equiv M$. Quotients of this type describe the toric varieties (see §5.2).

[2] The situation in the non-normal case is significantly more delicate, e.g. [5, §4].

As is standard in algebraic geometry, one considers the convex cone $A(X)$ spanned by ample line bundles in $N^1_{\mathbb{R}}(X)$ (where $N^1_{\mathbb{R}}(X) = NS(X) \otimes_{\mathbb{Z}} \mathbb{R}$ are the numerical equivalence classes with real coefficients). By definition, the GIT quotient only makes sense on the cone $f^{-1}(A(X)) \cap NS^G_{\mathbb{Q}}(X)$, where $f : NS^G_{\mathbb{R}}(X) \to N^1_{\mathbb{R}}(X)$ is the natural forgetful map. Furthermore, there might exist ample G-linearized bundles $\mathcal{L}$ for which there are no semistable points (and thus the quotient is empty). It follows that one has to restrict to the *cone of* G*-ample line bundles*:

$$C^G(X) \subseteq f^{-1}(A(X)) \subset NS^G_{\mathbb{R}}(X) \tag{3.4}$$

spanned by classes of ample G-linearized line bundles $\mathcal{L}$ such that $X^{ss}(\mathcal{L}) \neq \emptyset$ (i.e. $\mathcal{L}$ is ample and G-effective, or simply G-ample, cf. [105, p. 701], [32, Def. 3.1.1, Def. 3.2.1]). A first result of VGIT is then ([105, (2.3)]):

Proposition 3.5. *The cone* $C^G(X)$ *is convex and it is the intersection of* $f^{-1}(A(X))$ *with a rational polyhedron in* $NS^G_{\mathbb{R}}(X)$.

The boundary of the cone corresponds to either ample line bundles for which $X^s(\mathcal{L}) = \emptyset$ (and thus the quotient $X/\!\!/G$ has less than expected dimension) or nef but not ample line-bundles ([32, Prop. 3.2.8]). In particular, note that if $X^s(\mathcal{L}) \neq \emptyset$ for some ample $\mathcal{L}$, the interior of the cone $C^G(X)$ is not empty. Note also that $C^G(X)$ is not necessarily rational polyhedral (it is so when regarded as a subset of $f^{-1}(A(X))$). This phenomenon has to do with the pathological behavior of the ample cone of X. For example, one can take $G = \{1\}$ (the trivial group) and X one of the varieties with "round" ample cone (then $C^G(X) = A(X)$ is round, but the proposition is still valid). This shows that the structural results of VGIT have to do more with G than X.

3.1.2. GIT equivalence and the partition of $C^G(X)$ into chambers.

We have discussed above that the parameter space for linearizations is $C^G(X) \cap NS^G_{\mathbb{Q}}(X)$. Then, one defines a coarsening of the algebraic equivalence of linearizations (compare Prop. 3.1) as follows ([32, Def. 3.4.1]).

Definition 3.6. For G and X as before, we say that the linearizations $\mathcal{L}$ and $\mathcal{L}'$ are *GIT equivalent* if $X^{ss}(\mathcal{L}) = X^{ss}(\mathcal{L}')$ (which also implies that $X^s(\mathcal{L}) = X^s(\mathcal{L}')$ and $X/\!\!/_{\mathcal{L}}G \cong X/\!\!/_{\mathcal{L}'}G$).

A first result of VGIT is that there are only finitely many GIT equivalence classes, i.e. there are only finitely many open subsets of X that can be realized as $X^{ss}(\mathcal{L})$ for some linearization (cf. [32, Thm. 1.3.9]). One then gets that the GIT equivalence induces a stratification of $C^G(X)$ into locally closed strata (for the Euclidean topology on $NS^G_{\mathbb{R}}(X)$; see also §3.2.1 below). To fix the terminology, which is slightly different between [105] and [32], we define:

Definition 3.7. A *GIT cell* σ is a maximal connected locally closed subset of $C^G(X)$ such that any two (classes of) linearizations $\mathcal{L}, \mathcal{L}' \in \sigma \cap NS^G_{\mathbb{Q}}(X)$ are GIT equivalent.

A chamber is a cell of maximal dimension ($= \dim_{\mathbb{R}} \mathrm{NS}^{G}_{\mathbb{R}}(X)$). *A wall* is a codimension 1 cell separating two adjacent chambers.

As noted above the stratification is finite and furthermore it is rational polyhedral in $C^G(X)$. In conclusion, one obtains one of the first major results of VGIT (cf. [105, (2.3), (2.4)] and [32, Thm. 0.2.3]; see also [98]):

Theorem 3.8. *The following hold:*

i) *There exist only finitely many cells (and thus also chambers and walls).*
ii) *The closure of a cell (in particular, of a chamber) is a convex rational polyhedral cone in* $C^G(X)$.
iii) *The cell closures form a fan covering of* $C^G(X)$.

Remark 3.9. The walls in Dolgachev-Hu [32] are defined to be the top dimensional strata for which $X^{ss}(\mathcal{L}) \neq X^s(\mathcal{L})$. The chambers of [32] are the chambers of Def. 3.7 with the additional condition $X^{ss}(\mathcal{L}) = X^s(\mathcal{L})$. While for many group actions, the two notions of chambers coincide, there exist examples with $X^s(\mathcal{L}) \subsetneq X^{ss}(\mathcal{L})$ for all $\mathcal{L} \in C^G(X)$ (see Ressayre's appendix to [32]) or only for some open regions in the interior of $C^G(X)$ (see [77], and §3.4 below).

3.2. The structure of wall crossings

3.2.1. Semi-continuity for the semistable loci and birational transformations.

A first step towards understanding the change of GIT quotient as the linearization moves from one chamber to the other is to notice the following semicontinuity property of the semistable and stable loci (cf. [105, Lem. 4.1], [32, §3.4], [98, Prop. 4]).

Lemma 3.10. *Let* $\mathcal{L}_0 \in C^G(X)$ *be the class of a* G*-ample linearization. Then for every nearby linearization* $\mathcal{L} \in C^G(X) \cap \mathrm{NS}^G_{\mathbb{Q}}(X) \cap B_\epsilon(\mathcal{L}_0)$ *(where* $B_\epsilon(\mathcal{L}_0)$ *denotes the Euclidean ball of radius* $0 < \epsilon \ll 1$ *centered at* $\mathcal{L}_0$*):*

$$X^s(\mathcal{L}_0) \subseteq X^s(\mathcal{L}) \subseteq X^{ss}(\mathcal{L}) \subseteq X^{ss}(\mathcal{L}_0). \tag{3.11}$$

Note that the inclusion of semistable loci from (3.11) forces the reverse inclusion of the stable loci. Recall that a point is stable w.r.t. $\mathcal{L}$ iff the stabilizer G_x is finite (an invariant property w.r.t. $\mathcal{L}$) and $G \cdot x$ is a closed orbit in $X^{ss}(\mathcal{L})$. Since $X^{ss}(\mathcal{L}_0)$ is an enlargement of $X^{ss}(\mathcal{L})$, the orbit $G \cdot x$ might cease to be closed in $X^{ss}(\mathcal{L}_0)$ due to the inclusion of smaller orbits in the boundary of $G \cdot x$ (see §2.1.2). The toy example $\mathbb{C}\setminus\{0\} = \mathbb{C}^* \cdot 1 \subset \mathbb{C} = \mathbb{C}\setminus\{0\} \cup \{0\}$ clearly illustrates this point. In other words, (3.11) says that in a family of linearizations $\mathcal{L}_t$ approaching a special linearization $\mathcal{L}_0$ the set of semistable points can jump up, forcing the quotient corresponding to $\mathcal{L}_0$ to become "smaller". More precisely, in the context of 3.10, the functorial properties of the GIT quotient give a diagram:

$$
\begin{array}{ccc}
X^s(\mathcal{L})/G & \hookleftarrow & X^s(\mathcal{L}_0)/G \\
\downarrow & & \downarrow \\
X/\!\!/_{\mathcal{L}}G & \xrightarrow{\varphi} & X/\!\!/_{\mathcal{L}_0}G
\end{array}
\tag{3.12}
$$

In particular, if $X^s(\mathcal{L}_0) \neq \emptyset$, the morphism φ is clearly birational (with $X^s(\mathcal{L}_0)/G$ a common dense open subset). Furthermore, φ is easily described in naive terms. Namely, if $\pi_{\mathcal{L}}$ and $\pi_{\mathcal{L}_0}$ are the two quotient maps and $G \cdot x_0$ is an orbit in $X^{ss}(\mathcal{L}_0) \setminus X^{ss}(\mathcal{L})$, then we have the following contraction

$$\varphi(\pi_{\mathcal{L}}(x)) = \pi_{\mathcal{L}_0}(x_0)$$

for all $x \in X^{ss}(\mathcal{L})$ with closed orbit (and thus separated points in $X/\!\!/_{\mathcal{L}}G$) such that $\overline{G \cdot x} \supset G \cdot x_0$ in $X^{ss}(\mathcal{L}_0)$.

Remark 3.13. One should note that φ is not always birational, it can be a fibration (but then $\mathcal{L}_0$ belongs to the boundary of $C^G(X)$). Also, it might happen that φ is an isomorphism, even though $X^{ss}(\mathcal{L}) \neq X^{ss}(\mathcal{L}_0)$.

3.2.2. The wall crossing transformations in VGIT are often flips. After the finiteness result of Thm. 3.8, the second set of major results of VGIT is concerned with structural results on the birational morphism φ of (3.12). By the general theory of birational transformations, one distinguishes three possibilities for the morphism φ:

(1) φ is a divisorial contraction;
(2) φ is a small contraction (the exceptional loci have codimension at least 2);
(3) φ is a fibration.

It turns out that the possibilities (1) and (3) are quite special and relatively rare (e.g. (3) only occurs at the boundary of $C^G(X)$). Thus, the interesting case is (2). Again, by results in birational geometry, it is known that the small contractions force the target space to have bad singularities (e.g. [72, p. 33]). The solution to this issue in birational geometry is the *flip*, i.e. one replaces a small contraction $f : X \to Y$ by another small contraction $f^+ : X^+ \to Y$ with certain properties (e.g. [72, Def. 2.8]). In particular, if it exists, X^+ is uniquely determined and has good singularities.

The discussion above suggests that one has to consider not only the morphism $\varphi : X/\!\!/_{\mathcal{L}}G \to X/\!\!/_{\mathcal{L}_0}G$ as the stability changes when passing from a chamber to the wall, but also the morphism coming from the other adjacent chamber. In other words, instead of considering the collapsing morphisms as the linearization approaches the wall, it is better to consider the *wall crossing* behavior. More specifically, one considers two (classes of) linearizations $\mathcal{L}_+, \mathcal{L}_- \in C^G(X)$ such that the GIT semistability only changes at some $\mathcal{L}_0 \in (\mathcal{L}_-, \mathcal{L}_+)$ on the interval joining $\mathcal{L}_-$ and $\mathcal{L}_+$

(in $C^G(X) \subset NS^G_{\mathbb{R}}(X)$). The discussion of §3.2.1 automatically gives the following diagram similar to that for flips from birational geometry:

$$\begin{array}{ccc} X/\!\!/_{-}G & \overset{g}{\dashrightarrow} & X/\!\!/_{+}G\,, \\ & \underset{\varphi_-}{\searrow} \quad \underset{\varphi_+}{\swarrow} & \\ & X/\!\!/_{0}G & \end{array} \tag{3.14}$$

where φ_+, φ_- are birational morphism as in (3.12) and g is the induced birational map (and, for simplicity, we omit $\mathcal{L}$ from notation). Indeed (3.14) is a flip diagram if one defines a more general notion of flip as follows ([105, p. 693]):

Definition 3.15. Let $f_- : X_- \to X$ be a small proper birational morphism. Let D be a $\mathbb{Q}$-Cartier divisor on X_- such that D is relatively negative to f_-. Then, a D-flip is a small proper birational morphism $f_+ : X_- \to X$ such that

i) the Weil divisor g_*D is $\mathbb{Q}$-Cartier, where $g : X_- \dashrightarrow X_+$ is the induced birational map;
ii) g_*D is f_+-ample.

We conclude by stating one of the main results of VGIT: the wall crossing birational map g is a flip. More precisely, one has:

Theorem 3.16. [105, Thm. (3.3)] *With notation as in* (3.14), *if both* $X/\!\!/_{-}G$ *and* $X/\!\!/_{+}G$ *are not empty, then* φ_- *and* φ_+ *are proper and birational. If they are both small, then the rational map* $g : X/\!\!/_{-}G \dashrightarrow X/\!\!/_{+}G$ *is a flip with respect to* $\mathcal{O}(1)$ *on* $X/\!\!/_{+}G$ *(the relative bundle of the projective morphism* $X/\!\!/_{+}G \to X/\!\!/_{0}G$*).*

Remark 3.17. As already hinted, we note that it is possible to have φ_- to be a divisorial contraction and φ_+ to be an isomorphism (see §3.4). However, this is a relatively rare occurrence; most often the wall crossing behavior is flip-like as described in the theorem.

3.2.3. Explicit description of the flips. The final set of results of VGIT gives a rather explicit description of the morphisms $\varphi_\pm$ and the flip g.

We describe first the morphisms $\varphi_\pm$ set-theoretically. In the set-up of (3.14), the relationships (3.11) among the semistable loci can be made more precise:

$$X^s(\mathcal{L}_0) = X^s(\mathcal{L}_-) \cap X^s(\mathcal{L}_+) \subseteq X^{ss}(\mathcal{L}_-) \cap X^{ss}(\mathcal{L}_+) \subsetneq X^{ss}(\mathcal{L}_0). \tag{3.18}$$

Let $V = X^{ss}(\mathcal{L}_+) \cap X^{ss}(\mathcal{L}_-)$ be the open G-invariant subset of $X^{ss}(\mathcal{L}_0)$ of points that remain semistable when we cross the wall. Clearly, the morphisms $\varphi_\pm$ and the rational map g are isomorphisms over the common open subset $U := V/\!\!/G$. The exceptional locus for φ_- is then

$$E_- := (X^{ss}(\mathcal{L}_-) \setminus X^{ss}(\mathcal{L}_+)) /\!\!/_{\mathcal{L}_-} G \subset X/\!\!/_{\mathcal{L}_-} G$$

and similarly for φ_+. It also easy to see that $\varphi_-(E_-) = \varphi_+(E_+) = Z$, where Z is the closed subset of $X/\!\!/_0 G$ defined by

$$Z = (X^{ss}(\mathcal{L}_0) \setminus V) \subset X/\!\!/_{\mathcal{L}_0} G.$$

In particular, note that Thm. 3.16 says that $\dim E_+ + \dim E_- = \dim X/\!\!/_0 G - 1$. However, one needs to be careful as Z can have several connected components, called *strata*, corresponding to various possibilities for the stabilizers of closed orbits in $X^{ss}(0) \setminus V$.

To understand the scheme theoretic structure of the morphisms $\varphi_\pm$, one defines the following ideal sheaves:

$$\mathcal{I}_- := \langle H^0\left(X, \mathcal{L}_+^N\right)^G \rangle,$$

and similarly $\mathcal{I}_+$, where N is a sufficiently large and divisible integer. These ideal sheaves descend to ideal sheaves on the quotients $X/\!\!/_\pm G$ which then define the exceptional loci $E_\pm$ mentioned above. Similarly, the ideal sheaf corresponding to Z is $(\mathcal{I}_+ + \mathcal{I}_-)/\!\!/_0 G$. The following results of Thaddeus then describes the flip g in terms of blow-ups and blow-downs as in the familiar situation of the flip connecting the two small resolution of the three-dimensional quadric cone.

Theorem 3.19 ([105, Thm. 3.5]). *The pullbacks of $(\mathcal{I}_+ + \mathcal{I}_-)/\!\!/_0 G$ by the morphisms $\varphi_\pm : X/\!\!/_\pm G \to X/\!\!/_0 G$ are exactly $\mathcal{I}_\pm/\!\!/_\pm G$. The blow-ups of $X/\!\!/_\pm G$ at $\mathcal{I}_\pm/\!\!/_\pm G$, and of $X/\!\!/_0 G$ at $(\mathcal{I}_+ + \mathcal{I}_-)/\!\!/_0 G$, are all naturally isomorphic to the irreducible component of the fibered product $X/\!\!/_- G \times_{X/\!\!/_0 G} X/\!\!/_+ G$ dominating $X/\!\!/_0 G$.*

Without further simplifying assumptions these blow-ups are quite complicated, for instance they typically involve blow-ups of non-reduced or reducible schemes. To get more standard blow-ups, one makes the following assumptions:

(1) X is smooth (otherwise the singularities of the quotient will reflect the singularities of X);
(2) the stabilizers $G_x \cong \mathbb{G}_m$ for points $x \in X^{ss}(\mathcal{L}_0) \setminus V$ (to guarantee a reasonable structure for the blow-up locus Z).

We can now state a final major result of VGIT (cf. [105, Thm. 5.6], [32, Thm. 4.2.7]):

Theorem 3.20. *Assume that X is smooth and $z \in Z$ corresponds to a closed orbit $G \cdot x$ such that $G_x \cong \mathbb{G}_m$. Then over a neighborhood of z in Z, the exceptional divisors $E_\pm$ of $\varphi_\pm$ are fibrations, locally trivial in the étale (or analytic) topology, with fiber weighted projective spaces.*

The identification of the weights of the weighted projective fibers in the theorem above can be done using Luna slice theorem (see §4.2).

3.3. Some concluding remarks on VGIT

We conclude with a summary of the main results of VGIT:

(1) There are finitely many possibilities for the GIT quotients $X/\!/_{\mathcal{L}}G$ as one varies the linearization $\mathcal{L}$. The set of linearizations is partitioned into rational polyhedral chambers parameterizing GIT equivalent linearizations (Thm. 3.8).
(2) The semistable loci satisfy a semi-continuity property (Lem. 3.10). This induces morphisms between quotients for nearby linearizations.
(3) The birational change of the GIT quotient as the linearization moves from one chamber to another by passing a wall is a flip (Thm. 3.16), which can be understood in terms of blow-ups and blow-downs (Thm. 3.19).
(4) Under some genericity assumptions, the flips of (3) can be described quite explicitly (Thm. 3.20).

These results have opened the door to a multitude of applications to the construction and geometry of moduli spaces (for a sample see section 6). More surprisingly, VGIT had strong influences in birational geometry. While the birational geometry seen in VGIT is only a particular case, Hu and Keel [55] have shown that in some sense, it is the most well behaved part of birational geometry. This is discussed in section 5.

3.4. An example of VGIT

We choose to illustrate the variation of GIT quotients with a somewhat non-standard example (taken from [77]). We use this example since it exemplifies in a simple geometric situation several aspects of VGIT. Furthermore, it illustrates one of the key strengths of VGIT construction of moduli spaces: its flexibility. Specifically, one might be interested in two birational models for a moduli space (e.g. obtained by some special constructions). If those two models can be realized as GIT quotients $X/\!/_{\mathcal{L}_1}G$ and $X/\!/_{\mathcal{L}_2}G$ for two choices of linearization, then, using VGIT, one gets an explicit understanding of the birational relationship between them. For the connection of the VGIT example presented here to other types of constructions of moduli spaces see Prop. 3.22, §6.2, and [77].

Our example is probably the simplest VGIT analogue of the plane curve example of [94, §4.2]. Namely, we consider the moduli space of pairs (C, L) consisting of a plane curve C of degree d and a line L in $\mathbb{P}^2$. The natural GIT set-up for the study of this moduli space is that of the group $G = SL(3)$ acting diagonally on the parameter space

$$X = \mathbb{P}(H^0(\mathbb{P}^2, \mathcal{O}_{\mathbb{P}^2}(d))) \times \mathbb{P}(H^0(\mathbb{P}^2, \mathcal{O}_{\mathbb{P}^2}(1))) \cong \mathbb{P}^N \times \mathbb{P}^2,$$

with $N = \binom{d+2}{2} - 1$. In this situation, the space of linearizations $NS^G(X)$ is identified with $\mathrm{Pic}(X) \cong \mathbb{Z} \times \mathbb{Z}$. To fix the notation, let π_1, π_2 be the two projections from X to $\mathbb{P}^N$ and $\mathbb{P}^2$ respectively. For $(a, b) \in \mathbb{Z} \times \mathbb{Z}$, we define $\mathcal{O}(a, b) = \pi_1^*\mathcal{O}(a) \otimes \pi_2^*\mathcal{O}(b)$ and say that it has *slope* $t = \frac{b}{a} \in \mathbb{Q} \cup \{\infty\}$. Since replacing a linearization by a multiple does change the GIT quotient, the quotient $X/\!/_{\mathcal{L}}G$ and the (semi)stable set $X^{s(s)}(\mathcal{L})$

depend only on the slope t of $\mathcal{L}$; we call a point $x \in X^{s(s)}(\mathcal{L})$ t-*(semi)stable* and denote the quotient by $\mathcal{M}(t)$.

The nef cone $\overline{A(X)} \subset NS^G_{\mathbb{R}}(X)$ of X is the upper quadrant $a, b \geqslant 0$. One then identifies the subcone given by closure of the G-ample cone ([77, (2.5)]) as:

$$\overline{C^G(X)} = \left\{0 \leqslant t = \frac{b}{a} \leqslant \frac{d}{2}, a \geqslant 0\right\} \subset \overline{A(X)} \subset NS^G_{\mathbb{R}}(X) \cong \mathbb{R} \times \mathbb{R}. \tag{3.21}$$

The two extremal rays of the cone exemplify the two types of failure of G-ampleness. Namely, $t = 0$ corresponds to a semi-ample, but not ample linearization (giving the projection $\pi_1 : X \to \mathbb{P}^N$), while $t = \frac{d}{2}$ corresponds to an ample linearization for which there is no stable point. In this case, the GIT quotient still makes sense for these two boundary walls. One then gets:

(1) $\mathcal{M}(0) = \mathbb{P}^N /\!/ G$ is the GIT quotient for degree d curves; the VGIT map $\mathcal{M}(\epsilon) \to \mathcal{M}(0)$ is the forgetful map $(C, L) \to C$ (generically a $\mathbb{P}^2$-fibration).
(2) $\mathcal{M}\left(\frac{d}{2}\right)$ is isomorphic to the GIT quotient for d unordered points in $\mathbb{P}^1$; the VGIT map $\mathcal{M}\left(\frac{d}{2} - \epsilon\right) \to \mathcal{M}\left(\frac{d}{2}\right)$ is the forgetful map $(C, L) \to C \cap L \subset L \cong \mathbb{P}^1$ (generically a weighted projective fibration).

This description of the quotient at the boundary walls of $\overline{C^G(X)}$ implies that the variation of the slope $t \in [0, \frac{d}{2}]$ interpolates between two conditions of stability: the stability of degree d curves in $\mathbb{P}^2$ (at $t = 0$) and the stability of d-tuples of points in $\mathbb{P}^1$ (at $t = \frac{d}{2}$). In other words, as t increases, C is allowed to be more singular, but the intersection $C \cap L$ should satisfy stronger transversality conditions. This can be made more precise, by relating the GIT approach to the construction of the moduli of pairs to an approach based on MMP as in [45]:

Proposition 3.22. *Let (C, L) be a degree d pair. If the pair $\left(\mathbb{P}^2, \frac{3}{d+t}(C + tL)\right)$ is log canonical, then (C, L) is t-semistable.*

Remark 3.23. We emphasize that the implication in the above proposition is only one direction. For some discussion of the relationship between GIT and MMP stability see [45, §10] and [65].

The finiteness result of VGIT (Thm. 3.8) says that there are only finitely many non-isomorphic GIT quotients $\mathcal{M}(t)$ and that the cone $\overline{C^G(X)}$ is partitioned into subcones given by the closures of GIT chambers. In this simple situation, there will be a finite number of critical slopes $t_0 = 0 < t_1 < \cdots < t_k = \frac{d}{2}$ (for which the stability changes) corresponding to the walls. The subcones partitioning $\overline{C^G(X)}$ will be spanned by rays of slopes t_i and t_{i+1}.

For concreteness, we restrict here to pairs of degree 3. Then, the critical slopes are $t_0 = 0$, $t_1 = \frac{3}{5}$, $t_2 = 1$, and $t_3 = \frac{3}{2}$. Furthermore, using the numerical criterion (see §4.1), one easily computes the stability of degree 3 pairs:

Proposition 3.24. *Let (C, L) be a degree 3 pair. If L passes through a singular point of C, then the pair (C, L) is t-unstable for all $t > 0$. Otherwise, (C, L) is t-(semi)stable for*

$t \in (\alpha, \beta)$ *(resp.* $t \in [\alpha, \beta]$*), where* α *and* β *are given by*

$$\alpha = \begin{cases} 0 & \textit{if C has at worst nodes} \\ \frac{3}{5} & \textit{if C has an } A_2 \textit{ singularity} \\ 1 & \textit{if C has an } A_3 \textit{ singularity} \\ \frac{3}{2} & \textit{if C has a } D_4 \textit{ singularity} \end{cases} \quad \textit{and } \beta = \begin{cases} \frac{3}{5} & \textit{if L is inflectional to C} \\ 1 & \textit{if L is tangent to C} \\ \frac{3}{2} & \textit{if L is transversal to C} \end{cases}.$$

In this example, $X^{ss}(t) = X^s(t)$ for $t \in \left((0, \frac{3}{2}) \setminus \{\frac{3}{5}, 1\}\right) \cap \mathbb{Q}$, i. e. the quotients corresponding to chambers are geometric quotients. This is no longer the case starting with degree 4 (compare Rem. 3.9). The quotients corresponding to the walls will contain a unique closed orbit of a strictly semistable pair (C, L). For instance for the wall $t = \frac{3}{5}$, there exists a unique strictly semistable pair (C_0, L_0) with closed orbit which corresponds to $\alpha = \beta = t = \frac{3}{5}$: i.e. a pair consisting of a cuspidal cubic C_0 and an inflectional line L_0 (in a smooth point of C_0). Such a pair is unique up to projective equivalence, and it has a $\mathbb{G}_m$ stabilizer.

As for the birational transformations, note that there exist 7 quotients: four corresponding to the walls $\mathcal{M}(0) \cong \mathbb{P}^1$, $\mathcal{M}(\frac{3}{5})$, $\mathcal{M}(1)$, $\mathcal{M}(\frac{3}{2}) \cong \{\text{pt.}\}$ and three geometric quotients corresponding to the chambers $\mathcal{M}(\epsilon)$, $\mathcal{M}(1-\epsilon)$, and $\mathcal{M}(1+\epsilon)$. The boundary morphisms $\mathcal{M}(\epsilon) \to \mathcal{M}(0)$ and $\mathcal{M}(1+\epsilon) \to \mathcal{M}(\frac{3}{2})$ were already described above in terms of natural forgetful maps. It remains to describe the morphisms at the interior walls. For $t = 1$, one gets $\mathcal{M}(1-\epsilon) \to \mathcal{M}(1)$ is a divisorial contraction, and $\mathcal{M}(1+\epsilon) \cong \mathcal{M}(1)$. The most interesting case is $t = \frac{3}{5}$, which gives a flip as in Theorems 3.16 and 3.20. The center Z is a point corresponding to the unique closed orbit strictly semistable for $t = \frac{3}{5}$ given by the pair (C_0, L_0) identified in the previous paragraph. The exceptional divisor E_+ parameterizes pairs (C, L) such that C is a cuspidal cubic and L is a line, not passing through the cusp and not inflectional (see the stability condition from above and (3.18)). Similarly, E_- corresponds to pairs (C, L) such that C is an irreducible (but not cuspidal) cubic and L is an inflectional line. All the orbits of pairs parameterized by E_+ and E_- have in their closure the orbit of (C_0, L_0). Thus, the inclusion of (C_0, L_0) in $X^{ss}(\frac{3}{5})$ forces the collapse of $E_\pm \cong \mathbb{P}^1$ to $Z = \{\text{pt.}\}$ (via $\varphi_\pm$ as in (3.14)). The local structure of the flip at $t = \frac{3}{5}$ is discussed in §4.2.

4. Tools for the analysis of GIT quotients

In this section we discuss two essential tools for the study of GIT quotients: *the numerical criterion* and *Luna's slice theorem*. The numerical criterion gives an efficient way of finding the semistable and stable loci for a group G acting on a projective variety X by reducing the problem to the study of the induced actions for 1-parameter subgroups of G. On the other hand, the slice theorem gives a local description of the quotient by reducing the action of G to the action of the stabilizer G_x for a closed orbit $G \cdot x \subset X^{ss}$. These tools are well-known and widely used. Here, we focus on the application of these tools in a VGIT situation.

4.1. The Hilbert-Mumford numerical criterion

The numerical criterion is the main tool available for the analysis of the stable and semistable loci in a GIT situation (e.g. see [94, §4.2] for the simple case of plane curves, and [93] for the important case of $\overline{M}_g$). Also, the numerical criterion can be used to prove most of the results of VGIT ([32]).

Let $\mathcal{L} \in \mathrm{Pic}^G(X)$. For $x \in X$ and $\lambda : \mathbb{G}_m \to G$ a 1-parameter subgroup (1-PS), one defines $-\mu^{\mathcal{L}}(x,\lambda)$ to be the weight of the induced action of $\mathbb{G}_m$ on the fiber $\mathcal{L}_{x_0} \cong \mathbb{A}^1$, where $x_0 = \lim_{s\to 0} \lambda(s)\cdot x$. This numerical function $\mu^{\mathcal{L}}(x,\lambda)$ is used to check the (semi)stability of points in X via the following criterion.

Theorem 4.1 (Hilbert-Mumford Numerical Criterion). *Let $\mathcal{L}$ be an ample* G*-linearized line bundle. Then $x \in X$ is stable (resp. semistable) with respect to $\mathcal{L}$ if and only if $\mu^{\mathcal{L}}(x,\lambda) > 0$ (resp. $\mu^{\mathcal{L}}(x,\lambda) \geqslant 0$) for every nontrivial* 1*-PS λ of* G.

The numerical function $\mu^{\mathcal{L}}(x,\lambda)$ satisfies several properties (e.g. [94, pg. 49]). The most relevant one in the VGIT context is that for fixed x and λ

$$\mu^{\mathcal{L}}(x,\lambda) : \mathrm{Pic}^G(X) \to \mathbb{Z}$$

is a group homomorphism (N.B. in fact, $\mu^{\mathcal{L}}$ depends only on the numerical equivalence class of $\mathcal{L}$ in $\mathrm{NS}^G(X)$). In particular, for two linearizations $\mathcal{L}_-$ and $\mathcal{L}_+$, considering the linearizations on the segment joining them (in $\mathrm{NS}^G_{\mathbb{Q}}(X)$)

$$\mathcal{L}(t) = \mathcal{L}_-^{(1-t)} \otimes \mathcal{L}_+^{t}, \quad \text{for } t \in [0,1] \cap \mathbb{Q},$$

one gets

$$\mu^{\mathcal{L}(t)}(x,\lambda) = (1-t)\cdot \mu^{\mathcal{L}_-}(x,\lambda) + t\cdot \mu^{\mathcal{L}_+}(x,\lambda). \tag{4.2}$$

This equation essentially says that the (semi)stability for $\mathcal{L}(t)$ is an interpolation between the (semi)stability of $\mathcal{L}_-$ and $\mathcal{L}_+$ (see §3.4 for a geometric example). Furthermore, the VGIT results (esp. Thm. 3.8) say that the semistability conditions change only at finitely many $t_1, \dots, t_n \in [0,1] \cap \mathbb{Q}$ (corresponding to intersection of the segment joining $\mathcal{L}_-$ and $\mathcal{L}_+$ with walls in $\overline{C^G(X)}$). Thus, a typical analysis of the stability in a VGIT situations involves the following steps:

(1) describe the (semi)stability for $\mathcal{L}_\pm$;
(2) identify the critical points $t_1, \dots, t_n$;
(3) describe the (semi)stability for $\mathcal{L}(t_i)$ and $\mathcal{L}(t_i \pm \epsilon)$ (for $0 < \epsilon \ll 1$); in particular, identify the blow-up locus $Z(t_i) \subset X/\!\!/_{\mathcal{L}(t_i)}G$ and the closed orbits in $X^{ss}(\mathcal{L}(t_i))$ parameterized by $Z(t_i)$ (see §3.2.3 for notations).

As sketched below, these steps can be accomplished in a somewhat algorithmic way by using the numerical criterion. To understand the flip structure of theorems 3.16

and 3.20 (and in particular the exceptional loci $E_i^{\pm}$) a more appropriate tool is the slice theorem described in §4.2.

4.1.1. A GIT analysis for a fixed linearization $\mathcal{L}$ via the numerical criterion typically has two parts: a combinatorial part and a geometric part. The combinatorial part consists in fixing a maximal torus $T \subset G$ and analyzing the stability for the induced T action on X w.r.t. the induced T linearization $\mathcal{L}_T$. This is what we describe in some detail below. The geometric part consists of extrapolating from T-stability to G-stability. In practice, this is equivalent to interpreting the T-stability in intrinsic geometric terms. More specifically, note

$$X^{ss}(\mathcal{L}) = \bigcap_{T' \text{ max. torus}} X^{ss}(\mathcal{L}_{T'}) = \bigcap_{g \in G} X^{ss}(\mathcal{L}_{g \cdot T \cdot g^{-1}}) = \bigcap_{g \in G} g \cdot X^{ss}(\mathcal{L}_T).$$

Thus, x is G-semistable iff all its translates $g \cdot x$ are T-semistable. Since $x \in X$ represents a geometric object and $g \cdot x$ is an isomorphic object, one typically aims to show that unstable with respect to T is equivalent to some list of bad geometric features for the object corresponding to x. A key idea, due to Kempf [64], is that for an unstable object x there exists an essentially unique maximally destabilizing 1-PS λ. This 1-PS λ, determines a parabolic subgroup $P_\lambda \subset G$, which is the stabilizer of a flag of linear subspaces (see [94, §2.2], [64], [31, §9.5]). The geometric properties that force x to be unstable will be related to this flag (e.g. for plane curves C, the destabilizing flag will typically consists of a singular point $p \in C$ and a special tangent line L through p). We emphasize however that, in general, the geometric analysis is quite delicate and it can only be done case by case. For instance, for hypersurfaces the combinatorial part is quite easy, but the geometric analysis for higher dimensional hypersurfaces was completed only for cubics threefolds and fourfolds ([7], [77]).

Now we consider a fixed maximal torus $T \subset G$ and an ample linearization $\mathcal{L}$ (considered as a T-linearization). As described below, the T-stability is essentially a combinatorial question. As usual, let $M = \mathrm{Hom}(T, \mathbb{G}_m) \cong \mathbb{Z}^n$ (where n the dimension of T) be the lattice of characters and $N = \mathrm{Hom}(\mathbb{G}_m, T) = M^*$ the dual lattice of 1-PS. In particular, there is a natural perfect pairing:

$$\langle \cdot, \cdot \rangle : M \times N \to \mathbb{Z}.$$

The linearization $\mathcal{L}$ gives (after possibly replacing it with a power) a T-equivariant embedding $X \hookrightarrow \mathbb{P}(V)$, with T acting linearly on V. In particular, one has an eigenspace decomposition

$$V = \bigoplus_{\chi \in M} V_\chi, \tag{4.3}$$

where $V_\chi = \{v \in V \mid t \cdot v = \chi(t)v\}$. Only finitely many characters, say $\chi_1, \ldots, \chi_k$ are relevant to the decomposition (4.3). An element $x \in X$ has a lift $\tilde{x} \in V$, which decomposes $\tilde{x} = \sum_{i=1}^{k} v_i$ with $v_i \in V_{\chi_i}$. Then the numerical function is simply

$$\mu^{\mathcal{L}}(x,\lambda) = \max_{v_i \neq 0} \langle \lambda, \chi_i \rangle.$$

This leads to two basic observations:

i) for fixed x and $\mathcal{L}$, $\mu^{\mathcal{L}}(x,\lambda)$ is a piecewise linear function in $\lambda \in N \cong \mathbb{Z}^n$ (here λ is 1-PS varying in a fixed torus T);
ii) the stability of x actually depends only on the combinatorial object, *the state of* x: $\mathrm{st}^{\mathcal{L}}(x) = \{\chi_i \mid v_i \neq 0\} \subset \{\chi_1, \dots, \chi_k\} \subset M \cong \mathbb{Z}^n$.

To emphasize the second point, we let

$$\mu(\mathrm{st}^{\mathcal{L}}(x), \lambda) := \max_{\chi \in \mathrm{st}^{\mathcal{L}}(x)} \langle \chi, \lambda \rangle \; (= \mu^{\mathcal{L}}(x,\lambda))$$

(see also [64, p. 306]). Since $\mu^{\mathcal{L}}(x,\lambda^{-1}) = -\min_{v_i \neq 0} \langle \lambda, \chi_i \rangle$, it follows easily that *x is stable is equivalent to the origin in $M_{\mathbb{R}}$ being contained in the interior of the convex hull of* $\mathrm{st}^{\mathcal{L}}(x)$. For example, if X is the parameter space for degree d hypersurfaces in $\mathbb{P}^n$, $G = \mathrm{SL}(n+1)$, T is the standard torus, then V is space of degree d polynomials. The eigenspaces V_χ are spanned by the degree d monomials. The states are then the usual diagrams of Mumford ([93, p. 10]).

4.1.2. Now consider two linearization $\mathcal{L}^+$ and $\mathcal{L}^-$ each assumed very ample. Each of them will give an embedding in a (different) projective space and an eigenspace decomposition as in (4.3). For every $x \in X$, we will get two states $\mathrm{st}^+(x) \subset \{\chi_1^+, \dots, \chi_{k^+}^+\}$ and $\mathrm{st}^-(x) \subset \{\chi_1^-, \dots, \chi_{k^-}^-\}$ both being subsets of the character space M. For instance, in the situation of §3.4, for a pair (C, L) given by equations (c, l), for appropriate choices, the positive state is the subset of monomials occurring in c and the negative one is the subset of monomials in l.

For a line bundle $\mathcal{L}(t)$ as in (4.2), we get

$$\mu^{\mathcal{L}(t)}(x,\lambda) = (1-t)\cdot \mu(\mathrm{st}^-(x),\lambda) + t \cdot \mu(\mathrm{st}^+(x),\lambda). \tag{4.4}$$

We now make the trivial observation that there are only finitely many positive and negative states possible. Denote them by $\mathrm{st}_1^+, \dots, \mathrm{st}_{n^+}^+$ and $\mathrm{st}_n^-, \dots, \mathrm{st}_{n^-}^-$. Furthermore, since $\mu(\mathrm{st}_i^+, \lambda)$ is piecewise linear function in λ for each $i \in \{1, \dots, n^+\}$, there exists a finite polyhedral decomposition $\mathcal{P}$ (with some unbounded regions) of the space of 1-PS M such that all $\mu(\mathrm{st}_i^+, \lambda)$ and $\mu(\mathrm{st}_j^-, \lambda)$ are simultaneously piecewise linear with respect to $\mathcal{P}$. From the equation (4.4) the same will be true about the functions $\mu^{\mathcal{L}(t)}(x,\lambda)$ simultaneously for all $t \in [0,1]$. It follows easily that to test for the positivity of the numerical function $\mu^{\mathcal{L}(t)}(x,\lambda)$ it suffices to select a finite number of 1-PS $\lambda_1, \dots, \lambda_s$ that can be determined based on the decomposition $\mathcal{P}$.

Since a change of stability at $t \in (0,1)$ corresponds to $\mu^{\mathcal{L}(t)}(x,\lambda) = 0$ and $\mu^{\mathcal{L}(t-\epsilon)}(x,\lambda) \neq 0$ for some λ, one gets that all the critical t (i.e. corresponding to the walls) should be of the form

$$t_{ij}^{k} = \frac{\mu(st_i^-, \lambda_k)}{\mu(st_i^-, \lambda_k) - \mu(st_j^+, \lambda_k)}$$

with $\mu(st_i^-, \lambda_k)$ and $\mu(st_j^+, \lambda_k)$ of opposite signs. The index i runs over the index set $\{1, \dots, n_-\}$ for the negative states. Similarly, $j \in \{1, \dots, n_+\}$, and $k \in \{1, \dots, s\}$. In any case, one obtains a computable finite set of possible critical t. In other words, the set $\{\mathcal{L}(t_{ij}^k)\}$ contains all the possible walls in the interval that joins $\mathcal{L}^-$ and $\mathcal{L}^+$. Not all these values are actually achieved because of the following two necessary conditions for $\mathcal{L}(t_{ij}^k)$ to be realizable as a wall:

a) the states st_i^- and st_j^+ (corresponding to t_{ij}^k) should be geometrically realizable for some $x \in X$;
b) for x with fixed states st_i^- and st_j^+, one should have $\mu^{\mathcal{L}(t)}(x, \lambda) \geqslant 0$ for all λ (it suffices to check for λ_i).

The algorithm outlined above is not very efficient since the number of states $n_\pm$ is of order $2^{k_\pm}$, where $k_\pm$ is the number of characters (e.g. degree d monomials) occurring in the decomposition (4.3) for $\mathcal{L}_\pm$. However, various improvements are well-known (see [18], [90]). The author has implemented for [77] such an improved algorithm that computes all the stability conditions for pairs (C, L) as in §3.4 for reasonable degrees d (say $d \leqslant 15$). However, the geometric analysis is feasible only for much lower degrees (say $d \leqslant 6$; compare [102]).

4.1.3. Finally, we note that if x is semistable there exists a λ such that $\mu(x, \lambda) = 0$ and $x_0 = \lim_{s \to 0} \lambda(s) \cdot x$ is still semistable. The limit point x_0 will be stabilized by a subgroup $\mathbb{G}_m \subset G$ corresponding to the 1-PS λ. Thus, the analysis outlined above identifies also the relevant closed orbits $G \cdot x_0$ (parametrized by $Z(t)$ at a critical value t) as well as their stabilizers G_{x_0} (or more precisely a maximal torus in $G_{x_0}^0$).

4.2. The Luna slice theorem

A good tool for understanding the flips occurring in VGIT (cf. Thm. 3.16 and Thm. 3.20) is Luna's slice theorem ([85], [94, Appendix 1.D]).

Notation 4.5. Let H be a subgroup of G. For variety W with a left H action, we denote by $G *_H W := (G \times W)/H$, where the action of H on the product is given by $h \cdot (g, w) = (gh^{-1}, hw)$. Note that $G *_H W$ has a natural left G action.

Theorem 4.6 (Luna Slice Theorem). *Given $x \in X^{ss}$ with closed orbit $G \cdot x$ and X smooth, there exits a G_x-invariant normal slice $V_x \subset X^{ss}$ (smooth and affine) to $G \cdot x$ such that we have the following commutative diagram with Cartesian squares:*

$$\begin{array}{ccccc}
G *_{G_x} \mathcal{N}_x & \xleftarrow{\textit{étale}} & G *_{G_x} V_x & \xrightarrow{\textit{étale}} & X^{ss} \\
\downarrow & & \downarrow & & \downarrow \\
\mathcal{N}_x / G_x & \xleftarrow{\textit{étale}} & (G *_{G_x} V_x)/G & \xrightarrow{\textit{étale}} & X /\!\!/ G
\end{array}$$

where $\mathcal{N}_x$ is the fiber at x of the normal bundle to the orbit $G \cdot x$.

Remark 4.7. Note that the stabilizer G_x of a closed orbit in X^{ss} is reductive (Matsushima criterion). Thus, we are still in a standard GIT situation.

The slice theorem says that a local model for the action of G on X near x is given by the affine quotient $\mathcal{N}_x /\!\!/ G_x$ (N.B. $\mathcal{N}_x$ is vector space endowed with the natural G_x-action). The theorem can be easily adapted to a VGIT situation. Namely, assume that G_x^0 is torus. Then, a local model for the variation of quotients $X^{ss} /\!\!/_{\mathcal{L}_-} G \xrightarrow{\varphi_-} X^{ss} /\!\!/_{\mathcal{L}_0} G$ around the point $x \in X^{ss}(\mathcal{L}_0) \setminus X^s(\mathcal{L}_0)$ with closed orbit is given by $\mathcal{N}_x /\!\!/_{\chi_-} G_x \to \mathcal{N}_x / G_x$ from a projective quotient $\mathcal{N}_x /\!\!/_{\chi_-} G_x$ to the affine quotient $\mathcal{N}_x / G_x$ for a suitable character χ_- (compare §5.2). More specifically, as in the standard case G_x acts on $\mathcal{N}_x$. Since G_x fixes x, G_x acts on the fiber $(\mathcal{L}_-)_x$ with a character χ_- (N.B. the character on $(\mathcal{L}_0)_x$ is trivial since $x \in X^{ss}(\mathcal{L}_0)$ and x is stabilized by G_x). In particular, if $G_x \cong \mathbb{G}_m$, the results of Thm. 3.20 are easy to see. Namely, G_x acts with positive, zero, negative weights on the vector space $\mathcal{N}_x$. With an appropriate choice of $\pm$, one gets that $\mathcal{N}_x^{ss}(\chi_-)$ is the complement of the positive weight subspace of $\mathcal{N}_x$. Thus, the exceptional divisor E_- for φ_- is a weighted projective bundle over the 0-weight direction in $\mathcal{N}_x$ (corresponding to $Z \subset X /\!\!/_0 G$, see §3.2.3) and the weights are the negative weights.

We now consider the example discussed in §3.4. Namely, we recall that for the wall corresponding to $t = \frac{3}{5}$, there exists a unique closed orbit of a strictly semistable pair $x = (C_0, L_0)$ where C_0 is a cubic with a cusp and L_0 is inflectional. The defining equations in $\mathbb{P}^2$ can be taken to be $(x_0 x_2^2 + x_1^3 = 0)$ and $(x_0 = 0)$ respectively. With respect to the chosen coordinates the stabilizer $G_x \cong \mathbb{G}_m$ is diagonal with weights $(5, -1, -4)$. The action of G_x on the normal slice $\mathcal{N}_x$ is determined via the normal bundle sequence:

$$(4.8) \qquad 0 \to \mathcal{T}_{G\cdot x} \to \mathcal{T}_{X|G\cdot x} \to \mathcal{N}_{G\cdot x/X} \to 0.$$

We have $x = (c, l) \in X = \mathbb{P}^9 \times \mathbb{P}^2 = |3L| \times |L|$, which gives the weights of $G_x \cong \mathbb{G}_m$ on $\mathcal{T}_{X,x}$ are $-18, -12, -9, -6, -3, 0, 3, 6, 6, 9, 9$. Similarly, the weights on $\mathcal{T}_{G\cdot x, x}$ are $0, \pm 3, \pm 6, \pm 9$ (N.B. $G = SL(3)$, $G \cdot x \cong G/G_x$). It follows that the G_x action on $\mathcal{N}_x$ has weights $-18, -12, 6, 9$. This gives

$$E_- \cong W\mathbb{P}(12,18) \cong W\mathbb{P}(2,3) \cong \mathbb{P}^1$$

and similarly $E_+ \cong \mathbb{P}^1$. In conclusion, the birational transformation that occurs at $t = \frac{3}{5}$ for degree 3 pairs has the effect of replacing the point Z in $X /\!\!/_0 G$ corresponding to the pair (C_0, L_0) with the rational curves $E_\pm$ in $X /\!\!/_\pm \mathcal{L}$ respectively, as described in §3.4.

5. GIT and birational geometry

A byproduct of VGIT is a treasure of well-behaved examples in birational geometry, both of local (flips) and global nature (rational polyhedral decomposition of certain cones). It is perhaps not surprising then that VGIT has applications (e.g. used in the proof of the weak factorization theorem [2, §2.5], [54]) and influences in birational geometry (e.g. Mori dream spaces). In this section, we review the connections between VGIT and birational geometry.

5.1. Singularities of GIT quotients

Singularities play a central role in birational geometry. We start our discussion of the relationship between GIT and birational geometry with a brief review of the singularities of quotients. First, using the categorical properties of the quotient, it follows that if X is normal, then $X/\!\!/G$ is also normal. Since the singularities of $X/\!\!/G$ will reflect the singularities of X, we assume for simplicity that X is smooth. The local structure of the quotient at $\pi(x)$ is described as the quotient $\mathcal{N}_x/G_x$ of a normal slice (which can be assumed smooth affine) modulo the stabilizer G_x (see §4.2). Thus, the determining factor for the type of the singularities of the quotient are the stabilizers G_x for $x \in X^{ss}$ with closed orbit. Without any assumptions on stabilizers, Hochster–Roberts [51] proved that $X/\!\!/G$ has Cohen-Macaulay singularities. Then, Boutot [20] strengthened this to $X/\!\!/G$ has rational singularities (see [72, Thm. 5.10] for the relationship to normal Cohen-Macaulay singularities).

As discussed before, the best GIT situation is when there are no strictly semistable points ($X^{ss} = X^s$). In this case, all the stabilizers are finite and $X/\!\!/G$ has only finite quotient (or orbifold) singularities. It follows that $X/\!\!/G$ is $\mathbb{Q}$-factorial with rational singularities ([72, 5.15]). Furthermore, $X/\!\!/G$ has log terminal singularities (e.g. [99, Thm. 2]). Thus, from the point of view of the birational geometry the quotient $X/\!\!/G$ essentially behaves as a smooth variety. A more delicate question, which is relevant in the context of moduli (see [47] and [44]), is whether the singularities of $X/\!\!/G$ are canonical. This is answered by the Reid-Tai criterion (e.g. [47, App. 1]). Namely, the finite quotient singularities are locally of type $\mathbb{C}^n/G$, for G a finite subgroup of $GL(n, \mathbb{C})$. One can further assume that G acts freely in codimension 1. Then the singularity is canonical iff for each $g \in G$: $\sum \frac{1}{2\pi i} \log \zeta_k \geqslant 1$, where ζ_k are the eigenvalues of g, and log is suitably normalized. In the particular case $G \subset SL(n, \mathbb{C})$ (thus $\prod \zeta_k = 1$), the singularities are Gorenstein (i.e. index 1) canonical singularities (cf. [107]).

Remark 5.1. For surfaces, the log terminal singularities are precisely the finite quotient singularities. The canonical singularities (i.e. du Val singularities) are those of type $\mathbb{C}^2/G$ for G a finite subgroup of $SL(2, \mathbb{C})$ (see [72, (4.18), (4.20)]).

Another situation of interest is when G (or every G_x) is a torus, then $X/\!\!/G$ has toric singularities (i.e. the type of singularities that occur for toric varieties). This is a well understood and well behaved class of singularities (see [28, §11.4]). The main issue in this case is that $X/\!\!/G$ typically fails to be $\mathbb{Q}$-factorial or even $\mathbb{Q}$-Gorenstein (if $X^{ss} \neq X^s$). Recall that a toric variety (the local model of the quotient here) is $\mathbb{Q}$-factorial iff the corresponding fan is simplicial. The more general condition

$\mathbb{Q}$-Gorenstein has a similar description (see [28, 11.4.12(a)]). Thus, one can easily construct examples of quotients $X/\!\!/G$ that are not $\mathbb{Q}$-Gorenstein by starting with a fan that does not satisfy the $\mathbb{Q}$-Gorenstein condition, then the associated toric variety has a GIT quotient description (see §5.2) that produces the desired example. On the other hand, assuming that $X/\!\!/G$ is $\mathbb{Q}$-Gorenstein, then $X/\!\!/G$ is automatically log terminal ([99, Thm. 2]). Furthermore, assuming $\mathbb{Q}$-Gorenstein, the conditions of terminal or canonical singularities can be described in terms of the associated fan ([28, 11.4.12(b)]).

In conclusion, the singularities of quotients $X/\!\!/_{\mathcal{L}}G$ are mild for linearizations $\mathcal{L}$ belonging to chambers, but the quotients corresponding to walls typically fail to be $\mathbb{Q}$-factorial (or even $\mathbb{Q}$-Gorenstein). Thus, in order to satisfy the usual assumptions of birational geometry, one needs to perform a flip (by passing to a nearby chamber).

Remark 5.2. For quotients corresponding to linearizations lying on the wall, one can apply Kirwan's (partial) desingularization procedure ([67]) to resolve the singularities corresponding to the closed orbits $G \cdot x$ with G_x^0 non-trivial. From the description of the local structure of the quotient (see §4.2), it is clear that Kirwan's resolution dominates the VGIT flip similarly to the case of the cone over the quadric surface (compare with Thm. 3.19).

5.2. GIT and toric varieties

GIT for torus actions on the affine space is essentially equivalent to the theory of toric varieties ([36]). Consequently, in this situation, GIT and VGIT can be described quite explicitly in combinatorial terms. This leads to numerous non-trivial examples in GIT and birational geometry. Here we only review the basics as a preparation for §5.3. For more details and examples, we refer the reader to [31, Ch. 12], [28, Chapters 5, 14,15], and [91, Ch. 6].

The quotient of an affine space $X = \mathbb{A}^n$ by the linear action of a torus $G = T \cong (\mathbb{G}_m)^r$ is an affine toric variety. Namely, it is standard that the action of the torus T on $\mathbb{A}^n$ can be diagonalized:

$$(t, x) \in T \times \mathbb{A}^n \to (\chi_1(t) \cdot x_1, \ldots, \chi_n(t) \cdot x_n) \in \mathbb{A}^n,$$

where $\chi_i \in M_T \cong \mathbb{Z}^r$ are characters of T and $(x_1, \ldots, x_n)$ are suitable coordinates on $\mathbb{A}^n$. Then, $X/G = \operatorname{Spec} R^G$ (see §2.1) is an affine toric variety. Namely, the ring of invariants is

$$R^G = k[x_1, \ldots, x_n]^T = k[S],$$

where S is the semigroup

$$S = \left\{ \sum_{i=1,n} \alpha_i \chi_i = 0,\ \alpha_i \in \mathbb{Z}_{\geqslant 0} \right\} = \ker\left(M_{\mathbb{A}^n} \to M_T\right) \cap (\mathbb{Z}_{\geqslant 0})^n.$$

Here $M_{\bullet}$ denotes the lattice of characters associated to a toric variety, and the toric structure on $\mathbb{A}^n$ is determined by the diagonalization of the action of T.

More interesting quotients of torus actions on affine spaces are obtained by considering projective quotients (see §2.2). In this situation, the only line bundle is the trivial bundle $\mathcal{O}_X$, but there is a choice of linearization corresponding to a choice $\chi \in M_T$ of non-trivial character of T (compare Prop. 3.1). Similarly to the affine situation, one has $X/\!/_\chi G = \mathrm{Proj}(R^G_\chi)$, where the ring R^G_χ and the grading are defined with respect to χ as follows: the degree $d \geqslant 0$ part of R^G_χ is $k[S_d]$, where

$$S_d := \left\{ \sum_{i=1,n} \alpha_i \chi_i = d\chi,\ \alpha_i \in \mathbb{Z}_{\geqslant 0} \right\}.$$

Equivalently, R^G_χ is the usual ring of invariants for the action of $G = T$ on $\mathbb{A}^{n+1} \cong \mathbb{A}^n \times \mathbb{A}^1$ defined by the given action of T on $X = \mathbb{A}^n$ and by χ^{-1} on $\mathbb{A}^1$ (N.B. recall that a projective quotient $X/\!/G$ is defined by considering the affine cone over X, here $\mathbb{A}^{n+1}$; the linearization is a lift of the action on X to the affine cone). It follows easily that $X/\!/_\chi G$ is a toric variety, which is projective over the affine toric variety $X/G \cong \mathrm{Spec}\, k[S_0]$ (compare Rem. 2.5 and [28, Prop. 14.1.12]). Furthermore, the variation of GIT quotients that occurs when one changes the linearization on $\mathcal{O}_X$ by means of a character of the torus T is well understood (in combinatorial terms) and reviewed in other places (see [61], [28, Ch. 14, Ch. 15]).

Remark 5.3. The quotient $X/\!/_\chi G$ is a projective variety if S_0 is trivial, which is equivalent to saying that the convex hull of the characters $\chi_i \in M_T \cong \mathbb{Z}^r$ does not contain the origin (see [31, Thm. 12.2]). For example, the weighted projective spaces are GIT quotients of $\mathbb{A}^n$ by a $\mathbb{G}_m$ action with positive weights.

Conversely, any toric variety (projective over affine) can be described as a GIT quotient. In fact, Cox [27] showed that this can be done in an essentially canonical way. Namely, we recall that a toric variety V (assumed without a torus factor) with associated torus T_V can be described by means of a fan Δ living in $N_\mathbb{R}$, where $N = \mathrm{Hom}(\mathbb{G}_m, T_V)$ is the lattice of 1-parameter subgroups. Let $M = N^*$ be the character lattice. The group of Weil divisors $A_{n-1}(V)$ modulo rational equivalence is described by the exact sequence ([36, pg. 63]):

$$0 \to M \to \mathbb{Z}^{\Delta(1)} \to A_{n-1}(V) \to 0$$

where $\mathbb{Z}^{\Delta(1)}$ is the free abelian group generated by the rays of Δ that correspond to T_V-invariants Weil divisors. Applying $\mathrm{Hom}(\cdot, \mathbb{G}_m)$ to the above sequence gives:

$$1 \to T \to (\mathbb{G}_m)^{\Delta(1)} \to T_V \to 1 \tag{5.4}$$

Then the toric variety V has the following presentation as a GIT quotient:

Theorem 5.5 ([27, Thm. 2.1]). *Let V be a toric variety determined by a (non-degenerate) fan Δ. Let $X = \mathbb{A}^{\Delta(1)}$ and $T = \mathrm{Hom}(A_{n-1}(X), \mathbb{G}_m)$ acting on X by means of the inclusion $T \subset (\mathbb{G}_m)^{\Delta(1)}$. Then $V \cong X/\!/G$. Moreover, the quotient is a geometric quotient (i.e. $X^s = X^{ss}$) iff V is a simplicial toric variety.*

Remark 5.6. A toric variety has finite abelian quotient singularities iff the corresponding fan Δ is simplicial.

5.3. Mori dream spaces

Two of the main results of VGIT are: the G-ample cone $C^G(X)$ is partitioned into finitely many rational polyhedral chambers (Thm. 3.8) and the quotients corresponding to these chambers are related by flips (Thm. 3.16). Hu and Keel [55] have noticed that these are intrinsic and very desirable facts about the birational geometry of the quotients $X/\!\!/G$.

We recall that cone decompositions occur naturally in birational geometry. Namely, let V be a $\mathbb{Q}$-factorial projective variety. As usual, $N^1_{\mathbb{R}}(V)$ denotes the vector space of numerical equivalence classes of divisors (with $\mathbb{R}$ coefficients). Inside $N^1_{\mathbb{R}}(V)$, there are two natural cones:

$$\mathrm{Nef}(V) \subseteq \mathrm{Mov}(V) \subset N^1_{\mathbb{R}}(V),$$

the nef cone (the closure of *the ample cone*) and (the closure of) *the movable cone*. In addition to $\mathrm{Nef}(V)$, the movable cone $\mathrm{Mov}(V)$ contains several other subcones that are obtained from other birational models V' of V. Specifically, if $f : V \dashrightarrow V'$ is a birational map such that V' is $\mathbb{Q}$-factorial and f is an isomorphism in codimension 1, there is a natural identification $f^* : N^1_{\mathbb{R}}(V') \cong N^1_{\mathbb{R}}(V)$. Then, $f^*(\mathrm{Nef}(V'))$ is a top dimensional subcone of $\mathrm{Mov}(V)$. Furthermore, for non-isomorphic birational models, the associated cones have disjoint interiors. Simply note that if D is the pull-back of an ample divisor on V', then the ring of sections

$$R(V, D) = \oplus H^0(V, \mathcal{O}(nD))$$

is finitely generated and $V' \cong \mathrm{Proj}(R(V, D))$.

Inspired by the VGIT situation, but also for intrinsic reasons, the ideal situation would be that the movable cone decomposes in finitely many rational polyhedral chambers of type $f^*(\mathrm{Nef}(V'))$, called *Mori chambers*. Consequently, Hu and Keel [55] have defined the notion of *Mori dream space* to capture this situation.

Definition 5.7. A a projective $\mathbb{Q}$-factorial variety V is a *Mori dream space* if

i) $\mathrm{Pic}(V)$ is a finitely generated abelian group (thus $\mathrm{Pic}(V) \otimes \mathbb{Q} = N^1_{\mathbb{Q}}(V)$);
ii) $\mathrm{Nef}(V)$ is the affine hull of finitely many semi-ample line bundles;
iii) there are finitely many $f_i : V \dashrightarrow V_i$ which are isomorphisms in codimension one such that each movable divisor D on X is the pullback of some semiample divisor from some model V_i.

As the name suggests, for a Mori dream space V there is a satisfactory understanding of all birational models of V (see [55, Prop. 1.11]). The theory of variations of GIT quotients produces examples of Mori dream spaces. Namely, generalizing the situation for toric varieties, projective quotients of type $V = X/\!\!/_{\chi}T$ (for an affine variety X and a torus T acting on X) are Mori dream spaces (see [55, Cor. 2.4] for a precise statement). Varying the choice of linearization through the character χ, one obtains other birational models for V. In fact, under an appropriate identification of the lattice of characters with $N^1_{\mathbb{Q}}(V)$, the Mori chambers in Mov(V) coincide with the GIT chambers of Thm. 3.8 (see [55, Thm. 2.3]).

More surprisingly, all Mori dream spaces are GIT quotients of the type described above. Again inspired by the case of toric varieties (see Thm. 5.5), one defines for V a $\mathbb{Q}$-factorial variety with finitely generated Pic(V), *the Cox ring*

$$\mathrm{Cox}(V) := \bigoplus H^0(V, L_1^{n_1} \otimes \ldots, \otimes L_r^{n_r}),$$

where $L_1, \ldots, L_r$ is a basis of Pic(V). Then, the main result of Hu and Keel [55] is that Mori dream spaces, finitely generated Cox rings, and VGIT for torus actions on affine varieties are essentially equivalent notions.

Theorem 5.8 ([55]). *Let V be a $\mathbb{Q}$-factorial projective variety. Then V is a Mori dream space iff the Cox ring* Cox(V) *is finitely generated. In this case, V is the quotient of the affine variety $X = \mathrm{Spec}(\mathrm{Cox}(V))$ by the torus $T = \mathrm{Hom}(N^1(V), \mathbb{G}_m)$. The Mori decomposition coincides with the decomposition coming from VGIT by varying the linearization by characters of* T.

The case of toric varieties discussed in §5.2 corresponds to the case Cox(V) is the polynomial ring (and thus $X \cong \mathbb{A}^n$). We refer the reader to [28, §15, esp. Thm. 15.1.10] for a combinatorial description of the nef and moving cones in the case of toric varieties. Other examples of Mori dream spaces include the Fano (and log Fano) varieties (see [19, §1.3]).

For some recent surveys on Mori dream spaces and Cox rings see [87] and [76] respectively. For a more detailed survey on the relationship between birational geometry and VGIT see [53].

6. Applications of birational geometry of GIT quotients to moduli

Throughout the history of the subject, GIT and moduli spaces were highly interconnected. As mentioned in the introduction, some of the major achievements of GIT are the constructions of the moduli spaces of curves, of abelian varieties, of vector bundles, and of sheaves. Naturally, many of the applications of VGIT have as starting point these successful GIT stories. In particular, a very active and successful area is the application of VGIT to the computation of various cohomological groups associated to certain moduli spaces of vector bundles. A good example in this sense (and one of the motivations of [105]) is Thaddeus' proof of the Verlinde formula ([104]). On the other hand, as mentioned in section 5 (esp. §5.3), some of the applications and influences of VGIT are quite unexpected and not necessarily concerned with moduli spaces. Moreover, the interplay between VGIT and birational geometry, as well as the recent progress in birational geometry have made the study of the birational geometry of moduli spaces a central theme of modern moduli theory.

In this final section, we review two research topics in moduli spaces in which GIT and birational geometry are the central characters. These two applications are not necessarily the most representative applications of GIT or VGIT, but we believe they illustrate the main points of this survey. Namely, GIT is a useful tool for the construction of moduli spaces and then VGIT enhances the standard GIT constructions by giving them flexibility.

6.1. GIT and the birational geometry of $\overline{M}_{g,n}$

One of the great successes of GIT is its use to prove that the moduli space of curves is a projective variety. We recall that Deligne–Mumford [30] have constructed a smooth compactification for the moduli space of curves as a stack $\overline{\mathcal{M}}_g$ (a smooth proper Deligne-Mumford stack). Subsequently, Gieseker and Mumford (e.g. [93]) have constructed the associated coarse scheme $\overline{M}_g$ of $\overline{\mathcal{M}}_g$ via GIT by using asymptotic stability for the Hilbert scheme (or Chow variety) of ν-canonical embedded curves ($\nu \gg 0$). Subsequently, Mumford and Knudsen [69] have shown the projectivity of the moduli of pointed stable curves without actually constructing $\overline{M}_{g,n}$ via GIT. This was accomplished only recently by [17] and [103] (see [89] for a survey on GIT and $\overline{M}_g$). It is worth mentioning that it is possible to construct and prove the projectivity of $\overline{M}_g$ (and $\overline{M}_{g,n}$) completely avoiding GIT (see for example Kollár [70]). These methods are also applicable to moduli of higher dimensional varieties (e.g. [4]), for which there are very few results obtained via GIT (see however [41] and [106]).

The birational geometry of $\overline{M}_g$ is a topic of great interest in algebraic geometry ever since the seminal paper of Harris and Mumford [47]. Initially, the main interest was the Kodaira dimension of M_g and various conjectures on the cones of effective curves and effective divisors on $\overline{M}_g$ (e.g. [46]), esp. the Fulton conjecture (see [39], and [33] for a recent survey). Recently, partially inspired by the developments of VGIT, the focus shifted somewhat to the search for various log canonical models for the moduli spaces of curves, pointed curves (see the survey [35] in this volume), or even stable maps (e.g. [26]). Many of the log canonical models of M_g that were constructed so far are obtained via GIT. Here we briefly review two standard examples $M_{0,n}$ and M_g from a GIT perspective (for some results on the birational geometry of $M_{g,1}$ obtained via VGIT, see [59]).

6.1.1. Birational geometry of $M_{0,n}$ The compactification $\overline{M}_{0,n}$ of the moduli space of ordered n points on $\mathbb{P}^1$ is easily seen to be birational to $\mathbb{P}^{n-3}$. Kapranov [60] (and Keel [62] in a slightly different way) has given an explicit construction for $\overline{M}_{0,n}$ as a sequence of blow-ups of certain linear configurations in $\mathbb{P}^{n-3}$. Moreover, Kapranov gives a description of $\overline{M}_{0,n}$ as the Chow quotient of a Grassmanian by a torus, which can be then related to a VGIT construction (e.g. [40, Thm. 1.2]). A remaining key question is to understand the cone of effective curves (or equivalently the nef cone) of $\overline{M}_{0,n}$. A conjectural description of the cone of (numerically) effective curves on $\overline{M}_{g,n}$ is given by Fulton's conjecture (see [63] included in this volume). Moreover, a positive answer to this conjecture for $\overline{M}_{0,n}$ would imply the conjecture for all $\overline{M}_{g,n}$ (cf. [39]). GIT and VGIT can be used to understand a (small) slice of the nef cone of $\overline{M}_{0,n}$ (see [6]). Furthermore, recently discovered connections to conformal blocks and more creative GIT constructions (e.g. [38], [37]) might shed more light on the Fulton conjecture and related questions.

More importantly from our perspective, the study of the birational geometry of $\overline{M}_{0,n}$ led to the discovery by Hu and Keel [55] of the close connection between GIT and birational geometry (see §5.3). Still, the questions asked there ([55, §3]): *is* $\overline{M}_{0,n}$ *a Mori dream space?*, does not seem to have an answer yet (except for $n \leqslant 6$, in which case $\overline{M}_{0,n}$ is a log Fano variety and thus a Mori dream space; see also [25]).

6.1.2. Hassett – Keel program for M_g Another topic of great interest recently is the search for a canonical model for the moduli space of curves M_g. We recall that for large g ($g \geqslant 23$), the moduli of curves M_g is of general type, thus it is a reasonable question to ask for a canonical model M_g^{can}. Since $(\overline{M}_g, \Delta)$ (where Δ denotes the boundary divisor) is a log canonical model, a promising approach is to study the canonical model via the interpolation:

$$\overline{M}_g(\alpha) = \mathrm{Proj}(\oplus_n H^0(\overline{M}_g, n(K_{\overline{M}_g} + \alpha\Delta)).$$

Note that for $\alpha = 1$ one gets $\overline{M}_g$ (for all g), while for $g \geqslant 23$ and $\alpha = 0$ one gets M_g^{can}. Using general results in birational geometry (esp. [19]), one can show that there exists finitely many isomorphism classes $\overline{M}_g(\alpha)$, which are related by birational transformations, giving a picture similar to a VGIT situation. Moreover, it is expected that most of the resulting spaces $\overline{M}_g(\alpha)$ have modular interpretation (giving some alternate compactifications for M_g). The study of the spaces $\overline{M}_g(\alpha)$ and of their modular interpretation is the so called *Hassett–Keel program*. Note that even though the ultimate goal of this study is to understand geometrically M_g^{can}, and thus one needs $g \geqslant 23$, it still makes sense to study $\overline{M}_g(\alpha)$ for small genus. Results for large α were obtained by Hassett and Hyeon ([50], [49]) and for low genus by Hassett, Hyeon and Lee ([48], [57]). A prediction for the critical slopes is given by [11] (which is closely related to the discussion of §4.1 on finding the critical values in a VGIT situation).

Since the subject is well surveyed in other parts (see esp. [35] in this volume and [89]), we close by making some brief comments relevant to our survey on applications of GIT to moduli spaces. While Mumford and Gieseker [93] used asymptotic stability on the Hilbert scheme (or Chow variety) of ν-canonically embedded curves for $\nu \geqslant 5$ to construct $\overline{M}_g$, the most powerful tool so far to construct various $\overline{M}_g(\alpha)$ was to use GIT on Hilbert schemes for ν-canonically embedded curves for small ν. Specifically, if $\mathrm{Hilb}^m_{g,\nu}$ denotes the main component of the scheme of m-Hilbert points for ν-canonical curves, then the associated GIT quotients $\mathrm{Hilb}^m_{g,\nu} /\!/ SL(N+1)$

(with $C \xrightarrow{\omega_C^{\otimes \nu}} \mathbb{P}^N$) tend to produce examples of $\overline{M}_g(\alpha)$. For example, for $\nu = 3,4$ and asymptotic linearizations ($m \gg 0$) one gets the moduli of pseudo-stable curves $\overline{M}_g^{ps} \cong \overline{M}_g\left(\frac{9}{11}\right)$ (the first modification of $\overline{M}_g$ as α decreases), see [100] and [58]. Similarly, the case $\nu = 2$ and asymptotic linearizations produces the first instance of flip in the Hassett–Keel program (see [49]). Finally, only recently it was proved that for $\nu = 1,2$ and small m is the generic smooth genus g curve semistable when viewed as a point of $\mathrm{Hilb}^m_{g,\nu}$ (see [12]). For a more detailed discussion of the role played by GIT in the Hassett–Keel program see [35, §2.4] and [13].

Probably, the two main open questions related to the Hassett–Keel program and GIT are:

(1) Describe the GIT stability for the Hilbert scheme of canonically embedded curves (see also [89, §7.4]).
(2) A uniform GIT procedure that gives all the spaces $\overline{M}_g(\alpha)$ as instances of a VGIT problem.

While the behavior of the spaces $\overline{M}_g(\alpha)$ is as coming from VGIT, the second question seems purely speculative at this point. Also the answer to the first question seems far off with the current techniques. In fact, the questions are interesting even for low genera. Namely, for $g = 3$ a canonical (non-hyperelliptic) curve is plane quartic and thus a hypersurface. It follows that there is a unique GIT quotient for canonical genus 3 curves, which was well understood for a long time (e.g. [94, p. 80]). More recently, a complete analysis of the Hassett–Keel program in this case was done by Hyeon–Lee [57]; in particular, the GIT quotient for plane quartics was shown to be the final non-trivial space $\overline{M}_3(\alpha)$. For genus 4, a canonical curve is a (2,3) complete intersection. A natural parameter space is then a projective bundle $\mathbb{P}(E)$ over the space of quadrics (N.B. $\mathbb{P}(E)$ is birational to the corresponding Hilbert scheme $\mathrm{Hilb}_{4,1}$). Since $\mathrm{Pic}(\mathbb{P}(E)) \cong \mathbb{Z}^2$, GIT for canonical genus 4 curves leads naturally to a VGIT situation. The resulting VGIT problem and the connection to the Hassett–Keel program are discussed in [24] (see also [23] and [34]).

6.2. GIT and Hodge theory

As mentioned several times in this survey, there are usually several constructions for a moduli space, including GIT. Another standard construction for a moduli space is via Hodge theory (for a discussion of a closely related topic see [88] in this volume). For instance, the moduli space of elliptic curves can be described as the quotient $\mathfrak{h}/\mathrm{SL}(2,\mathbb{Z})$ of the Siegel upper half space by the modular group. As mentioned in the introduction there is also a natural GIT construction for the moduli space of elliptic curves. Consequently, one obtains:

$$\overline{M}_1 \cong \mathbb{P}^1 \cong (\mathfrak{h}/\mathrm{SL}(2,\mathbb{Z}))^* \cong \mathbb{P}\,\mathrm{Sym}^3 V^* /\!\!/ \mathrm{SL}(3,\mathbb{C}), \tag{6.1}$$

where * denotes the Satake–Baily–Borel compactification ([16]). This result is somewhat surprising given the different nature of the objects under consideration: $(\mathfrak{h}/\mathrm{SL}(2,\mathbb{Z}))^*$ is of analytic and arithmetic nature, while $\mathbb{P}\,\mathrm{Sym}^3 V^* /\!/ \mathrm{SL}(3,\mathbb{C})$ is purely algebraic.

It turns out that (6.1) is not an isolated result, but there is a series of similar results: for some of the $M_{0,n}$ with $n \leqslant 12$ ([29]), for M_g for $g \leqslant 4$ and $g = 6$ ([74], [82], [75], [15]), for low degree K3 surfaces ([102], [81, §8.2]), and for moduli of cubic surfaces ([8]), cubic threefolds ([7, 9], [84]), and cubic fourfolds ([78, 79] and [83]). More precisely, a similar isomorphism to (6.1) only holds for the cases of $M_{0,n}$ considered by Deligne–Mostow [29], and the moduli spaces of cubic surfaces ([8]). In all the other cases mentioned above the GIT construction and the Hodge theoretic construction differ, but in a rather minimal way: roughly speaking, a Heegner divisor associated to a hyperplane arrangement $\mathcal{H}$ inside the period domain $\mathcal{D}/\Gamma$ has to be "flipped" (i.e. blown-up to normal crossings and then contracted in the opposite direction). Looijenga [80, 81] has made this statement precise. Namely, for moduli spaces birational to arithmetic quotients $\mathcal{D}/\Gamma$ of Type IV domains or complex balls, Looijenga gave a comparison theorem ([81, Thm. 7.6]) to GIT quotients, which says that under appropriate hypotheses (satisfied by the geometric examples mentioned above):

$$\overline{(\mathcal{D}/\Gamma)}_{\mathcal{H}} \cong X /\!/_{\mathcal{L}} G, \tag{6.2}$$

where $\overline{(\mathcal{D}/\Gamma)}_{\mathcal{H}}$ is a birational modification of $\mathcal{D}/\Gamma$ associated to a hyperplane arrangement $\mathcal{H}$ (determined by the particular geometric situation). To understand (6.2), we recall that the Satake–Baily–Borel compactification $(\mathcal{D}/\Gamma)^*$ is a projective variety which can be defined as the Proj of the ring of Γ-automorphic forms on $\mathcal{D}$. Similarly, $\overline{(\mathcal{D}/\Gamma)}_{\mathcal{H}}$ is Proj of a ring of meromorphic forms with poles along $\mathcal{H}$, and thus a projective variety with a tautological polarization. Then, (6.2) is essentially equivalent to saying that a linearization $\mathcal{L}$ and a hyperplane arrangement $\mathcal{H}$ can be chosen so that the natural polarizations of $X /\!/_{\mathcal{L}} G$ and $\overline{(\mathcal{D}/\Gamma)}_{\mathcal{H}}$ agree on a open set with high codimension complement.

We note that there are numerous consequences of a result of type (6.2) for a moduli space. On one hand, the algebraic description as a GIT quotient $X /\!/_{\mathcal{L}} G$ can be used to prove properness statements about the period map (e.g. [102], [79], [83]). Conversely, the description $\overline{(\mathcal{D}/\Gamma)}_{\mathcal{H}}$ comes equipped with a rich arithmetic structure which can be then interpreted geometrically (e.g. results about the Neron–Severi group of K3 surfaces).

A particularly interesting case is that of genus 3 curves. Namely, there exists a natural GIT compactification $\overline{M}_3^{\mathrm{GIT}}$ obtained by viewing the smooth non-hyperelliptic curves of genus 3 as plane quartics, and a ball quotient description $(\mathcal{B}_6/\Gamma_6)^*$ due to Kondo [74]. These two birational models of $\overline{M}_3$ are closely related by results of the type described above. Concretely, there exists a common partial resolution

$\widehat{M}_3$ which can be viewed either as a partial Kirwan desingularization of $\overline{M}_3^{GIT}$ or as a Looijenga type arithmetic modification of $(\mathcal{B}_6/\Gamma_6)^*$ (see [82] and [14]). On the other hand, in the context of the Hassett–Keel program (see §6.1.2), one studies the log canonical models $\overline{M}_3(\alpha)$ (see [57]). It turns out that $\overline{M}_3^{GIT}$ and $(\mathcal{B}_6/\Gamma_6)^*$ actually occur as the last two non-trivial log canonical models in genus 3. Specifically, the following holds:

i) $\overline{M}_3^{GIT} \cong \overline{M}_3(\frac{17}{28})$;
ii) $\widehat{M}_3 \cong \overline{M}_3(\frac{7}{10} - \epsilon)$;
iii) $(\mathcal{B}_6/\Gamma_6)^* \cong \overline{M}_3(\frac{7}{10})$.

Similar results hold for genus 4 curves as well (see [75] and [23]).

Another interesting example, where GIT and Hodge theory and also VGIT occur, is [77]. Specifically, one considers the moduli space of degree d pairs (C, L) as described in §3.4. In the particular case $d = 5$, we prove that a special instance of the VGIT quotient (specifically $\mathcal{M}(1)$ in the notation of §3.4) is isomorphic to the Baily-Borel compactification of an arithmetic quotient $\mathcal{D}/\Gamma$ of type IV. On the other hand, another instance of the quotient (i.e. $\mathcal{M}(\frac{5}{2} - \epsilon)$) is closely related to the deformation space of a certain class of singularities (namely N_{16}). The interest in [77] is to study this deformation space, but structural results are only known for the space of $\mathcal{D}/\Gamma$. The VGIT set-up described in §3.4 connects these two spaces. In other words, in many situations, VGIT allows one to extract information from a known space (here $\mathcal{M}(1) \cong (\mathcal{D}/\Gamma)^*$) and translate it into information about a target space (here $\mathcal{M}(\frac{5}{2} - \epsilon)$).

References

[1] D. Abramovich, Q. Chen, D. Gillam, Y. Huang, M. Olsson, M. Satriano and, S. Sun. Logarithmic Geometry and Moduli, to appear in this volume (arXiv:1006.5870v1). ← 260
, in: Handbook of moduli, Vol. I (Gavril Farkas and Ian Morrison eds.), 1–63. *Advanced Lectures in Mathematics*, **24**, Higher Education Press & International Press, Beijing-Boston, 2012.

[2] D. Abramovich, K. Karu, K. Matsuki, and J. Włodarczyk. Torification and factorization of birational maps. *J. Amer. Math. Soc.*, **15**(3), 531–572, 2002 (electronic). MR 1896232 (2003c:14016) ← 281

[3] D. Abramovich and A. Vistoli. Compactifying the space of stable maps. *J. Amer. Math. Soc.*, **15**(1), 27–75, 2002 (electronic). MR 1862797 (2002i:14030) ← 260

[4] V. Alexeev. Moduli spaces $M_{g,n}(W)$ for surfaces, in: Higher-dimensional complex varieties (Trento, 1994), 1–22, de Gruyter, Berlin, 1996. MR 1463171 (99b:14010) ← 260, 286

[5] V. Alexeev. Complete moduli in the presence of semiabelian group action. *Ann. of Math. (2)*, **155**(3), 611–708, 2002. MR 1923963 (2003g:14059) ← 267

[6] V. Alexeev and D. Swinarski. Nef divisors on $\overline{M}_{0,n}$ from GIT, 2008. (arXiv:0812.0778) ← 287

[7] D. Allcock. The moduli space of cubic threefolds. *J. Algebraic Geom.*, **12**(2), 201–223, 2003. MR 1949641 (2003k:14043) ← 277, 289

[8] D. Allcock, J. A. Carlson, and D. Toledo. The complex hyperbolic geometry of the moduli space of cubic surfaces. *J. Algebraic Geom.*, **11**(4), 659–724, 2002. MR 1910264 (2003m:32011) ← 289

[9] D. Allcock, J. A. Carlson, and D. Toledo. The moduli space of cubic threefolds as a ball quotient, to appear in: Memoirs of the A.M.S. ← 289

[10] J. Alper. Good moduli spaces for Artin stacks, 2008. (arXiv:0804.2242) ← 266

[11] J. Alper, M. Fedorchuk, and D. I. Smyth. Singularities with G_m-action and the log minimal model program for $\overline{M}_g$, 2010. (arXiv:1010.3751v1) ← 287

[12] J. Alper, M. Fedorchuk, and D. I. Smyth. Finite Hilbert stability of (bi)canonical curves, 2011. (arXiv:1109.4986) ← 288

[13] J. Alper and D. Hyeon. GIT constructions of log canonical models of M_g, 2011. (arXiv:1109.2173) ← 288

[14] M. Artebani. A compactification of M_3 via K3 surfaces. *Nagoya Math. J.*, **196**, 1–26, 2009. MR 2591089 (2011a:14070) ← 290

[15] M. Artebani and S. Kondō. The moduli of curves of genus six and K3 surfaces. *Trans. Amer. Math. Soc.*, **363**(3), 1445–1462, 2011. ← 289

[16] W. L. Baily, Jr. and A. Borel. Compactification of arithmetic quotients of bounded symmetric domains. *Ann. of Math. (2)*, **84**, 442–528, 1966. MR 0216035 (35 #6870) ← 289

[17] E. Baldwin and D. Swinarski. A geometric invariant theory construction of moduli spaces of stable maps. *Int. Math. Res. Pap. IMRP (2008)*, no. **1**, Art. ID rpn 004, 104. MR 2431236 (2009f:14018) ← 286

[18] D. Bayer and I. Morrison. Standard bases and geometric invariant theory. *J. Symbolic Comput.*, **6** (2–3), 209–217, 1988. ← 279

[19] C. Birkar, P. Cascini, C. D. Hacon, and J. McKernan. Existence of minimal models for varieties of log general type. *J. Amer. Math. Soc.*, **23**(2), 405–468, 2010. MR 2601039 (2011f:14023) ← 261, 285, 287

[20] J. F. Boutot. Singularités rationnelles et quotients par les groupes réductifs. *Invent. Math.*, **88**(1), 65–68, 1987. MR 877006 (88a:14005) ← 281

[21] M. Brion and C. Procesi. Action d'un tore dans une variété projective, in: Operator algebras, unitary representations, enveloping algebras, and invariant theory (Paris, 1989), 509–539. *Progr. Math.*, Vol. **92**, Birkhäuser Boston, Boston, MA, 1990. MR 1103602 (92m:14061) ← 261

[22] L. Caporaso. A compactification of the universal Picard variety over the moduli space of stable curves. *J. Amer. Math. Soc.*, **7**(3), 589–660, 1994. MR 1254134 (95d:14014) ← 266

[23] S. Casalaina-Martin, D. Jensen, and R. Laza. The geometry of the ball quotient model of the moduli space of genus four curves, in: Compact moduli spaces and vector bundles, 107–136, *Contemp. Math.*, **564**, Amer. Math. Soc., Providence, RI, 2012. ← 288, 290

[24] S. Casalaina-Martin, D. Jensen, and R. Laza. Log canonical models and variation of GIT for genus four canonical curves, (arXiv:1109.4986), 2012. ← 288

[25] A. M. Castravet. The Cox ring of $\overline{M}_{0,6}$. *Trans. Amer. Math. Soc.*, **361**(7), 3851–3878, 2009. MR 2491903 (2009m:14037) ← 287

[26] D. Chen and I. Coskun. Towards Mori's program for the moduli space of stable maps, to appear in *Amer. J. Math.* (arXiv:0905.2947). ← 286

[27] D. A. Cox. The homogeneous coordinate ring of a toric variety. *J. Algebraic Geom.*, 4(1), 17–50, 1995. MR 1299003 (95i:14046) ← 283

[28] D. A. Cox, J. Little, and H. Schenck. Toric Varieties, to appear in: *Graduate Studies in Mathematics series*, 2010. ← 281, 282, 283, 285

[29] P. Deligne and G. D. Mostow. Monodromy of hypergeometric functions and nonlattice integral monodromy. *Inst. Hautes Études Sci. Publ. Math.*, no. **63**, 5–89, 1986. MR 849651 (88a:22023a) ← 289

[30] P. Deligne and D. Mumford. The irreducibility of the space of curves of given genus. *Inst. Hautes Études Sci. Publ. Math.* , no. **36**, 75–109, 1969. MR 0262240 (41 #6850) ← 260, 286

[31] I. V. Dolgachev. Lectures on invariant theory. *London Mathematical Society Lecture Note Series*, vol. **296**, Cambridge University Press, Cambridge, 2003. MR 2004511 (2004g:14051) ← 262, 277, 282, 283

[32] I. V. Dolgachev and Y. Hu. Variation of geometric invariant theory quotients. *Inst. Hautes Études Sci. Publ. Math.* , no. **87**, 5–56, 1998. MR 1659282 (2000b:14060) ← 260, 261, 262, 266, 267, 268, 269, 272, 276

[33] G. Farkas. Birational aspects of the geometry of $\overline{M}_g$. *Surveys in differential geometry*, vol. **XIV**; Geometry of Riemann surfaces and their moduli spaces, *Surv. Differ. Geom.*, vol. **14**, 57–110, Int. Press, Somerville, MA, 2009. MR 2655323 ← 286

[34] M. Fedorchuk. *The final log canonical model of the moduli space of stable curves of genus four*, 2011. (arXiv:1106.5012) ← 288

[35] M. Fedorchuk and D. I. Smyth. Alternate compactifications of moduli spaces of curves, in: Handbook of moduli, Vol. I (Gavril Farkas and Ian Morrison eds.), 331–414. *Advanced Lectures in Mathematics*, **24**, Higher Education Press & International Press, Beijing-Boston, 2012. ← 261, 286, 287, 288

[36] W. Fulton. Introduction to toric varieties. *Annals of Mathematics Studies*, vol. **131**, Princeton University Press, Princeton, NJ, 1993, The William H. Roever Lectures in Geometry. MR 1234037 (94g:14028) ← 282, 283

[37] N. Giansiracusa. Conformal blocks and rational normal curves, to appear in: *J. Algebraic Geom.* (arXiv:1012.4835) ← 287

[38] N. Giansiracusa and M. Simpson. GIT compactifications of $M_{0,n}$ from conics. *Int. Math. Res. Not. IMRN*, no. **14**, 3315–3334, 2011. MR 2817681 ← 287

[39] A. Gibney, S. Keel, and I. Morrison. Towards the ample cone of $\overline{M}_{g,n}$. *J. Amer. Math. Soc.*, **15**(2), 273–294, 2002. MR 1887636 (2003c:14029) ← 286, 287

[40] A. Gibney and D. Maclagan. Equations for Chow and Hilbert quotients. *Algebra Number Theory*, **4**(7), 855–885, 2010. MR 2776876 ← 286

[41] D. Gieseker. Global moduli for surfaces of general type. *Invent. Math.*, **43**(3), 233–282, 1977. MR 0498596 (58 #16687) ← 260, 286

[42] D. Gieseker. On the moduli of vector bundles on an algebraic surface. *Ann. of Math. (2)*, **106**(1), 45–60, 1977. MR 466475 (81h:14014) ← 261

[43] M. Goresky and R. MacPherson. On the topology of algebraic torus actions, in: Algebraic groups Utrecht 1986, *Lecture Notes in Math.*, vol. **1271**, 73–90, Springer, Berlin, 1987. MR 911135 (89a:14064) ← 266

[44] V. A. Gritsenko, K. Hulek, and G. K. Sankaran. Moduli of K3 Surfaces and Irreducible Symplectic Manifolds. [In Volume I.] ← 281

[45] P. Hacking. Compact moduli of plane curves. *Duke Math. J.*, **124**(2), 213–257, 2004. MR 2078368 (2005f:14056) ← 274

[46] J. Harris and I. Morrison. Slopes of effective divisors on the moduli space of stable curves. *Invent. Math.*, **99**(2), 321–355, 1990. MR 1031904 (91d:14009) ← 286

[47] J. Harris and D. Mumford. On the Kodaira dimension of the moduli space of curves, with an appendix by William Fulton. *Invent. Math.*, **67**(1), 23–88, 1982. MR 664324 (83i:14018) ← 281, 286

[48] B. Hassett. Classical and minimal models of the moduli space of curves of genus two, in: Geometric methods in algebra and number theory. *Progr. Math.*, vol. **235**, 169–192, Birkhäuser Boston, Boston, MA, 2005. MR 2166084 (2006g:14047) ← 287

[49] B. Hassett and D. Hyeon. Log minimal model program for the moduli space of curves: The first flip, To appear in *Ann. of Math.*, 2008. (arXiv:0806.3444) ← 261, 287, 288

[50] B. Hassett and D. Hyeon. Log canonical models for the moduli space of curves: the first divisorial contraction. *Trans. Amer. Math. Soc.*, **361**(8), 4471–4489, 2009. MR 2500894 (2009m:14039) ← 261, 287

[51] M. Hochster and J. L. Roberts. Rings of invariants of reductive groups acting on regular rings are Cohen-Macaulay. *Advances in Math.* **13**, 115–175, 1974. MR 0347810 (50 #311) ← 281

[52] B. Howard, J. Millson, A. Snowden, and R. Vakil. The equations for the moduli space of n points on the line. *Duke Math. J.*, **146**(2), 175–226, 2009. MR 2477759 (2009m:14070) ← 260

[53] Y. Hu. Geometric invariant theory and birational geometry, in: Third International Congress of Chinese Mathematicians. Part 1, 2. *AMS/IP Stud. Adv. Math.*,**42**, pt. 1, vol. 2, 155–175, Amer. Math. Soc., Providence, RI, 2008. MR 2409630 (2009i:14063) ← 285

[54] Y. Hu and S. Keel. A GIT proof of Włodarczyk weighted factorization theorem, 1999. (arXiv:math/9904146) ← 281

[55] Y. Hu and S. Keel. *Mori dream spaces and GIT*, Michigan Math. J. **48** (2000), 331–348, Dedicated to William Fulton on the occasion of his 60th birthday. MR 1786494 (2001i:14059) ← 261, 273, 284, 285, 287

[56] J. E. Humphreys. Linear algebraic groups. *Graduate Texts in Mathematics*, no. **21**, Springer-Verlag, New York, 1975. MR 0396773 (53 #633) ← 263

[57] D. Hyeon and Y. Lee. Log minimal model program for the moduli space of stable curves of genus three. *Math. Res. Lett.*, **17**(4), 625–636, 2010. MR 2661168 ← 287, 288, 290

[58] D. Hyeon and I. Morrison. Stability of tails and 4-canonical models. *Math. Res. Lett.*, **17**(4), 721–729, 2010. MR 2661175 (2011f:14077) ← 288

[59] D. Jensen. Birational contractions of $\overline{M}_{3,1}$ and $\overline{M}_{4,1}$, to appear in: *Trans. Am. Math. Soc.* (arXiv:1010.3377) ← 286

[60] M. M. Kapranov. Chow quotients of Grassmannians. I, I. M. Gel′fand Seminar. *Adv. Soviet Math.*, vol. **16**, 29–110, Amer. Math. Soc., Providence, RI, 1993. MR 1237834 (95g:14053) ← 286

[61] M. M. Kapranov, B. Sturmfels, and A. V. Zelevinsky. Quotients of toric varieties. *Math. Ann.*, **290**(4), 643–655, 1991. MR 1119943 (92g:14050) ← 283

[62] S. Keel. Intersection theory of moduli space of stable n-pointed curves of genus zero. *Trans. Amer. Math. Soc.*, **330**(2), 545–574, 1992. MR 1034665 (92f:14003) ← 286

[63] S. Keel and J. M[c]Kernan. Contractible extremal rays on $\overline{M}_{0,n}$, in: Handbook of moduli, Vol. II (Gavril Farkas and Ian Morrison eds.), 115–130. *Advanced Lectures in Mathematics*, **25**, Higher Education Press & International Press, Beijing-Boston, 2012. ← 287

[64] G. R. Kempf. Instability in invariant theory, *Ann. of Math. (2)*, **108**(2), 299–316, 1978. MR 506989 (80c:20057) ← 277, 278

[65] H. Kim and Y. Lee, *Log canonical thresholds of semistable plane curves*, Math. Proc. Cambridge Philos. Soc. **137**(2), 273–280, 2004. MR 2090618 (2005m:14055) ← 274

[66] F. C. Kirwan. Cohomology of quotients in symplectic and algebraic geometry. *Mathematical Notes*, vol. **31**, Princeton University Press, Princeton, NJ, 1984. MR 766741 (86i:58050) ← 261

[67] F. C. Kirwan. Partial desingularisations of quotients of nonsingular varieties and their Betti numbers. *Ann. of Math. (2)*, **122**(1), 41–85, 1985. MR 799252 (87a:14010) ← 261, 282

[68] F. C. Kirwan. Quotients by non-reductive algebraic group actions, in: Moduli spaces and vector bundles. *London Math. Soc. Lecture Note Ser.*, vol. **359**, 311–366, Cambridge Univ. Press, Cambridge, 2009. MR 2537073 (2011a:14092) ← 262

[69] F. F. Knudsen and D. Mumford. The projectivity of the moduli space of stable curves. I. Preliminaries on "det" and "Div". *Math. Scand.*, **39**(1), 19–55, 1976. MR 0437541 (55 #10465) ← 286

[70] J. Kollár. Projectivity of complete moduli. *J. Differential Geom.*, **32**(1), 235–268, 1990. MR 1064874 (92e:14008) ← 286

[71] J. Kollár. Moduli of varieties of general type, in: Handbook of moduli, Vol. II (Gavril Farkas and Ian Morrison eds.), 131–158. *Advanced Lectures in Mathematics*, **25**, Higher Education Press & International Press, Beijing-Boston, 2012. ← 260

[72] J. Kollár and S. Mori. Birational geometry of algebraic varieties. *Cambridge Tracts in Mathematics*, vol. **134**, Cambridge University Press, Cambridge, 1998. MR 1658959 (2000b:14018) ← 270, 281

[73] J. Kollár and N. I. Shepherd-Barron. Threefolds and deformations of surface singularities, *Invent. Math.*, **91**(2), 299–338, 1988. MR 922803 (88m:14022) ← 260

[74] S. Kondō. A complex hyperbolic structure for the moduli space of curves of genus three. *J. Reine Angew. Math.*, **525**, 219–232, 2000. MR 1780433 (2001j:14039) ← 289

[75] S. Kondō. The moduli space of curves of genus 4 and Deligne-Mostow's complex reflection groups, in; Algebraic geometry 2000, Azumino (Hotaka), *Adv. Stud. Pure Math.*, vol. **36**, 383–400, Math. Soc. Japan, Tokyo, 2002. MR 1971521 (2004h:14033) ← 289, 290

[76] A. Laface and M. Velasco. A survey on Cox rings. *Geom. Dedicata*, **139**, 269–287, 2009. MR 2481851 (2010m:14065) ← 285

[77] R. Laza. Deformations of singularities and variation of GIT quotients., *Trans. Amer. Math. Soc.*, **361**(4), 2109–2161, 2009. MR 2465831 (2009k:14006) ← 269, 273, 274, 277, 279, 290

[78] R. Laza. The moduli space of cubic fourfolds. *J. Algebraic Geom.*, **18**(3), 511–545, 2009. MR 2496456 (2010c:14039) ← 289

[79] R. Laza. The moduli space of cubic fourfolds via the period map. *Ann. of Math. (2)*, **172**(1), 673–711, 2010. MR 2680429 ← 289

[80] E. Looijenga. Compactifications defined by arrangements. I. The ball quotient case. *Duke Math. J.*, **118**(1), 151–187, 2003. MR 1978885 (2004i:14042a) ← 261, 289

[81] E. Looijenga. Compactifications defined by arrangements. II. Locally symmetric varieties of type IV. *Duke Math. J.*, **119**(3), 527–588, 2003. MR 2003125 (2004i:14042b) ← 261, 289

[82] E. Looijenga. Invariants of quartic plane curves as automorphic forms. *Algebraic geometry, Contemp. Math.*, vol. **422**, 107–120, Amer. Math. Soc., Providence, RI, 2007. MR 2296435 (2008b:14045) ← 289, 290

[83] E. Looijenga. The period map for cubic fourfolds. *Invent. Math.*, **177**(1), 213–233, 2009. MR 2507640 (2010h:32013) ← 289

[84] E. Looijenga and R. Swierstra. The period map for cubic threefolds. *Compos. Math.*, **143**(4), 1037–1049, 2007. MR 2339838 (2008f:32015) ← 289

[85] D. Luna. Adhérences d'orbite et invariants. *Invent. Math.*, **29**(3), 231–238, 1975. MR 0376704 (51 #12879) ← 279

[86] M. Maruyama. Moduli of stable sheaves. I. *J. Math. Kyoto Univ.*, **17**(1), 91–126, 1977. MR 0450271 (56 #8567) ← 261

[87] J. McKernan. Mori dream spaces. *Jpn. J. Math.*, **5**(1), 127–151, 2010. MR 2609325 ← 285

[88] J. S. Milne. Shimura varieties and moduli, in: Handbook of moduli, Vol. II (Gavril Farkas and Ian Morrison eds.), 467–548. *Advanced Lectures in Mathematics,* **25**, Higher Education Press & International Press, Beijing-Boston, 2012. (arXiv: 1105.0887). ← 288

[89] I. Morrison. GIT constructions of moduli spaces of stable curves and maps, *Surveys in differential geometry*, Vol. **XIV**. Geometry of Riemann surfaces and their moduli spaces, *Surv. Differ. Geom.*, Vol. **14**, 315–369. Int. Press, Somerville, MA, 2009. MR 2655332 ← 265, 286, 287, 288

[90] I. Morrison and D. Swinarski. *Gröbner techniques for low-degree Hilbert stability*, Exp. Math. **20**(1), 34–56, 2011. MR 2802723 ← 279

[91] S. Mukai. An introduction to invariants and moduli. *Cambridge Studies in Advanced Mathematics*, vol. **81**, Cambridge University Press, Cambridge, 2003, Translated from the 1998 and 2000 Japanese editions by W. M. Oxbury. MR 2004218 (2004g:14002) ← 262, 282

[92] S. Mukai. Geometric realization of T-shaped root systems and counterexamples to Hilbert's fourteenth problem, in: Algebraic transformation groups and algebraic varieties. *Encyclopaedia Math. Sci.*, vol. **132**, 123–129, Springer, Berlin, 2004. MR 2090672 (2005h:13008) ← 262

[93] D. Mumford. Stability of projective varieties. *Enseignement Math. (2)*, **23**(1–2), 39–110, 1977. MR 0450272 (56 #8568) ← 260, 265, 276, 278, 286, 287

[94] D. Mumford, J. Fogarty, and F. Kirwan. Geometric invariant theory, third ed. *Ergebnisse der Mathematik und ihrer Grenzgebiete (2)*, vol. **34**, Springer-Verlag, Berlin, 1994. MR 1304906 (95m:14012) ← 260, 262, 263, 264, 265, 267, 273, 276, 277, 279, 288

[95] M. Nagata. On the fourteenth problem of Hilbert. *Proc. Internat. Congress Math.* 1958, 459–462, Cambridge Univ. Press, New York, 1960. MR 0116056 (22 #6851) ← 262

[96] P. E. Newstead. Introduction to moduli problems and orbit spaces. *Tata Institute of Fundamental Research Lectures on Mathematics and Physics*, vol. **51**, Tata Institute of Fundamental Research, Bombay, 1978. MR 546290 (81k:14002) ← 262

[97] V. L. Popov and È. B. Vinberg. Invariant theory, in: Algebraic geometry, 4 (Russian), Itogi Nauki i Tekhniki, Akad. Nauk SSSR Vsesoyuz. Inst. Nauchn. i Tekhn. Inform., Moscow, 1989, 137–314, 315. ← 262

[98] N. Ressayre. The GIT-equivalence for G-line bundles. *Geom. Dedicata*, **81**(1–3), 295–324, 2000. MR 1772211 (2001e:14047) ← 269

[99] H. Schoutens. Log-terminal singularities and vanishing theorems via non-standard tight closure. *J. Algebraic Geom.*, **14**(2), 357–390, 2005. MR 2123234 (2006e:13005) ← 281, 282

[100] D. Schubert. A new compactification of the moduli space of curves. *Compositio Math.*, **78**(3), 297–313, 1991. MR 1106299 (92d:14018) ← 288

[101] C. S. Seshadri. Space of unitary vector bundles on a compact Riemann surface. *Ann. of Math. (2)*, **85**, 303–336, 1967. MR 0233371 (38 #1693) ← 261

[102] J. Shah. A complete moduli space for K3 surfaces of degree 2. *Ann. of Math. (2)*, **112**(3), 485–510, 1980. MR 595204 (82j:14030) ← 265, 266, 279, 289

[103] D. Swinarski. GIT stability of weighted pointed curves, to appear in: *Trans. Am. Math. Soc.* (arXiv: 0801.1288). ← 286

[104] M. Thaddeus. Stable pairs, linear systems and the Verlinde formula. *Invent. Math.*, **117**(2), 317–353, 1994. MR 1273268 (95e:14006) ← 285

[105] M. Thaddeus. Geometric invariant theory and flips. *J. Amer. Math. Soc.*, **9**(3), 691–723, 1996. MR 1333296 (96m:14017) ← 260, 261, 266, 267, 268, 269, 271, 272, 285

[106] E. Viehweg. Quasi-projective moduli for polarized manifolds. *Ergebnisse der Mathematik und ihrer Grenzgebiete (3) [Results in Mathematics and Related Areas (3)]*, vol. **30**, Springer-Verlag, Berlin, 1995. MR 1368632 (97j:14001) ← 261, 286

[107] K. Watanabe. Certain invariant subrings are Gorenstein. I, II. *Osaka J. Math.*, **11**, 1–8, 1974; *ibid.*, **11**, 379–388, 1974. MR 0354646 (50 #7124) ← 281

Stony Brook University, Department of Mathematics, Stony Brook, NY 11794
E-mail address: rlaza@math.sunysb.edu

Good degenerations of moduli spaces

Jun Li

Abstract. We survey the construction of good degenerations of moduli spaces associated to a simple degeneration of smooth projective varieties.

Contents

This research work was partially supported by the NSF grant NSF-0601002.

1. Introduction

Degeneration is a classical tool in algebraic geometry; for instance, it was already used by the Italian school in studying enumeration problems. Nowadays, degeneration is an indispensable tool in studying moduli problems in algebraic geometry, exemplified by its applications to moduli of stable curves, of stable maps, and of stable sheaves.

In this note, we will discuss a class of degenerations, which we call *good degenerations*, that is well suited to study geometry and topology of the moduli spaces. This construction is inspired by the pioneering work of Gieseker on degeneration of the moduli of bundles, and of Floer-Donaldson theory in gauge theory.

In this paper, we will focus on constructing good degenerations of moduli associated to a simple degeneration X/C of smooth (projective) schemes over a smooth curve $0 \in C$. We say

$$\pi : X \longrightarrow C$$

is a simple degeneration of smooth projective scheme *if* X *is smooth;* π *is projective;* π *has smooth fiber over* $t \neq 0 \in C$, *and the central fiber* X_0 *has normal crossing singularity such that the singular locus* D *of* X_0 *is smooth.* We denote by Y the normalization of X_0 and $\tilde{D} = Y \times_{X_0} D \subset Y$. We call $(Y, \tilde{D})$ the relative pair associated with Y.

Given a specified moduli problem, we seek to find a *good degeneration* that is a scheme (or a stack) $\mathcal{M}_{X/C}$, proper over C, such that

I. for $t \neq 0 \in C$, $\mathcal{M}_{X_t} := \mathcal{M}_{X/C} \times_C t$ is the moduli space of the given moduli problem of X_t;
II. the local geometry of the total space $\mathcal{M}_{X/C}$ is comparable to the local geometry of $\mathcal{M}_{X_t}$, $t \neq 0 \in C$; the central fiber $\mathcal{M}_{X_0}$ is "virtually" a local complete intersection in $\mathcal{M}_{X/C}$;
III. there is a relative moduli space $\mathcal{M}_{(Y,\tilde{D})}$ (of the specified moduli problem) of the relative pair $(Y, \tilde{D})$ and a finite morphism $\mathcal{M}_{(Y,\tilde{D})} \to \mathcal{M}_{X_0}$ that separates the branches of $\mathcal{M}_{X_0}$ due to the "virtual" local complete intersection closed immersion $\mathcal{M}_{X_0} \subset \mathcal{M}_{X/C}$;
IV. the space $\mathcal{M}_{X/C}$ has a modular description.

We comment that property-I makes $\mathcal{M}_{X/C}$ a degeneration of the moduli spaces of X_t; property-II assures that the local geometry of the total family resembles that of its general fibers, and that the central fiber is "virtually" a local complete intersection. This allows one to study the geometry of the general fibers via the central fiber. Property-III is crucial. Aiming at studying the geometry of $\mathcal{M}_{X_t}$, it allows one to transform problems on $\mathcal{M}_{X_0}$ to that of $\mathcal{M}_{(Y,\tilde{D})}$. Finally, the modular interpretation of $\mathcal{M}_{X/C}$ (and $\mathcal{M}_{(Y,\tilde{D})}$) makes working with $\mathcal{M}_{(Y,\tilde{D})}$ manageable.

Gieseker's construction of his degeneration of the moduli of rank two vector bundles on curves is the pioneering work in this subject.

1.1. Gieseker degeneration

We recall the Gieseker degeneration of moduli of rank two stable bundles on smooth curves [12]. Let $X \to C$ be a simple degeneration of smooth curves, and assume that the central fiber X_0 is a nodal curve with a single node D. The relative moduli of stable sheaves provides us a coarse moduli scheme $M_{X/C}(2,1)$ whose fiber over $t \in C$ is the moduli space $M_{X_t}(2,1)$. Since X_0 is singular, the total space of $M_{X/C}(2,1)$ is not smooth at stable sheaves $\mathcal{E}$ on X_0 that are not locally free. This makes it difficult to study the geometry of $M_{X_t}(2,1)$, like its Chern classes, via the family $M_{X/C}(2,1)$.

The more difficult part is the failure of property-III for the family $M_{X/C}(2,1)$. Given a vector bundle $\mathcal{E}$ on X_0, we can reconstruct $\mathcal{E}$ by gluing its pull back $\tilde{\mathcal{E}}$ on Y (the normalization of X_0) along $\tilde{D} \subset Y$. The gluing data, in case $\mathcal{E}$ has rank two, is parameterized by the group $GL(2)$, which is manageable. However, when $\mathcal{E}$ is non-locally free at D, reconstructing $\mathcal{E}$ from a sheaf on Y requires data more involved than $GL(2)$. This non-uniformity of gluing data breaks down the known construction that transforms the geometry of $M_{X_0}(2,1)$ to "the moduli of relative stable sheaves" $M_{(Y,\tilde{D})}(2,1)$, if the later can be constructed.

Instead Gieseker [12] considered a different degeneration scheme. Using Hilbert embeddings of curves induced by vector bundles on curves and studying their stability properties (by Gieseker and Morrison [13]), he constructed a new degeneration $M^{Gi}_{X/C}(2,1)$ whose fiber over $t \neq 0 \in C$ is the usual moduli space $M_{X_t}(2,1)$; whose total space is smooth; whose central fiber has normal crossing singularities, and points in this space have modular interpretation. It is the first elaborate example of *good degenerations* of moduli spaces known to the author.

It is instructive to recall the modular description of points in the central fiber of Gieseker's good degeneration. Elements of the central fiber $M^{Gi}_{X_0}(2,1)$ consists of stable vector bundles on semistable model $X[n]_0$ of X_0, which is the nodal curve by inserting an n-chain of rational curves to the node D of X_0. Let $M_{X_0}(2,1)^{l.f.} \subset M_{X_0}(2,1)$ be the open subset of locally free stable sheaves; then $M_{X_0}(2,1)^{l.f.}$ is open dense in $M^{Gi}_{X_0}(2,1)$. The complement

$$M^{Gi}_{X_0}(2,1) - M_{X_0}(2,1)^{l.f.}$$

consists of stable locally free sheaves on $X[1]_0$ or $X[2]_0$.

Since deformations of nodal curves and of vector bundles on curves are unobstructed, the total space $M^{Gi}_{X/C}(2,1)$ is smooth; one also sees that $M^{Gi}_{X/C}(2,1) \to C$ is a local complete intersection. Because elements in $M^{Gi}_{X_0}(2,1)$ are vector bundles on nodal curves that stabilize to X_0, $M^{Gi}_{X_0}(2,1)$ has a modular interpretation, satisfying property-IV of a good degeneration.

It is informative to see how $M^{Gi}_{X_0}(2,1)$ is related to the moduli of relative stable bundles $M^{Gi}_{(Y,\tilde{D})}(2,1)$. First, as $M^{Gi}_{X_0}(2,1)$ has local complete intersection singularities, property-III suggests that $M^{Gi}_{(Y,\tilde{D})}(2,1)$ is the normalization of $M^{Gi}_{X_0}(2,1)$, which is the case. Not going into details, as we will later in the paper, we merely state that elements in $M^{Gi}_{(Y,\tilde{D})}(2,1)$ are locally free sheaves on Y or on Y with extra $\mathbb{P}^1$ attached to $\tilde{D}$, plus framing data along the two marked points, where the marked points are $\tilde{D}$ in case no extra $\mathbb{P}^1$ are attached, and are points on the extra $\mathbb{P}^1$'s attached.

Summarizing, Gieseker's construction tells us that to form a good degeneration, we need to enlarge the base variety X_0 in case the moduli objects are degenerate along the singular divisor $D \subset X_0$, and replace the degenerate moduli objects by regular objects on new base schemes that are after inserting chains of ruled varieties at D.

1.2. From gauge theory

Another source of inspiration comes from four manifold gauge theory. In differential geometry, one considers a family of smooth four-manifolds degenerating to a singular one by stretching a four-manifold M along the normal direction of a (real) hypersurface $\Sigma \subset M$. The limit is a union of two four-manifolds M_+ and M_- with ends isomorphic to $[0,\infty) \times \Sigma$, and the two ends $\Sigma \times \{\infty\}$ are identified. We use $M_- \cup_\Sigma M_+$ to record this union with ends identified.

We let g_t, $t \in [1,\infty)$, be a family of Riemannian metrics on M so that as $t \to \infty$, g_t converges to a metric on $M_- \cup M_+$ with cylindrical metrics at the end of $M_- \cup M_+$. We fix a principal bundle E on M, and form the space $\mathcal{M}^{ASD}_t$ of gauge-equivalence classes of ASD connections on E with respect to the metric g_t. The task is to find a good limit $\lim_{t\to\infty} \mathcal{M}^{ASD}_t$.

The Floer-Donaldson theory constructs $\mathcal{M}^{ASD}_\infty$ as follows [9, 23]. It consists of equivalence classes of ASD connections $A_\bullet = (A_-, A_1, \ldots, A_n, A_+)$ on

$$(1.1)\qquad M_- \cup_\Sigma \Sigma \times \mathbb{R} \cup_\Sigma \ldots \cup_\Sigma \Sigma \times \mathbb{R} \cup_\Sigma M_+, \quad \text{a chain of } n \text{ copies inserted,}$$

such that

(a) $A_\pm$ are connections on $M_\pm$ and A_i are connections on the i-th cylinder in the chain;
(b) the limit of $A_\bullet$ at the ends that meet in (1.1) are compatible, and
(c) $(A_-, A_1, \ldots, A_n, A_+) \cong (A'_-, A'_1, \ldots, A'_n, A'_+)$ if $A_\pm = A'_\pm$ and $A_i = \ell_i^* A'_i$ for some translations $\ell_i : \Sigma \times \mathbb{R} \to \Sigma \times \mathbb{R}$ along $\mathbb{R}$ direction.

Key to our construction is the equivalence relation (c) via translations of $\Sigma \times \mathbb{R}$. It was to understand this equivalence via translations ℓ that led the author to realize that to construct good degenerations one needs to work with the quotient of expanded degenerations, which eventually leads to the notion of stack of expanded degenerations.

1.3. Admissible covers

The moduli space of admissible covers constructed by Harris and Mumford is another example of constructing good compactifications [15]. By using moduli of pointed curves, where the points are the branched points of a branched cover, the theory of admissible covers in effect inserts rational curves in case the branch points collide or appear at the nodes. This line of idea was further developed in the work of Fantechi and Pandharipande [11].

1.4. Good degenerations

We now describe briefly the construction of good degenerations using the stack of expanded degenerations. For brevity, we use the case of moduli of stable morphisms as an example.

Let X/C be a simple degeneration of smooth varieties with $D \subset X_0$ the singular locus. For simplicity, we assume D is connected, $Y = Y_- \cup Y_+$ is a union of two connected components, and label $D_\pm = \tilde{D} \cap Y_\pm$. We fix a relatively ample line bundle H on X/C; for integers g, k, d we form the moduli $\overline{\mathcal{M}}_{g,k}(X_t, d)$ of k-pointed genus g stable morphisms $f\colon \Sigma \to X_t$ of degree $d = \deg f^*H$. We seek to find a good degeneration of $\overline{\mathcal{M}}_{g,k}(X_t, d)$.

The logical degeneration is to take the moduli $\overline{\mathcal{M}}_{g,k}(X_0, d)$ of stable morphisms to X_0 of degree d. However, this makes fulfilling property-III difficult. Indeed, let

$$f : \Sigma \longrightarrow X_0 = Y_- \cup Y_+$$

be a morphism. In case no irreducible components of Σ are mapped to $D \subset X_0$, f decomposes into a pair of stable morphisms $f_\pm = f|_{\Sigma_\pm} : \Sigma_\pm \to Y_\pm$, $\Sigma_\pm = f^{-1}(Y_\pm)$.

This decomposition breaks down when some irreducible components of Σ are mapped to D. We call such maps degenerate along $D \subset X_0$. More than the difficulty in decomposing the maps, which is preferable in light of property-III, the deformations of stable maps degenerate along D encounter extra obstructions not present for stable maps to smooth varieties, thus making property-II less likely to hold.

Following Gieseker's degeneration of moduli of bundles, we consider stable morphisms to $X[n]_0$ that are regular along the singular loci of $X[n]_0$, where $X[n]_0$ is by inserting a chain of n-copies of the ruled variety

$$\Delta = \mathbf{P}_D(1 \oplus N_{D_+/Y_+})$$

(over D) to D in X_0. (We will construct $X[n]_0$ in the next section.)

Hinted by Floer-Donaldson theory, two stable maps $f_i : \Sigma_i \to X[n]_0$ are equivalent if there is an isomorphism $\sigma : X[n]_0 \to X[n]_0$ preserving the projections $X[n]_0 \to X_0$, and an isomorphism $\sigma' : \Sigma_1 \to \Sigma_2$ preserving the marked points of Σ_i, such that $\sigma \circ f_1 = f_2 \circ \sigma'$.

Since $X[n]_0 \to X_0$ is by contracting the chain of ruled variety Δ, automorphisms of $X[n]_0/X_0$ are dilations along fibers of $\Delta \to D$ fixing two sections of Δ/D. This way, the automorphisms of $X[n]_0/X_0$ are the algebraic version of the translations used in Floer-Donaldson theory.

The self-equivalences of a stable morphism $f\colon \Sigma \to X[n]_0$ form a group, which we denote by $\mathrm{Aut}_{\mathfrak{X}}(f)$. We call f stable (as a morphism to $\mathfrak{X}$, to be defined later) if $\mathrm{Aut}_{\mathfrak{X}}(f)$ is finite. There is one more technical requirement: f must be pre-deformable, which we will address later. (Pre-deformable is the high dimensional version of admissible cover of Harris and Mumford [15].)

Putting together, the central fiber of the good degeneration of $\overline{\mathcal{M}}_{g,k}(X_t, d)$ constructed in [26] has set-theoretic description

$$\left\{ f:\Sigma \to X[n]_0 \,\middle|\, \begin{array}{l} n \geqslant 0,\ f \text{ is non-degenerate, pre-deformable along} \\ \text{the singular loci of } X[n]_0, \text{ and } \mathrm{Aut}_{\mathfrak{X}}(f) \text{ is finite.} \end{array} \right\} \Big/ \cong .$$

Constructing the stack structure of this set-theoretic description of the central fiber, and fitting it into the family $\coprod_{t\in C-0} \overline{\mathcal{M}}_{g,k}(X_t, d)$, is achieved by working with the stack $\mathfrak{X} \to \mathfrak{C}$ of expanded degenerations. Here the stack $\mathfrak{C}$ is over C; $\mathfrak{X}$ is a family over $\mathfrak{C}$. The only member of $\mathfrak{C}$ over $t \neq 0 \in C$ is t; the family $\mathfrak{X}$ over t is X_t. The members of $\mathfrak{C}$ over $0 \in C$ consists of points $\{o_n : n \geqslant 0\}$; fiber of $\mathfrak{X}$ over o_n is $X[n]_0$. Since for $m \leqslant n$, $X[n]_0$ specializes to $X[m]_0$, the stack $\mathfrak{C}$ is non-separated, and for $m \leqslant n$, $\{o_n\}$ lies in the closure of $\{o_m\}$. Finally, the automorphism group of o_n is G_m^n, where $G_m = GL(1, k)$; its lift to $X[n]_0$ is the group of automorphisms of $X[n]_0$ preserving the projection $X[n]_0 \to X_0$.

Under this formulation, the set specified is the set of all morphisms f to $\mathfrak{X} \times_C 0$ that are non-degenerate, pre-deformable, and have finite automorphism groups $\mathrm{Aut}_{\mathfrak{X}}(f)$. Continuing this line of thinking, the good degeneration of $\overline{\mathcal{M}}_{g,k}(X_t, d)$ is the moduli of stable morphisms to $\mathfrak{X}$ that are non-degenerate, pre-deformable, and have finite automorphism groups $\mathrm{Aut}_{\mathfrak{X}}$.

1.5. Known constructions of good degenerations

The Gieseker degeneration of moduli of rank two stable bundles on curves has spurred further development in several research subjects in algebraic geometry.

The first fruitful area is on moduli of stable bundles on curves. In rank one case, following D. Gieseker's idea, L. Caporaso constructed a compactification of the Picard scheme over the universal family of the moduli of stable curves [4].

In high rank case, D.S. Nagaraj and C.S. Seshadri generalized Gieseker's construction to the case $(r, d) = 1$ [35]; they constructed the Gieseker degeneration using a refined version of moduli of torsion free sheaves on singular curves; the connection of Gieseker degeneration with generalized parabolic bundles was established in this work.

Circumventing the technical study of Hilbert stability by D. Gieseker and I. Morrison's in the rank two case [13], the author used Simpson stability of vector bundles on semistable curves to construct a good moduli of stable bundles (of $(r, d) = 1$) on the universal curve of the stack of semistable curves [28]. Using Hilbert stability, A. Schmitt constructed the moduli of semistable vector bundles over the universal curve of the stack of semistable curves without coprime assumption on rank and degree [39]. Later, I. Kausz studied the stack of Gieseker type (i.e. those appear in the Gieseker degeneration) vector bundles on the universal curve of the stack of semistable curves [20].

The approach presented in this note by Y-H. Kiem and the author [21] gave an alternative construction of the Gieseker degeneration of the moduli of high rank vector bundles; they used the stack of expanded degenerations developed in author's work on good degenerations of moduli of stable morphisms [26, 27]. X-T. Sun [45] proved that the degenerations constructed in [35] and in [21] are identical.

I add that the stack of expanded degenerations allows the authors of [21] to use different stability conditions to study in details the flips arising from the decomposition of the moduli spaces, thus proving the Newstead and Ramanan conjecture in the case of rank three. Their proof benefited from the works [19, 34].

Works on degenerations of moduli of G-bundles include G. Faltings [10], D.S. Nagaraj and C.S. Seshadri [34], and X-T. Sun [42, 44].

Gieseker degenerations have been applied to study factorization of generalized theta functions. The representative works include [7, 8, 36, 41, 43].

The stack of expanded degenerations was constructed by the author in his work on constructing good degenerations of moduli of stable morphisms to simple degenerations of projective varieties, and on proving a degeneration formula of Gromov-Witten invariants [26, 27]. Prior constructions in differential geometry were accomplished by A-M. Li and Y-B. Ruan [25], and by E. Ionel and T. Parker [18]. These constructions were inspired by the Floer-Donaldson theory in four manifold gauge theory [9, 23].

Degeneration formula of Gromov-Witten invariants [27] played a pivotal role in studying Gromov-Witten invariants. A partial list of representative works include [3, 17, 29, 32, 37],

A good degeneration of the Hilbert schemes of curves of a simple degeneration of smooth threefolds has been constructed by B. Wu in his thesis. It has been generalized to include Grothendieck's Quot scheme and of the moduli of stable pair of Le Potier on arbitrary simple degeneration by B. Wu and the author [31]. Applications of this degeneration formula include [24, 33].

1.6. Organization of the paper

We will review the construction of the stack of expanded degenerations and of expanded relative pairs. They are the foundation toward constructing good degenerations of moduli spaces. Subsequently, we will review the construction of good degeneration of moduli of stable morphisms, of vector bundles over curves and over the universal family of the moduli of stable curves, and of the Hilbert scheme of subschemes. The discussion of their structures and some open problems will be addressed after.

Acknowledgement. I thank B. Wu for carefully going through the draft of this article. I thank the referee for suggestions which made the final version more reader friendly.

1.7. List of symbols

$\mathfrak{X} \to \mathfrak{C}$: the stack of expanded degenerations of X/C; objects are $X[n]_0$; $\mathfrak{C}$ the base stack of $\mathfrak{X}$ (cf. Def. 2.12, 2.19).

$\mathfrak{X}_0^\dagger \to \mathfrak{C}_0^\dagger$: the stack of expanded central fibers with marked singular divisors; $\mathfrak{C}_0^\dagger$ is its base stack (cf. Prop. 2.37, 2.44).

$\mathfrak{D}_\pm \subset \mathfrak{Y}$: the stack of expanded relative pairs of $D_- \cup D_+ \subset Y$, objects are $D[n_-, n_+]_{\pm,0} \subset Y[n_-, n_+]_0$ (cf. Section 2.4).

$\mathfrak{A}$ and $\mathfrak{A}_\diamond$: $\mathfrak{A}_\diamond$ is the base stack of $\mathfrak{Y}$, with tautological $\mathfrak{Y} \to \mathfrak{A}_\diamond$ (cf. Def. 2.33); $\mathfrak{A}$ is the base stack such that $\mathfrak{C} = C \times_{\mathbf{A}^1} \mathfrak{A}$ (cf. Def. 2.11).

$\mathfrak{D}_+ \subset \mathfrak{Y}_+$: the stack of expanded relative pairs of $D_+ \subset Y_+$; objects are $D[n]_0 \subset Y[n]_0$; its base stack is $\mathfrak{A}_+ = \mathfrak{A}$ (cf. Prop. 2.48).

$\mathfrak{X}^\beta \to \mathfrak{C}^\beta$: the stack of weighted expanded degeneration of total weights β; its base stack is $\mathfrak{C}^\beta$ (cf. Prop. 2.53, 2.56).

$\mathfrak{X}_0^{\dagger,\beta}$: the stack of weighted expanded central fibers with marked singular divisors and total weights β; its base stack is $\mathfrak{C}_0^{\dagger,\beta}$ (cf. (2.58)).

$\mathfrak{Y}_+^{\delta_+,\delta_0}$: the stack of weighted expanded relative pairs of $D_+ \subset Y_+$; its base stack is $\mathfrak{A}_0^{\delta_+,\delta_0}$ (cf. (2.62)).

2. The stack of expanded degenerations

We work with a fixed algebraically closed field $\mathbf{k}$ of characteristic 0. We denote $G_m = GL(1, \mathbf{k})$. Let $\pi\colon X \to C$ be a flat, projective family over a smooth pointed affine curve $0 \in C$.

Definition 2.1. *We call $\pi\colon X \to C$ a simple degeneration if X is smooth; π is projective; π is smooth away from the central fiber X_0; X_0 has normal crossing singularities; the singular locus $D \subset X_0$ is smooth; and the preimage of each connected component of D in the normalization of X_0 has two connected components.*

Associated with a simple degeneration is the relative pair of the central fiber.

Definition 2.2. *Let Y be the normalization of X_0, and let $\tilde{D} \subset Y$ be the preimage of $D \subset X_0$ via $Y \to X_0$. We call $(Y, \tilde{D})$ the relative pair associated with X_0.*

Two simple situations occur frequently in applications. One is when Y has two connected components and D is connected. In this case we usually denote $(Y, \tilde{D})$ as the union of two pairs (Y_-, D_-) and (Y_+, D_+). The other case is when Y is irreducible and $\tilde{D}$ is smooth.

From now on, we pick a disjoint union decomposition $\tilde{D} = D_- \cup D_+$ so that both D_- and D_+ are isomorphic to D under the projection $Y \to X_0$. (We caution that unless otherwise stated, D can be non-connected.) In this section, we construct the stack of expanded degenerations of $X \to C$ and its universal family $\mathfrak{X} \to \mathfrak{C}$. Some aspects of the stack $\mathfrak{X}$ are new; however, the proofs of the results listed in this section are mostly elementary, and can be found in [26].

The stack of expanded degenerations consists of an Artin stack $\mathfrak{C}$ with a universal family $\mathfrak{X} \to \mathfrak{C}$ together with projections fitting into a commutative square

$$\begin{array}{ccc} \mathfrak{X} & \xrightarrow{p} & X \\ \downarrow & & \downarrow \pi \\ \mathfrak{C} & \longrightarrow & C \end{array} \tag{2.3}$$

Away from $0 \in C$, $\mathfrak{C} \times_C C^* = C^*$, where $C^* = C - 0$, and the square above over C^* is a Cartesian product; over $0 \in C$, $\mathfrak{C} \times_C 0$ consists of points $\{o_n, n \geqslant 0\}$, each has automorphism group G_m^n, and $\mathfrak{X} \times_{\mathfrak{C}} o_n$ is by inserting an n-chain of ruled variety Δ at the singular locus $D \subset X_0$.

2.1. The stack $\mathfrak{C}$

We begin with constructing a stack $\mathfrak{A}/\mathbf{A}^1$; the stack $\mathfrak{C}$ will be the fiber product of $\mathfrak{A}$ with C over $\mathbf{A}^1$.

We consider $\mathbf{A}^{n+1}$ with the group action

$$(t_1, \cdots, t_{n+1})^\sigma = (\sigma_1 t_1, \sigma_1^{-1}\sigma_2 t_2, \cdots, \sigma_{n-1}^{-1}\sigma_n t_n, \sigma_n^{-1} t_{n+1}), \quad \sigma \in G_m^n.$$

This group action generates equivalence relations on $\mathbf{A}^{n+1}$.

We need another class of equivalences. Let $I \subset [n+1] = \{1, \ldots, n+1\}$ be any subset, $I^\circ = [n+1] - I$ its complement; we let

$$\mathrm{ind}_I\colon [m+1] \to I \subset [n+1] \quad \text{and} \quad \mathrm{ind}_{I^\circ}\colon [n-m] \to I^\circ \subset [n+1]$$

be the unique order-preserving surjective maps, (thus $|I| = m+1$). We let

$$\mathbf{A}_I^{n+1} = \{(t) \in \mathbf{A}^{n+1} \mid t_i = 0,\ i \in I^\circ\} \subset \mathbf{A}^{n+1}, \tag{2.4}$$

and let

$$\mathbf{A}_{U(I)}^{n+1} = \{(t) \in \mathbf{A}^{n+1} \mid t_i \neq 0,\ i \in I^\circ\} \subset \mathbf{A}^{n+1}. \tag{2.5}$$

The set I defines a canonical isomorphism

$$\tilde{\tau}_I : \mathbf{A}^{m+1} \times G_m^{n-m} \xrightarrow{\cong} \mathbf{A}_{U(I)}^{n+1} \tag{2.6}$$

via the rule

$$(t'_1, \cdots, t'_{m+1}; \sigma_1, \cdots, \sigma_{n-m}) \mapsto (t_1, \cdots, t_{n+1}), \quad \begin{cases} t_k = t'_l, & \text{if } k = \mathrm{ind}_I(l); \\ t_k = \sigma_l, & \text{if } k = \mathrm{ind}_{I^\circ}(l). \end{cases} \tag{2.7}$$

Restricting to $(\sigma_1, \ldots, \sigma_{n-m}) = (1)$, it defines a tautological embedding τ_I.

$$\tau_I : \mathbf{A}^{m+1} \longrightarrow \mathbf{A}^{n+1}, \tag{2.8}$$

via $\tau_I(t'_1, \cdots, t'_{m+1}) = \tilde{\tau}_I(t'_1, \cdots, t'_{m+1}, 1, \ldots, 1)$. We call such τ_I the standard embeddings of $\mathbf{A}^{m+1}$ in $\mathbf{A}^{n+1}$.

Given two subsets $I, I' \subset [n+1]$ of same cardinalities, the $\tilde{\tau}_I$ and $\tilde{\tau}_{I'}$ define an isomorphism

$$\tilde{\tau}_{I,I'} = \tilde{\tau}_I \circ \tilde{\tau}_{I'}^{-1} : \mathbf{A}^{n+1}_{U(I')} \longrightarrow \mathbf{A}^{n+1}_{U(I)}. \tag{2.9}$$

Next, we let $\mathbf{A}^{n+1} \to \mathbf{A}^{n+2}$ be the closed immersion τ_I using $I = [n+1] \subset [n+2]$. Let $G_m^n \to G_m^{n+1}$ be the group homomorphism defined via $(\sigma_1, \ldots, \sigma_n) \mapsto (\sigma_1, \ldots, \sigma_n, 1)$. Then using this homomorphism, and viewing $\mathbf{A}^{n+1}$ as scheme over $\mathbf{A}^1$ via $(t) \mapsto t_1 \ldots t_{n+1}$, the morphism

$$\tau_I : \mathbf{A}^{n+1} \to \mathbf{A}^{n+2} \tag{2.10}$$

is a G_m^n equivariant $\mathbf{A}^1$-morphism, with G_m^n acting on $\mathbf{A}^1$ trivially. Further, the equivalences $\tilde{\tau}_{I,I'}$ of $\mathbf{A}^{n+1}$ are the restrictions of the same equivalences $\tilde{\tau}_{I,I'}$, by considering I, I' as subsets in $[n+2]$ via the tautological inclusion $[n+1] \subset [n+2]$.

Definition 2.11. *We define $\mathfrak{A}_n$ be the quotient of $\mathbf{A}^{n+1}$ by G_m^n and the equivalences $\tilde{\tau}_{I,I'}$ for all pairs $I, I' \subset [n+1]$ with $|I| = |I'|$. The morphism* (2.10) *defines an open immersion $\mathfrak{A}_n \to \mathfrak{A}_{n+1}$. We define $\mathfrak{A}$ be the direct limit $\mathfrak{A} = \varinjlim_n \mathfrak{A}_n$.*

Note that the tautological $\mathbf{A}^{n+1} \to \mathfrak{A}_n$ is a surjective smooth chart; thus the collection $\{\mathbf{A}^{n+1} \to \mathfrak{A}\}_{n \geqslant 0}$ forms a smooth atlas of $\mathfrak{A}$.

Now let $0 \in C$ be the pointed smooth affine curve given. Without loss of generality, we assume that there is an étale morphism $C \to \mathbf{A}^1$ so that the inverse image of $0 \in \mathbf{A}^1$ is the distinguished point $0 \in C$. We define

$$\mathfrak{C} = C \times_{\mathbf{A}^1} \mathfrak{A}. \tag{2.12}$$

It is clear that $\mathfrak{C}$ does not depend on the choice of $C \to \mathbf{A}^1$, and is covered by smooth charts

$$C[n] := C \times_{\mathbf{A}^1} \mathbf{A}^{n+1} \longrightarrow \mathfrak{C} = C \times_{\mathbf{A}^1} \mathfrak{A}.$$

Let $o_n \in \mathfrak{A}$ be the image of $0 \in \mathbf{A}^{n+1}$ under the tautological $\mathbf{A}^{n+1} \to \mathfrak{A}$. By abuse of notation, we denote by the same $o_n \in \mathfrak{C}$ the lift of $o_n \in \mathfrak{A}$ and $0 \in C$. By construction, o_n has automorphism group G_m^n; and $\overline{o_k} = \{o_{k'} : k' \geqslant k\}$.

2.2. The stack $\mathfrak{X}$

We begin with describing $\mathfrak{X} \times_C 0$. We keep the decomposition $\tilde{D} = D_- \cup D_+$ specified at the beginning of this section. Let $N_\pm$ be the normal line bundle of $D_\pm$ in Y. Since π is a simple degeneration, $N_- \otimes N_+ \cong \mathcal{O}_D$. (Here and later we implicitly identify $D_\pm$ with D using $D_- \cup D_+ = \tilde{D} \to D$.)

We introduce the ruled variety

$$\Delta = \mathbf{P}(N_+ \oplus 1);$$

it is the $\mathbf{P}^1$-bundle over D of the direct sum of N_+ and the trivial line bundle 1. We let $D_+ = \mathbf{P}(1)$ and $D_- = \mathbf{P}(N_+)$; we call them the two tautological sections of

Δ/D. For any $\sigma \in G_m$, the automorphism of $N_+ \oplus 1$ via $(a,b)^\sigma = (\sigma \cdot a, b)$ defines a D-automorphism

$$(2.13) \qquad \sigma : \Delta \longrightarrow \Delta, \quad [a,b]^\sigma = [\sigma a, b].$$

that fixes D_- and $D_+ \subset \Delta$. We call this the tautological G_m-action on Δ.

From the construction, the normal bundle of $D_\pm$ in Δ is $N_\pm$. We then take n copies of Δ, indexed by $\Delta_1, \cdots, \Delta_n$, and form a new scheme $X[n]_0$ according to the following rule: we identify $D_- \subset Y$ with $D_+ \subset \Delta_1$, ($D_- \cong D_+$ is via the isomorphism $D_\pm \to D$;) identify $D_- \subset \Delta_i$ with $D_+ \subset \Delta_{i+1}$, and identify $D_- \subset \Delta_n$ with $D_+ \subset Y$. We denote

$$X[n]_0 = Y \sqcup \Delta_1 \sqcup \cdots \sqcup \Delta_n \sqcup (Y),$$

where $\sqcup$ means that we identify $D_- \subset \Delta_i$ with $D_+ \subset \Delta_{i+1}$. (Here $Y = \Delta_0 = \Delta_{n+1}$, single copy of Y with two indices, and putting the further right Y in parenthesis indicating that it is the same Y appearing in the further left.) We denote $D_i \subset X[n]_0$ be the D_- in Δ_{i-1}, which is also the $D_+ \subset \Delta_i$; the singular locus of $X[n]_0$ is the union of $D_1, \ldots, D_{n+1}$.

Figure 1. The two ends are the same Y, in the middle a chain of n Δ's are inserted; the D_- of Y is glued to D_+ of Δ_1, which is named D_1.

Because the inserted Δ_i intersects the remainder components along D_i and $D_{i+1} \subset \Delta_i$, the tautological G_m-action on Δ_i (cf. (2.13)) lifts to an automorphism of $X[n]_0$ that acts trivially on all Δ_j, $j \neq i$. We let G_m^n acts on $X[n]_0$ so that its i-th factor acts on $X[n]_0$ via the mentioned lift of the tautological G_m-action on Δ_i. Let $p\colon X[n]_0 \longrightarrow X_0$ be the projection contracting all inserted components $\Delta_1, \cdots, \Delta_n$. With the trivial action on X, the G_m^n-action on $X[n]_0$ preserves p.

We now construct the family $\mathfrak{X} \to \mathfrak{C}$ associated with $X \to C$. For each n, we form $X \times_C C[n]$, as a family over $C[n]$. It is smooth away from the fibers over $C[n] \times_C 0$, and is a G_m^n-scheme over $C[n]$, after lifting the G_m^n-actions on $C[n]$ and the trivial action on X. Let $0 \in C[n]$ be the preimage of $0 \in \mathbf{A}^{n+1}$ in $C[n]$. We continue to denote $C^* = C \backslash 0$; we let $C[m]^* = C[m] \times_C C^*$.

Lemma 2.14. *There is a unique small resolution $X[n] \to X \times_C C[n]$, coupled with the projection $p\colon X[n] \to X$ induced from $X \times_C C[n] \to X$, characterized by:*

(1) *the total space $X[n]$ is smooth;*
(2) *the central fiber $(X[n] \times_{C[n]} 0, p)$ is the $(X[n]_0, p)$ constructed;*

(3) *let* $\tau_I \colon C[m] \to C[n]$ *be a standard subfamily induced by* $\tau_I : \mathbf{A}^{m+1} \to \mathbf{A}^{n+1}$ *(cf. (2.8)); then the induced family* $(\tau_I^* X[n], \tau_I^* p)$ *is isomorphic to* $(X[m], p)$ *as families over* $C[m]$*, extending the identity map*

$$\tau_I^* X[n]|_{C[m]^*} = X[m]|_{C[m]^*} = X \times_C C[m]^*;$$

(4) *let* $\mathbf{A}_l^{n+1}$ *be (2.4) (using* $I = \{l\}$*), let* $L_l = C[n] \times_{\mathbf{A}^{n+1}} \mathbf{A}_l^{n+1}$*, and let* $\iota_l \colon L_l \to C[n]$ *be the inclusion; then the induced family* $\iota_l^* X[n]$ *is a smoothing of* $X[n]_0$ *along its* l*-th singular divisor* D_l.

Because of (2), we will view $X[n]_0$ as the central fiber $X[n] \times_{C[n]} 0$.

Lemma 2.15. *The* G_m^n *action on* $C[n]$ *with the trivial action on* X *lifts to a unique* G_m^n*-action on* $X[n]$*. The induced* G_m^n *action on* $X[n]_0$ *is the action described before Lemma 2.14. For* $I, I' \subset [n+1]$ *of identical cardinalities, the equivalence* $\tilde{\tau}_{I,I'}$ *in (2.9) lifts to a* C*-isomorphism*

$$\tilde{\tau}_{I,I',X} : X[n]|_{C[n]_{U(I')}} \cong X[n]|_{C[n]_{U(I)}}, \tag{2.16}$$

where $C[n]_{U(I)} = C[n] \times_{\mathbf{A}^{n+1}} \mathbf{A}_{U(I)}^{n+1}$.

As an illustration, let $C = \mathbf{A}^1$, and $X/\mathbf{A}^1$ be a smoothing of a nodal curve $X_0 = Y_1 \sqcup Y_2$ with a single node D. Then $C[1] = \mathbf{A}^2$; the central fiber $X[1]_0 = Y_1 \sqcup \Delta \sqcup Y_2$, $\Delta = \mathbb{P}^1$, has two singular divisors $D_1 = Y_1 \cap \Delta$ and $D_2 = \Delta \cap Y_2$. Restricting $X[1]$ to the first coordinate line $\mathbf{A}_1^2$, we obtain a family that smoothes $D_1 \subset X[1]_0$ but not D_2; restricting to the second coordinate line $\mathbf{A}_2^2$ the family smoothes D_2 but not D_1.

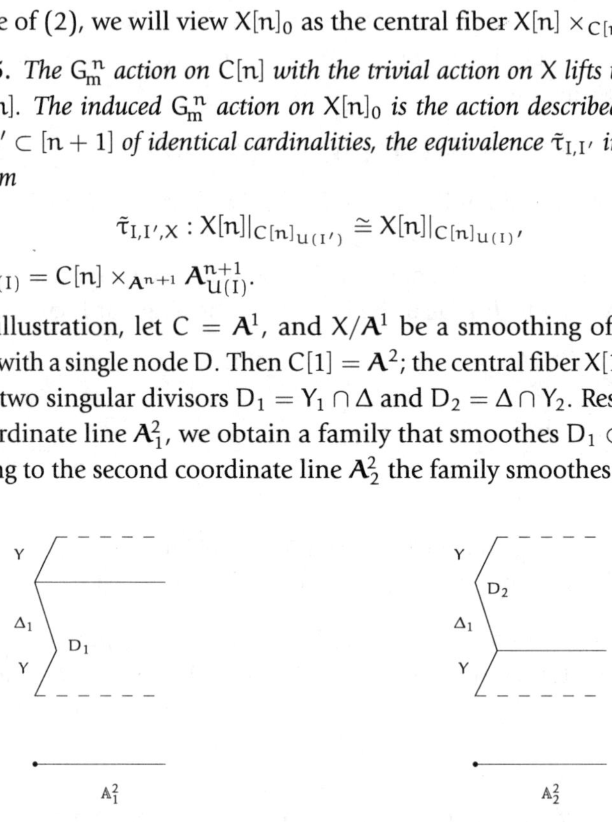

Figure 2. D_1 is smoothed over $\mathbf{A}_1^2$; D_2 is smoothed over $\mathbf{A}_2^2$.

Definition 2.17. *We define* $\mathfrak{X}_n$ *be* $X[n]$ *quotient by* G_m^n *and the equivalences* $\tilde{\tau}_{I,I'}$ *for all* $I, I \subset [n+1]$ *of* $|I| = |I'|$*. We let* $\mathfrak{p} \colon \mathfrak{X} \to X$ *be the morphism induced by the tautological projection* $p : X[n] \to X$.

The quotient is an Artin stack; it is over C since the G_m^n and the equivalence $\tilde{\tau}_{I,I',X}$ are defined over C.

Using the inclusion $[n+1] \subset [n+2]$, the induced $\mathbf{A}^{n+1} \to \mathbf{A}^{n+2}$ in (2.10), and $\tau_{+1} : C[n] \to C[n+1]$, we have tautological immersion

$$\mathfrak{X}_n \longrightarrow \mathfrak{X}_{n+1}. \tag{2.18}$$

One checks that via the homomorphism $G_m^n \to G_m^{n+1}$ defined by $(t) \mapsto (t,1)$, $\tau_{+1}: C[n] \to C[n+1]$ is G_m^n equivariant and the equivalence relations $\tilde{\tau}_{I,I',X}$ extend to equivalence relations $\tilde{\tau}_{I,I',X}$ by viewing $I, I' \subset [n+1]$ as subsets in $[n+2]$. Clearly, the immersions (2.18) preserve their projections to X, are open immersions, and form a direct system in $\{n \geqslant 0\}$.

Definition 2.19. *We define* $\mathfrak{X} = \varinjlim_n \mathfrak{X}_n$*; we define* $\mathfrak{p}: \mathfrak{X} \to X$ *be the induced projection.*

Theorem 2.20. *The morphisms* $X[n] \to C[n]$ *induce a representable* C*-morphism* $\mathfrak{X} \to \mathfrak{C}$*. It fits into the commutative square* (2.3).

We call $(\mathfrak{X} \to \mathfrak{C}, \mathfrak{p})$ the family of expanded degenerations of $X \to C$.

2.3. Working definition of $\mathfrak{X}$

Using the stack $(\mathfrak{X} \to \mathfrak{C}, \mathfrak{p})$, we define an expanded degeneration over a scheme $S \to C$ be the pull back family $\mathfrak{X} \times_{\mathfrak{C}} S$ for a C-morphism $S \to \mathfrak{C}$. We denote by $\mathfrak{X}(S)$ the collection of all S-families of expanded degenerations.

To define families in $\mathfrak{X}(S)$ using traditional terminology, we introduce effective family and effective equivalence of expanded degenerations.

Definition 2.21. *Let* S *be a* C*-scheme. An effective* S*-family in* $\mathfrak{X}(S)$ *is a pull back family* $\xi^*X[n] = X[n] \times_{C[n]} S$ *for a* C*-morphism* $\xi: S \to C[n]$ *with the projection* $p: \xi^*X[n] \to X$ *induced by* $X[n] \to X$.

Two effective families $\xi_1^*X[n_1]$ *and* $\xi_2^*X[n_2]$ *are effectively equivalent if there are standard embeddings* $\tau_i: C[n_i] \to C[n]$ *as in Lemma 2.14-(2) so that* $\tau_i \circ \xi_1 = (\tau_2 \circ \xi_2)^\sigma$ *for a* $\sigma: S \to G_m^n$.

Proposition 2.22. *An* S*-family of expanded degenerations of a* C*-scheme* S *consists of* $(\mathcal{X}, p)$*, where* $\mathcal{X} \to S$ *is a family of schemes and* $p: \mathcal{X} \to X$ *a morphism, an open covering* $\{S_\alpha\}$ *of* S *and isomorphisms*

$$(\mathcal{X}_\alpha = \mathcal{X} \times_S S_\alpha, p_\alpha = p|_{\mathcal{X}_\alpha}) \cong \xi_\alpha^* X[n_\alpha],$$

where $\xi_\alpha: S_\alpha \to C[n_\alpha]$*, such that restricting to the intersections* $S_\alpha \cap S_\beta$*, the identity map of* $\mathcal{X}|_{S_\alpha \cap S_\beta}$ *is induced by an effective equivalence of* $\xi_\alpha^*X[n_\alpha]|_{S_\alpha \cap S_\beta}$ *and* $\xi_\beta^*X[n_\beta]|_{S_\alpha \cap S_\beta}$.

An arrow of two S*-families* $(\mathcal{X}, p)$ *and* $(\mathcal{X}', p')$ *of expanded degenerations is an* S*-isomorphism* $\rho: \mathcal{X} \to \mathcal{X}'$ *that locally is an effective equivalence, using the local effective representatives of* $\mathcal{X}$ *and* $\mathcal{X}'$.

Remark 2.23. *We comment that this formulation is necessary when* D *is not connected. Otherwise, the automorphisms of* $X[n]_0$ *will be bigger than* G_m^n.

In case D is connected, the description is more direct.

Proposition 2.24. *Suppose* D *is connected. An* S*-family of expanded degenerations of a* C*-scheme* S *consists of* $(\mathcal{X}, p)$*, where* $\mathcal{X} \to S$ *is a family of schemes and* $p: \mathcal{X} \to X$ *a morphism, such that there is an open covering* $\{S_\alpha\}$ *of* S *so that for each* α, $\mathcal{X}_\alpha = \mathcal{X} \times_S S_\alpha$

together with $p_\alpha = p|_{\mathcal{X}_\alpha}$ *is isomorphic to* $X[n] \times_{C[n]} S_\alpha \to S_\alpha$ *with its projection to* X, *for an* n *and a morphism* $S_\alpha \to C[n]$.

An arrow of two S-families $(\mathcal{X}, p)$ *and* $(\mathcal{X}', p')$ *of expanded degenerations is an S-isomorphism* $\rho : \mathcal{X} \to \mathcal{X}'$ *that commutes with the projections* $p : \mathcal{X} \to X$ *and* $p' : \mathcal{X}' \to X$.

In [26], the author used this working definition to construct the collection of families of expanded degenerations in case D is connected. It was shown there that the description in Proposition 2.24 satisfies the characterization given in Proposition 2.22.

In [26], the author used the working definition of $\mathfrak{X}(S)$ to construct good degeneration of moduli of stable morphisms. Though the stack $\mathfrak{X} \to \mathfrak{C}$ was mentioned in [26], the current formulation is influenced by discussing this construction with several experts in the subject.

2.4. The stack $\mathfrak{D}_\pm \subset \mathfrak{Y}$

The stack of expanded pair $(Y, D_\pm)$ is pivotal to the decomposition of the central fiber of a good degeneration. It is the first step toward constructing the moduli space $\mathcal{M}_{(Y,\tilde{D})}$ mentioned in the property-III in the introduction.

We caution that in case $D \cong D_- \cong D_+$ has more than one irreducible components, different splitting of $\tilde{D} = D_- \cup D_+$ will result in different bookkeeping; nevertheless not affecting the application of the good degenerations.

We now construct the stack

$$\mathfrak{D}_\pm \subset \mathfrak{Y} \longrightarrow \mathfrak{A}_\diamond \tag{2.25}$$

of expanded pairs of $(Y, D_\pm)$.

We fix the convention on indexing $\mathbf{A}^{n_-+n_+}$ and $G_m^{n_-+n_+}$. In this paper, whenever we see product of $n_- + n_+$ copies, we index the individual factor by indices $-n_-, \dots, -1, 1, \dots, n_+$. (Note that index 0 is skipped.) Thus the $(-n_-)$-th coordinate line of $\mathbf{A}^{n_-+n_+}$ is $(t, 0, \dots, 0)$, and the n_+-th coordinate line is $(0, \dots, 0, t)$. The same convention applies to indexing factors of $G_m^{n_-+n_+}$. We let $G_m^{n_-+n_+}$ act on $\mathbf{A}^{n_-+n_+}$ via the traditional convention

$$(t_{-n_-}, \dots, t_{-1}, t_1, \dots, t_{n_+})^\sigma = (\sigma_{-n_-} t_{-n_-}, \dots, \sigma_{-1} t_{-1}, \sigma_1 t_1, \dots, \sigma_{n_+} t_{n_+}).$$

We then construct

$$D[n_-, n_+]_\pm \subset Y[n_-, n_+] \longrightarrow \mathbf{A}^{n_-+n_+}, \quad p : Y[n_-, n_+] \longrightarrow Y, \tag{2.26}$$

inductively by the rule:

(1) $(Y[0,0], D[0,0]_\pm) = (Y, D_\pm)$;
(2) $Y[n_-, n_+ + 1]$ is the blow-up of $Y[n_-, n_+] \times \mathbf{A}^1$ along $D[n_-, n_+]_+ \times 0$, and $D[n_-, n_+ + 1]_\pm$ is the proper transform of $D[n_-, n_+]_\pm \times \mathbf{A}^1$;
(3) $Y[n_- + 1, n_+]$ is the blow-up of $\mathbf{A}^1 \times Y[n_-, n_+]$ along $0 \times D[n_-, n_+]_-$, and $D[n_- + 1, n_+]_\pm$ is the proper transform of $\mathbf{A}^1 \times D[n_-, n_+]_\pm$;
(4) $p : Y[n_-, n_+] \to Y$ is the one induced by the identity $Y \to Y$.

Following the convention, the extra copy of $\mathbf{A}^1$ added to the right in item (2) is the $(n_+ + 1)$-th factor of $\mathbf{A}^{n_- + (n_+ + 1)}$; the copy $\mathbf{A}^1$ added to the left in item (3) is the $(-n_- - 1)$-th copy in $\mathbf{A}^{(n_- + 1) + n_+}$.

The central fiber $D[n_-, n_+]_{\pm,0} \subset Y[n_-, n_+]_0$ of (2.26) is easily described. We let $N_\pm$ be the normal line bundle of $D_\pm$ in Y; let $\Delta = \mathbf{P}(N_+ \oplus 1)$ with distinguished divisors $D_+ = \mathbf{P}(1)$ and $D_- = \mathbf{P}(N)$. Then

$$Y[n_-, n_+]_0 = \Delta_{-n_-} \sqcup \cdots \sqcup \Delta_{-1} \sqcup Y \sqcup \Delta_1 \sqcup \cdots \sqcup \Delta_{n_+}, \quad n_-, n_+ \geqslant 0,$$

where the square cup "$\sqcup$" means that we identify the divisor $D_- \subset \Delta_i$ with $D_+ \subset \Delta_{i+1}$, understanding that $\Delta_0 = Y$, and $\Delta_i = \Delta$ for $i \neq 0$.

The central fiber $D[n_-, n_+]_{-,0}$ of $D[n_-, n_+]$ is the divisor D_+ in Δ_{-n_-}; the central fiber $D[n_-, n_+]_{+,0}$ is the divisor $D_- \subset \Delta_{n_+}$.

We let $p : Y[n_-, n_+]_0 \to Y$ be induced by $p : Y[n_-, n_+] \to Y$ (cf. item (4)); it is by contracting all $\Delta_{i \neq 0}$. The scheme $Y[n_-, n_+]_0$ has simple normal crossings singularities.

We call

$$(2.27) \qquad (Y[n_-, n_+]_0, D[n_-, n_+]_{\pm,0}) \quad \text{with} \quad p : Y[n_-, n_+]_0 \to Y$$

and the $G_m^{n_- + n_+}$-action an expanded relative pair of $(Y, D_\pm)$.

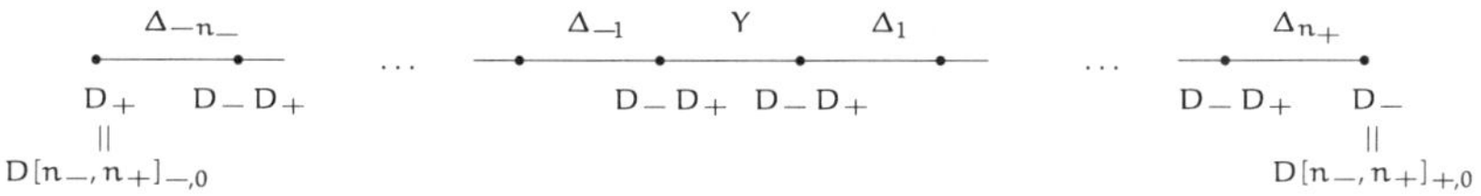

Figure 3. The Y_-, Δ's, and Y_+ glue to form $Y[n_-, n_+]_0$; the two end divisors are the new relative divisor of $Y[n_-, n_+]_0$.

The family $Y[n_-, n_+] \to \mathbf{A}^{n_- + n_+}$ has the following additional properties:

(5) let $\ell_l \to \mathbf{A}^{n_- + n_+}$ be the l-th coordinate line of $\mathbf{A}^{n_- + n_+}$, $-n_- \leqslant l \leqslant n_+$, $l \neq 0$, then the restriction $Y[n_-, n_+] \times_{\mathbf{A}^{n_- + n_+}} \ell_l$ is a smoothing of the divisor $D_l = \Delta_{l-1} \cap \Delta_l$ if $l > 0$, of $D_l = \Delta_l \cap \Delta_{l+1}$ if $l < 0$.

(Notice that $Y[n_-, n_+]_0$ has singular divisors D_l, $-n_- \leqslant l \leqslant n_+$ and $l \neq 0$.)

The family (2.26) is $G_m^{n_- + n_+}$-equivariant; the pair (2.27) is $G_m^{n_- + n_+}$-equivariant since $G_m^{n_- + n_+}$ acts on $Y[n_-, n_+]/\mathbf{A}^{n_- + n_+}$ and $0 \in \mathbf{A}^{n_- + n_+}$ is a fixed point. The k-th factor of the G_m in $G_m^{n_- + n_+}$ acts trivially on all Δ_i except Δ_k; on Δ_k the action is the tautological G_m-action of (2.13).

Like the stack $\mathfrak{X} \to \mathfrak{C}$, the stack (2.25) we aim to construct will be the limit of the quotients of (2.26) by $G_m^{n_- + n_+}$ and another class of equivalences associated to subsets

$$(2.28) \qquad I \subset [-n_-, n_+] - \{0\}.$$

(We define its complement $I^\circ = [-n_-, n_+] - I \cup \{0\}$.)

Given an I as in (2.28), we define $\mathbf{A}_{\mathrm{I}}^{n_-+n_+}$ and $\mathbf{A}_{\mathrm{U(I)}}^{n_-+n_+} \subset \mathbf{A}^{n_-+n_+}$ be as in (2.4) and (2.5). Like (2.6), for any I as in (2.28), letting $m_\pm = |\mathrm{I} \cap \mathbb{Z}_\pm|$, we have isomorphism

$$\tilde{\tau}_{\mathrm{I}} : \mathbf{A}^{m_-+m_+} \times \mathrm{G}_{\mathrm{m}}^{(n_--m_-)+(n_+-m_+)} \longrightarrow \mathbf{A}_{\mathrm{U(I)}}^{n_-+n_+}, \tag{2.29}$$

and for any I′ as in (2.28) with

$$m_\pm = |\mathrm{I} \cap \mathbb{Z}_\pm| = |\mathrm{I}' \cap \mathbb{Z}_\pm|, \tag{2.30}$$

the pair (I, I′) defines an isomorphism

$$\tilde{\tau}_{\mathrm{I,I'}} = \tilde{\tau}_{\mathrm{I}} \circ \tilde{\tau}_{\mathrm{I'}}^{-1} : \mathbf{A}_{\mathrm{U(I')}}^{n_-+n_+} \longrightarrow \mathbf{A}_{\mathrm{U(I)}}^{n_-+n_+}. \tag{2.31}$$

As before, we let

$$\tau_{\mathrm{I}} : \mathbf{A}^{m_-+m_+} \longrightarrow \mathbf{A}^{n_-+n_+} \tag{2.32}$$

be the restriction of $\tilde{\tau}_{\mathrm{I}}$ to $\mathbf{A}^{m_-+m_+} \times \{1\}$, where 1 is the identity element.

Following the construction, one checks that for any I as in (2.28), we have a canonical isomorphism

$$\tau_{\mathrm{I,Y}} : \mathrm{Y}[m_-, m_+] \longrightarrow \tau_{\mathrm{I}}^* \mathrm{Y}[n_-, n_+],$$

lifting the τ_{I} in (2.32); for any pair (I, I′) of subsets in (2.28) satisfying (2.30), we have canonical isomorphism

$$\tilde{\tau}_{\mathrm{I,I',Y}} : \mathrm{Y}[n_-, n_+]|_{\mathbf{A}_{\mathrm{U(I')}}^{n_-+n_+}} \longrightarrow \mathrm{Y}[n_-, n_+]|_{\mathbf{A}_{\mathrm{U(I)}}^{n_-+n_+}},$$

lifting the $\tilde{\tau}_{\mathrm{I,I'}}$ in (2.31).

Definition 2.33. *We define* $\mathfrak{A}_{\diamond, n_-+n_+}$ *be the quotient of* $\mathbf{A}^{n_-+n_+}$ *by* $\mathrm{G}_{\mathrm{m}}^{n_-+n_+}$ *and the equivalences* $\tilde{\tau}_{\mathrm{I,I'}}$ *for all allowable pairs* (I, I′) *in* (2.30); *we define* $\mathfrak{A}_\diamond$ *be the limit of* $\mathfrak{A}_{\diamond, n_-+n_+}$, *taking* $n_-, n_+ \to \infty$. $\mathfrak{A}_\diamond$ *is an Artin stack.*

We define $\mathfrak{D}_{n_-+n_+,\pm} \subset \mathfrak{Y}_{n_-+n_+}$ *be the quotient of* $\mathrm{D}[n_-, n_+]_\pm \subset \mathrm{Y}[n_-, n_+]$ *by* $\mathrm{G}_{\mathrm{m}}^{n_-+n_+}$ *and the equivalences* $\tilde{\tau}_{\mathrm{I,I',Y}}$ *for all pairs* (I, I′) *satisfying* (2.30); *we define* $\mathfrak{D}_\pm \subset \mathfrak{Y}$ *be the limit of* $\mathfrak{D}_{n_-+n_+,\pm} \subset \mathfrak{Y}_{n_-+n_+}$, *taking* $n_-, n_+ \to \infty$. *We let* $\mathfrak{p} : \mathfrak{Y} \to \mathrm{Y}$ *be the projection induced by the tautological* $\mathrm{Y}[n_-, n_+] \to \mathrm{Y}$.

Theorem 2.34. *The projections* $\mathrm{Y}[n_-, n_+] \to \mathbf{A}^{n_-+n_+}$ *induce a representable morphism* $\mathfrak{D}_\pm \subset \mathfrak{Y} \to \mathfrak{A}_\diamond$.

We call $\mathfrak{D}_\pm \subset \mathfrak{Y} \to \mathfrak{A}_\diamond$ with $\mathfrak{p} : \mathfrak{Y} \to \mathrm{Y}$ the stack of expanded relative pairs of $(\mathrm{Y}, \mathrm{D}_\pm)$. Using $(\mathfrak{D}_\pm \subset \mathfrak{Y} \to \mathfrak{A}_\diamond, \mathfrak{p})$, we define the collection $\mathfrak{Y}(\mathrm{S})$ of expanded families of pair $(\mathrm{Y}, \mathrm{D}_\pm)$ over a scheme S be

$$\mathfrak{D}_\pm \times_{\mathfrak{A}_\diamond} \mathrm{S} \subset \mathfrak{Y} \times_{\mathfrak{A}_\diamond} \mathrm{S}, \quad \mathrm{S} \to \mathfrak{A}_\diamond.$$

In case D is connected, we have

Proposition 2.35. *Suppose* D *is connected. An* S*-family of expanded relative pair of* $(Y, D_\pm)$ *consists of* $(\mathcal{D}_\pm \subset \mathcal{Y}, p)$, *where* $\mathcal{Y} \to S$ *is a family of schemes,* $\mathcal{D}_\pm \subset \mathcal{Y}$ *is a pair of Cartier divisors,* $p : \mathcal{Y} \to Y$ *a morphism, such that there is an open covering* $\{S_\alpha\}$ *of* S *so that for each* α, $\mathcal{D}_{\pm\alpha} = \mathcal{D}_\pm \times_S S_\alpha \subset \mathcal{Y}_\alpha = \mathcal{Y} \times_S S_\alpha$, *together with* $p_\alpha = p|_{\mathcal{Y}_\alpha}$, *is* S_α*-isomorphic to* $D[n_-, n_+]_\pm \times_{\mathbf{A}^{n_-+n_+}} S_\alpha \subset Y[n_-, n_+] \times_{\mathbf{A}^{n_-+n_+}} S_\alpha$ *with its projection to* Y, *for an* (n_-, n_+) *and a morphism* $S_\alpha \to \mathbf{A}^{n_-+n_+}$.

An arrow from $\mathcal{D}_\pm \subset \mathcal{Y}$ *to* $\mathcal{D}'_\pm \subset \mathcal{Y}'$ *of two* S*-families of expanded pairs of* $(Y, D_\pm)$ *is an* S*-isomorphism* $\rho : \mathcal{Y} \to \mathcal{Y}'$ *that commutes with the projections* $p : \mathcal{Y} \to Y$ *and* $p' : \mathcal{Y}' \to Y$, *and such that* $\rho(\mathcal{D}_\pm) = \mathcal{D}'_\pm$.

2.5. Decomposition of degenerations

This subsection sets the foundation for property-III of good degenerations; in case $Y = Y_- \cup Y_+$, we refine this construction by introducing the stack of weighted decompositions in the next subsection.

We work out the decomposition of the stack

$$\mathfrak{X}_0 := \mathfrak{X} \times_C 0$$

by introducing the stack of node-marking objects in $\mathfrak{X}_0$.

Definition 2.36. *A node-marking of* $X[n]_0$ *is a marking of one of the singular divisor* D_k *of* $X[n]_0$. *A node-marking of a family* $\mathcal{X} \to S$ *in* $\mathfrak{X}_0(S)$ *is an* S*-morphism* $\eta : D \times S \to \mathcal{X}$ *so that for any closed* $s \in S$, $\eta(D \times s) \subset \mathcal{X}_s$ *is a node-marking of* $\mathcal{X}_s$.

An arrow between two $\mathcal{X}$ *and* $\mathcal{X}'$ *in* $\mathfrak{X}_0(S)$ *with node-markings* η *and* η' *is an arrow* $\rho : \mathcal{X} \to \mathcal{X}'$ *in* $\mathfrak{X}_0(S)$ *so that for any closed* $s \in S$, $\rho \circ \eta(D \times s) = \eta'(D \times s)$.

Proposition 2.37. *The collection of families in* $\mathfrak{X}_0$ *with node-markings form an Artin stack, denoted by* $\mathfrak{X}_0^\dagger$, *and called the stack of expanded central fibers with marked singular divisors. Forgetting the node-marking defines a morphism*

$$\phi : \mathfrak{X}_0^\dagger \longrightarrow \mathfrak{X}_0.$$

Proof. Let $X[n] \to \mathfrak{X}$ be a smooth chart. Then $X[n] \times_C 0 \to \mathfrak{X}_0$ is a smooth chart. By construction,

$$X[n] \times_C 0 = X[n] \times_{\mathbf{A}^{n+1}} (\mathbf{A}^{n+1} \times_{\mathbf{A}^1} 0).$$

Denoting $\mathbf{A}^{n+1}_{k^c} = \{(t) \in \mathbf{A}^{n+1} \mid t_k = 0\}$, (this is $\mathbf{A}^{n+1}_I$ for $I = [n+1] - k$,) we have the union

$$\mathbf{A}^{n+1} \times_{\mathbf{A}^1} 0 = \bigcup_{k=1}^{n+1} \mathbf{A}^{n+1}_{k^c}.$$

Further, $X[n] \times_{\mathbf{A}^{n+1}} \mathbf{A}^{n+1}_{k^c}$ has normal crossing singularity and its singular divisor is isomorphic to $D \times \mathbf{A}^{n+1}_{k^c}$; we denote this divisor by

$$\eta_k : D \times \mathbf{A}^{n+1}_{k^c} \longrightarrow X[n] \times_{\mathbf{A}^{n+1}} \mathbf{A}^{n+1}_{k^c}. \tag{2.38}$$

By Definition 2.36, (2.38) is a node-marking of $X[n] \times_{\mathbf{A}^{n+1}} \mathbf{A}^{n+1}_{k^c}$; thus

$$(X[n] \times_{\mathbf{A}^{n+1}} \mathbf{A}^{n+1}_{k^c}, \eta_k) \in \mathfrak{X}_0^\dagger(\mathbf{A}^{n+1}_{k^c}). \tag{2.39}$$

The disjoint union of (2.39) for all $1 \leqslant k \leqslant n+1$ form a smooth atlas of $\mathfrak{X}_0^\dagger$. This proves that $\mathfrak{X}_0^\dagger$ is an Artin stack. □

It will be useful to construct a stack $\mathfrak{C}_0^\dagger$ and an arrow $\mathfrak{C}_0^\dagger \to \mathfrak{C}$ so that

$$\begin{array}{ccc} \mathfrak{X}_0^\dagger & \longrightarrow & \mathfrak{X} \\ \downarrow & & \downarrow \\ \mathfrak{C}_0^\dagger & \longrightarrow & \mathfrak{C} \end{array} \tag{2.40}$$

is a Cartesian product.

We construct $\mathfrak{C}_0^\dagger$ as follows. For a pair of integers $1 \leqslant k \leqslant n+1$, we consider the G_m^n-scheme $\mathbf{A}_{k^c}^{n+1}$, where the G_m^n-action is induced from $\mathbf{A}_{k^c}^{n+1} \subset \mathbf{A}^{n+1}$ and the G_m^n action on $\mathbf{A}^{n+1}$. We define the closed immersion

$$\tau_{+1} : \mathbf{A}_{k^c}^{n+1} \longrightarrow \mathbf{A}_{k^c}^{n+2}, \quad (z) \mapsto (z, 1). \tag{2.41}$$

Under the homomorphism $G_m^n \to G_m^{n+1}$ via $(\sigma) \mapsto (\sigma, 1)$, τ_{+1} is G_m^n equivariant.

Like the case for $\mathfrak{C}$, we need to quotient out the second class of equivalences. In the following, for any $I \subset [n+1]$ and k an integer, we denote $I_{<k} = \{i \in I \mid i < k\}$; similarly for $I_{>k}$.

We consider $k \in I \subset [n+1]$ and $k' \subset I' \subset [n+1]$ such that

$$|I_{<k}| = |I'_{<k}| \quad \text{and} \quad |I_{>k}| = |I'_{>k}|. \tag{2.42}$$

Then the equivalence $\tilde{\tau}_{I,I'}$ of (2.9) (or (2.31)) restricted to $\mathbf{A}_{k^c}^{n+1} \cap \mathbf{A}_{U(I')}^{n+1}$ defines an isomorphism

$$\tau_{(I,k),(I',k')} : \mathbf{A}_{k^c}^{n+1} \cap \mathbf{A}_{U(I')}^{n+1} \longrightarrow \mathbf{A}_{k^c}^{n+1} \cap \mathbf{A}_{U(I)}^{n+1}. \tag{2.43}$$

Definition 2.44. *We define $\mathfrak{C}_{n,0}^\dagger$ be the disjoint union $\coprod_{k=1}^{n+1} \mathbf{A}_k^{n+1}/G_m^n$ (stacky quotients) quotient by the equivalences $\tau_{(I,k),(I',k')}$ for all pairs $k \in I$ and $k' \in I'$ satisfying* (2.42).

We define open immersions $\mathfrak{C}_{n,0}^\dagger \to \mathfrak{C}_{n+1,0}^\dagger$ using the inclusion τ_{+1}. We define $\mathfrak{C}_0^\dagger = \varinjlim_n \mathfrak{C}_{n,0}^\dagger$.

Proposition 2.45. *The morphisms $X[n] \times_{\mathbf{A}^{n+1}} \mathbf{A}_{k^c}^{n+1} \to \mathbf{A}_{k^c}^{n+1}$, where $X[n] \times_{\mathbf{A}^{n+1}} \mathbf{A}_{k^c}^{n+1}$ is with the node-marking* (2.39), *induce a morphism $\mathfrak{X}_0^\dagger \to \mathfrak{C}_0^\dagger$ that fits into the Cartesian product* (2.40).

As $\coprod \mathbf{A}_{k^c}^{n+1} \to \mathfrak{C}_{n,0}^\dagger$ is a smooth chart of $\mathfrak{C}_{n,0}^\dagger$, and the former is the normalization of $\mathbf{A}^{n+1} \times_{\mathbf{A}^1} 0$, the morphism $\mathfrak{C}_0^\dagger \to \mathfrak{C}_0$ is a normalization. It is fitting to call $\mathfrak{X}_0^\dagger \to \mathfrak{X}_0$ the decomposition of locally complete intersection singularity of $\mathfrak{C}_0$.

The final step of the decomposition is the following isomorphism result

Proposition 2.46. *There is a canonical isomorphism $\mathfrak{C}^\dagger \cong \mathfrak{A}_\diamond$ so that $\mathfrak{X}_0^\dagger$ is derived from $\mathfrak{Y}$ by identifying the stacks $\mathfrak{D}_-$ with $\mathfrak{D}_+$ via the isomorphisms $\mathfrak{D}_- \cong D \times \mathfrak{A}_\diamond \cong \mathfrak{D}_+$.*

Proof. We define $\mathbf{A}^{n_-+n_+} \to \mathbf{A}^{n+1}_{k^c}$, $k = n_- + 1$, $n = n_- + n_+$, via

$$(t_{-n_-}, \dots, t_{-1}, t_1, \dots, t_{n_+}) \mapsto (t_{-1}, \dots, t_{-n_-}, 0, t_{n_+}, \dots, t_1).$$

This is G_m^n equivariant via a homomorphism $G_m^n \to G_m^n$, and induces a morphism $\mathfrak{A}_\diamond \to \mathfrak{C}_0^\dagger$. The remainder of the proof is straightforward. □

We let Φ be the induced morphism. It fits into the commutative square

$$\begin{array}{ccc} \mathfrak{Y} & \stackrel{\Phi}{\longrightarrow} & \mathfrak{X}_0^\dagger \\ \downarrow & & \downarrow \\ \mathfrak{A}_\diamond & \longrightarrow & \mathfrak{C}_0^\dagger \end{array} \tag{2.47}$$

so that for $\zeta : \mathfrak{C}_0^\dagger \times D \to \mathfrak{X}_0^\dagger$ the node-marking,

$$\Phi|_{\mathfrak{D}_\pm} : \mathfrak{D}_\pm \longrightarrow \zeta(\mathfrak{C}_0^\dagger \times D)$$

are isomorphisms. Further, away from $\mathfrak{D}_- \cup \mathfrak{D}_+$, Φ is an isomorphism. We call Φ the decomposition morphism.

Finally, we comment that in case $Y = Y_- \cup Y_+$ is a disjoint union of two smooth varieties, then $Y[n_-, n_+]$ is a disjoint union of two connected components, and $\mathfrak{Y}$ is a disjoint union of two connected stacks.

The two connected components can easily be described. Let $D_\pm = Y_\pm \cap \tilde{D}$. For the pair $D_+ \subset Y_+$, by replacing Y by Y_+, understanding $D_- = \emptyset$ and keeping $n_- = 0$ in the construction (2.25), we obtain the stack $\mathfrak{D}_+ \subset \mathfrak{Y}_+$ over the base stack $\mathfrak{A}_+ = \mathfrak{A}$. Similarly we obtain $\mathfrak{D}_- \subset \mathfrak{Y}_-$ over $\mathfrak{A}_- = \mathfrak{A}$.

Further, using the obvious $\mathbf{A}^{n_-+n_+} \cong \mathbf{A}^{n_-} \times \mathbf{A}^{n_+}$, we obtain an isomorphism

$$\mathfrak{A}_\diamond \cong \mathfrak{A}_- \times \mathfrak{A}_+.$$

Proposition 2.48. *Let $Y = Y_1 \cup Y_+$ be a disjoint union of two smooth varieties; let $D_\pm = Y_\pm \cap \tilde{D}$. Then we have canonical isomorphism*

$$(\mathfrak{D}_- \subset \mathfrak{Y}_-) \times \mathfrak{A}_+ \cup (\mathfrak{D}_+ \subset \mathfrak{Y}_+) \times \mathfrak{A}_- \cong \mathfrak{D} \subset \mathfrak{Y},$$

as stacks over $\mathfrak{A}_\diamond \cong \mathfrak{A}_- \times \mathfrak{A}_+$. Further, the decomposition morphism Φ (in (2.47)) takes the form

$$\Phi : \mathfrak{Y}_- \times \mathfrak{A}_+ \cup \mathfrak{Y}_+ \times \mathfrak{A}_- \longrightarrow \mathfrak{X}_0^\dagger.$$

Over a closed point in $\mathfrak{C}_0^\dagger$, the decomposition morphism simply states that

$$X[n] = Y_-[n_-] \cup Y_+[n_+], \tag{2.49}$$

a union after identifying $D_-[n_-] \subset Y_-[n_-]$ with $D_+[n_+] \subset Y_+[n_+]$.

2.6. Decomposition of degenerations II

In this subsection, we suppose $Y = Y_- \cup Y_+$ is a disjoint union of two irreducible varieties. We will form the weighted decomposition stack that will reveal the *virtual local complete intersection* property of the good degenerations postulated in the introduction.

Let $\mathcal{M}_{X/C}$ be an example of good degeneration constructed using the family $\mathfrak{X} \to \mathfrak{C}$. By definition, it is a stack over $\mathfrak{X}$. We will show that there is a new stack $\mathfrak{C}^{\beta}$, étale over $\mathfrak{C}$, factoring through

$$\begin{array}{ccc} \mathcal{M}_{X/C} & \longrightarrow & \mathfrak{X} \\ \downarrow & & \downarrow \\ \mathfrak{C}^{\beta} & \longrightarrow & \mathfrak{C}, \end{array}$$

such that $\mathfrak{C}^{\beta} \to C$ is a local complete intersection morphism, and $\mathcal{M}_{X/C} \to \mathfrak{C}^{\beta}$ has local property similar to that of the general fibers of $\mathcal{M}_{X/C} \to C$.

The stack $\mathfrak{C}^{\beta}$ essentially is $\mathfrak{C}$ together with the data to identify the decomposition divisor of $X[n]_0 = Y_-[n_-] \cup Y_+[n_+]$ along which a moduli object ξ on $X[n]_0$ (in $\mathcal{M}_{X/C} \times_C 0$) decomposes. The data we use will be the topological data of the restrictions $\xi|_{Y_-[n_-]}$ and $\xi|_{Y_+[n_+]}$.

We fix an additive group Λ. Using $Y = Y_- \cup Y_+$, we index the irreducible components of $X[n]_0$ as $\Delta_0 = Y_-$, $\Delta_{n+1} = Y_+$, and other Δ_i are as usual, from 1 to n.

Definition 2.50. *A weight assignment of* $X[n]_0$ *is a function*

$$w : \{\Delta_0, \dots, \Delta_{n+1}, D_1, \dots, D_{n+1}\} \longrightarrow \Lambda,$$

where $w(\Delta_i)$ *and* $w(D_j)$ *are the weights of* Δ_i *and* D_j. *A weight assignment of* X_t, $t \neq 0$, *is a single value assignment* $w(X_t) \in \Lambda$. *A weight assignment* w *of* $\mathcal{X} \in \mathfrak{X}(S)$ *is a collection* $\{w_s \mid s \in S\}$ *of weight assignments* w_s *of* $\mathcal{X}_s$.

We define the weight assignment of $(Y_\pm[n_\pm], D_\pm[n_\pm])$ *be weight assignments* $w_\pm$ *of the irreducible components of* $Y_\pm[n_\pm]$, *of its* D_k*'s, and of* $D_\pm[n_\pm]$.

For any subchain $\Delta_l \cup \ldots \cup \Delta_{l'}$ we define the total weight

$$w(\Delta_l \cup \ldots \cup \Delta_{l'}) = \sum_{i=l}^{l'} w(\Delta_i) - \sum_{j=l+1}^{l'} w(D_j).$$

Let $s_0 \in S$ be an irreducible curve, and let w be a weight assignment of $\mathcal{X} \in \mathfrak{X}(S)$. Suppose $\mathcal{X}_{s_0} \cong X[n]_0$ and for a general $s \in S$, $\mathcal{X}_s \cong X[m]_0$ or is smooth. In case it is $X[m]_0$, then $m \leqslant n$, and by the construction of $X[n]$, there are

$$k_0 = 0 < k_1 < \ldots < k_{m+1} < k_{m+2} = n + 2$$

so that the $\Delta_i \subset \mathcal{X}_s$ specializes to the union of the chain $\Delta_{k_i} \cup \ldots \cup \Delta_{k_{i+1}-1} \subset \mathcal{X}_{s_0}$; equivalently, (only) the singular divisors $D_{k_i} \subset \mathcal{X}_{s_0}$, $i = 1, \ldots, m+1$, are not smoothed in the family $\mathcal{X}$.

Definition 2.51. *Let $\mathcal{X} \in \mathfrak{X}(S)$ and $s_0 \in S$ be an irreducible curve. We say the assignment w of $\mathcal{X}$ is continuous at s_0 if the following hold:*
(a). when $\mathcal{X}_{s_0} \cong X[n]_0$ and $\mathcal{X}_s \cong X[m]_0$, then for every $i = 1, \ldots, m+1$, we have

$$w_s(\Delta_i) = w_{s_0}(\Delta_{k_i} \cup \ldots \cup \Delta_{k_{i+1}-1}), \quad w_s(D_i) = w_{s_0}(D_{k_i});$$

(b). when $\mathcal{X}_s$ is smooth, then $w_s(\mathcal{X}_s) = w_{s_0}(\mathcal{X}_{s_0})$.

In general, a weight assignment of $\mathcal{X} \in \mathfrak{X}(S)$ for any S is continuous if for any irreducible curve $S_0 \to S$ the pull back family $\mathcal{X} \times_S S_0$ with the induced weight assignment is continuous.

Example 2.52. *Suppose* $\dim X/C = 1$. *In case there is a locally free sheaf $\mathcal{E}$ on $\mathcal{X}$, assigning each $\Delta_k \subset \mathcal{X}_s$ the degree of $|_{\Delta_k}$ and assigning each $D_l \subset \mathcal{X}_s$ zero is a continuous weight assignment taking values in $\mathbb{Z}$.*

Proposition 2.53. *Given a $\beta \in \Lambda$. We define $\mathfrak{X}^\beta(S)$ be the collections of pairs $(\mathcal{X}, w)$, where $\mathcal{X} \in \mathfrak{X}(S)$ and w is a continuous weight assignment of $\mathcal{X}$ of total weight β. Then $\mathfrak{X}^\beta$ is an Artin stack.*

By forgetting the weights, we obtain the forgetful morphism $\mathfrak{X}^\beta \to \mathfrak{X}$. We now construct a weighted stack $\mathfrak{C}^\beta$ together with a forgetful morphism $\mathfrak{C}^\beta \to \mathfrak{C}$ so that $\mathfrak{X}^\beta$ is the Cartesian product

$$\begin{array}{ccc} \mathfrak{X}^\beta & \longrightarrow & \mathfrak{X} \\ \downarrow & & \downarrow \\ \mathfrak{C}^\beta & \longrightarrow & \mathfrak{C} \end{array} \tag{2.54}$$

This is done by defining a weight assignment of a $t \in C[n]$ be a weight assignment of the the dual graph Γ_t of $X[n]_t$. The dual graph of $X[n]_0$ is a chain of $(n+2)$ edges connected by $(n+1)$ vertices. We index the edges by $0, \ldots, n+1$, associated to the irreducible components $\Delta_0, \ldots, \Delta_{n+1}$ of $X[n]_0$; we index the vertices by $1, \ldots, n+1$, associated to the divisors $D_1, \ldots, D_{n+1}$. A weight assignment Γ_t is a map $w_t : \Gamma_t \to \Lambda$ that assigns to vertices and edges of Γ_t values in Λ. With the associations given, a weight assignment of $t \in C[n]$ is identical to a weight assignment of $X[n]_t$.

We define a weight assignment w of $C[n]$ be a collection of weight assignments w_t of all $t \in C[n]$. We define continuous weight assignments of $C[n]$ similar to Definition 2.51.

Definition 2.55. *We define $\mathfrak{C}_n^\beta$ be the quotient of the disjoint union of $(C[n], w)$, where w is a continuous weight assignments of $C[n]$ of total weights β, by the G_m^n-action, and by the equivalence relations $\tilde{\tau}_{I,I'}$ in (2.9) (or (2.31)) that preserves the weight assignments. We define $\mathfrak{C}^\beta = \varinjlim_n \mathfrak{C}_n^\beta$.*

Proposition 2.56. *The weight assignment makes $\mathfrak{C}^\beta$ an Artin stack, induces a tautological morphism $\mathfrak{X}^\beta \to \mathfrak{C}^\beta$. The forgetful morphism $\mathfrak{C}^\beta \to \mathfrak{C}$ is étale and fits into the Cartesian square* (2.54).

Replacing $\mathfrak{X}/\mathfrak{C}$ by $\mathfrak{X}_0^\dagger/\mathfrak{C}_0^\dagger$, we obtain a pair

$$\mathfrak{X}_0^{\dagger,\beta} \longrightarrow \mathfrak{C}_0^{\dagger,\beta},$$

where closed points in $\mathfrak{X}_0^{\dagger,\beta}$ are $(X[n]_0, D_k, w)$, where $D_k \subset X[n]_0$ are node-markings and w are weight assignments of $X[n]_0$ of total weights β; $\mathfrak{C}_0^{\dagger,\beta}$ is defined accordingly, combining the construction of $\mathfrak{C}_0^\dagger$ and $\mathfrak{C}^\beta$.

The pair $\mathfrak{X}_0^{\dagger,\beta} \longrightarrow \mathfrak{C}_0^{\dagger,\beta}$ is a disjoint union of open and closed substacks indexed by the set of splittings of β. We let

$$\Lambda_\beta^{\mathrm{spl}} = \{\delta = (\delta_\pm, \delta_0) \mid \delta_\pm, \delta_0 \in \Lambda,\ \delta_- + \delta_+ - \delta_0 = \beta\}. \tag{2.57}$$

For each $\delta \in \Lambda_\beta^{\mathrm{spl}}$, we define $\mathfrak{X}_0^{\dagger,\delta}(\mathbf{k}) \subset \mathfrak{X}_0^{\dagger,\beta}(\mathbf{k})$ be the subset of $(X[n]_0, D_k, w)$ such that

$$w(\Delta_0 \cup \ldots \cup \Delta_{k-1}) = \delta_-, \quad w(\Delta_k \cup \ldots \cup \Delta_{n+1}) = \delta_+, \quad w(D_k) = \delta_0.$$

Clearly, it is both open and closed in $\mathfrak{X}_0^{\dagger,\beta}(\mathbf{k})$, thus defining an open and closed substack $\mathfrak{X}_0^{\dagger,\delta} \longrightarrow \mathfrak{X}_0^{\dagger,\beta}$.

Accordingly, we can form the stack $\mathfrak{C}_0^{\dagger,\delta}$ and the morphism $\mathfrak{C}_0^{\dagger,\delta} \to \mathfrak{C}_0^{\dagger,\beta}$ that fits into a Cartesian product

$$\begin{array}{ccc} \mathfrak{X}_0^{\dagger,\delta} & \longrightarrow & \mathfrak{X}_0^{\dagger,\beta} \\ \downarrow & & \downarrow \\ \mathfrak{C}_0^{\dagger,\delta} & \longrightarrow & \mathfrak{C}_0^{\dagger,\beta} \end{array} \tag{2.58}$$

We let

$$\Phi_\delta : \mathfrak{C}_0^{\dagger,\delta} \longrightarrow \mathfrak{C}^\beta \tag{2.59}$$

be the lower horizontal morphism composed with the forgetful morphism $\mathfrak{C}_0^{\dagger,\beta} \to \mathfrak{C}^\beta$. The weighted decomposition takes the form

Proposition 2.60. *There are canonical choices of line bundles with sections* (L_δ, s_δ) *on* $\mathfrak{C}^\beta$, *indexed by* $\delta \in \Lambda_\beta^{\mathrm{spl}}$, *such that*

(1) *let* $t \in \Gamma(\mathcal{O}_{\mathbf{A}^1})$ *be the standard coordinate function and* $\pi : \mathfrak{C}^\beta \to \mathbf{A}^1$ *be the tautological projection, then*

$$\bigotimes_{\delta \in \Lambda_\beta^{\mathrm{spl}}} L_\delta \cong \mathcal{O}_{\mathfrak{C}^\beta} \quad \textit{and} \quad \prod_{\delta \in \Lambda_\beta^{\mathrm{spl}}} s_\delta = \pi^* t;$$

(2) *the morphism* Φ_δ *factors through* $s_\delta^{-1}(0) \subset \mathfrak{C}^\beta$ *and effects an isomorphism* $\mathfrak{C}_0^{\dagger,\delta} \cong s_\delta^{-1}(0)$.

The proof of this decomposition is essentially proved in [27]. Note that this Proposition states that $\mathfrak{C}_0^\beta \subset \mathfrak{C}^\beta$ is a complete intersection substack, and the disjoint union of $\mathfrak{C}_0^{\dagger,\delta}$ is its normalization.

Corollary 2.61. *The morphism $\mathfrak{C}^\beta \to C$ is a complete intersection morphism.*

We complete the weighted decomposition by introducing the stack of weighted relative pairs. For an $\delta \in \Lambda_\beta^{\mathrm{spl}}$, we define the stack $\mathfrak{Y}_+^{\delta_+,\delta_0}$ so that $\mathfrak{Y}_+^{\delta_+,\delta_0}(S)$ consists of data $(\mathcal{Y}_+, \mathcal{D}_+, w_+)$, where $(\mathcal{Y}_+, \mathcal{D}_+) \in \mathfrak{Y}_+(S)$ and w_+ are weight assignments of $(\mathcal{Y}_+, \mathcal{D}_+)$ so that for any closed $s \in S$, $w_{+,s}(\mathcal{D}_{+,s}) = \delta_0$ and the total weights $w_{+,s}(\mathcal{Y}_{+,s}) = \delta_+$. By replacing "+" by "−", we obtain $\mathfrak{Y}_-^{\delta_-,\delta_0}$.

We let $\mathfrak{A}_-^{\delta_+,\delta_0}$ and $\mathfrak{A}_+^{\delta_-,\delta_0}$ be the stacks defined similarly so that we have Cartesian products

$$\begin{array}{ccc} \mathfrak{Y}_\pm^{\delta_\pm,\delta_0} & \longrightarrow & \mathfrak{Y}_\pm \\ \downarrow & & \downarrow \\ \mathfrak{A}_\pm^{\delta_\pm,\delta_0} & \longrightarrow & \mathfrak{A}_\pm \end{array} \tag{2.62}$$

By gluing the two relative divisors $\mathcal{D}_-$ and $\mathcal{D}_+$ of $(\mathcal{Y}_\pm, \mathcal{D}_\pm, w_\pm) \in \mathfrak{Y}_\pm^{\delta_\pm,\delta_0}(S)$ and combining the weights w_- and w_+, we obtain the following commutative square of morphisms

$$\begin{array}{ccc} \mathfrak{Y}_-^{\delta_-,\delta_0} \times \mathfrak{Y}_+^{\delta_+,\delta_0} & \longrightarrow & \mathfrak{X}_0^{\dagger,\delta} \\ \downarrow & & \downarrow \\ \mathfrak{A}_-^{\delta_-,\delta_0} \times \mathfrak{A}_+^{\delta_+,\delta_0} & \xrightarrow{\Psi_\delta} & \mathfrak{C}_0^{\dagger,\delta} \end{array}$$

Proposition 2.63. *The morphism Ψ_δ is an isomorphism.*

We will apply this construction to the case of degenerations of moduli of vector bundles, of stable maps, and of Hilbert scheme of subschemes, in the next section.

3. Examples of good degenerations of moduli spaces

Given a moduli problem, and a simple degeneration of smooth projective scheme X/C, one constructs a good degeneration of the moduli space of the specified moduli objects on X/C by working out the moduli space of the specified non-degenerate moduli objects of the family $\mathfrak{X}/\mathfrak{C}$. Following the construction, its central fiber is the moduli space of non-degenerate moduli objects of the family $\mathfrak{X}_0/\mathfrak{C}_0$. Following the decomposition scheme, it is finitely mapped onto by the moduli space of the non-degenerate moduli objects of $\mathfrak{X}_0^\dagger/\mathfrak{A}_0^\dagger$, which is isomorphic to the moduli of non-degenerate relative moduli objects of $\mathfrak{Y}/\mathfrak{A}_\diamond$.

Working out this program produces a good degeneration of the the moduli spaces of the specified moduli objects on the family X/C.

3.1. Moduli of stable morphisms

Let X/C be a simple degeneration of smooth projective schemes. For simplicity, we assume X_t are connected. We fix an ample line bundle H on X, integers χ, k and $d \geqslant 0$. We will review the construction of the good degeneration of moduli of stable morphisms constructed by the author in [26].

We begin with the notion of non-degenerate, predeformable and stable morphisms to $\mathfrak{X}$. Following [37], we will work with stable morphisms of not necessarily connected domains. Consequently, in this subsection nodal curves are proper, but not necessarily connected.

Given $X[n] \to C[n]$ and an integer $1 \leqslant l \leqslant n+1$, we denote

$$X[n]_{l^c} = X[n] \times_{\mathbf{A}^{n+1}} \mathbf{A}^{n+1}_{l^c},$$

(recall $\mathbf{A}^{n+1}_{l^c} = \{(t) \mid t_l = 0\}$,) and denote its singular divisor by $\mathbf{D}_l$; $\mathbf{D}_l$ is isomorphic to $D \times \mathbf{A}^n$, and $\mathbf{D}_l \cap X[n]_0$ is the D_l in $X[n]_0$. We denote $\pi_n : X[n] \to \mathbf{A}^{n+1}$ the projection.

Definition 3.1. *Let $\pi : \mathcal{C} \to S$ with $p_i : S \to \mathcal{C}$ be a flat family of k-pointed nodal curves. Let $S \to C[n]$ be a morphism and $f : \mathcal{C} \to \mathfrak{X} = X[n] \times_{C[n]} S$ be an S-morphism. We say f is non-degenerate along $\mathbf{D}_l$ if $f^{-1}(\mathbf{D}_l \times_{C[n]} S)$ is purely codimension one in $f^{-1}(X[n] \times_{\mathbf{A}^{n+1}} \mathbf{A}^{n+1}_{l^c})$, and is disjoint from $p_1(S), \ldots, p_k(S)$. We say f is non-degenerate if it is non-degenerate along all $\mathbf{D}_l$.*

We next recall the notion of pre-deformable morphisms. We fix an $1 \leqslant l \leqslant n+1$. Let $W \subset X[n]$ be any affine open subset. We let $\hat{W}$ be the formal completion of W along $\mathbf{D}_l$. We let $\hat{\pi}_n : \hat{W} \to \mathbf{A}^{n+1}$ be the tautological projection. We then pick $w_1, w_2 \in \Gamma(\mathcal{O}_{\hat{W}})$ so that

$$w_1 w_2 = \hat{\pi}_n^*(t_l),$$

(t_l is the l-th coordinate function of $\mathbf{A}^{n+1}$,) and $(w_i = 0) \subset \hat{W}$ defines the two branches of $\hat{W} \cap X[n]_{l^c}$.

Let $f : \mathcal{C} \to \mathfrak{X}$ as in Definition 3.1 be an S-morphism that is non-degenerate along $\mathbf{D}_l$. Let

$$v \in f^{-1}(\mathbf{D}_l \times_{C[n]} S)$$

be any closed point. We pick (W, w_1, w_2) as before so that $f(v) \in W$; we pick $v \in V \subset f^{-1}(W)$ and $U \subset S$ be affine open subsets so that $V_{\text{node}} = f^{-1}(\mathbf{D}_l \times_{C[n]} S) \cap V$ is connected, $\pi(V) \subset U$ and $\pi(V_{\text{node}})$ is closed in U. We then let $\hat{V}$ be the formal completion of V along V_{node}; let $\hat{U}$ be the formal completion of U along $\pi(V_{\text{node}}) \subset U$. Let $\hat{\pi} : \hat{V} \to \hat{U}$ be the projection.

We let $\operatorname{Spec} \mathbf{k}[\![z_1, z_2]\!] \to \operatorname{Spec}[\![u]\!]$, $u \mapsto z_1 z_2$, be the versal deformation of the node $(z_1 z_2 = 0)$. We let $\Gamma(\mathcal{O}_{\hat{S}})^*$ be invertible elements in $\Gamma(\mathcal{O}_{\hat{S}})$.

Definition 3.2. *Let the notation be as before. We say $f : \mathcal{C} \to \mathfrak{X}$ is predeformable along $\mathbf{D}_l \cap W$ near $v \in f^{-1}(\mathbf{D}_l \times_{C[n]} S)$ if there are U, V and W as stated, and there is an* *integer* $m \geqslant 1$, *two* $\epsilon_1, \epsilon_2 \in \Gamma(\mathcal{O}_{\hat{U}})^*$, *a morphism* $\hat{U} \to \operatorname{Spec} \mathbf{k}[\![u]\!]$ *and* ϕ *that fit into the following Cartesian product*

$$\begin{array}{ccc} \hat{V} & \xrightarrow{\phi} & \operatorname{Spec} \mathbf{k}[\![z_1, z_2]\!] \\ \downarrow{\scriptstyle \hat{\pi}} & & \downarrow \\ \hat{U} & \longrightarrow & \operatorname{Spec} \mathbf{k}[\![u]\!] \end{array}$$

so that the associated morphism $\hat{f} : \hat{V} \to \hat{W}$ *has the form, (possibly after switching* w_1 *and* w_2*,)*

$$\hat{f}^*(w_i) = \hat{\pi}^*(\epsilon_i) \cdot \phi^* z_i^m, \ i = 1, 2.$$

We say $f : \mathcal{C} \to \mathcal{X}$ *is pre-deformable if it is pre-deformable near every point in* $f^{-1}(\mathbf{D}_l \times_{C[n]} S)$ *for all* $1 \leqslant l \leqslant n+1$.

Lastly, for any $f : C \to X[n]_0$ we define the automorphism group $\operatorname{Aut}_{\mathfrak{X}}(f)$ be pairs of isomorphisms $h : C \to C$ and $\sigma \in G_m^n$ so that $f \circ h = \sigma \circ f$, where h fix the marked points of C and σ are the automorphisms of $X[n]_0$ induced by its G_m^n action. We say f has no connected trivial component if no connected component of C is mapped to a point in $X[n]_0$.

Definition 3.3. *Given any C-scheme S, we define*

$$\overline{\mathcal{M}}_{\chi,k}(\mathfrak{X}, d)^\bullet(S)$$

be the collection of all $f : (\mathcal{C}, p_i) \to \mathcal{X}$, *where* $(\mathcal{C}, p_i)$ *are flat S-family of* k*-pointed nodal curves (not necessarily with connected fibers,* p_i *are sections of marked points), of Euler characteristics* $\chi(\mathcal{O}_{\mathcal{C}_s}) = \chi$ *for* $s \in S$ *closed,* $\mathcal{X} \in \mathfrak{X}(S)$, *such that* f *are non-degenerate;* f *are predeformable, and for any closed* $s \in S$, $f_s = f|_{\mathcal{C}_s} : (\mathcal{C}_s, p_{i,s}) \to \mathcal{X}_s$ *has no connected trivial components and the automorphism group* $\operatorname{Aut}_{\mathfrak{X}}(f_s)$ *is finite.*

An arrow between two such families $(f, (\mathcal{C}, p_i) \to \mathcal{X})$ *and* $(f' : (\mathcal{C}', p_i') \to \mathcal{X}')$ *in* $\overline{\mathcal{M}}_{\chi,k}(\mathfrak{X}, d)^\bullet(S)$ *consists of an S-isomorphism* $h_1 : (\mathcal{C}, p_i) \to (\mathcal{C}', p_i')$ *and an arrow* $h_2 : \mathcal{X} \to \mathcal{X}'$ *in* $\mathfrak{X}(S)$ *so that* $f' \circ h_1 = h_2 \circ f$.

It is easy to check that $\overline{\mathcal{M}}_{\chi,k}(\mathfrak{X}, d)^\bullet$ is a groupoid over C. The first main theorem [26] is the following existence theorem.

Theorem 3.4. *The groupoid* $\overline{\mathcal{M}}_{\chi,k}(\mathfrak{X}, d)^\bullet$ *is a Deligne-Mumford stack, separated, C-proper, and of finite type.*

The proof consists of two parts: one is to show that the $\overline{\mathcal{M}}_{\chi,k}(\mathfrak{X}, d)^\bullet$ is bounded and is étale covered by a Deligne-Mumford stack; the second is to prove the properness and separatedness. We sketch the proof by a series of Lemmas.

Following the standard method of studying stable morphisms, one checks that the groupoid $\overline{\mathcal{M}}_{\chi,k}(\mathfrak{X}, d)^\bullet$ is a stack. And since $\mathfrak{X} \times_C t = X_t$ for $t \neq 0 \in C$,

$$\overline{\mathcal{M}}_{\chi,k}(\mathfrak{X}, d)^\bullet \times_C t = \overline{\mathcal{M}}_{\chi,k}(X_t, d)^\bullet,$$

where the later is the moduli of stable morphisms of not necessarily connected domains, and without trivial connected components.

To show that $\overline{\mathcal{M}}_{\chi,k}(\mathfrak{X}, d)^\bullet$ is a Deligne-Mumford stack we will construct its étale atlas by Deligne-Mumford stacks. For degenerations constructed using $\mathfrak{X}$, the following method works in general.

For an integer n, we form the relative moduli $\overline{\mathcal{M}}_{\chi,k}(X[n]/C[n], d)^\bullet$ of stable morphisms to $X[n]/C[n]$ of degree d.[1] It is proper over $C[n]$ and is G_m^n equivariant since $X[n]$ is proper over $C[n]$ and is G_m^n equivariant. We let

$$\overline{\mathcal{M}}_{\chi,k}(X[n]/C[n], d)^\bullet_{st} \subset \overline{\mathcal{M}}_{\chi,k}(X[n]/C[n], d)^\bullet \tag{3.5}$$

be the set of all stable morphisms $f : C \to X[n]_t$, $t \in C[n]$, so that f is non-degenerate, pre-deformable and $\mathrm{Aut}_{\mathfrak{X}}(f)$ is finite.

It is easy to see that being non-degenerate and having $\mathrm{Aut}_{\mathfrak{X}}(f)$ finite are open conditions. In [26], the author showed that within the open subset of non-degenerate morphisms in $\overline{\mathcal{M}}_{\chi,k}(X[n]/C[n], d)^\bullet$, the condition pre-deformable is a closed condition. Thus (3.5) is a locally closed substack. Further, since all these conditions are G_m^n invariant, (3.5) is a G_m^n-equivariant locally closed substack.

Lastly, $\mathrm{Aut}_{\mathfrak{X}}(f)$ finite ensures that the G_m^n action on $\overline{\mathcal{M}}_{\chi,k}(X[n]/C[n], d)^\bullet_{st}$ has finite stabilizers. Thus

$$\overline{\mathcal{M}}_{\chi,k}(X[n]/C[n], d)^\bullet_{st}/G_m^n$$

is a Deligne-Mumford stack.

By the construction of $\overline{\mathcal{M}}_{\chi,k}(\mathfrak{X}, d)^\bullet$, the universal family of $\overline{\mathcal{M}}_{\chi,k}(X[n]/C[n], d)^\bullet_{st}$ induces a morphism

$$\overline{\mathcal{M}}_{\chi,k}(X[n]/C[n], d)^\bullet_{st} \longrightarrow \overline{\mathcal{M}}_{\chi,k}(\mathfrak{X}, d)^\bullet. \tag{3.6}$$

Use the G_m^n-equivariance of $\overline{\mathcal{M}}_{\chi,k}(X[n]/C[n], d)^\bullet_{st}$ and its universal family, we conclude that (3.6) is G_m^n-equivariant with trivial G_m^n-action on $\overline{\mathcal{M}}_{\chi,k}(\mathfrak{X}, d)^\bullet$. Therefore, (3.6) descends to its quotient by G_m^n. Taking union over all n, we obtain

$$\coprod_{n\geqslant 0} \overline{\mathcal{M}}_{\chi,k}(X[n]/C[n], d)^\bullet_{st}/G_m^n \longrightarrow \overline{\mathcal{M}}_{\chi,k}(\mathfrak{X}, d)^\bullet. \tag{3.7}$$

Lemma 3.8. *The morphism is surjective and étale.*

Proof. The proof is a direct check. □

This proves that $\overline{\mathcal{M}}_{\chi,k}(\mathfrak{X}, d)^\bullet$ is a Deligne-Mumford stack. It is of finite type because of the following boundedness Lemma.

Lemma 3.9. *Given* (χ, k, d), *there is a constant* N *so that whenever* $f : \mathcal{C} \to X[n]_0$ *is a stable map in* $\overline{\mathcal{M}}_{\chi,k}(\mathfrak{X}, d)^\bullet(k)$, *then* $n \leqslant N$.

[1] Here by relative moduli we mean that its S-families are S-morphisms $f : \mathcal{C} \to X[n] \times_{C[n]} S$, where $\mathcal{C}$ are S-families of nodal curves.

Proof. Let $f : \mathcal{C} \to X[n]_0$ be a stable morphism in $\overline{\mathcal{M}}_{\chi,k}(\mathfrak{X}, d)^\bullet(\mathbf{k})$. We let $\mathcal{C}_i = f^{-1}(\Delta_i)$, with the reduced scheme structure. We define the weight assignment of $X[n]_0$ be

$$\omega_f(\Delta_i) = \deg(\mathcal{C}_i, f^*H) - 2\chi(\mathcal{O}_{\mathcal{C}_i}) + \#(B \cap \mathcal{C}_i) + \#\tau(\mathcal{C}_i), \quad \omega_f(D_i) = 0,$$

where $B \subset \mathcal{C}$ is the set of marked points of f; $\tau(\mathcal{C}_i)$ is the set of smooth points of $\mathcal{C}_i$ that are nodes of $\mathcal{C}$. (We agree $\omega_f(\Delta_i) = 0$ if $\mathcal{C}_i = \emptyset$.)

One checks that since f is stable, $\omega_f(\Delta_i)$ takes integer values, and is $\geqslant 1$ unless $f^{-1}(\Delta_i) = \emptyset$, in which case it is 0. One also checks that $\sum_i \omega_f(\Delta_i) = d - 2\chi + k$. Thus taking $N = d - 2\chi + k$ will do the job. □

We next show that $\overline{\mathcal{M}}_{\chi,k}(\mathfrak{X}, d)^\bullet$ is C-proper using the valuative criterion. Along the way, the key is to modify a family of morphisms whose general member f_η is in $\overline{\mathcal{M}}_{\chi,k}(\mathfrak{X}, d)^\bullet$ while its special member f_{η_0} is not. Our strategy is to identify the degeneracy of f_{η_0}; (assuming $f_{\eta_0} : \mathcal{C}_{\eta_0} \to X[n]_0$,) embed $X[n]$ in $X[n+1]$ using the standard embedding, and then use the group action to construct a new specialization that decreases the degeneracy of f_{η_0}. This process terminates due to the boundedness.

Lemma 3.10. *Let $S = \operatorname{Spec} R \to C$ be a C-scheme, where R is a discrete valuation domain with residue field $\mathbf{k}$, with η and η_0 be its generic and closed point. Let $f_\eta \in \overline{\mathcal{M}}_{\chi,k}(\mathfrak{X}, d)^\bullet(\eta)$ be a stable morphism over η. Then possibly after a base change $\tilde{S} \to S$, $f_\eta \times_S \tilde{S}$ extends to a family $\tilde{f} \in \overline{\mathcal{M}}_{\chi,k}(\mathfrak{X}, d)^\bullet(\tilde{S})$.*

Proof. Let f_η be $f_\eta : \mathcal{C}_\eta \to X[n]$ with the marked points implicitly understood. We prove the case where $\eta \to C$ does not factor though $0 \in C$.

Since $\eta \to C$ does not factor through $0 \in C$, we can assume without loss of generality that f_η is represented by $f_\eta : \mathcal{C}_\eta \to X$. Then f_η is an ordinary stable morphism over η. Therefore, possibly after a base change of S, which we still denote by S for brevity, f_η extends to an S-family of ordinary stable morphisms $f_1 : \mathcal{C}_1 \to X$.

Let η_0 be the closed point of S. In case $S \to C$ sends η_0 to $C - 0$, then f_1 is already a family of stable morphisms in $\overline{\mathcal{M}}_{\chi,k}(\mathfrak{X}, d)^\bullet(S)$; we are done. We now suppose η_0 is mapped to $0 \in C$.

Definition 3.11. *We say f_η admits a quasi-stable extension to $X[n]$ if after a finite base change of S, which we continue to denote by S, we can find an S-family of morphisms $f_n : \mathcal{C}_n \to X[n] \times_{C[n]} S$ over an $\iota : S \to C[n]$ so that $\iota(\eta_0) = 0 \in C[n]$ and that (1). $f_\eta \cong f_{n,\eta}$ as elements in $\overline{\mathcal{M}}_{\chi,k}(\mathfrak{X}, d)^\bullet(\eta)$, where $f_{n,\eta} = f_n \times_S \eta$; (2). $\operatorname{Aut}_{\mathfrak{X}}(f_{n,\eta_0})$ is finite.*

Quasi-stable extensions exist since f_1 is one. On the other hand, the proof of 3.9 shows that $n \leqslant N$ for the N given in Lemma 3.9.

Now we let n be the largest possible integer so that there is a quasi-stable extension f_n. We show that f_n is a family of stable morphisms. Suppose $f_{n,\eta_0} : \mathcal{C}_{n,\eta_0} \to X[n]_0$ is non-degenerate, then an easy argument shows that f_n is pre-deformable. Hence $f_n \in \overline{\mathcal{M}}_{\chi,k}(\mathfrak{X}, d)^\bullet(S)$ as desired.

Suppose f_{n,η_0} is degenerate. Then for some D_l, $f^{-1}_{n,\eta_0}(D_l)$ is one-dimensional. (In case $f^{-1}_{n,\eta_0}(D_l)$ contains a marked points of $\mathcal{C}_{n,\eta_0}$, it must contains the irreducible component that contains this marked point.) We let A_0 be $f^{-1}_{n,\eta_0}(D_l)$ with the reduced scheme structure; we let $A_- \cup A_+$ be the closure of $\mathcal{C}_{n,\eta_0} - A_0$ with reduced scheme structure so that $f_{n,\eta_0}(A_-) \subset \cup_{i<l-1}\Delta_i$ and $f_{n,\eta_0}(A_+) \subset \cup_{i\geqslant l}\Delta_i$.

We let $I_- = [n+2]-\{l\}$ and $I_+ = [n+2]-\{l+1\}$. Using the standard immersions $\tau_{I_\pm,X} : X[n] \to X[n+1]$ in item (3) Lemma 3.1, we obtain morphisms

$$f_\pm := \tau_{I_\pm} \circ f : \mathcal{C}_n \longrightarrow X[n+1].$$

Let $\mathbf{D}_i \subset X[n+1]$ be the divisor introduced before Lemma 3.1. Since $\tau^{-1}_{I_-}(\mathbf{D}_{l+1}) = D_l \subset X[n]_0$ and $\tau^{-1}_{I_+}(\mathbf{D}_l) = D_l \subset X[n]_0$, we have $f_-(A_0) \subset \mathbf{D}_{l+1}$ and $f_+(A_0) \subset \mathbf{D}_l$.

We pick a curve $T \subset \mathcal{C}_n$, flat over S, so that its closed point is a general closed point in A_0. We let $\lambda_l : G_m \longrightarrow G_m^{n+1}$ be the l-th factor in G_m^{n+1}. Then after replacing S by its finite base change (still denoted by S), we can find a morphism $\xi_\eta : \eta \to G_m$ so that

$$(\lambda_l \circ \xi_\eta) \cdot f_{n,\eta} : \mathcal{C}_\eta = \mathcal{C}_{n,\eta} \longrightarrow X[n+1] \tag{3.12}$$

(where $\cdot$ means the group action) restricting to $T \times_S \eta \subset \mathcal{C}_{n,\eta}$ specializes to a point in $\Delta_l - D_l \cup D_{l+1} \subset X[n+1]_0$.

Because $X[n+1] \to C[n+1]$ is proper, after a finite base change, (3.12) extends to an S-family of stable morphisms

$$f_{n+1} : \mathcal{C}_{n+1} \longrightarrow X[n+1].$$

By a direct inspection, one shows that

$$\mathcal{C}_{n+1} - f^{-1}_{n+1}(\Delta_l) \cong \mathcal{C}_n - f^{-1}_n(D_l) = \mathcal{C}_n - A_0,$$

extending the identity $\mathcal{C}_{n+1} \times_S \eta = \mathcal{C}_n \times_S \eta$, and under this isomorphism

$$f_{n+1}|_{\mathcal{C}_{n+1}-f^{-1}_{n+1}(\Delta_l)} = f_n|_{\mathcal{C}_n - A_0} : \mathcal{C}_n - A_0 \longrightarrow X[n+1]_0 - \Delta_l \cong X[n]_0 - D_l.$$

Since T specializes to a general point in A_0, f_{n+1} is quasi-stable. This contradicts to the maximal assumption on n.

Finally, if $S \to C$ does factor through $0 \in C$, we decompose f_n into a union of relative morphisms, and apply the same trick. We will not go into details here. This proves the Lemma. □

The proof of separatedness follows from the separatedness of the moduli of ordinary stable morphisms. We will outline the proof for the case of degeneration of Hilbert schemes later in this section. Combined, we have proved Theorem 3.4.

Next we discuss the decomposition of the central fiber

$$\overline{\mathfrak{M}}_{\chi,k}(\mathfrak{X}_0, d)^\bullet = \overline{\mathfrak{M}}_{\chi,k}(\mathfrak{X}, d)^\bullet \times_C 0.$$

The first step is to construct the moduli of stable morphisms to the stack $\mathfrak{X}_0^\dagger/\mathfrak{C}_0^\dagger$.

We form the groupoid of families of stable morphisms

$$\overline{\mathcal{M}}_{X,k}(\mathfrak{X}_0^\dagger, d)^\bullet(S),$$

like in Definition 3.3 with $\mathfrak{X} \to \mathfrak{C}$ replaced by $\mathfrak{X}_0^\dagger \to \mathfrak{C}_0^\dagger$.

Proposition 3.13. *The groupoid $\overline{\mathcal{M}}_{X,k}(\mathfrak{X}_0^\dagger, d)^\bullet$ is a Deligne-Mumford stack, separated, proper and of finite type. The morphism $\mathfrak{X}_0^\dagger \to \mathfrak{X}$ induces a local closed morphism (the top line) that fits into the Cartesian product*

$$\begin{array}{ccc} \overline{\mathcal{M}}_{X,k}(\mathfrak{X}_0^\dagger, d)^\bullet & \longrightarrow & \overline{\mathcal{M}}_{X,k}(\mathfrak{X}, d)^\bullet \\ \downarrow & & \downarrow \\ \mathfrak{C}_0^\dagger & \longrightarrow & \mathfrak{C} \end{array}$$

The next step is to form the moduli of relative stable morphisms to $D_\pm \subset Y$. Recall that a partition of an integer is $\mu = (\mu_1, \dots, \mu_{\ell(\mu)})$, $0 < \mu_1 \leqslant \dots \leqslant \mu_{\ell(\mu)}$.

Definition 3.14. *Given two partitions of integers (μ^-, μ^+), we call a morphism*

$$f : \mathcal{C} \longrightarrow Y[n_-, n_+]_0, \ y_1, \dots, y_k; x_1^-, \dots, x_{\ell(\mu^-)}^-, x_1^+, \dots, x_{\ell(\mu^+)}^+ \in \mathcal{C} \text{ marked points}$$

a relative stable morphism to $\mathfrak{D}_\pm \subset \mathfrak{Y}$ of contacts (μ^-, μ^+) if f is non-degenerate and pre-deformable along $D_{-n_-}, \dots, D_{n_+}$; $\mathrm{Aut}_{\mathfrak{Y}}(f)$ is finite, and as divisors

$$f^{-1}(D[n_-, n_+]_{\pm,0}) = \sum \mu_i^\pm x_i^\pm.$$

Here as usual, y_i and $x_j^\pm$ being marked points ensure that they are distinct smooth points of $\mathcal{C}$. The automorphism group $\mathrm{Aut}_{\mathfrak{Y}}(f)$ consists of pairs of $h : \mathcal{C} \to \mathcal{C}$ fixing the marked points and $\sigma \in G_m^{n_-+n_+}$ so that $f \circ h = \sigma \cdot f$.

Given (μ^-, μ^+), we form the groupoid of families of relative stable morphisms of contacts (μ^-, μ^+):

$$\overline{\mathcal{M}}_{X,k}(\mathfrak{Y}, d, (\mu^-, \mu^+))^\bullet,$$

like in Definition 3.3 with stable morphisms replaced by relative stable morphisms to $\mathfrak{D}_\pm \subset \mathfrak{Y}$ of contacts (μ^-, μ^+).

Theorem 3.15. *The groupoid $\overline{\mathcal{M}}_{X,k}(\mathfrak{Y}, d, (\mu^-, \mu^+))^\bullet$ is a Deligne-Mumford stack, separated, proper and of finite type.*

Given a relative stable morphism f as in Definition 3.14, we associate it to

$$\big((x_1^-, \dots, x_{\ell(\mu^-)}^-), (x_1^+, \dots, x_{\ell(\mu^+)}^+)\big) \in D^{\ell(\mu^-)} \times D^{\ell(\mu^+)}.$$

This assignment extends to a pair of evaluation morphisms

$$(\mathrm{ev}_-, \mathrm{ev}_+) : \overline{\mathcal{M}}_{X,k}(\mathfrak{Y}, d, (\mu^-, \mu^+))^\bullet \longrightarrow D^{\ell(\mu^-)} \times D^{\ell(\mu^+)},$$

which provides a bridge between $\overline{\mathcal{M}}_{X,k}(\mathfrak{Y}, d, (\mu^-, \mu^+))^\bullet$ and $\overline{\mathcal{M}}_{X,k}(\mathfrak{X}_0^\dagger, d)^\bullet$.

Proposition 3.16. *The decomposition* (2.47) *defines a canonical morphism*

$$\coprod_\mu \Phi_\mu : \coprod_\mu \overline{\mathfrak{M}}_{\chi+\ell(\mu),k}(\mathfrak{Y}, d, (\mu,\mu))^\bullet \times_{D^{\ell(\mu)}\times D^{\ell(\mu)}} D^{\ell(\mu)} \longrightarrow \overline{\mathfrak{M}}_{\chi,k}(\mathfrak{X}_0^\dagger, d)^\bullet.$$

This morphism is surjective, a local closed immersion, and finite. Restricting to each summand associated to μ, Φ_μ *is a fiber bundle to its image with fibers isomorphic to* $\mathrm{Aut}(\mu)$. *Finally, the images of* Φ_μ *are disjoint for different* μ.

Proof. For any stable map $f : \mathcal{C} \to X[n]_0$ in $\overline{\mathfrak{M}}_{\chi,k}(\mathfrak{X}_0^\dagger, d)^\bullet$ with the node-marking $D_l \subset X[n]_0$, we first resolve $X[n]_0$ along D_l (via (2.47)) to get a relative pair

$$D[n_-, n_+]_{\pm,0} \subset Y[n_-, n_+]_0, \quad n_- = l-1,\ n_+ = n-l+1,$$

where $D[n_-, n_+]_{\pm,0}$ is the preimage of $D_l \subset X[n]_0$ under $Y[n_-, n_+]_0 \to X[n]_0$. Let $\mathcal{C}_{[l]}$ be $\mathcal{C}$ after resolving the nodes $f^{-1}(D_l)$ of $\mathcal{C}$. We define

$$f_{[l]} : \mathcal{C}_{[l]} \longrightarrow Y[n_-, n_+]$$

be that induced by f.

We assign the relative divisor to $f_{[l]}$. For any $x \in f^{-1}(D_l)$, since f is pre-deformable, f has bi-contact order $\nu(x) > 0$ with D_l at x. We let $x_1, \ldots, x_\ell$ be the points in $f^{-1}(D_l)$ so ordered that $\mu_i = \nu(x_i)$ is increasing. We then let $x_i^\pm \in \mathcal{C}_{[l]}$ be the preimages of x_i (under $\mathcal{C}_{[l]} \to \mathcal{C}$) so that $f_{[l]}(x_i^\pm) \subset D[n_-, n_+]_{\pm,0}$. Let $y_1, \ldots, y_k$ be the marked points of f; they lift uniquely to $\mathcal{C}_{[l]}$. Let $\mu = (\nu(x_1), \ldots, \nu(x_\ell))$.

The data

$$f_{[l]} : \mathcal{C}_{[l]} \longrightarrow Y[n_-, n_+], \quad y_1, \ldots, y_k; x_1^\pm, \ldots, x_{\ell(\mu)}^\pm \in \mathcal{C}_{[l]}$$

define a relative stable morphism in $\overline{\mathfrak{M}}_{\chi+\ell,n}(\mathfrak{Y}, d, (\mu,\mu))^\bullet$.

Given f, there are exactly $|\mathrm{Aut}(\mu)|$ many different reordering of $x_1, \ldots, x_\ell$ keeping $\nu(x_i)$ increasing. Here $\mathrm{Aut}(\mu)$ is the collection of permutations σ of $[\ell(\mu)]$ so that $\mu_i = \mu_{\sigma(i)}$ for all $i \in [\ell(\mu)]$.

By construction, $f_{[l]}$ lies in

$$\overline{\mathfrak{M}}_{\chi+\ell,k}(\mathfrak{Y}, d, (\mu,\mu))^\bullet \times_{D^{\ell(\mu)}\times D^{\ell(\mu)}} D^{\ell(\mu)}, \tag{3.17}$$

where the arrow $\overline{\mathfrak{M}}_{\chi+\ell,k}(\mathfrak{Y}, d, (\mu,\mu))^\bullet \to D^{\ell(\mu)} \times D^{\ell(\mu)}$ is via $(\mathrm{ev}_-, \mathrm{ev}_+)$; the arrow from $D^{\ell(\mu)}$ to the base scheme is the diagonal morphism.

Conversely, given $(f, \mathcal{C}, y_i, x_i^\pm)$ lie in the product (3.17), we can form $\bar{\mathcal{C}}$ by gluing x_i^- with x_i^+ for all i, and define $\bar{f} : \bar{\mathcal{C}} \to X[n]_0$ be induced by f, $n = n_- + n_+$, using that $X[n]_0$ is by identifying $D[n_-, n_+]_{-,0}$ with $D[n_-, n_+]_{+,0}$ in $Y[n_-, n_+]_0$.

Working out the family version of the above two constructions we prove the Proposition. □

In case the moduli stacks $\overline{\mathfrak{M}}_{\chi+\ell(\mu),k}(\mathfrak{Y}, d, (\mu,\mu))^\bullet$ are not smooth, this decomposition is not entirely satisfactory: the structure of $\overline{\mathfrak{M}}_{\chi,k}(\mathfrak{X}_0^\dagger, d)^\bullet$ is not completely revealed by this morphism. The refined version of this decomposition is more easily stated in case $Y = Y_- \cup Y_+$ is a union of two smooth varieties.

We now suppose $Y = Y_- \cup Y_+$, and let $D_\pm = \tilde{D} \cap Y_\pm$. We use the weighted decomposition stack $\mathfrak{C}^\beta$ introduced in the last section.

We let $\Lambda = \mathbb{Z}[u, v]$ be the ring of polynomials with integer coefficients. For a stable morphism $f : \mathcal{C} \to X[n]_0$, with $B \subset \mathcal{C}$ its marked points, we define a weight assignment of $X[n]_0$ by the rule: for $\Delta_i \subset X[n]_0$, we denote $\mathcal{C}_i = f^{-1}(\Delta_i)_{\text{red}}$; we define

$$w(\Delta_i) = \chi(\mathcal{O}_{\mathcal{C}_i}) + |B \cap \mathcal{C}_i| \cdot u + \deg(f^*H|_{\mathcal{C}_i}) \cdot v, \quad w(D_j) = |f^{-1}(D_j)|.$$

One checks that this rule gives a continuous weight assignment of the universal family $f : \mathcal{C} \to \mathcal{X}$ of $\overline{\mathcal{M}}_{\chi,k}(\mathfrak{X}, d)^\bullet$, of total weight

$$\beta = \chi + ku + dv \in \mathbb{Z}[u, v].$$

This makes $\mathcal{X}$ a family in $\mathfrak{X}^\beta$, thus induces a morphism

$$\pi_\beta : \overline{\mathcal{M}}_{\chi,k}(\mathfrak{X}, d)^\bullet \longrightarrow \mathfrak{C}^\beta.$$

Now let $\delta \in \Lambda_\beta^{\text{spl}}$ be in the set of splitting of β (cf. (2.57)). We let $\mathfrak{C}_0^{\dagger,\delta} \to \mathfrak{C}^\beta$ be as in (2.59); we define

$$\overline{\mathcal{M}}(\mathfrak{X}_0^\dagger, \delta)^\bullet = \overline{\mathcal{M}}_{\chi,k}(\mathfrak{X}, d)^\bullet \times_{\mathfrak{C}^\beta} \mathfrak{C}_0^{\dagger,\delta}.$$

(Here we skip the subscript (χ, k) because these data are included in δ.) Applying Proposition 2.60, we have

Proposition 3.18. *Let* (L_δ, s_δ) *and the notation be as in Proposition 2.60. Then*

(1) $\otimes_{\delta \in \Lambda_\beta^{\text{spl}}} \pi_\beta^* L_\delta \cong \mathcal{O}_{\overline{\mathcal{M}}_{\chi,k}(\mathfrak{X},d)^\bullet}$, *and* $\prod_{\delta \in \Lambda_\beta^{\text{spl}}} \pi_\beta^* s_\delta = \pi_\beta^* \pi^* t$;

(2) *as closed substacks,* $\overline{\mathcal{M}}(\mathfrak{X}_0^\dagger, \delta)^\bullet = (\pi_\beta^* s_\delta = 0)$.

We can split $\overline{\mathcal{M}}(\mathfrak{X}_0^\dagger, \delta)^\bullet$ further. Let $\delta_\pm = \chi_\pm + k_\pm u + d_\pm v \in \mathbb{Z}[u, v]$ and $\delta_0 = \ell$. Let μ be a partition of an integer of length $\ell(\mu) = \ell$. We denote by

$$\Phi_{\delta,\mu} : \overline{\mathcal{M}}_{\chi_-,k_-}(\mathfrak{Y}_-, d_-, \mu)^\bullet \times_{D^\ell} \overline{\mathcal{M}}_{\chi_+,k_+}(\mathfrak{Y}_+, d_+, \mu)^\bullet \longrightarrow \overline{\mathcal{M}}(\mathfrak{X}_0^\dagger, \delta)^\bullet \tag{3.19}$$

be the gluing morphism analogous to the one in Proposition 3.16. It is a local closed immersion. We denote by $\operatorname{Im}(\Phi_{\delta,\mu})$ be the image stack, and denote by $\Phi'_{\delta,\mu}$ the morphism (3.19) with $\overline{\mathcal{M}}(\mathfrak{X}_0^\dagger, \delta)^\bullet$ replaced by $\operatorname{Im}(\Phi_{\delta,\mu})$.

Lemma 3.20. *For each* μ, $\operatorname{Im}(\Phi_{\delta,\mu})$ *is both closed and open in* $\overline{\mathcal{M}}(\mathfrak{X}_0^\dagger, \delta)^\bullet$. *For distinct* μ *and* μ', $\operatorname{Im}(\Phi_{\delta,\mu})$ *and* $\operatorname{Im}(\Phi_{\delta,\mu'})$ *are disjoint. Finally, the morphism* $\Phi'_{\delta,\mu}$ *is a Galois covering of Galois group* $\operatorname{Aut}(\mu)$.

Proof. The proof is straightforward. We only comment that $\operatorname{Aut}(\mu)$ acts on the domain of $\Phi_{\delta,\mu}$ by permuting the index of the marked points $(x_1^\pm, \ldots, x_\ell^\pm)$. □

Because of the Lemma, there are finitely many distinct partitions $\mu^1, \ldots, \mu^r$ so that $\overline{\mathcal{M}}(\mathfrak{X}_0^\dagger, \delta)^\bullet$ is homeomorphic to the disjoint union of $\operatorname{Im}(\Phi_{\delta,\mu^i})$. We now let μ be one of $\mu^1, \ldots, \mu^r$. Since $\operatorname{Im}(\Phi_{\delta,\mu})$ is both open and closed in $\overline{\mathcal{M}}(\mathfrak{X}_0^\dagger, \delta)^\bullet$, we can define $\overline{\mathcal{M}}(\mathfrak{X}_0^\dagger, \delta, \mu)^\bullet$ be $\mathrm{Im}(\Phi_{\delta,\mu})$ endowed with the open stack structure of $\overline{\mathcal{M}}(\mathfrak{X}_0^\dagger, \delta)^\bullet$. Then

$$\overline{\mathcal{M}}(\mathfrak{X}_0^\dagger, \delta)^\bullet = \coprod_{i=1}^{r} \overline{\mathcal{M}}(\mathfrak{X}_0^\dagger, \delta, \mu^i)^\bullet.$$

Let μ be one of $\mu^1, \dots, \mu^r$; let $\mu = (\mu_1, \dots, \mu_\ell)$. As we will see momentarily that in case $\mu_\ell \geqslant 2$, then in general $\overline{\mathcal{M}}(\mathfrak{X}_0^\dagger, \delta, \mu)^\bullet \neq \mathrm{Im}(\Phi_{\delta,\mu})$. The distinction can be seen using the section s_δ. Since $\Phi'_{\delta,\mu}$ is a Galois cover, if we let $\overline{\mathcal{M}}(\mathfrak{X}_0^\dagger, \delta, \mu)^{\bullet\wedge}$ be the formal completion of $\overline{\mathcal{M}}_{X,k}(\mathfrak{X}, d)^\bullet$ along $\overline{\mathcal{M}}(\mathfrak{X}_0^\dagger, \delta, \mu)^\bullet$, then there is an $\mathrm{Aut}(\mu)$-Galois cover

$$\overline{\mathcal{M}}(\mathfrak{X}_0^\dagger, \delta, \mu)^{\bullet\sim} \longrightarrow \overline{\mathcal{M}}(\mathfrak{X}_0^\dagger, \delta, \mu)^{\bullet\wedge}$$

so that $\Phi_{\delta,\mu}$ lifts to a morphism that is a homeomorphism

$$\tilde{\Phi}_{\delta,\mu} : \overline{\mathcal{M}}(\mathfrak{Y}_-, \delta_-, \mu)^\bullet \times_{D^\ell} \overline{\mathcal{M}}(\mathfrak{Y}_+, \delta_+, \mu)^\bullet \longrightarrow \overline{\mathcal{M}}(\mathfrak{X}_0^\dagger, \delta, \mu)^{\bullet\sim}. \tag{3.21}$$

(Here we use $\overline{\mathcal{M}}(\mathfrak{Y}_\pm, \delta_\pm, \mu)^\bullet = \overline{\mathcal{M}}_{X_\pm,k_-}(\mathfrak{Y}_\pm, d_\pm, \mu)^\bullet$.)

Proposition 3.22. *There are line bundles with sections (P_i, ξ_i) on $\overline{\mathcal{M}}(\mathfrak{X}_0^\dagger, \delta, \mu)^{\bullet\sim}$, $i = 1, \dots, \ell$, such that $\otimes_{i=1}^{\ell} P^{\otimes \mu_i} \cong \pi_\beta^* L_\delta$, and $\prod_{i=1}^{\ell} \xi_i^{\mu_i} = \pi_\beta^* s_\delta$.*

For the proof, we refer to [27]. We comment that this Proposition reveals in full strength of a good degeneration: even in case the moduli space is singular, there is a precise formulation that decomposes the central fiber of a good degeneration as union of virtual "Cartier divisors", and with a universal multiplicity modification, these virtual "Cartier divisor" are the image stack of the gluing morphisms from the moduli of relative moduli objects. We will see this again when we do good degeneration of Hilbert scheme of subschemes.

We comment that applying these decomposition Propositions, the author was able to workout the relation of various virtual cycles of the relevant moduli stacks, and proved that their virtual cycles fits into the identity

$$[\overline{\mathcal{M}}_{X,k}(\mathfrak{X}_0, d)^\bullet]^{\mathrm{vir}} = \sum_{\delta,\mu} \frac{\mu!}{|\mathrm{Aut}(\mu)|} \Delta^! \left([\overline{\mathcal{M}}(\mathfrak{Y}_-, \delta_-, \mu)^\bullet]^{\mathrm{vir}} \times [\overline{\mathcal{M}}(\mathfrak{Y}_+, \delta_+, \mu)^\bullet]^{\mathrm{vir}} \right),$$

where $\mu! = \prod \mu_i$, and $\Delta^!$ is the Gysin homomorphism associates to the square

$$\begin{array}{ccc}
\overline{\mathcal{M}}(\mathfrak{Y}_-, \delta_-, \mu)^\bullet \times_{D^\ell} \overline{\mathcal{M}}(\mathfrak{Y}_+, \delta_+, \mu)^\bullet & \longrightarrow & \overline{\mathcal{M}}(\mathfrak{Y}_-, \delta_-, \mu)^\bullet \times \overline{\mathcal{M}}(\mathfrak{Y}_+, \delta_+, \mu)^\bullet \\
\downarrow & & \downarrow {\scriptstyle (ev_-, ev_+)} \\
D^{\ell(\mu)} & \xrightarrow{\ \Delta\ } & D^{\ell(\mu)} \times D^{\ell(\mu)}
\end{array}$$

This degeneration is an indispensable tool in studying Gromov-Witten invariants. For an incomplete list of references, see the introduction.

3.2. Moduli of vector bundles over universal curves

The relative moduli space of stable vector bundles over the universal curves $\mathcal{C}_g$ on $\mathcal{M}_g$ (of the moduli of smooth genus g curves, a Deligne-Mumford stack,) is a basic moduli space whose existence follows directly from the existence of the moduli of stable vector bundles over smooth curves. An important problem is to compactify it as a relative moduli space of bundles over the universal curve of the Deligne-Mumford stack $\overline{\mathcal{M}}_g$ of stable curves. Because moduli of stable bundles over smooth curves are smooth, one hopes to construct the compactification that is smooth, and its projection to $\overline{\mathcal{M}}_g$ is a locally complete intersection morphism.

In the case of rank 1, this was achieved by Caporaso [4]. In high rank case, using relative GIT theory Pandharipande constructed the moduli of relative stable sheaves on the universal curves of $\overline{\mathcal{M}}_g$ [38]. However, this compactification is unlikely to have smooth total space. A good solution should be a construction along the line of Gieseker's construction. The author constructed, in case $(r, d) = 1$, this compactification using Simpson bundles over the universal curves of the stack of semistable curves [28]. Using Hilbert stability, Schmitt constructed a similar compactification using GIT [39]; his construction works for all r and d, is projective over $\overline{\mathcal{M}}_g$. Both construction have smooth total spaces and normal crossing fibers over $\overline{\mathcal{M}}_g$.

In this section we will review the construction given in [28] using the stack of semistable curves and Simpson stability [40] of vector bundles. As one can see, it is a stacky substitute of the work of [4, 12, 13, 26].

We fix a genus $g \geqslant 2$, a rank $r > 1$ and relative prime $(r, \chi) = 1$. We intend to construct a compactification of the relative moduli of stable vector bundles on $\mathcal{C}_g \to \mathcal{M}_g$ of rank r and Euler characteristic χ. To avoid technical issues in finding relative ample line bundles, we will replace χ by a large prime in the form $\chi + cr$, $c \in \mathbb{Z}$. This change of χ will not affect the relative moduli space over the stack of smooth curves $\mathcal{M}_g$; it will affect the choice of compactification we are about to construct.

We briefly outline our construction. We begin with the (Artin) stack of all connected semistable[2] curves $\mathcal{M}_g^{ss}$ and its universal curve $\mathcal{C}_g^{ss}$. We form the relative stack $\mathfrak{V}_{r,\chi}$ of locally free sheaves on $\mathcal{C}_g^{ss}/\mathcal{M}_g^{ss}$, of rank r and Euler characteristic χ. We then introduce a stability on locally free sheaves on nodal curves, using Simpson stability of sheaves. After showing that the stability does not depend on the choices of $\mathbb{Q}$-ample line bundles we used, we can form the substack $\mathfrak{V}_{r,\chi}^{st}$ of all stable locally free sheaves. It is an open substack of $\mathfrak{V}_{r,\chi}$; it is a proper, separated and smooth Artin stack; its projection to $\overline{\mathcal{M}}_g$ is a locally complete intersection morphism.

As we will see, the pair $\mathcal{M}_g^{ss} \to \overline{\mathcal{M}}_g$ is parallel to the pair $\mathfrak{X} \to X$ in that the fiber over a nodal $[C] \in \overline{\mathcal{M}}_g$ consists of all semistable curves whose stabilizations are C.

[2]A semistable curve is a connected, nodal curve such that each of its smooth rational component meets other components in at least two nodes.

In this sense, the construction given in this subsection fits into the general scheme outlined in Section one.

We give more details of this construction.

Definition 3.23. *For any scheme S and $\mathcal{X} \in \mathcal{M}_g^{ss}(S)$, we define $\mathfrak{V}_{r,\chi}(\mathcal{X})$ be the set of all rank r locally free sheaves $\mathcal{E}$ on $\mathcal{X}$ so that the Euler characteristic of $\mathcal{E}$ restricted to fibers of $\mathcal{X}/S$ are χ. Given two members $\mathcal{E}$ and $\mathcal{E}'$ over $\mathcal{X}$ and $\mathcal{X}' \in \mathcal{M}^{ss}(S)$, respectively, an arrow from $\mathcal{E}$ to $\mathcal{E}'$ consists of a pair (ϕ, σ), where $\sigma : \mathcal{X} \to \mathcal{X}'$ is an arrow in $\mathcal{M}_g^{ss}(S)$ and ϕ is an isomorphism $\phi : \sigma^*\mathcal{E}' \to \mathcal{E}$.*

The following existence theorem holds by tautology.

Proposition 3.24. *The groupoid $\mathfrak{V}_{r,\chi}$ is a stack. Further there is a tautological morphism $\mathfrak{V}_{r,\chi} \to \mathcal{M}_g^{ss}$ sending $\mathcal{E} \in \mathfrak{V}_{r,\chi}(\mathbf{k})$ to the base curve X of $\mathcal{E}$.*

Our next step is to use stability to pick an open substack of $\mathfrak{V}_{r,\chi}$. Let X be a semistable curve in $\mathcal{M}_g^{ss}$ and let $\rho : X \to X^{st}$ be its stabilization. Since $g \geqslant 2$, the dualizing sheaf $\omega_{X^{st}}$ is ample. We let R(X) be the set of all irreducible components of X contracted under ρ. We pick a map $\eta : R(X) \to \mathbb{Z}_+$. We then let H_η be a $\mathbb{Q}$-line bundle on X so that its restriction to a non-contracted component $D \subset X$ is $\rho^*\omega_{X^{st}}|_D$, and its restriction to a contracted component $D \subset X$ has degree $\eta(D)\epsilon$, where $\epsilon \sim 0^+$ is a sufficiently small positive rational number. Clearly, H_η is $\mathbb{Q}$-ample on X. For any sheaf $\mathcal{F}$ on X, we denote by $P_{\mathcal{F},\eta}$ the polynomial $P_{\mathcal{F},\eta}(n) = \chi(\mathcal{F} \otimes H_\eta^{\otimes n})$ for n so that $H_\eta^{\otimes n}$ is a line bundle. By Riemann-Roch theorem, $P_{\mathcal{F},\eta}$ is well-defined, and linear in n. We denote by $\mathrm{l.c.}P_{\mathcal{F},\eta}$ the leading coefficient of $P_{\mathcal{F},\eta}$ in n.

Definition 3.25. *Let $\mathcal{E}$ be a locally free sheaf on X. We say $\mathcal{E}$ is weakly stable (with respect to H_η) if for any proper subsheaf $0 \neq \mathcal{F} \subset \mathcal{E}$ we have*

$$\frac{\chi(\mathcal{F})}{\mathrm{l.c.}P_{\mathcal{F},\eta}} < \frac{\chi(\mathcal{E})}{\mathrm{l.c.}P_{\mathcal{E},\eta}}. \tag{3.26}$$

We say $\mathcal{E}$ is stable if for each rational component $D \in R(X)$ the restriction $\mathcal{E}|_D$ is not $\mathcal{O}_D^{\oplus r}$.

We remark that the weakly stability is the Simpson stability associated to the ample divisor H_η.

Our next step is to take $\mathfrak{V}_{r,\chi}^{st}$ the open substack of $\mathfrak{V}_{r,\chi}$ of all stable locally free sheaves. There is one technical issue. In defining the stability we need to use the polarization H_η, which depends on the function $\eta : R(X) \to \mathbb{Z}_+$ and ϵ sufficiently small. There is unlikely a choice of $\mathbb{Q}$-line bundle over the universal family $\mathcal{C}_g^{ss}$ whose restriction to each fiber is H_η for some η.

What is true is that for each $[X] \in \mathcal{M}_g^{ss}$, we can find an open $[X] \in \mathcal{U} \subset \mathcal{M}_g^{ss}$ so that the universal curve over $\mathcal{U}$ has a single $\mathbb{Q}$-ample line bundle in the form prescribed. Thus by using that Simpson stability is an open condition, we can form the subset

$$\mathfrak{V}_{r,\chi}^{st}(\mathcal{U}) \subset \mathfrak{V}_{r,\chi} \times_{\mathcal{M}_g^{ss}} \mathcal{U} \tag{3.27}$$

of stable sheaves in $\mathfrak{V}_{r,\chi} \times_{\mathcal{M}_g^{ss}} \mathcal{U}$ stable with respect to the $\mathbb{Q}$-ample line bundle chosen. Because being stable is an open condition, the above inclusion is an open substack.

Fortunately, the stability so defined does not depend on the choice of η, when χ is a sufficiently large prime.

Lemma 3.28. *Let $\chi > 2rg$ be a prime. Then for every semistable curve X the stability of vector bundles $\mathcal{E}$ over X of rank r and Euler characteristic χ do not depend on the choice of $\eta : R(X) \to \mathbb{Z}_+$ and sufficiently small ϵ.*

Proof. Let X be fixed and let η_1 and η_2 be two such maps. Suppose $\mathcal{E}$ is a locally free sheaf on X as stated so that $\mathcal{E}$ is H_{η_1}-stable but not H_{η_2} -stable. Then there must be a proper subsheaf $0 \neq \mathcal{F} \subset \mathcal{E}$ so that (3.26) holds with η replaced by η_1 and does not hold with η replaced by η_2. Since $\mathcal{E}$ is H_{η_1}-stable, $\chi > rg$, and ϵ is sufficiently small; further, it is proved in [21] that the restriction of $\mathcal{E}$ to each contracted component $D \in R(X)$ has

$$\mathcal{E}|_D \cong \mathcal{O}_D^{\oplus a} \oplus \mathcal{O}_D(1)^{\oplus b}, \tag{3.29}$$

and there is a non-contracted component $D \subset X$ so that $\mathcal{F}|_D$ is not a torsion sheaf. From the second property, we know that $\mathrm{l.c.}P_{\mathcal{F},\eta_j}$ is very close to the positive integer

$$\ell(\mathcal{F}) := \mathrm{l.c.}\chi(\mathcal{F} \otimes \rho^* \omega_{X^{st}}^{\otimes n}) \in \mathbb{Z}_+.$$

Since $\chi(\mathcal{F})$ and $\chi(\mathcal{E})$ are independent of η_j, we must have

$$\frac{\chi(\mathcal{F})}{\ell(\mathcal{F})} = \frac{\chi}{2r(g-1)}.$$

Further, since $\mathcal{F} \subset \mathcal{E}$ is a subsheaf, $\ell(\mathcal{F}) \leqslant 2r(g-1)$. Hence because χ is prime, the above identity is possible only if $\ell(\mathcal{F}) = 2r(g-1)$, namely the quotient $\mathcal{E}/\mathcal{F}$ is only supported on the contracted components of X. Then applying the fact that the restriction of $\mathcal{E}$ to contracted components are non-negative, we conclude $\chi(\mathcal{E}/\mathcal{F}) > 0$. This contradicts to the identity above, proving that the stability is independent of the choice of η. □

We also have the following boundedness result.

Lemma 3.30. *Let $\mathcal{E}$ be a vector bundle over a semistable curve $X \in \mathcal{M}_g^{ss}$. Suppose ϵ is sufficiently small and $\mathcal{E}$ is H_η-stable as defined in (3.26), then the number of the contracted components of X is no more than rg, where $r = \mathrm{rank}\,\mathcal{E}$.*

Proof. We only sketch the proof here. Let $D \in R(X)$. Using stability, one obtains the isomorphism (3.29). Let Σ be a chain of contracted rational curves in X; let $p_1, p_2 \in \Sigma$ be the two points on Σ that connects Σ to the rest of X. Then applying stability condition, one derives the vanishing

$$H^0(\Sigma, \mathcal{E}|_\Sigma(-p_1 - p_2)) = 0. \tag{3.31}$$

This combined with (3.29) ensure that the number of rational curves in Σ is no more than r. Finally, we get the bound $\leqslant rg$ since X has at most g many nodes. □

Because of these Lemmas, in the following, we will speak of stable bundles $\mathcal{E}$ over $X \in \mathcal{M}_g^{ss}$ without explicitly referencing the choice of H_η and sufficiently small ϵ. Therefore, we can define the sub-groupoid

$$\mathfrak{V}_{r,\chi}^{st} \subset \mathfrak{V}_{r,\chi}$$

to be the one so that for the open $\mathcal{U} \subset \mathcal{M}_g^{ss}$ in (3.27), $\mathfrak{V}_{r,\chi}^{st} \times_{\mathcal{M}_g^{ss}} \mathcal{U} = \mathfrak{V}_{r,\chi}^{st}(\mathcal{U})$. By the previous two Lemmas, $\mathfrak{V}_{r,\chi}^{st}$ is well-defined and is open in $\mathfrak{V}_{r,\chi}$. The previous Lemma also shows that $\mathfrak{V}_{r,\chi}^{st}$ is bounded.

We define the automorphism group of $\mathcal{E} \in \mathfrak{V}_{r,\chi}^{st}(k)$, denoted by $\mathrm{Aut}(\mathcal{E})$, be the set of all arrows $\mathcal{E} \to \mathcal{E}$ in the stack $\mathfrak{V}_{r,\chi}(k)$. By definition, a $\xi \in \mathrm{Aut}(\mathcal{E})$ is a pair $\xi = (\sigma, \psi)$ of an automorphism $\sigma : X \to X$ and an isomorphism $\psi : \mathcal{E} \to \sigma^*\mathcal{E}$. Let X^{st} be the stabilization of X, and let $\sigma^{st} : X^{st} \to X^{st}$ be the automorphism induced by σ. Then assigning $\xi \mapsto \sigma^{st}$ defines a homomorphism

$$\mathrm{Aut}(\mathcal{E}) \to \mathrm{Aut}(X^{st}). \tag{3.32}$$

Note that for any $c \in G_m$, multiplication by c (dilation) is an element in $\mathrm{Aut}(\mathcal{E})$ that lies in the kernel of (3.32).

Lemma 3.33. *For every $\mathcal{E} \in \mathfrak{V}_{r,\chi}^{st}(k)$ over X, the kernel of (3.32) is the tautological subgroup G_m of dilations.*

Proof. Since $\mathcal{E}$ is H_η-stable for ϵ sufficiently small, it is proved in [21] that the push forward $\rho_*\mathcal{E}$, where $\rho : X \to X^{st}$, is a pure dimension one sheaf on X^{st}; using that χ is a large prime, one checks that $\rho_*\mathcal{E}$ is Simpson stable with respect to the polarization $\omega_{X^{st}}$.

However, the isomorphism ψ induces an automorphism of $\rho_*\mathcal{E}$. Since the later is Simpson stable, it must be a multiplication by a $c \in G_m$. Hence $c^{-1} \cdot \psi$ is the identity on $\mathcal{E}|_{X-R(X)}$. Then one uses the structure result (3.29) and the vanishing (3.31) to show that $\sigma : X \to X$ must be the identity map, thus $c^{-1} \cdot \psi$ must be the identity as well. This proves the Lemma. □

Theorem 3.34. *The stack $\mathfrak{V}_{r,\chi}^{st}$ is a smooth Artin stack, separated, proper over $\overline{\mathcal{M}}_g$, and of finite type. The morphism $\mathfrak{V}_{r,\chi}^{st} \to \mathcal{M}_g^{ss}$ is smooth, and the morphism $\mathfrak{V}_{r,\chi}^{st} \to \overline{\mathcal{M}}_g$ is a local complete intersection morphism.*

Proof. Since deformation of locally free sheaves on nodal curves are unobstructed, the morphism $\mathfrak{V}_{r,\chi}^{st} \to \mathcal{M}_g^{ss}$ is smooth. Thus since $\mathcal{M}_g^{ss} \to \overline{\mathcal{M}}_g$ is a local complete intersection morphism, $\mathfrak{V}_{r,\chi}^{st} \to \overline{\mathcal{M}}_g$ is a local complete intersection morphism. Because of the bound proved in Lemma 3.30, the stack $\mathfrak{V}_{r,\chi}^{st}$ is of finite type over $\overline{\mathcal{M}}_g$.

The proof that $\mathfrak{V}_{r,\chi}^{st}$ is separated and proper follows the same idea as the proof of the moduli of stable morphisms. We will comment on this issue for the degeneration of Hilbert scheme of subschemes on a simple degeneration. □

We comment that the moduli space $\mathfrak{V}^{st}_{r,\chi}$ constructed has a decomposition along the divisors $\Delta_\delta \subset \overline{\mathcal{M}}_g - \mathcal{M}_g \subset \overline{\mathcal{M}}_g$. The structure result of this decomposition will be clear after we work out the corresponding decomposition theorem for a good degeneration of curves. This will be accomplished in the next subsection.

We let $V^{st}_{r,\chi} \to \overline{\mathcal{M}}_g$ be the coarse moduli space of $\mathfrak{V}^{st}_{r,\chi}$; it is over $\mathcal{M}^{ss}_g$. For any $[X] \in \overline{\mathcal{M}}_g$, the fiber $V^{st}_{r,\chi} \times_{\overline{\mathcal{M}}_g} [X]$ is an algebraic space. The morphisms $V^{st}_{r,\chi} \to \mathcal{M}^{ss}_g$ and $V^{st}_{r,\chi} \to \overline{\mathcal{M}}_g$ have similar properties stated in Theorem 3.34.

Lastly, we comment that following the work of Schmitt [39], for a vector bundle V on a semistable curve X whose restriction to every not-stable rational curve $R \subset X$ has positive degree, we can use $\det E \otimes \omega_X^{\otimes m}$, m a fixed large integer, as a polarization to study the stability of V. This way, the assumption χ sufficiently large and $(r, \chi) = 1$ in the previous discussion is redundant. We refer the detailed treatment to [39].

3.3. Flips of degeneration of moduli of bundles

In this subsection, we will exhibit the special feature of decomposition property-III of good degenerations of vector bundles. We will see that decomposing the central fiber of a good degeneration will need stability depend on the marked-node. This class of stabilities results flips of related spaces. We will exhibit such structure by going through a construction worked out by Kiem and the author [21].

Let X/C be a simple degeneration of smooth curves of $g \geqslant 2$. Though the construction should work for any stable curve X_0, following [21], we assume throughout this subsection that X_0 is irreducible.

We form the stack of expanded degenerations $\mathfrak{X} \to \mathfrak{C}$ as before. For a pair of relatively prime integers $(r, \chi) = 1$, we form the groupoid

$$\mathcal{M}_{\mathfrak{X}/\mathfrak{C}}(r, \chi)$$

of stable locally free sheaves on $\mathfrak{X}/\mathfrak{C}$ of rank r and Euler characteristic χ: for any C-scheme S, $\mathcal{M}_{\mathfrak{X}/\mathfrak{C}}(r, \chi)(S)$ consists of locally free sheaves $\mathcal{E}$ on $\mathcal{X} \in \mathfrak{X}(S)$ so that for any closed $s \in S$, $\mathcal{E}_s = \mathcal{E}|_{\mathcal{X}_s}$ is Simpson stable (cf. Definition 3.25)[3]; an arrow from $\mathcal{E}$ and $\mathcal{E}'$ on $\mathcal{X}$ consists of an arrow $\rho : \mathcal{X} \to \mathcal{X}'$ in $\mathfrak{X}(S)$ together with an isomorphism $\mathcal{E} \cong \rho^*\mathcal{E}'$.

Theorem 3.35. *The groupoid $\mathcal{M}_{\mathfrak{X}/\mathfrak{C}}(r, \chi)$ is a smooth Artin stack, proper and separated over C, and of finite type. Further, $\mathcal{M}_{\mathfrak{X}/\mathfrak{C}}(r, \chi) \to C$ is smooth away from $0 \in C$ and its fiber over $0 \in C$ has normal crossing singularities.*

By replacing $\mathfrak{X}/\mathfrak{C}$ with $\mathfrak{X}_0^\dagger/\mathfrak{C}_0^\dagger$, we obtain a similarly defined groupoid

$$\mathcal{M}_{\mathfrak{X}_0^\dagger/\mathfrak{C}_0^\dagger}(r, \chi).$$

By forgetting the node-marking of $\mathfrak{X}_0^\dagger$, the morphism $\mathfrak{X}_0^\dagger \to \mathfrak{X}_0$ defines a morphism

$$\mathcal{M}_{\mathfrak{X}_0^\dagger/\mathfrak{C}_0^\dagger}(r, \chi) \longrightarrow \mathcal{M}_{\mathfrak{X}/\mathfrak{C}}(r, \chi) \times_C 0. \tag{3.36}$$

[3]When X_0 is irreducible, the stability condition does not depend on η if $(r, \chi) = 1$.

Theorem 3.37. *The groupoid $\mathcal{M}_{\mathfrak{X}_0^\dagger/\mathfrak{C}_0^\dagger}(r,\chi)$ is a smooth Artin stack, proper and separated, and of finite type. The morphism (3.36) is the normalization morphism.*

Due to the issue of stability, the decomposition realizing $\mathcal{M}_{\mathfrak{X}_0^\dagger/\mathfrak{C}_0^\dagger}(r,\chi)$ as gluing of "stable" bundles on $\mathfrak{D}_\pm \subset \mathfrak{Y}$ over $\mathfrak{A}_\diamond$ takes a more involved form compared with the moduli of stable morphisms. This is achieved after introducing a finer stability condition and GPB bundles.

We introduce more notations. We denote by $q \in X_0$ its singular point; denote Y the normalization of X_0 and q_- and $q_+ \in Y$ the two preimages of q under $Y \to X_0$. ($q_\pm$ are the $D_\pm$ mentioned before.) For simplicity, we denote $X_n = X[n]_0$, which is inserting a chain of n-$\mathbb{P}^1$'s at q. For X_n, we denote by $\Delta_1, \cdots, \Delta_n$ the n rational curves inserted, ordered so that if we view Y as an irreducible component of X_n, (when $n \geqslant 2$,) $\Delta_1 \cap Y = q_- = q_1$, $\Delta_i \cap \Delta_{i+1} = q_{i+1}$ and $\Delta_n \cap X = q_+ = q_{n+1}$; we let $R = \Delta_1 \cup \ldots \cup \Delta_n$; thus for $n \geqslant 1$, $X_n = R \cup Y$. By our convention, G_m^n acts on X_n, with the i-th factor of G_m^n acts trivially on all but Δ_i.

We denote by $X_n^\dagger$ be X_n with one of its node marked; we denote the marked node by $q^\dagger$. A contraction $X_n \to X_m$ (resp. $X_m^\dagger \to X_n^\dagger$) is by contracting some of the rational curves Δ_i (resp. so that it sends the marked-node of $X_m^\dagger$ to the marked-node of $X_n^\dagger$).

Let $X_n^\dagger = (X_n, q^\dagger)$ and let $\mathcal{E}$ be a rank r locally free sheaf of $\mathcal{O}_{X_n}$-modules with $\chi(\mathcal{E}) = \chi$. We say $\mathcal{E}$ is admissible if the restriction $\mathcal{E}|_{\Delta_i}$ to each Δ_i has no negative degree factor. We let ϵ be a sufficiently small positive rational, and define a $\mathbb{Q}$-polarization H on X_n whose degree along the component Y (resp. Δ_i) be $1 - n\epsilon$ (resp. ϵ).

Now let $\mathcal{F}$ be any subsheaf of $\mathcal{E}$. We define the rank of $\mathcal{F} \subset \mathcal{E}$ at $q^\dagger \in X_n^\dagger$ to be

$$r^\dagger(\mathcal{F}) = \dim \operatorname{Im}\{\mathcal{F}|_{q^\dagger} \to \mathcal{E}|_{q^\dagger}\}.$$

For real α we define the α-slope (implicitly depending on ϵ) of $\mathcal{F} \subset \mathcal{E}$ to be

$$\mu(\mathcal{F},\alpha) = (\chi(\mathcal{F}) - \alpha\, r^\dagger(\mathcal{F}))/\operatorname{rank}\mathcal{F} \in \mathbb{Q}, \tag{3.38}$$

where

$$\operatorname{rank}\mathcal{F} = (1 - n\,\epsilon)\operatorname{rank}\mathcal{F}|_Y + \sum_{i=1}^{n} \epsilon \operatorname{rank}\mathcal{F}|_{\Delta_i}.$$

We define the automorphism group $\operatorname{Aut}_{\mathfrak{X}}(\mathcal{E})$ to be the group of pairs (σ, f) so that $\sigma \in G_m^n$ and f is an isomorphism $\mathcal{E} \cong \sigma^*\mathcal{E}$.

Definition 3.39. *Let $\mathcal{E}$ be a rank r locally free sheaf over $X_n^\dagger$. For $\alpha \in [0,1)$, we say $\mathcal{E}$ is α-semistable (resp. weakly α-stable) if for any proper subsheaf $\mathcal{F} \subset \mathcal{E}$ and ϵ sufficiently small we have*

$$\mu(\mathcal{F},\alpha) \leqslant \mu(\mathcal{E},\alpha) \quad (\text{resp.} <).$$

We say $\mathcal{E}$ is α-stable if $\mathcal{E}$ is weakly α-stable and $\deg \mathcal{E}|_{\Delta_i} > 0$ *for all i.*

We remark that when $\alpha = 0$, the α-stability does not involve $r^\dagger$, and hence can be defined for X_n, without the marked-node. In this case, the 0-stability coincides with the Simpson stability of sheaves.

We quote the basic facts about α-stable sheaves on $X_n^\dagger$. To avoid complications arising from strictly semistable sheaves, we will restrict ourselves to $\alpha \in [0,1) - \Lambda_r$:

$$\Lambda_r = \{\alpha \in [0,1) \mid \alpha = \frac{r_0\chi - r\chi_0}{r(r_0 - r^\dagger)},\ 0 < r_0 < r,\ 0 \leqslant r^\dagger \leqslant r,\ 2r_0 - r \leqslant \chi \leqslant 2r_0\}.$$

Here $\chi_0, r_0, r^\dagger$ are integers. Clearly, Λ_r is a discrete subset of $[0,1)$. When $(\chi, r) = 1$, $0 \notin \Lambda_r$.

Lemma 3.40 ([21]). *Let $(r,\chi) = 1$ and $\chi > r$. Let $\alpha \in [0,1) - \Lambda_r$. Then for any rank r α-semistable sheaf $\mathcal{E}$ on $X_n^\dagger$ of Euler characteristic $\chi(\mathcal{E}) = \chi$, we have*
(a) in case $n \geqslant 1$, the restriction $\mathcal{E}|_{D_i}$ has no negative degree factors and $\mathcal{E}$ has no nontrivial section that vanishes on $Y \subset X_n$;
(b) for any (partial) contraction $\pi: X_m^\dagger \to X_n^\dagger$ the pull back $\pi^*\mathcal{E}$ is weakly α-stable;
(c) suppose $\mathcal{E}$ is α-stable. Then $n \leqslant r$ and $\mathrm{Aut}_{\mathfrak{X}}(\mathcal{E}) = G_m$.

For $(r,\chi) = 1$ and an $\alpha \in [0,1) - \Lambda_r$, we form the groupoid

$$\mathcal{M}^\alpha_{\mathfrak{X}_0^\dagger/\mathfrak{C}_0^\dagger}(r,\chi)$$

of α-stable sheaves on $\mathfrak{X}_0^\dagger$, like we did in constructing $\mathcal{M}_{\mathfrak{X}/\mathfrak{C}}(r,\chi)$ and $\mathcal{M}_{\mathfrak{X}_0^\dagger/\mathfrak{C}_0^\dagger}(r,\chi)$.

Theorem 3.41. *For $(r,\chi) = 1$ and $\alpha \in [0,1) - \Lambda_r$, the groupoid $\mathcal{M}^\alpha_{\mathfrak{X}_0^\dagger/\mathfrak{C}_0^\dagger}(r,\chi)$ is a smooth Artin stack with stabilizer G_m at every point. Let $M^\alpha_{\mathfrak{X}_0^\dagger/\mathfrak{C}_0^\dagger}(r,\chi)$ be its coarse moduli space. Then $M^\alpha_{\mathfrak{X}_0^\dagger/\mathfrak{C}_0^\dagger}(r,\chi)$ is a smooth, proper and separated algebraic space.*

We next introduce the notion of generalized parabolic bundles (in short GPB). It was introduced by Bhosle in [2]. Let Y^+ be the pair $(Y, q_- + q_+)$. A rank r GPB on Y^+ is a pair $\mathcal{V}^G = (\mathcal{V}, \mathcal{V}^0)$ of a rank r locally free sheaf $\mathcal{V}$ on Y and an r-dimensional subspace $\mathcal{V}^0 \subset \mathcal{V}|_{q_-+q_+}$. (We denote by $\mathcal{V}|_{q_-+q_+}$ the vector space $\mathcal{V}|_{q_-} \oplus \mathcal{V}|_{q_+}$.) For any subsheaf $\mathcal{F} \subset \mathcal{V}$ we denote $\mathcal{F}|_{q_-+q_+} \cap \mathcal{V}^0 = \mathrm{Im}\{\mathcal{F}|_{q_-+q_+} \to \mathcal{V}|_{q_-+q_+}\} \cap \mathcal{V}^0$; we define

$$r^+(\mathcal{F}) = \dim\big(\mathcal{F}|_{q_-+q_+} \cap \mathcal{V}^0\big).$$

Definition 3.42. *Let $\alpha \in [0,1)$. A GPB $\mathcal{V}^G = (\mathcal{V}, \mathcal{V}^0)$ is α-stable if for any proper subbundle $\mathcal{F} \subset \mathcal{V}$ we have $\mu^G(\mathcal{F},\alpha) < \mu^G(\mathcal{V},\alpha)$, where*

$$\mu^G(\mathcal{F},\alpha) = (\chi(\mathcal{F}) + (1-\alpha)r^+(\mathcal{F}))/r(\mathcal{F}).^4$$

Let $\mathfrak{G}^\alpha_{Y^+}(r,\chi)$ be the groupoid of all isomorphism classes of α-stable rank r GPBs on Y^+ of Euler characteristics χ. It is known that $\mathfrak{G}^\alpha_{Y^+}(r,\chi)$ is a smooth Artin stack, proper and separated, and of finite type. Its coarse moduli space $G^\alpha_{Y^+}(r,\chi)$ is projective.

---

[4]The α-stability is the $(1-\alpha)$-stability of GPB introduced in [2].

In [21], a decomposition of $\mathcal{M}^{\alpha}_{\mathfrak{X}_0^\dagger/\mathfrak{C}_0^\dagger}(r,\chi)$ of the form

$$\Phi_\alpha : \mathcal{M}^{\alpha}_{\mathfrak{X}_0^\dagger/\mathfrak{C}_0^\dagger}(r,\chi) \longrightarrow \mathfrak{G}^{\alpha}_{Y^+}(r,\chi'), \quad \chi' = \chi + r \tag{3.43}$$

was constructed. We describe this morphism. We begin with locally free sheaves on $R = \Delta_1 \cup \ldots \cup \Delta_n$. (Recall our convention that $q_i = \Delta_{i-1} \cap \Delta_i$, $q_1 = q_- \in \Delta_1$ and $q_+ = q_{n+1} \in \Delta_n$.) Let $\mathcal{F}$ be any admissible locally free sheaf on R (meaning that $\mathcal{F}|_{\Delta_i}$ has no negative degree factor). Inductively, we define vector spaces $W_i \subset \mathcal{F}|_{q_i}$ by $W_1 = \{0\}$ and

$$W_i = \{s(q_i) \mid s \in H^0(\Delta_i, \mathcal{F}), s(q_{i-1}) \in W_{i-1}\} \subset \mathcal{F}|_{q_i}.$$

We define $T_\rightarrow = W_{n+1}$. It is also defined via

$$T_\rightarrow = \{s(q_{n+1}) \mid s(q_1) = 0 \text{ and } s \in H^0(R, \mathcal{F})\}. \tag{3.44}$$

We call $T_\rightarrow$ the *transfer* of $0 \in \mathcal{F}|_{q_1}$ along R.

If we reverse the order of R by putting $\tilde{\Delta}_i = \Delta_{n-i+1}$, we call the resulting transfer $T_\leftarrow \subset \mathcal{F}|_{q_1}$ the *reverse transfer* of $0 \in \mathcal{F}|_{q_{n+1}}$. Notice that we have a well-defined homomorphism $\mathcal{F}_{q_{n+1}} \rightarrow \mathcal{F}|_{q_1}/T_\leftarrow$ by assigning to each element $c \in \mathcal{F}|_{q_{n+1}}$ the class of $[s(q_1)] \in \mathcal{F}|_{q_1}/T_\leftarrow$ for some $s \in H^0(R, \mathcal{F})$ such that $s(q_{n+1}) = c$. The kernel of this homomorphism is precisely the transfer $T_\rightarrow$. Hence we have a canonical isomorphism

$$\xi : \mathcal{F}|_{q_{n+1}}/T_\rightarrow \xrightarrow{\cong} \mathcal{F}|_{q_1}/T_\leftarrow. \tag{3.45}$$

There is another way to see this isomorphism. Let $H^0(R, \mathcal{F}^\vee) \otimes \mathcal{O}_R \rightarrow \mathcal{F}^\vee$ and

$$\varphi : \mathcal{F} \longrightarrow H^0(R, \mathcal{F}^\vee)^\vee \otimes \mathcal{O}_R$$

be the canonical homomorphism. Then $\ker(\varphi|_{q_1}) = T_\leftarrow$, $\ker(\varphi|_{q_{n+1}}) = T_\rightarrow$ and the isomorphism (3.45) is induced by

$$\mathcal{F}|_{q_1} \longrightarrow H^0(R, \mathcal{F}^\vee)^\vee \longleftarrow \mathcal{F}|_{q_{n+1}}.$$

Definition 3.46. *We say a locally free sheaf $\mathcal{F}$ on R is regular if there are integers a_i so that to each $1 \leqslant i \leqslant n$,*

$$\mathcal{F}|_{\Delta_i} = \mathcal{O}_{\Delta_i}^{r-a_i} \oplus \mathcal{O}_{\Delta_i}(1)^{a_i} \quad \textit{and} \quad \dim W_{i+1} = \dim W_i + a_i.$$

Now let $\mathcal{E}$ be a rank r locally free sheaf on $X_n^\dagger$ so that its restriction to the chain of rational curves $R \subset X_n$ is regular. Let $n_- + n_+ = n$ be integers so that after resolving the singularity of $X_n^\dagger$ at $q^\dagger$ we obtain Y_{n_-,n_+}, which is Y with a chain of n_- (resp. n_+) rational curves attached to $q_- \in Y$ (resp, $q_+ \in Y$). Let

$$\rho : Y_{n_-,n_+} \longrightarrow X_n^\dagger \quad \text{and} \quad \pi : Y_{n_-,n_+} \longrightarrow Y$$

be the tautological projections. Let q'_- and q'_+ be the two preimages $\rho^{-1}(q^\dagger)$, obviously indexed.

First, $\tilde{\mathcal{E}} := \rho^*\mathcal{E}$ is a locally free sheaf on Y_{n_-,n_+} and $\mathcal{E}$ can be reconstructed from $\tilde{\mathcal{E}}$ and the isomorphism $\phi : \tilde{\mathcal{E}}|_{q'_-} \equiv \mathcal{E}|_{q^\dagger} \equiv \tilde{\mathcal{E}}|_{q'_+}$ via

$$0 \longrightarrow \mathcal{E} \longrightarrow \rho_*\tilde{\mathcal{E}} \longrightarrow (\tilde{\mathcal{E}}|_{q'_-} \oplus \tilde{\mathcal{E}}|_{q'_+})/\Gamma_\phi \longrightarrow 0,$$

where $\Gamma_\phi \subset \tilde{\mathcal{E}}|_{q'_-} \oplus \tilde{\mathcal{E}}|_{q'_+}$ is the graph of ϕ. Here we view the last non-zero term in the sequence as a $\mathbf{k}(q^\dagger)$ vector space, which is naturally a sheaf of $\mathcal{O}_{X_n}$-modules. In this way the locally free sheaf $\mathcal{E}$ on $X_n^\dagger$ is equivalent to the GPB $(\tilde{\mathcal{E}}, \Gamma_\phi)$ over Y_{n_-,n_+}.

We now show how the GPB $(\tilde{\mathcal{E}}, \Gamma_\phi)$ associates to a GPB over Y^+. First let $\mathcal{V} = (\pi_*\tilde{\mathcal{E}}^\vee)^\vee$. Since the restriction of $\tilde{\mathcal{E}}$ to R is regular, $\pi_*\tilde{\mathcal{E}}^\vee$, and hence $\mathcal{V}$, are locally free on Y. Clearly, we have $\chi(\mathcal{V}) = \chi(\tilde{\mathcal{E}}) = \chi(\mathcal{E}) + r$.

Next, by construction we have canonical $\pi^*(\pi_*\tilde{\mathcal{E}}^\vee) \to \tilde{\mathcal{E}}^\vee$ and its dual

$$\tilde{\mathcal{E}} \longrightarrow \pi^*\mathcal{V} = \pi^*(\pi_*\tilde{\mathcal{E}}^\vee)^\vee. \tag{3.47}$$

Restricting to q'_- and q'_+, we obtain

$$h_1 : \tilde{\mathcal{E}}|_{q'_-} \longrightarrow \pi^*\mathcal{V}|_{q'_-} = \mathcal{V}|_{q_-} \quad \text{and} \quad h_2 : \tilde{\mathcal{E}}|_{q'_+} \longrightarrow \pi^*\mathcal{V}|_{q'_+} = \mathcal{V}|_{q_+}.$$

We then define $\mathcal{V}^0 \subset \mathcal{V}|_{q_-} \oplus \mathcal{V}|_{q_+}$ to be the image of $\Gamma_\phi \subset \tilde{\mathcal{E}}|_{q'_-} \oplus \tilde{\mathcal{E}}|_{q'_+}$ under the homomorphism $h_1 \oplus h_2$. It was proved in [21] that $\dim \mathcal{V}^0 = r$. Thus $(\mathcal{V}, \mathcal{V}^0)$ is a GPB on Y^+.

Proposition 3.48. *Let $\mathcal{E}$ be an α-stable locally free sheaf on $X_n^\dagger$. Then $\mathcal{E}|_R$ is regular and its associated GPB $\mathcal{V}^G = (\mathcal{V}, \mathcal{V}^0)$ is α-stable.*

A corollary of this Proposition is the morphism Φ_α in (3.43).

We comment that the α-stability of $\mathcal{F}$ on $X_n^\dagger$ is independent of the marked-node $q^\dagger$ when $\alpha = 0$. Thus

$$\mathcal{M}^0_{\mathfrak{X}_0^\dagger/\mathfrak{C}_0^\dagger}(r,\chi) = \mathcal{M}_{\mathfrak{X}_0^\dagger/\mathfrak{C}_0^\dagger}(r,\chi) \longrightarrow \mathcal{M}_{\mathfrak{X}_0/\mathfrak{C}_0}(r,\chi)$$

is the normalization of the central fiber of the good degeneration $\mathcal{M}_{\mathfrak{X}/\mathfrak{C}}(r,\chi)$. After studying wall-crossing of $\mathcal{M}^\alpha_{\mathfrak{X}_0^\dagger/\mathfrak{C}_0^\dagger}(r,\chi)$ for α varying from 0 to 1, we obtain $\mathcal{M}^1_{\mathfrak{X}_0^\dagger/\mathfrak{C}_0^\dagger}(r,\chi)$ from $\mathcal{M}^0_{\mathfrak{X}_0^\dagger/\mathfrak{C}_0^\dagger}(r,\chi)$ via a sequence of flips. We then apply the morphism Φ_1 to relate $\mathcal{M}^1_{\mathfrak{X}_0^\dagger/\mathfrak{C}_0^\dagger}(r,\chi)$ to $\mathfrak{G}^1_{Y^+}(\chi+r,r)$, which is explicit. Since the 1-stability of a GPB $(\mathcal{V},\mathcal{V}^0)$ is independent of the choice of $\mathcal{V}^0$, forgetting $\mathcal{V}^0$ defines a fiber bundle morphism

$$\mathfrak{G}^1_{Y^+}(\chi+r,r) \longrightarrow \mathcal{M}_Y(r,\chi+r),$$

to the moduli of stable bundles on Y. This process connects the geometry of $\mathcal{M}_{X_t}(r,\chi)$ with that of $\mathcal{M}_Y(r,\chi+r)$. Gieseker used this to prove the Newstead and Ramanan conjecture for $\mathcal{M}_X(2,1)$. Carrying Gieseker's construction to rank three case, Kiem and the author proved the rank three version of this conjecture.

Theorem 3.49 ([12, 21]). *Let X be a genus g smooth curve, then $c_i(\mathcal{M}_X(2,1)) = 0$ for $i > 2g-2$; $c_i(\mathcal{M}_X(3,1)) = 0$ for $i > 6g-5$.*

We refer the details of this construction to the paper [21]. Part of the construction outlined here can be achieved using other methods, like GIT. For instance, X-T. Sun proved that the degenerations constructed in [NS2] and in [KL] are identical. However, using the stack $\mathfrak{X}/\mathfrak{C}$ and its variants allow us to study in details the geometry of the degenerations, eventually leading to the proof of the vanishing theorem of Chern classes for moduli of rank three stable bundles.

3.4. Degenerations of Hilbert schemes

The last case we will study is the degeneration of relative Hilbert schemes of subschemes. The dimension three case was worked out by B. Wu in his thesis. The general version, including the degenerations of Grothendick's Quot-scheme, and of PT stable pairs, has been worked out by B. Wu and the author [31].

Let X/C be a good degeneration of smooth projective schemes. We assume that the normalization Y of X_0 consists to two connected components: $Y = Y_- \cup Y_+$. We let $D_\pm = \tilde{D} \cap Y_\pm$.

Fixing a relative ample line bundle H on X/C and a polynomial $P(m)$, we seek to find a good degeneration of the Hilbert scheme of subschemes Z of X_t, $t \in C - 0$, so that the Hilbert polynomials of $\mathcal{O}_Z = \mathcal{O}_{X_t}/\mathcal{I}_Z$ are P.

Like before, we form the stack $\mathfrak{X}/\mathfrak{C}$ of expanded degenerations of X/C; we introduce the notion of admissible ideal sheaves and construct the good degeneration as the moduli of admissible ideal sheaves on $\mathfrak{X}/\mathfrak{C}$ with given Hilbert polynomial P.

We begin with the notion of admissible ideal sheaves of $X[n]$ and of relative pairs $(Y_+[n], D_+[n])$ and $(Y_-[n], D_-[n])$.

Definition 3.50. *Let $V \subset W$ be a closed subscheme in a scheme W. We say an ideal sheaf $\mathcal{I}_Z \subset \mathcal{O}_W$ is normal to V if the canonical $\mathcal{I}_Z \otimes_{\mathcal{O}_W} \mathcal{O}_V \to \mathcal{O}_V$ is injective. Let S be any scheme. We call an ideal sheaf $\mathcal{I}_Z \subset \mathcal{O}_{W\times S}$ an S-flat family of ideal sheaves if*

$$\mathcal{I}_Z \otimes_{\mathcal{O}_{W\times S}} \mathcal{O}_{W\times s} \longrightarrow \mathcal{O}_{W\times s} \tag{3.51}$$

is injective for all $s \in S$.

It is proved in [31] that for a flat family of ideal sheaves of $\mathcal{O}_W$, the property of normal to the closed subscheme $V \subset W$ is an open condition.

Definition 3.52. *We call an ideal sheaf $\mathcal{I}_Z \subset \mathcal{O}_{X[n]_0}$ (resp. $\mathcal{I}_Z \subset \mathcal{O}_{Y_+[n]_0}$) admissible if it is normal to all $D_l \subset X[n]_0$ (resp. $D_l \subset Y_+[n]_0$). A relative ideal sheaf on $(Y_+[n]_0, D_+[n]_0)$ is an admissible ideal sheaf $\mathcal{I}_Z \subset \mathcal{O}_{Y_+[n]_0}$ normal to $D_+[n]_0$.*

We use Hilbert polynomials to keep track of the topological data of ideal sheaves. Fixing the H over X/C given before, for any family $(\mathcal{X}, p) \in \mathfrak{X}(S)$, $p : \mathcal{X} \to X$ the tautological projection, and any S-flat coherent sheaf $\mathcal{F}$ on $\mathcal{X}$, we denote $\mathcal{F}(m) = \mathcal{F} \otimes p^*H^{\otimes m}$; for any closed $s \in S$, we define $P_{\mathcal{F}_s}$ be the polynomial

$$P_{\mathcal{F}_s}(m) = \chi(\mathcal{F}(m) \otimes_{\mathcal{O}_\mathcal{X}} \mathcal{O}_{\mathcal{X}_s}).$$

For $\mathcal{F} = \mathcal{I}_Z$ an ideal sheaf on $X[n]_0$, we will work with the Hilbert polynomial $P_Z = P_{\mathcal{O}_Z}$ of the structure sheaf $\mathcal{O}_Z = \mathcal{O}_{X[n]_0}/\mathcal{I}_Z$.

Similarly, pulling back H to (Y_+, D_+) we obtain ample line bundle H_+. Using the projection $p : Y_+[n]_0 \to Y$, we define for any admissible ideal sheaf $\mathcal{I}_Z$ of $(Y_+[n], D_+[n])$, $\mathcal{O}_Z(m) = \mathcal{O}_Z \otimes p^* H_+^{\otimes m}$, and a pair of polynomials

$$P_Z(m) = \chi(\mathcal{O}_Z(m)), \quad P_{Z\cap D}(m) = \chi(\mathcal{O}_{Z\cap D_+[n]_0}(m)).$$

Stability of (relative) ideal sheaves are defined by requiring the automorphism groups be finite. Let $\mathcal{I}_Z$ be an ideal sheaf of $X[n]_0$. We define $\mathrm{Aut}_{\mathfrak{X}}(\mathcal{I}_Z)$ be $\sigma \in \mathbb{G}_m^n$ so that $\sigma^*\mathcal{I}_Z = \mathcal{I}_Z$ as ideal sheaves of $\mathcal{O}_{X[n]_0}$, where σ acts on $X[n]_0$ as before. We define $\mathrm{Aut}_{\mathfrak{Y}_\cdot}(\mathcal{I}_Z)$ for a relative ideal sheaf on $(Y_\cdot[n], D_\cdot[n])$ similarly.

Definition 3.53. *We call an ideal sheaf $\mathcal{I}_Z$ of $X[n]_0$ stable if in addition to being admissible, $\mathrm{Aut}_{\mathfrak{X}}(\mathcal{I}_Z)$ is finite. We define stable relative ideal sheaves $\mathcal{I}_Z$ of $(Y_\cdot[n]_0, D_\cdot[n]_0)$ similarly.*

The good degeneration of the Hilbert scheme of ideal sheaves is the moduli of stable ideal sheaves of the family $\mathfrak{X}/\mathfrak{C}$. We define the groupoid $\mathcal{I}^P_{\mathfrak{X}/\mathfrak{C}}$ of families of stable ideal sheaves: given any scheme S over C, we define $\mathcal{I}^P_{\mathfrak{X}/\mathfrak{C}}(S)$ be the set of all $(\mathcal{I}_{\mathcal{Z}}, \mathcal{X}, p)$, where $(\mathcal{X}, p) \in \mathfrak{X}(S)$ and $\mathcal{I}_{\mathcal{Z}}$ is an S-flat family of stable ideal sheaves on $\mathcal{X}$ of Hilbert polynomials P. An arrow between $(\mathcal{I}_{\mathcal{Z}_1}, \mathcal{X}_1, p_1)$ and $(\mathcal{I}_{\mathcal{Z}_2}, \mathcal{X}_2, p_2) \in \mathcal{I}^P_{\mathfrak{X}/\mathfrak{C}}(S)$ is an arrow $\sigma : \mathcal{X}_1 \to \mathcal{X}_2$ in $\mathfrak{X}(S)$ so that $\mathcal{I}_{\mathcal{Z}_1} = \sigma^*\mathcal{I}_{\mathcal{Z}_2}$ as subsheaves of $\mathcal{O}_{\mathcal{X}_1}$. For $\rho : S \to T$, the map $\mathcal{I}^P_{\mathfrak{X}/\mathfrak{C}}(\rho) : \mathcal{I}^P_{\mathfrak{X}/\mathfrak{C}}(T) \to \mathcal{I}^P_{\mathfrak{X}/\mathfrak{C}}(S)$ is defined by pull back.

Theorem 3.54. *$\mathcal{I}^P_{\mathfrak{X}/\mathfrak{C}}$ is a Deligne-Mumford stack, separated and proper over C; it is of finite type.*

Proof. We sketch the proof. Let $\mathcal{I}_Z$ be an ideal sheaf of $\mathcal{O}_{X[n]_0}$. We define a numerical quantity that measures the failure of $\mathcal{I}_Z$ being admissible.

Let $X[n]_0^{sm}$ be the smooth locus of $X[n]_0$; let Z^{min} (resp. Z_k^{min}) be the schematic closure of $Z \cap X[n]_0^{sm}$ in $X[n]_0$ (resp. $Z \cap (\Delta_k - D_k - D_{k+1})$ in Δ_k). It is clear that $\mathcal{I}_Z$ is admissible if and only if

$$Z = Z^{min} \quad \text{and} \quad Z_{k-1}^{min} \cap D_k = Z_k^{min} \cap D_k, \quad \text{for all } 1 \leqslant k \leqslant n+1. \tag{3.55}$$

To each D_k we define the error term err_Z^k. Since Z^{min} is a closed subscheme of Z, we have restriction homomorphism $\mathcal{O}_Z \to \mathcal{O}_{Z^{min}}$. We let $\mathcal{T}$ be its kernel. We write $\mathcal{T}$ as a direct sum of $\mathcal{T}_k$, $k = 1, \ldots, n+1$, where $\mathcal{T}_k$ supports (set theoretically) on D_k. We define polynomial

$$\mathrm{err}_Z^k = 2P_{\mathcal{T}_k} + P_{Z_{k-1}^{min}\cap D_k} + P_{Z_k^{min}\cap D_k} - 2P_{Z_{k-1}^{min}\cap Z_k^{min}}.$$

Because the leading coefficients of $P_{\mathcal{T}_k}$ and of $P_{Z_{k-i}^{min}\cap D_k} - P_{Z_{k-1}^{min}\cap Z_k^{min}}$, $i = 0,1$, are non-negative, they vanish simultaneously if and only if $\mathrm{err}_Z^k = 0$, which is equivalent to (3.55). Thus $\mathcal{I}_Z$ is admissible along D_k if and only if $\mathrm{err}_Z^k = 0$.

We now prove the properness using the valuative criterion. Let $S = \operatorname{Spec} R \to C$, where R be a discrete valuation $\mathbf{k}$-algebra. Let $\mathcal{Z} \subset \mathcal{X}$ with $\mathcal{X} \in \mathfrak{X}(S)$ be closed subscheme flat over S such that $\mathcal{I}_{\mathcal{Z}_\eta} \in \mathfrak{I}^P_{\mathfrak{X}/\mathfrak{C}}(\eta)$, $\mathrm{Aut}_{\mathfrak{X}}(\mathcal{I}_{\mathcal{Z}_{\eta_0}})$ is finite, and $\mathrm{err}_{\mathcal{Z}_{\eta_0}} \neq 0$. We will show that we can find a finite base change $S' = \operatorname{Spec} R' \to S$, an $\mathcal{X}' \in \mathfrak{X}(S')$ and an S'-flat closed subscheme $\mathcal{Z}' \subset \mathcal{X}'$ such that

(1) for η' the generic point of S', $\mathcal{X}'_{\eta'} \cong \mathcal{X}_\eta \times_\eta \eta'$ and under this isomorphism $\mathcal{Z}'_{\eta'} = \mathcal{Z}_\eta \times_\eta \eta'$;
(2) $\mathrm{Aut}_{\mathfrak{X}}(\mathcal{I}_{\mathcal{Z}'_{\eta'_0}})$ is finite;
(3) for η'_0 the closed point of S', $\mathrm{err}_{\mathcal{Z}'_{\eta'_0}} \prec \mathrm{err}_{\mathcal{Z}_{\eta_0}}$.

Here we say two polynomials P_1 and P_2 with positive leading coefficients having $P_1 \prec P_2$ if either $\deg P_1 < \deg P_2$, or $\deg P_1 = \deg P_2$ and the leading coefficients $\mathrm{l.c.}\, P_1 < \mathrm{l.c.}\, P_2$.

Since R is a discrete valuation ring, we can assume that $\mathcal{X}/S$ is given by $\xi : S \to C[n]$. In this note, we assume S is flat over C and $\eta_0 \mapsto 0 \in C[n]$. (We skip the other cases). We pick k so that $\mathrm{err}^k_{\mathcal{Z}_0}$ has the same (polynomial) degree as $\mathrm{err}_{\mathcal{Z}_0}$. We let u be a uniformizing parameter of S, and let $\pi_n : C[n] \to \mathbf{A}^{n+1}$ be the projection. Since S is flat over C and $\eta_0 \mapsto 0 \in C[n]$, we can write

$$\pi_n \circ \xi = (c_1 u^{e_1}, \ldots, c_{n+1} u^{e_{n+1}}), \quad c_i \in R^*,\ e_i \in \mathbb{Z}_+. \tag{3.56}$$

(R^* are the invertible elements in R.) We let $\tau_k : C[n] \times G_m \to C[n+1]$ be induced from $\mathbf{A}^{n+1} \times G_m \to \mathbf{A}^{n+2}$:

$$(t_1, \cdots, t_{n+1}, \sigma) \mapsto (t_1, \cdots, t_{k-1}, \sigma, \sigma^{-1} t_k, t_{k+1}, \cdots, t_{n+1}).$$

We introduce ξ_k

$$\xi_k = \tau_k \circ (\xi \times (\cdot)^{e_k}) : S \times G_m \longrightarrow C[n] \times G_m \longrightarrow C[n+1],$$

where $(\cdot)^{e_k} : G_m \to G_m$ sends $\sigma \mapsto \sigma^{e_k}$, we let $\mathcal{X}' = \xi_k^* X[n+1]$ be the pull back family, which is over $S \times G_m$. Then

$$\mathcal{X}' \cong \xi^* X[n] \times G_m = \mathcal{X} \times G_m,$$

because of the canonical isomorphism $\tau_k^* X[n+1] \cong X[n] \times G_m$ as families over $C[n] \times G_m$.

Let $\mathcal{X}' = \mathcal{X} \times G_m \to \mathcal{X}$ be the projection, and let $\mathcal{Z}' = \mathcal{Z} \times_{\mathcal{X}} \mathcal{X}' \subset \mathcal{X}'$ be the pull back family. The family $\mathcal{Z}'$ induces a $C[n+1]$-morphism

$$f_k : S \times G_m \longrightarrow \mathrm{Hilb}^P_{X[n+1]/C[n+1]}.$$

Since $\mathrm{Hilb}^P_{X[n+1]/C[n+1]}$ is proper over $C[n+1]$, there exists a regular surface V, proper over $C[n+1]$, a birational inclusion $h : S \times G_m \hookrightarrow V$, and a morphism $V \to \mathrm{Hilb}^P_{X[n+1]/C[n+1]}$, making the two triangles below commutative:

$$\begin{array}{ccc} V & \longrightarrow & \mathrm{Hilb}^P_{X[n+1]/C[n+1]} \\ {\scriptstyle h}\uparrow & \overset{f_k}{\nearrow} & \downarrow \\ S \times G_m & \xrightarrow{\ \xi_k\ } & C[n+1] \end{array}$$

Let $\tilde{\mathfrak{X}} = V \times_{C[n+1]} X[n+1]$ and let $\tilde{\mathcal{Z}} \subset \tilde{\mathfrak{X}}$ be the pull back of the universal family of the relative Hilbert scheme $\mathrm{Hilb}^P_{X[n+1]/C[n+1]}$.

The key is to study the family $\tilde{\mathcal{Z}}$ over $V \times_C 0$. First, we see that

$$\pi_{n+1} \circ \xi_k : S \times G_m \to \mathbf{A}^{n+2}$$

factors through a proper $V' \to \mathbf{A}^{n+2}$, where V' is by embedding $S \times G_m \subset S \times \mathbf{A}^1$ after adding 0 to $\mathbf{A}^1 - 0 \cong G_m$, blowing up $(\eta_0, 0) \in S \times \mathbf{A}^1$ and removing the proper transform of $S \times 0$. By the explicit expression (3.56), we see that $\pi_{n+1} \circ \xi_k$ factors through the proper $V' \to \mathbf{A}^{n+2}$, and $V' \times_{\mathbf{A}^1} 0 = \Sigma_1' \cup \Sigma_2'$, where $\Sigma_1' = \eta_0 \times \mathbf{A}^1$ and Σ_2' is the exceptional divisor of $V' \to S \times \mathbf{A}^1$.

Again, by the explicit construction of V', it is easy to see that the composite $V \to C[n+1] \to \mathbf{A}^{n+2}$ factors through $V' \to \mathbf{A}^{n+2}$. Let $\Sigma_i \subset V$ be the proper transform of Σ_i'; let $E = V \times_{C[n+1]} 0$. Then $V \times_C 0 = E \cup \Sigma_1 \cup \Sigma_2$ and E is a union of rational curves. Since $V \times_C 0$ is connected, we can find a chain of rational curves $R \subset E$ so that $\Sigma_1 \cup R \cup \Sigma_2$ is connected. We index $R = R_1 \cup \cdots \cup R_m$ so that $q_1 = \Sigma_1 \cap R_1$, $R_i \cap R_{i+1}$ and $q_{m+1} = R_m \cap \Sigma_2$ are non-empty intersection points.

For $x \in R$, let $\tilde{\mathcal{Z}}_x \subset X[n+1]_0$ be the fiber of $\tilde{\mathcal{Z}}$ over x. The technical part of the proof is to show that

(a) For any $x \in R$, $\tilde{\mathcal{Z}}_x|_{X[n+1]_0 - \Delta_k} \equiv \mathcal{Z}_{\eta_0}|_{X[n]_0 - D_k}$;
(b) $\mathcal{I}_{\tilde{\mathcal{Z}}_{q_1}}$ is normal to D_k, and $\mathcal{I}_{\tilde{\mathcal{Z}}_{q_{m+1}}}$ is normal to D_{k+1};
(c) there is an $x \in R$ so that (the leading coefficient) $\text{l.c.}\,\mathrm{err}^k_{\tilde{\mathcal{Z}}_x} < \text{l.c.}\,\mathrm{err}^k_{\mathcal{Z}_{\eta_0}}$.

With this proved, (we will skip this part since it is a technical argument on ideal sheaves,) we find a discrete valuation ring R' and morphism $S' = \operatorname{Spec} R' \to V$ so that $\eta_0' \mapsto x$ and $S' \to V \to S$ is flat. Pulling back the family $\tilde{Z} \subset \tilde{\mathfrak{X}}$ we obtain an S'-flat family $\mathcal{Z}' \subset \mathfrak{X}'$ that satisfies (a)-(c) listed.

Finally, we note that the leading terms of the collection of possible non-zero polynomials err_Z are of the form $a_r \frac{m^r}{r!}$, $a \in \mathbb{Z}_+$. Because of our strict definition of $\prec$, there are no infinite descending sequence of such polynomials. Therefore, this process of decreasing $\mathrm{err}_{\mathcal{Z}_{\eta_0}}$ terminates at finite steps. This proves the properness criterion.

We now sketch the proof of separateness. Let

$$(\mathcal{I}_{\mathcal{Z}_1}, \mathfrak{X}_1) \text{ and } (\mathcal{I}_{\mathcal{Z}_2}, \mathfrak{X}_2) \in \mathfrak{I}^P_{\mathfrak{X}/\mathfrak{C}}(S)$$

be two families of stable ideal sheaves, where $S = \operatorname{Spec} R$ flat over C as before, such that there is a $\rho_\eta : \mathfrak{X}_{1,\eta} \to \mathfrak{X}_{2,\eta}$ in $\mathfrak{X}(\eta)$ such that $\mathcal{I}_{\mathcal{Z}_{1,\eta}} = \rho_\eta^* \mathcal{I}_{\mathcal{Z}_{2,\eta}}$.

We express $\mathfrak{X}_i$ as $\xi_i^* X[n_i]$ induced by $\xi_i : S \to C[n_i]$, $\xi_i(\eta_0) = 0$. We express

$$\pi_{n_i} \circ \xi_i = \left(c_{i,1} u^{e_{i,1}}, \ldots, c_{i,n_i+1} u^{e_{i,n_i+1}}\right), \quad c_{i,j} \in R^*, \ e_{i,j} \in \mathbb{Z},$$

as in (3.56). Because $\mathfrak{X}_{1,\eta} = \mathfrak{X}_{2,\eta} \in \mathfrak{X}(\eta)$,

$$n := \sum^{n_1} e_{1,j} = \sum^{n_2} e_{2,j}. \tag{3.57}$$

We then define $\xi_i' : S \to C[n]$ by

$$\pi_{[n]} \circ \xi_i' = (c_{i,1}u^{e_{i,1}}, 1 \ldots, 1, c_{i,2}u^{e_{i,2}}, 1, \ldots, 1, c_{i,n_i+1}u^{e_{i,n_1+1}}, 1, \ldots, 1),$$

where after each $c_{i,1}u^{e_{i,j}}$ we repeat $e_{i,j} - 1$ many 1's; define $\bar{\xi}_i : S \to C[n]$ by

$$\pi_{[n]} \circ \bar{\xi}_i' = (c_{i,1}u, u \ldots, u, c_{i,2}u, u, \ldots, u, c_{i,n_i+1}u, u, \ldots, u),$$

where after each $c_{i,1}u$ we repeat $e_{i,j} - 1$ many u's.

We let $\mathcal{X}_i' = \xi_i'^* X[n]$, and let $\bar{\mathcal{X}}_i = \bar{\xi}_i^* X[n]$. They have the following properties:

(1) there is a $\sigma_{i,\eta} : \eta \to G_m^n$ so that $\bar{\xi}_i = (\xi_i')^{\sigma_{i,\eta}}$;
(2) the families $\xi_i'^* X[n] = \mathcal{X}_i \in \mathfrak{X}(S)$;
(3) there is a $\sigma : S \to G_m^n$ so that $\bar{\xi}_1 = (\bar{\xi}_2)^\sigma$;
(4) the isomorphisms $\bar{\mathcal{X}}_{i,\eta} \cong \mathcal{X}_{i,\eta}$ induced by (1) and (2) extend to morphisms $h_i : \bar{\mathcal{X}}_i \to \mathcal{X}_i$;
(5) the restriction of h_i to η_0, $h_{i,\eta_0} : \bar{\mathcal{X}}_{i,\eta_0} \to \mathcal{X}_{i,\eta_0}$, is a contraction of all components $\Delta_j \subset \bar{\mathcal{X}}_{i,\eta_0}$ except $\Delta_0, \Delta_{e_{i,1}}, \Delta_{e_{i,1}+e_{i,2}} \ldots, \Delta_{e_{i,1}+\ldots+e_{i,n_i+1}}$.

We now prove $e_{1,j} = e_{2,j}$ for all j. Indeed, using isomorphism $\bar{\mathcal{X}}_{i,\eta} \cong \mathcal{X}_{i,\eta}$ stated in (4) we define $\bar{\mathcal{Z}}_{i,\eta} = \bar{\mathcal{X}}_{i,\eta} \times_{\mathcal{X}_{i,\eta}} \mathcal{Z}_{i,\eta} \subset \bar{\mathcal{X}}_{i,\eta}$. Let $\bar{\mathcal{Z}}_i \subset \bar{\mathcal{X}}_i$ be the S-flat completion of $\bar{\mathcal{Z}}_{i,\eta}$. Such completion exists since the relative Hilbert scheme $\mathrm{Hilb}^P_{\bar{\mathcal{X}}_i/S}$ is proper over S.

Since $\mathcal{Z}_{i,\eta_0} \subset \mathcal{X}_{i,\eta_0}$ is stable, in particular admissible, one checks that $\bar{\mathcal{X}}_i \times_{\mathcal{X}_i} \mathcal{Z}_i$, where $\bar{\mathcal{X}}_i \to \mathcal{X}_i$ is the h_i given in (4), is flat over S. Then by the separatedness of the relative Hilbert scheme, $\bar{\mathcal{Z}}_i = \bar{\mathcal{X}}_i \times_{\mathcal{X}_i} \mathcal{Z}_i$.

Then since $\bar{\mathcal{Z}}_{1,\eta} = h^* \mathcal{Z}_{2,\eta}$ under the isomorphism $h : \bar{\mathcal{X}}_1 \to \bar{\mathcal{X}}_2$ given in (3), we must have $\bar{\mathcal{Z}}_1 = h^* \bar{\mathcal{Z}}_2$. In particular, this implies $e_{1,1} = e_{2,1}$, $e_{1,1} + e_{1,2} = e_{2,1} + e_{2,2}$, etc. Thus combined with identity (3.57), we conclude $n_1 = n_2$ and $e_{1,j} = e_{2,j}$ for all j. This implies that the arrow $\mathcal{X}_{1,\eta} \cong \mathcal{X}_{2,\eta}$ in $\mathfrak{X}(\eta)$ extends to an arrow $\mathcal{X}_1 \cong \mathcal{X}_2$ in $\mathfrak{X}(S)$. By the separatedness of the Hilbert scheme, we get $(\mathcal{I}_{\mathcal{Z}_1}, \mathcal{X}_1) \cong (\mathcal{I}_{\mathcal{Z}_2}, \mathcal{X}_2)$ in $\mathfrak{I}^P_{\mathfrak{X}/\mathfrak{C}}(S)$. This proves that $\mathfrak{I}^P_{\mathfrak{X}/\mathfrak{C}}$ is separated.

The proof that $\mathfrak{I}^P_{\mathfrak{X}/\mathfrak{C}}$ is of finite type is similar to that of the good degeneration of moduli of stable morphisms. The technical details are more involved; we refer the details to [31]. This completes the proof of the Theorem. □

3.5. Virtual complete intersection revisited

We now show how good degenerations are virtual local complete intersections over the base stack. We use the good degeneration of ideal sheaves of subschemes worked out as an example. We let $\mathfrak{I}^P_{\mathfrak{X}/\mathfrak{C}}$ be as before, and denote the central fiber

$$\mathfrak{I}^P_{\mathfrak{X}_0/\mathfrak{C}_0} = \mathfrak{I}^P_{\mathfrak{X}/\mathfrak{C}} \times_C 0.$$

As mentioned in Subsection 2.6, the virtual local complete intersection is over the weighted stack $\mathfrak{C}^\beta$, in the notation of Subsection 2.6. Following the notation developed in Subsection 2.6, we let $\Lambda = \mathbb{Q}[m]$ and form the weighted stack $\mathfrak{C}^P$ ($\beta = P$

in $\mathfrak{C}^{\beta}$) of total weight P, where P is the Hilbert polynomial used in the previous Subsection.

For each $\mathcal{I}_Z \in \mathfrak{I}^P_{\mathfrak{X}/\mathfrak{C}}(\mathbf{k})$, an ideal sheaf of $\mathcal{O}_{X[n]_0}$, it assigns a weight w to $X[n]_0$ by assigning each irreducible $\Delta_l \subset X[n]_0$ (resp. divisor $D_l \subset X[n]_0$) the Hilbert polynomial $P_{Z\cap\Delta_l}$ (resp. $P_{Z\cap D_l}$). Since $\mathcal{I}_Z$ is admissible, this rule applied to any family $(\mathcal{I}_{\mathcal{Z}}, \mathcal{X}) \in \mathfrak{I}^P_{\mathfrak{X}/\mathfrak{C}}(S)$ defines a continuous weight assignment of the family $\mathcal{X}/S$. Under this choice of weight assignment, the morphism $\mathfrak{I}^P_{\mathfrak{X}/\mathfrak{C}} \to \mathfrak{C}$ factors through

$$\pi_P : \mathfrak{I}^P_{\mathfrak{X}/\mathfrak{C}} \longrightarrow \mathfrak{C}^P. \tag{3.58}$$

We now form the set of splittings Λ^{spl}_P of P: it is the set of triples $\delta = (\delta_\pm, \delta_0)$ in Λ so that $\delta_- + \delta_+ - \delta_0 = P$. For any $\delta \in \Lambda^{spl}_P$, we form the moduli of stable relative ideal sheaves of $\mathfrak{D}_\pm \subset \mathfrak{Y}_\pm$ over $\mathfrak{A}_\pm \cong \mathfrak{A}$:

$$\mathfrak{I}^{\delta_-,\delta_0}_{\mathfrak{Y}_-/\mathfrak{A}_-} \quad \text{and} \quad \mathfrak{I}^{\delta_+,\delta_0}_{\mathfrak{Y}_+/\mathfrak{A}_+}.$$

For any scheme S, we define $\mathfrak{I}^{\delta_-,\delta_0}_{\mathfrak{Y}_-/\mathfrak{A}_-}(S)$ be the collection of $(\mathcal{I}_{\mathcal{Z}}, \mathcal{Y}, \mathcal{D})$, where $(\mathcal{Y}, \mathcal{D}) \in \mathfrak{Y}_-(S)$ and $\mathcal{I}_{\mathcal{Z}}$ is an S-flat family of stable relative ideal sheaves on the pair $\mathcal{D} \subset \mathcal{Y}$ such that for any closed $s \in S$, $P_{\mathcal{Z}_s} = \delta_-$ and $P_{\mathcal{Z}_s\cap\mathcal{D}} = \delta_0$. (Here $P_{\mathcal{Z}_s}$ and $P_{\mathcal{Z}_s\cap\mathcal{D}}$ are the Hilbert polynomials of $\mathcal{O}_{\mathcal{Z}_s}$ and $\mathcal{O}_{\mathcal{Z}_s\cap\mathcal{D}}$ with respect to the pull back of H via $\mathcal{Y} \to Y \to X$.) We form $\mathfrak{I}^{\delta_+,\delta_0}_{\mathfrak{Y}_+/\mathfrak{A}_+}$ similarly.

Theorem 3.59. *The groupoid $\mathfrak{I}^{\delta_\pm,\delta_0}_{\mathfrak{Y}_\pm/\mathfrak{A}_\pm}$ are Deligne-Mumford stacks, proper and separated, and of finite type.*

Using $\delta \in \Lambda^{spl}_P$, we form the stack $\mathfrak{C}^{\dagger,\delta}_0$, according to the rule specified in Section two. We define

$$\mathfrak{I}^{\delta}_{\mathfrak{X}^\dagger_0/\mathfrak{C}^\dagger_0} = \mathfrak{I}^P_{\mathfrak{X}/\mathfrak{C}} \times_{\mathfrak{C}} \mathfrak{C}^{\dagger,\delta}_0.$$

It parameterizes stable ideal sheaves $\mathcal{I}_Z$ on $X[n]_0$ with a node-marking $D_k \subset X[n]_0$ so that the Hilbert polynomials of Z restricted to $\cup_{i<k}\Delta_i$, to $\cup_{i\geqslant k}\Delta_i$ and to D_k are δ_-, δ_+ and δ_0, respectively.

For each $\delta \in \Lambda^{spl}_P$, like the case of stable morphisms, we have the gluing morphism that factors through $\mathfrak{I}^{\delta}_{\mathfrak{X}^\dagger_0/\mathfrak{C}^\dagger_0}$ (it originally maps to $\mathfrak{I}^P_{\mathfrak{X}/\mathfrak{C}} \times_C 0$):

$$\Phi_\delta : \mathfrak{I}^{\delta_-,\delta_0}_{\mathfrak{Y}_-/\mathfrak{A}_-} \times_{\mathrm{Hilb}^{\delta_0}_D} \mathfrak{I}^{\delta_+,\delta_0}_{\mathfrak{Y}_+/\mathfrak{A}_+} \longrightarrow \mathfrak{I}^{\delta}_{\mathfrak{X}^\dagger_0/\mathfrak{C}^\dagger_0}. \tag{3.60}$$

Using the collection of pairs of line bundles and sections (L_δ, s_δ) for $\delta \in \Lambda^{spl}_P$ constructed in Proposition 2.60, and let π_P be as in (3.58), we have

Theorem 3.61. *Let (L_δ, s_δ) and the notation be as in Proposition 2.60. Then*

(1) $\otimes_{\delta\in\Lambda^{spl}_P}\pi_P^*L_\delta \cong \mathcal{O}_{\mathfrak{I}^P_{\mathfrak{X}/\mathfrak{C}}}$, *and* $\prod_{\delta\in\Lambda^{spl}_P}\pi_P^*s_\delta = \pi_P^*\pi^*t$;

(2) *as closed substacks,* $\mathfrak{I}^{\delta}_{\mathfrak{X}^\dagger_0/\mathfrak{A}^\dagger_0} = (\pi_P^*s_\delta = 0)$;

(3) *The gluing morphism Φ_δ is an isomorphism of Deligne-Mumford stacks.*

When $X \to C$ is a family of surfaces, each $\mathfrak{I}^{\delta}_{\mathfrak{X}_0^\dagger/\mathfrak{A}_0^\dagger}$ is smooth. Thus $\mathfrak{I}^{P}_{\mathfrak{X}/\mathfrak{C}} \to C$ is a complete intersection morphism. In case $X \to C$ is a family of threefold, using the Atiyah class constructed in [16], we obtain a perfect relative obstruction theory of $\mathfrak{I}^{P}_{\mathfrak{X}/\mathfrak{C}} \to \mathfrak{C}^{P}$. (Indeed, a semi-perfect relative obstruction theory in the sense of [5].) Restricting to over $(s_\delta = 0) \subset \mathfrak{C}^{P}$, we obtain perfect relative obstruction theory of $\mathfrak{I}^{\delta}_{\mathfrak{X}_0^\dagger/\mathfrak{C}_0^\dagger} \to (s_\delta = 0)$. If we call morphisms with perfect relative perfect obstruction theories "virtual" smooth, then $\mathfrak{I}^{P}_{\mathfrak{X}/\mathfrak{C}} \to C$ is a "virtual" local complete intersection morphism.

4. Further comments

Comparing the construction of good degenerations via the stack of expanded degenerations with the research on moduli spaces, four issues come to mind that are worthy of further research. The first is the projectivity of the coarse moduli space of the degeneration; the second is the property-III via obstruction theory; the third is on good degeneration of moduli of sheaves, and the last is to generalize this to the case of non-simple degenerations.

4.1. Projective coarse moduli

Let $X \to C$ be a simple degeneration of smooth projective schemes. Given a moduli problem, we denote by $\mathcal{M}_{\mathfrak{X}/\mathfrak{C}}$ the good degeneration constructed using the stack $\mathfrak{X} \to \mathfrak{C}$. Let $M_{\mathfrak{X}/\mathfrak{C}}$ be its coarse moduli space, which exists in the category of algebraic spaces.

Conjecture 4.1. *Suppose the specified moduli problem for any projective scheme with normal crossing singularities always has projective coarse moduli space. Then the coarse moduli space $M_{\mathfrak{X}/\mathfrak{C}}$ is projective over C.*

This conjecture holds true for the moduli of bundles over curves by the work of D.S. Nagara and C.S. Seshadri [35] and X-T. Sun [45]; and by the GIT construction of A. Schmitt [39].

For other moduli spaces, like the case of moduli of G-bundles over curves proved by G. Faltings [10], one possible approach is to construct a line bundle on the moduli stack $\mathcal{M}_{\mathfrak{X}/\mathfrak{C}}$ via universal construction and show that it descends to an ample line bundle on its coarse moduli space $M_{\mathfrak{X}/\mathfrak{C}}$. Because the coarse moduli space for the specified moduli problem for a projective scheme with at worst normal crossing singularity is projective, there should be a standard candidate for such line bundle.

4.2. Property-III via obstruction theory

We continue the notation of the previous subsection. The property-III is simple when the moduli space $\mathcal{M}_{\mathfrak{X}/\mathfrak{C}}$ is smooth: $\mathcal{M}_{\mathfrak{X}_0^\dagger/\mathfrak{A}_0^\dagger} \to \mathcal{M}_{\mathfrak{X}_0/\mathfrak{A}_0}$ is the normalization. In general, it can be stated in terms of the relative obstruction theory of $\mathcal{M}_{\mathfrak{X}/\mathfrak{C}}/\mathfrak{C}$.

We take the good degeneration of Hilbert schemes constructed as an illustration. This example for $\dim X/C = 3$ gives the degeneration of Donaldson-Thomas invariants of ideal sheaves. Let

$$S \longrightarrow \mathfrak{I}^P_{\mathfrak{X}/\mathfrak{C}}$$

be a quasi-projective and étale chart, with $\pi_S : \mathcal{X} \to S$ and $\mathcal{I}_{\mathcal{Z}} \subset \mathcal{O}_{\mathcal{X}}$ its tautological family. Viewing $\mathcal{I}_{\mathcal{Z}}$ as a flat family of rank one torsion free sheaves of fixed determinant, we obtain an arrow in the derived category $D(S)$ that defines a relative obstruction theory of $S/\mathfrak{C}$:

$$R^{\cdot}\mathrm{Hom}_{\pi_S}(\mathcal{I}_{\mathcal{Z}}, \mathcal{I}_{\mathcal{Z}})_0^{\vee} \longrightarrow \mathbb{L}^{\geqslant -1}_{S/\mathfrak{C}},$$

where the subscript 0 means the complex of traceless part. Because $\dim X/C = 3$, it is a perfect obstruction theory [16, 46]. By the recipe of virtual cycle construction [1, 30], we obtain cone-cycle

$$[C_S] \in Z_* h^1/h^0(F^{\bullet}_S), \quad F^{\bullet}_S = R^{\cdot}\mathrm{Hom}_{\pi_S}(\mathcal{I}_{\mathcal{Z}}, \mathcal{I}_{\mathcal{Z}})_0.$$

Here we follow the construction of [1]; $h^1/h^0(F^{\bullet}_S)$ is the bundle stack of the two-term perfect complex $F^{\bullet}_S$. Because this construction is canonical, the cycles $[C_S]$ for all étale charts S of $\mathfrak{I}^P_{\mathfrak{X}/\mathfrak{C}}$ patch to form a cycle

$$[C] \in Z_* h^1/h^0(F^{\bullet}), \quad F^{\bullet} = R^{\cdot}\mathrm{Hom}_{\pi}(\mathcal{I}_{\mathcal{Z}}, \mathcal{I}_{\mathcal{Z}})_0,$$

where $\mathcal{I}_{\mathcal{Z}} \subset \mathcal{O}_{\mathcal{X}}$, $\pi : \mathcal{X} \to \mathfrak{I}^P_{\mathfrak{X}/\mathfrak{C}}$, is the universal family of $\mathfrak{I}^P_{\mathfrak{X}/\mathfrak{C}}$. By virtual cycle construction, let

$$\eta^! : Z_* h^1/h^0(F^{\bullet}) \longrightarrow A_* \mathfrak{I}^P_{\mathfrak{X}/\mathfrak{C}}$$

be the Gysin map by intersecting with the zero section of $h^1/h^0(F^{\bullet})$ (cf. [22]), and for $c \in C$ let

$$c^! : A_* \mathfrak{I}^P_{\mathfrak{X}/\mathfrak{C}} \longrightarrow A_* \mathfrak{I}^P_{X_c}, \quad \mathfrak{I}^P_{X_c} = \mathfrak{I}^P_{\mathfrak{X}/\mathfrak{C}} \times_C c,$$

be the Gysin map intersecting with $c \in C$. Then for $c \neq 0 \in C$,

$$[\mathfrak{I}^P_{X_c}]^{\mathrm{vir}} = c^!(\eta^![C]) \in A_* \mathfrak{I}^P_{X_c}. \tag{4.2}$$

We now suppose that X/C is a family of Calabi-Yau threefolds. Then the class (4.2) lies in $A_0 \mathfrak{I}^P_{X_c}$. By the numerical equivalence property of Gysin map, we have

$$\deg[\mathfrak{I}^P_{X_c}]^{\mathrm{vir}} = \deg c^!(\eta^![C]) = \deg 0^!(\eta^![C]).$$

Because $0^!$ is equal to the localized first Chern class $c_1^{\mathrm{loc}}(\mathcal{O}_{\mathbf{A}^1}, t)$, applying item (1) in Theorem 3.61, we obtain

$$0^!(\eta^![C]) = c_1^{\mathrm{loc}}(\mathcal{O}_{\mathbf{A}^1}, t)([C]) = \sum_{\delta \in \Lambda_P^{\mathrm{spl}}} c_1^{\mathrm{loc}}(\pi_P^* L_\delta, \pi_P^* s_\delta)([C]).$$

Applying item (2) of the same theorem, we conclude that each individual term in the summation is equal to the virtual cycle $[\mathfrak{I}^{\delta}_{\mathfrak{X}_0^{\dagger}/\mathfrak{A}_0^{\dagger}}]^{\mathrm{vir}}$. Finally, by comparing the obstruction theories of the domain and target of Φ_δ in (3.60), we conclude that

$$c_1^{\mathrm{loc}}(\pi_P^* L_\delta, \pi_P^* s_\delta)([C]) = \triangle^!\big([ev_{-*}[\mathfrak{I}^{\delta_-,\delta_0}_{\mathfrak{Y}_-/\mathfrak{A}_\diamond}]^{\mathrm{vir}} \times ev_{+*}[\mathfrak{I}^{\delta_+,\delta_0}_{\mathfrak{Y}_+/\mathfrak{A}_\diamond}]^{\mathrm{vir}}\big),$$

where $\triangle : \mathrm{Hilb}_D^{\delta_0} \to \mathrm{Hilb}_D^{\delta_0} \times \mathrm{Hilb}_D^{\delta_0}$ is the diagonal morphism.

In conclusion, the decomposition Theorem 3.61 proves the degeneration formula of Donaldson-Thomas invariants

$$\deg [\mathfrak{I}^P_{X_c}]^{\mathrm{vir}} = \sum_{\delta \in \Lambda_P^{\mathrm{spl}}} \deg \triangle^! \big([\mathrm{ev}_{-*}[\mathfrak{I}^{\delta_-,\delta_0}_{\mathfrak{Y}_-/\mathfrak{A}_-}]^{\mathrm{vir}} \times \mathrm{ev}_{+*}[\mathfrak{I}^{\delta_+,\delta_0}_{\mathfrak{Y}_+/\mathfrak{A}_+}]^{\mathrm{vir}} \big).$$

4.3. Good degeneration of moduli of sheaves

The difficulty in this case is the interplay between degeneration and the stability condition. Examining the proof for good degenerations of Hilbert schemes, it is reasonable to expect that using the stack $\mathfrak{X}/\mathfrak{C}$ and the moduli of (possibly Simpson) stable sheaves that are normal to the singular loci of the expanded degenerations will provide a good degeneration of the moduli of sheaves. The difficult part is the decomposition. Seeing the example of degenerations of bundles on curves, one ponders the shape of a reasonable theory for the case of, say, surfaces, or even threefolds. Due to the surge of interest in studying moduli of sheaves on Calabi-Yau threefolds, progress along this direction will be very useful.

Another related question is whether one can construct good degenerations of moduli of derived objects. As such moduli spaces are beginning to play an important role in the study of the geometry of Calabi-Yau threefolds, the construction of their good degenerations will undoubtedly provide new tools to their study.

4.4. Non-simple degenerations

The first case of non-simple degeneration is assuming that the central fiber of the family X/C has normal crossing singularity. In this case, the first difficulty is whether a workable expanded degenerations still exist. In case of the moduli of stable morphisms, the recent work of Q-L. Chen [6], and of Gross and Siebert [14] on using log-morphisms to construct moduli of relative stable morphisms give hope that a "good" degeneration in the category of log-morphisms is within reach. In general, we are hopeful that a breakthrough will occur in this direction in the near future. Such progress will provide new tools to study moduli spaces via degenerations.

References

[1] K. Behrend and B. Fantechi. Intrinsic normal cone. *Invent. Math.*, **128** (1), 45–88, 1997. ← 347

[2] U. N. Bhosle. Generalized parabolic bundles and applications. II. *Proc. Indian Acad. Sci. Math. Sci.*, **106**(4), 403–420, 1996. ← 337

[3] J. Bryan and R. Pandharipande. The local Gromov-Witten theory of curves. *J. Amer. Math. Soc.*, **21**(1), 101–136, 2008. ← 305

[4] L. Caporaso. A compactification of the universal Picard variety over the moduli space of stable curves. *J. Amer. Math. Soc.*, **7**(3), 589–660, 1994. ← 304, 331

[5] H-L. Chang and J. Li. Semi-Perfect Obstruction theory and DT Invariants of Derived Objects. Preprint. (arXiv:1105.3261) ← 346

[6] Q-L. Chen. Logarithmic stable maps to Deligne-Faltings pairs I. Preprint. (arXiv:1008.3090) ← 348

[7] G. Daskalopoulos and R. Wentworth. Local degeneration of the moduli space of vector bundles and factorization of rank two theta functions. I. *Math. Ann.*, **297**, 417–466, 1993. ← 305

[8] G. Daskalopoulos and R. Wentworth. Factorization of rank two theta functions. II: Proof of the Verlinde formula. *Math. Ann.*, **304**, 21–51, 1996. ← 305

[9] S. Donaldson. Floer homology groups in Yang-Mills theory. *Cambridge tracts in mathematics*, **147**, Cambridge University Press, Cambridge, 2002. ← 302, 305

[10] G. Faltings. Moduli-stacks for bundles on semistable curves. *Math. Ann.*, **304**, 489–515, 1996. ← 305, 346

[11] B. Fantechi and R. Pandharipande.Stable maps and branch divisors. *Compositio Math.*, **130**(3), 345–364, 2002. ← 303

[12] D. Gieseker. A degeneration of the moduli space of stable bundles. *J. Differential Geom.*, **19**(1), 173–206, 1984. ← 301, 331, 339

[13] D. Gieseker and I. Morrison. Hilbert stability of rank-two bundles on curves. *J. Differential Geom.*, **19**(1), 1–29, 1984. ← 301, 304, 331

[14] M. Gross and B. Siebert. Logarithmic Gromov-Witten invariants. Preprint. (arXiv:1102.4322) ← 348

[15] J. Harris and D. Mumford. The Kodaira dimension of the moduli space of curves of genus $\geqslant 23$. *Invent. Math.*, **90**(2), 359–387, 1987. ← 302, 304

[16] D. Huybrechts and R. P. Thomas. Deformation-obstruction theory for complexes via Atiyah and Kodaira–Spencer classes. *Math. Ann.*, **346**, 545–569, 2010. ← 346, 347

[17] E. Ionel. Topological recursive relations in $H^{2g}(\mathcal{M}_{g,n})$. *Invent. Math.*, **148**(3), 627–658, 2002. ← 305

[18] E. Ionel and T. Parker. The symplectic sum formula for Gromov-Witten invariants. *Ann. of Math.* (2), **159**(3), 935–1025, 2004. ← 305

[19] I. Kausz. A modular compactification of the general linear group. *Doc. Math.*, **5**, 553–594, 2000. ← 305

[20] I. Kausz. A Gieseker type degeneration of moduli stacks of vector bundles on curves. *Trans. Amer. Math. Soc.*, **357**(12), 489–4955, 2005. ← 305

[21] Y-H. Kiem and J. Li. Vanishing of the top Chern classes of the moduli of vector bundles. *J. Differential Geom.*, **76**(1), 45–115, 2007. ← 305, 333, 334, 335, 337, 338, 339, 340

[22] A. Kresch. Cycle groups for Artin stacks. *Invent. Math.*, **138**(3), 495–536, 1999. ← 347

[23] P. B. Kronheimer and T.S. Morwka. Embedded surfaces and the structure of Donaldson's polynomial invariants. *J. Diff. Geom.*, **41**(3), 573–734, 1995. ← 302, 305

[24] M. Levine and R. Pandharipande. Algebraic cobordism revisited. *Invent. Math.*, **176**(1), 63–130, 2009. ← 305

[25] A-M. Li and Y. Ruan. Symplectic surgery and Gromov-Witten invariants of Calabi-Yau 3-folds. *Invent. Math.*, **145**(1), 151–218, 2001. ← 305

[26] J. Li. Stable morphisms to singular schemes and relative stable morphisms. *J. Differential Geom.*, **57**(3), 509–578, 2001. ← 304, 305, 307, 312, 322, 323, 324, 331

[27] J. Li. A degeneration formula of GW-invariants. *J. Differential Geom.*, **60**(2), 199–293, 2002. ← 305, 321, 330

[28] J. Li. Moduli spaces associated to a singular variety and the moduli of bundles over universal curves, in: Vector bundles and representation theory (Columbia, MO, 2002), 57–74. *Contemp. Math.*, **322**, Amer. Math. Soc., Providence, RI, 2003. ← 304, 331

[29] J. Li, C-C. Liu, K-F. Liu, and J. Zhou. A mathematical theory of the topological vertex. *Geom. Topol.*, **13**(1), 527–621, 2009. ← 305

[30] J. Li and G. Tian. Virtual moduli cycles and Gromov-Witten invariants of algebraic varieties. *J. Amer. Math. Soc.*, **11**(1), 119–174, 1998. ← 347

[31] J. Li and B. Wu. Good degeneration of Quot schems and coherent systems. In preparation. ← 305, 340, 344

[32] D. Maulik and R. Pandharipande. A topological view of Gromov-Witten theory. *Topology*, **45**(5), 887–918, 2006. ← 305

[33] D. Maulik, R. Pandharipande, and R. Rhomas. Curves on K3 surfaces and modular forms. Preprint. (arXiv:1001.2719). ← 305

[34] D. S. Nagaraj and C. S. Seshadri. Degenerations of the moduli spaces of vector bundles on curves. I. *Proc. Indian Acad. Sci. Math. Sci.*, **107**(2), 101-137, 1997. ← 305

[35] D. S. Nagaraj and C. S. Seshadri. Degenerations of the moduli spaces of vector bundles on curves. II. Generalized Gieseker moduli spaces. *Proc. Indian Acad. Sci. Math. Sci.*, **109**(2), 165–201, 1999. ← 304, 305, 346

[36] M. S. Narasimhan and T. R. Ramadas. Factorisation of generalised theta function I. *Invent. Math.*, **114**(3), 565–623, 1993. ← 305

[37] A. Okounkov and R. Pandharipande. Virasoro constraints for target curves. *Invent. Math.*, **163**(1), 47–108, 2006. ← 305, 322

[38] R. Pandharipande. A compactification over Mg of the universal moduli space of slopesemistable vector bundles. *J. Amer. Math. Soc.*, **9**(2), 425–471, 1996. ← 331

[39] A. Schmitt. The Hilbert Compactification of the Universal Moduli Space of Semistable Vector Bundles over Smooth Curves. *J. Differential Geom.*, **66**(2), 169–209, 2004. ← 305, 331, 335, 346

[40] C. Simpson. Moduli of representations of the fundamental group of a smooth projective variety. II. *Inst. Hautes Études Sci. Publ. Math.*, **80**, 5–79, 1994. ← 331

[41] X-T. Sun. Degeneration of moduli spaces and generalized theta functions. *J. Algebraic Geom.*, **9**(3), 459–527, 2000. ← 305

[42] X-T. Sun. Degeneration of SL(n)-bundles on a reducible curve, in: Algebraic geometry in East Asia (Kyoto, 2001), 229–243, World Sci. Publ., River Edge, NJ, 2002. ← 305

[43] X-T. Sun. Factorization of generalized theta functions in the reducible case. *Ark. Mat.*, **41**(1), 165–202, 2003. ← 305

[44] X-T. Sun. Moduli spaces of SL(r)-bundles on singular irreducible curves. *Asian J. Math.*, **7**(4), 609–625, 2003. ← 305

[45] X-T. Sun. Remarks on Gieseker's degeneration and its normalization. *AMS/IP Studies in Advanced Mathematics*, **42**, 177–191, 2008. ← 305, 346

[46] R. Thomas. A holomorphic Casson invariant for Calabi-Yau 3-folds and bundles on K3 fibrations. *J. Diff. Geom.*, **54**, 367–438, 2000. ← 347

Department of Mathematics, Stanford University, Stanford, USA
E-mail address: `jli@math.stanford.edu`

Localization in Gromov-Witten theory and Orbifold Gromov-Witten theory

Chiu-Chu Melissa Liu

Abstract. In this expository article, we explain how to use localization to compute Gromov-Witten invariants of smooth toric varieties and orbifold Gromov-Witten invariants of smooth toric Deligne-Mumford stacks.

Contents

2000 *Mathematics Subject Classification.* Primary 14N35; Secondary 14H10.

Key words and phrases. Gromov-Witten invariants, toric varieties, virtual localization, orbifold Gromov-Witten invariants, toric DM stacks.

1. Introduction

Let X be a smooth projective variety over $\mathbb{C}$. Naively, Gromov-Witten invariants count parametrized algebraic curves of X; more precisely, they are intersection numbers on moduli spaces of stable maps to X. Let $\overline{\mathcal{M}}_{g,n}(X,\beta)$ be the Kontsevich's moduli space of n-pointed, genus g, degree β stable maps $f:(C,x_1,\dots,x_n)\to X$, where $\beta=f_*[C]\in H_2(X;\mathbb{Z})$. It is a proper Deligne-Mumford stack with a perfect obstruction theory of virtual dimension

$$d^{\mathrm{vir}}=\int_\beta c_1(T_X)+(\dim X-3)(1-g)+n, \tag{1.1}$$

where $\int$ stands for the pairing between the (rational) homology and cohomology. Given $i \in \{1, \ldots, n\}$, there is an evaluation map $ev_i : \overline{\mathcal{M}}_{g,n}(X, \beta) \to X$ which sends a moduli point $[f : (C, x_1, \ldots, x_n) \to X] \in \overline{\mathcal{M}}_{g,n}(X, \beta)$ to $f(x_i) \in X$, and there is a line bundle $\mathbb{L}_i$ over $\overline{\mathcal{M}}_{g,n}(X, \beta)$ whose fiber at the moduli point $[f : (C, x_1, \ldots, x_n) \to X]$ is the cotangent line $T^*_{x_i}C$ at the i-th marked point x_i. Gromov-Witten invariants of X are defined to be

$$(1.2) \qquad \langle \tau_{a_1}(\gamma_1), \ldots, \tau_{a_n}(\gamma_n)\rangle^X_{g,\beta} := \int_{[\overline{\mathcal{M}}_{g,n}(X,\beta)]^{\mathrm{vir}}} \prod_{i=1}^n (ev_i^*\gamma_i \psi_i^{a_i}) \in \mathbb{Q}$$

where $\gamma_1, \ldots, \gamma_n \in H^*(X; \mathbb{Q})$, $\psi_i = c_1(\mathbb{L}_i) \in H^2(\overline{\mathcal{M}}_{g,n}(X, \beta); \mathbb{Q})$, and

$$[\overline{\mathcal{M}}_{g,n}(X, \beta)]^{\mathrm{vir}} \in H_{2d^{\mathrm{vir}}}(X; \mathbb{Q})$$

is the virtual fundamental class (Li-Tian [34], Behrend-Fantechi [5]).

When X is a toric variety, the torus action on X induces torus actions on moduli spaces of stable maps to X. By virtual localization (Graber-Pandharipande [19], see also Behrend [4] and Kresch [33]),

$$(1.3) \qquad \int_{[\overline{\mathcal{M}}_{g,n}(X,\beta)]^{\mathrm{vir}}} \prod_{i=1}^n (ev_i^*\gamma_i \psi_i^{a_i}) = \sum_F \int_{[F]^{\mathrm{vir}}} \frac{i_F^* \prod_{i=1}^n (ev_i^*\gamma_i^T (\psi_i^T)^{a_i})}{e_T(N_F^{\mathrm{vir}})},$$

where

- T is the torus acting on X and on $\overline{\mathcal{M}}_{g,n}(X, \beta)$,
- the sum on the right hand side of (1.3) is over connected components of the set of T-fixed points in $\overline{\mathcal{M}}_{g,n}(X, \beta)$,
- $\gamma_i^T \in H_T^*(X; \mathbb{Q})$ is a T-equivariant lift of γ_i,
- $\psi_i^T \in H_T^2(\overline{\mathcal{M}}_{g,n}(X, \beta); \mathbb{Q})$ is a T-equivariant lift of ψ_i.
- $i_F^* : H_T^*(\overline{\mathcal{M}}_{g,n}(X, \beta); \mathbb{Q}) \to H_T^*(F; \mathbb{Q})$ is induced by the inclusion map $i_F : F \to \overline{\mathcal{M}}_{g,n}(X, \beta)$,
- $e_T(N_F^{\mathrm{vir}})$ is the T-equivariant Euler class of the virtual normal bundle N_F^{vir} of F in $\overline{\mathcal{M}}_{g,n}(X, \beta)$.

Up to a finite morphism, each connected component F is a product of moduli spaces of stable curves (with marked points). F is a proper smooth DM stack, and $[F]^{\mathrm{vir}}$ is the usual fundamental class $[F] \in H_*(F; \mathbb{Q})$. The right hand side of (1.3) can be expressed in terms of Hodge integrals, which are intersection numbers on moduli spaces of stable curves. The terminology "virtual localization" was introduced in [19] and the term "Hodge integral" was introduced in [14] precisely to study the virtual localization formula in [19]. Algorithms of computing Hodge integrals are known; a brief review of the relevant results will be given in Section 3.1. This gives an algorithm of evaluating Gromov-Witten invariants for any smooth projective toric varieties, in all genera and all degrees. Indeed, this algorithm was first described by Kontsevich for genus zero Gromov-Witten invariants of $\mathbb{P}^r$ in 1994 [32], before the construction of virtual fundamental class and the proof of virtual localization. The moduli spaces $\overline{\mathcal{M}}_{0,n}(\mathbb{P}^r, d)$ of genus zero stable maps to $\mathbb{P}^r$ are proper *smooth*

DM stacks, so there exists a fundamental class $[\overline{\mathcal{M}}_{0,n}(\mathbb{P}^r, d)] \in H_*(\overline{\mathcal{M}}_{0,n}(\mathbb{P}^r, d); \mathbb{Q})$, and one may apply the classical Atiyah-Bott localization formula [3] in this case. H. Spielberg derived a formula for genus 0 Gromov-Witten invariants of smooth toric varieties in his thesis [41, 42].

For a noncompact smooth toric variety X, Gromov-Witten invariants are usually not defined, but one may use the right hand side of (1.3) to define T-equivariant Gromov-Witten invariants of X. They are elements in the fractional field of $H^*(BT; \mathbb{Q})$, the rational equivariant cohomology ring of the classifying space BT of T.

Chen-Ruan developed Gromov-Witten theory for symplectic orbifolds [9]. The algebraic counterpart, the orbifold Gromov-Witten theory for smooth Deligne-Mumford (DM) stacks, was developed by Abramovich-Graber-Vistoli [1, 2]. Orbifold Gromov-Witten invariants of a smooth DM stack $\mathcal{X}$ are defined as intersection numbers on moduli spaces of twisted stable maps to $\mathcal{X}$. When $\mathcal{X}$ is a smooth toric DM stack, the torus action on $\mathcal{X}$ induces torus actions on moduli spaces of twisted stable maps to $\mathcal{X}$. By virtual localization, orbifold Gromov-Witten invariants of a smooth toric DM stack can be expressed in terms of Hurwitz-Hodge integrals, which are intersections numbers of moduli spaces of twisted stable maps to $\mathcal{B}G = [pt/G]$, the classifying space of a finite group G. Algorithms for computing Hurwitz-Hodge integrals are known; a brief review of the relevant results will be given in Section 7.5.

The goal of this article is to provide details of the localization calculations described above. In Section 2, we review equivariant intersection theory and localization. In Section 3, we give a brief review of Gromov-Witten theory. In Section 4, we give a brief review of smooth toric varieties, and introduce toric graphs. In Section 5, we use virtual localization to derive a formula for Gromov-Witten invariants of smooth toric varieties in terms of Hodge integrals. Most of Section 5 is straightforward generalization of the $\mathbb{P}^r$ case discussed in [32] (genus 0) and [19, Section 4], [4, Section 4] (higher genus); see also [20, Chapter 27]. Smooth DM stacks, orbifold Gromov-Witten theory, and smooth toric DM stacks are reviewed in Section 6, Section 7, and Section 8, respectively. In Section 9, we use virtual localization to derive a formula of orbifold Gromov-Witten invariants of smooth toric DM stacks in terms of abelian Hurwitz-Hodge integrals. Our main reference of Section 9 is P. Johnson's thesis [24], which contains detailed localization computations for 1-dimensional toric DM stacks. D. Ross's recent preprint [40] contains localization computations for 3-dimensional Calabi-Yau toric DM stacks.

Acknowledgments

I wish to thank Dan Abramovich, Lev Borisov, Dan Edidin, Ezra Getzler, Tom Graber, Paul Johnson, Zhengyu Zong for helpful communications, and the referee for his or her comments. Special thanks go to Tom Graber and Paul Johnson for their help with orbifold Gromov-Witten theory and virtual localization.

2. Equivariant intersection theory and localization

In this section, we review equivariant intersection theory and localization. In Section 2.1 – Section 2.5 we discuss equivariant cohomology of topological spaces and localization on smooth manifolds. In Section 2.7 we give a brief summary of equivariant intersection theory on schemes and Deligne-Mumford stacks in terms of equivariant Chow groups and equivariant operational Chow cohomology groups [12]. We state the virtual localization formula in Section 2.8.

In this paper, we consider cohomology groups, Chow groups, and operational Chow groups with rational coefficients. We write $H^*(\bullet)$, $A_*(\bullet)$, $A^*(\bullet)$ instead of $H^*(\bullet;\mathbb{Q})$, $A_*(\bullet;\mathbb{Q})$, $A^*(\bullet;\mathbb{Q})$.

2.1. Equivariant cohomology

Let G be a Lie group, and let EG be a contractible topological space on which G acts freely on the *right*. The quotient $BG = EG/G$ is a classifying space of principal G-bundles, and the natural projection $EG \to BG$ is a universal principal G-bundle; EG and BG are defined up to homotopy equivalences.

A G-space is a topological space together with a continuous *left* G-action. Given a G-space X, define a *right* G-action on $EG \times X$ by

$$(p,x)\cdot g = (p\cdot g, g^{-1}\cdot x). \tag{2.1}$$

The *homotopy orbit space* X_G is defined to be the quotient of $EG\times X$ by the *free* G-action (2.1). The G-equivariant cohomology of X is defined to be the ordinary cohomology of the homotopy orbit space X_G:

$$H^*_G(X) := H^*(X_G).$$

In particular, the G-equivariant cohomology of a point pt is the ordinary cohomology of the classifying space BG:

$$H^*_G(\mathrm{pt}) = H^*(BG)$$

Example 2.2 (Classifying space of $\mathbb{C}^*$-bundles). *The Lie group $\mathbb{C}^*$ acts on $\mathbb{C}^\infty - \{0\}$ on the right by*

$$v\cdot\lambda = \lambda v,\quad \lambda\in\mathbb{C}^*,\quad v\in\mathbb{C}^\infty-\{0\}.$$

$\mathbb{C}^\infty - \{0\}$ is contractible, and the $\mathbb{C}^*$-action on $\mathbb{C}^\infty - \{0\}$ is free. Therefore (up to homotopy equivalence)

$$E\mathbb{C}^* = \mathbb{C}^\infty - \{0\},\quad B\mathbb{C}^* = (\mathbb{C}^\infty - \{0\})/\mathbb{C}^* = \mathbb{P}^\infty,$$

where $\mathbb{P}^\infty$ is the infinite dimensional complex projective space. Let $\mathcal{O}_{\mathbb{P}^\infty}(-1)$ be the tautological line bundle over $\mathbb{P}^\infty$, and let u be the first Chern class of $\mathcal{O}_{\mathbb{P}^\infty}(-1)$:

$$u := c_1(\mathcal{O}_{\mathbb{P}^\infty}(-1)) \in H^2(\mathbb{P}^\infty) = H^2_{\mathbb{C}^*}(\mathrm{pt}).$$

Then $H^*_{\mathbb{C}^*}(\mathrm{pt}) = H^*(B\mathbb{C}^*) = \mathbb{Q}[u]$.

In this paper we will consider action by an algebraic torus $T = (\mathbb{C}^*)^l$. Let $\pi_i : BT = (B\mathbb{C}^*)^l \to B\mathbb{C}^*$ be the projection to the i-th factor, and let $u_i = \pi_i^* u \in H^2(BT)$. Then

$$H_T^*(pt) = H^*(BT) = \mathbb{Q}[u_1, \ldots, u_l].$$

Example 2.3. *Let $\widetilde{T} = (\mathbb{C}^*)^{r+1}$ act on the r-dimensional complex projective space $\mathbb{P}^r$ by*

$$(\tilde{t}_0, \ldots, \tilde{t}_r) \cdot [z_0, \ldots, z_r] = [\tilde{t}_0 z_0, \ldots, \tilde{t}_r z_r], \quad (\tilde{t}_0, \ldots, \tilde{t}_r) \in \widetilde{T}, \quad [z_0, \ldots, z_r] \in \mathbb{P}^r.$$

For $i = 0, \ldots, r$, let $p_i : B\widetilde{T} = (B\mathbb{C}^*)^{r+1} \to B\mathbb{C}^*$ be the projection to the i-th factor. Then $\mathbb{P}^r_{\widetilde{T}}$ can be identified with the total space of the $\mathbb{P}^r$-bundle

$$\mathbb{P}\big(\oplus_{i=0}^r p_i^*(\mathcal{O}_{\mathbb{P}^\infty}(-1))\big) \to BT.$$

In general, let $E \to X$ be a rank $(r+1)$ complex vector bundle over a topological space X, and let $\pi : \mathbb{P}(E) \to X$ be the projectivization of E, which is an $\mathbb{P}^r$-bundle over X. Then the cohomology $H^*(\mathbb{P}(E))$ of the total space $\mathbb{P}(E)$ is an $H^*(X)$-algebra generated by H with a single relation

$$H^{r+1} + c_1(E)H^r + \cdots + c_{r+1}(E) = 0,$$

where $c_i(E)$ is the i-th Chern class of E, and H is of degree 2.

In our case $E = \oplus_{i=0}^r p_i^* \mathcal{O}_{\mathbb{P}^\infty}(-1)$, so the total Chern class of E is given by

$$c(E) = \prod_{i=0}^r (1 + \tilde{u}_i), \quad \tilde{u}_i = p_i^*(c_1(\mathcal{O}_{\mathbb{P}^\infty}(-1))).$$

We have

$$H_{\widetilde{T}}^*(\mathbb{P}^r) = H^*(\mathbb{P}^r_{\widetilde{T}}) \cong \mathbb{Q}[H, \tilde{u}_0, \ldots, \tilde{u}_r] / \langle \prod_{i=0}^r (H + \tilde{u}_i) \rangle,$$

where $\mathbb{Q}[H, \tilde{u}_0, \ldots, \tilde{u}_r]$ is the ring of polynomials in $H, \tilde{u}_0, \ldots, \tilde{u}_r$ with coefficients in $\mathbb{Q}$, and $\langle \prod_{i=0}^r (H + \tilde{u}_i) \rangle$ is the principal ideal generated by $\prod_{i=0}^r (H + \tilde{u}_i)$.

The trivial fiber bundle $EG \times X \to EG$ with base EG, fiber X descends to a fiber bundle $X_G \to BG$ with base BG, fiber X. The inclusion of a fiber, $i_X : X \to X_G$, induces a ring homomorphism $i_X^* : H_G^*(X) = H^*(X_G) \to H^*(X)$.

2.2. Equivariant vector bundles and equivariant characteristic classes

Let G be a Lie group. A continuous map $f : X \to Y$ between G-spaces is called *G-equivariant* if $f(g \cdot x) = g \cdot f(x)$ for all $g \in G$ and $x \in X$.

Let $p : V \to X$ be a (real or complex) vector bundle over a G-space X. We say $p : V \to X$ is a G-equivariant vector bundle over X if the following properties hold.

- V is a G-space.
- p is G-equivariant.

- For every $g \in G$, define $\tilde{\phi}_g : V \to V$ by $v \mapsto g \cdot v$, and $\phi_g : X \to X$ by $x \mapsto g \cdot x$. Then $\tilde{\phi}_g$ is a vector bundle map covering ϕ_g:

$$\begin{array}{ccc} V & \xrightarrow{\tilde{\phi}_g} & V \\ p\downarrow & & p\downarrow \\ X & \xrightarrow{\phi_g} & X \end{array}$$

Example 2.4. *When X is a point, a complex vector bundle over X is a complex vector space V, and a G-equivariant vector bundle over X is a representation* $\rho : G \to GL(V)$.

Let $\pi : V \to X$ be a G-equivariant vector bundle over a G-space X. Then V_G is a vector bundle over X_G. Let c be a characteristic class of vector bundles (for example, Chern classes c_k and Chern characters ch_k for complex vector bundles, or the Euler class e for oriented real vector bundles). We define the corresponding G-equivariant class c^G by

$$c^G(V) := c(V_G) \in H^*(X_G) = H^*_G(X).$$

If V is a G-equivariant complex vector bundle over X then we call $c_k^G(V) \in H_G^{2k}(X)$ (resp. $ch_k^G(V) \in H_G^{2k}(X)$) the G-equivariant k-th Chern class (resp. the G-equivariant k-th Chern character) of V. If V is a G-equivariant oriented real bundle of rank r over X then we call $e^G(V) \in H^r_G(X)$ the G-equivariant Euler class of V.

Example 2.5. *For any* $a \in \mathbb{Z}$, *let* $\mathbb{C}_a$ *be the 1-dimensional representation of* $\mathbb{C}^*$ *with character* $t \mapsto t^a$. *Then* $\mathbb{C}_a$ *can be viewed as a* $\mathbb{C}^*$*-equivariant vector bundle over a point. We have*

$$(\mathbb{C}_a)_{\mathbb{C}^*} = \{(u, v) \in (\mathbb{C}^\infty - \{0\}) \times \mathbb{C}\}/(u, v) \sim (tu, t^{-a}v) \cong \mathcal{O}_{\mathbb{P}^\infty}(-a)$$

$$c_1^{\mathbb{C}^*}(\mathbb{C}_a) = au \in H^2_{\mathbb{C}^*}(pt) = \mathbb{Z}u.$$

2.3. Push-forward

Let X, Y be compact oriented manifolds of dimension r, s, respectively, and let $[X] \in H_r(X)$, $[Y] \in H_s(Y)$ be the fundamental classes. A continuous map $f : X \to Y$ induces a group homomorphism

$$f_* : H^k(X) \to H^{k+s-r}(Y)$$

characterized by

$$(f_*\alpha) \cap [Y] = f_*(\alpha \cap [X]) \in H_{r-k}(Y).$$

In particular, if $s \geqslant r$ then $f_*1 \in H^{s-r}(Y)$ is the Poincaré dual of $f_*[X] \in H_r(Y)$. The push-forward map $f_* : H^*(X) \to H^*(Y)$ is a homomorphism of $H^*(Y)$-modules:

$$f_*(\alpha \cup f^*\beta) = (f_*\alpha) \cup \beta, \quad \alpha \in H^*(X),\ \beta \in H^*(Y).$$

If $g : Y \to Z$ is a continuous map and Z is a compact oriented manifold then $g_* \circ f_* = (g \circ f)_* : H^*(X) \to H^*(Z)$.

Example 2.6. (1) *Let* V *be a rank* q *oriented real vector bundle over* Y*, and let* X *be the transversal intersection of a section* $s : Y \to V$ *and the zero section. Let* $f : X \to Y$ *be the inclusion. Then* $f_* 1 = e(V) \in H^q(Y)$.

(2) *Let* $p_X : X \to pt$ *be the constant map to a point. Then* $p_{X*} : H^*(X) \to H^*(pt) \cong \mathbb{Q}$ *can be identified with* $\int_X$.

Suppose that a Lie group G acts on X and on Y, and let $[X]^G \in H_r^G(X)$, $[Y]^G \in H_s^G(X)$ be G-equivariant fundamental class, where $H_*^G(X)$ is the G-equivariant homology groups with rational coefficients, constructed from G-invariant cycles in X. A G-equivariant map $f : X \to Y$ induces a group homomorphism $f_* : H_k^G(X) \to H_k^G(Y)$. It also induces

$$f_* : H_G^k(X) \to H_G^{k+s-r}(Y)$$

characterized by

$$f_*(\alpha^G) \cap [Y]^G = f_*(\alpha^G \cap [X]^G) \in H_{k-r}^G(X).$$

In particular, if $s \geqslant r$ then $f_* 1 \in H_G^{s-r}(Y)$ is the equivariant Poincaré dual of $f_*[X]^G \in H_r^G(Y)$.

We have the following commutative diagram:

$$\begin{array}{ccc} H_G^k(X) & \xrightarrow{f_*} & H_G^{k+s-r}(Y) \\ i_X^* \downarrow & & \downarrow i_Y^* \\ H^k(X) & \xrightarrow{f_*} & H^{k+s-r}(Y) \end{array}$$

where i_X^*, i_Y^* are defined as in the last paragraph of Section 2.1. If $s \geqslant r$ then $f_* 1 \in H^{s-r}(Y)$ is the equivariant Poincaré dual of $f_*[X]^G \in H_r^G(Y)$.

The push-forward map $f_* : H_G^*(X) \to H_G^*(Y)$ is a homomorphism of $H_G^*(Y)$-modules:

$$f_*(\alpha^G \cup f^*\beta^G) = (f_*\alpha^G) \cup \beta^G, \quad \alpha^G \in H_G^*(X),\ \beta^G \in H_G^*(Y).$$

If G acts on another compact oriented manifold Z, and $g : Y \to Z$ is a G-equivariant map, then $g_* \circ f_* = (g \circ f)_* : H_k^G(X) \to H_k^G(Z)$, $H_G^k(X) \to H_G^{k+\dim Z - \dim X}(Z)$.

Example 2.7. (1) *Let* V *be a G-equivariant rank* q *oriented real vector bundle over* Y*, and let* X *be the transversal intersection of a G-equivariant section* $s : Y \to V$ *and the zero section. Let* $f : X \to Y$ *be the inclusion. Then* $f_* 1 = e^G(V) \in H_G^q(Y)$.

(2) *Let* $p_X : X \to pt$ *be the constant map to a point. Then the pullback map* $(p_X)_* : H_G^*(X) \to H_G^*(pt) = H^*(BG)$ *is denoted by* $\int_{[X]^G}$.

2.4. Localization

Suppose that $T = (\mathbb{C}^*)^l$ acts on a compact oriented manifold M, and suppose that each connected component of the T fixed points set $M^T \subset M$ is a compact orientable submanifold of M. Let $F_1, \ldots, F_N$ be the connected components of M^T. Then $(F_j)_T = F_j \times BT$, so

$$H_T^*(F_j) = H^*(F_j \times BT) \cong H^*(F_j) \otimes_{\mathbb{Q}} H^*(BT) = H^*(F_j) \otimes_{\mathbb{Q}} R_T$$

where $R_T = H^*(BT) = \mathbb{Q}[u_1, \ldots, u_l]$. Let $Q_T = \mathbb{Q}(u_1, \ldots, u_l)$ be the fractional field of R_T. The equivariant Euler class $e^T(N_j)$ of the normal bundle N_j of F_j in M is invertible in $H^*(F_j) \otimes_{\mathbb{Q}} Q_T$. The inclusion $i_j : F_j \to M$ induces a homomorphism $(i_j)_* : H_T^*(F_j) \to H_T^*(M)$ of R_T-modules and can be extended to

$$(i_j)_* : H_T^*(F_j) \otimes_{R_T} Q_T \to H_T^*(M) \otimes_{R_T} Q_T.$$

Theorem 2.8 (Atiyah-Bott localization formula [3]).
If $\alpha^T \in H_T^*(X)$ *then*

$$\alpha^T = \sum_{j=1}^{N} (i_j)_* \frac{i_j^* \alpha^T}{e^T(N_j)}. \tag{2.9}$$

Corollary 2.10 (integration formula [3, Equation (3.8)]). *If* $\alpha \in H_T^*(X)$ *then*

$$\int_{[X]^T} \alpha^T = \sum_{j=1}^{N} \int_{[F_j]^T} \frac{i_j^* \alpha^T}{e^T(N_j)}. \tag{2.11}$$

Each term of the right hand side is a rational function in $u_1, \ldots, u_l$, while the left hand side is a *polynomial* in $u_1, \ldots, u_l$. If $\alpha \in H_T^k(X)$ then $\int_{[X]^T} \alpha^T \in H_T^{k-\dim X}(pt)$. In particular,

- If $k = \dim X$ then $\int_{[X]^T} \alpha^T \in \mathbb{Q}$.
- If $k < \dim X$ or if $k - \dim X$ is odd, then $\int_{[X]^T} \alpha^T = 0$.

$H_T^{2m}(pt)$ is the space of polynomials in $u_1, \ldots, u_l$ with $\mathbb{Q}$ coefficients, homogeneous of degree m.

We also have $i_{j_1}^* i_{j_2 *} = 0$ if $j_1 \neq j_2$. Therefore the inclusion $i : M^T = \cup_{j=1}^N F_j \to M$ induces an isomorphism

$$i_* : H_T^*(M^T) \otimes_{R_T} Q_T = \bigoplus_{j=1}^{N} H^*(F_j) \otimes_{\mathbb{Q}} Q_T \to H_T^*(M) \otimes_{R_T} Q_T.$$

Example 2.12. *Let* $\widetilde{T} = (\mathbb{C}^*)^{r+1}$ *act on* $\mathbb{P}^r$ *as in Example 2.3. Then the fixed points set consists of* $(r+1)$ *isolated points.*

$$(\mathbb{P}^r)^{\widetilde{T}} = \{p_0 = [1,0,\ldots,0], \quad p_1 = [0,1,0,\ldots,0], \quad p_r = [0,\ldots,0,1]\}.$$

Let D_j *be the* $\widetilde{T}$*-invariant divisor defined by* $x_j = 0$. *Then* x_j *is a* $\widetilde{T}$*-equivariant section of the* $\widetilde{T}$*-equivariant line bundle* $\mathcal{O}_{\mathbb{P}^r}(D_j)$. $\{x_k \mid k \neq j\}$ *defines a* T*-equivariant section* s_j *of the rank* r *vector bundle* $\oplus_{k\neq j}\mathcal{O}_{\mathbb{P}^r}(D_j)$. *The section* s_j *intersects the zero section transversally at a single point* p_j. *Let* $i_j : p_j \to \mathbb{P}^r$ *be the inclusion, and let* $h_j = (c_1)_{\widetilde{T}}(\mathcal{O}(D_j)) \in H_{\widetilde{T}}^2(\mathbb{P}^r)$. *Then*

$$(i_j)_* 1 = \prod_{k \neq j} h_k \in H_{\widetilde{T}}^{2r}(\mathbb{P}^r).$$

We have

$$i_j^* h_k = c_1^{\widetilde{T}}(\mathcal{O}_{\mathbb{P}^r}(D_k)_{p_j}) = \widetilde{u}_k - \widetilde{u}_j \in H^2_{\widetilde{T}}(p_j).$$

$D_0 \cap D_1 \cap \cdots \cap D_r$ *is empty, so*

$$h_0 h_1 \cdots h_r = 0.$$

For a fixed $k \in \{1, \ldots, r\}$, $(i_j)^*(h_k - h_0) = \widetilde{u}_k - \widetilde{u}_0$ *for all* $j \in \{0, \ldots, r\}$. *By localization,* $h_k - h_0 = \widetilde{u}_k - \widetilde{u}_0$. *Define*

$$H = h_0 - \widetilde{u}_0 = h_1 - \widetilde{u}_1 = \cdots = h_r - \widetilde{u}_r.$$

Then

$$(i_j)_* 1 = \prod_{k \neq j} (H + \widetilde{u}_k), \quad i_j^* H = -\widetilde{u}_j, \quad \prod_{j=0}^{l} (H + \widetilde{u}_j) = 0,$$

where the last identity agrees with the relation derived in Example 2.3.

Definition 2.13 (equivariant integration on noncompact spaces). *Suppose that* X *is a* noncompact *oriented manifold, but* X^T *is a finite union of compact, orientable submanifolds* $F_1, \ldots, F_N$. *We define* $\int_{[X]^T} : H_T^*(M) \to \mathbb{Q}(u_1, \ldots, u_l)$ *by*

$$\int_{[X]^T} \alpha^T = \sum_{j=1}^{N} \int_{[F_j]^T} \frac{i_j^* \alpha^T}{e^T(N_j)}.$$

In the above Definition 2.13, $\int_{[X]^T} \alpha^T$ can be nonzero even if $\alpha^T \in H_T^k(X)$ and $k < \dim M$. The following is a simple example.

Example 2.14. *Let* $T = (\mathbb{C}^*)^2$ *act on* $\mathbb{C}^2$ *by* $(t_1, t_2) \cdot (z_1, z_2) = (t_1 z_1, t_2 z_2)$ *for* $(t_1, t_2) \in T$, $(z_1, z_2) \in \mathbb{C}^2$. *Then*

$$\int_{[\mathbb{C}^2]^T} 1 = \frac{1}{e^T(T_0 \mathbb{C}^2)} = \frac{1}{u_1 u_2}.$$

2.5. Equivariant Riemann-Roch

Let X be a compact complex manifold with a holomorphic T-action. The constant map $X \to pt$ also induces an additive map between equivariant K-theories:

$$\pi_! : K_T(X) \to K_T(pt), \quad \mathcal{E} \mapsto \sum_i (-1)^i H^i(X, \mathcal{E})$$

where $\mathcal{E}$ is a T-equivariant holomorphic vector bundle over X, and $H^i(X, \mathcal{E})$ are the sheaf cohomology groups, which are representations of T.

A representation of T is determined by its T-equivariant Chern character ch^T. We can compute $ch^T(\pi_! \mathcal{E})$ by Grothendieck-Riemann-Roch (GRR) theorem and the Atiyah-Bott localization formula. Applying GRR to the fibration $\pi : X_T \to BT$, we have

$$ch^T(\pi_! \mathcal{E}) = \int_{[X]^T} ch^T(\mathcal{E}) td^T(TX)$$

where $td^T(TX)$ is the T-equivariant Todd class of the tangent bundle TX of X. By the integration formula (2.11),

$$\int_{[X]^T} ch^T(\mathcal{E})td^T(TX) = \sum_{j=1}^{N} \int_{[F_j]^T} \frac{i_j^*\left(ch^T(\mathcal{E})td^T(TX)\right)}{e^T(N_j)}.$$

We now specialize to the case where F_j are isolated points. We write $p_1, \ldots, p_N$ instead of $F_1, \ldots, F_N$. Let $r = \dim_{\mathbb{C}} X$, and let

$$x_{j,1}, \ldots, x_{j,r} \in H_T^2(pt) = \bigoplus_{i=1}^{l} \mathbb{Q}u_i$$

be the weights of the T-action on the tangent space $T_{p_j}X$ of X at p_j. Then

$$i_j^* td^T(TX) = \prod_{k=1}^{r} \frac{x_{j,k}}{1-e^{-x_{j,k}}}, \quad e^T(N_j) = e^T(T_{p_j}X) = \prod_{k=1}^{r} x_{j,k}.$$

Let $m = \mathrm{rank}_{\mathbb{C}}\mathcal{E}$, and let

$$y_{j,1}, \ldots, y_{j,m} \in H_T^2(pt)$$

be the weights of the T-action on the fiber $\mathcal{E}_{p_j}$ of $\mathcal{E}$ at p_j. Then

$$i_j^* ch^T(\mathcal{E}) = \sum_{l=1}^{m} e^{y_{j,l}}.$$

Therefore

$$ch^T(\pi_! \mathcal{E}) = \sum_{j=1}^{N} \frac{\sum_{l=1}^{m} e^{y_{j,l}}}{\prod_{k=1}^{r}(1-e^{-x_{j,k}})}. \tag{2.15}$$

Example 2.16. *Suppose that $T = (\mathbb{C}^*)^l$ acts on on $\mathbb{P}^1$, and let $L \to \mathbb{P}^1$ be a T-equivariant line bundle. The weights of the T-actions on $T_0\mathbb{P}^1$, $T_\infty\mathbb{P}^1$, L_0, L_∞ are given by u, $-u$, w, $w - au$, respectively, where $u, w \in H_T^2(pt; \mathbb{Q})$ and $a \in \mathbb{Z}$ is the degree of L. Then*

$$ch^T(H^0(\mathbb{P}^1, L) - H^1(\mathbb{P}^1, L)) = \int_{[\mathbb{P}^1]^T} ch^T(L)td^T(T\mathbb{P}^1)$$

$$= \frac{e^w}{1-e^{-u}} + \frac{e^{w-au}}{1-e^u} = \begin{cases} \sum_{i=0}^{a} e^{w-iu}, & a \geqslant 0 \\ -\sum_{i=1}^{-a-1} e^{w+iu} & a < 0 \end{cases}$$

Indeed,

$$H^0(\mathbb{P}^1, L) = \begin{cases} \sum_{i=0}^{a} e^{w-iu}, & a \geqslant 0, \\ 0, & a < 0. \end{cases} \qquad H^1(\mathbb{P}^1, L) = \begin{cases} 0, & a \geqslant 0, \\ \sum_{i=1}^{-a-1} e^{w+iu}, & a < 0. \end{cases}$$

2.6. Basic intersection theory in algebraic geometry

We refer to [16] for intersection theory on schemes, and to [44] for intersection theory on Deligne-Mumford stacks.

Given a scheme or a Deligne-Mumford stack M over $\mathbb{C}$, let $A_*(M) = \oplus_k A_k(M)$ be the Chow groups of M with rational coefficients, and let $A^*(M) = \oplus_k A^k(M)$ be the operational Chow cohomology groups (see [16]) with rational coefficients. There is a cap product

$$A^k(M) \times A_l(M) \to A_{l-k}(M), \quad (\alpha, \beta) \mapsto \alpha \cap \beta,$$

and a group homomorphism $\deg : A_0(M) \to \mathbb{Q}$. If M is a scheme and $p \in M$ is a smooth point then $\deg[p] = 1$; if M is a Deligne-Mumford stack and $p \in M$ is a smooth point with automorphism group $\mathrm{Aut}(p)$ (which is a finite group) then $\deg[p] = 1/|\mathrm{Aut}(p)|$. (In this paper, $|S|$ denotes the cardinality of a finite set S.) We extend deg to $A_*(X)$ by sending $A_k(X)$ to zero for $k \neq 0$.

If M is a proper smooth scheme or a proper smooth Deligne-Mumford stack of dimension r, then there is a fundamental class

$$[M] \in A_r(M).$$

We define $\int_M : A^*(M) \to \mathbb{Q}$ by

$$\int_M \alpha = \deg(\alpha \cap [M]) \in \mathbb{Q}.$$

2.7. Equivariant intersection theory in algebraic geometry

We have discussed equivariant intersection theory on topological spaces, and localization of equivariant cohomology on manifolds. We now discuss equivariant intersection theory on schemes and Deligne-Mumford stacks, and virtual localization. This is similar to the discussion in Section 2.1 – 2.5, so we will just give a brief summary in the case $G = T = (\mathbb{C}^*)^l$. We refer to [12] for equivariant intersection theory on schemes and algebraic spaces.

Suppose that $T = (\mathbb{C}^*)^l$ acts on a scheme or a Deligne-Mumford stack M over $\mathbb{C}$. The T-equivariant operational Chow cohomology of M is defined to be the ordinary operational Chow cohomology of the quotient stack $[M/T]$:

$$A_T^*(M) := A^*([M/T]).$$

In particular,

$$A_T^*(pt) = A^*([pt/T]) = \mathbb{Q}[u_1, \ldots, u_l].$$

The T-equivariant Chow groups $A_*^T(M)$ is constructed from T-invariant cycles in M. We refer to [12] for the construction.

A T-equivariant vector bundle $V \to M$ corresponds to a vector bundle $[V/T] \to [M/T]$. Define the T-equivariant Chern classes and Chern characters of E by

$$c_k^T(V) := c_k([V/T]) \in A_T^k(M), \quad \mathrm{ch}_k^T(V) := \mathrm{ch}_k([V/T]) \in A_T^k(M).$$

Now suppose that M is a proper Deligne-Mumford stack with a T-equivariant perfect obstruction theory of virtual dimension m. In particular, locally there exists a two term complex of T-equivariant vector bundles $E \to F$ over M, where $\mathrm{rank} F - \mathrm{rank} E = m$, such that we have an exact sequence of T-equivariant sheaves:

$$0 \to \mathcal{T}^1 \to F^\vee \to E^\vee \to \mathcal{T}^2 \to 0.$$

The perfect obstruction theory defines a T-equivariant virtual fundamental class

$$[M]^{\mathrm{vir},T} \in A_m^T(M)$$

which defines

$$\int_{[M]^{\mathrm{vir},T}} : A_T^k(M) \to A_T^{k-m}(\mathrm{pt}).$$

In particular, $\int_{[M]^{\mathrm{vir},T}}$ sends $A_T^k(M)$ to 0 if $k < m$.

2.8. Virtual localization

Let M^T denote the substack of T-fixed points in M. Let $F_1, \ldots, F_N$ be the connected components of M^T. We assume that each F_j is a proper Deligne-Mumford substack. Given any $\xi \in M^T$, let T^1 and T^2 be the tangent and obstruction spaces at ξ. Then T acts on T^1 and T^2. Let $T^{i,f} \subset T^i$ be the maximal subspace where T acts trivially. Then $T^i = T^{i,f} \oplus T^{i,m}$. We call $T^{i,f}$ and $T^{i,m}$ the fixed and moving parts of T^i, respectively. Then $T^{i,f}$ defines a perfect obstruction theory on each F_j and a virtual fundamental class $[F_j]^{\mathrm{vir},T} \in A_*^T(F_j)$. The virtual normal bundle N_j^{vir} of F_j in M is $N_j^{\mathrm{vir}} = T_j^{1,m} - T_j^{2,m}$. The T-equivariant Euler class $e_T(N_j^{\mathrm{vir}}) \in A_T^*(F_j)$ is invertible in

$$A_T^*(F_j) \otimes_{R_T} Q_T = A^*(F_j) \otimes_{\mathbb{Q}} Q_T$$

where $R_T = \mathbb{Q}[u_1, \ldots, u_l]$, $Q_T = \mathbb{Q}(u_1, \ldots, u_l)$. Let $i_j : F_j \to M$ be the inclusion. Assuming the existence of a T-equivariant embedding from M into a *smooth* Deligne-Mumford stack, Graber and Pandharipande proved the following localization formula [19] (see also K. Behrend [4], A. Kresch [33]):

Theorem 2.17 (virtual localization).

$$[X]_T^{\mathrm{vir}} = \sum_{j=1}^{N} (i_j)_* \frac{[F_j]^{\mathrm{vir},T}}{e^T(N_j^{\mathrm{vir}})}. \tag{2.18}$$

Corollary 2.19 (integration formula). *If* $\alpha^T \in A_T^*(M)$ *then*

$$\int_{[M]^{\mathrm{vir},T}} \alpha^T = \sum_{j=1}^{N} \int_{[F_j]^{\mathrm{vir},T}} \frac{i_j^* \alpha^T}{e^T(N_j^{\mathrm{vir}})}. \tag{2.20}$$

Definition 2.21 (equivariant integration on non-proper Deligne-Mumford stack). *Suppose that* X *is a* non-proper *Deligne-Mumford stack with a perfect obstruction theory,* *and* X^T *is a finite union of* proper *Deligne-Mumford stacks* $F_1, \ldots, F_N$. *We define* $\int_{[X]^{vir,T}}$: $A_T^*(M) \to \mathbb{Q}(u_1, \ldots, u_l)$ *by*

$$\int_{[X]^{vir,T}} \alpha^T = \sum_{j=1}^{N} \int_{[F_j]^{vir,T}} \frac{i_j^* \alpha^T}{e^T(N_j)}.$$

3. Gromov-Witten theory

In this section, we give a brief review of Gromov-Witten theory. We work over $\mathbb{C}$.

3.1. Moduli of stable curves and Hodge integrals

An n-pointed, genus g prestable curve is a connected algebraic curve C of arithmetic genus g together with n ordered marked points $x_1, \ldots, x_n \in C$, where C has at most nodal singularities, and $x_1, \ldots, x_n$ are distinct smooth points. An n-pointed, genus g prestable curve $(C, x_1, \ldots, x_n)$ is *stable* if its automorphism group is finite, or equivalently,

$$\mathrm{Hom}_{\mathcal{O}_C}(\Omega_C(x_1 + \cdots + x_n), \mathcal{O}_C) = 0.$$

Let $\overline{\mathcal{M}}_{g,n}$ be the moduli space of n-pointed, genus g stable curves, where n, g are nonnegative integers. We assume that $2g - 2 + n > 0$, so that $\overline{\mathcal{M}}_{g,n}$ is nonempty. Then $\overline{\mathcal{M}}_{g,n}$ is a proper smooth Deligne-Mumford stack of dimension $3g - 3 + n$ [11, 30, 28, 29]. The tangent space of $\overline{\mathcal{M}}_{g,n}$ at a moduli point $[(C, x_1, \ldots, x_n)] \in \overline{\mathcal{M}}_{g,n}$ is given by

$$\mathrm{Ext}^1_{\mathcal{O}_C}(\Omega_C(x_1 + \cdots + x_n), \mathcal{O}_C).$$

Since $\overline{\mathcal{M}}_{g,n}$ is a proper Deligne-Mumford stack, we may define

$$\int_{\overline{\mathcal{M}}_{g,n}} : A^*(\overline{\mathcal{M}}_{g,n}) \to \mathbb{Q}.$$

We now introduce some classes in $A^*(\overline{\mathcal{M}}_{g,n})$. There is a forgetful morphism $\pi : \overline{\mathcal{M}}_{g,n+1} \to \overline{\mathcal{M}}_{g,n}$ given by forgetting the $(n+1)$-th marked point (and contracting the unstable irreducible component if there is one):

$$[(C, x_1, \ldots, x_n, x_{n+1})] \mapsto [(C^{st}, x_1, \ldots, x_n)]$$

where $(C^{st}, x_1, \ldots, x_n)$ is the stabilization of the prestable curve $(C, x_1, \ldots, x_n)$. The morphism $\pi : \overline{\mathcal{M}}_{g,n+1} \to \overline{\mathcal{M}}_{g,n}$ can be identified with the universal curve over $\overline{\mathcal{M}}_{g,n}$.

- (λ classes) Let ω_π be the relative dualizing sheaf of $\pi : \overline{\mathcal{M}}_{g,n+1} \to \overline{\mathcal{M}}_{g,n}$. The Hodge bundle $\mathbb{E} = \pi_* \omega_\pi$ is a rank g vector bundle over $\overline{\mathcal{M}}_{g,n}$ whose fiber over the moduli point $[(C, x_1, \ldots, x_n)] \in \overline{\mathcal{M}}_{g,n}$ is $H^0(C, \omega_C)$, the space of sections of the dualizing sheaf ω_C of the curve C. The λ classes are defined by

$$\lambda_j = c_j(\mathbb{E}) \in A^j(\overline{\mathcal{M}}_{g,n}).$$

- (ψ classes) The i-th marked point x_i gives rise a section $s_i : \overline{\mathcal{M}}_{g,n} \to \overline{\mathcal{M}}_{g,n+1}$ of the universal curve. Let $\mathbb{L}_i = s_i^* \omega_\pi$ be the line bundle over $\overline{\mathcal{M}}_{g,n}$ whose fiber over the moduli point $[(C, x_1, \ldots, x_n)] \in \overline{\mathcal{M}}_{g,n}$ is the cotangent line $T^*_{x_i} C$ of C at x_i. The ψ classes are defined by

$$\psi_i = c_1(\mathbb{L}_i) \in A^1(\overline{\mathcal{M}}_{g,n}).$$

Hodge integrals are top intersection numbers of λ classes and ψ classes:

$$\int_{\overline{\mathcal{M}}_{g,n}} \psi_1^{a_1} \cdots \psi_n^{a_n} \lambda_1^{k_1} \cdots \lambda_g^{k_g} \in \mathbb{Q}. \tag{3.1}$$

By definition, (3.1) is zero unless

$$a_1 + \cdots + a_n + k_1 + 2k_2 + \cdots + gk_g = 3g - 3 + n.$$

Using Mumford's Grothendieck-Riemann-Roch calculations in [37], Faber proved, in [13], that general Hodge integrals can be uniquely reconstructed from the ψ integrals (also known as *descendant integrals*):

$$\int_{\overline{\mathcal{M}}_{g,n}} \psi_1^{a_1} \cdots \psi_n^{a_n}. \tag{3.2}$$

The descendant integrals can be computed recursively by Witten's conjecture which asserts that the ψ integrals (3.2) satisfy a system of differential equations known as the KdV equations [45]. The KdV equations and the string equation determine all the ψ integrals (3.2) from the initial value $\int_{\overline{\mathcal{M}}_{0,3}} 1 = 1$. For example, from the initial value $\int_{\overline{\mathcal{M}}_{0,3}} 1 = 1$ and the string equation, one can derive the following formula of genus 0 descendant integrals:

$$\int_{\overline{\mathcal{M}}_{0,n}} \psi_1^{a_1} \cdots \psi_n^{a_n} = \frac{(n-3)!}{a_1! \cdots a_n!} \tag{3.3}$$

where $a_1 + \cdots + a_n = n - 3$ [32, Section 3.3.2].

Witten's conjecture was first proved by Kontsevich in [31]. By now, Witten's conjecture has been reproved many times (Okounkov-Pandharipande [38], Mirzakhani [35], Kim-Liu [27], Kazarian-Lando [26], Chen-Li-Liu [8], Kazarian [25], Mulase-Zhang [36], etc.).

3.2. Moduli of stable maps

Let X be a nonsingular projective or quasi-projective variety over $\mathbb{C}$, and let $\beta \in H_2(X; \mathbb{Z})$. An n-pointed, genus g, degree β prestable map to X is a morphism $f : (C, x_1, \ldots, x_n) \to X$, where $(C, x_1, \ldots, x_n)$ is an n-pointed, genus g prestable curve, and $f_*[C] = \beta$. Two prestable maps

$$f : (C, x_1, \ldots, x_n) \to X, \quad f' : (C', x'_1, \ldots, x'_n) \to X$$

are isomorphic if there exists an isomorphism $\phi : (C, x_1, \ldots, x_n) \to (C', x_1', \ldots, x_n')$ of pointed prestable curves such that $f = f' \circ \phi$. A prestable map is *stable* if its automorphism group is finite. The notion of stable maps was introduced by Kontsevich [32].

The moduli space $\overline{\mathcal{M}}_{g,n}(X, \beta)$ of n-pointed, genus g, degree β stable maps to X is a Deligne-Mumford stack which is proper when X is projective [6].

3.3. Obstruction theory and virtual fundamental classes

The tangent space T^1 and the obstruction space T^2 at a moduli point $[f : (C, x_1, \ldots, x_n) \to X] \in \overline{\mathcal{M}}_{g,n}(X, \beta)$ fit in the *tangent-obstruction exact sequence*:

$$\begin{aligned} 0 \to &\mathrm{Ext}^0_{\mathcal{O}_C}(\Omega_C(x_1 + \cdots + x_n), \mathcal{O}_C) \to H^0(C, f^*T_X) \to T^1 \\ \to &\mathrm{Ext}^1_{\mathcal{O}_C}(\Omega_C(x_1 + \cdots + x_n), \mathcal{O}_C) \to H^1(C, f^*T_X) \to T^2 \to 0 \end{aligned} \tag{3.4}$$

where

- $\mathrm{Ext}^0_{\mathcal{O}_C}(\Omega_C(x_1 + \cdots + x_n), \mathcal{O}_C)$ is the space of infinitesimal automorphisms of the domain $(C, x_1, \ldots, x_n)$,
- $\mathrm{Ext}^1_{\mathcal{O}_C}(\Omega_C(x_1 + \cdots + x_n), \mathcal{O}_C)$ is the space of infinitesimal deformations of the domain $(C, x_1, \ldots, x_n)$,
- $H^0(C, f^*T_X)$ is the space of infinitesimal deformations of the map f, and
- $H^1(C, f^*T_X)$ is the space of obstructions to deforming the map f.

T^1 and T^2 form sheaves $\mathcal{T}^1$ and $\mathcal{T}^2$ on the moduli space $\overline{\mathcal{M}}_{g,n}(X, \beta)$.

Let X be a nonsingular projective variety. We say X is *convex* if $H^1(C, f^*TX) = 0$ for all genus 0 stable maps f. Projective spaces $\mathbb{P}^n$, or more generally, generalized flag varieties G/P, are examples of convex varieties. When X is convex and $g = 0$, the obstruction sheaf $\mathcal{T}^2 = 0$, and the moduli space $\overline{\mathcal{M}}_{0,n}(X, \beta)$ is a *smooth* Deligne-Mumford stack.

In general, $\overline{\mathcal{M}}_{g,n}(X, \beta)$ is a *singular* Deligne-Mumford stack equipped with a *perfect obstruction theory*: there is a two term complex of locally free sheaves $E \to F$ on $\overline{\mathcal{M}}_{g,n}(X, \beta)$ such that

$$0 \to \mathcal{T}^1 \to F^\vee \to E^\vee \to \mathcal{T}^2 \to 0$$

is an exact sequence of sheaves. (See [5] for the complete definition of a perfect obstruction theory.) The *virtual dimension* d^{vir} of $\overline{\mathcal{M}}_{g,n}(X, \beta)$ is the rank of the virtual tangent bundle $T^{\mathrm{vir}} = F^\vee - E^\vee$.

$$d^{\mathrm{vir}} = \int_\beta c_1(T_X) + (\dim X - 3)(1 - g) + n \tag{3.5}$$

Suppose that $\overline{\mathcal{M}}_{g,n}(X, \beta)$ is *proper*. (Recall that if X is projective then $\overline{\mathcal{M}}_{g,n}(X, \beta)$ is proper for any g, n, β.) Then there is a *virtual fundamental class*

$$[\overline{\mathcal{M}}_{g,n}(X, \beta)]^{\mathrm{vir}} \in A_{d^{\mathrm{vir}}}(\overline{\mathcal{M}}_{g,n}(X, \beta)).$$

The virtual fundamental class has been constructed by Li-Tian [34], Behrend-Fantechi [5] in algebraic Gromov-Witten theory. The virtual fundamental class allows us to define

$$\int_{[\overline{\mathcal{M}}_{g,n}(X,\beta)]^{\mathrm{vir}}} : A^*(\overline{\mathcal{M}}_{g,n}(X,\beta)) \longrightarrow \mathbb{Q}, \quad \alpha \mapsto \deg(\alpha \cap [\overline{\mathcal{M}}_{g,n}(X,\beta)]^{\mathrm{vir}}).$$

3.4. Gromov-Witten invariants

Let X be a nonsingular projective variety. Gromov-Witten invariants are rational numbers defined by applying

$$\int_{[\overline{\mathcal{M}}_{g,n}(X,\beta)]^{\mathrm{vir}}} : A^*(\overline{\mathcal{M}}_{g,n}(X,\beta)) \to \mathbb{Q}$$

to certain classes in $A^*(\overline{\mathcal{M}}_{g,n}(X,\beta))$.

Let $ev_i : \overline{\mathcal{M}}_{g,n}(X,\beta) \to X$ be the evaluation at the i-th marked point: ev_i sends $[f : (C, x_1, \ldots, x_n) \to X] \in \overline{\mathcal{M}}_{g,n}(X,\beta)$ to $f(x_i) \in X$. Given $\gamma_1, \ldots, \gamma_n \in A^*(X)$, define

$$\langle \gamma_1, \ldots, \gamma_n \rangle^X_{g,\beta} = \int_{[\overline{\mathcal{M}}_{g,n}(X,\beta)]^{\mathrm{vir}}} ev_1^*\gamma_1 \cup \cdots \cup ev_n^*\gamma_n \in \mathbb{Q}. \tag{3.6}$$

These are known as the *primary* Gromov-Witten invariants of X. More generally, we may also view $[\overline{\mathcal{M}}_{g,n}(X,\beta)]^{\mathrm{vir}}$ as a class in $H_{2d}(\overline{\mathcal{M}}_{g,n}(X,\beta))$. Then (3.6) is defined for ordinary cohomology classes $\gamma_1, \ldots, \gamma_n \in H^*(X)$, including odd cohomology classes which do not come from $A^*(\overline{\mathcal{M}}_{g,n}(X,\beta))$.

Let $\pi : \overline{\mathcal{M}}_{g,n+1}(X,\beta) \to \overline{\mathcal{M}}_{g,n}(X,\beta)$ be the universal curve. For $i = 1, \ldots, n$, let $s_i : \overline{\mathcal{M}}_{g,n}(X,\beta) \to \overline{\mathcal{M}}_{g,n+1}(X,\beta)$, be the section which corresponds to the i-th marked point. Let $\omega_\pi \to \overline{\mathcal{M}}_{g,n+1}(X,\beta)$ be the relative dualizing sheaf of π, and let $\mathbb{L}_i = s_i^*\omega_\pi$ be the line bundle over $\overline{\mathcal{M}}_{g,n}(X,\beta)$ whose fiber at the moduli point $[f : (C, x_1, \ldots, x_n) \to X] \in \overline{\mathcal{M}}_{g,n}(X,\beta)$ is the cotangent line $T^*_{x_i}C$ at the i-th marked point x_i. The ψ-classes are defined to be

$$\psi_i := c_1(\mathbb{L}_i) \in A^1(\overline{\mathcal{M}}_{g,n}(X,\beta)), \quad i = 1, \ldots, n.$$

We use the same notation ψ_i to denote the corresponding classes in the ordinary cohomology group $H^2(\overline{\mathcal{M}}_{g,n}(X,\beta))$.

The *descendant* Gromov-Witten invariants are defined by

$$\langle \tau_{a_1}(\gamma_1) \cdots \tau_{a_n}(\gamma_n) \rangle^X_{g,\beta} := \int_{[\overline{\mathcal{M}}_{g,n}(X,\beta)]^{\mathrm{vir}}} ev_1^*\gamma_1 \cup \psi_1^{a_1} \cup \cdots \cup ev_n^*\gamma_n \cup \psi_n^{a_n} \in \mathbb{Q}. \tag{3.7}$$

Suppose that $\gamma_i \in H^{d_i}(X)$. Then (3.7) is zero unless

$$\sum_{i=1}^n d_i + 2\sum_{i=1}^n a_i = 2\Big(\int_\beta c_1(TX) + (\dim X - 3)(1-g) + n\Big). \tag{3.8}$$

Remark 3.9. *Note that*

$$\psi_i \neq \pi^*\psi_i.$$

where the ψ_i *on the left hand side is an element in* $H^2(\overline{\mathcal{M}}_{g,n+1}(X,\beta))$*, whereas the* ψ_i *on the right hand side is an element in* $H^2(\overline{\mathcal{M}}_{g,n}(X,\beta))$*. Indeed, let* $D_i \subset \overline{\mathcal{M}}_{g,n+1}(X,\beta)$ *be the divisor associated to the section* s_i*. Then* $\psi_i = \pi^*\psi_i + [D_i]$ [45, Section 2b].

4. Toric varieties

In this section, we review geometry and topology of nonsingular toric varieties. We refer to [17] for the theory of toric varieties. We also introduce the toric graph of a nonsingular toric variety that satisfies some mild assumptions. In Section 5, we will see that the toric graph contains all the information needed for computing Gromov-Witten invariants and equivariant Gromov-Witten invariants of the toric variety.

4.1. Basic notation

Let X be a smooth toric variety of dimension r. Then X contains the algebraic torus $T = (\mathbb{C}^*)^r$ as a dense open subset, and the action of T on itself extends to X. Let $N = \mathrm{Hom}(\mathbb{C}^*, T) \cong \mathbb{Z}^r$ be the lattice of 1-parameter subgroups of T, and let $M = \mathrm{Hom}(T, \mathbb{C}^*)$ be the lattice of irreducible characters of T. Then $M = \mathrm{Hom}(N, \mathbb{Z})$ is the dual lattice of N. Let $N_{\mathbb{R}} = N \otimes_{\mathbb{Z}} \mathbb{R}$ and $M_{\mathbb{R}} = M \otimes_{\mathbb{Z}} \mathbb{R}$, so that they are dual real vector spaces of dimension r.

The toric variety X is defined by a fan $\Sigma \subset N_{\mathbb{R}}$. Let $\Sigma(k)$ be the set of k-dimensional cones in Σ. A k-dimensional cone $\sigma \in \Sigma(k)$ corresponds to an $(r-k)$-dimensional orbit closure $V(\sigma)$ of the T-action on X. We make the following assumption:

Assumption 4.1.

- $\Sigma(r)$ *is nonempty, so that* X *contains at least one fixed point.*
- *Each* $(r-1)$ *dimensional cone* $\tau \in \Sigma(r-1)$ *is contained in at least one top dimensional cone* $\sigma \in \Sigma(r)$.

We introduce some notation:

- Let $\{e_1, \ldots, e_r\}$ be a $\mathbb{Z}$-basis of N, and let $\{u_1, \ldots, u_r\}$ be the dual $\mathbb{Z}$-basis of $M = \mathrm{Hom}(N, \mathbb{Z})$: $\langle u_i, e_i \rangle = \delta_{ij}$.
- Given linearly independent vectors $w_1, \ldots, w_k \in N$, define

  $$\mathrm{Cone}(\{w_1, \ldots, w_k\}) = \{t_1 w_1 + \cdots t_k w_k \mid t_1, \ldots, t_k \in \mathbb{R}_{\geqslant 0}\}.$$

  We define $\mathrm{Cone}(\emptyset) = \{0\}$.

Example 4.2 ($\mathbb{P}^r$). $N = \oplus_{i=1}^r \mathbb{Z} e_i$. *Let*

$$v_i = e_i, \quad 1 \leqslant i \leqslant r, \quad v_0 = -e_1 - \cdots - e_r.$$

The projective space $\mathbb{P}^r$ *is a nonsingular projective toric variety of dimension* r*, defined by the fan*

$$\Sigma = \{\mathrm{Cone}(S) \mid S \subset \{v_0, \ldots, v_r\}, |S| \leqslant r\}.$$

Example 4.3 ($\mathcal{O}_{\mathbb{P}^1}(-1) \oplus \mathcal{O}_{\mathbb{P}^1}(-1)$). $N = \mathbb{Z}e_1 \oplus \mathbb{Z}e_2 \oplus \mathbb{Z}e_3$. *Define*

$$v_1 = e_1, \quad v_2 = e_2, \quad v_3 = e_3, \quad v_4 = e_1 + e_2 - e_3.$$

Given $1 \leqslant i_1 < \cdots < i_k \leqslant 4$, *define*

$$\sigma_{i_1 \cdots i_k} = \mathrm{Cone}(\{v_{i_1}, \ldots, v_{i_k}\}).$$

The total space of $\mathcal{O}_{\mathbb{P}^1}(-1) \oplus \mathcal{O}_{\mathbb{P}^1}(-1)$ *is a nonsingular quasi-projective toric variety of dimension* 3, *defined by the fan*

$$\Sigma = \{\{0\}, \sigma_1, \sigma_2, \sigma_3, \sigma_4, \sigma_{12}, \sigma_{13}, \sigma_{23}, \sigma_{24}, \sigma_{34}, \sigma_{123}, \sigma_{234}\}.$$

4.2. One-skeleton

The set of T fixed points in X is given by

$$\{p_\sigma := V(\sigma) : \sigma \in \Sigma(r)\}.$$

The set of 1-dimensional T orbit closures in X is given by

$$\{\ell_\tau := V(\tau) : \tau \in \Sigma(r-1)\}.$$

Under our assumption, each ℓ_τ is either an affine line $\mathbb{C}$ or a projective line $\mathbb{P}^1$. We define

$$\Sigma(r-1)_c = \{\tau \in \Sigma(r-1)_c : \ell_\tau \cong \mathbb{P}^1\}.$$

Note that $\Sigma(r-1)_c = \Sigma(r-1)$ if X is proper. We define the 1-skeleton of X to be the union of 1-dimensional orbit closures:

$$X^1 := \bigcup_{\tau \in \Sigma(r-1)} \ell_\tau. \tag{4.4}$$

We define the set of flags in Σ to be

$$\begin{aligned} F(\Sigma) &= \{(\tau, \sigma) \in \Sigma(r-1) \times \Sigma(r) \mid \tau \subset \sigma\} \\ &= \{(\tau, \sigma) \in \Sigma(r-1) \times \Sigma(r) \mid p_\sigma \in \ell_\tau\}. \end{aligned}$$

Example 4.5 ($\mathbb{P}^r$). *We use the notation in Example 4.2. Define*

$$\begin{aligned} \sigma_i &= \mathrm{Cone}\{v_j \mid j \neq i\}, \quad i = 0, \ldots, r, \\ \tau_{ij} &= \sigma_i \cap \sigma_j \in \Sigma(r-1), \quad 0 \leqslant i < j \leqslant r. \end{aligned}$$

Then

$$\begin{aligned} &\Sigma(r) = \{\sigma_i \mid i = 0, \ldots, r\} \\ &\Sigma(r-1) = \Sigma(r-1)_c = \{\tau_{ij} \mid 0 \leqslant i < j \leqslant r\} \\ &F(\Sigma) = \{(\tau_{ij}, \sigma_i) \mid 0 \leqslant i < j \leqslant n\} \cup \{(\tau_{ij}, \sigma_j) \mid 0 \leqslant i < j \leqslant r\}. \end{aligned}$$

Example 4.6 ($\mathcal{O}_{\mathbb{P}^1}(-1) \oplus \mathcal{O}_{\mathbb{P}^1}(-1)$). *We use the notation in Example 4.3.*

$$\Sigma(3) = \{\sigma_{123}, \sigma_{234}\}, \quad \Sigma(2) = \{\sigma_{12}, \sigma_{13}, \sigma_{23}, \sigma_{24}, \sigma_{34}\}, \quad \Sigma(2)_c = \{\sigma_{23}\}$$

$$F(\Sigma) = \{(\sigma_{12}, \sigma_{123}), (\sigma_{13}, \sigma_{123}), (\sigma_{23}, \sigma_{123}), (\sigma_{23}, \sigma_{234}), (\sigma_{24}, \sigma_{234}), (\sigma_{34}, \sigma_{234})\}$$

4.3. Toric graph

The sets $\Sigma(r)$, $\Sigma(r-1)$ and $F(\Sigma)$ define a connected graph Υ. Each top dimensional cone $\sigma \in \Sigma(r)$ corresponds to a vertex $v(\sigma)$ in Υ. Each $(r-1)$ dimensional cone $\tau \in \Sigma(r-1)$ corresponds to an edge $e(\tau)$ in Υ; $e(\tau)$ is a ray if $\ell_\tau \cong \mathbb{C}$, and is a line segment if $\ell_\tau \cong \mathbb{P}^1$. The vertex $v(\sigma)$ is contained in the edge $e(\tau)$ if and only if the fixed point p_σ is contained in the (affine or projective) line ℓ_τ.

Given any top dimensional cone $\sigma \in \Sigma(r)$, define the following subset of $\Sigma(r-1)$:

$$E_\sigma = \{\tau \in \Sigma(r-1) \mid \tau \subset \sigma\} = \{\tau \in \Sigma(r-1) \mid p_\sigma \in \ell_\tau\}.$$

Then $|E_\sigma| = n$. Therefore Υ is an r-valent graph.

Given a flag $(\tau, \sigma) \in F(\Sigma)$, let $w(\tau, \sigma) \in M = \mathrm{Hom}(T, \mathbb{C}^*)$ be the weight of T-action on $T_{p_\sigma}\ell_\tau$, the tangent line to ℓ_τ at the fixed point p_σ, namely,

$$w(\tau, \sigma) := c_1^T(T_{p_\sigma}\ell_\tau) \in H_T^2(p_\sigma; \mathbb{Z}) \cong M.$$

This gives rise to a map $w : F(\Sigma) \to M$ satisfying the following properties.

(1) Given any $\sigma \in \Sigma(r)$, the set $\{w(\tau, \sigma) \mid \tau \in E_\sigma\}$ form a $\mathbb{Z}$-basis of M. These are the weights of the tangent space $T_{p_\sigma}X$ to X at the fixed point p_σ.

(2) Any $\tau \in \Sigma(r-1)_c$ is contained in two top dimensional cones $\sigma, \sigma' \in \Sigma(r)$.
   (a) $w(\tau, \sigma) + w(\tau, \sigma') = 0$.
   (b) Let $E_\sigma = \{\tau_1, \ldots, \tau_r\}$, where $\tau_r = \tau$. For any $\tau_i \in E_\sigma$ there exists a unique $\tau_i' \in E_{\sigma'}$ and $a_i \in \mathbb{Z}$ such that

$$w(\tau_i', \sigma') = w(\tau_i, \sigma) - a_i w(\tau, \sigma).$$

In particular, $\tau_r' = \tau_r = \tau$ and $a_r = 2$.

Let τ be as in (2). The normal bundle of $\ell_\tau \cong \mathbb{P}^1$ in X is given by

$$N_{\ell_\tau/X} \cong L_1 \oplus \cdots \oplus L_{n-1}$$

where L_i is a degree a_i T-equivariant line bundle over ℓ_τ such that the weights of the T-actions on the fibers $(L_i)_{p_\sigma}$ and $(L_i)_{p_{\sigma'}}$ are $w(\tau_i, \sigma)$ and $w(\tau_i', \sigma')$, respectively.

Example 4.7 ($\mathbb{P}^r$). *In notation in Example 4.5,*

$$w(\sigma_{0j}, \sigma_0) = -w(\sigma_{0j}, \sigma_j) = u_j, \quad j = 1, \ldots, r,$$
$$w(\sigma_{ij}, \sigma_i) = -w(\sigma_{ij}, \sigma_j) = u_j - u_i, \quad 1 \leqslant i < j \leqslant r.$$

Example 4.8 ($\mathcal{O}_{\mathbb{P}^1}(-1) \oplus \mathcal{O}_{\mathbb{P}^1}(-1)$). *In notation in Example 4.3 and Example 4.6,*

$$w(\sigma_{23}, \sigma_{123}) = u_1, \quad w(\sigma_{13}, \sigma_{123}) = u_2, \quad w(\sigma_{12}, \sigma_{123}) = u_3,$$
$$w(\sigma_{23}, \sigma_{234}) = -u_1, \quad w(\sigma_{24}, \sigma_{234}) = u_1 + u_3, \quad w(\sigma_{34}, \sigma_{234}) = u_1 + u_2.$$

We now give another interpretation of the weight $w(\tau, \sigma)$ associated to a flag $(\tau, \sigma) \in F(\Gamma)$. There is a unique $\rho \in \Sigma(1)$ such that $\rho \subset \sigma$ and $\rho \not\subset \tau$. $D_\rho := V(\rho)$ is a T-invariant divisor which intersects the T-invariant (affine or projective) line ℓ_τ

transversally at the T-fixed point p_σ. Then $w(\tau,\sigma)$ is the weight of the T-action on $\mathcal{O}(D_\rho)_{p_\sigma}$, i.e.

$$w(\tau,\sigma)=c_1^T(\mathcal{O}(D_\rho)_{p_\sigma}).$$

The formal completion $\hat{X}$ of X along X^1, together with the T-action, can be reconstructed from the graph Υ and $w: F(\Sigma)\to M$. We call (Υ, w) the *toric graph* defined by Σ.

4.4. Induced torus action

Suppose that there is a group homomorphism $\phi: T'\to T$ from another torus $T'\cong(\mathbb{C}^*)^s$ to $T\cong(\mathbb{C}^*)^r$. Then T' acts on X by

$$t'\cdot x=\phi(t')\cdot x,\quad t'\in T,\ x\in X.$$

The group homomorphism $\phi: T'\to T$ induces group homomorphisms

$$\phi_*: N'=\mathrm{Hom}(\mathbb{C}^*,T')\longrightarrow \mathrm{Hom}(\mathbb{C}^*,T)$$
$$\phi^*: M=\mathrm{Hom}(T,\mathbb{C}^*)\longrightarrow \mathrm{Hom}(T',\mathbb{C}^*)$$

An important example is the big torus $\widetilde{T}=(\mathbb{C}^*)^s$ coming from the geometric quotient, where $s=|\Sigma(1)|\geqslant r$. Let I_Σ be the ideal of $\mathbb{C}[z_1,\dots,z_s]$ generated by

$$\{\prod_{\rho_i\not\subset\sigma} z_i:\sigma\in\Sigma\},$$

and let $Z(I_\Sigma)$ be the closed subscheme of $\mathbb{C}^s$ defined by I_Σ. Then

$$X=(\mathbb{C}^s-Z(I_\Sigma))/(\mathbb{C}^*)^{s-r}.$$

Let $\Sigma(1)=\{\rho_1,\dots,\rho_s\}$. For each ρ_α there exists a unique primitive vector $v_\alpha\in N$ such that $\rho_\alpha\cap N=\mathbb{Z}_{\geqslant 0}v_\alpha$. The group homomorphism

$$\phi_*:\widetilde{N}=\bigoplus_{\alpha=1}^{s}\mathbb{Z}\widetilde{e}_\alpha\longrightarrow N=\bigoplus_{i=1}^{r}\mathbb{Z}e_i,\quad \widetilde{e}_\alpha\mapsto v_\alpha$$

induces group homomorphisms

$$\phi:\widetilde{T}=(\mathbb{C}^*)^s\longrightarrow T=(\mathbb{C}^*)^r,\quad (\tilde{t}_1,\dots,\tilde{t}_s)\mapsto(\prod_{\alpha=1}^{s}\tilde{t}_\alpha^{\langle u_1,v_\alpha\rangle},\dots,\prod_{\alpha=1}^{s}\tilde{t}_\alpha^{\langle u_r,v_\alpha\rangle})$$

and

$$\phi^*: M=\bigoplus_{i=1}^{r}\mathbb{Z}u_i\longrightarrow\widetilde{M}=\bigoplus_{\alpha=1}^{s}\mathbb{Z}\widetilde{u}_\alpha,\quad u_i\mapsto\sum_{\alpha=1}^{s}\langle u_i,v_\alpha\rangle\widetilde{u}_\alpha.$$

Example 4.9 ($\mathbb{P}^r$). $\mathbb{P}^r=(\mathbb{C}^{r+1}-\{0\})/\mathbb{C}^*$. *The group homomorphism*

$$\phi_*:\widetilde{N}=\bigoplus_{i=0}^{r}\mathbb{Z}\widetilde{e}_i\longrightarrow N=\bigoplus_{i=1}^{r}\mathbb{Z}e_i,\quad \widetilde{e}_i\mapsto v_i,$$

induces group homomorphisms

$$\phi:\widetilde{T}=(\mathbb{C}^*)^{r+1}\to T=(\mathbb{C}^*)^r,\quad(\tilde{t}_0,\dots,\tilde{t}_r)\mapsto(\tilde{t}_1\tilde{t}_0^{-1},\ \ ,\tilde{t}_r\tilde{t}_0^{-1}).$$

and

$$\phi^* : M = \bigoplus_{i=1}^{r} \mathbb{Z}u_i \longrightarrow \widetilde{M} = \bigoplus_{i=0}^{r} \mathbb{Z}\widetilde{u}_i, \quad u_i \mapsto \widetilde{u}_i - \widetilde{u}_0, \quad i = 1, \ldots, r.$$

We have

$$\phi^* \circ w(\tau_{ij}, \sigma_i) = -w(\tau_{ij}, \sigma_j) = \widetilde{u}_j - \widetilde{u}_i, \quad 0 \leqslant i < j \leqslant r.$$

4.5. Cohomology and equivariant cohomology

Let X be a smooth toric variety of dimension r defined by a fan Σ. Let $\Sigma(1) = \{\rho_1, \ldots, \rho_s\}$, and let $v_\alpha \in N$ be the unique primitive vector such that $\rho_\alpha \cap N = \mathbb{Z}_{\geqslant 0} v_\alpha$. Let $D_\alpha = V(\rho_\alpha)$.

Given $\sigma \in \Sigma(k)$, the scheme theoretic intersection of toric subvarieties D_α and $V(\sigma)$ is given by

$$D_\alpha \cap V(\sigma) = \begin{cases} V(\gamma) & \text{if } \sigma \text{ and } v_\alpha \text{ span the cone } \gamma \in \Sigma(k+1), \\ \emptyset & \text{if } \sigma \text{ and } v_\alpha \text{ do not span a cone in } \Sigma. \end{cases} \tag{4.10}$$

Now assume that X is projective, so that $V(\sigma)$ is projective for all $\sigma \in \Sigma$. Given a k-dimensional cone $\sigma \in \Sigma(k)$, let $[V(\sigma)] \in H^{2k}(X)$ be the Poincaré dual of the homology class represented by $V(\sigma)$, and let $[V(\sigma)]^T \in H_T^{2k}(X)$ be the equivariant Poincaré dual of the T-equivariant homology class represented by the T-invariant subvariety $V(\sigma)$. Then $A^k(X) = H^{2k}(X)$ is generated, as a $\mathbb{Q}$-vector space, by $\{[V(\sigma)] \mid \sigma \in \Sigma(k)\}$, and $A_T^k(X) = H_T^{2k}(X)$ is generated by $\{[V(\sigma)]^T \mid \sigma \in \Sigma(k)\}$. We have

(i) If $v_{i_1}, \ldots, v_{i_k}$ do not span a cone of Σ then

$$[D_{i_1}] \cup \cdots \cup [D_{i_k}] = 0 \in H^{2k}(X),$$
$$[D_{i_1}]^T \cup \cdots \cup [D_{i_k}]^T = 0 \in H_T^{2k}(X).$$

(ii) For any $u \in M \subset H_T^2(X)$,

$$\sum_{\alpha=1}^{s} \langle u, v_\alpha \rangle [D_\alpha] = 0 \in H^2(X), \quad \sum_{\alpha=1}^{s} \langle u, v_\alpha \rangle [D_\alpha]^T = u \in H_T^2(X).$$

The above (i) follows from (4.10). To see (ii), let $\chi^u : T \to \mathbb{C}^*$ be the character which corresponds to $u \in M$. Then χ^u is a rational function on X which defines a T-invariant principal divisor $\sum_{\alpha=1}^{s} \langle u, v_\alpha \rangle [D_\alpha]^T$. Relations (i) and (ii) are essentially all the relations in $H^*(X)$ or $H_T^*(X)$.

Definition 4.11. (1) *Let I be the ideal in $\mathbb{Q}[X_1, \ldots, X_s]$ generated by the monomials* $\{X_{i_1} \cdots X_{i_k} \mid v_{i_1}, \ldots, v_{i_k}$ do not generate a cone in $\Sigma\}$.

(2) *Let J be the ideal in $\mathbb{Q}[X_1, \ldots, X_s]$ generated by* $\{\sum_{\alpha=1}^{s} \langle u, v_\alpha \rangle X_\alpha \mid u \in M\}$.

(3) *Let I' be the ideal in $R_T[X_1, \ldots, X_s] = \mathbb{Q}[X_1, \ldots, X_s, u_1, \ldots, u_r]$ generated by the monomials* $\{X_{i_1} \cdots X_{i_k} \mid v_{i_1, \ldots, i_k}$ do not generate a cone in $\Sigma\}$.

(4) *Let J' be the ideal in $R_T[X_1, \ldots, X_s] = \mathbb{Q}[X_1, \ldots, X_s, u_1, \ldots, u_r]$ generated by* $\{\sum_{\alpha=1}^{s} \langle u, v_\alpha \rangle X_\alpha - u \mid u \in M\}$.

With all the above definitions, the cohomology and equivariant cohomology rings of X can be describe explicitly as follows. (See for example [17, Section 5.2], [18, Lecture 14].)

Theorem 4.12.

$$H^*(X) \cong \mathbb{Q}[X_1, \dots, X_s]/(I + J).$$
$$H_T^*(X) \cong \mathbb{Q}[X_1, \dots, X_s, u_1, \dots, u_r]/(I' + J') \cong \mathbb{Q}[X_1, \dots, X_s]/I.$$

The isomorphism is given by $X_\alpha \mapsto [D_\alpha]$ *or* $[D_\alpha]^T$.

The ring $\mathbb{Q}[X_1, \dots, X_s]/I$ is known as the Stanley-Reisner ring. The ring homomorphism

$$i_X^* : H_T^*(X) = \mathbb{Q}[X_1, \dots, X_s, u_1, \dots, u_r]/(I' + J') \to H^*(X) = \mathbb{Q}[X_1, \dots, X_s]/(I + J)$$

is surjective. The kernel is the ideal generated by $u_1, \dots, u_r$. We say $\gamma^T \in H_T^*(X)$ is a T-equivariant lift of $\gamma \in H^*(X)$ if $i_X^*(\gamma^T) = \gamma$.

Example 4.13 ($\mathbb{P}^r$).

$$H^*(\mathbb{P}^r) \cong \mathbb{Q}[X_0, \dots, X_r]/\langle X_0 \cdots X_r, X_1 - X_0, \dots, X_r - X_0\rangle \cong \mathbb{Q}[X]/\langle X^{r+1}\rangle.$$
$$H_T^*(\mathbb{P}^r) \cong \mathbb{Q}[X_0, \dots, X_r, u_1, \dots, u_r]/\langle X_0 \cdots X_r, X_1 - X_0 - u_1, \dots, X_r - X_0 - u_r\rangle$$
$$= \mathbb{Q}[X, u_1, \dots, u_r]/\langle X(X + u_1) \cdots (X + u_r)\rangle.$$

5. Gromov-Witten invariants of smooth toric varieties

Let X be a nonsingular toric variety of dimension r. Then $T = (\mathbb{C}^*)^r$ acts on X, and acts on $\overline{\mathcal{M}}_{g,n}(X, \beta)$ by

$$t \cdot [f : (C, x_1, \dots, x_n) \to X] \mapsto [t \cdot f : (C, x_1, \dots, x_n) \to X]$$

where $(t \cdot f)(z) = t \cdot f(z)$, $z \in \mathbb{C}$. The evaluation maps $ev_i : \overline{\mathcal{M}}_{g,n}(X, \beta) \to X$ are T-equivariant and induce $ev_i^* : A_T^*(X) \to A_T^*(\overline{\mathcal{M}}_{g,n}(X, \beta))$.

5.1. Equivariant Gromov-Witten invariants

Suppose that $\overline{\mathcal{M}}_{g,n}(X, \beta)$ is proper, so that there are virtual fundamental classes

$$[\overline{\mathcal{M}}_{g,n}(X, \beta)]^{vir} \in A_{d^{vir}}(\overline{\mathcal{M}}_{g,n}(X, \beta)), \quad [\overline{\mathcal{M}}_{g,n}(X, \beta)]^{vir,T} \in A^T_{d^{vir}}(\overline{\mathcal{M}}_{g,n}(X, \beta)),$$

where

$$d^{vir} = \int_\beta c_1(TX) + (r - 3)(1 - g) + n.$$

Given $\gamma_i \in A^{d_i}(X) = H^{2d_i}(X)$ and $a_i \in \mathbb{Z}_{\geqslant 0}$, define $\langle \tau_{a_1}(\gamma_1) \cdots \tau_{a_n}(\gamma_n)\rangle_{g,\beta}^X$ as in Section 3.4:

$$\langle \tau_{a_1}(\gamma_1) \cdots \tau_{a_n}(\gamma_n)\rangle_{g,\beta}^X = \int_{[\overline{\mathcal{M}}_{g,n}(X,\beta)]^{vir}} \prod_{i=1}^n \left(ev_i^*\gamma_i \cup \psi_i^{a_i}\right) \in \mathbb{Q}. \tag{5.1}$$

By definition, (5.1) is zero unless $\sum_{i=1}^{n} d_i = d^{\mathrm{vir}}$. In this case,

$$(5.2) \qquad \langle \tau_{a_1}(\gamma_1)\cdots\tau_{a_N}(\gamma_n)\rangle_{g,\beta}^{X} = \int_{[\overline{\mathcal{M}}_{g,n}(X,\beta)]^{\mathrm{vir},T}} \prod_{i=1}^{n} \left(\mathrm{ev}_i^*\gamma_i^T \cup (\psi_i^T)^{a_i}\right).$$

Here $\gamma_i^T \in A_T^{d_i}(X)$ is any T-equivariant lift of $\gamma_i \in A^{d_i}(X)$, and $\psi_i^T \in A_T^1(\overline{\mathcal{M}}_{g,n}(X,\beta))$ is any T-equivariant lift of $\psi_i \in A^1(\overline{\mathcal{M}}_{g,n}(X,\beta))$.

In this section, fix a choice of ψ_i^T as follows. A stable map $f:(C,x_1,\ldots,x_n)\to X$ induces $\mathbb{C}$-linear maps $T_{x_i}C \to T_{f(x_i)}X$ for $i=1,\ldots,n$ which in turn give rise to maps $\mathbb{L}_i^\vee \to \mathrm{ev}_i^*TX$. The T-action on X induces a T-action on TX, so that TX is a T-equivariant vector bundle over X, and ev_i^*TX is a T-equivariant vector bundle over $\overline{\mathcal{M}}_{g,n}(X,\beta)$. Let T act on $\mathbb{L}_i$ such that $\mathbb{L}_i^\vee \to \mathrm{ev}_i^*TX$ is T-equivariant, and define

$$\psi_i^T = c_1^T(\mathbb{L}_i) \in A_T^1(\overline{\mathcal{M}}_{g,n}(X,\beta)), \quad i=1,\ldots,n.$$

Then ψ_i^T is a T-equivariant lift of $\psi_i = c_1(\mathbb{L}_i) \in A^1(\overline{\mathcal{M}}_{g,n}(X,\beta))$.

Given $\gamma_i^T \in A_T^{d_i}(X)$, we define equivariant Gromov-Witten invariants

$$(5.3) \qquad \begin{aligned} \langle \tau_{a_1}(\gamma_1^T),\cdots,\tau_{a_n}(\gamma_n^T)\rangle_{g,\beta}^{X_T} &:= \int_{[\overline{\mathcal{M}}_{g,n}(X,\beta)]^{\mathrm{vir},T}} \prod_{i=1}^{n} \left(\mathrm{ev}_i^*\gamma_i^T (\psi_i^T)^{a_i}\right) \\ &\in \mathbb{Q}[u_1,\ldots,u_l](\sum_{i=1}^{n} d_i - d^{\mathrm{vir}}). \end{aligned}$$

where $\mathbb{Q}[u_1,\ldots,u_l](k)$ is the space homogeneous polynomials of degree k with rational coefficients in $u_1,\ldots,u_l$. In particular,

$$\langle \tau_{a_1}(\gamma_1^T),\cdots,\tau_{a_n}(\gamma_n^T)\rangle_{g,\beta}^{X_T} = \begin{cases} 0, & \sum_{i=1}^{n} d_i < d^{\mathrm{vir}}, \\ \langle \tau_{a_1}(\gamma_1),\cdots,\tau_{a_n}(\gamma_n)\rangle_{g,\beta}^{X} \in \mathbb{Q}, & \sum_{i=1}^{n} d_i = d^{\mathrm{vir}}. \end{cases}$$

where $\gamma_i = i_X^*\gamma_i^T \in A^{d_i}(X)$. (Recall that $i_X: X \to X_T$ is the inclusion of a fiber of $X_T \to BT$.)

In this section, we will compute the equivariant Gromov-Witten invariants (5.3) by localization. Section 5.2 – Section 5.4 below are mostly straightforward generalizations of the $\mathbb{P}^r$ case discussed in [32] (genus 0), and [19, Section 4], [4, Section 4] (higher genus). H. Spielberg derived a formula for genus 0 Gromov-Witten invariants of smooth toric varieties [41, 42]. See also [20, Chapter 27].

Let $\overline{\mathcal{M}}_{g,n}(X,\beta)^T \subset \overline{\mathcal{M}}_{g,n}(X,\beta)$ be the substack of T fixed points, and let $i: \overline{\mathcal{M}}_{g,n}(X,\beta)^T \to \overline{\mathcal{M}}_{g,n}(X,\beta)$ be the inclusion. Let N^{vir} be the virtual normal bundle of substack $\overline{\mathcal{M}}_{g,n}(X,\beta)^T$ in $\overline{\mathcal{M}}_{g,n}(X,\beta)$; in general, N^{vir} has different ranks on different connected components of $\overline{\mathcal{M}}_{g,n}(X,\beta)^T$. By virtual localization,

(5.4)
$$\int_{[\overline{\mathcal{M}}_{g,n}(X,\beta)]^{\mathrm{vir},T}} \prod_{i=1}^{n} \left(\mathrm{ev}_i^*\gamma_i^T \cup (\psi_i^T)^{a_i}\right) = \int_{[\overline{\mathcal{M}}_{g,n}(X,\beta)^T]^{\mathrm{vir},T}} \frac{i^*\prod_{i=1}^{n}\left(\mathrm{ev}_i^*\gamma_i^T \cup (\psi_i^T)^{a_i}\right)}{e^T(N^{\mathrm{vir}})}.$$

Indeed, we will see that $\overline{\mathcal{M}}_{g,n}(X,\beta)^T$ is proper even when $\overline{\mathcal{M}}_{g,n}(X,\beta)$ is not. When $\overline{\mathcal{M}}_{g,n}(X,\beta)$ is not proper, we *define*

(5.5)

$$\langle\tau_{a_1}(\gamma_1^T),\ldots,\tau_{a_n}(\gamma_n^T)\rangle_{g,\beta}^X=\int_{[\overline{\mathcal{M}}_{g,n}(X,\beta)^T]^{\mathrm{vir},T}}\frac{i^*\prod_{i=1}^n\left(ev_i^*\gamma_i^T\cup(\psi_i^T)^{a_i}\right)}{e^T(N^{\mathrm{vir}})}\in Q_T.$$

When $\overline{\mathcal{M}}_{g,n}(X,\beta)$ is not proper, the right hand side of (5.5) is a rational function (instead of a polynomial) in $u_1,\ldots,u_r$. It can be nonzero when $\sum d_i<d^{\mathrm{vir}}$, and does not have a nonequivariant limit (obtained by setting $u_i=0$) in general.

5.2. Torus fixed points and graph notation

In this subsection, we describe the T-fixed points in $\overline{\mathcal{M}}_{g,n}(X,\beta)$. Following Kontsevich [32], given a stable map $f:(C,x_1,\ldots,x_n)\to X$ such that

$$[f:(C,x_1,\ldots,x_n)\to X]\in\overline{\mathcal{M}}_{g,n}(X,\beta)^T,$$

we will associate a decorated graph $\vec{\Gamma}$.

We first give a formal definition.

Definition 5.6. *A decorated graph* $\vec{\Gamma}=(\Gamma,\vec{f},\vec{d},\vec{g},\vec{s})$ *for* n*-pointed, genus* g*, degree* β *stable maps to* X *consists of the following data.*

(1) Γ *is a compact, connected 1 dimensional CW complex. We denote the set of vertices (resp. edges) in* Γ *by* $V(\Gamma)$ *(resp.* $E(\Gamma)$*). Let*

$$F(\Gamma)=\{(e,v)\in E(\Gamma)\times V(\Gamma)\mid v\in e\}$$

be the set of flags in Γ.

(2) *The* label map $\vec{f}:V(\Gamma)\cup E(\Gamma)\to\Sigma(r)\cup\Sigma(r-1)_c$ *sends a vertex* $v\in V(\Gamma)$ *to a top dimensional cone* $\sigma_v\in\Sigma(r)$*, and sends an edge* $e\in E(\Gamma)$ *to an* $(r-1)$*-dimensional cone* $\tau_e\in\Sigma(r-1)_c$*. Moreover,* $\vec{f}$ *defines a map from the graph* Γ *to the graph* Υ*: if* (e,v) *is a flag in* Γ *then* $(\mathbf{e}(\tau_e),\mathbf{v}(\sigma_v))$ *is a flag in* Υ*, or equivalently,* $(\tau_e,\sigma_v)\in F(\Sigma)$.

(3) *The* degree map $\vec{d}:E(\Gamma)\to\mathbb{Z}_{>0}$ *sends an edge* $e\in E(\Gamma)$ *to a positive integer* d_e.

(4) *The* genus map $\vec{g}:V(\Gamma)\to\mathbb{Z}_{\geqslant 0}$ *sends a vertex* $v\in V(\Gamma)$ *to a nonnegative integer* g_v.

(5) *The* marking map $\vec{s}:\{1,2,\ldots,n\}\to V(\Gamma)$ *is defined if* $n>0$.

The above maps satisfy the following two constraints:

(i) *(topology of the domain)* $\sum_{v\in V(\Gamma)}g_v+|E(\Gamma)|-|V(\Gamma)|+1=g$.

(ii) *(topology of the map)* $\sum_{e\in E(\Gamma)}d_e[\ell_{\tau_e}]=\beta$.

Let $G_{g,n}(X,\beta)$ *be the set of all decorated graphs* $\vec{\Gamma}=(\Gamma,\vec{f},\vec{d},\vec{g},\vec{s})$ *satisfying the above constraints.*

We now describe the geometry and combinatorics associated to a stable map $f:(C,x_1,\dots,x_n)\to X$ that represents a T fixed point in $\overline{\mathcal{M}}_{g,n}(X,\beta)$.

For any $t\in T$, there exists an automorphism $\phi_t:(C,x_1,\dots,x_n)$ such that $t\cdot f(z)=f\circ\phi_t(z)$ for any $z\in C$. Let C' be an irreducible component of C, and let $f'=f|_{C'}:C'\to X$. There are two possibilities:

Case 1: f' is a constant map, and $f(C')=\{p_\sigma\}$, where p_σ is a fixed point in X associated to some $\sigma\in\Sigma(r)$

Case 2: $C'\cong\mathbb{P}^1$ and $f(C')=\ell_\tau$, where ℓ_τ is a T-invariant $\mathbb{P}^1$ in X associated to some $\tau\in\Sigma(r-1)_c$.

We define a decorated graph $\vec{\Gamma}$ associated to $f:(C,x_1,\dots,x_n)\to X$ as follows.

(1) (Vertices) We assign a vertex v to each connected component C_v of $f^{-1}(X^T)$.
  (a) (label) $f(C_v)=\{p_\sigma\}$ for some top dimensional cone $\sigma\in\Sigma(r)$; we define $\vec{f}(v)=\sigma_v=\sigma$.
  (b) (genus) C_v is a curve or a point. If C_v is a curve then we define $\vec{g}(v)=g_v$ to be the arithmetic genus of C_v; if C_v is a point then we define $\vec{g}(v)=g_v=0$.
  (c) (marking) For $i=1,\dots,n$, define $\vec{s}(i)=v$ if $x_i\in C_v$.

(2) (Edges) For any $\tau\in\Sigma(r-1)$, let $O_\tau\cong\mathbb{C}^*$ be the 1-dimensional orbit whose closure is ℓ_τ. Then

$$X^1\setminus X^T=\bigsqcup_{\tau\in\Sigma(r-1)}O_\tau$$

where the right hand side is a disjoint union of connected components. We assign an edge e to each connected component $O_e\cong\mathbb{C}^*$ of $f^{-1}(X^1\setminus X^T)$.
  (a) (label) Let $C_e\cong\mathbb{P}^1$ be the closure of O_e. Then $f(C_e)=\ell_\tau$ for some τ in $\Sigma(r-1)_c$; we define $\vec{f}(e)=\tau_e=\tau$.
  (b) (degree) We define $\vec{d}(e)=d_e$ to be the degree of the map $f|_{C_e}:C_e\cong\mathbb{P}^1\to\ell_\tau\cong\mathbb{P}^1$.

(3) (Flags) The set of flags in the graph Γ is defined by

$$F(\Gamma)=\{(e,v)\in E(\Gamma)\times V(\Gamma)\mid C_e\cap C_v\neq\emptyset\}.$$

The above (1), (2), (3) define a decorated graph $\vec{\Gamma}=(\Gamma,\vec{f},\vec{d},\vec{g},\vec{s})$ satisfying the constraints (i) and (ii) in Definition 5.6. Therefore $\vec{\Gamma}\in G_{g,n}(X,\beta)$. This gives a map from $\overline{\mathcal{M}}_{g,n}(X,\beta)^T$ to the discrete set $G_{g,n}(X,\beta)$. Let $\mathcal{F}_{\vec{\Gamma}}\subset\overline{\mathcal{M}}_{g,n}(X,\beta)^T$ denote the preimage of $\vec{\Gamma}$. Then

$$\overline{\mathcal{M}}_{g,n}(X,\beta)^T=\bigsqcup_{\vec{\Gamma}\in G_{g,n}(X,\beta)}\mathcal{F}_{\vec{\Gamma}}$$

where the right hand side is a disjoint union of connected components. We next describe the fixed locus $\mathcal{F}_{\vec{\Gamma}}$ associated to each decorated graph $\vec{\Gamma}\in G_{g,n}(X,\beta)$. For later convenience, we introduce some definitions.

Definition 5.7. *Given a vertex* $v \in V(\Gamma)$, *we define*

$$E_v = \{e \in E(\Gamma) \mid (e, v) \in F(\Gamma)\},$$

the set of edges emanating from v, *and define* $S_v = \vec{s}^{-1}(v) \subset \{1, \ldots, n\}$. *The valency of* v *is given by* $\mathrm{val}(v) = |E_v|$. *Let* $n_v = |S_v|$ *be the number of marked points contained in* C_v. *We say a vertex is* stable *if* $2g_v - 2 + \mathrm{val}(v) + n_v > 0$. *Let* $V^S(\Gamma)$ *be the set of stable vertices in* $V(\Gamma)$. *There are three types of unstable vertices:*

$$\begin{aligned} V^1(\Gamma) &= \{v \in V(\Gamma) \mid g_v = 0, \mathrm{val}(v) = 1, n_v = 0\}, \\ V^{1,1}(\Gamma) &= \{v \in V(\Gamma) \mid g_v = 0, \mathrm{val}(v) = n_v = 1\}, \\ V^2(\Gamma) &= \{v \in V(\Gamma) \mid g_v = 0, \mathrm{val}(v) = 2, n_v = 0\}. \end{aligned}$$

Then $V(\Gamma)$ *is the disjoint union of* $V^1(\Gamma)$, $V^{1,1}(\Gamma)$, $V^2(\Gamma)$, *and* $V^S(\Gamma)$.

The set of stable flags is defined to be

$$F^S(\Gamma) = \{(e, v) \in F(\Gamma) \mid v \in V^S(\Gamma)\}.$$

Given a decorated graph $\vec{\Gamma} = (\Gamma, \vec{f}, \vec{d}, \vec{g}, \vec{s})$, the curves C_e and the maps $f|_{C_e} : C_e \to \ell_{\tau_e} \subset X$ are determined by $\vec{\Gamma}$. If $v \notin V^S(\Gamma)$ then C_v is a point. If $v \in V^S(\Gamma)$ then C_v is a curve, and $y(e, v) := C_e \cap C_v$ is a node of C for $e \in E_v$.

$$\big(C_v, \{y(e, v) : e \in E_v\} \cup \{x_i \mid i \in S_v\}\big)$$

is a $(\mathrm{val}(v) + n_v)$-pointed, genus g_v curve, which represents a point in $\overline{\mathcal{M}}_{g_v, \mathrm{val}(v)+n_v}$. We call this moduli space $\overline{\mathcal{M}}_{g_v, E_v \cup S_v}$ instead of $\overline{\mathcal{M}}_{g_v, \mathrm{val}(v)+n_v}$ because we would like to label the marked points on C_v by $E_v \cup S_v$ instead of $\{1, 2, \ldots, \mathrm{val}(v) + n_v\}$. Then

$$\mathcal{M}_{\vec{\Gamma}} = \prod_{v \in V^S(\Gamma)} \overline{\mathcal{M}}_{g_v, E_v \cup S_v}.$$

The automorphism group $A_{\vec{\Gamma}}$ for any point $[f : (C, x_1, \ldots, x_n) \to X] \in \mathcal{F}_{\vec{\Gamma}}$ fits in the following short exact sequence of groups:

$$1 \to \prod_{e \in E(\Gamma)} \mathbb{Z}_{d_e} \to A_{\vec{\Gamma}} \to \mathrm{Aut}(\vec{\Gamma}) \to 1$$

where $\mathbb{Z}_{d_e}$ is the automorphism group of the degree d_e morphism

$$f|_{C_e} : C_e \cong \mathbb{P}^1 \to \ell_{\tau_e} \cong \mathbb{P}^1,$$

and $\mathrm{Aut}(\vec{\Gamma})$ is the automorphism group of the decorated graph $\vec{\Gamma} = (\Gamma, \vec{f}, \vec{d}, \vec{g}, \vec{s})$. There is a morphism $i_{\vec{\Gamma}} : \mathcal{M}_{\vec{\Gamma}} \to \overline{\mathcal{M}}_{g,n}(X, \beta)$ whose image is the fixed locus $\mathcal{F}_{\vec{\Gamma}}$ associated to $\vec{\Gamma} \in G_{g,n}(X, \beta)$. The morphism $i_{\vec{\Gamma}}$ induces an isomorphism $[\mathcal{M}_{\vec{\Gamma}}/A_{\vec{\Gamma}}] \cong \mathcal{F}_{\vec{\Gamma}}$.

5.3. Virtual tangent and normal bundles

Given a decorated graph $\vec{\Gamma} \in G_{g,n}(X, \beta)$ and a stable map $f : (C, x_1, \ldots, x_n) \to X$ which represents a point in the fixed locus $\mathcal{F}_{\vec{\Gamma}}$ associated to $\vec{\Gamma}$, let

$$B_1 = \mathrm{Hom}(\Omega_C(x_1 + \cdots + x_n), \mathcal{O}_C), \quad B_2 = H^0(C, f^*TX)$$
$$B_4 = \mathrm{Ext}^1(\Omega_C(x_1 + \cdots + x_n), \mathcal{O}_C), \quad B_5 = H^1(C, f^*TX)$$

T acts on B_1, B_2, B_4, B_5. Let B_i^m and B_i^f be the moving and fixed parts of B_i. We have the following exact sequences:

$$0 \to B_1^f \to B_2^f \to T^{1,f} \to B_4^f \to B_5^f \to T^{2,f} \to 0 \tag{5.8}$$

$$0 \to B_1^m \to B_2^m \to T^{1,m} \to B_4^m \to B_5^m \to T^{2,m} \to 0 \tag{5.9}$$

The irreducible components of C are

$$\{C_v \mid v \in V^S(\Gamma)\} \cup \{C_e \mid e \in E(\Gamma)\}.$$

The nodes of C are

$$\{y_v = C_v \mid v \in V^2(\Gamma)\} \cup \{y(e,v) \mid (e,v) \in F^S(\Gamma)\}$$

5.3.1. Automorphisms of the domain Given any $(e, v) \in F(\Gamma)$, let $y(e, v) = C_e \cap C_v$, and define

$$w_{(e,v)} := e^T(T_{y(e,v)}C_e) = \frac{w(\tau_e, \sigma_v)}{d_e} \in H_T^2(y(e,v)) = M \otimes_{\mathbb{Z}} \mathbb{Q}.$$

We have

$$B_1^f = \bigoplus_{\substack{e \in E(\Gamma) \\ (e,v),(e,v') \in F(\Gamma)}} \mathrm{Hom}(\Omega_{C_e}(y(e,v) + y(e,v')), \mathcal{O}_{C_e})$$
$$= \bigoplus_{\substack{e \in E(\Gamma) \\ (e,v),(e,v') \in F(\Gamma)}} H^0(C_e, TC_e(-y(e,v) - y(e,v'))$$
$$B_1^m = \bigoplus_{v \in V^1(\Gamma), (e,v) \in F(\Gamma)} T_{y(e,v)}C_e$$

5.3.2. Deformations of the domain Given any $v \in V^S(\Gamma)$, define a divisor $\mathbf{x}_v$ of C_v by

$$\mathbf{x}_v = \sum_{i \in S_v} x_i + \sum_{e \in E_v} y(e,v).$$

Then

$$B_4^f = \bigoplus_{v \in V^S(\Gamma)} \mathrm{Ext}^1(\Omega_{C_v}(\mathbf{x}_v), \mathcal{O}_C) = \bigoplus_{v \in V^S(\Gamma)} T\overline{\mathcal{M}}_{g_v, E_v \cup S_v}$$

$$B_4^m = \bigoplus_{v \in V^2(\Gamma), E_v = \{e, e'\}} T_{y_v}C_e \otimes T_{y_v}C_{e'} \oplus \bigoplus_{(e,v) \in F^S(\Gamma)} T_{y(e,v)}C_v \otimes T_{y(e,v)}C_e$$

where

$$e^T(T_{y_\nu}C_e \otimes T_{y_\nu}C_{e'}) = w_{(e,\nu)} + w_{(e',\nu)}, \quad \nu \in V^2(\Gamma)$$
$$e^T(T_{y(e,\nu)}C_\nu \otimes T_{y(e,\nu)}C_e) = w_{(e,\nu)} - \psi_{(e,\nu)}, \quad \nu \in V^S(\Gamma)$$

5.3.3. Unifying stable and unstable vertices From the discussion in Section 5.3.1 and Section 5.3.2,

$$\begin{aligned}\frac{e^T(B_1^m)}{e^T(B_4^m)} = &\prod_{\nu\in V^1(\Gamma),(e,\nu)\in F(\Gamma)} w_{(e,\nu)} \prod_{\nu\in V^2(\Gamma),E_\nu=\{e,e'\}} \frac{1}{w_{(e,\nu)}+w_{(e',\nu)}}\\ &\cdot \prod_{\nu\in V^S(\Gamma)} \frac{1}{\prod_{e\in E_\nu}(w_{(e,\nu)}-\psi_{(e,\nu)})}.\end{aligned} \tag{5.10}$$

Recall that

$$\mathcal{M}_{\vec{\Gamma}} = \prod_{\nu\in V^S(\Gamma)} \overline{\mathcal{M}}_{g_\nu,E_\nu\cup S_\nu}.$$

To unify the stable and unstable vertices, we use the following convention for the empty sets $\overline{\mathcal{M}}_{0,1}$ and $\overline{\mathcal{M}}_{0,2}$. Let w_1, w_2 be formal variables.

(i) $\overline{\mathcal{M}}_{0,1}$ is a -2 dimensional space, and

$$\int_{\overline{\mathcal{M}}_{0,1}} \frac{1}{w_1-\psi_1} = w_1. \tag{5.11}$$

(ii) $\overline{\mathcal{M}}_{0,2}$ is a -1 dimensional space, and

$$\int_{\overline{\mathcal{M}}_{0,2}} \frac{1}{(w_1-\psi_1)(w_2-\psi_2)} = \frac{1}{w_1+w_2} \tag{5.12}$$

$$\int_{\overline{\mathcal{M}}_{0,2}} \frac{1}{w_1-\psi_1} = 1. \tag{5.13}$$

(iii) $\mathcal{M}_{\vec{\Gamma}} = \prod_{\nu\in V(\Gamma)} \overline{\mathcal{M}}_{g_\nu,E_\nu\cup S_\nu}.$

With the above conventions (i), (ii), (iii), we may rewrite (5.10) as

$$\frac{e^T(B_1^m)}{e^T(B_4^m)} = \prod_{\nu\in V(\Gamma)} \frac{1}{\prod_{e\in E_\nu}(w_{(e,\nu)}-\psi_{(e,\nu)})}. \tag{5.14}$$

The following lemma shows that the conventions (i) and (ii) are consistent with the stable case $\overline{\mathcal{M}}_{0,n}$, $n \geqslant 3$.

Lemma 5.15. *For any positive integer* n *and formal variables* $w_1, \ldots, w_n$, *we have*

(a) $\displaystyle\int_{\overline{\mathcal{M}}_{0,n}} \frac{1}{\prod_{i=1}^n (w_i-\psi_i)} = \frac{1}{w_1\cdots w_n}\left(\frac{1}{w_1}+\cdots\frac{1}{w_n}\right)^{n-3}.$

(b) $\displaystyle\int_{\overline{\mathcal{M}}_{0,n}} \frac{1}{w_1-\psi_1} = w_1^{2-n}.$

Proof. (a) The cases $n = 1$ and $n = 2$ follow from the definitions (5.11) and (5.12), respectively. For $n \geqslant 3$, we have

$$\int_{\overline{\mathcal{M}}_{0,n}} \frac{1}{\prod_{i=1}^n (w_i - \psi_i)} = \frac{1}{w_1 \cdots w_n} \int_{\overline{\mathcal{M}}_{0,n}} \frac{1}{\prod_{i=1}^n (1 - \frac{\psi_i}{w_i})}$$
$$= \frac{1}{w_1 \cdots w_n} \sum_{a_1 + \cdots + a_n = n-3} w_1^{-a_1} \cdots w_n^{-a_n} \int_{\overline{\mathcal{M}}_{0,n}} \psi_1^{a_1} \cdots \psi_n^{a_n}$$

where

$$\int_{\overline{\mathcal{M}}_{0,n}} \psi_1^{a_1} \cdots \psi_n^{a_n} = \frac{(n-3)!}{a_1! \cdots a_n!}.$$

So

$$\int_{\overline{\mathcal{M}}_{0,n}} \frac{1}{\prod_{i=1}^n (w_i - \psi_i)} = \frac{1}{w_1 \cdots w_n} (\frac{1}{w_1} + \cdots \frac{1}{w_n})^{n-3}.$$

(b) The cases $n = 1$ and $n = 2$ follow from the definitions (5.11) and (5.13), respectively. For $n \geqslant 3$, we have

$$\int_{\overline{\mathcal{M}}_{0,n}} \frac{1}{w_1 - \psi_1} = \frac{1}{w_1} \int_{\overline{\mathcal{M}}_{0,n}} \frac{1}{1 - \frac{\psi_1}{w_1}} = \frac{1}{w_1} w_1^{3-n} = w_1^{2-n}$$

□

5.3.4. Deformation of the map Consider the normalization sequence

$$(5.16)\quad \begin{aligned} 0 \to \mathcal{O}_C \to & \bigoplus_{v \in V^S(\Gamma)} \mathcal{O}_{C_v} \oplus \bigoplus_{e \in E(\Gamma)} \mathcal{O}_{C_e} \\ \to & \bigoplus_{v \in V^2(\Gamma)} \mathcal{O}_{y_v} \oplus \bigoplus_{(e,v) \in F^S(\Gamma)} \mathcal{O}_{y(e,v)} \to 0. \end{aligned}$$

We twist the above short exact sequence of sheaves by f^*TX. The resulting short exact sequence gives rise a long exact sequence of cohomology groups

$$\begin{aligned} 0 \to B_2 \to & \bigoplus_{v \in V^S(\Gamma)} H^0(C_v) \oplus \bigoplus_{e \in E(\Gamma)} H^0(C_e) \\ \to & \bigoplus_{v \in V^2(\Gamma)} T_{f(y_v)} X \oplus \bigoplus_{(e,v) \in F^S(\Gamma)} T_{f(y(e,v))} X \\ \to B_5 \to & \bigoplus_{v \in V^S(\Gamma)} H^1(C_v) \oplus \bigoplus_{e \in E(\Gamma)} H^1(C_e) \to 0. \end{aligned}$$

where

$$H^i(C_v) = H^i(C_v, (f|_{C_v})^* TX) \cong H^i(C_v, \mathcal{O}_{C_v}) \otimes T_{p_{\sigma_v}} X,$$
$$H^i(C_e) = H^i(C_e, (f|_{C_e})^* TX)$$

for $i = 0, 1$. We have

$$H^0(C_v) = T_{p_{\sigma_v}} X$$
$$H^1(C_v) = H^0(C_v, \omega_{C_v})^\vee \otimes T_{p_{\sigma_v}} X.$$

Lemma 5.17. *Let* $\sigma \in \Sigma(r)$, *so that* p_σ *is a* T *fixed point in* X. *Define*

$$\begin{aligned} w(\sigma) &= e^T(T_{p_\sigma}X) \in H_T^{2r}(pt) \\ h(\sigma, g) &= \frac{e^T(\mathbb{E}^\vee \otimes T_{p_\sigma}X)}{e^T(T_{p_\sigma}X)} \in H_T^{2r(g-1)}(\overline{\mathcal{M}}_{g,n}). \end{aligned}$$

Then

$$w(\sigma) = \prod_{(\tau,\sigma)\in F(\Sigma)} w(\tau, \sigma). \tag{5.18}$$

$$h(\sigma, g) = \prod_{(\tau,\sigma)\in F(\Sigma)} \frac{\Lambda_g^\vee(w(\tau, \sigma))}{w(\tau, \sigma)} \tag{5.19}$$

where $\Lambda_g^\vee(u) = \sum_{i=0}^{g}(-1)^i \lambda_i u^{g-i}$.

Proof. $T_{p_\sigma}X = \bigoplus_{(\tau,\sigma)\in F(\Sigma)} T_{p_\sigma}\ell_\tau$, where $e^T(T_{p_\sigma}\ell_\tau) = w(\tau, \sigma)$. So

$$\begin{aligned} e^T(T_{p_\sigma}) &= \prod_{(\tau,\sigma)\in F(\Sigma)} w(\tau, \sigma), \\ \frac{e^T(\mathbb{E}^\vee \otimes T_{p_\sigma}\ell_\tau)}{e^T(T_{p_\sigma}\ell_\tau)} &= \prod_{(\tau,\sigma)\in F(\Sigma)} \frac{e^T(\mathbb{E}^\vee \otimes T_{p_\sigma}\ell_\tau)}{w(\tau, \sigma)}, \end{aligned}$$

where

$$e^T(\mathbb{E}^\vee \otimes T_{p_\sigma}\ell_\tau) = \sum_{i=0}^{g}(-1)^i c_i(\mathbb{E}) c_1^T(T_{p_\sigma}\ell_\tau)^{g-i} = \sum_{i=0}^{g}(-1)^i \lambda_i w(\tau, \sigma)^{g-i}.$$

□

The map $B_1 \to B_2$ sends $H^0(C_e, TC_e(-y(e,v) - y(e',v)))$ isomorphically to $H^0(C_e, (f|_{C_e})^* T\ell_{\tau_e})^f$, the fixed part of $H^0(C_e, (f|_{C_e})^* T\ell_{\tau_e})$.

Lemma 5.20. *Given* $d \in \mathbb{Z}_{>0}$ *and* $\tau \in \Sigma(r-1)_c$, *define* $\sigma, \sigma', \tau_i, \tau_i', a_i$ *as in Section 4.3, and let* $f_d : \mathbb{P}^1 \to \ell_\tau \cong \mathbb{P}^1$ *be the unique degree* d *map totally ramified over the two* T *fixed point* p_σ *and* $p_{\sigma'}$ *in* ℓ_τ. *Define*

$$h(\tau, d) = \frac{e^T(H^1(\mathbb{P}^1, f_d^* TX)^m)}{e^T(H^0(\mathbb{P}^1, f_d^* TX)^m)}.$$

Then

$$h(\tau, d) = \frac{(-1)^d d^{2d}}{(d!)^2 w(\tau, \sigma)^{2d}} \prod_{i=1}^{r-1} b(\frac{w(\tau, \sigma)}{d}, w(\tau_i, \sigma), da_i) \tag{5.21}$$

where

$$b(u, w, a) = \begin{cases} \prod_{j=0}^{a}(w - ju)^{-1}, & a \in \mathbb{Z}, a \geqslant 0, \\ \prod_{j=1}^{-a-1}(w + ju), & a \in \mathbb{Z}, a < 0. \end{cases} \tag{5.22}$$

Proof. We use the notation in Section 4.3. We have

$$N_{\ell_\tau/X} = L_1 \oplus \cdots \oplus L_{r-1}.$$

The weights of T-actions on $(L_i)_{p_\sigma}$ and $(L_i)_{p_\sigma}$ are $\mathbf{w}(\tau_i, \sigma)$ and $\mathbf{w}(\tau_i, \sigma) - a_i \mathbf{w}(\tau, \sigma)$, respectively. The weights of T-actions on $T_0\mathbb{P}^1$, $T_\infty \mathbb{P}^1$, $(f_d^* L_i)_0$, $(f_d^* L_i)_\infty$ are $u := \frac{\mathbf{w}(\tau,\sigma)}{d}$, $-u$, $w_i := \mathbf{w}(\tau_i, \sigma)$, $w_i - da_i u$, respectively. By Example 2.16,

$$\mathrm{ch}_T(H^0(\mathbb{P}^1, f_d^* L_i) - H^1(\mathbb{P}^1, f_d^* L_i)) = \begin{cases} \sum_{j=0}^{da_i} e^{w_i - ju}, & a_i \geqslant 0, \\ \sum_{j=1}^{-da_i - 1} e^{w_i + ju}, & a_i < 0. \end{cases}$$

Note that $w_i + ju$ is nonzero for any $j \in \mathbb{Z}$ since w_i and u are linearly independent for $i = 1, \ldots, n-1$. So

$$\frac{e^T\left(H^1(\mathbb{P}^1, f_d^* L_i)\right)}{e^T\left(H^0(\mathbb{P}^1, f_d^* L_i)\right)} = \frac{e^T\left(H^1(\mathbb{P}^1, f_d^* L_i)^m\right)}{e^T\left(H^0(\mathbb{P}^1, f_d^* L_i)^m\right)} = b(u, w_i, da_i)$$

where $b(u, w, a)$ is defined by (5.22). By Example 2.16,

$$\begin{aligned} \mathrm{ch}_T(H^0(\mathbb{P}^1, f_d^* T\ell_\tau) - H^1(\mathbb{P}^1, f_d^* T\ell_\tau)) &= \sum_{j=0}^{2d} e^{du - ju} \\ &= 1 + \sum_{j=1}^{d} (e^{j\mathbf{w}(\tau,\sigma)/d} + e^{-j\mathbf{w}(\tau,\sigma)/d}). \end{aligned}$$

So

$$\frac{e^T(H^1(\mathbb{P}^1, f_d^* T\ell_\tau)^m)}{e^T(H^0(\mathbb{P}^1, f_d^* T\ell_\tau)^m)} = \prod_{j=1}^{d} \frac{-d^2}{j^2 \mathbf{w}(\tau, \sigma)^2} = \frac{(-1)^d d^{2d}}{(d!)^2 \mathbf{w}(\tau, \sigma)^{2d}}.$$

Therefore,

$$\begin{aligned} \frac{e^T(H^1(\mathbb{P}^1, f_d^* TX)^m)}{e^T(H^0(\mathbb{P}^1, f_d^* TX)^m)} &= \frac{e^T(H^1(\mathbb{P}^1, f_d^* T\ell_\tau)^m)}{e^T(H^0(\mathbb{P}^1, f_d^* T\ell_\tau)^m)} \cdot \prod_{i=1}^{r-1} \frac{e^T(H^1(\mathbb{P}^1, f_d^* L_i)^m)}{e^T(H^0(\mathbb{P}^1, f_d^* L_i)^m)} \\ &= \frac{(-1)^d d^{2d}}{(d!)^2 \mathbf{w}(\tau, \sigma)^{2d}} \prod_{i=1}^{r-1} b(\frac{\mathbf{w}(\tau, \sigma)}{d}, \mathbf{w}(\tau_i, \sigma), da_i). \end{aligned}$$

□

Finally, $f(y_v) = p_{\sigma_v} = f(y(e, v))$, and

$$e^T(T_{p_{\sigma_v}} X) = \mathbf{w}(\sigma_v).$$

From the above discussion, we conclude that

$$\begin{aligned} \frac{e^T(B_5^m)}{e^T(B_2^m)} &= \prod_{v \in V^2(\Gamma)} \mathbf{w}(\sigma_v) \cdot \prod_{(e,v) \in F^S(\Gamma)} \mathbf{w}(\sigma_v) \cdot \prod_{v \in V^S(\Gamma)} \mathbf{h}(\sigma_v, g_v) \cdot \prod_{e \in E(\Gamma)} \mathbf{h}(\tau_e, d_e) \\ &= \prod_{v \in V(\Gamma)} \left(\mathbf{h}(\sigma_v, g_v) \cdot \mathbf{w}(\sigma_v)^{\mathrm{val}(v)}\right) \cdot \prod_{e \in E(\Gamma)} \mathbf{h}(\tau_e, d_e) \end{aligned}$$

where $\mathbf{w}(\sigma)$, $\mathbf{h}(\sigma, g)$, and $\mathbf{h}(\tau, d)$ are defined by (5.18), (5.19), (5.21), respectively.

5.4. Contribution from each graph

5.4.1. Virtual tangent bundle We have $B_1^f = B_2^f$, $B_5^f = 0$. So

$$T^{1,f} = B_4^f = \bigoplus_{v \in V^S(\Gamma)} T\overline{\mathcal{M}}_{g_v, E_v \cup S_v}, \quad T^{2,f} = 0.$$

We conclude that

$$[\prod_{v \in V^S(\Gamma)} \overline{\mathcal{M}}_{g_v, E_v \cup S_v}]^{\mathrm{vir}} = \prod_{v \in V^S(\Gamma)} [\overline{\mathcal{M}}_{g_v, E_v \cup S_v}].$$

5.4.2. Virtual normal bundle Let $N_{\vec{\Gamma}}^{\mathrm{vir}}$ be the pull back of the virtual normal bundle of $\mathcal{F}_{\vec{\Gamma}}$ in $\overline{\mathcal{M}}_{g,n}(X, \beta)$ under $i_{\vec{\Gamma}} : \mathcal{M}_{\vec{\Gamma}} \to \mathcal{F}_{\vec{\Gamma}}$. Then

$$\frac{1}{e^T(N_{\vec{\Gamma}}^{\mathrm{vir}})} = \frac{e^T(B_1^m) e^T(B_5^m)}{e^T(B_2^m) e^T(B_4^m)} = \prod_{v \in V(\Gamma)} \frac{\mathbf{h}(\sigma_v, g_v) \cdot \mathbf{w}(\sigma_v)^{\mathrm{val}(v)}}{\prod_{e \in E_v} (w_{(e,v)} - \psi_{(e,v)})} \cdot \prod_{e \in E(\Gamma)} \mathbf{h}(\tau_e, d_e)$$

5.4.3. Integrand Given $\sigma \in \Sigma(r)$, let

$$i_\sigma^* : A_T^*(X) \to A_T^*(p_\sigma) = \mathbb{Q}[u_1, \ldots, u_r]$$

be induced by the inclusion $i_\sigma : p_\sigma \to X$. Then

$$
\begin{aligned}
& i_{\vec{\Gamma}}^* \prod_{i=1}^{n} \left(ev_i^* \gamma_i^T \cup (\psi_i^T)^{a_i} \right) \\
(5.23) \quad = & \prod_{\substack{v \in V^{1,1}(E) \\ S_v = \{i\}, E_v = \{e\}}} i_{\sigma_v}^* \gamma_i^T (-w_{(e,v)})^{a_i} \cdot \prod_{v \in V^S(\Gamma)} \left(\prod_{i \in S_v} i_{\sigma_v}^* \gamma_i^T \prod_{e \in E_v} \psi_{(e,v)}^{a_i} \right)
\end{aligned}
$$

To unify the stable vertices in $V^S(\Gamma)$ and the unstable vertices in $V^{1,1}(\Gamma)$, we use the following convention: for $a \in \mathbb{Z}_{\geqslant 0}$,

$$(5.24) \qquad \int_{\overline{\mathcal{M}}_{0,2}} \frac{\psi_2^a}{w_1 - \psi_1} = (-w_1)^a.$$

In particular, (5.13) is obtained by setting $a = 0$. With the convention (5.24), we may rewrite (5.23) as

$$(5.25) \qquad i_{\vec{\Gamma}}^* \prod_{i=1}^{n} \left(ev_i^* \gamma_i^T \cup (\psi_i^T)^{a_i} \right) = \prod_{v \in V(\Gamma)} \left(\prod_{i \in S_v} i_{\sigma_v}^* \gamma_i^T \prod_{e \in E_v} \psi_{(e,v)}^{a_i} \right).$$

The following lemma shows that the convention (5.24) is consistent with the stable case $\overline{\mathcal{M}}_{0,n}$, $n \geqslant 3$.

Lemma 5.26. *Let n, a be integers, $n \geqslant 2$, $a \geqslant 0$. Then*

$$\int_{\overline{\mathcal{M}}_{0,n}} \frac{\psi_2^a}{w_1 - \psi_1} = \begin{cases} \dfrac{\prod_{i=0}^{a-1}(n-3-i)}{a!} w_1^{a+2-n}, & n = 2 \text{ or } 0 \leqslant a \leqslant n-3, \\ 0, & \text{otherwise.} \end{cases}$$

Proof. The case $n = 2$ follows from (5.24). For $n \geqslant 3$,

$$\int_{\overline{\mathcal{M}}_{0,n}} \frac{\psi_2^a}{w_1 - \psi_1} = \frac{1}{w_1}\int_{\overline{\mathcal{M}}_{0,n}} \frac{\psi_2^a}{1 - \frac{\psi_1}{w_1}} = w_1^{a+2-n}\int_{\overline{\mathcal{M}}_{0,n}} \psi_1^{n-3-a}\psi_2^a$$

$$= w_1^{a+2-n}\frac{(n-3)!}{(n-3-a)!a!} = \frac{\prod_{i=0}^{a-1}(n-3-i)}{a!} w_1^{a+2-n}.$$

□

5.4.4. Integral The contribution of

$$\int_{[\overline{\mathcal{M}}_{g,n}(X,\beta)^T]^{\mathrm{vir},T}} \frac{i^*\prod_{i=1}^n (ev_i^*\gamma_i^T \cup (\psi_i^T)^{a_i})}{e^T(N^{\mathrm{vir}})}$$

from the fixed locus $\mathcal{F}_{\vec{\Gamma}}$ is given by

$$\frac{1}{|A_{\vec{\Gamma}}|}\prod_{e\in E(\Gamma)} h(\tau_e, d_e) \prod_{v\in V(\Gamma)}\Big(w(\sigma_v)^{\mathrm{val}(v)}\prod_{i\in S_v} i_{\sigma_v}^*\gamma_i^T\Big)$$

$$\cdot \prod_{v\in V(\Gamma)}\int_{\overline{\mathcal{M}}_{g_v,E_v\cup S_v}} \frac{h(\sigma_v, g_v)\cdot\prod_{e\in E_v}\psi_{(e,v)}^{a_i}}{\prod_{e\in E_v}(w_{(e,v)} - \psi_{(e,v)})}$$

where $|A_{\vec{\Gamma}}| = |\mathrm{Aut}(\vec{\Gamma})|\cdot\prod_{e\in E(\Gamma)} d_e$.

5.5. Sum over graphs

Summing over the contribution from each graph $\vec{\Gamma}$ given in Section 5.4.4 above, we obtain the following formula.

Theorem 5.27.

$$\langle \tau_{a_1}(\gamma_1^T)\cdots\tau_{a_n}(\gamma_n^T)\rangle_{g,\beta}^{X_T}$$

$$(5.28)\quad = \sum_{\vec{\Gamma}\in G_{g,n}(X,\beta)} \frac{1}{|\mathrm{Aut}(\vec{\Gamma})|}\prod_{e\in E(\Gamma)}\frac{h(\tau_e, d_e)}{d_e}\prod_{v\in V(\Gamma)}\Big(w(\sigma_v)^{\mathrm{val}(v)}\prod_{i\in S_v} i_{\sigma_v}^*\gamma_i^T\Big)$$

$$\cdot\prod_{v\in V(\Gamma)}\int_{\overline{\mathcal{M}}_{g,E_v\cup S_v}}\frac{h(\sigma_v, g_v)\prod_{i\in S_v}\psi_i^{a_i}}{\prod_{e\in E_v}(w_{(e,v)} - \psi_{(e,v)})}.$$

where $h(\tau, d)$, $w(\sigma)$, $h(\sigma, g)$ are given by (5.21), (5.18), (5.19), respectively, and we have the following convention for the $v \notin V^S(\Gamma)$:

$$\int_{\overline{\mathcal{M}}_{0,1}}\frac{1}{w_1 - \psi_2} = w_1, \quad \int_{\overline{\mathcal{M}}_{0,2}}\frac{1}{(w_1-\psi_1)(w_2-\psi_2)} = \frac{1}{w_1 + w_2},$$

$$\int_{\overline{\mathcal{M}}_{0,2}}\frac{\psi_2^a}{w_1 - \psi_1} = (-w_1)^a, \quad a\in\mathbb{Z}_{\geqslant 0}.$$

Given $g\in\mathbb{Z}_{\geqslant 0}$, r weights $\vec{w} = \{w_1,\ldots,w_r\}$, r partitions $\vec{\mu} = \{\mu^1,\ldots,\mu^r\}$, and $a_1,\ldots,a_k\in\mathbb{Z}$, let $\ell(\mu^i)$ be the length of μ^i, and let $\ell(\vec{\mu}) = \sum_{i=1}^r \ell(\mu^i)$. We define

$$\langle\tau_{a_1},\ldots,\tau_{a_k}\rangle_{g,\vec{\mu},\vec{w}} = \int_{\overline{\mathcal{M}}_{g,\ell(\vec{\mu})+k}}\prod_{i=1}^r\Big(\frac{\Lambda_g^\vee(w_i)w_i^{\ell(\vec{\mu})-1}}{\prod_{j=1}^{\ell(\mu^i)}\frac{w_i}{\mu_j^i} - \psi_j^i}\Big)\prod_{b=1}^k\psi_b^{a_i}.$$

Given $v \in V(\Gamma)$, define $\vec{w}(v) = \{w(\tau, \sigma_v) \mid (\tau, \sigma_v) \in F(\Sigma)\}$. Given $v \in V(\Gamma)$, and $\tau \in E_{\sigma_v}$, let $\mu^{v,\tau}$ be a (possibly empty) partition defined by $\{d_e \mid e \in E_v, \vec{f}(e) = \tau\}$, and define $\vec{\mu}(v) = \{\mu^{v,\tau} \mid (\tau, \sigma_v) \in F(\Sigma)\}$. Then (5.28) can be rewritten as

(5.29)

$$\begin{aligned}&\langle \tau_{a_1}(\gamma_1^T)\cdots\tau_{a_n}(\gamma_n^T)\rangle_{g,\beta}^{X_T}\\&= \sum_{\vec{\Gamma}\in G_{g,n}(X,\beta)} \frac{1}{|\mathrm{Aut}(\vec{\Gamma})|} \prod_{e\in E(\Gamma)} \frac{h(\tau_e, d_e)}{d_e} \prod_{v\in V(\Gamma)} \Big(\prod_{i\in S_v} i_{\sigma_v}^* \gamma_i \langle \prod_{i\in S_v} \tau_{a_i}\rangle_{g_v,\vec{\mu}(v),\vec{w}(v)}\Big).\end{aligned}$$

Recall that

$$g = \sum_{v\in V(\Gamma)} g_v + |E(\Gamma)| - |V(\Gamma)| + 1$$

so

$$2g-2 = \sum_{v\in V(\Gamma)} (2g_v - 2 + \mathrm{val}(v)).$$

Given $\vec{\Gamma} = (\Gamma, \vec{f}, \vec{d}, \vec{g}, \vec{s})$, let $\vec{\Gamma}' = (\Gamma, \vec{f}, \vec{d}, \vec{s})$ be the decorated graph obtained by forgetting the genus map. Let $G_n(X, \beta) = \{\vec{\Gamma}' \mid \vec{\Gamma} \in \cup_{g\geqslant 0} G_{g,n}(X, \beta)\}$. Define

$$\langle \tau_{a_1}(\gamma_1^T), \cdots, \tau_{a_n}(\gamma_n^T) \mid u\rangle_\beta^{X_T} = \sum_{g\geqslant 0} u^{2g-2} \langle \tau_{a_1}(\gamma_1^T), \cdots, \tau_{a_n}(\gamma_n^T)\rangle_{g,\beta}^{X_T} \tag{5.30}$$

$$\langle \tau_{a_1}, \ldots, \tau_{a_k} \mid u\rangle_{\vec{\mu},\vec{w}} = \sum_{g\geqslant 0} u^{2g-2+\ell(\vec{\mu})} \langle \tau_{a_1}, \ldots, \tau_{a_k}\rangle_{g,\vec{\mu},\vec{w}}. \tag{5.31}$$

Then we have the following formula for the generating function (5.30).

Theorem 5.32.

$$\begin{aligned}\langle \tau_{a_1}(\gamma_1^T)\cdots\tau_{a_n}(\gamma_n^T) \mid u\rangle_\beta^{X_T} &= \sum_{\vec{\Gamma}'\in G_n(X,\beta)} \frac{1}{|\mathrm{Aut}(\vec{\Gamma})|} \prod_{e\in E(\Gamma)} \frac{h(\tau_e, d_e)}{d_e}\\&\quad\cdot \prod_{v\in V(\Gamma)} \Big(\prod_{i\in S_v} i_{\sigma_v}^* \gamma_i^T \langle \prod_{i\in S_v} \tau_{a_i} \mid u\rangle_{\vec{\mu}(v),\vec{w}(v)}\Big).\end{aligned} \tag{5.33}$$

6. Smooth Deligne-Mumford stacks

We work over $\mathbb{C}$. Let $\mathcal{X}$ be a smooth Deligne-Mumford (DM) stack. Let $\pi : \mathcal{X} \to X$ be the natural projection to the coarse moduli space X.

6.1. The inertia stack and its rigidification

The inertia stack $\mathcal{IX}$ associated to $\mathcal{X}$ is a smooth DM stack such that the following diagram is Cartesian:

$$\begin{array}{ccc}\mathcal{IX} & \longrightarrow & \mathcal{X}\\ \downarrow & & \downarrow \Delta\\ \mathcal{X} & \xrightarrow{\ \Delta\ } & \mathcal{X}\times\mathcal{X}\end{array}$$

where $\Delta : \mathcal{X} \to \mathcal{X} \times \mathcal{X}$ is the diagonal map. An object in the category $\mathcal{I}\mathcal{X}$ is a pair (x, g), where x is an object in the category $\mathcal{X}$ and $g \in \mathrm{Aut}_{\mathcal{X}}(x)$:

$$\mathrm{Ob}(\mathcal{I}\mathcal{X}) = \{(x, g) \mid x \in \mathcal{X}, g \in \mathrm{Aut}_{\mathcal{X}}(x)\}.$$

The morphisms between two objects in the category $\mathcal{I}\mathcal{X}$ are:

$$\mathrm{Hom}_{\mathcal{I}\mathcal{X}}((x_1, g_1), (x_2, g_2)) = \{h \in \mathrm{Hom}_{\mathcal{X}}(x_1, x_2) \mid h \circ g_1 = g_2 \circ h\}.$$

In particular,

$$\mathrm{Aut}_{\mathcal{I}\mathcal{X}}(x, g) = \{h \in \mathrm{Aut}_{\mathcal{X}}(x) \mid h \circ g = g \circ h\}.$$

The rigidified inertia stack $\bar{\mathcal{I}}\mathcal{X}$ satisfies

$$\mathrm{Ob}(\bar{\mathcal{I}}\mathcal{X}) = \mathrm{Ob}(\mathcal{I}\mathcal{X}), \quad \mathrm{Aut}_{\bar{\mathcal{I}}\mathcal{X}}(x, g) = \mathrm{Aut}_{\mathcal{I}\mathcal{X}}(x, g)/\langle g\rangle,$$

where $\langle g\rangle$ is the subgroup of $\mathrm{Aut}_{\mathcal{I}\mathcal{X}}(x, g)$ generated by g.

There is a more topological interpretation of the inertia stack $\mathcal{I}\mathcal{X}$. Let $L\mathcal{X} = \mathrm{Map}(S^1, \mathcal{X})$ be the stack of loops in $\mathcal{X}$. The rotation of S^1 induces an S^1-action on $L\mathcal{X}$. The stack $(L\mathcal{X})^{S^1}$ of S^1 fixed loops can be identified with the inertial stack $\mathcal{I}\mathcal{X}$. An object in $\mathcal{I}\mathcal{X}$ is a morphism $[\mathrm{pt}/\mathbb{Z}] \to \mathcal{X}$ of stacks, which is determined by $x \in \mathrm{Ob}(\mathcal{X})$ and the image of $1 \in \mathbb{Z}$ in $\mathrm{Aut}(x)$.

There is a natural projection $q : \mathcal{I}\mathcal{X} \to \mathcal{X}$ which sends (x, g) to x. There is a natural involution $\iota : \mathcal{I}\mathcal{X} \to \mathcal{I}\mathcal{X}$ which sends (x, g) to (x, g^{-1}). We assume that $\mathcal{X}$ is connected. Let

$$\mathcal{I}\mathcal{X} = \bigsqcup_{i \in I} \mathcal{X}_i$$

be disjoint union of connected components. There is a distinguished connected component $\mathcal{X}_0$ whose objects are (x, id_x), where $x \in \mathrm{Ob}(\mathcal{X})$, and $\mathrm{id}_x \in \mathrm{Aut}(x)$ is the identity element. The involution ι restricts to an isomorphism $\iota_i : \mathcal{X}_i \to \mathcal{X}_{\iota(i)}$. In particular, $\iota_0 : \mathcal{X}_0 \to \mathcal{X}_0$ is the identity functor.

Example 6.1 (classifying space). *Let G be a finite group. The stack $\mathcal{B}G = [\mathrm{pt}/G]$ is a category which consists of one object x, and $\mathrm{Hom}(x, x) = G$. The objects of its inertia stack $\mathcal{I}\mathcal{B}G$ are*

$$\mathrm{Ob}(\mathcal{I}\mathcal{B}G) = \{(x, g) \mid g \in G\}.$$

The morphisms between two objects are

$$\mathrm{Hom}((x, g_1), (x, g_2)) = \{g \in G \mid g_2 g = g g_1\} = \{g \in G \mid g_2 = g g_1 g^{-1}\}.$$

Therefore

$$\mathcal{I}\mathcal{B}G \cong [G/G]$$

where G acts on G by conjugation. We have

$$\mathcal{I}\mathcal{B}G = \bigsqcup_{c \in \mathrm{Conj}(G)} (\mathcal{B}G)_c$$

where $\mathrm{Conj}(G)$ is the set of conjugacy classes in G, and $(\mathcal{B}G)_c$ is the connected component associated to the conjugacy class $c \in \mathrm{Conj}(G)$.

In particular, when G *is abelian,* $\mathrm{Conj}(G) = G$, *and*

$$\mathcal{I}\mathcal{B}G = \bigsqcup_{g \in G} (\mathcal{B}G)_g$$

where $(\mathcal{B}G)_g = [g/G]$.

Given a positive integer r, let μ_r denote the group of r-th roots of unity. It is a cyclic subgroup of $\mathbb{C}^*$ of order r, generated by

$$\zeta_r := e^{2\pi\sqrt{-1}/r}.$$

Example 6.2. *Let* $\mathbb{C}^*$ *acts on* $\mathbb{C}^2 - \{0\}$ *by*

$$\lambda \cdot (x, y) = (\lambda^2 x, \lambda^3 y), \quad \lambda \in \mathbb{C}^*, \quad (x, y) \in \mathbb{C}^2 - \{0\}.$$

Let $\mathcal{X}$ *be the quotient stack:*

$$\mathcal{X} = [(\mathbb{C}^2 - \{0\})/\mathbb{C}^*] = \mathbb{P}[2,3].$$

Then the coarse moduli space is $X = \mathbb{P}^1$.

We have

$$\mathcal{I}\mathcal{X} = \bigsqcup_{i=0}^{3} \mathcal{X}_i$$

where

$$\begin{aligned}
\mathcal{X}_0 = \mathcal{X}, &\ \ \mathrm{Ob}(\mathcal{X}_0) = \{((x,y),1) \mid (x,y) \in \mathbb{C}^2 - \{0\}\},\\
\mathcal{X}_1 = \mathcal{B}\mu_2, &\ \ \mathrm{Ob}(\mathcal{X}_1) = \{((1,0),-1)\},\\
\mathcal{X}_2 = \mathcal{B}\mu_3, &\ \ \mathrm{Ob}(\mathcal{X}_2) = \{((0,1), e^{2\pi\sqrt{-1}/3})\},\\
\mathcal{X}_3 = \mathcal{B}\mu_3, &\ \ \mathrm{Ob}(\mathcal{X}_3) = \{((0,1), e^{4\pi\sqrt{-1}/3})\}.
\end{aligned}$$

We have

$$\iota_0 : \mathcal{X}_0 \to \mathcal{X}_0, \quad \iota_1 : \mathcal{X}_1, \to \mathcal{X}_1, \quad \iota_2 : \mathcal{X}_2 \to \mathcal{X}_3.$$

6.2. Age

Given any object (x, g) in $\mathcal{I}\mathcal{X}$, $g : T_x\mathcal{X} \to T_x\mathcal{X}$ is a linear isomorphism such that $g^r = \mathrm{id}$, where r is the order of g. The eigenvalues of $g : T_x\mathcal{X} \to T_x\mathcal{X}$ are $\zeta_r^{l_1}, \ldots, \zeta_r^{l_n}$, where $l_i \in \{0, 1, \ldots, r-1\}$, $n = \dim_{\mathbb{C}} \mathcal{X}$. Define

$$\mathrm{age}(x, g) := \frac{l_1 + \cdots + l_n}{r}.$$

Then $\mathrm{age} : \mathcal{I}\mathcal{X} \to \mathbb{Q}$ is constant on each connected component $\mathcal{X}_i$ of $\mathcal{I}\mathcal{X}$. Define $\mathrm{age}(\mathcal{X}_i) = \mathrm{age}(x, g)$ where (x, g) is any object in $\mathcal{X}_i$. Note that

$$\mathrm{age}(\mathcal{X}_i) + \mathrm{age}(\mathcal{X}_{\iota(i)}) = \dim_{\mathbb{C}} \mathcal{X} - \dim_{\mathbb{C}} \mathcal{X}_i.$$

Example 6.3. *Let* $\mathcal{X}_0$, $\mathcal{X}_1$, $\mathcal{X}_2$, $\mathcal{X}_3$ *be defined as in Example 6.2. Then*

$$\mathrm{age}(\mathcal{X}_0) = 0, \quad \mathrm{age}(\mathcal{X}_1) = \frac{1}{2}, \quad \mathrm{age}(\mathcal{X}_2) = \frac{1}{3}, \quad \mathrm{age}(\mathcal{X}_3) = \frac{2}{3}.$$

6.3. The orbifold cohomology group and operational Chow group

In [10], W. Chen and Y. Ruan introduced the orbifold cohomology group of a complex orbifold. See [1, Section 4.4] for a more algebraic version.

The rational Chen-Ruan orbifold cohomology group of $\mathcal{X}$ is defined to be

$$H^*_{\mathrm{orb}}(\mathcal{X}) := \bigoplus_{a \in \mathbb{Q}_{\geqslant 0}} H^a_{\mathrm{orb}}(\mathcal{X})$$

where

$$H^a_{\mathrm{orb}}(\mathcal{X}) = \bigoplus_{i \in I} H^{a - 2\mathrm{age}(\mathcal{X}_i)}(\mathcal{X}_i).$$

The Chen-Ruan orbifold cohomology H^*_{orb} is denoted by H^*_{CR} in some papers, for example [24].

The rational orbifold operational Chow group of $\mathcal{X}$ is defined to be

$$A^*_{\mathrm{orb}}(\mathcal{X}) := \bigoplus_{a \in \mathbb{Q}_{\geqslant 0}} A^a_{\mathrm{orb}}(\mathcal{X})$$

where

$$A^a_{\mathrm{orb}}(\mathcal{X}) = \bigoplus_{i \in I} A^{a - \mathrm{age}(\mathcal{X}_i)}(\mathcal{X}_i).$$

Suppose that $\mathcal{X}$ is proper, and let

$$\int_{\mathcal{X}} : A^*(\mathcal{X}) \to \mathbb{Q}$$

be defined as in Section 2.6. Similarly, we have

$$\int_{\mathcal{X}} : H^*(\mathcal{X}) \to \mathbb{Q}.$$

The orbifold Poincaré pairing is defined by

$$(\alpha, \beta)_{\mathrm{orb}} := \begin{cases} \int_{\mathcal{X}_i} \alpha \cup \iota_i^* \beta, & j = \iota(i), \\ 0, & j \neq \iota(i), \end{cases}$$

where $\alpha \in H^*(\mathcal{X}_i)$, $\beta \in H^*(\mathcal{X}_j)$.

Example 6.4. *Let $\mathcal{X} = \mathbb{P}[2,3]$, and let $\mathcal{X}_0, \mathcal{X}_1, \mathcal{X}_2, \mathcal{X}_3$ be defined as in Example 6.2. Let $H \in H^2(\mathcal{X}) = A^1(\mathcal{X})$ be the pull back of the hyperplane class of $H^2(\mathbb{P}^1) = A^1(\mathbb{P}^1)$ under the map $\mathcal{X} = \mathbb{P}[2,3] \to \mathbb{P}^1$ to the coarse moduli space. We have*

$$H^*_{\mathrm{orb}}(\mathcal{X}) = H^0_{\mathrm{orb}}(\mathbb{P}[2,3]) \oplus H^{\frac{2}{3}}_{\mathrm{orb}}(\mathcal{X}) \oplus H^1_{\mathrm{orb}}(\mathcal{X}) \oplus H^{\frac{4}{3}}_{\mathrm{orb}}(\mathcal{X}) \oplus H^2_{\mathrm{orb}}(\mathcal{X}),$$

$$A^*_{\mathrm{orb}}(\mathcal{X}) = A^0_{\mathrm{orb}}(\mathcal{X}) \oplus A^{\frac{1}{3}}_{\mathrm{orb}}(\mathcal{X}) \oplus A^{\frac{1}{2}}_{\mathrm{orb}}(\mathcal{X}) \oplus A^{\frac{2}{3}}_{\mathrm{orb}}(\mathcal{X}) \oplus A^1_{\mathrm{orb}}(\mathcal{X}),$$

where

$$\begin{aligned}
H^0_{\rm orb}(\mathcal{X}) &= A^0_{\rm orb}(\mathcal{X}) = H^0(\mathcal{X}_0) = A^0(\mathcal{X}_0) = \mathbb{Q}1,\\
H^{\frac{2}{3}}_{\rm orb}(\mathcal{X}) &= A^{\frac{1}{3}}_{\rm orb}(\mathcal{X}) = H^0(\mathcal{X}_2) = A^0(\mathcal{X}_2) = \mathbb{Q}1_{\frac{1}{3}},\\
H^{1}_{\rm orb}(\mathcal{X}) &= A^{\frac{1}{2}}_{\rm orb}(\mathcal{X}) = H^0(\mathcal{X}_1) = A^0(\mathcal{X}_1) = \mathbb{Q}1_{\frac{1}{2}},\\
H^{\frac{4}{3}}_{\rm orb}(\mathcal{X}) &= A^{\frac{2}{3}}_{\rm orb}(\mathcal{X}) = H^0(\mathcal{X}_3) = A^0(\mathcal{X}_3) = \mathbb{Q}1_{\frac{2}{3}},\\
H^{2}_{\rm orb}(\mathcal{X}) &= A^{1}_{\rm orb}(\mathcal{X}) = H^2(\mathcal{X}_0) = A^1(\mathcal{X}_0) = \mathbb{Q}H.
\end{aligned}$$

7. Orbifold Gromov-Witten theory

In [9], Chen-Ruan developed Gromov-Witten theory for symplectic orbifolds. The algebraic counterpart, the Gromov-Witten theory for smooth DM stacks, was developed by Abramovich-Graber-Vistoli [1, 2]. In this section, we give a brief review of algebraic orbifold Gromov-Witten theory, following [2].

7.1. Twisted curves and their moduli

An n-pointed, genus g twisted curve is a connected proper one-dimensional DM stack $\mathcal{C}$ together with n disjoint closed substacks $\mathfrak{x}_1, \dots, \mathfrak{x}_n$ of $\mathcal{C}$, such that

(1) $\mathcal{C}$ is étale locally a nodal curve;
(2) formally locally near a node, $\mathcal{C}$ is isomorphic to the quotient stack

$$[\mathrm{Spec}(\mathbb{C}[x,y]/(xy))/\mu_r],$$

where the action of $\zeta \in \mu_r$ is given by $\zeta \cdot (x,y) = (\zeta x, \zeta^{-1} y)$;
(3) each $\mathfrak{x}_i \subset \mathcal{C}$ is contained in the smooth locus of $\mathcal{C}$;
(4) each stack $\mathfrak{x}_i$ is an étale gerbe over Spec$\mathbb{C}$ *with a section* (hence trivialization);
(5) $\mathcal{C}$ is a scheme outside the twisted points $\mathfrak{x}_1, \dots, \mathfrak{x}_n$ and the singular locus;
(6) the coarse moduli space C is a nodal curve of arithmetic genus g.

Let $\pi : \mathcal{C} \to C$ be the projection to the coarse moduli space, and let $x_i = \pi(\mathfrak{x}_i)$. Then $x_1, \dots, x_n$ are distinct smooth points of C, and $(C, x_1, \dots, x_n)$ is an n-pointed, genus g prestable curve.

Let $\mathcal{M}^{\rm tw}_{g,n}$ be the moduli of n-pointed, genus g twisted curves. Then $\mathcal{M}^{\rm tw}_{g,n}$ is a smooth algebraic stack, locally of finite type [39].

7.2. Riemann-Roch theorem for twisted curves

Let $(\mathcal{C}, \mathfrak{x}_1, \dots, \mathfrak{x}_n)$ be an n-pointed, genus g twisted curve, and let $(C, x_1, \dots, x_n)$ be the coarse curve, which is an n-pointed, genus g prestable curve. Let $\mathcal{E} \to \mathcal{X}$ be a vector bundle over $\mathcal{X}$. Then $\mathfrak{x}_i \cong \mathcal{B}\mu_{r_i}$, and $\zeta_{r_i} \in \mu_{r_i}$ acts on $\mathcal{E}|_{\mathfrak{x}_i}$ with eigenvalues $\zeta_{r_i}^{l_1}, \dots, \zeta_{r_i}^{l_N}$, where $l_i \in \{0, 1, \dots, r_i - 1\}$ and $N = \mathrm{rank}\mathcal{E}$. Define

$$\mathrm{age}_{x_i}(\mathcal{E}) := \frac{l_1 + \cdots + l_N}{r_i} \in \mathbb{Q}.$$

The Riemann-Roch theorem for twisted curves says

$$\chi(\mathcal{E}) = \int_{\mathcal{C}} c_1(\mathcal{E}) + \mathrm{rank}(\mathcal{E})(1-g) - \sum_{i=1}^{n} \mathrm{age}_{x_i}(\mathcal{E}). \tag{7.1}$$

Given a real number x, let $\lfloor x \rfloor$ denote the largest integer which is less or equal to x, and let $\langle x \rangle = x - \lfloor x \rfloor$.

Example 7.2. *Let* $\mathcal{C} = \mathbb{P}[2,3]$, $\mathfrak{x}_1 = [0,1]$, $\mathfrak{x}_2 = [1,0]$. *Then* $(\mathcal{C}, \mathfrak{x}_1, \mathfrak{x}_2)$ *is a 2-pointed, genus 0 twisted curve. The coarse moduli curve is* $(C, x_1, x_2) = (\mathbb{P}^1, [0,1], [1,0])$. *Let* $\mathcal{L}_n = \mathcal{O}_{\mathcal{C}}(n\mathfrak{x}_2)$, *where* $n \in \mathbb{Z}$. *Then*

$$\int_{\mathcal{C}} c_1(\mathcal{E}) = \frac{n}{2}, \quad \mathrm{rank}(\mathcal{L}_n) = 1, \quad \mathrm{age}_{x_1}(\mathcal{L}_n) = \langle \frac{n}{2} \rangle, \quad \mathrm{age}_{x_2}(\mathcal{L}_n) = 0,$$

so

$$\chi(\mathcal{L}_n) = 1 + \lfloor \frac{n}{2} \rfloor.$$

Let $\pi : \mathcal{C} = \mathbb{P}[2,3] \to C = \mathbb{P}^1$ *be the projection to the coarse moduli space. Then* $\pi^* \mathcal{O}_{\mathbb{P}^1}(kx_2) = \mathcal{L}_{2k}$. *We have*

$$\chi(\mathcal{L}_{2k}) = k + 1 = \chi(\mathcal{O}_{\mathbb{P}^1}(kx_2))$$

as expected.

7.3. Moduli of twisted stable maps

Let $\mathcal{X}$ be a proper smooth DM stack with a projective coarse moduli space X, and let β be an effective curve class in X. An n-pointed, genus g, degree β twisted stable map to $\mathcal{X}$ is a representable morphism $f : \mathcal{C} \to \mathcal{X}$, where the domain $\mathcal{C}$ is an n-pointed, genus g twisted curve, and the induced morphism $C \to X$ between the coarse moduli spaces is an n-pointed, genus g, degree β stable map to X.

Let $\overline{\mathcal{M}}_{g,n}(\mathcal{X}, \beta)$ be the moduli stack of n-pointed, genus g, degree β twisted stable maps to $\mathcal{X}$. Then $\overline{\mathcal{M}}_{g,n}(\mathcal{X}, \beta)$ is a proper DM stack.

For $j = 1, \ldots, n$, there are evaluation maps $\mathrm{ev}_j : \overline{\mathcal{M}}_{g,n}(\mathcal{X}, \beta) \to \mathcal{I}\mathcal{X}$. Given $\vec{i} = (i_1, \ldots, i_n)$, where $i_j \in I$, define

$$\overline{\mathcal{M}}_{g,\vec{i}}(\mathcal{X}, \beta) := \bigcap_{j=1}^{n} \mathrm{ev}_j^{-1}(\mathcal{X}_{i_j}).$$

Then $\overline{\mathcal{M}}_{g,\vec{i}}(\mathcal{X}, \beta)$ is a union of connected components of $\overline{\mathcal{M}}_{g,n}(\mathcal{X}, \beta)$, and

$$\overline{\mathcal{M}}_{g,n}(\mathcal{X}, \beta) = \bigsqcup_{\vec{i} \in I^n} \overline{\mathcal{M}}_{g,\vec{i}}(\mathcal{X}, \beta).$$

Remark 7.3. *In the definition of twisted curves in Section 7.1, if we replace (4) by*

(4)′ *each stack* $\mathfrak{x}_i$ *is an étale gerbes over* $\mathrm{Spec}\mathbb{C}$;

i.e. without a section, then the resulting moduli space is $\mathcal{K}_{g,n}(\mathcal{X}, \beta)$ *in [2], and the evaluation maps take values in the rigidified inertial stack* $\bar{\mathcal{I}}\mathcal{X}$ *instead of the initial stack* $\mathcal{I}\mathcal{X}$.

7.4. Obstruction theory and virtual fundamental classes

The tangent space T^1 and the obstruction space T^2 at a moduli point $[f : (\mathcal{C}, \mathfrak{x}_1, \ldots, \mathfrak{x}_n) \to \mathcal{X}] \in \overline{\mathcal{M}}_{g,n}(\mathcal{X}, \beta)$ fit in the *tangent-obstruction exact sequence*:

$$\begin{aligned} 0 \to & \mathrm{Ext}^0_{\mathcal{O}_\mathcal{C}}(\Omega_\mathcal{C}(\mathfrak{x}_1 + \cdots + \mathfrak{x}_n), \mathcal{O}_\mathcal{C}) \to H^0(\mathcal{C}, f^*T_\mathcal{X}) \to T^1 \\ \to & \mathrm{Ext}^1_{\mathcal{O}_\mathcal{C}}(\Omega_\mathcal{C}(\mathfrak{x}_1 + \cdots + \mathfrak{x}_n), \mathcal{O}_\mathcal{C}) \to H^1(\mathcal{C}, f^*T_\mathcal{X}) \to T^2 \to 0 \end{aligned} \tag{7.4}$$

where

- $\mathrm{Ext}^0_{\mathcal{O}_\mathcal{C}}(\Omega_\mathcal{C}(\mathfrak{x}_1 + \cdots + \mathfrak{x}_n), \mathcal{O}_\mathcal{C})$ is the space of infinitesimal automorphisms of the domain $(\mathcal{C}, \mathfrak{x}_1, \ldots, \mathfrak{x}_n)$,
- $\mathrm{Ext}^1_{\mathcal{O}_\mathcal{C}}(\Omega_\mathcal{C}(\mathfrak{x}_1 + \cdots + \mathfrak{x}_n), \mathcal{O}_\mathcal{C})$ is the space of infinitesimal deformations of the domain $(\mathcal{C}, \mathfrak{x}_1, \ldots, \mathfrak{x}_n)$,
- $H^0(\mathcal{C}, f^*T_\mathcal{X})$ is the space of infinitesimal deformations of the map f, and
- $H^1(\mathcal{C}, f^*T_\mathcal{X})$ is the space of obstructions to deforming the map f.

T^1 and T^2 form sheaves $\mathcal{T}^1$ and $\mathcal{T}^2$ on the moduli space $\overline{\mathcal{M}}_{g,\vec{i}}(\mathcal{X}, \beta)$. This defines a perfect obstruction theory of virtual dimension

$$d^{\mathrm{vir}}_{\vec{i}} = \int_\beta c_1(T_\mathcal{X}) + (\dim \mathcal{X} - 3)(1-g) + n - \sum_{j=1}^n \mathrm{age}(\mathcal{X}_{i_j})$$

on $\overline{\mathcal{M}}_{g,\vec{i}}(\mathcal{X}, \beta)$, which defines a virtual fundamental class

$$[\overline{\mathcal{M}}_{g,\vec{i}}(\mathcal{X}, \beta)]^{\mathrm{vir}} \in A_{d^{\mathrm{vir}}_{\vec{i}}}(\overline{\mathcal{M}}_{g,\vec{i}}(\mathcal{X}, \beta)).$$

The *weighted virtual fundamental class* is defined by

$$[\overline{\mathcal{M}}_{g,\vec{i}}(\mathcal{X}, \beta)]^w := \Big(\prod_{j=1}^n r_{i_j}\Big) [\overline{\mathcal{M}}_{g,\vec{i}}(\mathcal{X}, \beta)]^{\mathrm{vir}}.$$

7.5. Hurwitz-Hodge integrals

By Example 6.1, when $\mathcal{X} = \mathcal{B}G$ we have

$$\mathcal{I}\mathcal{B}G = \bigsqcup_{c \in \mathrm{Conj}(G)} (\mathcal{B}G)_c$$

where $\mathrm{Conj}(G)$ is the set of conjugacy classes of G. Give $\vec{c} = (c_1, \ldots, c_n) \in \mathrm{Conj}(G)^n$, let $\overline{\mathcal{M}}_{g,\vec{c}}(\mathcal{B}G) = \overline{\mathcal{M}}_{g,\vec{c}}(\mathcal{B}G, \beta = 0)$. Then $\overline{\mathcal{M}}_{g,\vec{c}}(\mathcal{B}G)$ is a union of connected components of $\overline{\mathcal{M}}_{g,n}(\mathcal{B}G) := \overline{\mathcal{M}}_{g,n}(\mathcal{B}G, 0)$, and

$$\overline{\mathcal{M}}_{g,n}(\mathcal{B}G) = \bigsqcup_{\vec{c} \in \mathrm{Conj}(G)^n} \overline{\mathcal{M}}_{g,\vec{c}}(\mathcal{B}G).$$

We now fix a genus g and n conjugacy classes $\vec{c} = (c_1, \ldots, c_n) \in \mathrm{Conj}(G)^n$. Let $\pi : \mathcal{U} \to \overline{\mathcal{M}}_{g,\vec{c}}(\mathcal{B}G)$ be the universal curve, and let $f : \mathcal{U} \to \mathcal{B}G$ be the universal map. Let $\rho : G \to GL(V)$ be an irreducible representation of G, where V is a finite dimensional vector space over $\mathbb{C}$. Then $\mathcal{E}_\rho := [V/G]$ is a vector bundle over $\mathcal{B}G = [pt/G]$. We have

$$\pi_* f^* \mathcal{E}_\rho = \begin{cases} \mathcal{O}_{\overline{\mathcal{M}}_{g,\vec{c}}(\mathcal{B}G)}, & \text{if } \rho : G \to GL(1, \mathbb{C}) \text{ is the trivial representation,} \\ 0, & \text{otherwise.} \end{cases}$$

The ρ-twisted Hurwitz-Hodge bundle $\mathbb{E}_\rho$ can be defined as the dual of the vector bundle $R^1\pi_* f^*\mathcal{E}_\rho$. If $\rho = 1$ is the trivial representation, then $\mathbb{E}_1 = \epsilon^*\mathbb{E}$, where $\epsilon : \overline{\mathcal{M}}_{g,\vec{c}}(\mathcal{B}G) \to \overline{\mathcal{M}}_{g,n}$, and $\mathbb{E} \to \overline{\mathcal{M}}_{g,n}$ is the Hodge bundle of $\overline{\mathcal{M}}_{g,n}$. So $\text{rank}\mathbb{E}_1 = g$. If ρ is a nontrivial irreducible representation, it follows from the Riemann-Roch theorem for twisted curves (see Section 7.2) that

$$\text{rank}\mathbb{E}_\rho = \text{rank}(\mathcal{E}_\rho)(g-1) + \sum_{j=1}^{n} \text{age}_{c_j}(\mathcal{E}_\rho), \tag{7.5}$$

where $\text{age}_{c_j}(\mathcal{E}_\rho)$ is given as follows. Choose $g \in c_j$. Let $r > 0$ be the order of g in G, let $N = \text{rank}\mathcal{E}_\rho = \dim V$. If the eigenvalues of $\rho(g) \in GL(V) = GL(N, \mathbb{C})$ are $\zeta_r^{l_1}, \ldots, \zeta_r^{l_N}$, where $l_1, \ldots, l_N \in \{0, 1, \ldots, r-1\}$, then

$$\text{age}_{c_j}(\mathcal{E}_\rho) = \frac{l_1 + \cdots + l_N}{r}.$$

The definition is independent of choice of $g \in c_j$. The map $\det \circ \rho : G \to GL(1, \mathbb{C})$ descends to a map $\det \circ \rho : \text{Conj}(G) \to GL(1, \mathbb{C})$. We have

$$\prod_{j=1}^{n} \det \circ \rho(c_j) = 1,$$

so

$$\sum_{j=1}^{n} \text{age}_{c_j}(\mathcal{E}_\rho) \in \mathbb{Z}.$$

Note that when G is abelian, any irreducible representation of G is 1-dimensional, so $\text{rank}(\mathcal{E}_\rho) = 1$ for any irreducible representation ρ of G.

- *Hodge classes.* Given an irreducible representation ρ of G, define

$$\lambda_i^\rho = c_i(\mathbb{E}_\rho) \in A^i(\overline{\mathcal{M}}_{g,\vec{c}}(\mathcal{B}G)), \quad i = 1, \ldots, \text{rank}\mathbb{E}_\rho.$$

- *Descendant classes.* There is a map $\epsilon : \overline{\mathcal{M}}_{g,\vec{c}}(\mathcal{B}G) \to \overline{\mathcal{M}}_{g,n}$. Define

$$\bar{\psi}_j = \epsilon^*\psi_j \in A^1(\overline{\mathcal{M}}_{g,\vec{c}}(\mathcal{B}G)), \quad j = 1, \ldots, n.$$

Hurwitz-Hodge integrals are top intersection numbers of Hodge classes λ_i^ρ and descendant classes $\bar{\psi}_j$:

$$\int_{\overline{\mathcal{M}}_{g,\vec{c}}(\mathcal{B}G)} \bar{\psi}_1^{a_1} \cdots \bar{\psi}_n^{a_n} (\lambda_1^{\rho_1})^{k_1} \cdots (\lambda_g^{\rho_g})^{k_g}. \tag{7.6}$$

In [46], J. Zhou described an algorithm of computing Hurwitz-Hodge integrals, as follows. By Tseng's orbifold quantum Riemann-Roch theorem [43], Hurwitz-Hodge integrals can be reconstructed from descendant integrals on $\overline{\mathcal{M}}_{g,\vec{c}}(\mathcal{B}G)$:

$$\int_{\overline{\mathcal{M}}_{g,\vec{c}}(\mathcal{B}G)} \bar{\psi}_1^{a_1}\cdots\bar{\psi}_n^{a_n}. \tag{7.7}$$

Jarvis-Kimura relate the descendant integrals on $\overline{\mathcal{M}}_{g,\vec{c}}(\mathcal{B}G)$ to those on $\overline{\mathcal{M}}_{g,n}$ [23]. We now state their result. Given $g \in \mathbb{Z}_{\geqslant 0}$ and $\vec{c} = (c_1,\dots,c_n) \in \mathrm{Conj}(G)^n$, let

$$V^G_{g,\vec{c}} := \{(a_1,b_1,\dots,a_g,b_g,e_1,\dots,e_n)\in G^{2g+n} \mid \prod_{i=1}^g [a_i,b_i] = \prod_{j=1}^n e_j,\ e_j\in c_j\}.$$

Then $\overline{\mathcal{M}}_{g,\vec{c}}(\mathcal{B}G)$ is nonempty iff $V^G_{g,\vec{c}}$ is nonempty.

Theorem 7.8 (Jarvis-Kimura [23, Proposition 3.4]). *Suppose that $2g-2+n>0$ and $V^G_{g,\vec{c}}$ is nonempty. Then*

$$\int_{\overline{\mathcal{M}}_{g,\vec{c}}(\mathcal{B}G)} \bar{\psi}_1^{a_1}\cdots\bar{\psi}_n^{a_n} = \frac{|V^G_{g,\vec{c}}|}{|G|}\int_{\overline{\mathcal{M}}_{g,n}} \psi_1^{a_1}\cdots\psi_n^{a_n}.$$

When G is abelian, each c_i is an element in G. $V^G_{g,\vec{c}}$ is nonempty iff $c_1\cdots c_n = 1$, and in this case $V^G_{g,\vec{c}} = G^{2g}$.

Corollary 7.9. *Let G be a finite abelian group. Suppose that $2g-2+n>0$, and $\vec{c}=(c_1,\dots,c_n)\in G^n$, where $c_1\cdots c_n = 1$. Then*

$$\int_{\overline{\mathcal{M}}_{g,\vec{c}}(\mathcal{B}G)} \bar{\psi}_1^{a_1}\cdots\bar{\psi}_n^{a_n} = |G|^{2g-1}\int_{\overline{\mathcal{M}}_{g,n}} \psi_1^{a_1}\cdots\psi_n^{a_n}.$$

7.6. Orbifold Gromov-Witten invariants

There is a morphism $\epsilon : \overline{\mathcal{M}}_{g,\vec{i}}(\mathcal{X},\beta)\to\overline{\mathcal{M}}_{g,n}(X,\beta)$. Define $\bar{\psi}_i = \epsilon^*\psi_i$. Let

$$\gamma_j \in A^{d_j}(\mathcal{X}_{i_j}) \subset A_{\mathrm{orb}}^{d_j+\mathrm{age}(\mathcal{X}_{i_j})}(\mathcal{X}).$$

Define orbifold Gromov-Witten invariants

$$\langle \bar{\tau}_{a_1}\gamma_1,\dots,\bar{\tau}_{a_n}\gamma_n\rangle^{\mathcal{X}}_{g,\beta} := \int_{[\overline{\mathcal{M}}_{g,\vec{i}}(\mathcal{X},\beta)]^w}\prod_{j=1}^n ev_j^*\gamma_j\bar{\psi}_j^{a_j} \tag{7.10}$$

which is zero unless

$$\sum_{j=1}^n (d_j + \mathrm{age}(\mathcal{X}_{i_j}) + a_j) = \int_\beta c_1(T_{\mathcal{X}}) + (1-g)(\dim\mathcal{X}-3)+n.$$

More generally, let

$$\gamma_j \in H^{d_j}(\mathcal{X}_{i_j}) \subset H_{\mathrm{orb}}^{d_j+2\mathrm{age}(\mathcal{X}_{i_j})}(\mathcal{X}),$$

and define orbifold Gromov-Witten invariants (7.10). Then it is zero unless

$$\sum_{j=1}^n (d_j + 2\mathrm{age}(\mathcal{X}_{i_j}) + 2a_j) = 2\left(\int_\beta c_1(T_{\mathcal{X}}) + (1-g)(\dim\mathcal{X}-3)+n\right).$$

8. Toric Deligne-Mumford stacks

In [7], Borisov, Chen, and Smith defined toric DM stacks in terms of stacky fans. Toric DM stacks are smooth DM stacks, and their coarse moduli spaces are simplicial toric varieties. A toric DM stack is called a *toric orbifold* if its generic stabilizer is trivial. Later, more geometric definitions of toric orbifolds and toric DM stacks are given by Iwanari [21, 22] and by Fantechi-Mann-Nironi [15], respectively.

8.1. Stacky fans

In this subsection, we recall the definition of stacky fans. Let N be a finitely generated abelian group, and let $N_{\mathbb{R}} = N \otimes_{\mathbb{Z}} \mathbb{R}$. We have a short exact sequence of abelian groups:

$$1 \to N_{\text{tor}} \to N \to \bar{N} = N/N_{\text{tor}} \to 1,$$

where N_{tor} is the subgroup of torsion elements in N. Then N_{tor} is a finite abelian group, and $\bar{N} \cong \mathbb{Z}^r$, where $r = \dim_{\mathbb{R}} N_{\mathbb{R}}$. The natural projection $N \to \bar{N}$ is denoted by $b \mapsto \bar{b}$.

Let Σ be a simplicial fan in $N_{\mathbb{R}}$ (see [17]), and let $\Sigma(1) = \{\rho_1, \dots, \rho_s\}$ be the set of 1-dimensional cones in the fan Σ. We assume that $\rho_1, \dots, \rho_s$ span $N_{\mathbb{R}}$, and fix $b_i \in N$ such that $\rho_i = \mathbb{R}_{\geqslant 0}\bar{b}_i$. A *stacky fan* $\mathbf{\Sigma}$ is defined as the data (N, Σ, β), where $\beta : \widetilde{N} := \oplus_{i=1}^{s} \mathbb{Z}\widetilde{b}_i \cong \mathbb{Z}^s \to N$ is a group homomorphism defined by $\widetilde{b}_i \mapsto b_i$. By assumption, the cokernel of β is finite.

We introduce some notation.

(1) $M = \text{Hom}(N, \mathbb{Z}) = \text{Hom}(\bar{N}, \mathbb{Z}) \cong (\mathbb{Z}^r)^*$.
(2) $\widetilde{M} = \text{Hom}(\widetilde{N}, \mathbb{Z}) \cong (\mathbb{Z}^s)^*$.
(3) Let $\Sigma(d)$ be the set of d-dimensional cones in Σ. Given $\sigma \in \Sigma(d)$, let $N_\sigma \subset N$ be the subgroup generated by $\{b_i \mid \rho_i \subset \sigma\}$, and let $\bar{N}_\sigma$ be the rank d sublattice of $\bar{N}$ generated by $\{\bar{b}_i \mid \rho_i \subset \sigma\}$. Let $M_\sigma = \text{Hom}(\bar{N}_\sigma, \mathbb{Z})$ be the dual lattice of $\bar{N}_\sigma$.

Given $\sigma \in \Sigma(d)$, the surjective group homomorphism $N_\sigma \to \bar{N}_\sigma$ induces an injective group homomorphism $\text{Hom}(\bar{N}_\sigma, \mathbb{Z}) \to \text{Hom}(N_\sigma, \mathbb{Z})$ which is indeed an isomorphism. So $\text{Hom}(N_\sigma, \mathbb{Z}) \cong M_\sigma \cong \mathbb{Z}^d$.

8.2. The Gale dual

The finite abelian group N_{tor} is of the form $\oplus_{j=1}^{l} \mathbb{Z}_{a_j}$. We choose a projective resolution of N:

$$0 \to \mathbb{Z}^l \xrightarrow{Q} \mathbb{Z}^{r+l} \to N \to 0.$$

Choose a map $B : \widetilde{N} \to \mathbb{Z}^{r+l}$ lifting $\beta : \widetilde{N} \to N$. Let $i_1 : \widetilde{N} \oplus \mathbb{Z}^l \to \widetilde{N}$ and $i_2 : \widetilde{N} \oplus \mathbb{Z}^l \to \mathbb{Z}^l$ be inclusions of the first and second factors, respectively. We have the following commutative diagram:

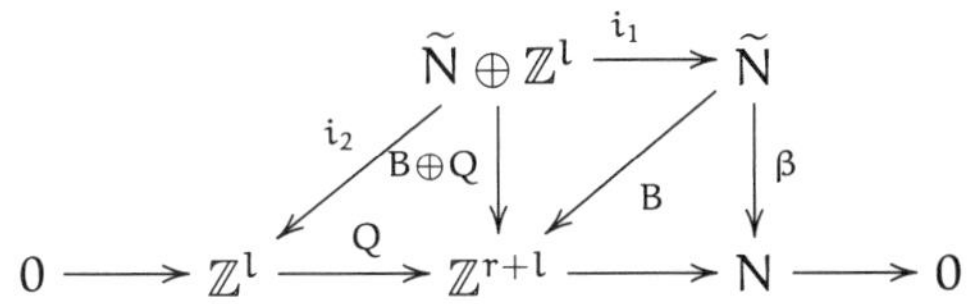

Define the dual group $DG(\beta)$ to be the the cokernel of $B^* \oplus Q^* : (\mathbb{Z}^{r+l})^* \longrightarrow \widetilde{M} \oplus (\mathbb{Z}^l)^*$. The Gale dual of the map $\beta : \widetilde{N} \to N$ is $\beta^\vee : \widetilde{M} \to DG(\beta)$.

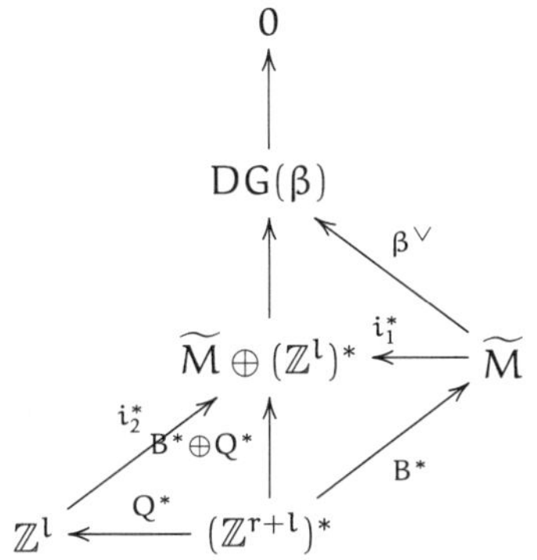

8.3. Construction of the toric DM stack

We follow [7, Section 3]. Applying $\mathrm{Hom}(-, \mathbb{C}^*)$ to $\beta^\vee : \widetilde{M} \to DG(\beta)$, one obtains

$$\phi : G_\Sigma := \mathrm{Hom}(DG(\beta), \mathbb{C}^*) \to \widetilde{T} := \mathrm{Hom}(\widetilde{M}, \mathbb{C}^*).$$

Let $G = \mathrm{Ker}\phi$. Then $G \cong \prod_{j=1}^{l} \mu_{a_j}$, where $\mu_{a_j} \subset \mathbb{C}^*$ is the group of a_j-th roots of unity, which is isomorphic to $\mathbb{Z}_{a_j}$. Let $\mathcal{B}G$ denote the quotient stack $[\{1\}/G]$. The algebraic torus $\widetilde{T}$ acts on $\mathbb{C}^s$ by

$$(\widetilde{t}_1, \ldots, \widetilde{t}_s) \cdot (z_1, \ldots, z_s) = (\widetilde{t}_1 z_1, \ldots, \widetilde{t}_s z_s), \quad (\widetilde{t}_1, \ldots, \widetilde{t}_s) \in \widetilde{T}, \quad (z_1, \ldots, z_s) \in \mathbb{C}^s.$$

Let G_Σ act on $\mathbb{C}^s$ by $g \cdot z := \phi(g) \cdot z$, where $g \in G_\Sigma$, $z \in \mathbb{C}^s$. Let $\mathcal{O}(\mathbb{C}^s) = \mathbb{C}[z_1, \ldots, z_s]$ be the coordinate ring of $\mathbb{C}^s$. Let I_Σ be the ideal of $\mathcal{O}(\mathbb{C}^s)$ generated by

$$\{\prod_{\rho_i \not\subset \sigma} z_i : \sigma \in \Sigma\}$$

and let $Z(I_\Sigma)$ be the closed subscheme of $\mathbb{C}^s$ defined by I_Σ. Then $U := \mathbb{C}^s - Z(I_\Sigma)$ is a quasi-affine variety over $\mathbb{C}$. The toric DM stack associated to the stacky fan $\mathbf{\Sigma}$ is defined to be the quotient stack

$$\mathcal{X}_{\mathbf{\Sigma}} := [U/G_\Sigma].$$

It is a smooth DM stack whose generic stabilizer is G, and its coarse moduli space is the toric variety X_Σ defined by the simplicial fan Σ. There is an open dense immersion

$$\iota : \mathcal{T} = [\widetilde{T}/G_{\mathbf{\Sigma}}] \hookrightarrow \mathcal{X}_{\mathbf{\Sigma}} = [U/G_{\mathbf{\Sigma}}],$$

where $\mathcal{T} \cong (\mathbb{C}^*)^r \times \mathcal{B}G$ is a DM torus. The action of $\mathcal{T}$ on itself extends to an action $a : \mathcal{T} \times \mathcal{X}_{\mathbf{\Sigma}} \to \mathcal{X}_{\mathbf{\Sigma}}$.

Example 8.1 (weighted projective spaces). *Let $w_1, \ldots, w_{r+1}$ be positive integers. The weighted projective space $\mathbb{P}[w_1, \ldots, w_{r+1}]$ is defined to be the quotient stack*

$$[(\mathbb{C}^{r+1} - \{0\})/\mathbb{C}^*],$$

where $\mathbb{C}^*$ acts on $\mathbb{C}^{r+1} - \{0\}$ by

$$\lambda \cdot (z_1, \ldots, z_{r+1}) = (\lambda^{w_1} z_1, \ldots, \lambda^{w_{r+1}} z_{r+1}).$$

$\mathbb{P}[w_1, \ldots, w_{r+1}]$ is a smooth DM stack and, if and only if g.c.d.$(w_1, \ldots, w_{r+1}) = 1$, an orbifold. We will show that it is indeed a toric DM stack defined by some stacky fan $\mathbf{\Sigma} = (\Sigma, N, \beta)$.

Let $e =$ g.c.d.$(w_1, \ldots, w_{r+1}) \in \mathbb{Z}_{>0}$, so that $(w_1, \ldots, w_{r+1}) = e(w_1', \ldots, w_{r+1}')$, where $w_1', \ldots, w_{r+1}'$ are positive integers such that g.c.d.$(w_1', \ldots, w_{r+1}') = 1$. Define

$$\widetilde{N} = \bigoplus_{i=1}^{r+1} \mathbb{Z}\widetilde{b}_i \cong \mathbb{Z}^{r+1}.$$

Define $\widetilde{b}_0 := \sum_{i=1}^{r+1} w_i' \widetilde{b}_i$, which is a primitive vector in the lattice $\widetilde{N}$, and define

$$\bar{N} = \widetilde{N}/\mathbb{Z}\widetilde{b}_0 \cong \mathbb{Z}^r.$$

Applying $\mathrm{Hom}(-, \mathbb{Z})$ to the surjective map $\widetilde{N} \to \bar{N}$, we obtain an injective map

$$i : M = \mathrm{Hom}(\bar{N}, \mathbb{Z}) \to \widetilde{M} = \mathrm{Hom}(\widetilde{N}, \mathbb{Z})$$

where M can be identified with the following rank r sublattice of $\widetilde{M}$:

$$M = \{\widetilde{m} \in \widetilde{M} \mid \langle \widetilde{m}, \widetilde{b}_0 \rangle = 0\}.$$

Let $\bar{b}_i \in \bar{N}$ be image of $\widetilde{b}_i$. Define $N = \bar{N} \oplus \mathbb{Z}/e\mathbb{Z}$, and let $b_i = (\bar{b}_i, 1)$. Define $\beta : \widetilde{N} \to N$ by $\beta(\widetilde{b}_i) = b_i$. A projective resolution of N is given by

$$0 \to \mathbb{Z} \xrightarrow{Q} \bar{N} \oplus \mathbb{Z} \to N = \bar{N} \oplus \mathbb{Z}/e\mathbb{Z} \to 0,$$

where $Q(1) = (0, e)$. The map $\beta : \widetilde{N} \to N$ can be lifted to $B : \widetilde{N} \to \bar{N} \oplus \mathbb{Z}$, $\widetilde{b}_i \mapsto (\bar{b}_i, 1)$. Let $\{\widetilde{b}_1^*, \ldots, \widetilde{b}_{r+1}^*\}$ be the $\mathbb{Z}$-basis of $\widetilde{M}$ dual to the $\mathbb{Z}$-basis $\{\widetilde{b}_1, \ldots, \widetilde{b}_{r+1}\}$ of $\widetilde{N}$. The map $B^* \oplus Q^* : M \oplus \mathbb{Z} \to \widetilde{M} \oplus \mathbb{Z}$ is given by

$$(m, 0) \mapsto (i(m), 0), \quad (0, 1) \mapsto (\sum_{i=1}^{r+1} \widetilde{b}_i^*, e).$$

The map $\widetilde{M} \oplus \mathbb{Z} \to DG(\beta) = \mathbb{Z}$ *is given by*

$$(\widetilde{b}_i^*, 0) \mapsto w_i \quad (0,1) \mapsto \sum_{j=1}^{r+1} w_j'.$$

Applying $\mathrm{Hom}(-, \mathbb{C}^*)$ *to* $[w_1 \ \cdots \ w_{r+1}] : \widetilde{M} = \mathbb{Z}^{r+1} \to DG(\beta) = \mathbb{Z}$, *we obtain*

$$\phi : G_\Sigma = \mathbb{C}^* \to \widetilde{T} = \mathrm{Hom}(\widetilde{M}, \mathbb{C}^*) = (\mathbb{C}^*)^{r+1}, \quad \lambda \mapsto (\lambda^{w_1}, \ldots, \lambda^{w_{r+1}}).$$

Therefore,

$$\mathcal{X}_\Sigma = (\mathbb{C}^{r+1} - \{0\})/G_\Sigma = \mathbb{P}[w_1, \ldots, w_{r+1}].$$

Example 8.2 (complete 1-dimensional toric orbifolds). *Suppose that* $\boldsymbol{\Sigma} = (\Sigma, N, \beta)$ *is a stacky fan which defines a 1-dimensional complete toric orbifold* $\mathcal{X}_{\boldsymbol{\Sigma}}$. *The coarse moduli space* X_Σ *must be* $\mathbb{P}^1$, *the unique 1-dimensional complete simplicial toric variety. So we have*

$$N = \mathbb{Z}, \quad \widetilde{N} = \mathbb{Z}^2, \quad v_1 = 1, \quad v_2 = 1 \quad b_1 = s_1, \quad b_2 = -s_2,$$

where s_1, s_2 *are positive integers. Let* $\boldsymbol{\Sigma}_{s_1,s_2}$ *denote the stacky fan*

$$(\Sigma, N = \mathbb{Z}, \beta = [\ s_1 \ \ -s_2\ ]),$$

and let $G_{s_1,s_2} = G_{\boldsymbol{\Sigma}_{s_1,s_2}}$. *There is a commutative diagram*

$$\begin{array}{ccccccccc}
1 & \longrightarrow & G_{s_1,s_2} & \xrightarrow{\phi_{s_1,s_2}} & \widetilde{T} = (\mathbb{C}^*)^2 & \xrightarrow{\pi_{s_1,s_2}} & T = \mathbb{C}^* & \longrightarrow & 1 \\
 & & \downarrow{\scriptstyle \hat{p}_{s_1,s_2}} & & \downarrow{\scriptstyle \widetilde{p}_{s_1,s_2}} & & \downarrow{\scriptstyle p} & & \\
1 & \longrightarrow & G_\Sigma = \mathbb{C}^* & \xrightarrow{\phi} & \widetilde{T} = (\mathbb{C}^*)^2 & \xrightarrow{\pi} & T = \mathbb{C}^* & \longrightarrow & 1
\end{array} \tag{8.3}$$

where the rows are short exact sequences of abelian groups. The arrows are group homomorphisms given explicitly as follows:

$$\widetilde{p}_{s_1,s_2}(\widetilde{t}_1, \widetilde{t}_2) = (\widetilde{t}_1^{s_1}, \widetilde{t}_2^{s_2}), \quad p(t) = t, \quad \pi_{s_1,s_2}(\widetilde{t}_1, \widetilde{t}_2) = \widetilde{t}_1^{s_1} \widetilde{t}_2^{-s_2}, \quad \pi(\widetilde{t}_1, \widetilde{t}_2) = \widetilde{t}_1 \widetilde{t}_2^{-1}.$$

$$G_{s_1,s_2} = \mathrm{Ker}(\pi_{s_1,s_2}) = \{(\widetilde{t}_1, \widetilde{t}_2) \in \widetilde{T} = (\mathbb{C}^*)^2 \mid \widetilde{t}_1^{s_1} \widetilde{t}_2^{-s_2} = 1\}$$
$$G_\Sigma = \mathrm{Ker}(\pi) = \{(\widetilde{t}_1, \widetilde{t}_2) \in \widetilde{T} = (\mathbb{C}^*)^2 \mid \widetilde{t}_1 \widetilde{t}_2^{-1} = 1\}$$

Following [24], *let* $\mathcal{C}_{s_1,s_2}$ *be the toric orbifold defined by the stacky fan* $\boldsymbol{\Sigma}_{s_1,s_2}$:

$$\mathcal{C}_{s_1,s_2} := \mathcal{X}_{\boldsymbol{\Sigma}_{s_1,s_2}} = \left[(\mathbb{C}^2 - \{(0,0)\})/G_{s_1,s_2}\right].$$

Note that Example 6.2 is a special case of this: $\mathbb{P}[2,3] = \mathcal{C}_{3,2}$. *More generally, when* s_1 *and* s_2 *are relatively prime,* $G_{s_1,s_2} \cong \mathbb{C}^*$ *and*

$$\mathcal{C}_{s_1,s_2} = [(\mathbb{C}^2 - \{(0,0)\})/\mathbb{C}^*] = \mathbb{P}[s_2, s_1]$$

where $\mathbb{C}^*$ *acts on* $\mathbb{C}^2$ *by* $\lambda \cdot (z_1, z_2) = (\lambda^{s_2} z_1, \lambda^{s_1} z_2)$. *In general,* $G_{s_1,s_2} \cong \mathbb{C}^* \times \mu_d$, *where* $d = \mathrm{g.c.d.}(s_1, s_2)$ *(see* [15, Example 7.29]*).*

The coarse moduli space of $\mathcal{C}_{s_1,s_2}$ *is the projective line:*

$$X_\Sigma = (\mathbb{C}^2 - \{(0,0)\})/\mathbb{C}^* = \mathbb{P}^1,$$

where $\mathbb{C}^*$ *acts on* $\mathbb{C}^2$ *by* $\lambda \cdot (z_1, z_2) = (\lambda z_1, \lambda z_2)$.

We have

$$\mathcal{IC}_{s_1,s_2} = \coprod_{\substack{\nu \in \mathbb{Z} \\ -s_2 < \nu < s_1}} \mathcal{C}_{s_1,s_2,\nu}$$

where

$$\mathcal{C}_{s_1,s_2,\nu} = \begin{cases} \mathcal{B}\mu_{s_1}, & 1 \leqslant \nu \leqslant s_1 - 1, \\ \mathcal{C}_{s_1,s_2}, & \nu = 0, \\ \mathcal{B}\mu_{s_2}, & 1 - s_2 \leqslant \nu \leqslant -1, \end{cases}$$

and

$$\mathrm{Ob}(\mathcal{C}_{s_1,s_2,\nu}) = \begin{cases} \{((0,1), \zeta_{s_1}^{\nu})\}, & 1 \leqslant \nu \leqslant s_1 - 1, \\ \{((x,y), 1) \mid (x,y) \in \mathbb{C}^2 - \{0\}\}, & \nu = 0, \\ \{((1,0), \zeta_{s_2}^{-\nu})\}, & 1 - s_2 \leqslant \nu \leqslant -1. \end{cases}$$

We have

$$\iota_0 : \mathcal{C}_{s_1,s_2,0} \to \mathcal{C}_{s_1,s_2,0},$$

and

$$\iota_\nu : \mathcal{C}_{s_1,s_2,\nu} \to \begin{cases} \mathcal{C}_{s_1,s_2,s_1-\nu}, & 1 \leqslant \nu \leqslant s_1 - 1, \\ \mathcal{C}_{s_1,s_2,s_2+\nu}, & 1 - s_2 \leqslant \nu \leqslant -1. \end{cases}$$

8.4. Rigidification

We define the *rigidification* of $\mathbf{\Sigma} = (N, \Sigma, \beta)$ to be the stacky fan $\mathbf{\Sigma}^{\mathrm{rig}} := (\bar{N}, \Sigma, \bar{\beta})$, where $\bar{\beta}$ is the composition of $\beta : \tilde{N} \to N$ with the projection $N \to \bar{N}$. Note that M, $\bar{N}_\sigma$, and M_σ defined in Section 8.1 depend only on $\mathbf{\Sigma}^{\mathrm{rig}}$. The generic stabilizer of the toric DM stack $\mathcal{X}_{\mathbf{\Sigma}^{\mathrm{rig}}}$ is trivial because $\bar{N} \cong \mathbb{Z}^n$ is torsion free. So $\mathcal{X}_{\mathbf{\Sigma}^{\mathrm{rig}}}$ is a toric orbifold. There is a morphism of stacky fans $\mathbf{\Sigma} \to \mathbf{\Sigma}^{\mathrm{rig}}$ which induces a morphism of toric DM stacks $\pi^{\mathrm{rig}} : \mathcal{X}_{\mathbf{\Sigma}} \to \mathcal{X}_{\mathbf{\Sigma}^{\mathrm{rig}}}$. The toric orbifold $\mathcal{X}_{\mathbf{\Sigma}^{\mathrm{rig}}}$ is called the *rigidification* of the toric DM stack $\mathcal{X}_{\mathbf{\Sigma}}$. The morphism $\pi^{\mathrm{rig}} : \mathcal{X}_{\mathbf{\Sigma}} \to \mathcal{X}_{\mathbf{\Sigma}^{\mathrm{rig}}}$ makes $\mathcal{X}_{\mathbf{\Sigma}}$ a G-gerbe over $\mathcal{X}_{\mathbf{\Sigma}^{\mathrm{rig}}}$.

$G_{\mathbf{\Sigma}^{\mathrm{rig}}} = G_{\mathbf{\Sigma}}/G$ is a subgroup of $\tilde{T}$. Let $T := \tilde{T}/G_{\mathbf{\Sigma}^{\mathrm{rig}}} \cong (\mathbb{C}^*)^r$. There is an open dense immersion

$$\iota^{\mathrm{rig}} : T = [\tilde{T}/G_{\mathbf{\Sigma}^{\mathrm{rig}}}] \hookrightarrow \mathcal{X}_{\mathbf{\Sigma}^{\mathrm{rig}}} = [U/G_{\mathbf{\Sigma}^{\mathrm{rig}}}].$$

8.5. Lifting the fan

Let $\mathbf{\Sigma} = (N, \Sigma, \beta)$ be a stacky fan, where $N \cong \mathbb{Z}^r$. Let U be defined as in Section 8.3. The open embedding $U \hookrightarrow \mathbb{C}^s$ is $\tilde{T}$-equivariant, and can be viewed as a morphism between smooth toric varieties. More explicitly, consider the s-dimensional cone

$$\tilde{\sigma}_0 = \mathrm{Cone}(\{\tilde{b}_1, \ldots, \tilde{b}_s\}) \subset \tilde{N}_{\mathbb{R}} = \tilde{N} \otimes_{\mathbb{Z}} \mathbb{R},$$

and let $\widetilde{\Sigma}_0 \subset \widetilde{N}_{\mathbb{R}}$ be the fan which consists of all the faces of $\widetilde{\sigma}_0$. Then $\mathbb{C}^s$ is the smooth toric variety defined by the fan $\widetilde{\Sigma}_0$. We define a subfan $\widetilde{\Sigma} \subset \widetilde{\Sigma}_0$ as follows. Given $\sigma \in \Sigma(d)$, such that $\sigma \cap \{\bar{b}_1, \ldots, \bar{b}_s\} = \{\bar{b}_{i_1}, \ldots, \bar{b}_{i_d}\}$, let

$$\widetilde{\sigma} = \mathrm{Cone}(\{\widetilde{b}_{i_1}, \ldots, \widetilde{b}_{i_d}\}) \subset \widetilde{N}_{\mathbb{R}}.$$

Then there is a bijection $\Sigma \to \widetilde{\Sigma}$ given by $\sigma \mapsto \widetilde{\sigma}$, and U is the smooth toric variety defined by $\widetilde{\Sigma}$.

For any d-dimensional cone $\widetilde{\sigma} \in \widetilde{\Sigma}$, let $I = \{i \mid \rho_i \subset \sigma\}$, and define

$$\begin{aligned}
U_{\widetilde{\sigma}} &= \mathrm{Spec}\mathbb{C}[\widetilde{\sigma}^\vee \cap \widetilde{M}] = \mathbb{C}^s - \{\prod_{i \notin I} z_i = 0\} \\
&= \{(z_1, \ldots, z_s) \in \mathbb{C}^s \mid z_i \neq 0 \text{ if } i \notin I\} \cong \mathbb{C}^d \times (\mathbb{C}^*)^{s-d}, \\
O_{\widetilde{\sigma}} &= \{(z_1, \ldots, z_s) \in \mathbb{C}^s \mid z_i = 0 \text{ iff } i \in I\} \cong (\mathbb{C}^*)^{s-d} \\
V(\widetilde{\sigma}) &= \{(z_1, \ldots, z_s) \in \mathbb{C}^s \mid z_i = 0 \text{ if } i \in I\} \cong \mathbb{C}^{s-d} \\
\widetilde{T}_{\widetilde{\sigma}} &= \{(\widetilde{t}_1, \ldots, \widetilde{t}_s) \in \widetilde{T} \mid \widetilde{t}_i = 1 \text{ for } i \notin I\} \cong (\mathbb{C}^*)^d.
\end{aligned}$$

Then

- $U_{\widetilde{\sigma}}$ is a Zariski open subset of U.
- $O_{\widetilde{\sigma}}$ is an orbit of the $\widetilde{T}$-action on U. The stabilizer of the $\widetilde{T}$-action on $O_{\widetilde{\sigma}}$ is $\widetilde{T}_{\widetilde{\sigma}}$, so $O_{\widetilde{\sigma}} = \widetilde{T}/\widetilde{T}_{\widetilde{\sigma}}$.
- $V(\widetilde{\sigma})$ is a closed subvariety of U.

Let $G_\sigma = \phi^{-1}(\widetilde{T}_{\widetilde{\sigma}})$ be the stabilizer of $G_{\mathbf{\Sigma}}$-action on $O_{\widetilde{\sigma}}$. Then G_σ is a finite abelian group. In particular, when $\sigma = \{0\}$ is the zero dimensional cone, $G_{\{0\}} = \mathrm{Ker}\phi = G$ is the generic stabilizer. Note that if $\sigma \subset \sigma'$ then $\widetilde{T}_{\widetilde{\sigma}} \subset \widetilde{T}_{\widetilde{\sigma}'}$ and $G_\sigma \subset G_{\sigma'}$.

We have $\widetilde{T}$-equivariant open embeddings

$$\widetilde{T} \hookrightarrow X_{\widetilde{\Sigma}} = U \hookrightarrow X_{\widetilde{\Sigma}_0} = \mathbb{C}^s.$$

We define

$$\mathcal{X}_\sigma := [U_{\widetilde{\sigma}}/G_{\mathbf{\Sigma}}], \quad \mathbf{V}(\sigma) = [V(\widetilde{\sigma})/G_{\mathbf{\Sigma}}], \quad \mathbf{O}_\sigma = [O_{\widetilde{\sigma}}/G_{\mathbf{\Sigma}}].$$

Then

- $\mathcal{X}_\sigma$ is an open substack of $\mathcal{X}$.
- $\mathbf{O}_\sigma$ is an orbit of the $\mathcal{T}$-action on $\mathcal{X}$.
- $\mathbf{V}(\sigma)$ is the closure of $\mathbf{O}_\sigma$.
- $\mathbf{V}(\sigma) \to \mathbf{V}(\sigma)^{\mathrm{rig}}$ is a G_σ-gerbe.

The $\widetilde{T}$-equivariant line bundles on $U_{\widetilde{\sigma}} = \mathrm{Spec}\mathbb{C}[\widetilde{\sigma}^\vee \cap \widetilde{M}]$ are in one-to-one correspondence with characters in $\mathrm{Hom}(\widetilde{T}_{\widetilde{\sigma}}, \mathbb{C}^*)$. Moreover, we have canonical isomorphisms

$$\mathrm{Hom}(\widetilde{T}_{\widetilde{\sigma}}, \mathbb{C}^*) \cong \widetilde{M}/(\widetilde{\sigma}^\perp \cap \widetilde{M}) \cong M_\sigma.$$

Given $\chi \in M_\sigma$, let $\mathcal{O}_{U_{\widetilde{\sigma}}}(\chi)$ denote the $\widetilde{T}$-equivariant line bundle on $U_{\widetilde{\sigma}}$ associated to $\chi \in M_\sigma$, and let $\mathcal{O}_{\mathcal{X}_\sigma}(\chi)$ denote the corresponding $\mathcal{T}$-equivariant line bundle on $\mathcal{X}_\sigma = [U_{\widetilde{\sigma}}/G_{\mathbf{\Sigma}}]$. Let $\widetilde{\chi} \in \widetilde{M}$ be any representative of the coset $\chi \in \widetilde{M}/(\widetilde{\sigma}^\perp \cap \widetilde{M}) \cong M_\sigma$.

The T-weights of $\Gamma(\mathcal{X}_\sigma, \mathcal{O}_{\mathcal{X}_\sigma}(\chi))$ are in one-to-one correspondence with points in $(\chi + \sigma^\vee) \cap M$.

More generally, a $\widetilde{T}$-equivariant coherent sheaf on U descends to a $\mathcal{T}$-equivariant coherent sheaf on $\mathcal{X} = [U/G_\Sigma]$; indeed, we may regard this as the definition of a $\mathcal{T}$-equivariant coherent sheaf on $\mathcal{X}$. Composing the map $T \to \mathcal{T} = [T/G]$ with the $\mathcal{T}$-action $a : \mathcal{T} \times \mathcal{X} \to \mathcal{X}$ on the toric DM stack $\mathcal{X}$, we obtain a T-action $\bar{a} : T \times \mathcal{X} \to \mathcal{X}$ on $\mathcal{X}$. Following Kresch [33], we define the $\mathcal{T}$-equivariant Chow groups of the stack $\mathcal{X}$ to be the Chow groups of the Artin stack $[\mathcal{X}/\mathcal{T}]$:

$$A^*_{\mathcal{T}}(\mathcal{X}) := A^*([\mathcal{X}/\mathcal{T}]), \quad A^*_{\mathcal{T}}(\mathcal{X}; \mathbb{Z}) := A^*([\mathcal{X}/\mathcal{T}]; \mathbb{Z}).$$

The identification of stacks

$$[\mathcal{X}/\mathcal{T}] = [U/\widetilde{T}]$$

implies that we may identify these Chow groups with the $\widetilde{T}$-equivariant Chow groups of U:

$$A^*_{\mathcal{T}}(\mathcal{X}) = A^*_{\widetilde{T}}(U), \quad A^*_{\mathcal{T}}(\mathcal{X}; \mathbb{Z}) = A^*_{\widetilde{T}}(U; \mathbb{Z}).$$

Note that we have an isomorphism of rational Chow groups

$$A^*_{\mathcal{T}}(\mathcal{X}) = A^*_T(\mathcal{X}).$$

As the following example shows, this isomorphism does not generally hold for integral Chow groups.

Example 8.4. *Let $\mathcal{X} = \mathbb{P}[w]$ be the zero dimensional weighted projective space, where w is an integer and $w > 1$. Then $\mathcal{X} = \mathcal{T} = \mathcal{B}\mu_w$ and $T = \{1\}$.*

$$A^1_{\mathcal{T}}(\mathcal{X}; \mathbb{Z}) = 0, \quad A^1_T(\mathcal{X}; \mathbb{Z}) = \mathbb{Z}/w\mathbb{Z}.$$

Let $\mathcal{V}$ be a $\mathcal{T}$-equivariant vector bundle over $\mathcal{X}$. Under the identification $A^*_T(\mathcal{X}) = A^*_{\mathcal{T}}(\mathcal{X})$ (or equivalently, $H^*_T(\mathcal{X}) = H^*_{\mathcal{T}}(\mathcal{X})$), the T-equivariant Chern classes of $\mathcal{V}$ are equal to the $\mathcal{T}$-equivariant Chern classes of $\mathcal{V}$:

$$c^T_k(\mathcal{V}) = c^{\mathcal{T}}_k(\mathcal{V}), \quad 0 \leqslant k \leqslant \operatorname{rank}\mathcal{V}.$$

Example 8.5. *Let $\mathcal{X} = \mathcal{C}_{s_1,s_2}$ be defined as in Example 8.2. Let $\mathfrak{p}_1 = [0,1]$ and $\mathfrak{p}_2 = [1,0]$ be the two T-fixed (stacky) points in $\mathcal{C}_{s_1,s_2}$. Then any T-equivariant line bundle on $\mathcal{C}_{s_1,s_2}$ is of the form*

$$\mathcal{L}_{c_1,c_2} = \mathcal{O}_{\mathcal{X}}(c_1\mathfrak{p}_1 + c_2\mathfrak{p}_2), \quad c_1, c_2 \in \mathbb{Z}.$$

We will compute

$$\operatorname{ch}^T\big(H^0(\mathcal{X}, \mathcal{L}_{c_1,c_2}) - H^1(\mathcal{X}, \mathcal{L}_{c_1,c_2})\big).$$

We have

$$N = \mathbb{Z}, \quad \Sigma = \{\{0\}, \quad \rho_1 = [0, \infty), \quad \rho_2 = (-\infty, 0]\}$$

Let

$$\mathcal{X}_1 = \mathcal{X}_{\rho_1}, \quad \mathcal{X}_2 = \mathcal{X}_{\rho_2}, \quad \mathcal{X}_{12} = \mathcal{X}_{\{0\}} = \mathcal{X}_1 \cap \mathcal{X}_2 = T = \mathbb{C}^*.$$

The cohomology groups $H^0(\mathcal{X}, \mathcal{L}_{c_1,c_2})$ *and* $H^1(\mathcal{X}, \mathcal{L}_{c_1,c_2})$ *are the kernel and cokernel of the following Čech complex:*

$$0 \to \Gamma(\mathcal{X}_1, \mathcal{L}_{c_1,c_2}) \oplus \Gamma(\mathcal{X}_2, \mathcal{L}_{c_1,c_2}) \xrightarrow{\delta} \Gamma(\mathcal{X}_{12}, \mathcal{L}_{c_1,c_2}) \to 0,$$

where $\delta(s_1, s_2) = s_1\big|_{\mathcal{X}_{12}} - s_2\big|_{\mathcal{X}_{12}}$. *Let* $u \in M$ *be the dual of the* $\mathbb{Z}$*-basis of* $v_1 \in N$. *Then*

$$\begin{aligned} \mathrm{ch}^T\big(\Gamma(\mathcal{X}_1, \mathcal{L}_{c_1,c_2})\big) &= \sum_{m\in\mathbb{Z}, s_1 m \geqslant -c_1} e^{mu} \\ \mathrm{ch}^T\big(\Gamma(\mathcal{X}_2, \mathcal{L}_{c_1,c_2})\big) &= \sum_{m\in\mathbb{Z}, -s_2 m \geqslant -c_2} e^{mu} \\ \mathrm{ch}^T\big(\Gamma(\mathcal{X}_{12}, \mathcal{L}_{c_1,c_2})\big) &= \sum_{m\in\mathbb{Z}} e^{mu}. \end{aligned}$$

Therefore,

$$\mathrm{ch}^T(H^0(\mathcal{X}, \mathcal{L}_{c_1,c_2})) = \begin{cases} \displaystyle\sum_{m\in\mathbb{Z}, -\frac{c_1}{s_1} \leqslant m \leqslant \frac{c_2}{s_2}} e^{mu}, & \frac{c_1}{s_1} + \frac{c_2}{s_2} \geqslant 0, \\ 0, & \frac{c_1}{s_1} + \frac{c_2}{s_2} < 0, \end{cases}$$

$$\mathrm{ch}^T(H^1(\mathcal{X}, \mathcal{L}_{c_1,c_2})) = \begin{cases} 0, & \frac{c_1}{s_1} + \frac{c_2}{s_2} \geqslant 0, \\ \displaystyle\sum_{m\in\mathbb{Z}, \frac{c_2}{s_2} < m < -\frac{c_1}{s_1}} e^{mu}, & \frac{c_1}{s_1} + \frac{c_2}{s_2} < 0. \end{cases}$$

More generally, suppose that a torus T' *(of any dimension) acts on the total space of* $\mathcal{L} = \mathcal{L}_{c_1,c_2}$, *such that*

$$c_1^{T'}(T_{p_1}\mathcal{C}_{s_1,s_2}) = \frac{-w_1}{s_1}, \quad c_1^{T'}(T_{p_2}\mathcal{C}_{s_1,s_2}) = \frac{w_1}{s_2}, \quad c_1^{T'}(\mathcal{L}_{p_1}) = w_2, \quad c_1^{T'}(\mathcal{L}_{p_2}) = w_3,$$

where $w_1, w_2, w_3 \in H^2(BT'; \mathbb{Q})$. *Then*

$$w_3 = w_2 + a w_1,$$

where

$$a = \frac{c_1}{s_1} + \frac{c_2}{s_2} \in \frac{\mathrm{g.c.d.}(s_1, s_2)}{s_1 s_2}\mathbb{Z}.$$

Let

$$\epsilon = \langle \frac{c_2}{s_2} \rangle \in \{0, \frac{1}{s_2}, \ldots, \frac{s_2 - 1}{s_2}\}.$$

Then

$$\mathrm{ch}^T(H^0(\mathcal{X}, \mathcal{L})) = \begin{cases} \displaystyle\sum_{m\in\mathbb{Z}, -\epsilon \leqslant m \leqslant a-\epsilon} e^{w_3 - (m+\epsilon)w_1} = \sum_{m=0}^{\lfloor a-\epsilon \rfloor} e^{w_3 - (m+\epsilon)w_1}, & a \geqslant 0, \\ 0, & a < 0, \end{cases}$$

$$\mathrm{ch}^T(H^1(\mathcal{X}, \mathcal{L})) = \begin{cases} 0, & a \geqslant 0, \\ \displaystyle\sum_{m\in\mathbb{Z}, \epsilon < m < \epsilon - a} e^{w_3 + (m-\epsilon)w_1} = \sum_{m=1}^{\lceil \epsilon - a - 1 \rceil} e^{w_3 + (m-\epsilon) + w_1}, & a < 0. \end{cases}$$

8.6. Toric graph

The coarse moduli space of the toric DM stack $\mathcal{X} = \mathcal{X}_{\mathbf{\Sigma}}$ defined by a stacky fan $\mathbf{\Sigma} = (N, \Sigma, \beta)$ is the simplicial toric variety $X = X_\Sigma$ defined by the simplicial fan $\Sigma \subset N_{\mathbb{R}}$. The definitions of the 1-skeleton X^1 and the flags in Σ in Section 4.3 for smooth toric varieties also work for simplicial toric varieties. The sets $\Sigma(r)$, $\Sigma(r-1)$ and $F(\Sigma)$ define a connected graph Υ. Let $T = (\mathbb{C}^*)^r$ be the torus acting on the coarse moduli X, and let $\mathcal{T}$ be the DM torus acting on $\mathcal{X}$. Then $\pi : \mathcal{X} \to X$ restricts to $\mathcal{T} \to T$, and $\mathcal{T} = T$ if and only if $\mathcal{X}$ is a toric orbifold.

Given $\sigma \in \Sigma(r)$ let $p_\sigma = V(\sigma)$ (resp. $\mathfrak{p}_\sigma = \mathcal{V}(\sigma)$) be the associated zero dimensional T-orbit (resp. $\mathcal{T}$-orbit) in X (resp. $\mathcal{X}$). Then $\mathfrak{p}_\sigma = [p_\sigma/G_\sigma] = \mathcal{B}G_\sigma$. Given $\tau \in \Sigma(r-1)$, let $\ell_\tau = V(\tau)$ (resp. $\mathfrak{l}_\tau = \mathcal{V}(\tau)$) be the associated one dimensional T-orbit closure (resp. $\mathcal{T}$-orbit closure) in X (resp. $\mathcal{X}$). Then $\mathfrak{l}_\tau$ is a 1-dimensional toric DM stack, and $\mathfrak{l}_\tau \to \mathfrak{l}_\tau^{\mathrm{rig}}$ is a G_τ-gerbe. Define a map $r : F(\Sigma) \to \mathbb{Z}_{>0}$ by

$$r(\tau, \sigma) = \frac{|G_\sigma|}{|G_\tau|}.$$

There there is a short exact sequence of abelian groups

$$1 \to G_\tau \longrightarrow G_\sigma \overset{\phi(\tau,\sigma)}{\longrightarrow} \mu_{r(\tau,\sigma)} \to 1,$$

where $\phi(\tau, \sigma) : G_\sigma \to \mathbb{C}^*$ is the character of the irreducible G_σ-representation $T_{\mathfrak{p}_\sigma}\mathfrak{l}_\tau$.

Given $\tau \in \Sigma(r-1)$, there are two cases:

(1) Suppose that $\tau \in \Sigma(r-1)_c$. Then τ is the intersection of two r-dimensional cones σ, σ'. We have $\ell_\tau \cong \mathbb{P}^1$ and $\mathfrak{l}_\tau^{\mathrm{rig}} \cong \mathcal{C}_{r(\tau,\sigma),r(\tau,\sigma')}$.

(2) Suppose that $\tau \notin \Sigma(r-1)_c$. Then there is a unique r-dimensional cone σ which contains τ. We have $\ell_\tau \cong \mathbb{C}$ and $\mathfrak{l}_\tau^{\mathrm{rig}} \cong [\mathbb{C}/\mu_{r(\tau,\sigma)}]$.

Given $(\tau, \sigma) \in F(\Gamma)$, let $\mathbf{w}(\tau, \sigma) \in M_\sigma$ be characterized by

$$\langle \mathbf{w}(\tau, \sigma), b_i \rangle = \begin{cases} 0 & \text{if } \rho_i \subset \tau, \\ 1 & \text{if } \rho_i \subset \sigma \text{ and } \rho_i \not\subset \tau. \end{cases}$$

This gives rise to a map $\mathbf{w} : F(\Sigma) \to M_{\mathbb{Q}}$ satisfying the following properties.

(1) $\mathbf{w}(\tau, \sigma)$ is the weight of T-action on $T_{\mathfrak{p}_\sigma}\mathfrak{l}_\tau$, the tangent line to $\mathfrak{l}_\tau$ at $\mathfrak{p}_\sigma$. In other words,

$$\mathbf{w}(\tau, \sigma) = c_1^{\mathcal{T}}(T_{\mathfrak{p}_\sigma}\mathfrak{l}_\tau) = H^2_{\mathcal{T}}(\mathfrak{p}_\sigma) = M_{\mathbb{Q}}.$$

(2) Given any $\sigma \in \Sigma(r)$, the set $\{\mathbf{w}(\tau, \sigma) \mid \tau \in E_\sigma\}$ form a $\mathbb{Z}$-basis of M_σ. These are the weights of the T-action on the tangent space $T_{\mathfrak{p}_\sigma}\mathcal{X}$ to $\mathcal{X}$ at the torus fixed (stacky) point $\mathfrak{p}_\sigma$.

(3) Any $\tau \in \Sigma(r-1)_c$ is contained in two top dimensional cones $\sigma, \sigma' \in \Sigma(r)$.
   (a) $r(\tau, \sigma)\mathbf{w}(\tau, \sigma) = -r(\tau, \sigma')\mathbf{w}(\tau, \sigma') \in M$.
   (b) $\mathfrak{l}_\tau^{\mathrm{rig}} \cong \mathcal{C}_{r(\tau,\sigma),r(\tau,\sigma')}$.

Let τ be as in (2). The normal bundle of $\mathfrak{l}_\tau$ in $\mathcal{X}$ is given by

$$N_{\mathfrak{l}_\tau/\mathcal{X}} \cong \mathcal{L}_1 \oplus \cdots \oplus \mathcal{L}_{r-1}$$

where $\mathcal{L}_i$ is a $\mathcal{T}$-equivariant line bundle over $\mathfrak{l}_\tau$ such that the weights of the T-actions on the fibers $(\mathcal{L}_i)_{\mathfrak{p}_\sigma}$ and $(\mathcal{L}_i)_{\mathfrak{p}_{\sigma'}}$ are $\mathbf{w}(\tau_i, \sigma) \in M_\sigma$ and $\mathbf{w}(\tau_i', \sigma') \in M_{\sigma'}$, respectively. We have

$$\mathbf{w}(\tau_i', \sigma') = \mathbf{w}(\tau_i, \sigma) - a_i r(\tau, \sigma)\mathbf{w}(\tau, \sigma) = \mathbf{w}(\tau_i, \sigma_i) + a_i r(\tau, \sigma')\mathbf{w}(\tau, \sigma')$$

where

$$a_i = \int_{\mathfrak{l}_\tau} c_1(\mathcal{L}_i) \in \mathbb{Q}.$$

8.7. Cohomology and equivariant cohomology

In this section, we recall the result of [7] on the Chow ring of toric Deligne-Mumford stacks. We also state the equivariant version.

Let $\mathcal{X} = \mathcal{X}_{\mathbf{\Sigma}}$ be the toric DM stack defined by a stacky fan $\mathbf{\Sigma} = (N, \Sigma, \beta)$, and let $X = X_\Sigma$ be the simplicial toric variety defined by the simplicial fan Σ. We assume that X is projective.

Definition 8.6. (1) *Let I be the ideal in $\mathbb{Q}[X_1, \ldots, X_s]$ generated by the monomials* $\{X_{i_1} \cdots X_{i_k} \mid v_{i_1}, \ldots, v_{i_k}$ do not generate a cone in $\Sigma\}$.

(2) *Let J be the ideal in $\mathbb{Q}[X_1, \ldots, X_s]$ generated by* $\{\sum_{\alpha=1}^{s} \langle u, \bar{b}_\alpha \rangle X_\alpha \mid u \in M\}$.

(3) *Let I' be the ideal in $R_T[X_1, \ldots, X_s] = \mathbb{Q}[X_1, \ldots, X_s, u_1, \ldots, u_r]$ generated by the monomials* $\{X_{i_1} \cdots X_{i_k} \mid v_{i_1}, \ldots, v_{i_k}$ do not generate a cone in $\Sigma\}$.

(4) *Let J' be the ideal in $R_T[X_1, \ldots, X_s] = \mathbb{Q}[X_1, \ldots, X_s, u_1, \ldots, u_r]$ generated by* $\{\sum_{\alpha=1}^{s} \langle u, \bar{b}_\alpha \rangle X_\alpha - u \mid u \in M\}$.

(5) $\deg(X_\alpha) = 2$, $\alpha = 1, \ldots, s$; $\deg(u_i) = 2$, $i = 1, \ldots, r$.

With all the above definitions, the cohomology and equivariant cohomology rings of $\mathcal{X}$ can be describe explicitly as follows.

Theorem 8.7. *We have the following isomorphisms of graded rings:*

$$H^*(\mathcal{X}) \cong \mathbb{Q}[X_1, \ldots, X_s]/(I + J).$$

$$H_T^*(\mathcal{X}) = H_{\mathcal{T}}^*(\mathcal{X}) \cong \mathbb{Q}[X_1, \ldots, X_s, u_1, \ldots, u_r]/(I' + J') \cong \mathbb{Q}[X_1, \ldots, X_s]/I.$$

The isomorphism is given by $X_\alpha \mapsto c_1(\mathcal{O}_\mathcal{X}(\mathcal{D}_\alpha))$ *or* $c_1^T(\mathcal{O}_\mathcal{X}(\mathcal{D}_\alpha))$.

The ring $\mathbb{Q}[X_1, \ldots, X_s]/I$ is known as the Stanley-Reisner ring. The ring homomorphism

$$i_\mathcal{X}^* : H_T^*(\mathcal{X}) = \mathbb{Q}[X_1, \ldots, X_s, u_1, \ldots, u_r]/(I' + J') \to H^*(\mathcal{X}) = \mathbb{Q}[X_1, \ldots, X_s]/(I + J)$$

is surjective. The kernel is the ideal generated by $u_1, \ldots, u_r$. We say $\gamma^T \in H_T^*(\mathcal{X})$ is a T-equivariant lift of $\gamma \in H^*(X)$ if $i_\mathcal{X}^*(\gamma^T) = \gamma$.

Example 8.8. *Let $\mathcal{C}_{s_1,s_2}$ be defined as in Example 8.2. Then*

$$H^*(\mathcal{C}_{s_1,s_2}) \cong \mathbb{Q}[X_1, X_2]/\langle s_1X_1 - s_2X_2, X_1X_2\rangle \cong \mathbb{Q}[X_1]/\langle X_1^2\rangle,$$
$$H_T^*(\mathcal{C}_{s_1,s_2}) \cong \mathbb{Q}[X_1, X_2]/\langle X_1X_2\rangle.$$

8.8. Orbifold cohomology and equivariant cohomology

In this section, we recall the results of [7] on orbifold Chow ring. We also state the equivariant version.

For any $\sigma \in \Sigma$, define

$$\mathrm{Box}(\sigma) = \{v \in N \mid \bar{v} = \sum_{\rho_i \subset \sigma} q_i \bar{b}_i,\ 0 \leqslant q_i < 1\}$$

Then there is an bijection between $\mathrm{Box}(\sigma)$ and $N(\sigma) = N/N_\sigma$. Define

$$\mathrm{Box}(\mathbf{\Sigma}) = \bigcup_{\sigma \in \Sigma} \mathrm{Box}(\sigma).$$

The inertia stack of $\mathcal{X} = \mathcal{X}_{\mathbf{\Sigma}}$ is

$$\mathcal{I}\mathcal{X} = \coprod_{v \in \mathrm{Box}(\mathbf{\Sigma})} \mathcal{X}(\mathbf{\Sigma}/\sigma(\bar{v}))$$

where $\sigma(\bar{v})$ is the minimal cone in $\mathbf{\Sigma}$ containing $\bar{v}$.

Given $v = \sum_\alpha q_\alpha \bar{v}_{\alpha=1}^r \in \mathrm{Box}(\mathbf{\Sigma})$, define

$$X^v := \prod_{\alpha=1}^{s} X_\alpha^{q_\alpha}.$$

As a $\mathbb{Q}$-vector spaces,

$$H^*_{\mathrm{orb}}(\mathcal{X}) = H^*(\mathcal{X}(\mathbf{\Sigma}/\sigma(\bar{v}))[\deg(X^v)].$$

Let $R_T = \mathbb{Q}[u_1, \ldots, u_r]$. As an R_T-module,

$$H^*_{\mathrm{orb},T}(\mathcal{X}) = H^*_{\mathrm{orb},\mathcal{T}}(\mathcal{X}) = \bigoplus_{v \in \mathrm{Box}(\mathbf{\Sigma})} H_T^*(\mathcal{X}(\mathbf{\Sigma}/\sigma(\bar{v}))[\deg(X^v)].$$

Example 8.9. *Let $\mathbf{\Sigma}_{s_1,s_2}$, $\mathcal{C}_{s_1,s_2}$, and $\{\mathcal{C}_{s_1,s_2,v}\}_{v=1-s_2}^{s_1-1}$ be defined as in Example 8.2. Then*

$$N = \bar{N} = \mathbb{Z}, \quad \mathrm{Box}(\mathbf{\Sigma}_{s_1,s_2}) = \{v \in \mathbb{Z} \mid 1 - s_2 \leqslant s_1 - 1\}.$$

$$\mathcal{X}(\mathbf{\Sigma}/\sigma(\bar{v})) = \mathcal{C}_{s_1,s_2,v}, \quad 1 - s_2 \leqslant v \leqslant s_1 - 1.$$

As a $\mathbb{Q}$-vector space,

$$H^*_{\mathrm{orb}}(\mathcal{C}_{s_1,s_2}) = H^*(\mathcal{C}_{s_1,s_2}) \oplus \bigoplus_{i=1}^{s_1-1} H^*(\mathcal{C}_{s_1,s_2,i})[\frac{2i}{s_1}] \oplus \bigoplus_{j=1}^{s_2-1} H^*(\mathcal{C}_{s_1,s_2,-j})[\frac{2j}{s_2}]$$
$$= \mathbb{Q}1 \oplus \mathbb{Q}H \oplus \bigoplus_{i=1}^{s_1-1} \mathbb{Q}1_{\frac{i}{s_1}} \oplus \bigoplus_{j=1}^{s_2-1} \mathbb{Q}1'_{\frac{j}{s_2}},$$

where $1_r, 1'_r \in H^{2r}_{\mathrm{orb}}(\mathcal{C}_{s_1,s_2})$.

We next describe the ring structure.

Definition 8.10. (1) *As a $\mathbb{Q}$-vector space, $\mathbb{Q}[N]^{\Sigma} = \oplus_{c\in N}\mathbb{Q}y^c$.*

(2) *As a R_T-module, $R_T[N]^{\Sigma} = \oplus_{c\in N}R_T y^c$.*

(3) *Define the multiplication on $\mathbb{Q}[N]^{\Sigma}$ and $R_T[N]^{\Sigma}$ by*

$$y^{c_1}\cdot y^{c_1} = \begin{cases} y^{c_1+c_2}, & \text{if there is } \sigma\in\Sigma \text{ such that } \bar{c}_1\in\sigma \text{ and } \bar{c}_2\in\sigma,\\ 0, & \text{otherwise.}\end{cases}$$

(4) *Given $c \in N$, let σ be the minimal cone in Σ containing $\bar{c}\in\bar{N}$. Then $\bar{c} = \sum_{\bar{b}_\alpha\in\sigma} m_\alpha \bar{b}_\alpha$ for some $m_\alpha\in\mathbb{Q}_{\geqslant 0}$. Define*

$$\deg(y^c) := 2\sum_{\bar{b}_\alpha\in\sigma} m_\alpha.$$

(5) *Let J be the ideal of $\mathbb{Q}[N]^{\Sigma}$ generated by $\{\sum_{\alpha=1}^{s}\langle u_i,\bar{b}_\alpha\rangle y^{b_\alpha} \mid i = 1,\dots,r\}$.*

(6) *Let J' be the ideal of $R_T[N]^{\Sigma}$ generated by $\{\sum_{\alpha=1}^{s}\langle u_i,\bar{b}_\alpha\rangle y^{b_\alpha} - u_i \mid i = 1,\dots,r\}$.*

Theorem 8.11. (1) *There is an isomorphisms of $\mathbb{Q}$-graded rings:*

$$H^*_{\mathrm{orb}}(\mathcal{X}) \cong \mathbb{Q}[N]^{\Sigma}/J.$$

(2) *There is an isomorphism of $\mathbb{Q}$-graded R_T-modules:*

$$H^*_{\mathrm{orb},T}(\mathcal{X}) = H^*_{\mathrm{orb},\mathcal{T}}(\mathcal{X}) \cong R_T[N]^{\Sigma}/J'.$$

Example 8.12. *Let $\mathcal{C}_{s_1,s_2}$ be defined as in Example 8.2. Then*

$$H^*_{\mathrm{orb}}(\mathcal{C}_{s_1,s_2}) \cong \mathbb{Q}[y_1,y_2]/\langle y_1y_2, s_1y_1^{s_1} - s_2y_2^{s_1}\rangle.$$

As a $\mathbb{Q}$-graded $\mathbb{Q}$-vector space,

$$H^*_{\mathrm{orb}}(\mathcal{C}_{s_1,s_2}) \cong \mathbb{Q}1 \oplus \mathbb{Q}H \bigoplus_{i=1}^{s_1-1}\mathbb{Q}y_1^i \oplus \bigoplus_{i=1}^{s_2-1}\mathbb{Q}y_2^j,$$

where

$$H = s_1y_1^{s_1} = s_2y_2^{s_2},\quad \deg(y_1) = \frac{2}{s_1},\quad \deg(y_2) = \frac{2}{s_2}.$$

Let 1_r and $1'_r$ be defined as in Example 8.9. Then

$$y_1^i = 1_{\frac{i}{s_1}},\quad y_2^j = 1'_{\frac{j}{s_2}}.$$

$$H^*_{\mathrm{orb},T}(\mathcal{C}_{s_1,s_2}) \cong \mathbb{Q}[y_1,y_2,u]/\langle y_1y_2, s_1y_1^{s_1} - s_2y_2^{s_1} - u\rangle$$

9. Orbifold Gromov-Witten invariants of smooth toric DM stacks

The main reference of this section is P. Johnson's thesis [24], which contains detailed localization computations for one-dimensional toric DM stacks.

Let $\mathcal{X}$ be a toric DM stack of dimension r defined by a stacky fan $\mathbf{\Sigma} = (N, \Sigma, \beta)$, and let $s = |\Sigma(1)| \geqslant r$. Let

$$\mathcal{I}\mathcal{X} = \bigsqcup_{i \in I} \mathcal{X}_i$$

be the inertia stack of $\mathcal{X}$, and let $\vec{i} = (i_1, \dots, i_n) \in I^n$. The torus T acts on $\mathcal{X}$, and acts on the moduli stack $\overline{\mathcal{M}}_{g,\vec{i}}(\mathcal{X}, \beta)$ by

$$t \cdot [f : (\mathcal{C}, \mathfrak{x}_1, \dots, \mathfrak{x}_n) \to \mathcal{X}] \mapsto [t \cdot f : (\mathcal{C}, \mathfrak{x}_1, \dots, \mathfrak{x}_n) \to \mathcal{X}]$$

where $(t \cdot f)(z) = t \cdot f(z)$, $z \in \mathcal{C}$. The evaluation maps $ev_j : \overline{\mathcal{M}}_{g,\vec{i}}(\mathcal{X}, \beta) \to \mathcal{X}_{i_j}$ are T-equivariant and induce $ev_j^* : A_T^*(\mathcal{X}_{i_j}) \to A_T^*(\overline{\mathcal{M}}_{g,\vec{i}}(\mathcal{X}, \beta))$.

9.1. Equivariant orbifold Gromov-Witten invariants

Suppose that $\overline{\mathcal{M}}_{g,\vec{i}}(\mathcal{X}, \beta)$ is proper, so that there are virtual fundamental classes

$$[\overline{\mathcal{M}}_{g,\vec{i}}(\mathcal{X}, \beta)]^{\mathrm{vir}} \in A_{d_{\vec{i}}^{\mathrm{vir}}}(\overline{\mathcal{M}}_{g,\vec{i}}(\mathcal{X}, \beta))$$
$$[\overline{\mathcal{M}}_{g,\vec{i}}(\mathcal{X}, \beta)]^{\mathrm{vir},T} \in A^T_{d_{\vec{i}}^{\mathrm{vir}}}(\overline{\mathcal{M}}_{g,\vec{i}}(\mathcal{X}, \beta)),$$

where $\vec{i} = (i_1, \dots, i_n) \in I^n$, and

$$d_{\vec{i}}^{\mathrm{vir}} = \int_\beta c_1(T\mathcal{X}) + (r-3)(1-g) + n - \sum_{j=1}^n \mathrm{age}(\mathcal{X}_{i_j}).$$

Recall that the weighted virtual fundamental class is given by

$$[\overline{\mathcal{M}}_{g,\vec{i}}(\mathcal{X}, \beta)]^w = \Big(\prod_{j=1}^n r_{i_j}\Big) [\overline{\mathcal{M}}_{g,\vec{i}}(\mathcal{X}, \beta)]^{\mathrm{vir}}.$$

Similarly,

$$[\overline{\mathcal{M}}_{g,\vec{i}}(\mathcal{X}, \beta)]^{w,T} = \Big(\prod_{j=1}^n r_{i_j}\Big) [\overline{\mathcal{M}}_{g,\vec{i}}(\mathcal{X}, \beta)]^{\mathrm{vir},T}$$

Given $\gamma_j \in A^{d_j}(\mathcal{X}_{i_j}) = H^{2d_i}(\mathcal{X}_{i_j}) = H_{\mathrm{orb}}^{2(d_j + \mathrm{age}(\mathcal{X}_{i_j}))}(\mathcal{X})$ and $a_j \in \mathbb{Z}_{\geqslant 0}$, define $\langle \bar{\tau}_{a_1}(\gamma_1) \cdots \bar{\tau}_{a_n}(\gamma_n) \rangle_{g,\beta}^{\mathcal{X}}$ as in Section 7.6:

$$\langle \bar{\tau}_{a_1}(\gamma_1) \cdots \bar{\tau}_{a_n}(\gamma_n) \rangle_{g,\beta}^{\mathcal{X}} = \int_{[\overline{\mathcal{M}}_{g,\vec{i}}(\mathcal{X}, \beta)]^w} \prod_{j=1}^n \Big(ev_j^* \gamma_j \cup \bar{\psi}_j^{a_j} \Big) \in \mathbb{Q}. \tag{9.1}$$

By definition, (9.1) is zero unless

$$\sum_{j=1}^n d_j = d_{\vec{i}}^{\mathrm{vir}}$$

or equivalently,

$$\sum_{j=1}^{n}(d_j+\mathrm{age}(\mathcal{X}_{i_j}))=\int_{\beta}c_1(T\mathcal{X})+(r-3)(1-g)+n.$$

In this case,

$$(9.2)\qquad \langle\bar{\tau}_{a_1}(\gamma_1)\cdots\bar{\tau}_{a_n}(\gamma_n)\rangle^{\mathcal{X}}_{g,\beta}=\int_{[\overline{\mathcal{M}}_{g,\vec{i}}(\mathcal{X},\beta)]^{w,T}}\prod_{j=1}^{n}(\mathrm{ev}_j^*\gamma_j^T\cup(\bar{\psi}_j^T)^{a_j})$$

where $\gamma_j^T\in A_T^{d_j}(X)$ is any T-equivariant lift of $\gamma_j\in A^{d_j}(X)$, and

$$\bar{\psi}_j^T\in A^1_T(\overline{\mathcal{M}}_{g,\vec{i}}(\mathcal{X},\beta))$$

is any T-equivariant lift of $\bar{\psi}_j\in A^1(\overline{\mathcal{M}}_{g,\vec{i}}(\mathcal{X},\beta))$.

Given $\gamma_j^T\in A_T^{d_j}(\mathcal{X}_{i_j})$, we define T-equivariant orbifold Gromov-Witten invariants

$$(9.3)\qquad \begin{aligned}\langle\bar{\tau}_{a_1}(\gamma_1^T),\cdots,\bar{\tau}_{a_n}(\gamma_n^T)\rangle^{\mathcal{X}_T}_{g,\beta}&:=\int_{[\overline{\mathcal{M}}_{g,\vec{i}}(\mathcal{X},\beta)]^{w,T}}\prod_{j=1}^{n}(\mathrm{ev}_i^*\gamma_i^T(\bar{\psi}_i^T)^{a_i})\\&\in\mathbb{Q}[u_1,\ldots,u_l](\sum_{j=1}^{n}d_j-d^{\mathrm{vir}}_{\vec{i}}).\end{aligned}$$

where $\mathbb{Q}[u_1,\ldots,u_l](k)$ is the space of degree k homogeneous polynomials in the variables $u_1,\ldots,u_l$ with rational coefficients. In particular,

$$\langle\bar{\tau}_{a_1}(\gamma_1^T),\cdots,\bar{\tau}_{a_n}(\gamma_n^T)\rangle^{\mathcal{X}_T}_{g,\beta}=\begin{cases}0, & \sum_{i=1}^{n}d_i<d^{\mathrm{vir}}_{\vec{i}},\\ \langle\bar{\tau}_{a_1}(\gamma_1),\cdots,\bar{\tau}_{a_n}(\gamma_n)\rangle^{\mathcal{X}}_{g,\beta}\in\mathbb{Q}, & \sum_{j=1}^{n}d_j=d^{\mathrm{vir}}_{\vec{i}}.\end{cases}$$

where $\gamma_j=i^*_{\mathcal{X}_{i_j}}\gamma_j^T\in A^{d_j}(\mathcal{X}_{i_j})$.

In this section, we will compute the T-equivariant orbifold Gromov-Witten invariants (9.3) by localization. Let $\overline{\mathcal{M}}_{g,\vec{i}}(\mathcal{X},\beta)^T\subset\overline{\mathcal{M}}_{g,\vec{i}}(\mathcal{X},\beta)$ be the substack of T fixed points, and let $i:\overline{\mathcal{M}}_{g,\vec{i}}(\mathcal{X},\beta)^T\to\overline{\mathcal{M}}_{g,\vec{i}}(\mathcal{X},\beta)$ be the inclusion. Let N^{vir} be the virtual normal bundle of substack $\overline{\mathcal{M}}_{g,\vec{i}}(\mathcal{X},\beta)^T$ in $\overline{\mathcal{M}}_{g,\vec{i}}(\mathcal{X},\beta)$; in general, N^{vir} has different ranks on different connected components of $\overline{\mathcal{M}}_{g,\vec{i}}(\mathcal{X},\beta)^T$. By virtual localization,

$$(9.4)\qquad \begin{aligned}&\int_{[\overline{\mathcal{M}}_{g,\vec{i}}(\mathcal{X},\beta)]^{w,T}}\prod_{j=1}^{n}(\mathrm{ev}_j^*\gamma_j^T\cup(\bar{\psi}_j^T)^{a_j})\\&=\int_{[\overline{\mathcal{M}}_{g,\vec{i}}(\mathcal{X},\beta)^T]^{w,T}}\frac{i^*\prod_{j=1}^{n}\left(\mathrm{ev}_j^*\gamma_j^T\cup(\bar{\psi}_j^T)^{a_j}\right)}{e^T(N^{\mathrm{vir}})}.\end{aligned}$$

Indeed, we will see that $\overline{\mathcal{M}}_{g,\vec{\iota}}(\mathcal{X},\beta)^T$ is proper even when $\overline{\mathcal{M}}_{g,\vec{\iota}}(\mathcal{X},\beta)$ is not. When $\overline{\mathcal{M}}_{g,\vec{\iota}}(\mathcal{X},\beta)$ is not proper, we *define*

$$(9.5)\qquad \langle \bar{\tau}_{a_1}(\gamma_1^T),\ldots,\bar{\tau}_{a_n}(\gamma_n^T)\rangle_{g,\beta}^{\mathcal{X}} = \int_{[\overline{\mathcal{M}}_{g,\vec{\iota}}(\mathcal{X},\beta)^T]^{w,T}} \frac{i^*\prod_{j=1}^n \left(\mathrm{ev}_j^*\gamma_j^T \cup (\bar{\psi}_j^T)^{a_j}\right)}{e^T(N^{\mathrm{vir}})} \in \mathbb{Q}(u_1,\ldots,u_r).$$

When $\overline{\mathcal{M}}_{g,\vec{\iota}}(\mathcal{X},\beta)$ is not proper, the right hand side of (9.5) is a rational function (instead of a polynomial) in $u_1,\ldots,u_r$. It can be nonzero when $\sum_{j=1}^n d_j < d_{\vec{\iota}}^{\mathrm{vir}}$, and does not have a nonequivariant limit (obtained by setting $u_i = 0$) in general.

9.2. Torus fixed points and graph notation

In this subsection, we describe the T-fixed points in $\overline{\mathcal{M}}_{g,\vec{\iota}}(\mathcal{X},\beta)$. Given a twisted stable map $f:(\mathcal{C},\mathfrak{x}_1,\ldots,\mathfrak{x}_n)\to\mathcal{X}$ such that

$$[f:(\mathcal{C},\mathfrak{x}_1,\ldots,\mathfrak{x}_n)\to\mathcal{X}]\in\overline{\mathcal{M}}_{g,\vec{\iota}}(\mathcal{X},\beta)^T,$$

we will associate a decorated graph $\vec{\Gamma}$. We first give a formal definition.

Definition 9.6. *A decorated graph* $\vec{\Gamma} = (\Gamma,\vec{f},\vec{d},\vec{g},\vec{s},\vec{k})$ *for* n*-pointed, genus* g*, degree* β *stable maps to* $\mathcal{X}$ *consists of the following data.*

(1) Γ *is a compact, connected 1 dimensional CW complex. We denote the set of vertices (resp. edges) in* Γ *by* $V(\Gamma)$ *(resp.* $E(\Gamma)$*). The set of flags of* Γ *is defined to be*

$$F(\Gamma) = \{(e,v)\in E(\Gamma)\times V(\Gamma) \mid v\in e\}.$$

(2) *The* label map $\vec{f}: V(\Gamma)\cup E(\Gamma)\to\Sigma(r)\cup\Sigma(r-1)_c$ *sends a vertex* $v\in V(\Gamma)$ *to a top dimensional cone* $\sigma_v\in\Sigma(r)$*, and sends an edge* $e\in E(\Gamma)$ *to an* $(r-1)$*-dimensional cone* $\tau_e\in\Sigma(r-1)_c$*. Moreover,* $\vec{f}$ *defines a map from the graph* Γ *to the graph* Υ*: if* $(e,v)\in F(\Gamma)$ *then* $(\tau_e,\sigma_v)\in F(\Sigma)$*.*
(3) *The* degree map $\vec{d}: E(\Gamma)\to\mathbb{Z}_{>0}$ *sends an edge* $e\in E(\Gamma)$ *to a positive integer* d_e*.*
(4) *The* genus map $\vec{g}: V(\Gamma)\to\mathbb{Z}_{\geqslant 0}$ *sends a vertex* $v\in V(\Gamma)$ *to a nonnegative integer* g_v*.*
(5) *The* marking map $\vec{s}:\{1,2,\ldots,n\}\to V(\Gamma)$ *is defined if* $n>0$*.*
(6) *The* twisting map $\vec{k}$ *sends an edge* $e\in E(\Gamma)$ *to an element* $k_e\in G_e := G_{\tau_e}$*, a flag* (e,v) *to an element* $k_{(e,v)}\in G_v := G_{\sigma_v}$*, a marking* $j\in\{1,\ldots,n\}$ *to an element* $k_j\in G_v$ *if* $\vec{\iota}(j)=v$*.*

The above maps satisfy the following two constraints:

(i) *(topology of the domain)* $\sum_{v\in V(\Gamma)} g_v + |E(\Gamma)| - |V(\Gamma)| + 1 = g.$

(ii) *(topology of the map)* $\sum_{e\in E(\Gamma)} d_e[\ell_{\tau_e}] = \beta.$

(iii) *(compatibility along an edge) Given any edge* $e \in E(\Gamma)$, *let* $v, v' \in V(\Gamma)$ *be its two ends. Then* $k_{(e,v)} \in G_v$ *and* $k_{(e,v')} \in G_{v'}$ *are determined by* $d_e \in \mathbb{Z}_{>0}$ *and* $k_e \in G_e$ [24, Lemma II.13].

(iv) *(compatibility at a vertex) Given* $v \in V(\Gamma)$, *let* E_v *and* S_v *be defined as in Definition 5.7. Then*

$$\prod_{e \in E_v} k_{(e,v)}^{-1} \prod_{j \in S_v} k_j = 1.$$

In particular, if $(e, v) \in F(\Gamma)$ *and* $v \in V^1(\Gamma)$ *then* $k_{(e,v)} = 1 \in G_v$.

(v) *(compatibility with* $\vec{i} = (i_1, \ldots, i_n)$*) Given* $j \in \{1, \ldots, n\}$, *if* $\vec{s}(j) = v$, *then the pair* (p_{σ_v}, k_j) *represent a point in* $\mathcal{X}_{i_j}$, *the connected component of* $\mathcal{IX}$ *labeled by* i_j.

Let $G_{g,\vec{i}}(\mathcal{X}, \beta)$ *be the set of all decorated graphs* $\vec{\Gamma} = (\Gamma, \vec{f}, \vec{d}, \vec{g}, \vec{s}, \vec{k})$ *satisfying the above constraints.*

Let $f : (\mathcal{C}, \mathfrak{x}_1, \ldots, \mathfrak{x}_n) \to \mathcal{X}$ be a twisted stable map which represents a T fixed point in $\overline{\mathcal{M}}_{g,\vec{i}}(\mathcal{X}, \beta)$. Let $\bar{f} : (C, x_1, \ldots, x_n) \to X$ be the corresponding stable map between coarse moduli spaces. Then $\bar{f} : (C, x_1, \ldots, x_n) \to X$ represents a T fixed point in $\overline{\mathcal{M}}_{g,n}(X, \beta)$, so we may define, as in Section 5.2, $\Gamma, \vec{f}, \vec{d}, \vec{g}, \vec{s}$, C_v for each vertex $v \in V(\Gamma)$, and C_e for each edge $e \in E(\Gamma)$. It remains to define the twisting map $\vec{k}$. Let $\mathcal{C}_v$ (resp. $\mathcal{C}_e$) be the preimage of C_v (resp. C_e) under the projection $\mathcal{C} \to C$.

- Given an edge $e \in E(\Gamma)$, the map $f_e := f|_{\mathcal{C}_e} : \mathcal{C}_e \to \mathfrak{l}_\tau$ is determined by the degree d_e of the map $\bar{f}_e := \bar{f}|_{C_e} : C_e = \mathbb{P}^1 \to \ell_\tau = \mathbb{P}^1$ and $k_e \in G_{\tau_e}$. Define $\vec{k}(e) = k_e$.
- Given $(e, v) \in F(\Gamma)$, let $\mathfrak{y}(e, v) = \mathcal{C}_e \cap \mathcal{C}_v$. Define $\vec{k}(e, v) = k_{(e,v)} \in G_v$ to be the image of the generator of the stabilizer of the stacky point $\mathfrak{y}(e, v)$ in the orbicurve $\mathcal{C}_e$.
- Under the evaluation map ev_j, the j-th marked point $\mathfrak{x}_j$ is mapped to $(\mathfrak{p}_\sigma, k)$ in the inertial stack $\mathcal{IX}$, where $\sigma \in V(\Sigma)$ and $k \in G_\sigma$. Then $\vec{f} \circ \vec{s}(j) = \sigma$. Define $\vec{k}(j) = k_j = k$.

Define

$$r_{(e,v)} = |\langle k_{(e,v)} \rangle|. \tag{9.7}$$

where $\langle k_{(e,v)} \rangle$ is the subgroup of G_v generated by $k_{(e,v)}$. Suppose that $v, v' \in V(\Gamma)$ are the two end points of the edge $e \in E(\Gamma)$. Then

$$\mathfrak{l}_\tau^{\mathrm{rig}} \cong \mathcal{C}_{r(\tau_e,\sigma_v), r(\tau_e,\sigma_{v'})}, \quad \mathcal{C}_e \cong \mathcal{C}_{r_{(e,v)}, r_{(e,v')}}.$$

To summarize, we have a map from $\overline{\mathcal{M}}_{g,\vec{i}}(\mathcal{X}, \beta)^T$ to the discrete set $G_{g,\vec{i}}(\mathcal{X}, \beta)$. Let $\mathcal{F}_{\vec{\Gamma}} \subset \overline{\mathcal{M}}_{g,\vec{i}}(\mathcal{X}, \beta)^T$ denote the preimage of $\vec{\Gamma}$. Then

$$\overline{\mathcal{M}}_{g,\vec{i}}(\mathcal{X}, \beta)^T = \bigsqcup_{\vec{\Gamma} \in G_{g,\vec{i}}(\mathcal{X},\beta)} \mathcal{F}_{\vec{\Gamma}}$$

where the right hand side is a disjoint union of connected components.

We now describe the fixed locus $\mathcal{F}_{\vec{\Gamma}}$ associated to each decorated graph $\vec{\Gamma} \in G_{g,\vec{i}}(\mathcal{X}, \beta)$. Given an edge $e \in E(\Gamma)$, the map $f_e : \mathcal{C}_e \to \mathfrak{l}_\tau$, where $\tau = \vec{f}(e)$, is determined by $\vec{\Gamma}$ up to isomorphism. The automorphism group of f_e is $G_e \times \mathbb{Z}_{\vec{d}(e)}$. The moduli space of f_e is

$$\mathcal{M}_e = \mathcal{B}(G_e \times \mathbb{Z}_{\vec{d}(e)}).$$

Given a stable vertex $v \in V^S(\Gamma)$, the map $f_v := f|_{\mathcal{C}_v} : \mathcal{C}_v \to \mathfrak{p}_\sigma = \mathcal{B}G_v$, where $\sigma = \vec{f}(v)$, represents a point in $\overline{\mathcal{M}}_{g_v, E_v \cup S_v}(\mathfrak{p}_\sigma)$, where E_v and S_v are defined as in Definition 5.7. For each $e \in E_v \subset E(\Gamma)$, there is an evaluation map

$$\mathrm{ev}_{(e,v)} : \overline{\mathcal{M}}_{g_v, E_v \cup S_v}(\mathfrak{p}_\sigma) \to \mathcal{I}\mathfrak{p}_{\sigma_v}.$$

For each $j \in S_v \subset \{1, \ldots, n\}$, there is an evaluation map

$$\mathrm{ev}_j : \overline{\mathcal{M}}_{g_v, E_v \cup S_v}(\mathfrak{p}_\sigma) \to \mathcal{I}\mathfrak{p}_{\sigma_v}.$$

We have

$$\mathcal{I}\mathfrak{p}_{\sigma_v} \cong \mathcal{I}\mathcal{B}G_v = \bigsqcup_{k \in G_v} (\mathcal{B}G_v)_k,$$

where $(\mathcal{B}G_v)_k$ are connected components of $\mathcal{I}\mathcal{B}G_v$ (see Example 6.1). The moduli space of f_v is

$$\overline{\mathcal{M}}_{g_v, \vec{i}_v}(\mathcal{B}G_v) := \bigcap_{e \in E_v} \mathrm{ev}_{(e,v)}^{-1}((\mathcal{B}G_v)_{k_{(e,v)}^{-1}}) \cap \bigcap_{j \in S_v} \mathrm{ev}_j^{-1}((\mathcal{B}G_v)_{k_j}).$$

To obtain a T fixed point $[f : (\mathcal{C}, \mathfrak{x}_1, \ldots, \mathfrak{x}_n) \to \mathcal{X}]$, we glue the the above maps f_v and f_e along the nodes. Let $V^2(\Gamma)$ and $F^S(\Gamma)$ be defined as in Definition 5.7. The nodes of $\mathcal{C}$ are

$$\{\mathfrak{y}_{(e,v)} = \mathcal{C}_e \cap \mathcal{C}_v \mid (e,v) \in F^S(\Gamma)\} \cup \{\mathfrak{y}_v = \mathcal{C}_v \mid v \in V^2(\Gamma), E_v = \{e_1, e_2\}\}.$$

We define $\widetilde{\mathcal{M}}_{\vec{\Gamma}}$ by the following 2-cartesian diagram

$$\begin{array}{ccc}
\widetilde{\mathcal{M}}_{\vec{\Gamma}} & \xrightarrow{f_E} & \prod_{e \in E(\Gamma)} \mathcal{M}_e \\
{\scriptstyle f_V}\downarrow & & \downarrow{\scriptstyle \mathrm{ev}_E} \\
\prod_{v \in V^S(\Gamma)} \overline{\mathcal{M}}_{g_v, \vec{i}_v}(\mathcal{B}G_v) & \xrightarrow{\mathrm{ev}_V} & \prod_{(e,v) \in F^S(\Gamma)} \bar{\mathcal{I}}\mathcal{B}G_v \times \prod_{v \in V^2(\Gamma)} \bar{\mathcal{I}}\mathcal{B}G_v
\end{array}$$

where ev_V and ev_E are given by evaluation at nodes, and $\bar{\mathcal{I}}\mathcal{B}G_v$ is the rigidified inertia stack. More precisely:

- For every stable flag $(e, v) \in F^S(\Gamma)$, let $\mathrm{ev}_{(e,v)}$ be the evaluation map at the node $\mathfrak{y}_{(e,v)}$, and let $\overline{\mathrm{ev}}_{(e,v)} = \iota \circ \mathrm{ev}_{(e,v)}$, where ι is the involution on $\mathcal{I}\mathcal{B}G_v$.
- For each $v \in V^2(\Gamma)$, let $E_V = \{e_1, e_2\}$ (we pick some ordering of the two edges in E_v), let $\mathrm{ev}_{(e_1,v)}$ be the evaluation map at the node $\mathfrak{y}_v$, and let $\mathrm{ev}_{(e_2,v)} = \iota \circ \mathrm{ev}_{(e_2,v)}$.

- Define

$$\mathrm{ev}_V = \prod_{(e,v)\in F^S(\Gamma)} \mathrm{ev}_{(e,v)}$$

$$\mathrm{ev}_E = \prod_{(e,v)\in F^S(\Gamma)} \overline{\mathrm{ev}}_{(e,v)} \times \prod_{\substack{v\in V^2(\Gamma)\\ (e,v)\in F(\Gamma)}} \mathrm{ev}_{(e,v)}.$$

The fixed locus associated to the decorated graph $\vec{\Gamma}$ is

$$\mathcal{F}_\Gamma = \widetilde{\mathcal{M}}_{\vec{\Gamma}}/\mathrm{Aut}(\vec{\Gamma}).$$

From the above definitions, up to some finite morphism, $\mathcal{F}_\Gamma$ can be identified with

$$\mathcal{M}_{\vec{\Gamma}} := \prod_{v\in V^S(\Gamma)} \overline{\mathcal{M}}_{g_v,\vec{i}_v}(\mathcal{B}G_v),$$

and

$$[\mathcal{F}_{\vec{\Gamma}}] = c_{\vec{\Gamma}}[\mathcal{M}_{\vec{\Gamma}}] \in A_*(\mathcal{M}_{\vec{\Gamma}})$$

where

$$c_{\vec{\Gamma}} = \frac{1}{|\mathrm{Aut}(\vec{\Gamma})|\prod_{e\in E(\Gamma)}(d_e|G_e|)} \cdot \prod_{(e,v)\in F^S(\Gamma)} \frac{|G_v|}{r_{(e,v)}} \cdot \prod_{v\in V^2(\Gamma)} \frac{|G_v|}{r_v}. \tag{9.8}$$

In the above equation:

- $\dfrac{|G_v|}{r_{(e,v)}} = |G_v/\langle k_{(e,v)}\rangle|$, where $G_v/\langle k_{(e,v)}\rangle$ is the automorphism group of $k_{(e,v)}^{-1}$ in the rigidified inertial stack $\bar{\mathcal{I}}\mathcal{B}G_v$.
- If $v \in V^2(\Gamma)$ and $E_v = \{e_1, e_2\}$, we define $r_v = r(e_1, v) = r(e_2, v)$.

9.3. Virtual tangent and normal bundles

Given a decorated graph $\vec{\Gamma} \in G_{g,\vec{i}}(\mathcal{X}, \beta)$ fixed locus $\mathcal{F}_{\vec{\Gamma}}$ and a twisted stable map $f : (\mathcal{C}, \mathfrak{x}_1, \dots, \mathfrak{x}_n) \to \mathcal{X}$ that represents a point of $\mathcal{F}_{\vec{\Gamma}}$. Let

$$B_1 = \mathrm{Hom}(\Omega_{\mathcal{C}}(\mathfrak{x}_1 + \cdots + \mathfrak{x}_n), \mathcal{O}_{\mathcal{C}}), \quad B_2 = H^0(\mathcal{C}, f^*T\mathcal{X})$$

$$B_4 = \mathrm{Ext}^1(\Omega_{\mathcal{C}}(\mathfrak{x}_1 + \cdots + \mathfrak{x}_n), \mathcal{O}_{\mathcal{C}}), \quad B_5 = H^1(\mathcal{C}, f^*T\mathcal{X})$$

T acts on B_1, B_2, B_3, B_4. Let B_i^m and B_i^f be the moving and fixed parts of B_i, respectively. Then

$$0 \to B_1^f \to B_2^f \to T^{1,f} \to B_4^f \to B_5^f \to T^{2,f} \to 0 \tag{9.9}$$

$$0 \to B_1^m \to B_2^m \to T^{1,m} \to B_4^m \to B_5^m \to T^{2,m} \to 0 \tag{9.10}$$

The irreducible components of $\mathcal{C}$ are

$$\{\mathcal{C}_v \mid v \in V^S(\Gamma)\} \cup \{\mathcal{C}_e \mid e \in E(\Gamma)\}.$$

Recall that the nodes of $\mathcal{C}$ are

$$\{\mathfrak{y}(e, v) = \mathcal{C}_e \cap \mathcal{C}_v \mid (e, v) \in F^S(\Gamma)\} \cup \{\mathfrak{y}_v = \mathcal{C}_v \mid v \in V^2(\Gamma)\}.$$

9.3.1. Automorphisms of the domain

$$\begin{aligned}
B_1^f &= \bigoplus_{\substack{e\in E(\Gamma)\\ (e,v),(e,v')\in F(\Gamma)}} \mathrm{Hom}(\Omega_{\mathcal{C}_e}(\mathfrak{y}(e,v)+\mathfrak{y}(e,v')),\mathcal{O}_{\mathcal{C}_e})\\
&= \bigoplus_{\substack{e\in E(\Gamma)\\ (e,v),(e,v')\in F(\Gamma)}} H^0(\mathcal{C}_e, T\mathcal{C}_e(-\mathfrak{y}(e,v)-\mathfrak{y}(e,v')))\\
B_1^m &= \bigoplus_{v\in V^1(\Gamma),(e,v)\in F(\Gamma)} T_{\mathfrak{y}(e,v)}\mathcal{C}_e
\end{aligned}$$

We define

$$w_{(e,v)} := e^T(T_{\mathfrak{y}(e,v)}\mathcal{C}_e) = \frac{r(\tau_e,\sigma_v)\mathbf{w}(\tau_e,\sigma_v)}{r_{(e,v)}d_e} \in H_T^2(\mathfrak{y}(e,v)) = M_{\mathbb{Q}}.$$

9.3.2. Deformations of the domain Given any $v\in V^S(\Gamma)$, define a divisor $\mathbf{x}_v$ of $\mathcal{C}_v$ by

$$\mathbf{x}_v = \sum_{i\in S_v}\mathfrak{x}_i + \sum_{e\in E_v}\mathfrak{y}(e,v).$$

Then

$$\begin{aligned}
B_4^f &= \bigoplus_{v\in V^S(\Gamma)} \mathrm{Ext}^1(\Omega_{\mathcal{C}_v}(\mathbf{x}_v),\mathcal{O}_{\mathcal{C}}) = \bigoplus_{v\in V^S(\Gamma)} T\overline{\mathcal{M}}_{g_v,\vec{i}_v}(\mathcal{B}G_v)\\
B_4^m &= \bigoplus_{v\in V^2(\Gamma),E_v=\{e,e'\}} T_{\mathfrak{y}_v}\mathcal{C}_e\otimes T_{\mathfrak{y}_v}\mathcal{C}_{e'} \oplus \bigoplus_{(e,v)\in F^S(\Gamma)} T_{\mathfrak{y}(e,v)}\mathcal{C}_v\otimes T_{\mathfrak{y}(e,v)}\mathcal{C}_e
\end{aligned}$$

where

$$\begin{aligned}
e^T(T_{\mathfrak{y}_v}\mathcal{C}_e\otimes T_{\mathfrak{y}_v}\mathcal{C}_{e'}) &= w_{(e,v)} + w_{(e',v)}, \quad v\in V^2(\Gamma)\\
e^T(T_{\mathfrak{y}(e,v)}\mathcal{C}_v\otimes T_{\mathfrak{y}(e,v)}\mathcal{C}_e) &= w_{(e,v)} - \frac{\bar{\psi}_{(e,v)}}{r_{(e,v)}}, \quad v\in V^S(\Gamma)
\end{aligned}$$

9.3.3. Unifying stable and unstable vertices From the discussion in Section 9.3.1 and Section 9.3.2,

$$\begin{aligned}
\frac{e^T(B_1^m)}{e^T(B_4^m)} &= \prod_{v\in V^1(\Gamma),(e,v)\in F(\Gamma)} w_{(e,v)} \prod_{v\in V^2(\Gamma),E_v=\{e,e'\}} \frac{1}{w_{(e,v)}+w_{(e',v)}}\\
&\quad\cdot \prod_{v\in V^S(\Gamma)} \frac{1}{\prod_{e\in E_v}(w_{(e,v)}-\bar{\psi}_{(e,v)}/r_{(e,v)})}.
\end{aligned} \tag{9.11}$$

Recall that

$$\mathcal{M}_{\vec{\Gamma}} = \prod_{v\in V^S(\Gamma)} \overline{\mathcal{M}}_{g_v,\vec{i}_v}(\mathcal{B}G_v).$$

$$c_{\vec{\Gamma}} = \frac{1}{|\mathrm{Aut}(\vec{\Gamma})|\prod_{e\in E(\Gamma)}(d_e|G_e|)} \prod_{(e,v)\in F^S(\Gamma)} \frac{|G_v|}{r_{(e,v)}} \prod_{v\in V^2(\Gamma)} \frac{|G_v|}{r_v}.$$

To unify the stable and unstable vertices, we use the following convention for the empty sets $\overline{\mathcal{M}}_{0,(1)}(\mathcal{B}G)$ and $\overline{\mathcal{M}}_{0,(c,c^{-1})}(\mathcal{B}G)$, where $1 \in G$ is the identity element, and $c \in G$. Let G be a finite abelian group. Let w_1, w_2 be formal variables.

- $\overline{\mathcal{M}}_{0,(1)}(\mathcal{B}G)$ is a -2 dimensional space, and

$$\int_{\overline{\mathcal{M}}_{0,(1)}(\mathcal{B}G)} \frac{1}{w_1 - \bar{\psi}_1} = \frac{w_1}{|G|} \tag{9.12}$$

- $\overline{\mathcal{M}}_{0,(c,c^{-1})}(\mathcal{B}G)$ is a -1 dimensional space, and

$$\int_{\overline{\mathcal{M}}_{0,(c,c^{-1})}(\mathcal{B}G)} \frac{1}{(w_1 - \bar{\psi}_1)(w_2 - \bar{\psi}_2)} = \frac{1}{(w_1 + w_2) \cdot |G|} \tag{9.13}$$

$$\int_{\overline{\mathcal{M}}_{0,(c,c^{-1})}(\mathcal{B}G)} \frac{1}{w_1 - \bar{\psi}_1} = \frac{1}{|G|} \tag{9.14}$$

From (9.12), (9.13), (9.14), we obtain the following identities for non-stable vertices:

(i) If $v \in V^1(\Gamma)$ and $(e,v) \in F(\Gamma)$, then $r_{(e,v)} = 1$, and

$$|G_v| \int_{\overline{\mathcal{M}}_{0,(1)}(\mathcal{B}G_v)} \frac{1}{w_{(e,v)} - \bar{\psi}_{(e,v)}} = w_{(e,v)}.$$

(ii) If $v \in V^2(\Gamma)$ and $E_v = \{e, e'\}$, let $c = \rho(e,v) = \rho(e',v)^{-1} \in G_v$, then

$$\frac{|G_v|}{r_v} \cdot \frac{|G_v|}{r_v} \cdot \int_{\overline{\mathcal{M}}_{0,(c,c^{-1})}(\mathcal{B}G_v)} \frac{1}{(w_{(e,v)} - \bar{\psi}_{(e,v)}/r_v)(w_{(e',v)} - \bar{\psi}_{(e',v)}/r_v)}$$

$$= \frac{|G_v|}{r_v} \cdot \frac{1}{w_{(e,v)} + w_{(e',v)}}.$$

(iii) If $v \in V^{1,1}(\Gamma)$ and $(e,v) \in F(\Gamma)$, then

$$\frac{|G_v|}{r_{(e,v)}} \int_{\overline{\mathcal{M}}_{0,(c,c^{-1})}(\mathcal{B}G_v)} \frac{1}{w_{(e,v)} - \bar{\psi}_1/r_{(e,v)}} = 1.$$

We then redefine $\mathcal{M}_{\vec{\Gamma}}$ and $c_{\vec{\Gamma}}$ as follows:

$$\mathcal{M}_{\vec{\Gamma}} = \prod_{v \in V(\Gamma)} \overline{\mathcal{M}}_{g_v, \vec{i}_v}(\mathcal{B}G_v), \quad [\mathcal{F}_{\vec{\Gamma}}] = c_{\vec{\Gamma}}[\mathcal{M}_{\vec{\Gamma}}], \tag{9.15}$$

$$c_{\vec{\Gamma}} = \frac{1}{|\mathrm{Aut}(\vec{\Gamma})| \prod_{e \in E(\Gamma)} (d_e |G_e|)} \prod_{(e,v) \in F(\Gamma)} \frac{|G_v|}{r_{(e,v)}}. \tag{9.16}$$

With the above conventions (9.12)–(9.16), we may rewrite (9.11) as

$$\frac{e^T(B_1^m)}{e^T(B_4^m)} = \prod_{v \in V(\Gamma)} \frac{1}{\prod_{e \in E_v} (w_{(e,v)} - \bar{\psi}_{(e,v)}/r_{(e,v)})}. \tag{9.17}$$

The following lemma shows that the conventions (9.12), (9.13), and (9.14) are consistent with the stable case $\overline{\mathcal{M}}_{0,(c_1,\ldots,c_n)}(\mathcal{B}G)$, $n \geqslant 3$.

Lemma 9.18. *Let* G *be a finite abelian group. Let* $\vec{c} = (c_1, \dots, c_n) \in G^n$, *where* $c_1 \cdots c_n = 1$. *Let* $w_1, \dots, w_n$ *be formal variables. Then*

(a) $$\int_{\overline{\mathcal{M}}_{0,\vec{c}}(\mathcal{B}G)} \frac{1}{\prod_{i=1}^{n}(w_i - \bar{\psi}_i)} = \frac{1}{|G| \cdot w_1 \cdots w_n}(\frac{1}{w_1} + \cdots \frac{1}{w_n})^{n-3}.$$

(b) $$\int_{\overline{\mathcal{M}}_{0,\vec{c}}(\mathcal{B}G)} \frac{1}{w_1 - \bar{\psi}_1} = \frac{w_1^{2-n}}{|G|}.$$

Proof. The unstable cases $n = 1$ and $n = 2$ follow from the definitions (9.12) and (9.13), respectively. The stable case ($n \geqslant 3$) follows from Corollary 7.9 and Lemma 5.15. □

9.3.4. Deformation of the map We first introduce some notation. Given $\sigma \in \Sigma(r)$ and $k \in G_\sigma$, let $(T_{p_\sigma}\mathcal{X})^k$ denote the subspace which is invariant under the action of k on $T_{p_\sigma}\mathcal{X}$. Then

$$(T_{p_\sigma}\mathcal{X})^k = (T_{p_\sigma}\mathcal{X})^{k^{-1}}.$$

Consider the normalization sequence

(9.19)
$$\begin{aligned} 0 \to \mathcal{O}_{\mathcal{C}} \to & \bigoplus_{v \in V^S(\Gamma)} \mathcal{O}_{\mathcal{C}_v} \oplus \bigoplus_{e \in E(\Gamma)} \mathcal{O}_{\mathcal{C}_e} \\ \to & \bigoplus_{v \in V^2(\Gamma)} \mathcal{O}_{\mathfrak{y}_v} \oplus \bigoplus_{(e,v) \in F^S(\Gamma)} \mathcal{O}_{\mathfrak{y}(e,v)} \to 0. \end{aligned}$$

We twist the above short exact sequence of sheaves by $f^*T\mathcal{X}$. The resulting short exact sequence gives rise a long exact sequence of cohomology groups

$$\begin{aligned} 0 \to B_2 \to & \bigoplus_{v \in V^S(\Gamma)} H^0(\mathcal{C}_v) \oplus \bigoplus_{e \in E(\Gamma)} H^0(\mathcal{C}_e) \\ \to & \bigoplus_{\substack{v \in V^2(\Gamma) \\ E_v = \{e, e'\}}} (T_{f(\mathfrak{y}_v)}\mathcal{X})^{k_{(e,v)}} \oplus \bigoplus_{(e,v) \in F^S(\Gamma)} (T_{f(\mathfrak{y}(e,v))}\mathcal{X})^{k_{(e,v)}} \\ \to B_5 \to & \bigoplus_{v \in V^S(\Gamma)} H^1(\mathcal{C}_v) \oplus \bigoplus_{e \in E(\Gamma)} H^1(\mathcal{C}_e) \to 0. \end{aligned}$$

where

$$H^i(\mathcal{C}_v) = H^i(\mathcal{C}_v, f_v^*T\mathcal{X}), \quad H^i(\mathcal{C}_e) = H^i(\mathcal{C}_e, f_e^*T\mathcal{X})$$

for $i = 0,1$.

$f(\mathfrak{y}_v) = p_{\sigma_v} = f(\mathfrak{y}(e,v))$. Given $(e,v) \in F(\Gamma)$, define

(9.20)
$$\mathbf{h}(e,v) = e^T((T_{p_\sigma}\mathcal{X})^{k_{(e,v)}}) = \prod_{(\tau,\sigma_v) \in F(\Sigma), \langle k_{(e,v)} \rangle \subset G_\tau} \mathbf{w}(\tau, \sigma_v).$$

The map $B_1 \to B_2$ sends $H^0(\mathcal{C}_e, T\mathcal{C}_e(-\mathfrak{y}(e,v) - \mathfrak{y}(e',v)))$ isomorphically to the fixed part $H^0(\mathcal{C}_e, f_e^*T\mathfrak{l}_{\tau_e})^f$ of $H^0(\mathcal{C}_e, f_e^*T\mathfrak{l}_{\tau_e})$.

It remains to compute

$$\mathbf{h}(v) := \frac{e^T(H^1(\mathcal{C}_v, f_v^*T\mathcal{X})^m)}{e^T(H^0(\mathcal{C}_v, f_v^*T\mathcal{X})^m)}, \quad \mathbf{h}(e) := \frac{e^T(H^1(\mathcal{C}_e, f_e^*T\mathcal{X})^m)}{e^T(H^0(\mathcal{C}_e, f_e^*T\mathcal{X})^m)}$$

We first introduce some notation.

- If $v \in V^S(\Gamma)$, then there is a cartesian diagram

$$\begin{array}{ccc} \widetilde{\mathcal{C}}_v & \xrightarrow{\tilde{f}_v} & \mathrm{pt} \\ \downarrow & & \downarrow \\ \mathcal{C}_v & \xrightarrow{f_v} & \mathcal{B}G_v. \end{array}$$

  Let $\widehat{G}_v$ denote the subgroup of G_v generated by the monodromies of the G_v-cover $\widetilde{\mathcal{C}}_v \to \mathcal{C}_v$. Then the number of connected components of $\widetilde{\mathcal{C}}_v$ is $|G_v/\widehat{G}_v|$, and each connected component is a $\widehat{G}_v$-cover of $\mathcal{C}_v$.
- Given $(\tau, \sigma) \in F(\Sigma)$, let $\phi(\tau, \sigma) \in G_\sigma^*$ be the irreducible character which corresponds to the 1-dimensional G_σ-representation $T_{p_\sigma} l_\tau$.
- Given an irreducible character ϕ of G_v, let $\mathbb{C}_\phi$ denote the 1-dimensional G_v-representation associated to ϕ. Define

$$\Lambda_\phi^\vee(u) = \sum_{i=0}^{\mathrm{rank}\mathbb{E}_\phi} (-1)^i \lambda_i^\phi u^{\mathrm{rank}\mathbb{E}_\phi - i},$$

  where $\lambda_i^\phi \in A^i(\overline{\mathcal{M}}_{g_v,\vec{i}_v}(\mathcal{B}G_v))$ are Hurwitz-Hodge classes associated to $\phi \in G_v^*$. Here $\mathrm{rank}\mathbb{E}_\rho$ is the rank of $\mathbb{E}_\rho \to \overline{\mathcal{M}}_{g_v,\vec{i}_v}(\mathcal{B}G_v)$. The rank of a Hurwitz-Hodge bundle $\mathbb{E}_\rho \to \overline{\mathcal{M}}_{g,\vec{c}}(\mathcal{B}G)$, where G is any finite group and $\rho \in G^*$, is given in Section 7.5.
- Given a G_v representation V, let V^{G_v} denote the subspace on which G_v acts trivially.

Lemma 9.21. *Suppose that* $v \in V^S(\Gamma)$ *and* $\vec{f}(v) = \sigma \in \Sigma(r)$. *Then*

$$h(v) = \frac{\prod_{(\sigma,\tau)\in E(\Gamma)} \Lambda^\vee_{\phi(\tau,\sigma)}(\mathbf{w}(\tau,\sigma))}{\prod_{(\sigma,\tau)\in E(\Gamma),\widehat{G}_v\subset G_\tau} \mathbf{w}(\tau,\sigma)} \tag{9.22}$$

Proof. We have

$$\begin{aligned} H^i(\mathcal{C}_v, f_v^* T\mathcal{X}) &= \left(H^i(\widetilde{\mathcal{C}}_v, \mathcal{O}_{\widetilde{\mathcal{C}}_v}) \otimes T_\sigma \mathcal{X}\right)^{G_v} \\ &\cong \bigoplus_{(\tau,\sigma)\in F(\Gamma)} \left(H^i(\widetilde{\mathcal{C}}_v, \mathcal{O}_{\widetilde{\mathcal{C}}_v}) \otimes \mathbb{C}_{\phi(\tau,\sigma)}\right)^{G_v}. \end{aligned}$$

The group homomorphism $G_v \to G_v/\widehat{G}_v$ induces an inclusion $(G_v/\widehat{G}_v)^* \to G_v^*$ of sets of irreducible characters, so $(G_v/\widehat{G}_v)^*$ can be viewed as a subset of G_v^*. $H^0(\widetilde{\mathcal{C}}_v, \mathcal{O}_{\widetilde{\mathcal{C}}_v})$ is the regular representation of $G_v/\widehat{G}_v$, so

$$H^0(\widetilde{\mathcal{C}}_v, \mathcal{O}_{\widetilde{\mathcal{C}}_v}) = \bigoplus_{\phi\in(G_v/\widehat{G}_v)^*} \mathbb{C}_\phi.$$

$\phi(\tau,\sigma) \in (G_v/\widehat{G}_v)^*$ iff $\widehat{G}_v \subset G_\tau$, so

$$e_T\Big(\big(H^0(\widetilde{\mathcal{C}}_v, \mathcal{O}_{\widetilde{\mathcal{C}}_v}) \otimes \mathbb{C}_{\phi(\tau,\sigma)}\big)^{G_v}\Big) = \begin{cases} w(\tau,\sigma), & \widehat{G}_v \subset G_\tau, \\ 1, & \widehat{G}_v \not\subset G_\tau. \end{cases}$$

Therefore,

$$e_T(H^0(\mathcal{C}_v, f_v^*T\mathcal{X})^m) = e_T(H^0(\mathcal{C}_v, f_v^*T\mathcal{X})) = \prod_{(\tau,\sigma)\in F(\Gamma), \widehat{G}_v \subset G_\tau} w(\tau,\sigma) \tag{9.23}$$

$$\big(H^1(\widetilde{\mathcal{C}}_v, \mathcal{O}_{\mathcal{C}_v}) \otimes \mathbb{C}_{\phi(\tau,\sigma)}\big)^{G_v} = \mathbb{E}^\vee_{\phi(\tau,\sigma)},$$

so

$$e_T(H^1(\mathcal{C}_v, f_v^*T\mathcal{X})^m) = e_T(H^1(\mathcal{C}_v, f_v^*T\mathcal{X})) = \prod_{(\tau,\sigma)\in F(\Gamma)} \Lambda^\vee_{\phi(\tau,\sigma)}(w(\tau,\sigma)). \tag{9.24}$$

Equation (9.22) follows from (9.23) and (9.24). □

Lemma 9.25. *Suppose that* $e \in E(\Gamma)$. *Let* $d = d_e \in \mathbb{Z}_{>0}$, *and let* $\tau = \vec{f}(e) \in \Sigma(r-1)_c$. *Define* $\sigma, \sigma', \tau_i, \tau_i', a_i$ *as in Section 4.3. Suppose that* $(e,v), (e,v') \in F(\Gamma)$, $\vec{f}(v) = \sigma$, $\vec{f}(v') = \sigma'$. *Then* $k_{(e,v)} \in G_\sigma$ *acts on* $T_{p_\sigma}\mathfrak{l}_\tau$ *by multiplication by* $e^{2\pi\sqrt{-1}\langle d/r(\tau,\sigma)\rangle}$, *and acts on* $T_{p_\sigma}\mathfrak{l}_{\tau_i}$ *by* $e^{2\pi\sqrt{-1}\epsilon_j}$, *where*

$$\langle\frac{d}{r(\tau,\sigma)}\rangle, \epsilon_1, \ldots, \epsilon_{r-1} \in \{0, \frac{1}{r_{(e,v)}}, \ldots, \frac{r_{(e,v)}-1}{r_{(e,v)}}\}.$$

Define

$$u = r(\tau,\sigma)w(\tau,\sigma) = -r(\tau,\sigma')w(\tau,\sigma').$$

Then

$$h(e) = \frac{(\frac{d}{u})^{\lfloor\frac{d}{r(\tau,\sigma)}\rfloor}}{\lfloor\frac{d}{r(\tau,\sigma)}\rfloor!} \frac{(-\frac{d}{u})^{\lfloor\frac{d}{r(\tau,\sigma')}\rfloor}}{\lfloor\frac{d}{r(\tau,\sigma')}\rfloor!} \prod_{i=1}^{r-1} b_i \tag{9.26}$$

where

$$b_i = \begin{cases} \displaystyle\prod_{j=0}^{\lfloor da_i - \epsilon_i\rfloor} (w(\tau_i,\sigma) - (j+\epsilon_i)\frac{u}{d})^{-1}, & a_i \geqslant 0, \\ \displaystyle\prod_{j=1}^{\lceil \epsilon_i - da_i - 1\rceil} (w(\tau_i,\sigma) + (j-\epsilon_i)\frac{u}{d}), & a_i < 0. \end{cases} \tag{9.27}$$

Proof. Let

$$w_i = w(\tau_i,\sigma), \quad i = 1, \ldots, r-1.$$

We have

$$N_{\mathfrak{l}_\tau/\mathcal{X}} = \mathcal{L}_1 \oplus \cdots \oplus \mathcal{L}_{r-1}.$$

- The weights of T-actions on $(\mathcal{L}_i)_{p_\sigma}$ and $(\mathcal{L}_i)_{p_{\sigma'}}$ are w_i and $w_i - a_i u$, respectively.
- The weights of T-action on $T_{p_\sigma}\mathfrak{l}_\tau$ and $T_{p_{\sigma'}}\mathfrak{l}_\tau$ are $\dfrac{u}{r(\tau,\sigma)}$ and $\dfrac{-u}{r(\tau,\sigma')}$, respectively.

- Let $\mathfrak{p}_\nu = f_e^{-1}(\mathfrak{p}_\sigma), \mathfrak{p}_{\nu'} = f_e^{-1}(\mathfrak{p}_{\sigma'})$ be the two torus fixed points in $\mathcal{C}_e$. Then the weights of T-action on $T_{\mathfrak{p}_\nu}\mathcal{C}_e$ and $T_{\mathfrak{p}_{\nu'}}\mathcal{C}_e$ are $\frac{\mathbf{u}}{dr_{(e,\nu)}}$ and $\frac{-\mathbf{u}}{dr_{(e,\nu')}}$, respectively.

By Example 8.5,

$$\mathrm{ch}_T(H^1(\mathcal{C}_e, f_e^*\mathcal{L}_i) - H^0(\mathcal{C}_e, f_e^*\mathcal{L}_i)) = \begin{cases} -\displaystyle\sum_{j=0}^{\lfloor da_i - \epsilon_i \rfloor} e^{\mathbf{w}_i - (j+\epsilon_i)\frac{\mathbf{u}}{d}}, & a_i \geqslant 0, \\ \displaystyle\sum_{j=1}^{\lceil \epsilon_i - da_i - 1 \rceil} e^{\mathbf{w}_i + (j-\epsilon_i)\frac{\mathbf{u}}{d}}, & a_i < 0. \end{cases}$$

Note that $\mathbf{w}_i - (j+\epsilon_i)\mathbf{u}$ and $\mathbf{w}_i + (j-\epsilon_i)\mathbf{u}$ are nonzero for any $j \in \mathbb{Z}$ since $\mathbf{w}_i$ and $\mathbf{u}$ are linearly independent for $i = 1, \dots, r-1$. So

$$\frac{e^T\left(H^1(\mathcal{C}_e, f_e^*\mathcal{L}_i)^m\right)}{e^T\left(H^0(\mathcal{C}_e, f_e^*\mathcal{L}_i)^m\right)} = \frac{e^T\left(H^1(\mathcal{C}_e, f_e^*\mathcal{L}_i)\right)}{e^T\left(H^0(\mathcal{C}_e, f_e^*\mathcal{L}_i)\right)} = \mathbf{b}_i$$

where $\mathbf{b}_i$ is defined by (9.27).

By Example 8.5 again,

$$\begin{aligned} &\mathrm{ch}_T(H^1(\mathcal{C}_e, f_e^*T\mathfrak{l}_\tau) - H^0(\mathcal{C}_e, f_e^*T\mathfrak{l}_\tau)) \\ = &\sum_{j\in\mathbb{Z}, -\langle \frac{d}{r(\tau,\sigma)}\rangle \leqslant j \leqslant \frac{d}{r(\tau,\sigma)} + \frac{d}{r(\tau,\sigma')} - \langle \frac{d}{r(\tau,\sigma)}\rangle} e^{\frac{\mathbf{u}}{r(\tau,\sigma)} - (j + \langle \frac{d}{r(\tau,\sigma)}\rangle)\frac{\mathbf{u}}{d}} \\ = &= 1 + \sum_{j=1}^{\lfloor \frac{d}{r(\tau,\sigma)} \rfloor} e^{j\frac{\mathbf{u}}{d}} + \sum_{j=1}^{\lfloor \frac{d}{r(\tau,\sigma')} \rfloor} e^{-j\frac{\mathbf{u}}{d}}. \end{aligned}$$

So

$$\begin{aligned} \frac{e^T(H^1(\mathcal{C}_e, f_e^*T\mathfrak{l}_\tau)^m)}{e^T(H^0(\mathcal{C}_e, f_e^*T\mathfrak{l}_\tau)^m)} &= \prod_{j=1}^{\lfloor \frac{d}{r(\tau,\sigma)} \rfloor} \frac{1}{j\frac{\mathbf{u}}{d}} \prod_{j=1}^{\lfloor \frac{d}{r(\tau,\sigma')} \rfloor} \frac{1}{-j\frac{\mathbf{u}}{d}} \\ &= \frac{(\frac{d}{\mathbf{u}})^{\lfloor \frac{d}{r(\tau,\sigma)} \rfloor}}{\lfloor \frac{d}{r(\tau,\sigma)} \rfloor !} \frac{(-\frac{d}{\mathbf{u}})^{\lfloor \frac{d}{r(\tau,\sigma')} \rfloor}}{\lfloor \frac{d}{r(\tau,\sigma')} \rfloor !} \end{aligned}$$

Therefore,

$$\begin{aligned} \frac{e^T(H^1(\mathcal{C}_e, f_e^*T\mathcal{X})^m)}{e^T(H^0(\mathcal{C}_e, f_e^*T\mathcal{X})^m)} &= \frac{e^T(H^1(\mathcal{C}_e, f_e^*T\mathfrak{l}_\tau)^m)}{e^T(H^0(\mathcal{C}_e, f_e^*T\mathfrak{l}_\tau)^m)} \cdot \prod_{i=1}^{r-1} \frac{e^T(H^1(\mathcal{C}_e, f_e^*\mathcal{L}_i)^m)}{e^T(H^0(\mathcal{C}_e, f_e^*\mathcal{L}_i)^m)} \\ &= \frac{(\frac{d}{\mathbf{u}})^{\lfloor \frac{d}{r(\tau,\sigma)} \rfloor}}{\lfloor \frac{d}{r(\tau,\sigma)} \rfloor !} \frac{(-\frac{d}{\mathbf{u}})^{\lfloor \frac{d}{r(\tau,\sigma')} \rfloor}}{\lfloor \frac{d}{r(\tau,\sigma')} \rfloor !} \prod_{i=1}^{r-1} \mathbf{b}_i \end{aligned}$$

□

From the above discussion, we conclude that

$$\frac{e^T(B_5^m)}{e^T(B_2^m)} = \prod_{v\in V^2(\Gamma), E_v=\{e,e'\}} \mathfrak{h}(e,v)\cdot \prod_{(e,v)\in F^S(\Gamma)} \mathfrak{h}(e,v)\cdot \prod_{v\in V^S(\Gamma)} \mathfrak{h}(v)\cdot \prod_{e\in E(\Gamma)} \mathfrak{h}(e)$$

where $\mathfrak{h}(e,v)$, $\mathfrak{h}(v)$, and $\mathfrak{h}(e)$ are defined by (9.20), (9.22), (9.26), respectively. To unify the stable and unstable vertices, we define

$$\mathfrak{h}(v) := \begin{cases} \dfrac{1}{\mathfrak{h}(e,v)}, & v\in V^1(\Gamma)\cup V^{1,1}(\Gamma), \quad E_v=\{e\}, \\ \dfrac{1}{\mathfrak{h}(e,v)} = \dfrac{1}{\mathfrak{h}(e',v)}, & v\in V^2(\Gamma), \quad E_v=\{e,e'\}. \end{cases}$$

Then

$$\frac{e^T(B_5^m)}{e^T(B_2^m)} = \prod_{v\in V(\Gamma)} \mathfrak{h}(v)\cdot \prod_{(e,v)\in F(\Gamma)} \mathfrak{h}(e,v)\cdot \prod_{e\in E(\Gamma)} \mathfrak{h}(e).$$

9.4. Contribution from each graph

9.4.1. Virtual tangent bundle We have $B_1^f = B_2^f$, $B_5^f = 0$. So

$$T^{1,f} = B_4^f = \bigoplus_{v\in V^S(\Gamma)} T\overline{\mathcal{M}}_{g_v,\vec{\iota}_v}(\mathcal{B}G_v), \quad T^{2,f}=0.$$

We conclude that

$$[\prod_{v\in V^S(\Gamma)} \overline{\mathcal{M}}_{g_v,\vec{\iota}_v}(\mathcal{B}G_v)]^{\mathrm{vir}} = \prod_{v\in V^S(\Gamma)} [\overline{\mathcal{M}}_{g_v,\vec{\iota}_v}(\mathcal{B}G_v)].$$

9.4.2. Virtual normal bundle Let $N_{\vec{\Gamma}}^{\mathrm{vir}}$ be the virtual bundle on $\mathcal{M}_{\vec{\Gamma}}$ which corresponds to the virtual normal bundle of $\mathcal{F}_{\vec{\Gamma}}$ in $\overline{\mathcal{M}}_{g,\vec{\iota}}(\mathcal{X},\beta)$. Then

$$\begin{aligned}\frac{1}{e_T(N_{\vec{\Gamma}}^{\mathrm{vir}})} &= \frac{e^T(B_1^m)e^T(B_5^m)}{e^T(B_2^m)e^T(B_4^m)} \\ &= \prod_{v\in V(\Gamma)} \frac{\mathfrak{h}(v)}{\prod_{e\in E_v}(w_{(e,v)}-\bar{\psi}_{(e,v)}/r_{(e,v)})} \prod_{(e,v)\in F(\Gamma)} \mathfrak{h}(e,v)\cdot \prod_{e\in E(\Gamma)} \mathfrak{h}(e)\end{aligned}$$

9.4.3. Integrand Given $\sigma\in\Sigma(r)$, let

$$i_\sigma^*: A_T^*(\mathcal{X}) \to A_T^*(\mathfrak{p}_\sigma) = \mathbb{Q}[u_1,\dots,u_r]$$

be induced by the inclusion $i_\sigma: \mathfrak{p}_\sigma \to \mathcal{X}$. Given $\vec{\Gamma}\in G_{g,\vec{\iota}}(\mathcal{X},\beta)$, let

$$i_{\vec{\Gamma}}^*: A_T^*(\overline{\mathcal{M}}_{g,\vec{\iota}}(\mathcal{X},\beta)) \to A_T^*(\mathcal{F}_{\vec{\Gamma}}) \cong A_T^*(\mathcal{M}_{\vec{\Gamma}})$$

be induced by the inclusion $i_{\vec{\Gamma}}: \mathcal{F}_{\vec{\Gamma}} \to \overline{\mathcal{M}}_{g,\vec{\iota}}(\mathcal{X},\beta)$. Then

$$\begin{aligned}& i_{\vec{\Gamma}}^* \prod_{i=1}^n (ev_i^*\gamma_i^T \cup (\bar{\psi}_i^T)^{a_i}) \\ (9.28)\quad &= \prod_{\substack{v\in V^{1,1}(E)\\ S_v=\{i\}, E_v=\{e\}}} i_{\sigma_v}^*\gamma_i^T(-w_{(e,v)})^{a_i}\cdot \prod_{v\in V^S(\Gamma)} \Big(\prod_{i\in S_v} i_{\sigma_v}^*\gamma_i^T \prod_{e\in E_v} \bar{\psi}_{(e,v)}^{a_i}\Big)\end{aligned}$$

To unify the stable vertices in $V^S(\Gamma)$ and the unstable vertices in $V^{1,1}(\Gamma)$, we use the following convention: for $a \in \mathbb{Z}_{\geqslant 0}$,

$$\int_{\overline{\mathcal{M}}_{0,(c,c^{-1})}(\mathcal{B}G)} \frac{\bar{\psi}_2^a}{w_1 - \bar{\psi}_1} = \frac{(-w_1)^a}{|G|}. \tag{9.29}$$

In particular, (9.14) is obtained by setting $a = 0$. With the convention (9.29), we may rewrite (9.28) as

$$i_{\vec{\Gamma}}^* \prod_{i=1}^{n} (ev_i^* \gamma_i^T \cup (\bar{\psi}_i^T)^{a_i}) = \prod_{v \in V(\Gamma)} \Big(\prod_{i \in S_v} i_{\sigma_v}^* \gamma_i^T \prod_{e \in E_v} \bar{\psi}_{(e,v)}^{a_i} \Big). \tag{9.30}$$

The following lemma shows that the convention (9.29) is consistent with the stable case $\overline{\mathcal{M}}_{0,(c_1,\dots,c_n)}(\mathcal{B}G)$, $n \geqslant 3$.

Lemma 9.31. *Let n, a be integers, $n \geqslant 2$, $a \geqslant 0$. Let $\vec{c} = (c_1, \dots, c_n) \in G^n$, where $c_1 \cdots c_n = 1$. Then*

$$\int_{\overline{\mathcal{M}}_{0,\vec{c}}(\mathcal{B}G)} \frac{\bar{\psi}_2^a}{w_1 - \bar{\psi}_1} = \begin{cases} \dfrac{\prod_{i=0}^{a-1}(n-3-i)}{a!|G|} w_1^{a+2-n}, & n = 2 \text{ or } 0 \leqslant a \leqslant n-3. \\ 0, & \text{otherwise.} \end{cases}$$

Proof. The case $n = 2$ follows from (9.29). For $n \geqslant 3$,

$$\int_{\overline{\mathcal{M}}_{0,\vec{c}}(\mathcal{B}G)} \frac{\psi_2^a}{w_1 - \bar{\psi}_1} = \frac{1}{w_1} \int_{\overline{\mathcal{M}}_{0,\vec{c}}(\mathcal{B}G)} \frac{\bar{\psi}_2^a}{1 - \frac{\bar{\psi}_1}{w_1}} = w_1^{a+2-n} \int_{\overline{\mathcal{M}}_{0,\vec{c}}(\mathcal{B}G)} \bar{\psi}_1^{n-3-a} \bar{\psi}_2^a$$

$$= w_1^{a+2-n} \cdot \frac{1}{|G|} \cdot \frac{(n-3)!}{(n-3-a)!a_!} = \frac{\prod_{i=0}^{a-1}(n-3-i)}{a!|G|} w_1^{a+2-n}.$$

□

9.4.4. Integral

Let

$$i^* : A_T^*(\overline{\mathcal{M}}_{g,\vec{i}}(\mathcal{X}, \beta)) \to A_T^*(\overline{\mathcal{M}}_{g,\vec{i}}(\mathcal{X}, \beta)^T)$$

be induced by the inclusion $i : \overline{\mathcal{M}}_{g,\vec{i}}(\mathcal{X}, \beta)^T \to \overline{\mathcal{M}}_{g,\vec{i}}(\mathcal{X}, \beta)$. The contribution of

$$\int_{[\overline{\mathcal{M}}_{g,\vec{i}}(\mathcal{X},\beta)^T]^{vir,T}} \frac{i^* \prod_{i=1}^n (ev_i^* \gamma_i^T \cup (\bar{\psi}_i^T)^{a_i})}{e^T(N^{vir})}$$

from the fixed locus $\mathcal{F}_{\vec{\Gamma}}$ is given by

$$c_{\vec{\Gamma}} \prod_{e \in E(\Gamma)} h(e) \prod_{(e,v) \in F(\Gamma)} h(e,v) \prod_{v \in V(\Gamma)} \Big(\prod_{i \in S_v} i_{\sigma_v}^* \gamma_i^T \Big)$$

$$\cdot \prod_{v \in V(\Gamma)} \int_{\overline{\mathcal{M}}_{g_v,\vec{i}_v}(\mathcal{B}G_v)} \frac{h(v) \cdot \prod_{e \in E_v} \bar{\psi}_{(e,v)}^{a_i}}{\prod_{e \in E_v} (w_{(e,v)} - \bar{\psi}_{(e,v)}/r_{(e,v)})}$$

where $c_{\vec{\Gamma}} \in \mathbb{Q}$ is defined by (9.16).

9.5. Sum over graphs

Summing over the contribution from each graph $\vec{\Gamma}$ given in Section 9.4.4 above, we obtain the following formula.

Theorem 9.32.

$$\begin{aligned}
&\langle \bar{\tau}_{a_1}(\gamma_1^T)\cdots\bar{\tau}_{a_n}(\gamma_n^T)\rangle_{g,\beta}^{\mathcal{X}_T} \\
&= \sum_{\vec{\Gamma}\in G_{g,\vec{\iota}}(\mathcal{X},\beta)} c_{\vec{\Gamma}} \prod_{e\in E(\Gamma)} h(e) \prod_{(e,v)\in F(\Gamma)} h(e,v) \prod_{v\in V(\Gamma)} \Big(\prod_{i\in S_v} i_{\sigma_v}^* \gamma_i^T\Big) \\
&\quad \cdot \prod_{v\in V(\Gamma)} \int_{\overline{\mathcal{M}}_{g,\vec{\iota}_v}(\mathcal{B}G_v)} \frac{h(v)\prod_{i\in S_v}\bar{\psi}_i^{a_i}}{\prod_{e\in E_v}(w_{(e,v)}-\bar{\psi}_{(e,v)}/r_{(e,v)})}.
\end{aligned} \tag{9.33}$$

where $h(e)$, $h(e,v)$, $h(v)$ *are given by* (9.26), (9.20), (9.22), *respectively, and we have the following convention for the* $v \notin V^S(\Gamma)$:

$$\int_{\overline{\mathcal{M}}_{0,(1)}(\mathcal{B}G)} \frac{1}{w_1-\bar{\psi}_2} = \frac{w_1}{|G|},$$

$$\int_{\overline{\mathcal{M}}_{0,(c,c^{-1})}(\mathcal{B}G)} \frac{1}{(w_1-\bar{\psi}_1)(w_2-\bar{\psi}_2)} = \frac{1}{|G|\cdot(w_1+w_2)},$$

$$\int_{\overline{\mathcal{M}}_{0,(c,c^{-1})}(\mathcal{B}G)} \frac{\bar{\psi}_2^a}{w_1-\bar{\psi}_1} = \frac{(-w_1)^a}{|G|}, \qquad a\in\mathbb{Z}_{\geqslant 0}.$$

References

[1] D. Abramovich, T. Graber, and A. Vistoli. Algebraic orbifold quantum products, in: Orbifolds in mathematics and physics (Madison, WI, 2001), 1–24, Contemp. Math., **310**, Amer. Math. Soc., Providence, RI, 2002. ← 356, 390, 391

[2] D. Abramovich, T. Graber, and A. Vistoli. Gromov-Witten theory of Deligne-Mumford stacks. *Amer. J. Math.*, **130**(5), 1337–1398, 2008. ← 356, 391, 392

[3] M. F. Atiyah and R. Bott. The moment map and equivariant cohomology, *Topology*, **23**(1), 1–28, 1984. ← 356, 361

[4] K. Behrend. Localization and Gromov-Witten invariants, in: Quantum cohomology, 3–38, *Lecture Notes in Math.*, **1776**, Springer, Berlin, 2002. ← 355, 356, 365, 376

[5] K. Behrend and B. Fantechi. Intrinsic normal cone. *Invent. Math.*, **128**(1), 45–88, 1997. ← 355, 368, 369

[6] K. Behrend and Y. Manin. Stacks of stable maps and Gromov-Witten invariants. *Duke Math. J.*, **85** (1), 1–60, 1996. ← 368

[7] L. Borisov, L. Chen, and G. Smith. The Orbifold Chow ring of a toric Deligne-Mumford stack. *J. Amer. Math. Soc.*, **18**(1), 193–215, 2005. ← 396, 397, 405, 406

[8] L. Chen, Y. Li, and K. Liu. Localization, Hurwitz numbers and the Witten conjecture. *Asian J. Math.*, **12**(4), 511–518, 2008. ← 367

[9] W. Chen and Y. Ruan. Orbifold Gromov-Witten theory, in: Orbifolds in mathematics and physics (Madison, WI, 2001), 25–85. *Contemp. Math.*, **310**, Amer. Math. Soc., Providence, RI, 2002. ← 356, 391

[10] W. Chen and Y. Ruan. A new cohomology theory of orbifold. *Comm. Math. Phys.*, **248**(1), 1–31, 2004. ← 390

[11] P. Deligne and D. Mumford. The irreducibility of the space of curves of given genus. *Inst. Hautes Études Sci. Publ. Math.*, no. **36**, 75–109, 1969. ← 366

[12] D. Edidin and W. Graham. Equivariant intersection theory. *Invent. Math.*, **131**(3), 595–634, 1998. ← 357, 364

[13] C. Faber. Algorithms for computing intersection numbers on moduli spaces of curves, with an application to the class of the locus of Jacobians, in: New trends in algebraic geometry (Warwick, 1996), 93–109. *London Math. Soc. Lecture Note Ser.*, **264**, Cambridge Univ. Press, Cambridge, 1999. ← 367

[14] C. Faber and R. Pandharipande. Hodge integrals and Gromov-Witten theory. *Invent. Math*, **139**(1), 173–199, 2000. ← 355

[15] B. Fantechi, E. Mann, and F. Nironi. Smooth toric Deligne-Mumford stacks, to appear in: *J. Reine Angew. Math.*, **648**, 201–244, 2010. ← 396, 399

[16] W. Fulton. Intersection theory. *Ergebnisse der Mathematik und ihrer Grenzgebiete*, Springer-Verlag, Berlin, 1984. ← 364

[17] W. Fulton. Introduction to toric varieties. *Annals of Mathematics Studies*, **131**, the William H. Roever Lectures in Geometry, Princeton University Press, Princeton, NJ, 1993. ← 370, 375, 396

[18] W. Fulton. Equivariant Cohomology in Algebraic Geometry. *Eilenberg lectures at Columbia University*, Spring 2007, notes by Dave Anderson are available at `http://www.math.washington.edu/$\sim $dandersn/eilenberg/` ← 375

[19] T. Graber and R. Pandharipande. Localization of virtual classes. *Invent. Math.*, **135**(2), 487–518, 1999. ← 355, 356, 365, 376

[20] K. Hori, S. Katz, A. Klemm, R. Pandharipande, R. Thomas, C. Vafa, R. Vakil, and E. Zaslow. Mirror Symmetry. *Clay Mathematics Monographs*, **1**, American Mathematical Society, Providence, RI; Clay Mathematics Institute, Cambridge, MA, 2003. ← 356, 376

[21] I. Iwanari. The category of toric stacks. *Compos. Math.*, **145**(3), 718–746, 2009. ← 396

[22] I. Iwanari. Logarithm geometry, minimal free resolutions and toric algebraic stacks. *Publ. Res. Inst. Math. Sci.*, **45**(4), 1095–1140, 2009. ← 396

[23] T. J. Jarvis and T. Kimura. Orbifold quantum cohomology of the classifying space of a finite group, in: Orbifolds in mathematics and physics (Madison, WI, 2001), 123–134. *Contemp. Math.*, **310**, Amer. Math. Soc., Providence, RI, 2002. ← 395

[24] P. Johnson. Equivariant Gromov-Witten theory of one dimensional stacks. arXiv:0903.1068, Ph.D. Thesis, University of Michigan, 2009. ← 356, 390, 399, 408, 411

[25] M. E. Kazarian. KP hierarchy for Hodge integrals. *Adv. Math.*, **221**(1), 1–21, 2009. ← 367

[26] M. E. Kazarian and S. K. Lando. An algebro-geometric proof of Witten's conjecture. *J. Amer. Math. Soc.*, **20**(4), 1079–1089, 2007. ← 367

[27] Y-S. Kim and K. Liu. Virasoro constraints and Hurwitz numbers through asymptotic analysis. *Pacific J. Math.*, **241**(2), 275–284, 2009. ← 367

[28] F. Knudsen. The projectivity of the moduli space of stable curves. II. The stacks $M_{g,n}$. *Math. Scand.*, **52**(2), 161–199, 1983. ← 366

[29] F. Knudsen. The projectivity of the moduli space of stable curves. III. The line bundles on $M_{g,n}$, and a proof of the projectivity of $\overline{M}_{g,n}$ in characteristic 0. *Math. Scand.*, **52**(2), 200–212, 1983. ← 366

[30] F. Knudsen and D. Mumford. The projectivity of the moduli space of stable curves. I. Preliminaries on "det" and "Div". *Math. Scand.*, **39**(1), 19–55, 1976. ← 366

[31] M. Kontsevich. Intersection theory on the moduli space of curves and the matrix Airy function. *Comm. Math. Phys.*, **147**(1), 1–23, 1992. ← 367

[32] M. Kontsevich. Enumeration of rational curves via torus actions, in: The moduli space of curves (Texel Island, 1994), 335–368. *Progr. Math.*, **129**, Birkhäuser Boston, Boston, MA, 1995. ← 355, 356, 367, 368, 376, 377

[33] A. Kresch. Cycle groups for Artin stacks. *Invent. Math.*, **138**(1), 495–536, 1999. ← 355, 365, 402

[34] J. Li and G. Tian. Virtual moduli cycles and Gromov-Witten invariants of algebraic varieties. *J. Amer. Math. Soc.*, **11**(1), 119–174, 1998. ← 355, 369

[35] M. Mirzakhani. Weil-Petersson volumes and intersection theory on the moduli space of curves. *J. Amer. Math. Soc.*, **20**(1), 1–23, 2007. ← 367

[36] M. Mulase and N. Zhang. Polynomial recursion formula for linear Hodge integrals. *Commun. Number Theory Phys.*, **4**(2), 267–293, 2010. ← 367

[37] D. Mumford. Towards an enumerative geometry of the moduli space of curves, in: Arithmetic and geometry, Vol. II, 271–328. *Progr. Math.*, **36**, Birkhäuser Boston, Boston, MA, 1983. ← 367

[38] A. Okounkov and R. Pandharipande. Gromov-Witten theory, Hurwitz numbers, and matrix models, in: Algebraic geometry—Seattle 2005. Part 1, 325–414. *Proc. Sympos. Pure Math.*, **80**, Part 1, Amer. Math. Soc., Providence, RI, 2009. ← 367

[39] M. C. Olsson. (Log) twisted curves. *Compos. Math.*, **143**(2), 476–494, 2007. ← 391

[40] D. Ross. Localization and gluing of orbifold amplitudes: the Gromov-Witten orbifold vertex. arXiv:1109.5995 ← 356

[41] H. Spielberg. A formula for the Gromov-Witten invariants of toric varieties. Thèse, Université Louis Pasteur (Strasbourg I), Strasbourg, 1999. ← 356, 376

[42] H. Spielberg. The Gromov-Witten invariants of symplectic manifolds. *C. R. Acad. Sci. Paris Sér. I Math.* **329**(8), 699–704, 1999. ← 356, 376

[43] H.-H. Tseng. Orbifold quantum Riemann-Roch, Lefschetz and Serre. *Geom. Topol.*, **14**(1), 1–81, 2010. ← 395

[44] A. Vistoli. Intersection theory on algebraic stacks and on their moduli spaces. *Invent. Math.*, **97**(3), 613–670, 1989. ← 364

[45] E. Witten. Two-dimensional gravity and intersection theory on moduli space. *Surveys in differential geometry (Cambridge, MA, 1990)*, 243–310, Lehigh Univ., Bethlehem, PA, 1991. ← 367, 370

[46] J. Zhou. On computations of Hurwitz-Hodge integrals. arXiv:0710.1679 ← 395

Department of Mathematics, Columbia University, 2990 Broadway, New York, NY 10027
E-mail address: ccliu@math.columbia.edu

From WZW models to modular functors

Eduard Looijenga

Abstract. In this survey paper we give a relatively simple and coordinate free description of the WZW model as a local system whose base is the $\mathbb{G}_m$-bundle associated to the determinant bundle on the moduli stack of pointed curves. We derive its main properties and show how it leads to a modular functor in the spirit of Segal. The approach presented here is almost purely algebro-geometric in character; it avoids the Boson-Fermion correspondence, operator product expansions as well as Teichmüller theory.

Contents

The tumultuous interaction between mathematicians and theoretical physicists that began more than two decades ago left some of us hardly time to take stock. It is telling for this era that it took physicists (Witten, mainly) to point out in the late eighties that there must exist a bridge between two, at the time hardly connected, mathematical land masses, *viz.* algebraic geometry and knot theory, and it is equally telling that it was only recently that this was materialized with mathematically rigorous underpinnings (and strictly speaking not even in the desired form yet). We are here referring on the algebro-geometric side to a subject that has its place in the present handbook, namely moduli spaces of vector bundles over curves, and on the other side to the knot invariants (like the Jones polynomial) that are furnished by Chern-Simons theory. The bridge metaphor is actually a bit misleading, because on either side the roads leading to it had yet to be constructed. Let us use the remainder of this introduction to survey very briefly the part this route that involves algebraic geometry, stopping short at the point were the crossing is made, then say which segment is covered by this paper and conclude in the customary manner by commenting on the various sections.

To set the stage, let C be a compact Riemann surface and G a (say, simply connected) complex algebraic group with simple Lie algebra $\mathfrak{g}$. Then there is a moduli stack $M(C, G)$ of G-principal bundles over C. This stack carries a natural ample line bundle $\Theta(C, G)$, which in fact generates its Picard group, and for which the vector space of sections of $\Theta(C, G)^{\otimes \ell}$, the so-called *Verlinde space of level* ℓ and here denoted by $\mathbb{H}_\ell(G)_C$, is finite dimensional for all ℓ. Its dimension is independent of C and indeed, if we vary C over a base S, then we get a vector bundle $\mathcal{H}_\ell(G)_{\mathcal{C}/S}$ over that base. Although we required G to be simply connected, one can make sense of this for reductive groups as well, although some care is needed. For instance, for $G = \mathbb{C}^\times$, we let $M(C, \mathbb{C}^\times)$ not be the full Picard variety $\mathrm{Pic}(C)$ of C, but pick the component $\mathrm{Pic}(C)^{g-1}$ parameterizing line bundles of degree $(g-1)$, as this is the one which carries a natural line bundle that can play the role of $\Theta(C, \mathbb{C}^\times)$ (and which is indeed known as the theta bundle). In that case $\mathbb{H}_\ell(G)_C$ is just the space of theta functions of degree ℓ. These theta functions satisfy a heat equation and it is our understanding that Mumford was the first to observe that this property may be interpreted as defining a flat connection for the associated projective space bundle. Hitchin [6] proved that this is also true for the case considered here: the projectivized Verlinde bundles come naturally with a flat connection. But if one aims for flat connections on the bundles themselves, then one should work on the total space of a $\mathbb{C}^\times$-bundle over S (which allows for nontrivial monodromy in a fiber). For the line bundle attached to this $\mathbb{C}^\times$-bundle we can take the determinant bundle of the direct image of the sheaf of relative differentials on $\mathcal{C}/S$. For many purposes—certainly for topological applications—it is desirable to allow for certain 'impurities' of the principal bundle, in the form of a parabolic structure. Such a structure is specified by giving on C a finite set of points $(x_i \in C)_{i \in I}$, and for each such point a finite dimensional irreducible representation V_i of G. It was shown by Scheinost-Schottenloher [13] that in this setting there are still corresponding Verlinde bundles that come with a flat connection after a pull-back to a $\mathbb{C}^\times$-bundle. There is an infinitesimal counterpart of the above construction via holomorphic conformal field theory where the group G enters only via its Lie algebra $\mathfrak{g}$, known as the *Wess-Zumino-Witten model*. This centers on the affine Lie algebra associated to $\mathfrak{g}$ and its representation theory and leads to similar constructs such as the Verlinde bundles with a projectively flat connection. Its mathematically rigorous treatment began with the fundamental paper by Tsuchiya-Ueno-Yamada [15] with subsequent extensions and refinements, mainly by Andersen-Ueno [1], [2]. It was however not a priori clear that this led to the same local system as the global approach. Indeed, this turned out to be not trivial at all: after partial results by Beauville-Laszlo and others, Laszlo-Sorger [11] proved that the Verlinde bundles can be identified and

Laszlo [10] showed that via this identification the two connections are the same as well, at least when no parabolic structure is present.

The bridge is now crossed as follows: a nonzero point of the determinant line over C can be topologically specified by means of the choice of Lagrangian sublattice in $H_1(C;\mathbb{Z})$. This enables us to understand the existence of the flat connection on the Verlinde bundles as telling us that these spaces only depend on the isotopy class of the complex structure of C. In particular, they naturally receive the structure of a projective representation of the mapping class group of the pointed surface. This puts these spaces into the topological realm and we thus arrive at an example of a topological quantum field theory, more precisely, at one of Segal's modular functors [14].

Let us now turn to the central goal of this paper, which is to define the Wess-Zumino-Witten connection and to derive its principal properties, to wit its flatness, factorization, the relation with the KZ-system, . . . , in short, to recover all the properties needed for defining the underlying (topological) modular functor as found in the papers above mentioned by Tsuchiya-Ueno-Yamada and Andersen-Ueno. For an audience of algebraic geometers knowing, or willing to accept, some rather basic facts about affine Lie algebras, our presentation is essentially self-contained. It is also shorter and possibly at several points more transparent than the literature we are aware of. This is to a large extent due to our consistent coordinate free approach, which not only has the advantage of making it unnecessary to constantly check for gauge invariance, but is also conceptually more satisfying. Cases in point are our definition of the WZW-connection and our treatment of the Fock representation (leading up to Corollary 2.7) which enables us to avoid resorting to the infinite wedge representation and allied techniques.

Let us take the occasion to point out that what makes the WZW-story still incomplete is an explanation of the duality property and the unitary structure that the associated modular functor should possess.

As promised, we finish with brief comments on the contents of the separate sections. The rather short Section 1 essentially elaborates on the notion of a projectively flat connection. Logically, this material should have its place later in the paper, but as it has some motivating content for what comes right after it, we felt it best to put it there. Section 2 introduces in a canonical way the Virasoro algebra and its Fock representation and the associated Segal-Suguwara construction in a relative setting. New is the last subsection about symplectic local systems, where we see the determinant bundle appear in a canonical fashion. The Lie algebra $\mathfrak{g}$ enters in Section 3. We found it helpful to present this material in an abstract algebraic setting, replacing for instance the ring of complex Laurent polynomials by a complete local field containing $\mathbb{Q}$ (or rather a direct sum of these), which is then also allowed to 'depend on parameters'. Our extension 3.3 of the Sugawara representation to a relative situation involving a Leibniz rule in the horizontal direction serves here as the origin of WZW-connection and its projective flatness. We keep that setting in Section 4, where the connection itself is defined. In the subsequent section we derive the coherence of the Verlinde sheaf and establish what is called the propagation of covacua. Special attention is paid to the genus zero case and it shown how the WZW-connection is then related to the one of Knizhnik-Zamolodchikov. Section 6 is devoted to the basic results associated to a double point degeneration such as local freeness, factorization and monodromy. Finally, in Section 7, we establish the conversion into a modular functor. Notice that the approach described here is elementary and does not resort to Teichmüller theory.

This paper is based on (but substantially supersedes) our arXiv preprint `math.AG/0507086`.

The author is grateful to two referees for helpful comments.

We find it convenient to work over an arbitrary algebraically closed field k of characteristic zero, but in Section 7, where we discuss the link with topological quantum field theory, we assume $k = \mathbb{C}$. As an intermediate base we use a regular k-algebra, denoted R.

1. Flat and projectively flat connections

A central notion of this article is that of a flat projective connection. Although it enters the scene much later in the paper, some of the work done in the first part is motivated by the particular way this notion appears here. So we start with a brief section discussing it.

We begin with recalling some basic facts. Let $\mathcal{H}$ be a rank r vector bundle over a smooth base S (in other words, is a locally a free $\mathcal{O}_S$-module of rank r). Then the Lie algebra $\mathcal{D}_1(\mathcal{H})$ of first order differential operators $\mathcal{H} \to \mathcal{H}$ fits in an exact sequence of coherent sheaves of Lie algebras

$$0 \to \mathcal{E}nd(\mathcal{H}) \to \mathcal{D}_1(\mathcal{H}) \xrightarrow{sb} \theta_S \otimes_{\mathcal{O}_S} \mathcal{E}nd(\mathcal{H}) \to 0,$$

where sb is the symbol map which assigns to $D \in \mathcal{D}_1(\mathcal{H})$ the k-derivation $\phi \in \mathcal{O}_S \mapsto [D, \phi] \in \mathcal{O}_S$. The local sections of $\mathcal{D}_1(\mathcal{H})$ whose symbol land in $\theta_S \otimes 1_{\mathcal{H}} \cong \theta_S$ make up a coherent subsheaf of Lie subalgebras $sb^{-1}(\theta_S) \subset \mathcal{D}_1(\mathcal{H})$ so that we have an exact sequence of coherent sheaves of Lie algebras

$$0 \to \mathcal{E}nd(\mathcal{H}) \to sb^{-1}(\theta_S) \to \theta_S \to 0.$$

A connection on $\mathcal{H}$ is then simply a section $X \in \theta_S \mapsto \nabla_X \in sb^{-1}(\theta_S)$ of $sb^{-1}(\theta_S) \to \theta_S$ and it is flat precisely if this section is a Lie homomorphism. This suggests that we define a flat connection on the associated projective space bundle $\mathbb{P}_S(\mathcal{H})$ as a Lie subalgebra $\hat{\mathcal{D}} \subset \mathcal{D}_1(\mathcal{H})$ with $sb(\hat{\mathcal{D}}) = \theta_S$ and $\hat{\mathcal{D}} \cap \mathcal{E}nd(\mathcal{H}) = \mathcal{O}_S \otimes 1_{\mathcal{H}} \cong \mathcal{O}_S$ so that we have an exact subsequence

$$0 \to \mathcal{O}_S \to \hat{\mathcal{D}} \to \theta_S \to 0$$

of the one displayed above. That this is a sensible definition follows from the observation that an $\mathcal{O}_S$-linear section ∇ of $\hat{\mathcal{D}} \to \theta_S$ defines a connection on $\mathcal{H}$ whose curvature form $R(\nabla)$ is a closed 2-form on S. Any other section ∇' differs from ∇ by an $\mathcal{O}_S$-linear map $\theta_S \to \mathcal{O}_S$, in other words, by a differential ω, and we have $R(\nabla') = R(\nabla) + d\omega$. So this indeed gives rise to a flat connection in $\mathbb{P}_S(\mathcal{H})$ and it is easily seen that this connection is independent of the choice of the section. Locally on S, $R(\nabla)$ is exact, and so we can always choose ∇ to be flat as a connection. Any other such local section that is flat is necessarily of the form $\nabla + d\phi$ with $\phi \in \mathcal{O}$ and conversely, any such local section has that property. The Lie algebra sheaf $\hat{\mathcal{D}}$ itself does not determine a connection on $\mathcal{H}$; this is most evident when $\mathcal{H}$ is a line bundle, for then we must have $\hat{\mathcal{D}}(\mathcal{H}) = \mathcal{D}_1(\mathcal{H})$.

In the above situation we let $\hat{\mathcal{D}}$ act on the determinant bundle $\det(\mathcal{H}) = \wedge^r_{\mathcal{O}_S}\mathcal{H}$ by means of the formula

$$\hat{D}(e_1 \wedge \cdots \wedge e_r) := \sum_{i=1}^{r} e_1 \wedge \cdots \wedge \hat{D}(e_i) \wedge \cdots \wedge e_r.$$

This is indeed well-defined, and identifies $\hat{\mathcal{D}}$ as a Lie algebra with the Lie algebra of first order differential operators $\mathcal{D}_1(\det(\mathcal{H}))$. Notice however that this identification makes $f \in \mathcal{O}_S \subset \hat{\mathcal{D}}$ act on $\det(\mathcal{H})$ as multiplication by rf.

Let us next observe that if λ is a line bundle on S and N is a positive integer, then a similar formula identifies $\mathcal{D}_1(\lambda)$ with $\mathcal{D}_1(\lambda^{\otimes N})$ (both as $\mathcal{O}_S$-modules and as k-Lie algebras), but induces multiplication by N on $\mathcal{O}_S$. This leads us to make the following

Definition 1.1. Let be given a smooth base variety S over which we are given a line bundle λ and a locally free $\mathcal{O}_S$-module $\mathcal{H}$ of finite rank. A *λ-flat connection* on $\mathcal{H}$ is homomorphism of $\mathcal{O}_S$-modules $u : \mathcal{D}_1(\lambda) \to \mathcal{D}_1(\mathcal{H})$ that is also a Lie homomorphism over k, commutes with the symbol maps (so these must land in θ_S) and takes scalars to scalars: $\mathcal{O}_S \subset \mathcal{D}_1(\lambda)$ is mapped to $\mathcal{O}_S \subset \mathcal{D}_1(\mathcal{H})$.

It follows from the preceding that such a homomorphism u determines a flat connection on the projectivization of $\mathcal{H}$. The map u preserves $\mathcal{O}_S$ and since this restriction is $\mathcal{O}_S$-linear, it is given by multiplication by some regular function w on S. If $D \in \theta_S$ is lifted to $\hat{D} \in \mathcal{D}_1(\lambda)$, then $u(\hat{D}) \in \mathcal{D}_1(\mathcal{H})$ is also a lift of D and so $D(w) = [u(\hat{D}), u(1)] = u([\hat{D}, 1]) = 0$. This shows that w must be locally constant; we call this the *weight* of u. So in the above discussion, $\hat{\mathcal{D}}$ comes with $\det(\mathcal{H})$-flat connection of weight r^{-1}.

It is clear that if the weight of u is constant zero, then u factors through θ_S, so that we get a flat connection in $\mathcal{H}$. This is also the case when $\lambda = \mathcal{O}_S$, for then $\mathcal{D}_1(\mathcal{O}_S)$ contains θ_S canonically as a direct summand (both as $\mathcal{O}_S$-module and as a sheaf of k-Lie algebras) and the flat connection is then given by the action of θ_S. This has an interesting consequence: if $\pi : \Lambda^\times \to S$ is the geometric realization of the $\mathbb{G}_m$-bundle defined by λ, then $\pi^*\lambda$ has a 'tautological' generating section and thus gets identified with $\mathcal{O}_{\Lambda^\times}$. Hence a λ-flat connection on $\mathcal{H}$ defines an ordinary flat connection on $\pi^*\mathcal{H}$. One checks that if w is the weight of u, then the connection is homogeneous of degree w along the fibers. So in case $k = \mathbb{C}$, $s \in S$ and $\tilde{s} \in \Lambda^\times$ lies over $s \in S$, then the multivalued map $(z, h) \in \mathbb{C}^\times \times H_s \mapsto (z\tilde{s}, z^w h) \in \Lambda_s^\times \times H_s$ is flat, and so the monodromy of the connection in $\Lambda_s^\times$ is scalar multiplication by $e^{2\pi\sqrt{-1}w}$.

We will also encounter a logarithmic version. Here we are given a closed subvariety $\Delta \subset S$ of lower dimension (usually a normal crossing hypersurface). Then the θ_S-stabilizer of the ideal defining Δ, denoted $\theta_S(\log\Delta)$, is a coherent $\mathcal{O}_S$-submodule of θ_S closed under the Lie bracket. If in Definition 1.1 we have u only defined on the preimage of $\theta_S(\log\Delta) \subset \theta_S$ in $\mathcal{D}_1(\lambda)$ (which we denote here by $\mathcal{D}_1(\lambda)(\log\Delta)$), then we say that we have a *logarithmic λ-flat connection relative to Δ* on $\mathcal{H}$.

2. The Virasoro algebra and its basic representation

Much of the material exposed in this section is a conversion of certain standard constructions (as can be found for instance in [8]) into a coordinate invariant and relative setting. But the way we introduce the Virasoro algebra is less standard and may be even new. A similar remark applies to part of the last subsection (in particular, Corollary 2.7), which is devoted to the Fock module attached to a symplectic local system.

In this section we fix an R-algebra $\mathcal{O}$ isomorphic to the formal power series ring $R[[t]]$. In other words, $\mathcal{O}$ comes with a principal ideal $\mathfrak{m}$ so that $\mathcal{O}$ is complete for the $\mathfrak{m}$-adic topology and the associated graded R-algebra $\oplus_{j=0}^{\infty}\mathfrak{m}^j/\mathfrak{m}^{j+1}$ is a polynomial ring over R in one variable. The choice of a generator t of the ideal $\mathfrak{m}$ identifies $\mathcal{O}$ with $R[[t]]$. We denote by L the localization of $\mathcal{O}$ obtained by inverting a generator of $\mathfrak{m}$. For $N \in \mathbb{Z}$, $\mathfrak{m}^N$ has the obvious meaning as an $\mathcal{O}$-submodule of L. The *$\mathfrak{m}$-adic topology* on L is the topology that has the collection of cosets $\{f + \mathfrak{m}^N\}_{f\in L, N\in\mathbb{Z}}$ as a basis of open subsets. We sometimes write $F^N L$ for $\mathfrak{m}^N$. We further denote by θ the L-module of continuous R-derivations from L into L and by ω the L-dual of θ. These L-modules come with filtrations (making them principal filtered L-modules): $F^N\theta$ consists of the derivations that take $\mathfrak{m}$ to $\mathfrak{m}^{N+1}$ and $F^N\omega$ consists of the L-homomorphisms $\theta \to L$ that take $F^0\theta$ to $\mathfrak{m}^N$. So in terms of the generator t above, $L = R((t))$, $\theta = R((t))\frac{d}{dt}$, $F^N\theta = R[[t]]t^{N+1}\frac{d}{dt}$, $\omega = R((t))dt$ and $F^N\omega = R[[t]]t^{N-1}dt$.

The residue map $\mathrm{Res} : \omega \to R$ which assigns to an element of $R((t))dt$ the coefficient of $t^{-1}dt$ is canonical, i.e., is independent of the choice of t. The R-bilinear map

$$r : L \times \omega \to R, \quad (f, \alpha) \mapsto \mathrm{Res}(f\alpha)$$

is a topologically perfect pairing of filtered R-modules: we have $r(t^k, t^{-l-1}dt) = \delta_{k,l}$ and so any R-linear $\phi : L \to R$ which is continuous (i.e., ϕ zero on $\mathfrak{m}^N$ for some N) is definable by an element of ω (namely by $\sum_{k>N} \phi(t^{-k})t^{k-1}dt$) and likewise for an R-linear continuous map $\omega \to R$.

A trivial Lie algebra

If we think of of the multiplicative group $L^\times$ of L as an algebraic group over R (or rather, as a group object in a category of ind schemes over R), then its Lie algebra, denoted here by $\mathfrak{l}$, is L, regarded as a R-module with trivial Lie bracket. It comes with a decreasing filtration $F^\bullet\mathfrak{l}$ (as a Lie algebra) defined by the valuation. The universal enveloping algebra $U\mathfrak{l}$ is clearly the symmetric algebra of $\mathfrak{l}$ as an R-module, $\mathrm{Sym}^\bullet_R(\mathfrak{l})$. The ideal $U_+\mathfrak{l} \subset U\mathfrak{l}$ generated by $\mathfrak{l}$ is also a right $\mathcal{O}$-module (since $\mathfrak{l}$ is). We complete it $\mathfrak{m}$-adically: given an integer $N \geqslant 0$, then an R-basis of the truncation $U_+\mathfrak{l}/(U\mathfrak{l} \circ F^N\mathfrak{l})$ is the collection $t^{k_1} \circ \cdots \circ t^{k_r}$ with $k_1 \leqslant k_2 \leqslant \cdots \leqslant k_r < N$. So elements of the completion

$$U_+\mathfrak{l} \to \overline{U}_+\mathfrak{l} := \varprojlim_N U_+\mathfrak{l}/U\mathfrak{l} \circ F^N\mathfrak{l}$$

are series of the form $\sum_{i=1}^{\infty} r_i t^{k_{i,1}} \circ \cdots \circ t^{k_{i,r_i}}$ with $r_i \in R$, $c \leqslant k_{1,i} \leqslant k_{2,i} \leqslant \cdots \leqslant k_{i,r_i}$ for some constant c. We put $\overline{U}\mathfrak{l} := R \oplus \overline{U}_+\mathfrak{l}$, which we could of course have defined just as well directly as

$$U\mathfrak{l} \to \overline{U}\mathfrak{l} := \varprojlim_N U\mathfrak{l}/U\mathfrak{l} \circ F^N\mathfrak{l}.$$

We will refer to this construction as the *$\mathfrak{m}$-adic completion on the right,* although in the present case there is no difference with the analogously defined $\mathfrak{m}$-adic completion on the left, as $\mathfrak{l}$ is commutative.

Any continuous derivation $D \in \theta$ defines an R-linear map $\omega \to L$ which is self-adjoint relative the residue pairing: $r(\langle D, \alpha\rangle, \beta) = r(\alpha, \langle D, \beta\rangle)$. We use that pairing to identify D with an element of the closure of $\mathrm{Sym}^2\,\mathfrak{l}$ in $\overline{U}\mathfrak{l}$. Let $C(D)$ be half this element, so that in terms of the above topological basis,

$$C(D) = \tfrac{1}{2} \sum_{i,j \in \mathbb{Z}} r(\langle D, t^{-i-1}dt\rangle, t^{-j-1}dt)t^i \circ t^j.$$

In particular for $D = D_k = t^{k+1}\frac{d}{dt}$, $C(D_k) = \frac{1}{2}\sum_{i+j=k} t^i \circ t^j$. Observe that the map $C : \theta \to \overline{U}\mathfrak{l}$ is continuous.

Oscillator and Virasoro algebra

The residue map defines a central extension of $\mathfrak{l}$, the *oscillator algebra* $\hat{\mathfrak{l}}$, which as an R-module is simply $\mathfrak{l} \oplus R$. If we denote the generator of the second summand by $\hbar$, then the Lie bracket is given by

$$[f + \hbar r, g + \hbar s] := \mathrm{Res}(g\, df)\hbar.$$

So $[t^k, t^{-l}] = k\delta_{k,l}\hbar$ and the center of $\hat{\mathfrak{l}}$ is $Re \oplus R\hbar$, where $e = t^0$ denotes the unit element of L viewed as an element of $\mathfrak{l}$. It follows that $U\hat{\mathfrak{l}}$ is an $R[e, \hbar]$-algebra. As an $R[\hbar]$-algebra it is obtained as follows: take the tensor algebra of $\mathfrak{l}$ (over R) tensored with $R[\hbar]$, $\otimes^{\bullet}_{R}\mathfrak{l} \otimes_R R[\hbar]$, and divide that out by the two-sided ideal generated by the elements $f \otimes g - g \otimes f - \mathrm{Res}(g df)\hbar$. The obvious surjection $\pi : U\hat{\mathfrak{l}} \to U\mathfrak{l} = \mathrm{Sym}^{\bullet}_{R}(\mathfrak{l})$ is the reduction modulo $\hbar$.

We filter $\hat{\mathfrak{l}}$ by letting $F^N\hat{\mathfrak{l}}$ be $F^N\mathfrak{l}$ for $N > 0$ and $F^N\mathfrak{l} + R\hbar$ for $N \leqslant 0$. This filtration is used to complete $U\hat{\mathfrak{l}}$ $\mathfrak{m}$-adically on the right:

$$U\hat{\mathfrak{l}} \to \overline{U\hat{\mathfrak{l}}} := \varprojlim_N U\hat{\mathfrak{l}}/U\hat{\mathfrak{l}} \circ F^N\mathfrak{l}.$$

Notice that this completion has the collection $t^{k_1} \circ \cdots \circ t^{k_r}$ with $r \geqslant 0$, $k_1 \leqslant k_2 \leqslant \cdots \leqslant k_r$, as topological $R[\hbar]$-basis. Since $\hat{\mathfrak{l}}$ is not abelian, the left and right $\mathfrak{m}$-adic topologies now differ. For instance, $\sum_{k \geqslant 1} t^k \circ t^{-k}$ does not converge in $\overline{U\hat{\mathfrak{l}}}$, whereas $\sum_{k\geqslant 1} t^{-k} \circ t^k$ does. The obvious surjection $\pi : \overline{U\hat{\mathfrak{l}}} \to \overline{U\mathfrak{l}}$ is still given by reduction modulo $\hbar$. We also observe that the filtrations of $\mathfrak{l}$ and $\hat{\mathfrak{l}}$ determine decreasing filtrations of their (completed) universal enveloping algebras, e.g., $F^N U\hat{\mathfrak{l}} = \sum_{r\geqslant 0} \sum_{n_1+\cdots+n_r \geqslant N} F^{n_1}\hat{\mathfrak{l}} \circ \cdots \circ F^{n_r}\hat{\mathfrak{l}}$.

Let us denote by $\mathfrak{l}_2$ the image of $\mathfrak{l} \otimes_R \mathfrak{l} \subset \hat{\mathfrak{l}} \otimes_R \hat{\mathfrak{l}} \to U\hat{\mathfrak{l}}$. Under the reduction modulo $\hbar$, $\mathfrak{l}_2$ maps onto $\mathrm{Sym}^2_R(\mathfrak{l}) \subset U\mathfrak{l}$ with kernel $R\hbar$. Its closure $\bar{\mathfrak{l}}_2$ in $\overline{U\hat{\mathfrak{l}}}$ maps onto the closure of $\mathrm{Sym}^2_R(\mathfrak{l})$ in $\overline{U\mathfrak{l}}$ with the same kernel.

The generator t defines a continuous R-linear map $D \in \theta \mapsto \hat{C}(D) \in \bar{\mathfrak{l}}_2$ characterized by

$$\hat{C}(D_k) := \tfrac{1}{2} \sum_{i+j=k} : t^i \circ t^j : .$$

We here adhered to the *normal ordering convention*, which prescribes that the factor with the highest index comes last and hence acts first (here the exponent serves as index). So $: t^i \circ t^j :$ equals $t^i \circ t^j$ if $i \leqslant j$ and $t^j \circ t^i$ if $i > j$. This map is clearly a lift of $C : \theta \to \mathrm{Sym}^2 \mathfrak{l}$, but is otherwise non-canonical.

Lemma 2.1. *We have*

(i) $[\hat{C}(D), f] = -\hbar D(f)$ *as an identity in* $\overline{U\hat{\mathfrak{l}}}$ *(where* $f \in \mathfrak{l} \subset \hat{\mathfrak{l}}$*) and*
(ii) $[\hat{C}(D_k), \hat{C}(D_l)] = -\hbar(l - k)\hat{C}(D_{k+l}) + \hbar^2 \tfrac{1}{12}(k^3 - k)\delta_{k+l,0}$.

Proof. For the first statement we compute $[\hat{C}(D_k), t^l]$. If we substitute $\hat{C}(D_k) = \frac{1}{2}\sum_{i+j=k} : t^i \circ t^j :$, then we see that only terms of the form $[t^{k+l} \circ t^{-l}, t^l]$ or $[t^{-l} \circ t^{k+l}, t^l]$ (depending on whether $k + 2l \leqslant 0$ or $k + 2l \geqslant 0$) can make a contribution and then have coefficient $\frac{1}{2}$ if $k + 2l = 0$ and 1 otherwise. In all cases the result is $-\hbar l t^{k+l} = -\hbar D_k(t^l)$.

Formula (i) implies that

$$[\hat{C}(D_k), \hat{C}(D_l)] = \lim_{N\to\infty} \sum_{|i|\leqslant N} \tfrac{1}{2}\left(D_k(t^i) \circ t^{l-i} + t^i \circ D_k(t^{l-i})\right)$$
$$= -\hbar \lim_{N\to\infty} \sum_{|i|\leqslant N} \left(it^{k+i} \circ t^{l-i} + t^i \circ (l-i)t^{k+l-i}\right).$$

This is up to a reordering equal to $-\hbar(l-k)\hat{C}(D_{k+l})$. The terms which do not commute and are in the wrong order are those for which $0 < k+i = -(l-i)$ (with coefficient i) and for which $0 < i = -(k+l-i)$ (with coefficient $(l-i)$). This accounts for the extra term $\hbar^2 \frac{1}{12}(k^3-k)\delta_{k+l,0}$. □

This lemma shows that $-\hbar^{-1}\hat{C}$ behaves better than $\hat{C}$ (but requires us of course to assume that $\hbar$ be invertible). In fact, it suggests to consider the set $\hat{\theta}$ of pairs $(D,u) \in \theta \times \hbar^{-1}\bar{l}_2$ for which $C(D) \in \mathrm{Sym}^2 l$ is the mod $\hbar$ reduction of $-\hbar u$, so that we have an exact sequence

$$0 \to R \to \hat{\theta} \to \theta \to 0$$

of R-modules. Then a non-canonical section of $\hat{\theta} \to \theta$ is given by $D \mapsto \hat{D} := (D, -\hbar^{-1}\hat{C}(D))$. In order to avoid confusion, we denote the generator of the copy of R by c_0.

Corollary-Definition 2.2. *This defines a central extension of Lie algebras, called the* Virasoro algebra *(of the* R*-algebra* L*). Precisely, if* $T : \hat{\theta} \to \overline{U\hat{l}}[\frac{1}{\hbar}]$ *is given by the second component, then* T *is injective, maps* $\hat{\theta}$ *onto a Lie subalgebra of* $\overline{U\hat{l}}[\frac{1}{\hbar}]$ *and sends* c_0 *to* 1*. If we transfer the Lie bracket on* $\overline{U\hat{l}}[\frac{1}{\hbar}]$ *to* $\hat{\theta}$*, then in terms of our non-canonical section,*

$$[\hat{D}_k, \hat{D}_l] = (l-k)\hat{D}_{k+l} + \frac{k^3-k}{12}\delta_{k+l,0}c_0.$$

Moreover, $\mathrm{ad}_{T(\hat{D})}$ *leaves* l *invariant (as a subspace of* $\overline{U\hat{l}}$*) and acts on that subspace by derivation with respect to* $D \in \theta$*.*

Remark 2.3. An alternative coordinate free definition of the Virasoro algebra, based on the algebra of pseudo-differential operators on L, can be found in [4].

Fock representation

It is clear that $F^0\hat{l} = R\hbar \oplus \mathcal{O}$ is an abelian subalgebra of $\hat{l}$. We let $F^0\hat{l} = \mathcal{O} \oplus R\hbar$ act on a free rank one module Rv_o by letting $\mathcal{O}$ act trivially and $\hbar$ as the identity. The induced representation of $\hat{l}$ over R,

$$\mathbb{F} := U\hat{l} \otimes_{U\hat{l} \circ F^0\hat{l}} Rv_o,$$

will be regarded as a $Ul[\hbar^{-1}]$-module. It comes with an increasing PBW (Poincaré-Birkhoff-Witt) filtration $W_\bullet\mathbb{F}$ by R-submodules, with $W_r\mathbb{F}$ being the image of $\oplus_{s\leqslant r}\hat{l}^{\otimes s} \otimes Rv_o$. Since the scalars $R \subset l$ are central in $\hat{l}$ and kill $\mathbb{F}$ (because $R \subset \mathcal{O}$), they act trivially in all of $\mathbb{F}$. As an R-module, $\mathbb{F}$ is free with basis the collection

$t^{-k_r} \circ \cdots \circ t^{-k_1} \otimes v_o$, where $r \geqslant 0$ and $1 \leqslant k_1 \leqslant k_2 \leqslant \cdots \leqslant k_r$ (for $r = 0$, read v_o). (In fact, $\mathrm{Gr}^W_\bullet \mathbb{F}$ can be identified as a graded R-module with the symmetric algebra $\mathrm{Sym}^\bullet(\mathfrak{l}/F^0\mathfrak{l})$.) This also shows that $\mathbb{F}$ is even a $\widehat{U\mathfrak{l}}[\hbar^{-1}]$-module. Thus $\mathbb{F}$ affords a representation of $\hat{\theta}$ over R, called its *Fock representation*.

It follows from Lemma 2.1 that for any $D \in \theta$ with lift $\hat{D} \in \hat{\theta}$,

$$T(\hat{D})t^{-k_r} \circ \cdots \circ t^{-k_1} \otimes v_o =$$
$$= \left(\sum_{i=1}^{r} t^{-k_r} \circ \cdots \circ D(t^{-k_i}) \circ \cdots \circ t^{-k_1} \right) \otimes v_o + t^{-k_r} \circ \cdots \circ t^{-k_1} \circ T(\hat{D})v_o.$$

Since $T(\hat{D})v_o = 0$ when $D \in F^0\theta$, it follows that $F^0\theta$ acts on $\mathbb{F}$ by coefficient-wise derivation. This observation has an interesting consequence. Consider the module of k-derivations $R \to R$ (denoted here simply by θ_R instead of the more accurate $\theta_{R/k}$) and the module $\theta_{L,R}$ of k-derivations of L that are continuous for the $\mathfrak{m}$-adic topology and preserve $R \subset L$. Since $L \cong R((t))$ as an R-algebra, every k-derivation $R \to R$ extends to one from L to L. So we have an exact sequence

$$0 \to \theta \to \theta_{L,R} \to \theta_R \to 0.$$

The following corollary essentially says that we have defined in the L-module $\mathbb{F}$ a Lie algebra $\hat{\theta}_{L,R}$ of first order (k-linear) differential operators which contains R as the degree zero operators and for which the symbol map (which is just the formation of the degree one quotient) has image $\theta_{L,R}$.

Corollary 2.4. *The actions on $\mathbb{F}$ of $F^0\theta_{L,R} = \mathfrak{m}\theta_{\mathcal{O},R} \subset \theta_{L,R}$ (given by coefficient-wise derivation, killing the generator v_o) and $\hat{\theta}$ coincide on $F^0\theta$ and generate a central extension of Lie algebras $\hat{\theta}_{L,R} \to \theta_{L,R}$ by Rc_o. Its defining representation on $\mathbb{F}$ (still denoted T) is faithful and has the property that for every lift $\hat{D} \in \hat{\theta}_{L,R}$ of $D \in \theta_{L,R}$ and $f \in \mathfrak{l}$ we have $[T(\hat{D}), f] = Df$ (in particular, it preserves every $U\hat{\mathfrak{l}}$-submodule of $\mathbb{F}$).*

Proof. The generator t can be used to define a section of $\theta_{L,R} \to \theta_R$: the set of elements of $\theta_{L,R}$ which kill t is a k-Lie subalgebra of $\theta_{L,R}$ which projects isomorphically onto θ_R. Now if $D \in \theta_{L,R}$, write $D = D_{\text{vert}} + D_{\text{hor}}$ with $D_{\text{vert}} \in \theta$ and $D_{\text{hor}}(t) = 0$ and define an R-linear operator $\hat{D}$ in $\mathbb{F}$ as the sum of $T(\hat{D}_{\text{vert}})$ and coefficient-wise derivation by D_{hor}. This map clearly has the properties mentioned.

As to its dependence on t: another choice yields a decomposition of the form $D = (D_{\text{hor}} + D_0) + (D_{\text{vert}} - D_0)$ with $D_0 \in F^0\theta$ and in view of the above $\hat{D}_0$ acts in $\mathbb{F}$ by coefficient-wise derivation. □

The Fock representation for a symplectic local system

In Section 4 we shall run into a particular type of finite rank subquotient of the Fock representation and it seems best to discuss the resulting structure here. We start out from the following data:

(i) a free R-module H of finite rank endowed with a symplectic form $\langle\ ,\ \rangle$: $H \otimes_R H \to R$, which is nondegenerate in the sense that the induced map $H \to H^*$, $a \mapsto \langle\ , a\rangle$ is an isomorphism of R-modules,

(ii) an R-submodule $\mathfrak{D} \subset \theta_R$ closed under the Lie bracket for which the inclusion is an equality over the generic point and a Lie action $D \mapsto \nabla_D$ of $\mathfrak{D}$ on H by k-derivations which preserves the symplectic form,

(iii) a Lagrangian R-submodule $F \subset H$.

Property (ii) means that $D \in \mathfrak{D} \mapsto \nabla_D \in \mathrm{End}_k(H)$ is R-linear, obeys the Leibniz rule: $\nabla_D(ra) = r\nabla_D(a) + D(r)a$ and satisfies $\langle \nabla_D a, b\rangle + \langle a, \nabla_D b\rangle = D\langle a, b\rangle$. In the cases of interest, $\mathfrak{D}$ will be the θ_R-stabilizer of a principal ideal in R (and often be all of θ_R). One might think of ∇ as a flat meromorphic connection on the symplectic bundle represented by H.

In this setting, a Heisenberg algebra is defined in an obvious manner: it is $\hat{H} := H \oplus R\hbar$ endowed with the bracket $[a + R\hbar, b + R\hbar] = \langle a, b\rangle\hbar$. We also have defined a Fock representation $\mathbb{F}(H, F)$ of $\hat{H}$ as the induced module of the rank one representation of $\hat{F} = F + R\hbar$ on R given by the coefficient of $\hbar$. Notice that if we grade $\mathbb{F}(H, F)$ with respect to the PBW filtration, we get a copy of the symmetric algebra of H/F over R. We aim to define a projective Lie action of $\mathfrak{D}$ on $\mathbb{F}(H, F)$.

We begin with extending the $\mathfrak{D}$-action to $\hat{H}$ by stipulating that it kills $\hbar$. This action clearly preserves the Lie bracket and hence determines one of $\mathfrak{D}$ on the universal enveloping algebra $U\hat{H}$. This does not however induce one in $\mathbb{F}(H, F)$, as ∇_D will not respect the right ideal in $U\hat{H}$ generated by $\hbar - 1$ and F. We will remedy this by means of a 'twist'.

We shall use the isomorphism $\sigma : H \otimes_R H \cong \mathrm{End}_R(H)$ of R-modules defined by associating to $a \otimes b$ the endomorphism $\sigma(a \otimes b) : x \in H \mapsto a\langle b, x\rangle \in H$. If we agree to identify an element in the tensor algebra of H, in particular, an element of H, as the operator in $U\hat{H}$ or $\mathbb{F}(H, F)$ given by left multiplication, then it is ready checked that for $x \in H$,

$$[a \circ b, x] = \sigma(a \otimes b + b \otimes a)(x).$$

We choose a Lagrangian supplement of F in H, i.e., a Lagrangian R-submodule $F' \subset H$ that is also a section of $H \to H/F$. Since F' is an abelian Lie subalgebra of $\hat{H}$, we have a natural map $\mathrm{Sym}^\bullet_R(F') \to \mathbb{F}(H, F)$. It is clearly an isomorphism of $\mathrm{Sym}^\bullet_R(F')$-modules. Now write ∇_D according to the Lagrangian decomposition $H = F' \oplus F$:

$$\nabla_D = \begin{pmatrix} \nabla_D^{F'} & \sigma'_D \\ \sigma_D & \nabla_D^F \end{pmatrix}.$$

Here the diagonal entries represent the induced connections on F' and F, whereas $\sigma_D \in \mathrm{Hom}_R(F', F)$ and $\sigma'_D \in \mathrm{Hom}_R(F, F')$. Since σ identifies $F \otimes_R F$ resp. $F' \otimes_R F'$ with $\mathrm{Hom}_R(F', F)$ resp. $\mathrm{Hom}_R(F, F')$, we can write $\sigma_D = \sigma(s_D)$ with $s_D \in F \otimes_R F$ and $\sigma'_D = \sigma(s'_D)$ and $s'_D \in F' \otimes_R F'$. These tensors are symmetric and represent the second fundamental form of $F' \subset H$ resp. $F \subset H$. Notice that if $a \in F$, then

$$[\nabla_D, a] = \nabla_D(a) = \nabla^F_D(a) + \sigma^F_{F'}(a) = \nabla^F_D(a) + \tfrac{1}{2}[s'_D, a]$$

and similarly, if $a' \in F'$, then $[\nabla_D, a'] = \nabla^{F'}_D(a') + \frac{1}{2}[s_D, a']$. This suggests we should assign to $D \in \mathfrak{D}$ the first order differential operator $T_{F'}(D)$ in $\mathbb{F}(H, F) \cong \mathrm{Sym}^\bullet F'$ defined by

$$T_{F'}(D) := \nabla^{F'}_D + \tfrac{1}{2}s_D + \tfrac{1}{2}s'_D.$$

Proposition 2.5. *The map $T_{F'} : \mathfrak{D} \to \mathrm{End}_k(\mathrm{Sym}^\bullet F')$ is R-linear and has the property that $[T_{F'}(D), a] = \nabla_D(a)$ for every $D \in \mathfrak{D}$ and $a \in \hat{H}$. Any other map $\mathfrak{D} \to \mathrm{End}_k(\mathrm{Sym}^\bullet F')$ enjoying these properties differs from $T_{F'}$ by a multiple of the identity operator, in other words, is of the form $D \mapsto T_{F'}(D) + \eta(D)$ for some $\eta \in \mathrm{Hom}_R(\mathfrak{D}, R)$.*

Proof. That $T_{F'}(D)$ has the stated property follows from the preceding. Let $\eta : \mathfrak{D} \to \mathrm{End}_k(\mathrm{Sym}^\bullet F')$ be the difference of two such maps. Then for every $D \in \mathfrak{D}$, $\eta(D) \in \mathrm{End}_R(\mathbb{F}(H, F))$ commutes with all elements of $\hat{H}$. Since $\mathbb{F}(H, F)$ is irreducible as a representation of $\hat{H}$, it follows that $\eta(D)$ is a scalar in R. □

Notice that if $u_1, \ldots, u_r \in \hat{H}$, then

$$T_{F'}(D)(u_r \circ \cdots \circ u_1 \otimes v_o) =$$
$$= \left(\sum_{i=1}^{r} u_r \circ \cdots \circ \nabla_D(u_i) \circ \cdots \circ u_1 + u_r \circ \cdots \circ u_1 \circ \tfrac{1}{2}s'_D\right) \otimes v_o.$$

So this looks like the operator $T_{\hat{D}}$ acting in $\mathbb{F}$ with s'_D playing the role of $-\hat{C}(D)$. Here is the key result about the 'curvature' of $T_{F'}$.

Lemma 2.6. *Given $D, E \in \mathfrak{D}$, then $[T_{F'}(D), T_{F'}(E)] - T_{F'}([D, E])$ is scalar multiplication by $1/2$ times the value on of the ∇^F-curvature on* $\det(F)$ *on the pair (D, E).*

Proof. The fact that ∇ preserves the Lie bracket is expressed by the following identities:

$$\nabla^F_D\nabla^F_E - \nabla^F_E\nabla^F_D - \nabla^F_{[D,E]} = \sigma_E\sigma'_D - \sigma_D\sigma'_E,$$
$$\nabla^{F'}_D\nabla^{F'}_E - \nabla^{F'}_E\nabla^{F'}_D - \nabla^{F'}_{[D,E]} = \sigma'_E\sigma_D - \sigma'_D\sigma_E,$$
$$\nabla^{\mathrm{Hom}(F',F)}_D(\sigma_E) - \nabla^{\mathrm{Hom}(F',F)}_E(\sigma_D) = \sigma_{[D,E]},$$
$$\nabla^{\mathrm{Hom}(F,F')}_D(\sigma'_E) - \nabla^{\mathrm{Hom}(F,F')}_E(\sigma'_D) = \sigma'_{[D,E]}.$$

The first two give the curvature of ∇^F and $\nabla^{F'}$ on the pair (D, E). The last two can also be written as operator identities in $\mathrm{Sym}^\bullet F'$:

$$[\nabla^{F'}_D, s_E] - [\nabla^{F'}_E, s_D] = s_{[D,E]},$$
$$[\nabla^{F'}_D, s'_E] - [\nabla^{F'}_E, s'_D] = s'_{[D,E]}.$$

If we feed these identities in:

$$\begin{aligned}
&[T_{F'}(D), T_{F'}(E)] - T_{F'}([D,E]) \\
&= [\nabla_D^{F'} + \tfrac{1}{2}s_D + \tfrac{1}{2}s'_D, \nabla_E^{F'} + \tfrac{1}{2}s_E + \tfrac{1}{2}s'_E] - (\nabla_{[D,E]}^{F'} + \tfrac{1}{2}s_{[D,E]} + \tfrac{1}{2}s'_{[D,E]}) \\
&= \big([\nabla_D^{F'}, \nabla_E^{F'}] - \nabla_{[D,E]}^{F'}\big) + \tfrac{1}{2}\big([\nabla_D^{F'}, s_E] - [\nabla_E^{F'}, s_D] - s_{[D,E]}\big)\big) \\
&\quad + \tfrac{1}{2}\big([\nabla_D^{F'}, s'_E] - [\nabla_E^{F'}, s'_D] - s'_{[D,E]}\big)\big) + \tfrac{1}{4}\big([s_D, s'_E] - [s_E, s'_D]\big)
\end{aligned}$$

(where we identified $\mathbb{F}(H,F)$ with $\mathrm{Sym}^\bullet F'$), we obtain

$$[T_{F'}(D), T_{F'}(E)] - T_{F'}([D,E]) = \big(\sigma'_E\sigma_D + \tfrac{1}{4}[s_D, s'_E]\big) - \big(\sigma'_D\sigma_E + \tfrac{1}{4}[s_{D'}, s_E]\big).$$

We must show that the right hand side is equal to $\frac{1}{2}\operatorname{Tr}(\sigma_E\sigma'_D - \sigma_D\sigma'_E)$, or perhaps more specifically, that $\sigma'_E\sigma_D + \frac{1}{4}[s_D, s'_E] = -\frac{1}{2}\operatorname{Tr}(\sigma_D\sigma'_E)$ (and similarly if we exchange D and E). This reduces to the following identity in linear algebra: if $\alpha \in F$ and $\beta \in F'$, then in $\mathrm{Sym}^\bullet F'$ we have

$$\sigma(\beta\otimes\beta)\sigma(\alpha\otimes\alpha) + \tfrac{1}{4}[\alpha\circ\alpha, \beta\circ\beta] = -\tfrac{1}{2}\operatorname{Tr}_{F'}(\sigma_{\alpha\otimes\alpha}\sigma_{\beta\otimes\beta}),$$

Indeed, a straightforward computation shows that

$$[\alpha\circ\alpha, \beta\circ\beta] = 2\langle\alpha,\beta\rangle(\alpha\circ\beta + \beta\circ\alpha) = 4\langle\alpha,\beta\rangle\beta\circ\alpha + 2\langle\alpha,\beta\rangle^2.$$

If we interpret $\langle\alpha,\beta\rangle\beta\circ\alpha$ as an operator in $\mathrm{Sym}^\bullet F'$, then applying it to $x \in F'$ yields $\langle\alpha,\beta\rangle\beta\langle\alpha,x\rangle = -\sigma(\beta\otimes\beta)\sigma(\alpha\otimes\alpha)(x)$. We also find that $\langle\alpha,\beta\rangle^2 = -\operatorname{Tr}_{F'}(\sigma(\alpha\otimes\alpha)\sigma(\beta\otimes\beta))$. □

If N is a free R-module of rank one, then by a *square root of* N we mean a free R-module Θ of rank one together with an isomorphism of $\Theta\otimes_R\Theta$ onto N.

Corollary 2.7. *Let* Θ *be a square root of* $\det_R(F)$. *Then the twisted Fock module* $\mathrm{Hom}_R(\Theta, \mathbb{F}(H,F))$ *comes with a natural action of* $\mathfrak{D}$ *by derivations.*

Proof. Given the Lagrangian supplement F' of F in H, then endow Θ with the unique $\mathfrak{D}$-module structure that makes the given isomorphism $\Theta\otimes_R\Theta \cong \det_R(F)$ one of $\mathfrak{D}$-modules: if $w \in \Theta$ is a generator and $\nabla_D^{\det F}(w\otimes w) = rw\otimes w$, then $\nabla_D^{\Theta}(w) = \frac{1}{2}rw$. This ensures that the $\mathfrak{D}$-action on $\mathrm{Hom}_R(\Theta, \mathrm{Sym}^\bullet F')$ preserves the Lie bracket. It remains to show that this action is independent of F'. This can be verified by a computation, but rather than carrying this out, we give an abstract argument that avoids this. It is based on the well-known fact that if H_o is a fixed symplectic k-vector space of finite dimension 2g, and $F_o \subset H_o$ is Lagrangian, then the set of Lagrangian supplements of F_o in H_o form in the Grassmannian of H_o an affine space over $\mathrm{Sym}_k^2 F_o$ (and hence is simply connected). Now by doing the preceding construction universally over the corresponding affine space over $\mathrm{Sym}_R^2 F$, we see that the flatness on the universal example immediately gives the independence. □

Remark 2.8. We will use this corollary mainly via the following reformulation. First we observe that the Lie algebra of first order k-linear differential operators $\Theta \to \Theta$

projects to θ_R (this is the symbol map) with kernel the scalars R. Denote by $\mathfrak{D}(\Theta)$ the preimage of $\mathfrak{D}$. This is clearly a Lie subalgebra. Then the above corollary can be understood as saying that there is a natural Lie action of $\mathfrak{D}(\Theta)$ on $\mathbb{F}(H, F)$ by first order differential operators, acting, in the terminology of Section 1, with weight 1. The image in $\mathrm{End}_k(\mathbb{F}(H, F))$ is the R-submodule of $\mathrm{End}_k(\mathbb{F}(H, F))$ generated by the $T_{F'}(D)$ and the identity operator. We may also use $\mathfrak{D}(\det_R(F))$ instead, although then the weight will be $\frac{1}{2}$. Note that our discussion of projectively flat connections in Section 1 now suggests a formulation in more geometric terms, namely that the pull-back of $\mathbb{F}(H, F)$ to the geometric realization of the $\mathbb{G}_m$-bundle over Spec(R) defined by $\det_R(F)$ acquires a flat meromorphic connection with fiber monodromy minus the identity.

Remark 2.9. The preceding follows the presentation of Boer-Looijenga [5] rather closely. The quadratic terms that enter in the definition of $T_{F'}$ are in a way a relict of the heat operator of which the theta functions associated to this symplectic local system are solutions (flat sections are expansions of theta functions relative to an unspecified lattice).

3. The Sugawara construction

In this section we show how the Virasoro algebra acts in the standard representions of a centrally extended loop algebra. This construction goes back to the physicist H. Sugawara (in 1968), but it was probably Graeme Segal who first noticed its relevance for the present context.

Most of the material below can for instance be found in [8] (Lecture 10) and [7] (Ch. 12), but our presentation slightly deviates from the standard sources in substance as well in form: we approach the Sugawara construction via the construction discussed in the previous section and put it in the (coordinate free) setting that makes it appropriate for the application we have in mind.

In this section, we fix a simple Lie algebra $\mathfrak{g}$ over k of finite dimension. We retain the data and the notation of Section 2.

Loop algebras

We identify $\mathfrak{g}\otimes\mathfrak{g}$ with the space of bilinear forms $\mathfrak{g}^*\times\mathfrak{g}^*\to k$, where $\mathfrak{g}^*$ denotes the k-dual of $\mathfrak{g}$, as usual. We form its space of $\mathfrak{g}$-covariants (relative to the adjoint action on both factors):

$$q : \mathfrak{g}\otimes\mathfrak{g} \to (\mathfrak{g}\otimes\mathfrak{g})_{\mathfrak{g}} =: \mathfrak{c}.$$

This space is known to be of dimension one and to consist of symmetric tensors. It has a canonical generator which is characterized by the property that it is represented by a $\mathfrak{g}$-invariant symmetric tensor $c \in \mathfrak{g}\otimes\mathfrak{g}$ with the property that $c(\theta\otimes\theta) = 2$ if $\theta \in \mathfrak{g}^*$ is a long root (relative to a choice of Cartan subalgebra $\mathfrak{h}$ of $\mathfrak{g}$; the roots then lie in the zero eigenspace of $\mathfrak{h}$ in $\mathfrak{g}^*$). This element is in fact invariant under the full automorphism group of the Lie algebra $\mathfrak{g}$, not just the inner ones. It is nondegenerate when viewed as a symmetric bilinear form on $\mathfrak{g}^*$ and so the inverse form on $\mathfrak{g}$ is defined. If we denote the latter by $\check{c}$, then the equivariant projection $q : \mathfrak{g} \otimes \mathfrak{g} \to \mathfrak{c}$ is given by $X \otimes Y \mapsto \check{c}(X, Y)c$.

It is well-known and easy to prove that c maps to the center of $U\mathfrak{g}$. This implies that c acts in any irreducible representation of $\mathfrak{g}$ by a scalar. In the case of the adjoint representation half this scalar is called the *dual Coxeter number* of $\mathfrak{g}$ and is denoted by $\check{h}$. So if we choose an orthonormal basis $\{X_\kappa\}_\kappa$ of $\mathfrak{g}$ relative to $\check{c}$ so that c takes the form $\sum_\kappa X_\kappa \otimes X_\kappa$, then

$$\sum_\kappa [X_\kappa, [X_\kappa, Y]] = 2\check{h}Y \quad \text{for all } Y \in \mathfrak{g}.$$

Let $L\mathfrak{g}$ stand for $\mathfrak{g} \otimes_k L$, but considered as a filtered R-Lie algebra (so we restrict the scalars to R) with $F^N L\mathfrak{g} = \mathfrak{g} \otimes_k \mathfrak{m}^N$. An argument similar to the one we used to prove that the pairing r is topologically perfect shows that the pairing

$$r_{\mathfrak{g}} : (\mathfrak{g} \otimes_k L) \times (\mathfrak{g} \otimes_k \omega) \to \mathfrak{c} \otimes_k R =: \mathfrak{c}_R$$

which sends $(Xf, Y\alpha)$ to $q(X \otimes Y)\operatorname{Res}(f\alpha)$ is topologically perfect (the basis dual to $(X_\kappa t^l)_{\kappa,l}$ is $(X_\kappa t^{-l-1}dt \otimes c)_{\kappa,l}$).

For an integer $N \geqslant 0$, the quotient $UL\mathfrak{g}/UL\mathfrak{g} \circ F^N L\mathfrak{g}$ is a free R-module (a set of generators is $X_{\kappa_1}t^{k_1} \circ \cdots \circ X_{\kappa_r}t^{k_r}$, $k_1 \leqslant \cdots \leqslant k_r < N$). We complete $UL\mathfrak{g}$ $\mathfrak{m}$-adically on the right:

$$\overline{U}L\mathfrak{g} := \varprojlim_N UL\mathfrak{g}/UL\mathfrak{g} \circ F^N L\mathfrak{g}.$$

A central extension of Lie algebras

$$0 \to \mathfrak{c}_R \to \widehat{L\mathfrak{g}} \to L\mathfrak{g} \to 0$$

is defined by endowing the sum $L\mathfrak{g} \oplus \mathfrak{c}_R$ with the Lie bracket

$$[Xf + cr, Yg + cs] := [X, Y]fg + r_{\mathfrak{g}}(Yg, Xdf).$$

Since the residue is zero on $\mathcal{O}$, the inclusion of $\mathcal{O}\mathfrak{g}$ in $\widehat{L\mathfrak{g}}$ is a homomorphism of Lie algebras. In fact, this is a canonical (and even unique) Lie section of the central extension over $\mathcal{O}\mathfrak{g}$, for it is just the derived Lie algebra of the preimage of $\mathcal{O}\mathfrak{g}$ in $\widehat{L\mathfrak{g}}$. The $\operatorname{Aut}(\mathfrak{g})$-invariance of c implies that the tautological action of $\operatorname{Aut}(\mathfrak{g})$ on $\mathfrak{g}$ extends to $\widehat{L\mathfrak{g}}$.

We filter $\widehat{L\mathfrak{g}}$ by setting $F^N\widehat{L\mathfrak{g}} = F^N L\mathfrak{g}$ for $N > 0$ and $F^N\widehat{L\mathfrak{g}} = F^N L\mathfrak{g} + \mathfrak{c}_R$ for $N \leqslant 0$. Then $U\widehat{L\mathfrak{g}}$ is a filtered $R[c]$-algebra whose reduction modulo c is $UL\mathfrak{g}$. The $\mathfrak{m}$-adic completion on the right

$$\overline{U\widehat{L\mathfrak{g}}} := \varprojlim_N U\widehat{L\mathfrak{g}}/(U\widehat{L\mathfrak{g}} \circ F^N L\mathfrak{g})$$

is still an $R[c]$-algebra and the obvious surjection $\overline{U\widehat{L\mathfrak{g}}} \to \overline{U}L\mathfrak{g}$ is the reduction modulo c. These (completed) enveloping algebras not only come with the (increasing) Poincaré-Birkhoff-Witt filtration, but also inherit a (decreasing) filtration from L.

Segal-Sugawara representation

Tensoring with $c \in \mathfrak{g} \otimes_k \mathfrak{g}$ defines the R-linear map

$$\mathfrak{l} \otimes_R \mathfrak{l} \to L\mathfrak{g} \otimes_R L\mathfrak{g}, \quad f \otimes g \mapsto c \cdot f \otimes g = \sum_\kappa X_\kappa f \otimes X_\kappa g,$$

which, when composed with $L\mathfrak{g} \otimes_R L\mathfrak{g} \subset \widehat{L\mathfrak{g}} \otimes_R \widehat{L\mathfrak{g}} \to U\widehat{L\mathfrak{g}}$, yields a map $\gamma : \mathfrak{l} \otimes_R \mathfrak{l} \to U\widehat{L\mathfrak{g}}$. Since $\gamma(f \otimes g - g \otimes f) = \sum_\kappa [X_\kappa f, X_\kappa g] = c \dim \mathfrak{g} \operatorname{Res}(g df)$, γ drops and extends naturally to an R-module homomorphism $\hat{\gamma} : \mathfrak{l}_2 \to U\widehat{L\mathfrak{g}}$ which sends $\hbar$ to $c \dim \mathfrak{g}$. This, in turn, extends continuously to a map from the closure $\bar{\mathfrak{l}}_2$ of $\mathfrak{l}_2$ in $\overline{U\hat{\mathfrak{l}}}$ to $\overline{U\widehat{L\mathfrak{g}}}$. As $\bar{\mathfrak{l}}_2$ contains the image of $\hat{C} : \theta \to \overline{U\hat{\mathfrak{l}}}$, and since c is $\operatorname{Aut}(\mathfrak{g})$-invariant, we get a R-homomorphism

$$\hat{C}_\mathfrak{g} := \hat{\gamma}\hat{C} : \theta \to (\overline{U\widehat{L\mathfrak{g}}})^{\operatorname{Aut}(\mathfrak{g})}.$$

We may also describe $\hat{C}_\mathfrak{g}$ in the spirit of Section 2: given $D \in \theta$, then the R-linear map

$$1 \otimes D : \mathfrak{g} \otimes_k \omega \to \mathfrak{g} \otimes_k L$$

is continuous and self-adjoint relative to $r_\mathfrak{g}$ and the perfect pairing $r_\mathfrak{g}$ allows us to identify it with an element of $\overline{U}L\mathfrak{g}$; this element produces our $\hat{C}_\mathfrak{g}(D)$. Thus the choice of the parameter t yields

$$\hat{C}_\mathfrak{g}(D_k) = \tfrac{1}{2} \sum_{\kappa, l} : X_\kappa t^{k-l} \circ X_\kappa t^l : \quad .$$

This formula can be used to define $\hat{C}_\mathfrak{g}$, but this approach does not exhibit its naturality.

Lemma 3.1. *For $X \in \mathfrak{g}$ and $f \in L$ we have*

$$[\hat{C}_\mathfrak{g}(D_k), Xf] = -(c + \check{h}) X D_k(f)$$

(an identity in $\overline{U\widehat{L\mathfrak{g}}}$) and upon a choice of a parameter t, then with the preceding notation

$$[\hat{C}_\mathfrak{g}(D_k), \hat{C}_\mathfrak{g}(D_l)] = (c + \check{h})(k - l)\hat{C}_\mathfrak{g}(D_{k+l}) + c(c + \check{h})\delta_{k+l,0}\frac{k^3 - k}{12} \dim \mathfrak{g}.$$

For the proof (which is a bit tricky, but not very deep), we refer to Lecture 10 of [8] (our $C_\mathfrak{g}(\hat{D}_k)$ is their T_k). This formula suggests that we make the central element $c + \check{h}$ of $\overline{U\widehat{L\mathfrak{g}}}$ invertible (its inverse might be viewed as a rational function on $\mathfrak{c}^*$), so that we can state this lemma in a more natural manner as follows.

Corollary 3.2 (Sugawara representation). *The map $\hat{D}_k \mapsto \frac{-1}{c+\check{h}} \hat{C}_\mathfrak{g}(D_k)$ induces a natural homomorphism of R-Lie algebras*

$$T_\mathfrak{g} : \hat{\theta} \to \left(\overline{U\widehat{L\mathfrak{g}}}[\tfrac{1}{c+\check{h}}]\right)^{\operatorname{Aut}(\mathfrak{g})}$$

which sends the central element $c_0 \in \hat{\theta}$ to $c(c + \check{h})^{-1} \dim \mathfrak{g}$. Moreover, if $\hat{D} \in \hat{\theta}$, then $\operatorname{ad}_{T_\mathfrak{g}(\hat{D})}$ leaves $L\mathfrak{g}$ invariant (as a subspace of $\overline{U\widehat{L\mathfrak{g}}}$) and acts on that subspace by derivation with respect to the image of $\hat{D}$ in θ.

A representation for $\widehat{L\mathfrak{g}}$

We fix $\ell \in k$ with $\ell \neq -\check{h}$. Let $F^1 L\mathfrak{g} \oplus \mathfrak{c}_R$ act on the free R-module of rank one Rv_ℓ via the projection onto the second factor $\mathfrak{c}_R = Rc$ with c acting as multiplication by ℓ. We regard $F^1 L\mathfrak{g} \oplus \mathfrak{c}_R$ as a subalgebra of $U\widehat{L\mathfrak{g}}$ so that we can form the induced module

$$\mathbb{F}_\ell(\mathfrak{g}, L) := U\widehat{L\mathfrak{g}} \otimes_{U(F^1 L\mathfrak{g} \oplus \mathfrak{c}_R)} Rv_\ell,$$

which we often simply denote by $\mathbb{F}_\ell(\mathfrak{g})$. We use v_ℓ also to denote its image in this module. As an R-module $\mathbb{F}_\ell(\mathfrak{g})$ is generated by $X_{\kappa_r} t^{-k_r} \circ \cdots \circ X_{\kappa_1} t^{-k_1} \otimes v_\ell$, where $r \geqslant 0$, $0 \leqslant k_1 \leqslant k_2 \leqslant \cdots \leqslant k_r$ and where $(X_\kappa)_\kappa$ is a given k-basis of $\mathfrak{g}$. If we let $\hat{\theta}$ act on $\mathbb{F}_\ell(\mathfrak{g})$ via $T_\mathfrak{g}$, then it follows from Corollary 3.2 that if $\hat{D} \in \hat{\theta}$ lifts $D \in \theta$, then

$$\begin{aligned} T_\mathfrak{g}(\hat{D}) X_{\kappa_r} t^{-k_r} \circ \cdots \circ X_{\kappa_1} t^{-k_1} \otimes v_\ell & \\ = \sum_{i=1}^{r} X_{\kappa_r} t^{-k_r} \circ \cdots X_{\kappa_i} D(t^{-k_i}) \circ \cdots \circ X_{\kappa_1} t^{-k_1} \otimes v_\ell & \\ + X_{\kappa_r} t^{-k_r} \circ \cdots \circ X_{\kappa_1} t^{-k_1} \circ T_\mathfrak{g}(\hat{D}) v_\ell. & \end{aligned}$$

Thus $\hat{\theta}$ is faithfully represented as a Lie algebra of R-linear endomorphisms of $\mathbb{F}_\ell(\mathfrak{g})$. If $D \in F^0\theta$, then clearly $T_\mathfrak{g}(\hat{D}) v_\ell = 0$ and hence we have the following counterpart of Corollary 2.4 (with the same proof). It tells us that $\hat{\theta}_{L,R}$ acts in $\mathbb{F}_\ell(\mathfrak{g})$ as a Lie algebra of first order differential operators, but with its degree zero part R acting with weight $(c + \check{h})^{-1} c \dim \mathfrak{g}$:

Corollary 3.3. *The Sugawara representation* $T_\mathfrak{g}$ *of* $\hat{\theta}$ *on* $\mathbb{F}_\ell(\mathfrak{g})$ *extends to* $\hat{\theta}_{L,R}$ *in such a manner that* $F^0\theta_{L,R}$ *acts by coefficientwise derivation (killing the generator* v_ℓ*), that* $[T_\mathfrak{g}(\hat{D}), Xf] = X(Df)$ *for* $X \in \mathfrak{g}$, $f \in L$ *and that* $T_\mathfrak{g}(\hat{D})$ *is* $\mathrm{Aut}(\mathfrak{g})$*-invariant. In particular, this action preserves every* $U\widehat{L\mathfrak{g}}$*-submodule of* $\mathbb{F}_\ell(\mathfrak{g})$.

Semi-local case

This refers to the situation where we allow the R-algebra L to be a finite direct sum of R-algebras isomorphic to $R((t))$: $L = \oplus_{i \in I} L_i$, where I is a nonempty finite index set and L_i as before. We then extend the notation employed earlier in the most natural fashion. For instance, $\mathcal{O}$, $\mathfrak{m}$, ω, $\mathfrak{l}$ are now the direct sums over I (as filtered objects) of the items suggested by the notation. If $r : L \times \omega \to R$ denotes the sum of the residue pairings of the summands, then r is still topologically perfect. However, we take for the oscillator algebra $\hat{\mathfrak{l}}$ not the direct sum of the $\hat{\mathfrak{l}}_i$, but rather the quotient of $\oplus_i \hat{\mathfrak{l}}_i$ that identifies the central generators of the summands with a single $\hbar$. We thus get a Virasoro extension $\hat{\theta}$ of θ by $c_0 R$ and a (faithful) oscillator representation of $\hat{\theta}$ in $\overline{U\hat{\mathfrak{l}}}$. The decreasing filtrations are the obvious ones. We shall denote by $\mathbb{F}$ the Fock representation $\mathbb{F}$ of $\hat{\mathfrak{l}}$ that ensures that the unit of every summand $\mathcal{O}_i$ acts the identity; it is then the induced representation of the rank one representation of $F^0\hat{\mathfrak{l}} = \mathcal{O} \oplus R\hbar$ in Rv_o.

In likewise manner we define $\widehat{L\mathfrak{g}}$ (a central extension of $\oplus_{i\in I}L\mathfrak{g}_i$ by c_R) and construct the associated Sugawara representation. The representation $\mathbb{F}_\ell(\mathfrak{g})$ of $\widehat{L\mathfrak{g}}$ is as before. We have defined $\hat{\theta}_{L,R}$ and Corollaries 3.2 and 3.3 continue to hold.

4. The WZW connection: algebraic aspects

From now on we place ourselves in the semi-local case, so $L = \oplus_{i\in I}L_i$ with I nonempty and finite and $L_i \cong R((t))$. For the sake of transparency, we begin with an abstract discussion that will lead us to the Fock representation of a symplectic local system.

Abstract spaces of covacua I

Let A be a R-subalgebra of L and let $\theta_{A/R}$ have the usual meaning as the Lie algebra of R-derivations $A \to A$. We denote by $A^\perp \subset L$ the annihilator of A relative to the residue pairing. We assume that:

(A_1) as an R-algebra, A is flat and of finite type and $A \cap \mathcal{O} = R$,

(A_2) the R-modules $L/(A+\mathcal{O})$ and $F := A^\perp \cap \mathcal{O}$ are free of finite rank and the residue pairing induces a perfect pairing $L/(A+\mathcal{O}) \otimes_R F \to R$.

(A_3) the universal continuous R-derivation $d : L \to \omega$ maps A to $A^\perp$ and the A-dual of the resulting A-homomorphism $\Omega_{A/R} \to A^\perp$ is an R-isomorphism $\mathrm{Hom}_A(A^\perp, A) \cong \theta_{A/R}$.

Remark 4.1. The example to keep in mind is the following. Since R is regular local k-algebra, it represents a smooth germ (S, o). Suppose we are given a family $\pi : \mathcal{C} \to S$ of smooth projective curves of genus g over this germ, endowed with pairwise disjoint sections $(x_i)_{i\in I}$. We let $\mathcal{O}_i$ be is the formal completion of $\mathcal{O}_\mathcal{C}$ along x_i, let L_i be obtained from $\mathcal{O}_i$ by inverting a generator for the ideal defining $x_i(S)$, and take for A the R-algebra of regular functions on $\mathcal{C}^\circ := \mathcal{C} - \cup_i x_i(S)$ (or rather its isomorphic image in $L = \oplus_i L_i$). It is a classical fact that the three properties A_1, A_2, A_3 are then satisfied. For instance, $L/(A+\mathcal{O})$ has according to Weil the interpretation of $R^1\pi_*\mathcal{O}_\mathcal{C}$ and hence is free of rank g. It is also classical that the annihilator of A in ω is precisely the image of the space relative rational differentials on $\mathcal{C}/S$ that are regular on $\mathcal{C}^\circ$ (so in this case $\Omega_{A/R} \to A^\perp$ is already an isomorphism before dualizing).

We put $H := A^\perp/A$. It follows from properties (A_1) and (A_2), that the natural map $F \to H$ is an embedding with image a Lagrangian subspace. Recall that $\theta_{A,R}$ denotes the Lie algebra of k-derivations $A \to A$ which preserve R. The kernel of the natural map $\theta_{A,R} \to \theta_R$ is $\theta_{A/R}$ and its image, is by definition the R-submodule of k-derivations $R \to R$ that extend to one of A. We denote this image by $\theta_R^A \subset \theta_R$ and refer to it as the module of *liftable derivations*. This module is clearly closed under the Lie bracket. We shall assume that we have equality in the generic point, so that θ_R^A is as our $\mathfrak{D}$. According to (A_3) any element of $\theta_{A/R}$ induces the zero map in H and so $\theta_{A,R}$ acts in H (as a k-Lie algebra) through $\theta_{A,R}$. It is clear that $\theta_{A,R} \subset \theta_{L,R}$.

(In the above example, H would represent the first De Rham cohomology module of $\mathcal{C}/S$, F the module of relative regular differentials, and we would have $\theta_R^A = \theta_R$, as every vector field germ on (S, o) lifts to rational vector field on $\mathcal{C}$ that is regular on $\mathcal{C}^\circ$. The Lie action is then that of covariant derivation of relative cohomology classes. The reason for us to allow $\theta_R^A \neq \theta_R$ is because we want to admit the central fibers of $\mathcal{C} \to S$ to have modest singularities; in that case θ_R^A is the θ_R-stabilizer of a principal ideal in R, the *discriminant* ideal of π.)

We write $\hat{\theta}_{A,R}$ for the preimage of $\theta_{A,R}$ in $\hat{\theta}_{L,R}$ and by $\hat{\theta}_R^A$ the quotient $\hat{\theta}_{A,R}/\theta_{A/R}$. These are extensions of $\theta_{A,R}$ resp. θ_R^A by c_0R. They can be split, but not canonically so.

Since $\mathrm{Ad}(A) \subset A^\perp$, the residue pairing vanishes on $A \times \mathrm{Ad}(A)$ and hence A is contained in $\hat{\mathfrak{l}}$ as an abelian Lie subalgebra. Let $\mathbb{F}_A := \mathbb{F}/A\mathbb{F}$ denote the space of A-covariants.

Theorem 4.2. *The following properties hold:*

(i) *The space of covariants $\mathbb{F}_A$ is naturally identified with the Fock representation $\mathbb{F}(H, F)$,*

(ii) *for every $D \in \theta_{A/R}$ there exists a lift $\hat{D} \in \hat{\theta}_{A/R}$ such that $T(\hat{D})$ lies in the closure of $A \circ \hat{\mathfrak{l}}$ in $\overline{U\hat{\mathfrak{l}}}$,*

(iii) *the representation of the Lie algebra $\hat{\theta}_{A,R}$ on $\mathbb{F}$ preserves the submodule $A\mathbb{F}$ and $\hat{\theta}_{A,R}$ acts in $\mathbb{F}_A$ through $\hat{\theta}_R^A$ by differential operators of degree $\leqslant 1$ (with c_0 acting as the identity),*

(iv) *if Θ is a square root of $\det_R(F)$, then the image of this action on $\mathbb{F}_A$ is equal to the image of the Lie algebra of first order differential operators $\theta_R^A(\Theta)$ (as described in Remark 2.8).*

Proof. The proof of the first assertion is straightforward and left to the reader.

Since $L/(A + \mathcal{O})$ is finitely generated as a R-module, we can choose a finite subset $M \subset L$ such that $L = A + \sum_{f \in M} Rf + \mathcal{O}$.

Now let $D \in \theta_{A/R}$. According to (A_3), we may view D as a L-linear map $\omega \to L$ which maps $A^\perp$ to A. This implies that $\hat{C}(\hat{D})$ lies in the closure of the image of $A \otimes_R \hat{\mathfrak{l}} + \hat{\mathfrak{l}} \otimes_R A$ in $\overline{U\hat{\mathfrak{l}}}$. It follows that $\hat{C}(\hat{D})$ has the form $\hbar r + \sum_{n \geqslant 1} f_n \circ g_n$ with $r \in R$, one of $f_n, g_n \in L$ being in A and the order of f_n smaller than that of g_n for almost all n. In view of the fact that the nonzero elements of A are of lower order than those of $\mathcal{O}$ and $f_n \circ g_n \equiv g_n \circ f_n \pmod{\hbar R}$, we can assume that all f_n lie in A and so we can arrange that $\hat{C}(\hat{D})$ lies in the closure of $A \circ \hat{\mathfrak{l}}$.

For (iii) we observe that if $D \in \theta_{A,R}$ and $f \in A$, then $[D, f] = Df$ lies in A. This shows that $T(\hat{D})$ preserves $A\mathbb{F}$ and hence acts in $\mathbb{F}_A$. When $D \in \theta_{A/R}$ and if we choose $\hat{D} \in \hat{\theta}_{A/R}$ as in (ii), then $T(\hat{D})$ is clearly zero in $\mathbb{F}_A$. Thus $\hat{\theta}_{A,R}$ acts in $\mathbb{F}_A$ through $\hat{\theta}_R^A$.

Property (iv) follows from the observation that the action of $\hat{\theta}_{A,R}$ on $\mathbb{F}_A \cong \mathbb{F}(H, F)$ evidently has the properties described in Proposition-Definition 2.5. □

Abstract spaces of covacua II

We continue with the setting of the previous subsection. With $\mathfrak{g}$ as before we have defined $\mathbb{F}_\ell(\mathfrak{g})$. We first consider the space of $A\mathfrak{g}$-covariants in $\mathbb{F}_\ell(\mathfrak{g})$,

$$\mathbb{F}_\ell(\mathfrak{g})_{A\mathfrak{g}} := \mathbb{F}_\ell(\mathfrak{g})/A\mathfrak{g}\mathbb{F}_\ell(\mathfrak{g}).$$

Proposition 4.3. *For $\hat{D} \in \hat{\theta}_{A/R}$, $T_\mathfrak{g}(\hat{D})$ lies in the closure of $A\mathfrak{g} \circ \widehat{L\mathfrak{g}}$ in $\overline{U\widehat{L\mathfrak{g}}}$. The Sugawara representation of the Lie algebra $\hat{\theta}_{A,R}$ on $\mathbb{F}_\ell(\mathfrak{g})$ preserves the submodule $A\mathfrak{g}\mathbb{F}_\ell(\mathfrak{g}) \subset \mathbb{F}_\ell(\mathfrak{g})$ and acts in the space of $A\mathfrak{g}$-covariants in $\mathbb{F}_\ell(\mathfrak{g})$, $\mathbb{F}_\ell(\mathfrak{g})_{A\mathfrak{g}}$, via $\hat{\theta}_R^A$; this representation is one by differential operators of degree $\leqslant 1$ (with c_0 acting as multiplication by $(c+\check{h})^{-1}c\dim\mathfrak{g}$).*

Proof. The proof is similar to the arguments used to prove Theorem 4.2. Since D maps $A^\perp$ to $A \subset L$, $1 \otimes D$ maps the submodule $\mathfrak{g} \otimes A^\perp$ of $\mathfrak{g} \otimes \omega$ to the submodule $\mathfrak{g} \otimes A = A\mathfrak{g}$ of $\mathfrak{g} \otimes L = L\mathfrak{g}$. It is clear that $\mathfrak{g} \otimes A^\perp$ and $A\mathfrak{g}$ are each others annihilator relative to the pairing $r_\mathfrak{g}$. This implies that $\hat{C}(\hat{D})$ lies in the closure of the image of $A\mathfrak{g}\otimes_k L\mathfrak{g}+L\mathfrak{g}\otimes_k A\mathfrak{g}$ in $\overline{U\widehat{L\mathfrak{g}}}$. It follows that $\hat{C}(\hat{D})$ has the form $cr+\sum_\kappa\sum_{n\geqslant 1} X_\kappa f_{\kappa,n} \circ X_\kappa g_{\kappa,n}$ with $r \in R$, one of $f_{\kappa_n}, g_{\kappa,n} \in L$ being in A and the order of f_{κ_n} smaller than that of $g_{\kappa,n}$ for almost all κ, n. Since the elements of A have order $\leqslant 0$ and $X_\kappa f_{\kappa,n} \circ X_\kappa g_{\kappa,n} \equiv X_\kappa g_{\kappa,n} \circ X_\kappa f_{\kappa,n} \pmod{cR}$, we can assume that all $f_{\kappa,n}$ lie in A and so the first assertion follows.

If $D \in \theta_{A,R}$, then for $X \in \mathfrak{g}$ and $f \in A$, we have $[D, Xf] = X(Df)$, which is an element of $A\mathfrak{g}$ (since $Df \in A$). This shows that $T_\mathfrak{g}(\hat{D})$ preserves $A\mathfrak{g}\mathbb{F}_\ell(\mathfrak{g})$. If $D \in \theta_{A/R}$, then it follows from the proven part that $T_\mathfrak{g}(\hat{D})$ maps $\mathbb{F}_\ell(\mathfrak{g})$ to $A\mathfrak{g}\mathbb{F}_\ell(\mathfrak{g})$ and hence induces the zero map in $\mathbb{F}_\ell(\mathfrak{g})_{A\mathfrak{g}}$. So $\hat{\theta}_{A,R}$ acts on $\mathbb{F}_\ell(\mathfrak{g})_{A\mathfrak{g}}$ via $\hat{\theta}_R^A$. □

For what follows we need to briefly review from [7] the theory of highest weight representations of a loop algebra such as $\widehat{L\mathfrak{g}}$. According to that theory, the natural analogues for $\widehat{L\mathfrak{g}}$ of the finite dimensional irreducible representations of the finite dimensional semi-simple Lie algebras are obtained as follows, assuming that I is a singleton. Fix an integer $\ell \geqslant 0$ and let V be a finite dimensional irreducible representation of $\mathfrak{g}$. Make V a k-representation of $F^0L\mathfrak{g}$ by letting c act as multiplication by ℓ and by letting $\mathfrak{g} \otimes_k \mathcal{O}$ act via its projection onto $\mathfrak{g}$. If we induce this up to $\widehat{L\mathfrak{g}}$ we get a representation $\widetilde{\mathbb{H}}_\ell(V)$ of $\widehat{L\mathfrak{g}}$ which clearly is a quotient of $\mathbb{F}_\ell(\mathfrak{g})$. Its irreducible quotient is denoted by $\mathbb{H}_\ell(V)$. This is integrable as an $\widehat{L\mathfrak{g}}$-module: if $Y \in \mathfrak{g}$ is nilpotent and $f \in L$, then Yf acts locally nilpotently in $\mathbb{H}_\ell(V)$ (which means that the latter is a union of finite dimensional Yf-invariant subspaces in which Yf acts nilpotently). We can be more precise if we fix a Cartan subalgebra $\mathfrak{h} \subset \mathfrak{g}$ and a system of positive roots $(\alpha_1, \dots, \alpha_r)$ in $\mathfrak{h}^*$. Let $\theta \in \mathfrak{h}^*$ the highest root, $\check{\theta} \in \mathfrak{h}$ the corresponding coroot and $X \in \mathfrak{g}$ a generator of the root space $\mathfrak{g}_\theta$.

Lemma 4.4. *If $\lambda \in \mathfrak{h}^*$ be the highest weight of V, then $\mathbb{H}_\ell(V)$ is zero unless $\lambda(\check{\theta}) \leqslant \ell$. Assuming this inequality, then $\mathbb{H}_\ell(V)$ can be obtained as the quotient of $U\widehat{L\mathfrak{g}}$ by the left ideal generated by $\mathfrak{g}\otimes_k \mathfrak{m}$, $c-\ell$ and $(Xf)^{1+\ell-\lambda(\check{\theta})}$, where we can take for f any $\mathcal{O}$-generator* *of* $F^{-1}\mathfrak{l}$. *In fact, the image of* V *in* $\mathbb{H}_\ell(V)$ *(which generates* $\mathbb{H}_\ell(V)$ *as a* $\widehat{L\mathfrak{g}}$*-representation) is annihilated by all expressions of the form* $Xf_N \circ \cdots \circ Xf_1$ *with* $f_k \in F^{-1}\mathfrak{l}$ *and* $N > \ell - \lambda(\check{\theta})$.

Proof. The first assertion is in the literature in the form of an Exercise (12.12 of [7]). As to the second statement: choose variables $u_1, \dots, u_N$ and observe that $f_u := f + \sum_k u_k f_k$ is an $\mathcal{O}$-generator of $F^{-1}\mathfrak{l}$ for generic u. So V is killed by $(Xf_u)^N$ for generic u and hence for all u. By taking the coefficient of $u_1 \cdots u_N$ (and using that the Xf_k's commute with each other), we find that $Xf_N \circ \cdots \circ Xf_1$ annihilates V. □

Let us call the k-span of an X as above a *highest root line*. Since the Cartan subalgebras of $\mathfrak{g}$ are all conjugate under the adjoint representation, the same is true for the highest root lines.

Definition 4.5. The *level* of a finite dimensional representation V of $\mathfrak{g}$ is the smallest integer ℓ for which some (or equivalently, any) highest root line $\mathfrak{n}$ has the property that $\mathfrak{n}^{\ell+1} \subset U\mathfrak{g}$ kills V. We denote it by $\ell(V)$.

It is clear that in terms of the above root data, the set P_ℓ of equivalence classes of irreducible representations of level $\leqslant \ell$ can be identified with the set of integral weights in a simplex, hence is finite. Notice that P_ℓ is invariant under dualization and more generally, under all outer automorphisms of $\mathfrak{g}$.

Returning to the general case in which I need not be a singleton, we put $\mathbb{H}_\ell(V) := \otimes_{i\in I}\mathbb{H}_\ell(V_i)$. So this is zero unless every V_i is of level $\leqslant \ell$. Inspired by the physicists terminology, the R-module $\mathbb{H}_\ell(V)_{A\mathfrak{g}}$ is called the space of *covacua* attached to A. The following proposition says that it is of finite rank and describes the WZW-connection.

Proposition 4.6 (Finiteness). *The space* $\mathbb{H}_\ell(V)$ *is finitely generated as a* $UA\mathfrak{g}$*-module (so that* $\mathbb{H}_\ell(V)_{A\mathfrak{g}}$ *is a finitely generated* R*-module). The Lie algebra* $\hat{\theta}^A_R$ *acts on* $\mathbb{H}_\ell(V)_{A\mathfrak{g}}$ *via the Sugawara representation with* c_0 *acting as multiplication by* $\frac{\ell}{\ell+\check{h}}\dim\mathfrak{g}$.

Proof. Choose a generator t_i of $\mathfrak{m}_i$. The issue being local on Spec(R), we may assume that after localizing R, there exists a finite set Φ of *negative* powers of these generators mapping to an R-basis set of $L/(\mathcal{O}+A)$. The nilpotent elements of $\mathfrak{g}$ span a nontrivial subspace that is invariant under the adjoint action and hence span all of $\mathfrak{g}$. Let $\Xi \subset \mathfrak{g}$ be a k-basis of $\mathfrak{g}$ consisting of nilpotent elements. Then for a pair $(X, f) \in \Xi \times \Phi$, Xf acts locally nilpotently in $\mathbb{H}_\ell(V)$ and so there exists a positive integer N such that the Nth power of any such element kills the image of $\otimes_{i\in I}V_i$ in $\mathbb{H}_\ell(V)$.

A PBW type of argument then shows that $\mathbb{H}_\ell(V)$ is the sum of the subspaces

$$A\mathfrak{g} \circ (X_r f_r)^{\circ n_r} \circ \cdots \circ (X_1 f_1)^{\circ n_1} \otimes (\otimes_{i\in I}V_i) \subset \mathbb{H}_\ell(V)$$

with $(X_i, f_i) \in \Xi \times \Phi$ pairwise distinct for $i = 1, \dots, r$, and $n_1 \geqslant \cdots \geqslant n_r \geqslant 0$. Since we get a nonzero element only when $n_1 < N$, we thus obtain a finite collection of R-module generators of $\mathbb{H}_\ell(V)_{A\mathfrak{g}}$. The remaining statements follow from 4.3. □

Remark 4.7. We expect the R-module $\mathbb{H}_\ell(V)_{A\mathfrak{g}}$ to be flat as well and this to be a consequence of a related property for the $UA\mathfrak{g}$-module $\mathbb{H}_\ell(V)$. Such a result, or rather an algebraic proof of it, might simplify the argument in [15] (see Section 6 for our version) which shows that the sheaf of covacua attached to a degenerating family of pointed curves is locally free.

Remark 4.8. It is clear from the definition that a system of $\mathfrak{g}$-equivariant isomorphisms $(\phi_i : V_i \cong V_i')_{i\in I}$ of finite dimensional irreducible representations induces an isomorphism $\phi_* : \mathbb{H}_\ell(V)_{A\mathfrak{g}} \cong \mathbb{H}_\ell(V')_{A\mathfrak{g}}$. By Schur's lemma, each ϕ_i is unique up to scalar in k and hence the same is true for ϕ_*. We may rigidify the situation by fixing in each representation V_i and V_i' involved a highest weight orbit for the closed connected subgroup of linear transformations whose Lie algebra is the image of $\mathfrak{g}$: if we require that every ϕ_i respects these orbits, then ϕ_i is unique.

We can also say something if we are given a $\sigma \in \mathrm{Aut}(\mathfrak{g})$. This turns every representation V of $\mathfrak{g}$ into another one (denoted ${}^\sigma V$) that has the same underlying vector space V, by letting $X \in \mathfrak{g}$ act as $\sigma(X)$ on V. The extension $\hat{\sigma}$ of σ to $\widehat{L\mathfrak{g}}$ does the same with $\mathbb{H}_\ell(V)$. It follows that we have an identification of $\widehat{L\mathfrak{g}}$-modules:

$$Y_r f_r \circ \cdots \circ Y_1 f_1 {}^\sigma\!\otimes (\otimes_{i\in I} v_i) \in \mathbb{H}_\ell({}^\sigma V) \mapsto$$
$$\sigma(Y_r) f_r \circ \cdots \circ \sigma(Y_1) f_1 \otimes (\otimes_{i\in I} v_i) \in {}^{\hat{\sigma}}\mathbb{H}_\ell(V).$$

Since σ preserves $A\mathfrak{g}$, this descends to an identification $\mathbb{H}_\ell(V^\sigma)_{A\mathfrak{g}} \cong \mathbb{H}_\ell(V)_{A\mathfrak{g}}$ of R-modules. It is clear from the definition above that this is also equivariant for the Segal-Sugawara representation and hence is an isomorphism of $\hat{\theta}_R^A$-modules.

Propagation principle

The following proposition is a bare version of what is known as the *propagation of vacua*; it essentially shows that trivial representations may be ignored (as long as some representations remain: if all are trivial, then we can get rid of all but one of them). If we do not care about the WZW-connection, then this is even true for nontrivial representations (a fact that can be found in Beauville [3]) so that we then essentially reduce the discussion to the case where I is a singleton.

Proposition 4.9. *Let* $J \subsetneq I$ *be such that* A *maps onto* $\oplus_{j\in J} L_j/\mathcal{O}_j$. *Denote by* $B \subset A$ *the kernel of the map* $A \to \oplus_{j\in J} L_j/\mathfrak{m}_j \cong R^J$ *(evidently an ideal) so that we have a surjective Lie homomorphism* $B\mathfrak{g} \to (R \otimes_k \mathfrak{g})^J$ *via which* $B\mathfrak{g}$ *acts on* $R \otimes_k (\otimes_{j\in J} V_j)$. *Then the map of* $B\mathfrak{g}$*-modules* $\mathbb{H}_\ell(V|I-J) \otimes_k (\otimes_{j\in J} V_j) \to \mathbb{H}_\ell(V)$ *induces an isomorphism on covariants:*

$$(\mathbb{H}_\ell(V|I-J) \otimes_k (\otimes_{j\in J} V_j))_{B\mathfrak{g}} \xrightarrow{\cong} \mathbb{H}_\ell(V)_{A\mathfrak{g}}.$$

If $\theta_R^{A,B} \subset \theta_R^A$ *denotes the module of* k*-derivations* $R \to R$ *that lift to* k*-derivations* $A \to A$ *which preserve* B *(or equivalently,* $\oplus_{j\in J}\mathfrak{m}_j$*), and* $\hat{\theta}_R^{A,B} \subset \hat{\theta}_R^A$ *stands for the corresponding extension, then the above isomorphism of covariants is compatible with the action of* $\hat{\theta}_R^{A,B}$ *on both sides, provided that the representations* V_j *are trivial for* $j \in J$.

Proof. For the first assertion it suffices to do the case when J is a singleton $\{o\}$. The hypotheses clearly imply that $\mathbb{H}_\ell(V|I-\{o\})\otimes V_o \to \mathbb{H}_\ell(V)_{A\mathfrak{g}}$ is onto. The kernel is easily shown to be $B\mathfrak{g}(\mathbb{H}_\ell(V|I-\{o\})\otimes V_o)$.

The second assertion follows in a straightforward manner from our definitions: if $\bar{D}\in\hat{\theta}_R^{A,B}$, then lift $\bar{D}$ to a k-derivation $D: A\to A$ which preserves B. This implies that D preserves each $\mathcal{O}_j$, $j\in J$. If we choose a parameter t_j for $\mathcal{O}_j$ so that $\mathcal{O}_j = R((t_j))$, then D takes in $\mathcal{O}_j$ the form $D^{(j)}_{\mathrm{hor}}+D^{(j)}_{\mathrm{vert}}$, with $D^{(j)}_{\mathrm{hor}}$ the extension of $\bar{D}$ which kills t_j and $D^{(j)}_{\mathrm{vert}} = c^{(j)}\partial/\partial t_j$ plus higher order terms with $c^{(j)}\in R$. The Sugawara action of $D^{(j)}_{\mathrm{vert}}$ on the subspace $V_j\subset\mathbb{H}_\ell(V_j)$ is up to a factor in R given by $\sum_\kappa t_j^{-1}X_\kappa\circ X_\kappa$. But if V_j is the trivial representation, then this is evidently zero. The second assertion now follows. □

Remark 4.10. Our discussion of the genus zero case will show that the isomorphism of covariants generally fails to be compatible relative to the $\hat{\theta}_R^{A,B}$-action.

Remark 4.11. Proposition 4.9 is sometimes used in the opposite direction: if $\mathfrak{m}_o\subset A$ is a principal ideal with the property that for a generator $t\in\mathfrak{m}_o$, the $\mathfrak{m}_o$-adic completion of A gets identified with $R((t))$, then let $\tilde{I}$ be the disjoint union of I and $\{o\}$, $\tilde{V}$ the extension of V to $\tilde{I}$ which assigns to o the trivial representation and $\tilde{A} := A[t^{-1}]$. With $(\tilde{I},\{o\})$ taking the role of (I,J), we then find that $\mathbb{H}_\ell(V)_{A\mathfrak{g}}\cong\mathbb{H}_\ell(\tilde{V})_{\tilde{A}\mathfrak{g}}$.

5. Bundles of covacua

Spaces of covacua in families

We specialize the discussion of Section 4 to a more concrete geometric situation. This leads us to sheafify many of the notions we introduced earlier and in such cases we shall modify our notation (or its meaning) accordingly. Suppose given a proper and flat morphism between k-varieties $\pi:\mathcal{C}\to S$ whose base S is smooth and connected and whose fibers are reduced connected curves that have complete intersection singularities only (but we do not assume that $\mathcal{C}$ is smooth over k). Since the family is flat, the arithmetic genus of the fibers is locally constant, hence constant, say equal to g. We also suppose given disjoint sections x_i of π, indexed by the finite nonempty set I whose union $\cup_{i\in I}x_i(S)$ lies in the smooth part of $\mathcal{C}$ and meets every irreducible component of a fiber. The last condition ensures that if $j:\mathcal{C}^\circ := \mathcal{C}-\cup_{i\in I}x_i(S)\subset\mathcal{C}$ is the inclusion, then πj is an affine morphism.

We denote by $(\mathcal{O}_i,\mathfrak{m}_i)$ the formal completion of $\mathcal{O}_\mathcal{C}$ along $x_i(S)$, by $\mathcal{L}_i$ the subsheaf of fractions of $\mathcal{O}_i$ with denominator a local generator of $\mathfrak{m}_i$ and by $\mathcal{O}$, $\mathfrak{m}$ and $\mathcal{L}$ the corresponding direct sums. But we keep on using ω, θ, $\hat{\theta}$ etc. for their sheafified counterparts. So these are now all $\mathcal{O}_S$-modules and the residue pairing is also one of $\mathcal{O}_S$-modules: $r:\mathcal{L}\times\omega\to\mathcal{O}_S$. We write $\mathcal{A}$ for $\pi_*j_*j^*\mathcal{O}_\mathcal{C}$ (a sheaf of $\mathcal{O}_S$-algebras that is also equal to the direct image of $\mathcal{O}_{\mathcal{C}^\circ}$ on S) and often identify this with its image in $\mathcal{L}$. We denote by $\theta_{\mathcal{A}/S}$ the sheaf of $\mathcal{O}_S$-derivations $\mathcal{A}\to\mathcal{A}$

and by $\omega_{\mathcal{A}/S}$ for the sheaf $\pi_* j_* j^* \omega_{\mathcal{C}/S}$ (which is also the direct image on S of the relative dualizing sheaf of $\mathcal{C}^\circ/S$; if $\mathcal{C}^\circ$ is smooth, this is simply the sheaf of relative differentials). So $\omega_{\mathcal{A}/S}$ is torsion free and embeds therefore in ω.

Lemma 5.1. *The properties* A_1, A_2 *and* A_3 *hold for the sheaf* $\mathcal{A}$. *Precisely,*

($\mathcal{A}_1$) *$\mathcal{A}$ is as a sheaf of $\mathcal{O}_S$-algebras flat and of finite type,*

($\mathcal{A}_2$) *$\mathcal{A} \cap \mathcal{O} = \mathcal{O}_S$ and $R^1\pi_*\mathcal{O}_\mathcal{C} = \mathcal{L}/(\mathcal{A} + \mathcal{O})$ is locally free of rank g,*

($\mathcal{A}_3$) *we have $\theta_{\mathcal{A}/S} = \mathrm{Hom}_\mathcal{A}(\omega_{\mathcal{A}/S}, \mathcal{A})$ and $\omega_{\mathcal{A}/S}$ is the annihilator of $\mathcal{A}$ with respect to the residue pairing.*

Proof. Property $\mathcal{A}_1$ is clear. It is also clear that $\mathcal{O}_S = \pi_*\mathcal{O}_\mathcal{C} \to \mathcal{A}\cap\mathcal{O}$ is an isomorphism. The long exact sequence defined by the functor π_* applied to the short exact sequence

$$0 \to \mathcal{O}_\mathcal{C} \to j_* j^* \mathcal{O}_\mathcal{C} \to \mathcal{L}/\mathcal{O} \to 0$$

tells us that $R^1\pi_*\mathcal{O}_\mathcal{C} = \mathcal{L}/(\mathcal{A} + \mathcal{O})$; in particular, the latter is locally free of rank g. Hence $\mathcal{A}_2$ holds as well.

In order to verify $\mathcal{A}_3$, we note that $\pi_*\omega_{\mathcal{C}/S}$ is the $\mathcal{O}_S$-dual of $R^1\pi_*\mathcal{O}_S$, and hence is locally free of rank g. The first part of $\mathcal{A}_3$ follows from the corresponding local property $\theta_{\mathcal{C}/S} = \mathrm{Hom}_{\mathcal{O}_\mathcal{C}}(\omega_{\mathcal{C}/S}, \mathcal{O}_\mathcal{C})$ by applying $\pi_* j^*$ to either side. This local property is known to hold for families of curves with complete intersection singularities. (A proof under the assumption that $\mathcal{C}$ is smooth—which does not affect the generality, since π is locally the restriction of that case and both sides are compatible with base change—runs as follows: if $j' : \mathcal{C}' \subset \mathcal{C}$ denotes the locus where π is smooth, then its complement is of codimension $\geqslant 2$ everywhere. Clearly, $\theta_{\mathcal{C}/S}$ is the $\mathcal{O}_\mathcal{C}$-dual of $\omega_{\mathcal{C}/S}$ on $\mathcal{C}'$ and since both are inert under $j'_* j'^*$, they are equal everywhere.)

The last assertion essentially restates the well-known fact that the polar part of a rational section of $\omega_{\mathcal{C}/S}$ must have zero residue sum, but can otherwise be arbitrary. More precisely, the image of $\omega_{\mathcal{A}/S}$ in $\omega/F^1\omega$ is the kernel of the residue map $\omega/F^1\omega \to \mathcal{O}_S$. The intersection $\omega_{\mathcal{A}/S} \cap F^1\omega$ is $\pi_*\omega_{\mathcal{C}/S}$ and is hence locally free of rank g. Since $(F^1\omega)^\perp = \mathcal{O}$, it follows that $(\omega_{\mathcal{A}/S})^\perp \cap \mathcal{O}$ and $\mathcal{L}/((\omega_{\mathcal{A}/S})^\perp + \mathcal{O})$ are locally free of rank 1 and g respectively. Since $\mathcal{A}$ has these properties also and is contained in $(\omega_{\mathcal{A}/S})^\perp$, we must have $\mathcal{A} = (\omega_{\mathcal{A}/S})^\perp$. □

For what follows one usually supposes that the fibers are stable I-pointed curves (meaning that every fiber of πj has only ordinary double points as singularities and has finite automorphism group) and is versal (so that the discriminant Δ_π of π is a reduced normal crossing divisor), but we shall not make these assumptions yet. Instead, we assume the considerable weaker property that the sections of the sheaf $\theta_S(\log \Delta_\pi)$ of vector fields on S tangent to Δ_π lift locally on S to vector fields on $\mathcal{C}$. (This is for instance the case if $\mathcal{C}$ is smooth and π is multi-transversal with respect to the (Thom) stratification of $\mathrm{Hom}(T\mathcal{C}, \pi^*TS)$ by rank [12].) Notice that we have a restriction homomorphism $\theta_S(\log \Delta_\pi) \otimes \mathcal{O}_{\Delta_\pi} \to \theta_{\Delta_\pi}$.

Let $\theta_{\mathcal{C},S} \subset \theta_{\mathcal{C}}$ denote the sheaf of derivations which preserve $\pi^*\mathcal{O}_S$. If we apply $\pi_* j_* j^*$ to the exact sequence $0 \to \theta_{\mathcal{C}/S} \to \theta_{\mathcal{C},S} \to \theta_{\mathcal{C},S}/\theta_{\mathcal{C}/S} \to 0$ and use our liftability assumption and the fact that πj is affine, we get the exact sequence

$$0 \to \theta_{\mathcal{A}} \to \theta_{\mathcal{A},S} \to \theta_S(\log \Delta_\pi) \to 0.$$

We defined $\hat{\theta}_{\mathcal{A},S}$ as the preimage of $\theta_{\mathcal{A},S}$ in $\hat{\theta}_{\mathcal{L},S}$ and $\hat{\theta}_S(\log \Delta_\pi)$ as the quotient $\hat{\theta}_{\mathcal{L},S}/\theta_{\mathcal{A}}$. These extend $\theta_{\mathcal{A},S}$ and θ_S by $c_0\mathcal{O}_S$. If we denote the *Hodge bundle*

$$\lambda := \lambda(\mathcal{C}/S) := \det(\pi_* \omega_{\mathcal{C}/S}),$$

then we see that $\hat{\theta}_S(\log \Delta_\pi)$ may be identified with the Lie sheaf $\mathcal{D}_1(\lambda)(\log \Delta_\pi)$ of first order differential operators $\lambda \to \lambda$ which preserve the subsheaf of sections vanishing on Δ_π.

Observe that $\mathcal{L}\mathfrak{g} = \mathfrak{g} \otimes_k \mathcal{L}$ is now a sheaf of Lie algebras over $\mathcal{O}_S$. The same applies to $\hat{\mathfrak{l}}$ and so we have a Virasoro extension $\hat{\theta}_S$ of θ_S by $c_0\mathcal{O}_S$. We have also defined $\mathcal{A}\mathfrak{g} = \mathfrak{g} \otimes_k \mathcal{A}$, which is a Lie subsheaf of $\mathcal{L}\mathfrak{g}$ as well as of $\widehat{\mathcal{L}\mathfrak{g}}$ and the Fock type $\widehat{\mathcal{L}\mathfrak{g}}$-module $\mathcal{F}_\ell(\mathfrak{g})$. The will also consider the sheaf of $\mathcal{A}\mathfrak{g}$-covariants in the latter,

$$\mathcal{F}_\ell(\mathfrak{g})_{\mathcal{C}/S} := \mathcal{F}_\ell(\mathfrak{g})_{\mathcal{A}\mathfrak{g}} = \mathcal{A}\mathfrak{g}\mathcal{F}_\ell(\mathfrak{g}) \backslash \mathcal{F}_\ell(\mathfrak{g}).$$

From Proposition 4.3 we get:

Corollary 5.2. *The representation of the Lie algebra* $\hat{\theta}_{\mathcal{A},S}$ *on* $\mathcal{F}_\ell(\mathfrak{g})$ *preserves* $\mathcal{A}\mathfrak{g}\mathcal{F}_\ell(\mathfrak{g})$ *and acts on* $\mathcal{F}_\ell(\mathfrak{g})_{\mathcal{C}/S}$ *via* $\hat{\theta}(\log \Delta_\pi)$ *with* c_0 *acting as multiplication by* $(\ell + \check{h})^{-1}\ell \dim \mathfrak{g}$. *This construction has a base change property along any smooth part* S' *of the discriminant in the sense that the residual action of* $\hat{\theta}(\log \Delta_\pi)$ *on* $\mathcal{F}_\ell(\mathfrak{g})_{\mathcal{C}_{S'}/S'} \cong \mathcal{F}_\ell(\mathfrak{g})_{\mathcal{C}/S} \otimes \mathcal{O}_{S'}$ *factors through* $\hat{\theta}_{S'}$.

The bundle of integrable representations $\mathcal{H}_\ell(V)$ over S is defined in the expected manner: it is obtained as a quotient of $\mathcal{F}_\ell(\mathfrak{g})$ in the way $\mathbb{H}_\ell(V)$ is obtained from $\mathbb{F}_\ell(\widehat{L\mathfrak{g}})$. We write $\mathcal{H}_\ell(V)_{\mathcal{C}/S}$ for $\mathcal{H}_\ell(V)_{\mathcal{A}\mathfrak{g}}$. The following theorem, which is mostly a summary of what we have done so far, is one of the main results of the theory.

Theorem 5.3 (WZW-connection). *The* $\mathcal{O}_S$*-module* $\mathcal{H}_\ell(V)_{\mathcal{C}/S}$ *is of finite rank; it is also locally free over* $S - \Delta_\pi$ *and the Lie action of* $\mathcal{D}_1(\lambda)(\log \Delta_\pi)$ *on* $\mathcal{H}_\ell(V)_{\mathcal{C}/S}$ *defines a logarithmic* λ*-flat connection relative to* Δ_π *of weight* $\frac{\ell}{2(\ell+\check{h})}\dim \mathfrak{g}$. *The same base change property holds along the smooth part of the discriminant as in Corollary 5.2. Furthermore, any* $\sigma \in \mathrm{Aut}(\mathfrak{g})$ *determines an isomorphism of* $\mathcal{D}_1(\lambda)(\log \Delta_\pi)$*-modules* $\mathcal{H}_\ell({}^\sigma V)_{\mathcal{C}/S} \cong \mathcal{H}_\ell(V)_{\mathcal{C}/S}$.

Proof. The first assertion follows from 4.6. The action of $\hat{\theta}$ factors (locally) through $\mathcal{D}_1(\sqrt{\lambda})(\log \Delta_\pi)$ for some square root $\sqrt{\lambda}$ of λ and has then weight $(\ell + \check{h})^{-1}\ell \dim \mathfrak{g}$. This amounts to an action of $\mathcal{D}_1(\lambda)(\log \Delta_\pi)$ of half that weight. The last assertion follows from Corollary 3.3. The rest is clear except perhaps the local freeness of $\mathcal{H}_\ell(V)_{\mathcal{C}/S}$ on $S - \Delta_\pi$. But this follows from the local existence of a connection in the $\mathcal{O}_S$-module $\mathcal{H}_\ell(V)_{\mathcal{C}}$. □

So if $\Lambda^\times \to S$ denotes the $\mathbb{G}_m$-bundle that is associated to λ, then we have a flat connection on the pull-back of $\mathcal{H}_\ell(V)_{\mathcal{C}/S}$ to $\Lambda^\times|S - \Delta_\pi$ with fiber monodromy scalar multiplication by a root of unity of order $\frac{\ell}{2(\ell+\check{h})}\dim\mathfrak{g}$.

Propagation principle continued

In the preceding subsection we made the assumption throughout that a union of sections of $\mathcal{C} \to S$ is given to ensure that its complement is affine over S. However, the propagation principle permits us to abandon that assumption. In fact, this leads us to let $\mathbb{V}$ stand for any map which assigns to every S-valued point x of $\mathcal{C}$ an irreducible $\mathfrak{g}$-representation $\mathbb{V}_x$ of level $\leqslant \ell$, subject to the condition that its *support*, $\mathrm{Supp}(\mathbb{V})$ (i.e., the union of the $x(S)$ for which $\mathbb{V}_x$ is generically not the trivial representation), is a trivial finite cover over S and contained in the locus where $\pi : \mathcal{C} \to S$ is smooth. We then might write $\mathcal{H}_\ell(\mathbb{V})$ for $\mathcal{H}_\ell(\mathbb{V}|_{\mathrm{Supp}(\mathbb{V})})$, but since $\mathcal{C} - \mathrm{Supp}(\mathbb{V})$ need not be affine over S, this does not yield the right notion of conformal block. We can find however, at least locally over S, additional pairwise disjoint sections of $\mathcal{C} \to S$ so that the complement $\mathcal{C}^\circ$ of their support and that of $\mathbb{V}$ is affine over S. Then we can form $\mathcal{H}_\ell(\mathbb{V}|\mathcal{C} - \mathcal{C}^\circ)$ and Proposition 4.9 shows that the resulting bundle of covacua $\mathcal{H}_\ell(\mathbb{V}|\mathcal{C} - \mathcal{C}^\circ)_{(\pi_*\mathcal{O}_{\mathcal{C}^\circ})\mathfrak{g}}$ with the projective connection is independent of the choices made. This suggests that we let $\mathcal{H}_\ell(\mathbb{V})$ resp. $\mathcal{H}_\ell(\mathbb{V})_{\mathcal{C}/S}$ stand for the sheaf associated to the presheaf

$$S \supset U \mapsto \varinjlim_{\tilde{S}} \mathcal{H}_\ell(\mathbb{V}|_{\tilde{S}}) \text{ resp. } \varinjlim_{\tilde{S}} \mathcal{H}_\ell(\mathbb{V}|_{\tilde{S}})_{\mathcal{C}_U/U},$$

where $\tilde{S}$ runs over the unions of pairwise disjoint sections as above. The latter, when twisted with the dual of $\det(\mathcal{C}/S)$, has, being a limit of presheaves with flat connections, a flat connection as well. It is clear that in this set-up there is also no need anymore to insist that the fibers of π be connected.

The genus zero case and the KZ-connection

We here assume C to be isomorphic to $\mathbb{P}^1$. Let $x_1, \dots, x_n \in C$ be distinct and contain $\mathrm{Supp}(\mathbb{V})$. Choose an affine coordinate z on C (which identifies C with $\mathbb{P}^1$) whose domain contains the x_i's and write z_i for $z(x_i)$. Notice that $t_\infty := z^{-1}$ may serve as a parameter for the local field at $z = \infty$. So if $\mathbb{H}_\ell(k)$ denotes the representation of $\widehat{\mathfrak{g}((z^{-1}))}$ attached to the trivial representation k of $\widehat{\mathfrak{g}((z^{-1}))}$, then by the propagation principle 4.9 we have $\mathbb{H}_\ell(\mathbb{V})_C = (V_1 \otimes \cdots \otimes V_n \otimes \mathbb{H}_\ell(k))_{\mathfrak{g}[z]}$, where $\mathfrak{g}[z]$ acts on V_i for $i \leqslant n$ via its evaluation at z_i. According to [7], the $\mathfrak{g}[z]$-homomorphism $U(\mathfrak{g}[z]) \to \mathbb{H}_\ell(k)$ is surjective and its kernel is the left ideal generated by $(zX)^{1+\ell}$, where $X \in \mathfrak{g}$ generates a highest root line. This implies that $\mathbb{H}_\ell(\mathbb{V})_{\mathbb{P}^1}$ can be identified with a quotient of the space of $\mathfrak{g}$-covariants $(V_1 \otimes \cdots \otimes V_n)_\mathfrak{g}$, namely its biggest quotient on which $(\sum_{i=1}^n z_i X^{(i)})^{1+\ell}$ acts trivially (where $X^{(i)}$ acts on V_i as X and on the other tensor factors V_j, $j \neq i$, as the identity). Now regard $z_1, \dots, z_n$ as variables. Our first observation is that a translation in $\mathbb{C}$ does not affect $\mathbb{H}_\ell(\mathbb{V})_C$:

if $a \in \mathbb{C}$, then the actions of $\sum_{i=1}^{n}(z_i + a)X^{(i)}$ and $\sum_{i=1}^{n} z_i X^{(i)}$ on $V_1 \otimes \cdots \otimes V_n$ differ the action of $aX \in \mathfrak{g}$. So we always arrange that $z_1 + \cdots + z_n = 0$. Consider in $\mathbb{C}^n$ the hyperplane S_{n-1} defined by $z_1 + \cdots + z_n = 0$ and denote by S_{n-1}° the open subset of pairwise distinct n-tuples. Then the trivial family over S_{n-1}°, $\mathcal{C} := \mathbb{P}^1 \times S_{n-1}^{\circ}$, comes with $n+1$ 'tautological' sections (including the one at infinity) so that we also have defined $\mathcal{C}^{\circ}$. This determines a sheaf $\mathcal{H}_\ell(\mathbb{V})_{\mathcal{C}/S_{n-1}^{\circ}}$ over S_{n-1}°. According to the preceding, we have an exact sequence

$$(V_1 \otimes \cdots \otimes V_n)_{\mathfrak{g}} \otimes_k \mathcal{O}_{S_{n-1}^{\circ}} \to (V_1 \otimes \cdots \otimes V_n)_{\mathfrak{g}} \otimes_k \mathcal{O}_S \to \mathcal{H}_\ell(\mathbb{V})_{\mathcal{C}/S_{n-1}^{\circ}} \to 0,$$

where the first map is given by $(\sum_{i=1}^{n} z_i X^{(i)})^{1+\ell}$. We identify its WZW connection, or rather, a natural lift of that connection to $V_1 \otimes \cdots \otimes V_n \otimes_k \mathcal{O}_{S_{n-1}^{\circ}}$. In order to compute the covariant derivative with respect to the vector field $\partial_i := \frac{\partial}{\partial z_i}$ on S_{n-1}°, we follow our recipe and lift it to $C \times S_{n-1}^{\circ}$ in the obvious way (with zero component along C). We continue to denote that lift by ∂_i and determine its (Sugawara) action on $\mathcal{H}_\ell(\mathbb{V})$. We first observe that ∂_i is tangent to all the sections, except the ith. Near that section we decompose it as $(\frac{\partial}{\partial z} + \partial_i) - \frac{\partial}{\partial z}$, where the first term is tangent to the ith section and the second term is vertical. The action of the former is easily understood: its lift to $V_1 \otimes \cdots \otimes V_n \otimes_k \mathcal{O}_{S_{n-1}^{\circ}}$ acts as derivation with respect to z_i. The vertical term, $-\frac{\partial}{\partial z}$, acts via the Sugawara representation, that is, it acts on the ith slot as $-\frac{1}{\ell+\check{h}}\sum_\kappa X_\kappa(z-z_i)^{-1} \circ X_\kappa$ and as the identity on the others, in other words, acts as $-\frac{1}{\ell+\check{h}}\sum_\kappa X_\kappa^{(i)}(z-z_i)^{-1} \circ X_\kappa^{(i)}$. This action does not induce one in $V_1 \otimes \cdots \otimes V_n \otimes_k \mathcal{O}_{S_{n-1}^{\circ}}$. To make it so, we add to this the action by an element of $\mathfrak{g}[\mathcal{C}^{\circ}]U\widehat{\mathcal{L}\mathfrak{g}}$ (which of course will act trivially in $\mathcal{H}_\ell(\mathbb{V})_{\mathcal{C}/S_{n-1}^{\circ}}$), namely

$$\frac{1}{\ell+\check{h}}\sum_\kappa X_\kappa(z-z_i)^{-1} \circ X_\kappa^{(i)} = \frac{1}{\ell+\check{h}}\sum_{j,\kappa} \frac{1}{z-z_i} X_\kappa^{(j)} \circ X_\kappa^{(i)}.$$

Doing this for every i, then the modification acts in $V_1 \otimes \cdots \otimes V_n \otimes_k \mathcal{O}_{S_{n-1}^{\circ}}$ as

$$\frac{1}{\ell+\check{h}}\sum_{j\neq i} \frac{1}{z_j - z_i} X_\kappa^{(j)} X_\kappa^{(i)}.$$

Let us regard the Casimir element c as an element of $\mathfrak{g} \otimes_k \mathfrak{g}$, and denote by $c^{(i,j)}$ its action in $V_1 \otimes \cdots \otimes V_n$ on the ith and jth factor (since c is symmetric, we have $c^{(i,j)} = c^{(j,i)}$, so that we need not worry about the order here). We conclude that the WZW-connection is induced by the connection on $V_1 \otimes \cdots \otimes V_n \otimes_k \mathcal{O}_{S_{n-1}^{\circ}}$ whose connection form is

$$\frac{1}{\ell+\check{h}}\sum_{i=1}^{n}\sum_{j\neq i} \frac{dz_i}{z_j - z_i} c^{(i,j)} = -\frac{1}{\ell+\check{h}}\sum_{1\leqslant i<j\leqslant n} \frac{d(z_i - z_j)}{z_i - z_j} c^{(i,j)}.$$

It commutes with the Lie action of $\mathfrak{g}$ on $V_1 \otimes \cdots \otimes V_n$ and so the connection passes to one on $(V_1 \otimes \cdots \otimes V_n)_{\mathfrak{g}} \otimes_k \mathcal{O}_{S_{n-1}^{\circ}}$. This lift of the WZW-connection is known as the *Knizhnik-Zamolodchikov connection*. It is not difficult to verify that it is flat (see glung [9]), so that we have not just a projectively flat connection, but a genuine one.

Proposition 5.4. *The map* $(V_1\otimes\cdots\otimes V_n)_{\mathfrak{g}}\otimes_k \mathcal{O}_{S^\circ_{n-1}} \to \mathcal{H}_\ell(\mathbb{V})_{\mathcal{C}/S^\circ_{n-1}}$ *is an isomorphism for* $n = 1,2$. *Hence for* $n = 1$ *(resp.* $n = 2$*),* $\mathcal{H}_\ell(\mathbb{V})_{\mathcal{C}/S^\circ_{n-1}}$ *is zero unless* V_0 *is the trivial representation (resp.* V_0 *and* V_1 *are each others dual), in which case it can be identified with* $\mathcal{O}_{S^\circ_{n-1}}$.

Proof. For $n = 1$ this is clear. For $n = 2$, the stalk of $\mathcal{H}_\ell(\mathbb{V})_{\mathcal{C}/S^\circ_1}$ at $(z, -z)$, $z \neq 0$, can be identified with the image in $(V_1 \otimes V_2)_{\mathfrak{g}}$ of the kernel of $(zX^{(1)} - zX^{(2)})^{1+\ell}$ acting in $V_1 \otimes V_2$. Since $X^{(1)} + X^{(2)}$ is zero in $(V_1 \otimes V_2)_{\mathfrak{g}}$ and $(X^{(1)})^{1+\ell}$ is zero in V_1, this $(V_1 \otimes V_2)_{\mathfrak{g}}$. □

Remark 5.5. A 3-pointed genus zero curve $(C \cong \mathbb{P}^1; x_1, x_2, x_3)$ has no moduli, and so we expect in this case an identification of $\mathbb{H}_\ell(\mathbb{V})_C$ also. Indeed, as is shown in [3], if V_1, V_2, V_3 are the associated irreducible $\mathfrak{g}$-representations of level $\leqslant \ell$, then $\mathbb{H}_\ell(\mathbb{V})_C$ is naturally identified with the biggest quotient of $V_1 \otimes V_2 \otimes V_3$ on which both $\mathfrak{g}$ and the endomorphisms $(z_1X^{(1)} + z_2X^{(2)} + z_3X^{(3)})^{1+\ell}$ act trivially for *all* values of (z_1, z_2, z_3). This last condition is of course equivalent to requiring that $X^p \otimes X^q \otimes X^r$ induces the zero map whenever $p + q + r > \ell$.

6. Factorization

In this section we consider the case when we are given a family $\pi_o : \mathcal{C}_o \to S_o$ of pointed curves of genus g with a smooth base germ $S_o = \mathrm{Spec}(R_o)$ (so R_o is a regular local ring) and for which we are given a section x_0 along which π_o has an ordinary double point. We assume that the fibers have no other singularities, in other words, that π_o is smooth outside x_0. After possibly making an étale base change of degree two we find a partial normalization $\nu : \tilde{\mathcal{C}}_o \to \mathcal{C}_o$ which separates the branches in the (strong) sense that ν is an isomorphism over the complement of $x_0(S_o)$ and x_0 has two disjoint lifts to $\mathcal{C}_o$ (which we shall denote by x_+ and x_-). In what follows we simply assume this to be already the case. There are two basic cases: the *nonseparating case*, where $\tilde{\mathcal{C}}_o/S_o$ is connected—in that case the fibers have genus $g-1$—and the *separating case*, where x_+ and x_- take values in different components $\tilde{\mathcal{C}}_\pm$ of $\tilde{C}_o$ such that the fiber genera $g_\pm$ of $\tilde{C}_\pm/S_o$ add up to g. Since the natural base of the WZW-connection is the $\mathbb{G}_m$-bundle defined by a determinant bundle (or a fractional power thereof), let us first recall what we get in the present case. The bundle of which we take the determinant is the direct image of the relative dualizing sheaf $\pi_{o*}\omega_{\mathcal{C}_o/S_o}$. This bundle contains the direct image of $\omega_{\tilde{\mathcal{C}}_o/S_o}$ and the two differ only at x_o: an element of $\omega_{\tilde{\mathcal{C}}_o/S_o,x_o}$ when pulled back under ν may have a simple pole at x_+ and x_- whose residues add up to zero. So we have a natural exact sequence

$$0 \to \nu_*\omega_{\tilde{\mathcal{C}}_o/S_o} \to \omega_{\mathcal{C}_o/S_o} \to \mathcal{O}_{S_o} \to 0,$$

where the last map is defined by taking the residue at x_+. If we take the direct image under π_o, we see that we have a natural injection $(\pi_o\nu)_*\omega_{\tilde{\mathcal{C}}_o/S_o} \to \pi_{o*}\omega_{\mathcal{C}_o/S_o}$. It is in fact an isomorphism in the separating case, whereas it has a cokernel naturally isomorphic to R_o in the nonseparating case. So after taking determinants we get in either case that $\lambda(\mathcal{C}_o/S_o) = \lambda(\tilde{\mathcal{C}}_o/S_o)$, where it is understood that in the separating case the right hand side equals $\lambda(\tilde{\mathcal{C}}_+/S_o) \otimes \lambda(\tilde{\mathcal{C}}_-/S_o)$.

We now also assume given a representation valued map $\mathbb{V}_o$ on the smooth part of $\mathcal{C}_o$ whose support is contained in a finite union of sections S_o so that we have defined $\mathcal{H}_\ell(\mathbb{V}_o)_{\mathcal{C}_o/S_o}$. A coarse version of the *factorization principle* expresses this R_o-module in terms of a space of covacua attached to the normalization $\tilde{\mathcal{C}}_o/S_o$. The more refined form describes it as a residue of a module of covacua on a smoothing of π_o and takes into account the flat connection.

Throughout this section $\Sigma_o \subset \mathcal{C}_o$ is a finite union of sections of $\mathcal{C}_o/S_o$ contained in the smooth part of $\mathcal{C}_o$, which contains the support of $\mathbb{V}_o$ and has the additional property that its complement $\mathcal{C}_o^\circ := \mathcal{C}_o - \Sigma_o$ is affine over S_o (this can always be arranged by adding some 'dummy' sections to the support of $\mathbb{V}_o$). We often identify Σ_o with its preimage in $\tilde{\mathcal{C}}_o$. Notice that $\tilde{\mathcal{C}}_o^\circ := \nu^{-1}\mathcal{C}_o^\circ = \tilde{\mathcal{C}}_o - \Sigma_o$ is also affine over S_o, being the normalization of an affine S_o-scheme. We write A_o resp. $\tilde{A}_o$ for their (coordinate) R_o-algebras.

Coarse version of the factorization property

Recall that P_ℓ denotes the set of isomorphism classes of irreducible representations of $\mathfrak{g}$ of level $\leqslant \ell$ and is invariant under dualization: if $\mu \in P_\ell$, then $\mu^* \in P_\ell$. Let V_μ be a $\mathfrak{g}$-representation in the equivalence class $\mu \in P_\ell$ and choose $\mathfrak{g}$-equivariant dualities

$$b_\mu : V_\mu \otimes V_{\mu^*} \to k,$$

where we assume that b_{μ^*} is the transpose of b_μ. Its transpose inverse $\check{b}_\mu \in V_\mu \otimes V_{\mu^*}$ then spans the line of $\mathfrak{g}$-invariants in $V_\mu \otimes V_{\mu^*}$.

Proposition 6.1. *Let $\tilde{\mathbb{V}}_{\mu,\mu^*}$ be the representation valued map on $\tilde{\mathcal{C}}_o$ which is constant equal to V_μ resp. V_{μ^*} on x_+ resp. x_- and is elsewhere equal to $\mathbb{V}_o$ (via the obvious identification defined by ν). Then the contractions $b_\mu : V_\mu \otimes V_{\mu^*} \to k$ define an isomorphism*

$$\oplus_{\mu \in P_\ell} \mathbb{H}_\ell(\tilde{\mathbb{V}}_{\mu,\mu^*})_{\tilde{\mathcal{C}}_o/S_o} \xrightarrow{\cong} \mathbb{H}_\ell(\mathbb{V}_o)_{\mathcal{C}_o/S_o}.$$

This is almost a formal consequence of:

Lemma 6.2. *Let M be a finite dimensional representation of $\mathfrak{g} \times \mathfrak{g}$ which is of level $\leqslant \ell$ relative to both factors. If M^δ denotes that same space viewed as $\mathfrak{g}$-module with respect to the diagonal embedding $\delta : \mathfrak{g} \to \mathfrak{g} \times \mathfrak{g}$, then the contraction $\oplus_{\mu \in P_\ell} M \otimes (V_\mu \boxtimes V_\mu^*) \to M$ that on each summand is defined by b_μ (the symbol $\boxtimes$ stands for the exterior tensor product of representations) induces an isomorphism between covariants:*

$$\oplus_{\mu \in P_\ell} \big(M \otimes (V_\mu \boxtimes V_\mu^*)\big)_{\mathfrak{g} \times \mathfrak{g}} \xrightarrow{\cong} M^\delta_{\mathfrak{g}}.$$

Proof. Without loss of generality we may assume that M is irreducible, or more precisely, equal to $V_\lambda \boxtimes V_{\lambda'}$ for some $\lambda, \lambda' \in P_\ell$. Then $M^\delta = V_\lambda \otimes V_{\lambda'}$. By Schur's lemma, $M^\delta_{\mathfrak{g}}$ is one-dimensional if $\lambda' = \lambda^*$ and trivial otherwise. That same lemma applied to $\mathfrak{g}\times\mathfrak{g}$ shows that $(M\otimes(V_\mu \boxtimes V^*_\mu))_{\mathfrak{g}\times\mathfrak{g}}$ is zero unless $(\lambda,\lambda') = (\mu^*,\mu)$, in which case it is one-dimensional and maps isomorphically to M^δ. □

Proof of 6.1. Evaluation in x_o resp. x_+, x_- define epimorphisms $A_o \to R_o$ resp. $\tilde{A}_o \to R_o \oplus R_o$ whose kernels may be identified by means of ν. We denote that common kernel by $\mathcal{J}$ and by B the algebra of regular functions on the smooth part of $\mathcal{C}^\circ_o$. This is also the algebra of regular functions on the complement of the two sections $x_\pm$ $\tilde{\mathcal{C}}^\circ_o$. If $\mathcal{J}\mathfrak{g}$ has the evident meaning, then the argument used to prove Proposition 4.6 shows that $M := \mathbb{H}_\ell(\mathbb{V}_o|\Sigma_o)_{\mathcal{J}\mathfrak{g}}$ is an R_o-module of finite rank. It underlies a representation of $\mathfrak{g}\times\mathfrak{g}$ of level $\leqslant \ell$ relative to both factors and is such that $M^\delta_{\mathfrak{g}} = \mathbb{H}_\ell(\mathbb{V}_o)_{A_o\mathfrak{g}} = \mathcal{H}_\ell(\mathbb{V}_o)_{\mathcal{C}_o/S_o}$. The assertion now follows from Lemma 6.2 and the argument used for the propagation principle which shows that $(M\otimes(V_\mu\boxtimes V^*_\mu))_{R_o\mathfrak{g}\times R_o\mathfrak{g}} = \mathbb{H}(\tilde{\mathbb{V}}_{\mu,\mu^*})_{B\mathfrak{g}} = \mathcal{H}_\ell(\tilde{\mathbb{V}}_{\mu,\mu^*})_{\tilde{\mathcal{C}}_o/S_o}$. □

A smoothing construction

In order to motivate the algebraic discussion that will follow, we choose generators $t_\pm$ of the ideals of the completed local R_o-algebras of $\tilde{\mathcal{C}}_o$ at $x_\pm$ and explain how they determine a *smoothing* of $\mathcal{C}_o/S_o$, that is, a way of making $\mathcal{C}_o$ the restriction over $S_o\times\{o\}$ of a flat morphism $\mathcal{C}\to S$, with $S := S_o\times_k \Delta$ (the spectrum of $R := R_o[[\tau]]$) which is smooth over $S - S_o$. The construction goes as follows: in the product $\tilde{\mathcal{C}}_o\times\Delta$, blow up $x_\pm\times\{o\}$ and let $\tilde{\mathcal{C}}$ be the formal neighborhood of the strict transform of $\tilde{\mathcal{C}}_o\times\{o\}$. So at the preimage of $x_\pm\times\{o\}$ we have on the strict transform of $\tilde{\mathcal{C}}\times\{o\}$ the formal S_o-chart $(t_\pm, \tau/t_\pm)$. Now let $\mathcal{C}$ be the quotient of $\tilde{\mathcal{C}}$ obtained by identifying these formal S_o-charts up to order: $(t_+, \tau/t_+) = (\tau/t_-, t_-)$, so that $(s_+, s_-) := (t_+, t_-)$ is now a formal S_o-chart of $\mathcal{C}$ on which we have $\tau = s_+s_-$ (in either domain τ represents the same regular function). We thus have defined a flat morphism $\mathcal{C}\to S_o\times\Delta = S$ (with τ as second component) with the stated properties.

Remark 6.3. If we were to work in the complex analytic category, then we could take for Δ the complex unit disk. The fiber of $\mathcal{C}/S$ over $(s,\tau)\in S_o\times\Delta$ is then obtained by removing from C_s the union of the two disks defined by $|t_\pm|\leqslant|\tau|$, and identifying the two closed annuli $|\tau| < |t_\pm| < 1$ by imposing the identity $t_+t_- = \tau$.

With a view toward a later application—namely, of extracting a topological quantum field theory from the WZW model—we note that there is even a limit if τ tends to zero if we keep its argument fixed. To see this, let us first observe that for $|\tau| < \frac{1}{2}$, the fiber is also obtained by removal of the union of the two open disks defined by $|t_\pm| < \sqrt{|\tau/2|}$, followed by the above identification of the two closed annuli $\sqrt{|\tau/2|}\leqslant|t_\pm|\leqslant\sqrt{|2\tau|}$. Now do a real oriented blow up $\hat{C}_s\to\tilde{C}_s$ of the points $x_\pm(s)\in\tilde{C}_s$. This means that the polar coordinates associated to $t_\pm$ are to be viewed as coordinates for the preimage of its domain on $\hat{C}_s$: $t_\pm = r_\pm\zeta_\pm$ with $|\zeta_\pm| = 1$ and $r_\pm\geqslant 0$ such that the exceptional set $\partial\hat{C}_s$ is defined by $r_\pm = 0$. Notice that $\partial\hat{C}_s$

is indeed the boundary of a surface; it has two components, each of which comes with a natural principal $U(1)$-action. If we write $\tau = \varepsilon\zeta$ accordingly with $|\zeta| = 1$ and $\varepsilon > 0$, then for $\sqrt{\varepsilon/2} \leqslant r_\pm \leqslant \sqrt{2\varepsilon}$, (r_+, ζ_+) must be identified with (r_-, ζ_-) precisely when $r_+r_- = \varepsilon$ and $\zeta_+\zeta_- = \zeta$. This has indeed a continuous extension over $\varepsilon = 0$, for then we just identify the two boundary circles corresponding to $r_\pm = 0$ by insisting that $\zeta_+\zeta_- = \zeta$. We thus obtain a family $\hat{\mathcal{C}} \to \hat{\Delta}$ over the real oriented blow up $\hat{\Delta} \to \Delta$ of Δ at its origin and whose fibers over $\partial\hat{\Delta}$ are as just described. The dependence of $\hat{\mathcal{C}}|\partial\hat{\Delta}$ is a priori on the coordinates $t_\pm$, but it is clear from the construction this dependence is in fact only via the (real) ray in $T_{x_+}\hat{C}_s \otimes T_{x_-}\hat{C}_s$ defined by $\frac{\partial}{\partial t_+}\big|_{x_+} \otimes_{\mathbb{C}} \frac{\partial}{\partial t_-}\big|_{x_-}$. The fibers of this family just differ by the way we identified the boundary circles and we thus see that the monodromy of the family is a positive Dehn twist defined by the welding circle. For later use we note that this construction takes place in the C^1-category: $\hat{\mathcal{C}}$ has a natural C^1-structure such that the projection to $\hat{\Delta}$ is C^1.

We should perhaps add that this has an algebro-geometric incarnation in terms of log structures and that $T_{x_+}\tilde{C}_s \otimes T_{x_-}\tilde{C}_s$ can be understood as the tangent space of the semi-universal deformation of the singular germ $(C_s, x(s))$ (equivalently, our data define a smooth point of the boundary divisor of some moduli stack $\overline{M}_{g,n}$ and $T_{x_+}\tilde{C}_s \otimes T_{x_-}\tilde{C}_s$ can be identified with its normal space).

We will denote by Σ the image of $\Sigma_o \times \Delta$ in both $\mathcal{C}$ and $\tilde{\mathcal{C}}$. In either case it is a union of sections over S. The representation valued map $\mathbb{V}_o$ on $\mathcal{C}_o$ is extended to $\mathcal{C}$ in the obvious way (so that its support is contained in Σ) and we denote this extension by $\mathbb{V}$. We let A stand for R-algebra of regular functions on $\mathcal{C}^o := \mathcal{C} - \Sigma$. Notice that $A_o = A/(\tau A)$ and that A embeds in $\tilde{A}_o[[\tau]]$.

The glueing tensor

Suppose that in the regular local algebra R we are given a subalgebra R_o and an element τ in the maximal ideal of R such that $R = R_o[[\tau]]$. Let L_+ and L_- be R-algebras, both isomorphic to $R((t))$. The 'ideal' in $L_\pm$ corresponding to $tR[[t]]$ is denoted by $\mathfrak{m}_\pm$. Let $L := L_+ \oplus L_-$ the direct sum as R-algebras. We assume given a closed R-subalgebra $\mathcal{O}_0 \subset L$ with the property that it can be topologically generated as a R_o-algebra by two generators s_+, s_- of the following type: there exist generators $t_\pm$ of $\mathfrak{m}_\pm$ such that $s_+ = (t_+, \tau/t_-)$ and $s_- = (\tau/t_+, t_-)$. So an element of $\mathcal{O}_0$ will then have the form

$$\begin{aligned}\sum_{m\geqslant 0, n\geqslant 0} a_{m,n} s_+^m s_-^n &= \sum_{m\geqslant 0, n\geqslant 0} a_{m,n}(t_+^{m-n}\tau^n, t_-^{n-m}\tau^m)\\ &= \sum_{k\geqslant 0}\Big(\sum_{m\geqslant 0} a_{m,k} t_+^{m-k}, \sum_{n\geqslant 0} a_{k,n} t_-^{n-k}\Big)\tau^k\\ &= \sum_{n>m\geqslant 0} a_{n,m}\tau^n s_+^{m-n} + \sum_{m\geqslant 0} a_{m,m}\tau^m + \sum_{m>n\geqslant 0} a_{n,m}\tau^m s_-^{n-m},\end{aligned}$$

with $a_{n,m} \in R_o$. Clearly, the coefficients $a_{m,n}$ can be arbitrary in R_o and the element in question is zero only when all $a_{m,n}$ are. So $\mathcal{O}_0$ is a copy of $R_o[[s_+, s_-]]$. The last identity shows that $\mathcal{O}_0$ is contained in the R-submodule generated by nonpositive powers of s_+ and s_-. A similar argument yields the following lemma and so the proof is left as an exercise.

Lemma 6.4. *Any continuous R_o-derivation of $\mathcal{O}_0$ which preserves $\tau \in \mathcal{O}_0$ extends uniquely to one of* L. *If we let $D_k^{\pm}$ stand for $t_{\pm}^{k+1}\frac{\partial}{\partial t_{\pm}}$, then it has there the form*

$$(D_0^+, 0) + \sum_{k\geqslant 0} \tau^k \Big(\sum_{m\geqslant 0} a_{m,k} D^+_{m-k}, \sum_{n\geqslant 0} a_{k,n} D^-_{n-k} \Big),$$

with $a_{m,n} \in R_o$.

We have defined $L\mathfrak{g}$ and its central extension $\widehat{L\mathfrak{g}}$. For $\mu \in P_\ell$, let $\mathbb{H}_\ell^{\pm}(V_\mu)$ denote the representation attached to V_μ of the central extension $\widehat{L_{\pm}\mathfrak{g}}$ of $L_{\pm}\mathfrak{g}$, so that the R-module $\mathbb{H}_\ell^+(V_\mu) \otimes_R \mathbb{H}_\ell^-(V_{\mu^*})$ is one of $\widehat{L\mathfrak{g}}$. These representations are defined over R_o (over k even) and so arise from a base change: $\mathbb{H}_\ell^{\pm}(V_\mu) = R \otimes_{R_o} \mathbb{H}_{o,\ell}^{\pm}(V_\mu)$ and likewise for their tensor product. The Casimir element c acts in V_μ as a scalar, a scalar we shall denote by c_μ. Observe that $c_{\mu^*} = c_\mu$. Its value is best expressed (and computed) in terms of a Cartan subalgebra $\mathfrak{h} \subset \mathfrak{g}$ and a system of positive roots relative to $\mathfrak{h}$: if we identify μ with its highest weight in $\mathfrak{h}^*$, then

$$c_\mu = c(\mu, \mu + 2\rho),$$

where ρ has the customary meaning as the half the sum of the positive roots. In particular, c_μ is a positive rational number (the denominator is in fact at most 3).

Lemma 6.5. *There exists a series $\varepsilon^\mu = \sum_{d=0}^{\infty} \varepsilon_d^\mu \tau^d \in \mathbb{H}_\ell^+(V_\mu) \otimes_{R_o} \mathbb{H}_\ell^-(V_{\mu^*})[[\tau]]$ (the* glueing tensor*) with constant term $\varepsilon_0^\mu = \check{b}_\mu$ that is annihilated by the image of $\mathcal{O}_0\mathfrak{g}$ in $\widehat{L\mathfrak{g}}$. Moreover, any continuous* R*-derivation* D *of $\mathcal{O}_0$ which preserves τ determines a $\hat{D} \in \hat{\theta}$ (relative to the Fock construction on the* R*-algebra* L*) with the property that ε^μ is an eigenvector of $T_{\mathfrak{g}}(\hat{D})$ with eigenvalue $-\frac{c_\mu}{2(\ell+\check{h})}$.*

Proof. We first observe the generators $t_\pm$ of $\mathfrak{m}_\pm$ define a grading on all the relevant objects on which we have defined the associated filtration F (e.g., the degree zero summand of $\mathbb{H}_\ell(V_\mu)$ is $R \otimes_k V_\mu$). It is known ([7], § 9.4) that the pairing b_μ : $V_\mu \times V_{\mu^*} \to k$ extends (in fact, in a unique manner) to a perfect R-pairing

$$b_\mu : \mathbb{H}_\ell^+(V_\mu) \times \mathbb{H}_\ell^-(V_{\mu^*}) \to R$$

with the property that $b_\mu(Xt_+^n u, u') + b_\mu(u, Xt_-^{-n}u') = 0$ for all $X \in \mathfrak{g}$ and $n \in \mathbb{Z}$. This formula implies that the restriction of b_μ to $\mathbb{H}_\ell^+(V_\mu)_{-d} \times \mathbb{H}_\ell^-(V_{\mu^*})_{-d'}$ is zero when $d \neq d'$ and is perfect when $d = d'$. So if $\varepsilon_d^\mu \in \mathbb{H}_\ell^+(V_\mu)_{-d} \otimes \mathbb{H}_\ell^-(V_{\mu^*})_{-d}$ denotes the latter's transpose inverse, then we have for all $n \in \mathbb{Z}$, $X \in \mathfrak{g}$ the following identity in $\mathbb{H}_\ell^+(V_\mu)_d \times \mathbb{H}_\ell^-(V_{\mu^*})_{-d-n}$:

$$(Xt_+^n \otimes 1)\varepsilon_{d+n}^\mu + (1 \otimes Xt_-^{-n})\varepsilon_d^\mu = 0.$$

This just says that $(Xt_+^n \otimes 1) + \tau^n(1 \otimes Xt_-^{-n})$ kills $\varepsilon^\mu := \sum_{d\geqslant 0} \varepsilon_d^\mu \tau^d$. Since $s_+^n = (t_+^n, \tau^n t_-^{-n})$, this amounts to saying that $Xs_+^n \in \mathcal{O}_0\mathfrak{g} \subset \widehat{L\mathfrak{g}}$ kills ε^μ. Likewise for Xs_-^n. Since any element of $\mathcal{O}_0$ lies in the R-submodule generated by the nonpositive powers of s_+ and s_-, it follows that ε^μ is killed by all of $\mathcal{O}_0\mathfrak{g}$.

The second statement is proved by a direct computation. If we use Lemma 6.4 to write D as an operator in L, then we find that it suffices to prove:

(i) $\tau^n T_\mathfrak{g}(\hat{D}^+_{m-n}) - \tau^m T_\mathfrak{g}(\hat{D}^-_{n-m})$ kills ε^μ for all $m, n \geqslant 0$, and

(ii) $T_\mathfrak{g}(\hat{D}^+_0)(\varepsilon^\mu) = -\frac{c_\mu}{2(\ell+\check{h})}\varepsilon^\mu$.

As to (i), if we substitute

$$T_\mathfrak{g}(\hat{D}^+_{m-n}) = -\frac{1}{2(\ell+\check{h})}\sum_{j\in\mathbb{Z}}\sum_\kappa : X_\kappa t_+^{m-n-j} \circ X_\kappa t_+^j :$$

and do likewise for $T_\mathfrak{g}(\hat{D}^-_{n-m})$, then this assertion follows easily.

For (ii) we first observe that $T_\mathfrak{g}(\hat{D}^+_0)$ preserves the grading of $\mathbb{H}^+_\ell(V_\mu)$ and acts on $\mathbb{H}^+_\ell(V_\mu)_0 = R\otimes_k V_\mu$ as $-(2\ell+2\check{h})^{-1}\sum_\kappa X_\kappa \circ X_\kappa$. This is just multiplication by $-\frac{c_\mu}{2(\ell+\check{h})}$. For an element $u \in \mathbb{H}^+_\ell(V_\mu)_{-d}$ of the form $u = Y_r t_+^{-k_r} \circ \cdots \circ Y_1 t_+^{-k_1} \circ v$ with $v \in V_\mu$, and $Y_\rho \in \mathfrak{g}$ (so that $d = k_r + \cdots + k_1$), we have

$$T_\mathfrak{g}(\hat{D}^+_0)(u) = -du + Y_r t_+^{-k_r} \circ \cdots \circ Y_1 t_+^{-k_1} \circ T_\mathfrak{g}(\hat{D}^+_0)(v) = (-d - \frac{c_\mu}{2(\ell+\check{h})})u.$$

Since $D_0^+(\tau^d) = d\tau^d$, it follows that $\varepsilon_d^\mu\tau^d$ is an eigenvector of $T_\mathfrak{g}(\hat{D}^+_0)$ with eigenvalue $-\frac{c_\mu}{2(\ell+\check{h})}$. □

Finer version of the factorization property

It is clear that our smoothing identifies the R-module $\mathbb{H}_\ell(\mathbb{V})$ with $\mathbb{H}_\ell(\mathbb{V}_o)[[\tau]]$. According to Proposition 4.6, $\mathcal{H}_\ell(\mathbb{V})_{\mathcal{C}/S} = \mathbb{H}_\ell(\mathbb{V})_{A\mathfrak{g}}$ is a finitely generated R-module. Since $A_o = A/\tau A$, reduction of $\mathbb{H}_\ell(\mathbb{V})_{A\mathfrak{g}}$ modulo τ yields $\mathbb{H}_\ell(\mathbb{V}_o)_{A_o\mathfrak{g}} = \mathcal{H}_\ell(\mathbb{V}_o)_{\mathcal{C}_o/S_o}$. Proposition 6.1 identifies the latter with $\oplus_{\mu\in P_\ell}\mathcal{H}_\ell(\tilde{\mathbb{V}}_{\mu,\mu^*})_{\tilde{\mathcal{C}}_o/S_o}$. It is our goal to extend this identification to one of the space of covacua $\mathcal{H}_\ell(\mathbb{V})_{\mathcal{C}/S}$ with the pull-back of $\oplus_{\mu\in P_\ell}\mathcal{H}_\ell(\tilde{\mathbb{V}}_{\mu,\mu^*})_{\mathcal{C}_o/S_o}$ along the projection $\pi_{S_o} : S \to S_o$ and to identify the connection on that pull-back. This will imply among other things that $\mathcal{H}_\ell(\mathbb{V})_{\mathcal{C}/S}$ is a free R-module.

Theorem 6.6. *The* R*-homomorphism defined by tensoring with the glueing tensor,*

$$E = (E_\mu)_\mu : \mathbb{H}_\ell(\mathbb{V}) \to \oplus_{\mu\in P_l}\mathbb{H}_\ell(\tilde{\mathbb{V}}_{\mu,\mu^*})[[\tau]],$$

$$U = \sum_{k\geqslant 0} u_k\tau^k \mapsto \Big(U\hat{\otimes}_R\varepsilon^\mu = \sum_{k,d\geqslant 0} u_k\otimes\varepsilon_d^\mu\tau^{k+d}\Big)_\mu,$$

is also a map of $A\mathfrak{g}$*-representations if we let* $A\mathfrak{g}$ *act on the right hand side via the inclusion* $A \subset \tilde{A}_o[[\tau]]$. *The resulting* R*-homomorphism of covariants,*

$$E_{\mathcal{C}/S} : \mathbb{H}_\ell(\mathbb{V})_{A\mathfrak{g}} \to \oplus_{\mu\in P_l}\mathbb{H}_\ell(\tilde{\mathbb{V}}_{\mu,\mu^*})_{\tilde{A}_o\mathfrak{g}}[[\tau]],$$

is an isomorphism (so that $\mathbb{H}_\ell(\mathbb{V})_{A_\mathfrak{g}}$ *is a free* R*-module). It is compatible with covariant differentiation with respect to* $\theta_S(\log S_o) = R[[\tau]] \otimes_{R_o} \theta_{R_o} + R[[\tau]]\tau\frac{d}{d\tau}$ *relative to the lift to* $\hat{\theta}_S(\log S_o)$ *of Lemma 6.5: it commutes with the action on* θ_{R_o}*, whereas* $\tau\frac{d}{d\tau}$ *respects each summand* $H_\ell(\tilde{\mathbb{V}}_{\mu,\mu^*})_{\tilde{A}_\mathfrak{g}}[[\tau]]$ *and acts there as the first order differential operator* $\tau\frac{d}{d\tau} + \frac{c_\mu}{2(\ell+\check{h})}$.

Proof. The first statement is immediate from Lemma 6.5. So the map on covariants is defined and is R-linear. If we reduce $E_{\mathcal{C}/S}$ modulo τ, we get the map

$$\mathbb{H}_\ell(\mathbb{V}_o)_{A_o\mathfrak{g}} \to \oplus_{\mu\in P_\ell}\mathbb{H}_\ell(\tilde{\mathbb{V}}_{\mu,\mu^*})_{\tilde{A}_o\mathfrak{g}}, \quad u \mapsto \sum_{\mu\in P_\ell} u\otimes\varepsilon_0^\mu,$$

and observe that this is just the inverse of the isomorphism of Proposition 6.1. Since the range of $E_{\mathcal{C}/S}$ is a free R-module, this implies that $E_{\mathcal{C}/S}$ is an isomorphism.

The commutativity with the action of θ_{R_o} is clear. According to Corollary 3.3 covariant derivation with respect to $\tau\frac{d}{d\tau}$ in $\mathbb{H}_\ell(\mathbb{V})_{\mathcal{C}/S}$ is defined by means of a k-derivation D of A which lifts $\tau\frac{d}{d\tau}$: if we write $D = \tau\frac{d}{d\tau} + \sum_{n\geqslant 0}\tau^n D^{(n)}$, where $D^{(n)}$ is a vector field on the smooth part of $\mathcal{C}/S$, then the covariant derivative is induced by $T_\mathfrak{g}(\hat{D}) = \tau\frac{d}{d\tau} + \sum_{n\geqslant 0}\tau^n T_\mathfrak{g}(D^{(n)})$ acting on $\mathbb{H}_\ell(\mathbb{V}_o)[[\tau]]$. From the last clause of Lemma 6.5 we get that when $U \in \mathbb{H}_\ell(\mathbb{V}_o)[[\tau]]$,

$$\begin{aligned} T_\mathfrak{g}(D)E_\mu(U) &= T_\mathfrak{g}(D)(U\varepsilon^\mu) \\ &= T_\mathfrak{g}(D)(U)\varepsilon^\mu - \frac{c_\mu}{2(\ell+\check{h})}U\varepsilon^\mu = E_\mu T_\mathfrak{g}(D)(U) - \frac{c_\mu}{2(\ell+\check{h})}E_\mu(U). \end{aligned}$$

On $\mathbb{H}_\ell(\tilde{\mathbb{V}}_{\mu,\mu^*})_{\tilde{A}_o\mathfrak{g}}[[\tau]]$, $T_\mathfrak{g}(D)$ acts as derivation by $\tau\frac{d}{d\tau}$ so the last clause follows. □

Corollary 6.7. *The monodromy of the WZW connection acting on* $\mathcal{H}_\ell(\mathbb{V})_{\mathcal{C}/S}$ *has finite order and acts in the summand* $\mathcal{H}_\ell(\tilde{\mathbb{V}}_{\mu,\mu^*})_{\tilde{\mathcal{C}}/S_o}[[\tau]]$ *as multiplication by the root of unity* $\exp(-\pi\sqrt{-1}\frac{c_\mu}{\ell+\check{h}})$.

Proof. The multivalued flat sections of $\mathcal{H}_\ell(\mathbb{V})_{\mathcal{C}/\Delta}$ decompose under $E_{\mathcal{C}/\Delta}$ as a direct sum labeled by P_ℓ. The summand corresponding to μ is the set of solutions of the differential equation $\tau\frac{d}{d\tau}U + \frac{c_\mu}{2(\ell+\check{h})}U = 0$. These are clearly of the form $u\tau^{-c_\mu/2(\ell+\check{h})}$ with $u \in \mathbb{H}_\ell(\tilde{\mathbb{V}}_{\mu,\mu^*})_{\tilde{A}_o\mathfrak{g}}$. If we let τ run over the unit circle, then we see that the monodromy is as asserted. Since $\frac{c_\mu}{\ell+\check{h}} \in \mathbb{Q}$, it has finite order. □

Remark 6.8. We use here the convention that the monodromy of the multivalued function z^α is $\exp(2\pi\alpha\sqrt{-1})$ (rather than $\exp(-2\pi\alpha\sqrt{-1})$). More pedantically: for us the monodromy is a *covariant* rather than a contra-variant functor from the fundamental groupoid to a linear category.

7. The modular functor attached to the WZW model

We show here that the results of Section 6 have topological counterparts that take the form of (what is called) a modular functor in topological quantum field theory.

Defining the functor

For what follows, the most natural setting would probably be that of quasi-conformal surfaces, but we have chosen to work with the more familiar notion of C^1-surfaces. This forced us however to introduce the auxiliary notion of an infinitesimal collar below.

The main objects will be *compact oriented* surfaces endowed with a C^1-structure, possibly with boundary, but where we assume that each boundary component comes with a principal action of the unit circle $U(1)$ that is compatible with the orientation it receives from the surface. In the rest of this paper, we will simply refer to such an object as a *surface.*

An *infinitesimal collar* of a surface is a inward pointing (nowhere zero) vector field defined on the boundary only with the property that it is locally trivial in the sense that we can find local C^1-diffeomorphism (r, u) of a neighborhood of the boundary onto $[0, \varepsilon) \times U(1)$ which is compatible with the $U(1)$-action on the boundary and takes the vector field to $\partial/\partial r|_{\{0\}\times U(1)}$. The choice of such a vector field determines a basis for each tangent space (the second tangent vector field being the derivative of the $U(1)$-action) and so we may think of this as a first order extension of the given $U(1)$ action. Suppose given such an infinitesimally collared surface Σ and two of its boundary components B_+, B_-. Let us call a *glueing map* for this pair an *anti-isomorphism* $\phi : B_- \to B_+$, that is, a C^1-diffeomorphism with the property that $\phi(ub) = u^{-1}\phi(b)$ for all $b \in B_-$ and $u \in U(1)$. We call it thus, because if we use it to identify B_- with B_+, then we get a new (infinitesimally collared) surface Σ_ϕ without the need of making any further choices: the C^1-structure must be such that the normal vector fields become each others antipode. Similarly, the topological quotient $\check{\Sigma}$ of Σ obtained by contracting each of its boundary components also acquires a C^1-structure: a function on $\check{\Sigma}$ is differentiable precisely when its lift to Σ is C^1 and is such that its derivative evaluated on the infinitesimal collar of a boundary component is the representation of a linear map in polar coordinates.

Definition 7.1. We call a conformal structure on the interior of the infinitesimally collared surface Σ *admissible* if it is compatible with the given C^1-structure as well as with the infinitesimal collaring: for every boundary component either the conformal structure extends to the boundary or extends across its image in $\check{\Sigma}$ and we demand that in the first case the infinitesimal collaring be perpendicular to the boundary, and that in the second (cuspidal) case it maps to a $U(1)$-orbit in the tangent space.

This somewhat unconventional definition is in part motivated by the following observation. A conformal structure on a manifold is just a Riemann metric given up to multiplication by a continuous function. More precisely, it is a section of the bundle of positive quadratic forms modulo positive scalars on the tangent bundle. As the fibers of this bundle have a convex structure, so has its space of sections. This also holds in the present case with the given boundary conditions, in particular the space of admissible conformal structures is contractible. And this is still true if we restrict ourselves to the admissible conformal structures that are cuspidal at a prescribed union of boundary components. This makes it a tractable notion from the point of view of homotopy.

Definition 7.2. A $\mathfrak{g}$-*marking* of a surface Σ consists of giving a map V that assigns to every boundary component of Σ a finite dimensional irreducible represention of $\mathfrak{g}$. We then denote the resulting set of data by (Σ, V). We say that the $\mathfrak{g}$-marking is of level $\leqslant \ell$ if V takes values in representations of level $\leqslant \ell$.

Let (Σ, V) be $\mathfrak{g}$-marked surface. We first suppose Σ endowed with an infinitesimal collaring. Choose an admissible purely cuspidal conformal structure C with respect to this infinitesimal collaring. Then $\check{\Sigma}$ acquires a conformal structure and hence (since $\check{\Sigma}$ is oriented) the structure of a compact Riemann surface, or equivalently, a nonsingular complex projective curve. We hope the reader forgives us for denoting that curve by C as well. It comes with an injection $\pi_0(\partial\Sigma) \to C$. If V takes values in representations of level $\leqslant \ell$, then we have defined the space of covacua $\mathbb{H}_\ell(\mathbb{V})_C$; otherwise we set $\mathbb{H}_\ell(\mathbb{V})_C = 0$. For another choice of purely cuspidal admissible conformal structure C', we can find a path of such structures $(C_t)_{0\leqslant t\leqslant 1}$ connecting C with C'. The projectively flat connection can be used to identify the corresponding projective spaces, and this identification is independent of the choice of path since they belong to the same homotopy class.

In order to lift this to the actual vector spaces, we need a 'rigging' of Σ as follows. Put $g := \dim H_1(\check{\Sigma};\mathbb{R})$ and denote by $\mathcal{L}(\Sigma) \subset \wedge^g H_1(\check{\Sigma};\mathbb{R})$ the set of $I \in \wedge^g H_1(\check{\Sigma};\mathbb{R})$ for which $L(I) := \ker(\wedge I : H_1(\check{\Sigma};\mathbb{R}) \to \wedge^{g+1}H_1(\check{\Sigma};\mathbb{R}))$ is a Lagrangian subspace (so that I is a generator of $\wedge^g L(I)$). Let us first assume that Σ is connected. It is known that if $g > 0$, then $\mathcal{L}(\Sigma)$ is connected, has infinite cyclic fundamental group with a canonical generator and is an orbit of the symplectic group $\mathrm{Sp}(H_1(\check{\Sigma};\mathbb{R}))$. For example, if $g = 1$, then $\mathcal{L}(\Sigma) = H_1(\check{\Sigma};\mathbb{R}) - \{0\} \cong \mathbb{R}^2 - \{0\}$. (If $g = 0$, then $\wedge^g H_1(\check{\Sigma};\mathbb{R}) = \mathbb{R}$ and so $\mathcal{L}(\Sigma)$ is canonically identified with $\mathbb{R} - \{0\}$.) An element of $\mathcal{L}(\Sigma)$ may arise if $\check{\Sigma}$ is given as the boundary of a compact oriented 3-manifold W: then the kernel of $H_1(\check{\Sigma};\mathbb{R}) \to H_1(W;\mathbb{R})$ is a Lagrangian sublattice and so an orientation of it yields an element of $\mathcal{L}(\Sigma)$.

Let $I \in \mathcal{L}(\Sigma)$. We note that every regular differential on C defines by integration a linear map $L(I) \to \mathbb{C}$ and the basic theory or Riemann surfaces tells us that we thus obtain a complex-linear isomorphism $H^0(C, \omega_C) \cong \mathrm{Hom}(L(I), \mathbb{C})$. Since this induces an isomorphism $\det H^0(C, \omega_C) \cong \mathrm{Hom}_{\mathbb{R}}(\det_{\mathbb{R}} L(I), \mathbb{C})$, the linear form that takes the value 1 in $I \in \det_{\mathbb{R}} L(I)$ yields a generator $I(C)$ of $\det H^0(C, \omega_C)$. Likewise the arc $(C_t)_{0\leqslant t\leqslant 1}$ lifts to a section $t \in [0,1] \mapsto I(C_t) \in \det H^0(C_t, \omega_{C_t})$ of the determinant bundle and this in turn yields via Theorem 5.3 an identification of $\mathbb{H}_\ell(\mathbb{V})_C$ with $\mathbb{H}_\ell(\mathbb{V})_{C'}$. As this identification is canonical, we now have attached to the triple (Σ, V, I) and the infinitesimal collaring of Σ a well-defined finite dimensional complex vector space $H_\ell(\Sigma, V, I)$. Actually, the infinitesimal collaring is irrelevant, for the infinitesimal collarings make up an affine space over the vector space of vector fields on $\partial\Sigma$ and hence form a contractible set.

For $g = 0$, we shall always take for $I \in \mathcal{L}(\Sigma)$ the canonical element that corresponds to 1 under the identification $\mathcal{L}(\Sigma) \cong \mathbb{R} - \{0\}$ so that we then have a well-defined vector space $H_\ell(\Sigma, V)$. Proposition 5.4 tells us what we get in some simple cases:

Proposition 7.3. *For Σ a disk (resp. a cylinder), $H_\ell(\Sigma, V)$ is zero unless V is the trivial representation (resp. the two representations attached to the boundary are each other's contra-gradient), in which case it is canonically equal to $\mathbb{C}$.*

Now drop the assumption that Σ be connected and let $\Sigma_1, \dots, \Sigma_r$ enumerate its distinct connected components. If $I \in \mathcal{L}(\Sigma)$ corresponds to $I_1 \otimes_{\mathbb{R}} \cdots \otimes_{\mathbb{R}} I_r$ with $I_k \in \mathcal{L}(\Sigma_k)$, then the tensor product $H_\ell(\Sigma_1, V_1, I_1) \otimes_{\mathbb{C}} \cdots \otimes_{\mathbb{C}} H_\ell(\Sigma_r, V_r, I_r)$ only depends on $I_1 \otimes_{\mathbb{R}} \cdots \otimes_{\mathbb{R}} I_r$ and so the preceding generalizes if we let $H_\ell(\Sigma, V, I)$ be this tensor product. We thus find:

Theorem 7.4. *Let (Σ, V) be a $\mathfrak{g}$-marked surface of level $\leqslant \ell$. Then we have naturally defined on the Lagrangian manifold $\mathcal{L}(\Sigma)$ a local system $\mathbb{H}_\ell(\Sigma, V)$ whose stalk at $I \in \mathcal{L}(\Sigma)$ is $H_\ell(\Sigma, V, I)$. This construction is functorial with respect to automorphisms of $\mathfrak{g}$ so that for every $\sigma \in \mathrm{Aut}(\mathfrak{g})$ we have a natural isomorphism $\mathbb{H}_\ell(\Sigma, {}^\sigma V) \cong \mathbb{H}_\ell(\Sigma, V)$.*

Proof. The last assertion follows from the last clause of Theorem 5.3, the rest is clear from the preceding discussion. □

Remark 7.5. The natural involution of $\mathfrak{g}$ with respect to a choice of root data takes every finite dimensional $\mathfrak{g}$-representation into one equivalent to its contra-gradient. So for such an involution σ we obtain an isomorphism between $H_\ell(\Sigma, V^*, I)$ and $H_\ell(\Sigma, V, I)$, but beware that this involution is only unique up to inner automorphism. However, one expects that there exists a canonical perfect pairing (which therefore does not involve a choice of σ) $H_\ell(\overline{\Sigma}, V^*, I) \otimes H_\ell(\Sigma, V, I) \to \mathbb{C}$, where $\overline{\Sigma}$ stands for Σ with the opposite orientation.

Action of the centrally extended mapping class group

Let $\Gamma(\Sigma)$ denote the group of orientation preserving isotopies of Σ which leave each of its components and each boundary component invariant (but not necessarily point-wise). This is isomorphic to the usual mapping class group of the pair consisting of $\check{\Sigma}$ and the finite subset of $\check{\Sigma}$ that appears as the image of $\pi_0(\partial\Sigma)$. The above lemma shows that if (Σ, V) is a $\mathfrak{g}$-marked surface, then for every $I \in \mathcal{L}(\Sigma)$, the mapping class $\phi \in \Gamma(\Sigma)$ gives rise an isomorphism $\phi_\# : H_\ell(\Sigma, V, I) \to H_\ell(\Sigma, V, \phi_* I)$. In other words, ϕ induces an automorphism of the local system $\mathbb{H}_\ell(\Sigma, V)$.

Assume Σ connected and $g > 0$ so that $\mathcal{L}(\Sigma)$ has infinite cyclic fundamental group. Fix a universal cover $\tilde{\mathcal{L}}(\Sigma) \to \mathcal{L}(\Sigma)$ and denote by $\tilde{H}_\ell(\Sigma, V)$ the space of sections of the pull-back of $\mathbb{H}(\Sigma, V)$ to this cover. The pairs $(\phi, \tilde{\phi}_\#)$ with ϕ a mapping class and $\tilde{\phi}_\# \in \mathrm{Aut}(\tilde{\mathcal{L}}(\Sigma))$ a lift of $\phi_\#$ define a central extension $\tilde{\Gamma}(\Sigma) \to \Gamma(\Sigma)$ of the mapping class group by $\mathbb{Z}$. We have arranged things in such a manner that this extension acts on $\tilde{\mathrm{H}}_\ell(\Sigma, V)$ with the central element $2(\ell + \check{h}) \in \mathbb{Z}$ acting trivially. The central extension is clearly one that already lives on the automorphism group of $H_1(\check{\Sigma})$ (an integral symplectic group of genus g). The latter is known to produce the universal central extension of the symplectic group. It has an abstract description in terms of a 2-cocyle, known as the Maslov index.

The glueing property

Let B_+ and B_- be distinct boundary components of Σ and $\phi : B_+ \to B_-$ a glueing map so that we have an associated quotient surface Σ_ϕ. We show that there is a natural embedding of $\mathcal{L}(\Sigma_\phi)$ in $\mathcal{L}(\Sigma)$. If B_+ and B_- lie on distinct components, then we have a natural identification between the symplectic lattices $H_1(\check{\Sigma})$ and $H_1((\Sigma_\phi)\check{})$ and so we also have a natural identification of $\mathcal{L}(\Sigma)$ with $\mathcal{L}(\Sigma_\phi)$. If B_+ and B_- lie on the same component, then their common image B in Σ_ϕ has the property that the image $\langle B \rangle$ of $H_1(B) \to H_1((\Sigma_\phi)\check{})$ is a primitive rank one sublattice. If we denote by $\langle B \rangle^\perp$ the annihilator of $\langle B \rangle$ with respect to the intersection pairing, then $\langle B \rangle \subset \langle B \rangle^\perp$ and the symplectic lattice $\langle B \rangle^\perp / \langle B \rangle$ is naturally identified with $H_1(\check{\Sigma})$. Since the intersection pairing identifies $H_1((\Sigma_\phi)\check{})/\langle B \rangle^\perp$ with the dual of $\langle B \rangle$, we see that we have a natural embedding of $\mathcal{L}(\Sigma_\phi)$ in $\mathcal{L}(\Sigma)$.

If we now combine the discussion in Remark 6.3 with Theorem 6.6, we obtain

Theorem 7.6 (Glueing property). *Suppose we have endowed Σ_ϕ with a $\mathfrak{g}$-marking, i.e., a map $V : \pi_0(\partial\Sigma_\phi) = \pi_0(\partial\Sigma) - \{\{B_+\}, \{B_-\}\} \to P_\ell$. For $\mu \in P_\ell$, denote by $V_{\mu,\mu^*} : \pi_0(\partial\Sigma) \to P_\ell$ the extension of V to a $\mathfrak{g}$-marking of Σ which assigns to B_+ resp. B_- the value λ resp. λ^*. Then the local system $\mathbb{H}_\ell(\Sigma_\phi, V)$ on $\mathcal{L}(\Sigma_\phi)$ can be naturally identified with the restriction of $\oplus_{\lambda \in P_\ell} \mathbb{H}_\ell(\Sigma, V_{\mu,\mu^*})$ with respect to the embedding of $\mathcal{L}(\Sigma_\phi)$ in $\mathcal{L}(\Sigma)$ defined above. Under this identification, the mapping class of Σ_ϕ obtained by the glueing maps $\{\zeta\phi\}_{\zeta \in U(1)}$ (a Dehn twist) acts on the summand $H_\ell(\Sigma, V_{\mu,\mu^*})$ as scalar multiplication by* $\exp\left(-\frac{\pi\sqrt{-1}c_\mu}{\ell + \check{h}}\right)$.

By repeated application of Theorem 7.6 in the opposite direction we can thus completely recover $\mathbb{H}_\ell(\Sigma, V)$ from a pair of pants decomposition of Σ: such a decomposition has a 3-holed sphere (a pair of pants) as its basic building block, a case that is taken care of by Remark 5.5. In particular we obtain a formula, at least in principle, for its dimension, known as the *Verlinde formula*. This process is nicely formalized by the notion of a fusion ring (see [3]). But if we wish to deal with the modular functor itself, then we are led to the representation theory of quantum groups. As we mentioned in the introduction, this has applications in knot theory via a threedimensional topological quantum field theory. For most of this we refer to the monograph of Turaev [16].

References

[1] J. E. Andersen and K. Ueno. Abelian conformal field theory and determinant bundles. *Internat. J. Math.*, **18**, 919–993, 2007. ← 428

[2] J. E. Andersen and K. Ueno. Geometric construction of modular functors from conformal field theory. *J. Knot Theory Ramifications*, **16**, 127–202, 2007. ← 428

[3] A. Beauville. Conformal blocks, fusion rules and the Verlinde formula, in: Proceedings of the Hirzebruch 65 Conference on Algebraic Geometry, 75–96. *Israel Math. Conf. Proc.*, **9**, Bar-Ilan Univ., Ramat Gan, 1996. ← 448, 454, 464

[4] A. A. Beilinson, Yu. I. Manin and V. V. Schechtman. Sheaves of the Virasoro and Neveu-Schwarz algebras, in: K-theory, arithmetic and geometry (Moscow, 1984–1986), 52–66. *Lecture Notes in Math.*, **1289**, Springer, Berlin, 1987. ← 435

[5] A. Boer and E. Looijenga. On the unitary nature of abelian conformal blocks. *J. Geom. Phys.*, **60**(2), 205–218, 2010. ← 440

[6] N. J. Hitchin. Flat connections and geometric quantization. *Comm. Math. Phys.*, **131**, 347–380, 1990. ← 428

[7] V. G. Kac. *Infinite-dimensional Lie algebras, 3rd ed.* Cambridge University Press, Cambridge ,1990. ← 440, 446, 447, 452, 458

[8] V. G. Kac and A. K. Raina. Bombay lectures on highest weight representations of infinite-dimensional Lie algebras. *Advanced Series in Mathematical Physics*, **2**, World Scientific Publishing Co., Inc., Teaneck, NJ, 1987. ← 432, 440, 442

[9] T. Kohno. Integrable connections related to Manin and Schechtman's higher braid groups, Ill. *J. Math.*, **34**, 476–484, 1990. ← 453

[10] Y. Laszlo. Hitchin's and WZW connections are the same. *J. Differential Geom.*, **49**, 547–576, 1998. ← 429

[11] Y. Laszlo and C. Sorger. The line bundles on the moduli of parabolic G-bundles over curves and their sections. *Ann. Sci. École Norm. Sup. (4)*, **30**, 499–525, 1997. ← 428

[12] E. J. N. Looijenga. Isolated singular points on complete intersections. *LMS Lecture Note Series*, **77**, Cambridge University Press, Cambridge, 1984. ← 450

[13] P. Scheinost and M. Schottenloher. Metaplectic quantization of the moduli spaces of flat and parabolic bundles. *J. Reine Angew. Math.*,**466**, 145– 219, 1995. ← 428

[14] G. Segal. The definition of conformal field theory, in: Topology, geometry and quantum field theory, 421–577. *London Math. Soc. Lecture Note Ser.*,**308**, Cambridge Univ. Press, Cambridge, 2004. ← 429

[15] A. Tsuchiya, K. Ueno, and Y. Yamada. Conformal field theory on universal family of stable curves with gauge symmetries, in: Integrable systems in quantum field theory and statistical mechanics, 459–566. *Adv. Stud. Pure Math.*, **19**, Academic Press, Boston, MA, 1989. ← 428, 448

[16] V. G. Turaev. Quantum invariants of knots and 3-manifolds. *de Gruyter Studies in Mathematics*, **18**, Walter de Gruyter & Co., Berlin, 1994. ← 464

Mathematisch Instituut, Universiteit Utrecht, Postbus 80.010, NL-3508 TA Utrecht, Nederland
E-mail address: E.J.N.Looijenga@uu.nl

Shimura varieties and moduli

J.S. Milne

Abstract. Connected Shimura varieties are the quotients of hermitian symmetric domains by discrete groups defined by congruence conditions. We examine their relation with moduli varieties.

Contents

Introduction

The hermitian symmetric domains are the complex manifolds isomorphic to bounded symmetric domains. The Griffiths period domains are the parameter spaces for polarized rational Hodge structures. A period domain is a hermitian symmetric domain if the universal family of Hodge structures on it is a variation of Hodge structures, i.e., satisfies Griffiths transversality. This rarely happens, but, as Deligne showed, every hermitian symmetric domain can be realized as the subdomain of a period domain on which certain tensors for the universal family are of type (p, p) (i.e., are Hodge tensors).

In particular, every hermitian symmetric domain can be realized as a moduli space for Hodge structures plus tensors. This all takes place in the analytic realm, because hermitian symmetric domains are not algebraic varieties. To obtain an algebraic variety, we must pass to the quotient by an arithmetic group. In fact, in order to obtain a moduli variety, we should assume that the arithmetic group is defined by congruence conditions. The algebraic varieties obtained in this way are the connected Shimura varieties.

The arithmetic subgroup lives in a semisimple algebraic group over $\mathbb{Q}$, and the variations of Hodge structures on the connected Shimura variety are classified in terms of auxiliary reductive algebraic groups. In order to realize the connected Shimura variety as a moduli variety, we must choose the additional data so that the variation of Hodge structures is of geometric origin.

The main result of the article classifies the connected Shimura varieties for which this is known to be possible. Briefly, in a small number of cases, the connected Shimura variety is a moduli variety for abelian varieties with polarization, endomorphism, and level structure (the PEL case); for a much larger class, the variety is a moduli variety for abelian varieties with polarization, Hodge class, and level structure (the PHL case); for all connected Shimura varieties except those of type E_6, E_7, and certain types D, the variety is a moduli variety for abelian *motives* with additional structure. In the remaining cases, the connected Shimura variety is not a moduli variety for abelian motives, and it is not known whether it is a moduli variety at all.

We now summarize the contents of the article.

§1. As an introduction to the general theory, we review the case of elliptic modular curves. In particular, we prove that the modular curve constructed analytically coincides with the modular curve constructed algebraically using geometric invariant theory.

§2. We briefly review the theory of hermitian symmetric domains. To give a hermitian symmetric domain amounts to giving a real semisimple Lie group H with trivial centre and a homomorphism u from the circle group to H satisfying certain conditions. This leads to a classification of hermitian symmetric domains in terms of Dynkin diagrams and special nodes.

§3. The group of holomorphic automorphisms of a hermitian symmetric domain is a real Lie group, and we are interested in quotients of the domain by certain discrete subgroups of this Lie group. In this section we review the fundamental theorems of Borel, Harish-Chandra, Margulis, Mostow, Selberg, Tamagawa, and others concerning discrete subgroups of Lie groups.

§4. The arithmetic locally symmetric varieties (resp. connected Shimura varieties) are the quotients of hermitian symmetric domains by arithmetic (resp. congruence) groups. We explain the fundamental theorems of Baily and Borel on the algebraicity of these varieties and of the maps into them.

§5. We review the definition of Hodge structures and of their variations, and state the fundamental theorem of Griffiths that motivated their definition.

§6. We define the Mumford-Tate group of a rational Hodge structure, and we prove the basic results concerning their behaviour in families.

§7. We review the theory of period domains, and explain Deligne's interpretation of hermitian symmetric domains as period subdomains.

§8. We classify certain variations of Hodge structures on locally symmetric varieties in terms of group-theoretic data.

§9. In order to be able to realize all but a handful of locally symmetric varieties as moduli varieties, we shall need to replace algebraic varieties and algebraic classes by more general objects. In this section, we prove Deligne's theorem that all Hodge classes on abelian varieties are absolutely Hodge, and have algebraic meaning, and we define abelian motives.

§10. Following Satake and Deligne, we classify the symplectic embeddings of an algebraic group that give rise to an embedding of the associated hermitian symmetric domain into a Siegel upper half space.

§11. We use the results of the preceding sections to determine which Shimura varieties can be realized as moduli varieties for abelian varieties (or abelian motives) plus additional structure.

Although the expert will find little that is new in this article, there is much that is not well explained in the literature. As far as possible, complete proofs have been included.

Notations

We use k to denote the base field (always of characteristic zero), and k^{al} to denote an algebraic closure of k. "Algebraic group" means "affine algebraic group scheme" and "algebraic variety" means "geometrically reduced scheme of finite type over a field". For a smooth algebraic variety X over $\mathbb{C}$, we let X^{an} denote the set $X(\mathbb{C})$ endowed with its natural structure of a complex manifold. The tangent space at a point p of space X is denoted by $T_p(X)$.

Vector spaces and representations are finite dimensional unless indicated otherwise. The linear dual of a vector space V is denoted by $V^\vee$. For a k-vector space V and commutative k-algebra R, $V_R = R \otimes_k V$. For a topological space S, we let V_S denote the constant local system of vector spaces on S defined by V. By a lattice in a real vector space, we mean a full lattice, i.e., the $\mathbb{Z}$-module generated by a basis for the vector space.

A *vector sheaf* on a complex manifold (or scheme) S is a locally free sheaf of $\mathcal{O}_S$-modules of finite rank. In order for $\mathcal{W}$ to be a vector subsheaf of a vector sheaf $\mathcal{V}$, we require that the maps on the fibres $\mathcal{W}_s \to \mathcal{V}_s$ be injective. With these definitions, vector sheaves correspond to vector bundles and vector subsheaves to vector subbundles.

The quotient of a Lie group or algebraic group G by its centre $Z(G)$ is denoted by G^{ad}. A Lie group or algebraic group is said to be *adjoint* if it is semisimple (in particular, connected) with trivial centre. An algebraic group is *simple* (resp. *almost simple*) if it connected noncommutative and every proper normal subgroup is trivial (resp. finite). An *isogeny* of algebraic groups is a surjective homomorphism with finite kernel. An algebraic group G is *simply connected* if it is semisimple and every isogeny $G' \to G$ with G' connected is an isomorphism. The inner automorphism of G defined by an element g is denoted by $\text{inn}(g)$. Let $\text{ad}: G \to G^{ad}$ be the quotient map. There is an action of G^{ad} on G such that $\text{ad}(g)$ acts as $\text{inn}(g)$ for all $g \in G(k^{al})$. For an algebraic group G over $\mathbb{R}$, $G(\mathbb{R})^+$ is the identity component of $G(\mathbb{R})$ for the real topology. For a finite extension of fields L/k and an algebraic group G over L, we write $(G)_{L/k}$ for algebraic group over k obtained by (Weil) restriction of scalars. As usual, $\mathbb{G}_m = GL_1$ and μ_N is the kernel of $\mathbb{G}_m \xrightarrow{N} \mathbb{G}_m$.

A *prime* of a number field k is a prime ideal in $\mathcal{O}_k$ (a finite prime), an embedding of k into $\mathbb{R}$ (a real prime), or a conjugate pair of embeddings of k into $\mathbb{C}$ (a complex prime). The ring of finite adèles of $\mathbb{Q}$ is $\mathbb{A}_f = \mathbb{Q} \otimes \left(\prod_p \mathbb{Z}_p\right)$.

We use ι or $z \mapsto \bar{z}$ to denote complex conjugation on $\mathbb{C}$ or on a subfield of $\mathbb{C}$, and we use $X \simeq Y$ to mean that X and Y isomorphic with a specific isomorphism — which isomorphism should always be clear from the context.

For algebraic groups we use the language of modern algebraic geometry, not the more usual language, which is based on Weil's Foundations. For example, if G and G' are algebraic groups over a field k, then by a homomorphism $G \to G'$ we mean a homomorphism defined over k, not over some universal domain. Similarly, a simple algebraic group over a field k need not be geometrically (i.e., absolutely) simple.

1. Elliptic modular curves

The first Shimura varieties, and the first moduli varieties, were the elliptic modular curves. In this section, we review the theory of elliptic modular curves as an introduction to the general theory.

Definition of elliptic modular curves

Let D be the complex upper half plane,

$$D = \{z \in \mathbb{C} \mid \Im(z) > 0\}.$$

The group $SL_2(\mathbb{R})$ acts transitively on D by the rule

$$\begin{pmatrix} a & b \\ c & d \end{pmatrix} z = \frac{az+b}{cz+d}.$$

A subgroup Γ of $SL_2(\mathbb{Z})$ is a congruence subgroup if, for some integer $N \geqslant 1$, Γ contains the principal congruence subgroup of level N,

$$\Gamma(N) \stackrel{\text{def}}{=} \{A \in SL_2(\mathbb{Z}) \mid A \equiv I \bmod N\}.$$

An elliptic modular curve is the quotient $\Gamma\backslash D$ of D by a congruence group Γ. Initially this is a one-dimensional complex manifold, but it can be compactified by adding a finite number of "cusps", and so it has a unique structure of an algebraic curve compatible with its structure as a complex manifold.[1] This curve can be realized as a moduli variety for elliptic curves with level structure, from which it is possible deduce many beautiful properties of the curve, for example, that it has a canonical model over a specific number field, and that the coordinates of the special points on the model generate class fields.

Elliptic modular curves as moduli varieties

For an elliptic curve E over $\mathbb{C}$, the exponential map defines an exact sequence

$$0 \to \Lambda \to T_0(E^{an}) \xrightarrow{\exp} E^{an} \to 0 \tag{1.1}$$

with

$$\Lambda \simeq \pi_1(E^{an}, 0) \simeq H_1(E^{an}, \mathbb{Z}).$$

The functor $E \rightsquigarrow (T_0E, \Lambda)$ is an equivalence from the category of complex elliptic curves to the category of pairs consisting of a one-dimensional $\mathbb{C}$-vector space and a lattice. Thus, to give an elliptic curve over $\mathbb{C}$ amounts to giving a two-dimensional $\mathbb{R}$-vector space V, a complex structure on V, and a lattice in V. It is known that D parametrizes elliptic curves plus additional data. Traditionally, to a point τ of D one attaches the quotient of $\mathbb{C}$ by the lattice spanned by 1 and τ. In other words, one fixes the real vector space and the complex structure, and varies the lattice. From the point of view of period domains and Shimura varieties, it is more natural to fix the real vector space and the lattice, and vary the complex structure.[2]

Thus, let V be a two-dimensional vector space over $\mathbb{R}$. A complex structure on V is an endomorphism J of V such that $J^2 = -1$. From such a J, we get a decomposition $V_{\mathbb{C}} = V_J^+ \oplus V_J^-$ of $V_{\mathbb{C}}$ into its $+i$ and $-i$ eigenspaces, and the isomorphism $V \to V_{\mathbb{C}}/V_J^-$ carries the complex structure J on V to the natural complex structure on $V_{\mathbb{C}}/V_J^-$. The map $J \mapsto V_{\mathbb{C}}/V_J^-$ identifies the set of complex structures on V with the

[1]We are using that the functor $S \rightsquigarrow S^{an}$ from smooth algebraic varieties over $\mathbb{C}$ to complex manifolds defines an equivalence from the category of *complete* smooth algebraic curves to that of *compact* Riemann surfaces.

[2]The choice of a trivialization of a variation of integral Hodge structures attaches to each point of the underlying space a fixed real vector space and lattice, but a varying Hodge structure — see below.

set of nonreal one-dimensional quotients of $V_{\mathbb{C}}$, i.e., with $\mathbb{P}(V_{\mathbb{C}}) \smallsetminus \mathbb{P}(V)$. This space has two connected components.

Now choose a basis for V, and identify it with $\mathbb{R}^2$. Let $\psi\colon V \times V \to \mathbb{R}$ be the alternating form

$$\psi(\left(\begin{smallmatrix} a \\ b \end{smallmatrix}\right), \left(\begin{smallmatrix} c \\ d \end{smallmatrix}\right)) = \det\left(\begin{smallmatrix} a & c \\ b & d \end{smallmatrix}\right) = ad - bc.$$

On one of the connected components, which we denote D, the symmetric bilinear form

$$(x, y) \mapsto \psi_J(x, y) \overset{\text{def}}{=} \psi(x, Jy)\colon V \times V \to \mathbb{R}$$

is positive definite and on the other it is negative definite. Thus D is the set of complex structures on V for which $+\psi$ (rather than $-\psi$) is a Riemann form. Our choice of a basis for V identifies $\mathbb{P}(V_{\mathbb{C}}) \smallsetminus \mathbb{P}(V)$ with $\mathbb{P}^1(\mathbb{C}) \smallsetminus \mathbb{P}^1(\mathbb{R})$ and D with the complex upper half plane.

Now let Λ be the lattice $\mathbb{Z}^2$ in V. For each $J \in D$, the quotient $(V, J)/\Lambda$ is an elliptic curve E with $H_1(E^{an}, \mathbb{Z}) \simeq \Lambda$. In this way, we obtain a one-to-one correspondence between the points of D and the isomorphism classes of pairs consisting of an elliptic curve E over $\mathbb{C}$ and an ordered basis for $H_1(E^{an}, \mathbb{Z})$.

Let E_N denote the kernel of multiplication by N on an elliptic curve E. Thus, for the curve $E = (V, J)/\Lambda$,

$$E_N(\mathbb{C}) = \tfrac{1}{N}\Lambda/\Lambda \simeq \Lambda/N\Lambda \approx (\mathbb{Z}/N\mathbb{Z})^2.$$

A level-N structure on E is a pair of points $\eta = (t_1, t_2)$ in $E(\mathbb{C})$ that forms an ordered basis for $E_N(\mathbb{C})$.

For an elliptic curve E over any field, there is an algebraically defined (Weil) pairing

$$e_N\colon E_N \times E_N \to \mu_N.$$

When the ground field is $\mathbb{C}$, this induces an isomorphism $\bigwedge^2 (E_N(\mathbb{C})) \simeq \mu_N(\mathbb{C})$. In the following, we fix a primitive Nth root ζ of 1 in $\mathbb{C}$, and we require that our level-N structures satisfy the condition $e_N(t_1, t_2) = \zeta$.

Identify $\Gamma(N)$ with the subgroup of SL(V) whose elements preserve Λ and act as the identity on $\Lambda/N\Lambda$. On passing to the quotient by $\Gamma(N)$, we obtain a one-to-one correspondence between the points of $\Gamma(N)\backslash D$ and the isomorphism classes of pairs consisting of an elliptic curve E over $\mathbb{C}$ and a level-N structure η on E. Let Y_N denote the algebraic curve over $\mathbb{C}$ with $Y_N^{an} = \Gamma(N)\backslash D$.

Let $f\colon E \to S$ be a family of elliptic curves over a scheme S, i.e., a flat map of schemes together with a section whose fibres are elliptic curves. A level-N structure on E/S is an ordered pair of sections to f that give a level-N structure on E_s for each closed point s of S.

Proposition 1.2. *Let* $f\colon E \to S$ *be a family of elliptic curves on a smooth algebraic curve* S *over* $\mathbb{C}$*, and let* η *be a level-*N *structure on* E/S*. The map* $\gamma\colon S(\mathbb{C}) \to Y_N(\mathbb{C})$ *sending*

$s \in S(\mathbb{C})$ *to the point of* $\Gamma(N)\backslash D$ *corresponding to* (E_s, η_s) *is regular, i.e., defined by a morphism of algebraic curves.*

Proof. We first show that γ is holomorphic. For this, we use that $\mathbb{P}(V_\mathbb{C})$ is the Grassmann manifold classifying the one-dimensional quotients of $V_\mathbb{C}$. This means that, for any complex manifold M and surjective homomorphism $\alpha\colon \mathcal{O}_M \otimes_\mathbb{R} V \to \mathcal{W}$ of vector sheaves on M with $\mathcal{W}$ of rank 1, the map sending $m \in M$ to the point of $\mathbb{P}(V_\mathbb{C})$ corresponding to the quotient $\alpha_m\colon V_\mathbb{C} \to \mathcal{W}_m$ of $V_\mathbb{C}$ is holomorphic.

Let $f\colon E \to S$ be a family of elliptic curves on a connected smooth algebraic variety S. The exponential map defines an exact sequence of sheaves on S^{an}

$$0 \longrightarrow R_1 f_* \mathbb{Z} \longrightarrow \mathcal{T}_0(E^{an}/S^{an}) \longrightarrow E^{an} \longrightarrow 0$$

whose fibre at a point $s \in S^{an}$ is the sequence (1.1) for E_s. From the first map in the sequence we get a surjective map

$$\mathcal{O}_{S^{an}} \otimes_\mathbb{Z} R_1 f_* \mathbb{Z} \twoheadrightarrow \mathcal{T}_0(E^{an}/S^{an}). \tag{1.3}$$

Let (t_1, t_2) be a level-N structure on E/S. Each point of S^{an} has an open neighbourhood U such that $t_1|_U$and $t_2|_U$ lift to sections $\tilde{t}_1$ and $\tilde{t}_2$ of $\mathcal{T}_0(E^{an}/S^{an})$ over U; now $N\tilde{t}_1$ and $N\tilde{t}_2$ are sections of $R_1 f_* \mathbb{Z}$ over U, and they define an isomorphism

$$\mathbb{Z}^2_U \to R_1 f_* \mathbb{Z}|_U.$$

On tensoring this with $\mathcal{O}_{U^{an}}$,

$$\mathcal{O}_{U^{an}} \otimes_\mathbb{Z} \mathbb{Z}^2_U \to \mathcal{O}_{U^{an}} \otimes R_1 f_* \mathbb{Z}|_U$$

and composing with (1.3), we get a surjective map

$$\mathcal{O}_{U^{an}} \otimes_\mathbb{R} V \twoheadrightarrow \mathcal{T}_0(E^{an}/S^{an})|U$$

of vector sheaves on U, which defines a holomorphic map $U \to \mathbb{P}(V_\mathbb{C})$. This maps into D, and its composite with the quotient map $D \to \Gamma(N)\backslash D$ is the map γ. Therefore γ is holomorphic.

It remains to show that γ is algebraic. We now assume that S is a curve. After passing to a finite covering, we may suSppose that N is even. Let $\bar{Y}_N$ (resp. $\bar{S}$) be the completion of Y_N (resp. S) to a smooth complete algebraic curve. We have a holomorphic map

$$S^{an} \xrightarrow{\gamma} Y_N^{an} \subset \bar{Y}_N^{an};$$

to show that it is regular, it suffices to show that it extends to a holomorphic map of compact Riemann surfaces $\bar{S}^{an} \to \bar{Y}_N^{an}$. The curve Y_2 is isomorphic to $\mathbb{P}^1 \smallsetminus \{0,1,\infty\}$. The composed map

$$S^{an} \xrightarrow{\gamma} Y_N^{an} \xrightarrow{\text{onto}} Y_2^{an} \approx \mathbb{P}^1(\mathbb{C}) \smallsetminus \{0,1,\infty\}$$

does not have an essential singularity at any of the (finitely many) points of $\bar{S}^{an} \setminus S^{an}$ because this would violate the big Picard theorem.[3] Therefore, it extends to a holomorphic map $\bar{S}^{an} \to \mathbb{P}^1(\mathbb{C})$, which implies that γ extends to a holomorphic map $\bar{\gamma}\colon \bar{S}^{an} \to \bar{Y}_N^{an}$, as required. □

Let $\mathcal{F}$ be the functor sending a scheme S of finite type over $\mathbb{C}$ to the set of isomorphism classes of pairs consisting of a family elliptic curves $f\colon E \to S$ over S and a level-N structure η on E. According to Mumford [44], Chapter 7, the functor $\mathcal{F}$ is representable when $N \geqslant 3$. More precisely, when $N \geqslant 3$ there exists a smooth algebraic curve S_N over $\mathbb{C}$ and a family of elliptic curves over S_N endowed with a level N structure that is universal in the sense that any similar pair on a scheme S is isomorphic to the pullback of the universal pair by a unique morphism $\alpha\colon S \to S_N$.

Theorem 1.4. *There is a canonical isomorphism* $\gamma\colon S_N \to Y_N$.

Proof. According to Proposition 1.2, the universal family of elliptic curves with level-N structure on S_N defines a morphism of smooth algebraic curves $\gamma\colon S_N \to Y_N$. Both sets $S_N(\mathbb{C})$ and $Y_N(\mathbb{C})$ are in natural one-to-one correspondence with the set of isomorphism classes of complex elliptic curves with level-N structure, and γ sends the point in $S_N(\mathbb{C})$ corresponding to a pair (E, η) to the point in $Y_N(\mathbb{C})$ corresponding to the same pair. Therefore, $\gamma(\mathbb{C})$ is bijective, which implies that γ is an isomorphism. □

In particular, we have shown that the curve S_N, constructed by Mumford purely in terms of algebraic geometry, is isomorphic by the obvious map to the curve Y_N, constructed analytically. Of course, this is well known, but it is difficult to find a proof of it in the literature. For example, Brian Conrad has noted that it is used without reference in [30].

Theorem 1.4 says that there exists a single algebraic curve over $\mathbb{C}$ enjoying the good properties of both S_N and Y_N.

2. Hermitian symmetric domains

The natural generalization of the complex upper half plane is a hermitian symmetric domain.

Preliminaries on Cartan involutions and polarizations

Let G be a connected algebraic group over $\mathbb{R}$, and let $\sigma_0\colon g \mapsto \bar{g}$ denote complex conjugation on $G_{\mathbb{C}}$ with respect to G. A *Cartan involution* of G is an involution θ of G (as an algebraic group over $\mathbb{R}$) such that the group

$$G^{(\theta)}(\mathbb{R}) = \{g \in G(\mathbb{C}) \mid g = \theta(\bar{g})\}$$

[3]Recall that this says that a holomorphic function on the punctured disk with an essential singularity at 0 omits at most one value in $\mathbb{C}$. Therefore a function on the punctured disk that omits two values has (at worst) a pole at 0, and so extends to a function from the whole disk to $\mathbb{P}^1(\mathbb{C})$.

is compact. Then $G^{(\theta)}$ is a compact real form of $G_{\mathbb{C}}$, and θ acts on $G(\mathbb{C})$ as $\sigma_0\sigma = \sigma\sigma_0$ where σ denotes complex conjugation on $G_{\mathbb{C}}$ with respect to $G^{(\theta)}$.

Consider, for example, the algebraic group GL_V attached to a real vector space V. The choice of a basis for V determines a transpose operator $g \mapsto g^t$, and $\theta: g \mapsto (g^t)^{-1}$ is a Cartan involution of GL_V because $GL_V^{(\theta)}(\mathbb{R})$ is the unitary group. The basis determines an isomorphism $GL_V \simeq GL_n$, and $\sigma_0(A) = \bar{A}$ and $\sigma(A) = (\bar{A}^t)^{-1}$ for $A \in GL_n(\mathbb{C})$.

A connected algebraic group G has a Cartan involution if and only if it has a compact real form, which is the case if and only if G is reductive. Any two Cartan involutions of G are conjugate by an element of $G(\mathbb{R})$. In particular, all Cartan involutions of GL_V arise, as in the last paragraph, from the choice of a basis for V. An algebraic subgroup G of GL_V is reductive if and only if it is stable under $g \mapsto g^t$ for some basis of V, in which case the restriction of $g \mapsto (g^t)^{-1}$ to G is a Cartan involution. Every Cartan involution of G is of this form. See [53], I, §4.

Let C be an element of $G(\mathbb{R})$ whose square is central (so $\mathrm{inn}(C)$ is an involution). A *C-polarization* on a real representation V of G is a G-invariant bilinear form $\varphi: V \times V \to \mathbb{R}$ such that the form $\varphi_C: (x, y) \mapsto \varphi(x, Cy)$ is symmetric and positive definite.

Theorem 2.1. *If* $\mathrm{inn}(C)$ *is a Cartan involution of* G, *then every finite dimensional real representation of* G *carries a* C*-polarization; conversely, if one* faithful *finite dimensional real representation of* G *carries a* C*-polarization, then* $\mathrm{inn}(C)$ *is a Cartan involution.*

Proof. An $\mathbb{R}$-bilinear form φ on a real vector space V defines a sesquilinear form $\varphi': (u, v) \mapsto \varphi_{\mathbb{C}}(u, \bar{v})$ on $V(\mathbb{C})$, and φ' is hermitian (and positive definite) if and only if φ is symmetric (and positive definite).

Let $G \to GL_V$ be a representation of G. If $\mathrm{inn}(C)$ is a Cartan involution of G, then $G^{(\mathrm{inn}\, C)}(\mathbb{R})$ is compact, and so there exists a $G^{(\mathrm{inn}\, C)}$-invariant positive definite symmetric bilinear form φ on V. Then $\varphi_{\mathbb{C}}$ is $G(\mathbb{C})$-invariant, and so

$$\varphi'(gu, (\sigma g)v) = \varphi'(u, v), \quad \text{for all } g \in G(\mathbb{C}),\ u, v \in V_{\mathbb{C}},$$

where σ is the complex conjugation on $G_{\mathbb{C}}$ with respect to $G^{(\mathrm{inn}\, C)}$. Now $\sigma g = \mathrm{inn}(C)(\bar{g}) = \mathrm{inn}(C^{-1})(\bar{g})$, and so, on replacing v with $C^{-1}v$ in the equality, we find that

$$\varphi'(gu, (C^{-1}\bar{g}C)C^{-1}v) = \varphi'(u, C^{-1}v), \quad \text{for all } g \in G(\mathbb{C}),\ u, v \in V_{\mathbb{C}}.$$

In particular, $\varphi(gu, C^{-1}gv) = \varphi(u, C^{-1}v)$ when $g \in G(\mathbb{R})$ and $u, v \in V$. Therefore, $\varphi_{C^{-1}}$ is G-invariant. As $(\varphi_{C^{-1}})_C = \varphi$, we see that φ is a C-polarization.

For the converse, one shows that, if φ is a C-polarization on a faithful representation, then φ_C is invariant under $G^{(\mathrm{inn}\, C)}(\mathbb{R})$, which is therefore compact. □

2.2. Variant. Let G be an algebraic group over $\mathbb{Q}$, and let C be an element of $G(\mathbb{R})$ whose square is central. A *C-polarization* on a $\mathbb{Q}$-representation V of G is a G-invariant bilinear form $\varphi: V \times V \to \mathbb{Q}$ such that $\varphi_{\mathbb{R}}$ is a C-polarization on $V_{\mathbb{R}}$. In order to show that a $\mathbb{Q}$-representation V of G is polarizable, it suffices to check that $V_{\mathbb{R}}$ is polarizable. We prove this when C^2 acts as +1 or −1 on V, which are the only cases we shall need. Let $P(\mathbb{Q})$ (resp. $P(\mathbb{R})$) denote the space of G-invariant bilinear forms on V (resp. on $V_{\mathbb{R}}$) that are symmetric when C^2 acts as +1 or skew-symmetric when it acts as −1. Then $P(\mathbb{R}) = \mathbb{R} \otimes_{\mathbb{Q}} P(\mathbb{Q})$. The C-polarizations of $V_{\mathbb{R}}$ form an open subset of $P(\mathbb{R})$, whose intersection with $P(\mathbb{Q})$ consists of the C-polarizations of V.

Definition of hermitian symmetric domains

Let M be a complex manifold, and let $J_p \colon T_pM \to T_pM$ denote the action of $i = \sqrt{-1}$ on the tangent space at a point p of M. A *hermitian metric* on M is a riemannian metric g on the underlying smooth manifold of M such that J_p is an isometry for all p.[4] A *hermitian manifold* is a complex manifold equipped with a hermitian metric g, and a *hermitian symmetric space* is a connected hermitian manifold M that admits a symmetry at each point p, i.e., an involution s_p having p as an isolated fixed point. The group Hol(M) of holomorphic automorphisms of a hermitian symmetric space M is a real Lie group whose identity component $\mathrm{Hol}(M)^+$ acts transitively on M.

Every hermitian symmetric space M is a product of hermitian symmetric spaces of the following types:

- Noncompact type — the curvature is negative[5] and $\mathrm{Hol}(M)^+$ is a noncompact adjoint Lie group; example, the complex upper half plane.
- Compact type — the curvature is positive and $\mathrm{Hol}(M)^+$ is a compact adjoint Lie group; example, the Riemann sphere.
- Euclidean type — the curvature is zero; M is isomorphic to a quotient of a space $\mathbb{C}^n$ by a discrete group of translations.

In the first two cases, the space is simply connected. A hermitian symmetric space is *indecomposable* if it is not a product of two hermitian symmetric spaces of lower dimension. For an indecomposable hermitian symmetric space M of compact or noncompact type, the Lie group $\mathrm{Hol}(M)^+$ is simple. See [27], Chapter VIII.

A *hermitian symmetric domain* is a connected complex manifold that admits a hermitian metric for which it is a hermitian symmetric space of noncompact type.[6] The hermitian symmetric domains are exactly the complex manifolds isomorphic to bounded symmetric domains (via the Harish-Chandra embedding; [53], II §4). Thus a connected complex manifold M is a hermitian symmetric domain if and only if

---

[4]Then g_p is the real part of a unique hermitian form on the complex vector space T_pM, which explains the name.

[5]This means that the sectional curvature $K(p, E)$ is < 0 for every $p \in M$ and every two-dimensional subspace E of T_pM.

[6]Usually a hermitian symmetric domain is defined to be a complex manifold *equipped* with a hermitian metric etc.. However, a hermitian symmetric domain in our sense satisfies conditions (A.1) and (A.2) of [31], and so has a canonical Bergman metric, invariant under all holomorphic automorphisms.

(a) it is isomorphic to a bounded open subset of $\mathbb{C}^n$ for some n, and
(b) for each point p of M, there exists a holomorphic involution of M (the *symmetry* at p) having p as an isolated fixed point.

For example, the bounded domain $\{z \in \mathbb{C} \mid |z| < 1\}$ is a hermitian symmetric domain because it is homogeneous and admits a symmetry at the origin ($z \mapsto -1/z$). The map $z \mapsto \frac{z-i}{z+i}$ is an isomorphism from the complex upper half plane D onto the open unit disk, and so D is also a hermitian symmetric domain. Its automorphism group is

$$\mathrm{Hol}(D) \simeq \mathrm{SL}_2(\mathbb{R})/\{\pm I\} \simeq \mathrm{PGL}_2(\mathbb{R})^+.$$

Classification in terms of real groups

2.3. Let U^1 be the circle group, $U^1 = \{z \in \mathbb{C} \mid |z| = 1\}$. For each point o of a hermitian symmetric domain D, there is a unique homomorphism $u_o\colon U^1 \to \mathrm{Hol}(D)$ such that $u_o(z)$ fixes o and acts on T_oD as multiplication by z ($z \in U^1$).[7] In particular, $u_o(-1)$ is the symmetry at o.

Example 2.4. Let D be the complex upper half plane and let $o = i$. Let $h\colon U^1 \to \mathrm{SL}_2(\mathbb{R})$ be the homomorphism $a + bi \mapsto \left(\begin{smallmatrix} a & b \\ -b & a \end{smallmatrix}\right)$. Then $h(z)$ fixes o, and it acts as z^2 on $T_o(D)$. For $z \in U^1$, choose a square root $\sqrt{z}$ in U^1, and let $u_o(z) = h(\sqrt{z})$ mod $\pm I$. Then $u_o(z)$ is independent of the choice of $\sqrt{z}$ because $h(-1) = -I$. The homomorphism $u_o\colon U^1 \to \mathrm{SL}_2(\mathbb{Z})/\{\pm I\} = \mathrm{Hol}(D)$ has the correct properties.

Now let D be a hermitian symmetric domain. Because $\mathrm{Hol}(D)$ is an adjoint Lie group, there is a unique real algebraic group H such that $H(\mathbb{R})^+ = \mathrm{Hol}(D)^+$. Similarly, U^1 is the group of $\mathbb{R}$-points of the algebraic torus $\mathbb{S}^1$ defined by the equation $X^2 + Y^2 = 1$. A point $o \in D$ defines a homomorphism $u\colon \mathbb{S}^1 \to H$ of real algebraic groups.

Theorem 2.5. *The homomorphism $u\colon \mathbb{S}^1 \to H$ has the following properties:*

SU1: *only the characters $z, 1, z^{-1}$ occur in the representation of $\mathbb{S}^1$ on $\mathrm{Lie}(H)_{\mathbb{C}}$ defined by u;*[8]

SU2: $\mathrm{inn}(u(-1))$ *is a Cartan involution.*

Conversely, if H is a real adjoint algebraic group with no compact factor and $u\colon \mathbb{S}^1 \to H$ satisfies the conditions (SU1,2), then the set D of conjugates of u by elements of $H(\mathbb{R})^+$ has a natural structure of a hermitian symmetric domain for which $u(z)$ acts on T_uD as multiplication by z; moreover, $H(\mathbb{R})^+ = \mathrm{Hol}(D)^+$.

Proof. The proof is sketched in [40], 1.21; see also [53], II, Proposition 3.2 □

---

[7] See, for example, [40], Theorem 1.9.

[8] The maps $\mathbb{S}^1 \xrightarrow{u} H_{\mathbb{R}} \xrightarrow{\mathrm{Ad}} \mathrm{Aut}(\mathrm{Lie}(H))$ define an action of $\mathbb{S}^1$ on $\mathrm{Lie}(H)$, and hence on $\mathrm{Lie}(H)_{\mathbb{C}}$. The condition means that $\mathrm{Lie}(H)_{\mathbb{C}}$ is a direct sum of subspaces on which $u(z)$ acts as z, 1, or z^{-1}.

Thus, the pointed hermitian symmetric domains are classified by the pairs (H, u) as in the theorem. Changing the point corresponds to conjugating u by an element of $H(\mathbb{R})$.

Classification in terms of root systems

We now assume that the reader is familiar with the classification of semisimple algebraic groups over an algebraically closed field in terms of root systems (e.g., [29]).

Let D be an indecomposable hermitian symmetric domain. Then the corresponding group H is simple, and $H_{\mathbb{C}}$ is also simple because H is an inner form of its compact form (by SU2).[9] Thus, from D and a point o, we get a simple algebraic group $H_{\mathbb{C}}$ over $\mathbb{C}$ and a nontrivial cocharacter $\mu \stackrel{\text{def}}{=} u_{\mathbb{C}}\colon \mathbb{G}_m \to H_{\mathbb{C}}$ satisfying the condition:

(*) $\qquad \mathbb{G}_m$ acts on $\mathrm{Lie}(H_{\mathbb{C}})$ through the characters $z,\ 1,\ z^{-1}$.

Changing o replaces μ by a conjugate. Thus the next step is to classify the pairs (G, M) consisting of a simple algebraic group over $\mathbb{C}$ and a conjugacy class of nontrivial cocharacters of G satisfying (*).

Fix a maximal torus T of G and a base S for the root system $R = R(G, T)$, and let R^+ be the corresponding set of positive roots. As each μ in M factors through some maximal torus, and all maximal tori are conjugate, we may choose $\mu \in M$ to factor through T. Among the μ in M factoring through T, there is exactly one such that $\langle \alpha, \mu \rangle \geqslant 0$ for all $\alpha \in R^+$ (because the Weyl group acts simply transitively on the Weyl chambers). The condition (*) says that $\langle \alpha, \mu \rangle \in \{1, 0, -1\}$ for all roots α. Since μ is nontrivial, not all of the $\langle \alpha, \mu \rangle$ can be zero, and so $\langle \tilde{\alpha}, \mu \rangle = 1$ where $\tilde{\alpha}$ is the highest root. Recall that the highest root $\tilde{\alpha} = \sum_{\alpha \in S} n_\alpha \alpha$ has the property that $n_\alpha \geqslant m_\alpha$ for any other root $\sum_{\alpha \in S} m_\alpha \alpha$; in particular, $n_\alpha \geqslant 1$. It follows that $\langle \alpha, \mu \rangle = 0$ for all but one simple root α, and that for that simple root $\langle \alpha, \mu \rangle = 1$ and $n_\alpha = 1$. Thus, the pairs (G, M) are classified by the simple roots α for which $n_\alpha = 1$ — these are called the *special* simple roots. On examining the tables, one finds that the special simple roots are as in the following table:

type	$\tilde{\alpha}$	special roots	#
A_n	$\alpha_1 + \alpha_2 + \cdots + \alpha_n$	$\alpha_1, \ldots, \alpha_n$	n
B_n	$\alpha_1 + 2\alpha_2 + \cdots + 2\alpha_n$	α_1	1
C_n	$2\alpha_1 + \cdots + 2\alpha_{n-1} + \alpha_n$	α_n	1
D_n	$\alpha_1 + 2\alpha_2 + \cdots + 2\alpha_{n-2} + \alpha_{n-1} + \alpha_n$	$\alpha_1, \alpha_{n-1}, \alpha_n$	3
E_6	$\alpha_1 + 2\alpha_2 + 2\alpha_3 + 3\alpha_4 + 2\alpha_5 + \alpha_6$	α_1, α_6	2
E_7	$2\alpha_1 + 2\alpha_2 + 3\alpha_3 + 4\alpha_4 + 3\alpha_5 + 2\alpha_6 + \alpha_7$	α_7	1
E_8, F_4, G_2		none	0

[9] If $H_{\mathbb{C}}$ is not simple, say, $H_{\mathbb{C}} = H_1 \times H_2$, then $H = (H_1)_{\mathbb{C}/\mathbb{R}}$, and every inner form of H is isomorphic to H itself (by Shapiro's lemma), which is not compact because $H(\mathbb{R}) = H_1(\mathbb{C})$.

Mnemonic: the number of special simple roots is one less than the connection index $(P(R)\colon Q(R))$ of the root system.[10]

To every indecomposable hermitian symmetric domain we have attached a special node, and we next show that every special node arises from a hermitian symmetric domain. Let G be a simple algebraic group over $\mathbb{C}$ with a character μ satisfying (*). Let U be the (unique) compact real form of G, and let σ be the complex conjugation on G with respect to U. Finally, let H be the real form of G such that $\mathrm{inn}(\mu(-1))\circ\sigma$ is the complex conjugation on G with respect to H. The restriction of μ to $U^1\subset\mathbb{C}^\times$ maps into $H(\mathbb{R})$ and defines a homomorphism u satisfying the conditions (SU1,2) of (2.5). The hermitian symmetric domain corresponding to (H,u) gives rise to (G,μ). Thus there are indecomposable hermitian symmetric domains of all possible types except E_8, F_4, and G_2.

Let H be a real simple group such that there exists a homomorphism $u\colon\mathbb{S}^1\to H$ satisfying (SV1,2). The set of such u's has two connected components, interchanged by $u\leftrightarrow u^{-1}$, each of which is an $H(\mathbb{R})^+$-conjugacy class. The u's form a single $H(\mathbb{R})$-conjugacy class except when s is moved by the opposition involution ([19], 1.2.7, 1.2.8). This happens in the following cases: type A_n and $s\neq\frac{n}{2}$; type D_n with n odd and $s=\alpha_{n-1}$ or α_n; type E_6 (see p. 527 below).

Example: the Siegel upper half space

A *symplectic space* (V,ψ) over a field k is a finite dimensional vector space V over k together with a nondegenerate alternating form ψ on V. The *symplectic group* $S(\psi)$ is the algebraic subgroup of GL_V of elements fixing ψ. It is an almost simple simply connected group of type C_{n-1} where $n=\frac{1}{2}\dim_k V$.

Now let $k=\mathbb{R}$, and let $H=S(\psi)$. Let D be the space of complex structures J on V such that $(x,y)\mapsto\psi_J(x,y)\overset{\text{def}}{=}\psi(x,Jy)$ is symmetric and positive definite. The symmetry is equivalent to J lying in $S(\psi)$. Therefore, D is the set of complex structures J on V for which $J\in H(\mathbb{R})$ and ψ is a J-polarization for H.

The action,

$$g,J\mapsto gJg^{-1}\colon H(\mathbb{R})\times D\to D,$$

of $H(\mathbb{R})$ on D is transitive ([40], §6). Each $J\in D$ defines an action of $\mathbb{C}$ on V, and

$$(2.6)\quad \psi(Jx,Jy)=\psi(x,y)\text{ all }x,y\in V\implies\psi(zx,zy)=|z|^2\psi(x,y)\text{ all }x,y\in V.$$

Let $h_J\colon\mathbb{S}\to GL_V$ be the homomorphism such that $h_J(z)$ acts on V as multiplication by z, and let $V_\mathbb{C}=V^+\oplus V^-$ be the decomposition of $V_\mathbb{C}$ into its $\pm i$ eigenspaces for J. Then $h_J(z)$ acts on V^+ as z and on V^- as $\bar{z}$, and so it acts on

$$\mathrm{Lie}(H)_\mathbb{C}\subset\mathrm{End}(V)_\mathbb{C}\simeq V_\mathbb{C}^\vee\otimes V_\mathbb{C}=(V^+\oplus V^-)^\vee\otimes(V^+\oplus V^-),$$

through the characters $z^{-1}\bar{z}$, 1, $z\bar{z}^{-1}$.

[10] It is possible to prove this directly. Let $S^+=S\cup\{\alpha_0\}$ where α_0 is the negative of the highest root — the elements of S^+ correspond to the nodes of the completed Dynkin diagram ([7], VI 4, 3). The group P/Q acts on S^+, and it acts simply transitively on the set {simple roots} $\cup\{\alpha_0\}$ ([19], 1.2.5).

For $z \in U^1$, (2.6) shows that $h_J(z) \in H$; choose a square root $\sqrt{z}$ of z in U^1, and let $u_J(z) = h_J(\sqrt{z}) \bmod \pm 1$. Then u_J is a well-defined homomorphism $U^1 \to H^{ad}(\mathbb{R})$, and it satisfies the conditions (SU1,2) of Theorem 2.5. Therefore, D has a natural complex structure for which $z \in U^1$ acts on $T_J(D)$ as multiplication by z and $\mathrm{Hol}(D)^+ = H^{ad}(\mathbb{R})^+$. With this structure, D is the (unique) indecomposable hermitian symmetric domain of type C_{n-1}. It is called the *Siegel upper half space* (of degree, or genus, n).

3. Discrete subgroups of Lie groups

The algebraic varieties we are concerned with are quotients of hermitian symmetric domains by the action of discrete groups. In this section, we describe the discrete groups of interest to us.

Lattices in Lie groups

Let H be a connected real Lie group. A *lattice* in H is a discrete subgroup Γ of finite covolume, i.e., such that H/Γ has finite volume with respect to an H-invariant measure. For example, the lattices in $\mathbb{R}^n$ are exactly the $\mathbb{Z}$-submodules generated by bases for $\mathbb{R}^n$, and two such lattices are commensurable[11] if and only if they generate the same $\mathbb{Q}$-vector space. Every discrete subgroup commensurable with a lattice is itself a lattice.

Now assume that H is semisimple with finite centre. A lattice Γ in H is *irreducible* if $\Gamma \cdot N$ is dense in H for every noncompact closed normal subgroup N of H. For example, if Γ_1 and Γ_2 are lattices in H_1 and H_2, then the lattice $\Gamma_1 \times \Gamma_2$ in $H_1 \times H_2$ is not irreducible because $(\Gamma_1 \times \Gamma_2) \cdot (1 \times H_2) = \Gamma_1 \times H_2$ is not dense. On the other hand, $SL_2(\mathbb{Z}[\sqrt{2}])$ can be realized as an irreducible lattice in $SL_2(\mathbb{R}) \times SL_2(\mathbb{R})$ via the embeddings $\mathbb{Z}[\sqrt{2}] \to \mathbb{R}$ given by $\sqrt{2} \mapsto \sqrt{2}$ and $\sqrt{2} \mapsto -\sqrt{2}$.

Theorem 3.1. *Let H be a connected semisimple Lie group with no compact factors and trivial centre, and let Γ be a lattice H. Then H can be written (uniquely) as a direct product $H = H_1 \times \cdots \times H_r$ of Lie subgroups H_i such that $\Gamma_i \overset{\text{def}}{=} \Gamma \cap H_i$ is an irreducible lattice in H_i and $\Gamma_1 \times \cdots \times \Gamma_r$ has finite index in Γ*

Proof. See [42], 4.24. □

Theorem 3.2. *Let D be a hermitian symmetric domain, and let $H = \mathrm{Hol}(D)^+$. A discrete subgroup Γ of H is a lattice if and only if $\Gamma \backslash D$ has finite volume. Let Γ be a lattice in H; then D can be written (uniquely) as a product $D = D_1 \times \cdots \times D_r$ of hermitian symmetric domains such that $\Gamma_i \overset{\text{def}}{=} \Gamma \cap \mathrm{Hol}(D_i)^+$ is an irreducible lattice in $\mathrm{Hol}(D_i)^+$ and $\Gamma_1 \backslash D_1 \times \cdots \times \Gamma_r \backslash D_r$ is a finite covering of $\Gamma \backslash D$.*

---

[11]Recall that two subgroup S_1 and S_2 of a group are *commensurable* if $S_1 \cap S_2$ has finite index in both S_1 and S_2. Commensurability is an equivalence relation.

Proof. Let u_o be the homomorphism $\mathbb{S}^1 \to H$ attached to a point $o \in D$ (see 2.3), and let θ be the Cartan involution $\mathrm{inn}(u_o(-1))$. The centralizer of u_o is contained in $H(\mathbb{R}) \cap H^{(\theta)}(\mathbb{R})$, which is compact. Therefore D is a quotient of $H(\mathbb{R})$ by a *compact* subgroup, from which the first statement follows. For the second statement, let $H = H_1 \times \cdots \times H_r$ be the decomposition of H defined by Γ (see 3.1). Then $u_o = (u_1, \ldots, u_r)$ where each u_i is a homomorphism $\mathbb{S}^1 \to H_i$ satisfying the conditions SU1,2 of Theorem 2.5. Now $D = D_1 \times \cdots \times D_r$ with D_i the hermitian symmetric domain corresponding to (H_i, u_i). This is the required decomposition. □

Proposition 3.3. *Let* $\varphi\colon H \to H'$ *be a surjective homomorphism of Lie groups with compact kernel. If* Γ *is a lattice in* H, *then* $\varphi(\Gamma)$ *is a lattice in* H'.

Proof. The proof is elementary (it requires only that H and H' be locally compact topological groups). □

Arithmetic subgroups of algebraic groups

Let G be an algebraic group over $\mathbb{Q}$. When $r\colon G \to GL_n$ is an injective homomorphism, we let

$$G(\mathbb{Z})_r = \{g \in G(\mathbb{Q}) \mid r(g) \in GL_n(\mathbb{Z})\}.$$

Then $G(\mathbb{Z})_r$ is independent of r up to commensurability ([4], 7.13), and we sometimes omit r from the notation. A subgroup Γ of $G(\mathbb{Q})$ is *arithmetic* if it is commensurable with $G(\mathbb{Z})_r$ for some r.

Theorem 3.4. *Let* $\varphi\colon G \to G'$ *be a surjective homomorphism of algebraic groups over* $\mathbb{Q}$. *If* Γ *is an arithmetic subgroup of* $G(\mathbb{Q})$, *then* $\varphi(\Gamma)$ *is an arithmetic subgroup of* $G'(\mathbb{Q})$.

Proof. See [4], 8.11. □

An arithmetic subgroup Γ of $G(\mathbb{Q})$ is obviously discrete in $G(\mathbb{R})$, but it need not be a lattice. For example, $\mathbb{G}_m(\mathbb{Z}) = \{\pm 1\}$ is an arithmetic subgroup of $\mathbb{G}_m(\mathbb{Q})$ of infinite covolume in $\mathbb{G}_m(\mathbb{R}) = \mathbb{R}^\times$.

Theorem 3.5. *Let* G *be a reductive algebraic group over* $\mathbb{Q}$, *and let* Γ *be an arithmetic subgroup of* $G(\mathbb{Q})$.

(a) *The quotient* $\Gamma\backslash G(\mathbb{R})$ *has finite volume if and only if* $\mathrm{Hom}(G, \mathbb{G}_m) = 0$; *in particular,* Γ *is a lattice if* G *is semisimple.*

(b) *(Godement compactness criterion) The quotient* $\Gamma\backslash G(\mathbb{R})$ *is compact if and only if* $\mathrm{Hom}(G, \mathbb{G}_m) = 0$ *and* $G(\mathbb{Q})$ *contains no unipotent element other than* 1.

Proof. See [4], 13.2, 8.4.[12] □

[12] Statement (a) was proved in particular cases by Siegel and others, and in general by Borel and Harish-Chandra [6]. Statement (b) was conjectured by Godement, and proved independently by Mostow and Tamagawa [43] and by Borel and Harish-Chandra [6].

Let k be a subfield of $\mathbb{C}$. An automorphism α of a k-vector space V is said to be *neat* if its eigenvalues in $\mathbb{C}$ generate a torsion free subgroup of $\mathbb{C}^\times$. Let G be an algebraic group over $\mathbb{Q}$. An element $g \in G(\mathbb{Q})$ is *neat* if $\rho(g)$ is neat for one faithful representation $G \hookrightarrow GL(V)$, in which case $\rho(g)$ is neat for every representation ρ of G defined over a subfield of $\mathbb{C}$. A subgroup of $G(\mathbb{Q})$ is *neat* if all its elements are. See [4], §17.

Theorem 3.6. *Let G be an algebraic group over $\mathbb{Q}$, and let Γ be an arithmetic subgroup of $G(\mathbb{Q})$. Then, Γ contains a neat subgroup of finite index. In particular, Γ contains a torsion free subgroup of finite index.*

Proof. In fact, the neat subgroup can be defined by congruence conditions. See [4], 17.4. □

Definition 3.7. A semisimple algebraic group G over $\mathbb{Q}$ is said to be of *compact type* if $G(\mathbb{R})$ is compact, and it is said to be of *of noncompact type* if it does not contain a nontrivial connected normal algebraic subgroup of compact type.

Thus a simply connected or adjoint group over $\mathbb{Q}$ is of compact type if all of its almost simple factors are of compact type, and it is of noncompact type if *none* of its almost simple factors is of compact type. In particular, an algebraic group may be of neither type.

Theorem 3.8 (Borel density theorem). *Let G be a semisimple algebraic group over $\mathbb{Q}$. If G is of noncompact type, then every arithmetic subgroup of $G(\mathbb{Q})$ is dense in the Zariski topology.*

Proof. See [4], 15.12. □

Proposition 3.9. *Let G be a simply connected algebraic group over $\mathbb{Q}$ of noncompact type, and let Γ be an arithmetic subgroup of $G(\mathbb{Q})$. Then Γ is irreducible as a lattice in $G(\mathbb{R})$ if and only if G is almost simple.*

Proof. $\Rightarrow$: Suppose $G = G_1 \times G_2$, and let Γ_1 and Γ_2 be arithmetic subgroups in $G_1(\mathbb{Q})$ and $G_2(\mathbb{Q})$. Then $\Gamma_1 \times \Gamma_2$ is an arithmetic subgroup of $G(\mathbb{Q})$, and so Γ is commensurable with it, but $\Gamma_1 \times \Gamma_2$ is not irreducible.

$\Leftarrow$: Let $G(\mathbb{R}) = H_1 \times \cdots \times H_r$ be a decomposition of the Lie group $G(\mathbb{R})$ such that $\Gamma_i \overset{\text{def}}{=} \Gamma \cap H_i$ is an irreducible lattice in H_i (cf. Theorem 3.1). There exists a finite Galois extension F of $\mathbb{Q}$ in $\mathbb{R}$ and a decomposition $G_F = G_1 \times \cdots \times G_r$ of G_F into a product of algebraic subgroups G_i over F such that $H_i = G_i(\mathbb{R})$ for all i. Because Γ_i is Zariski dense in G_i (Borel density theorem), this last decomposition is stable under the action of $\mathrm{Gal}(F/\mathbb{Q})$, and hence arises from a decomposition over $\mathbb{Q}$. This contradicts the almost simplicity of G unless $r = 1$. □

The rank, $\mathrm{rank}(G)$, of a semisimple algebraic group over $\mathbb{R}$ is the dimension of a maximal split torus in G, i.e., $\mathrm{rank}(G) = r$ if G contains an algebraic subgroup isomorphic to $\mathbb{G}_m^r$ but not to $\mathbb{G}_m^{r+1}$.

Theorem 3.10 (Margulis superrigidity theorem). *Let* G *and* H *be algebraic groups over* $\mathbb{Q}$ *with* G *simply connected and almost simple. Let* Γ *be an arithmetic subgroup of* $G(\mathbb{Q})$, *and let* $\delta\colon \Gamma \to H(\mathbb{Q})$ *be a homomorphism. If* $\operatorname{rank}(G_{\mathbb{R}}) \geqslant 2$, *then the Zariski closure of* $\delta(\Gamma)$ *in* H *is a semisimple algebraic group (possibly not connected), and there is a unique homomorphism* $\varphi\colon G \to H$ *of algebraic groups such that* $\varphi(\gamma) = \delta(\gamma)$ *for all* γ *in a subgroup of finite index in* Γ.

Proof. This the special case of [33], Chapter VIII, Theorem B, p. 258, in which $K = \mathbb{Q} = l$, $S = \{\infty\}$, $\mathbf{G} = G$, $\mathbf{H} = H$, and $\Lambda = \Gamma$. □

Arithmetic lattices in Lie groups

For an algebraic group G over $\mathbb{Q}$, $G(\mathbb{R})$ has a natural structure of a real Lie group, which is connected if G is simply connected (Theorem of Cartan).

Let H be a connected semisimple real Lie group with no compact factors and trivial centre. A subgroup Γ in H is *arithmetic* if there exists a simply connected algebraic group G over $\mathbb{Q}$ and a surjective homomorphism $\varphi\colon G(\mathbb{R}) \to H$ with compact kernel such that Γ is commensurable with $\varphi(G(\mathbb{Z}))$. Such a subgroup is a lattice by Theorem 3.5(a) and Proposition 3.3.

Example 3.11. Let $H = SL_2(\mathbb{R})$, and let B be a quaternion algebra over a totally real number field F such that $H \otimes_{F,v} \mathbb{R} \approx M_2(\mathbb{R})$ for exactly one real prime v. Let G be the algebraic group over $\mathbb{Q}$ such that $G(\mathbb{Q}) = \{b \in B \mid \operatorname{Norm}_{B/\mathbb{Q}}(b) = 1\}$. Then $H \otimes_{\mathbb{Q}} \mathbb{R} \approx M_2(\mathbb{R}) \times \mathbb{H} \times \mathbb{H} \times \cdots$ where $\mathbb{H}$ is usual quaternion algebra, and so there exists a surjective homomorphism $\varphi\colon G(\mathbb{R}) \to SL_2(\mathbb{R})$ with compact kernel. The image under φ of any arithmetic subgroup of $G(\mathbb{Q})$ is an arithmetic subgroup Γ of $SL_2(\mathbb{R})$, and every arithmetic subgroup of $SL_2(\mathbb{R})$ is commensurable with one of this form. If $F = \mathbb{Q}$ and $B = M_2(\mathbb{Q})$, then $G = SL_{2\mathbb{Q}}$ and $\Gamma \backslash SL_2(\mathbb{R})$ is noncompact (see §1); otherwise B is a division algebra, and $\Gamma \backslash SL_2(\mathbb{R})$ is compact by Godement's criterion (3.5b).

For almost a century, $PSL_2(\mathbb{R})$ was the only simple Lie group known to have non arithmetic lattices, and when further examples were discovered in the 1960s they involved only a few other Lie groups. This gave credence to the idea that, except in a few groups of low rank, all lattices are arithmetic (Selberg's conjecture). This was proved by Margulis in a very precise form.

Theorem 3.12 (Margulis arithmeticity theorem). *Every irreducible lattice in a semisimple Lie group is arithmetic unless the group is isogenous to* $SO(1,n) \times$ *(compact) or* $SU(1,n) \times$ *(compact).*

Proof. For a discussion of the theorem, see [42], §5B. For proofs, see [33], Chapter IX, and [67], Chapter 6. □

Theorem 3.13. *Let* H *be the identity component of the group of automorphisms of a hermitian symmetric domain* D, *and let* Γ *be a discrete subgroup of* H *such that* $\Gamma \backslash D$

has finite volume. If rank $H_i \geqslant 2$ *for each factor* H_i *in (3.1), then there exists a simply connected algebraic group* G *of noncompact type over* $\mathbb{Q}$ *and a surjective homomorphism* $\varphi\colon G(\mathbb{R}) \to H$ *with compact kernel such that* Γ *is commensurable with* $\varphi(G(\mathbb{Z}))$. *Moreover, the pair* (G, φ) *is unique up to a unique isomorphism.*

Proof. The group Γ is a lattice in H by Theorem 3.2. Each factor H_i is again the identity component of the group of automorphisms of a hermitian symmetric domain (Theorem 3.2), and so we may suppose that Γ is irreducible. The existence of the pair (G, φ) just means that Γ is arithmetic, which follows from the Margulis arithmeticity theorem (3.12).

Because Γ is irreducible, G is almost simple (see 3.9). As G is simply connected, this implies that $G = (G^s)_{F/\mathbb{Q}}$ where F is a number field and G^s is a geometrically almost simple algebraic group over F. If F had a complex prime, $G_{\mathbb{R}}$ would have a factor $(G')_{\mathbb{C}/\mathbb{R}}$, but $(G')_{\mathbb{C}/\mathbb{R}}$ has no inner form except itself (by Shapiro's lemma), and so this is impossible. Therefore F is totally real.

Let (G_1, φ_1) be a second pair. Because the kernel of φ_1 is compact, its intersection with $G_1(\mathbb{Z})$ is finite, and so there exists an arithmetic subgroup Γ_1 of $G_1(\mathbb{Q})$ such $\varphi_1|\Gamma_1$ is injective. Because $\varphi(G(\mathbb{Z}))$ and $\varphi_1(\Gamma_1)$ are commensurable, there exists an arithmetic subgroup Γ' of $G(\mathbb{Q})$ such that $\varphi(\Gamma') \subset \varphi_1(\Gamma_1)$. Now the Margulis superrigidity theorem 3.10 shows that there exists a homomorphism $\alpha\colon G \to G_1$ such that

$$\varphi_1(\alpha(\gamma)) = \varphi(\gamma) \tag{3.14}$$

for all γ in a subgroup Γ'' of Γ' of finite index. The subgroup Γ'' of $G(\mathbb{Q})$ is Zariski-dense in G (Borel density theorem 3.8), and so (3.14) implies that

$$\varphi_1 \circ \alpha(\mathbb{R}) = \varphi. \tag{3.15}$$

Because G and G_1 are almost simple, (3.15) implies that α is an isogeny, and because G_1 is simply connected, this implies that α is an isomorphism. It is unique because it is uniquely determined on an arithmetic subgroup of G. □

Congruence subgroups of algebraic groups

As in the case of elliptic modular curves, we shall need to consider a special class of arithmetic subgroups, namely, the congruence subgroups.

Let G be an algebraic group over $\mathbb{Q}$. Choose an embedding of G into GL_n, and define

$$\Gamma(N) = G(\mathbb{Q}) \cap \{A \in GL_n(\mathbb{Z}) \mid A \equiv 1 \bmod N\}.$$

A *congruence subgroup*[13] of $G(\mathbb{Q})$ is any subgroup containing $\Gamma(N)$ as a subgroup of finite index. Although $\Gamma(N)$ depends on the choice of the embedding, this definition does not — in fact, the congruence subgroups are exactly those of the form $K \cap G(\mathbb{Q})$ for K a compact open subgroup of $G(\mathbb{A}_f)$.

---

[13]Subgroup defined by congruence conditions.

For a surjective homomorphism $G \to G'$ of algebraic groups over $\mathbb{Q}$, the homomorphism $G(\mathbb{Q}) \to G'(\mathbb{Q})$ need not send congruence subgroups to congruence subgroups. For example, the image in $PGL_2(\mathbb{Q})$ of a congruence subgroup of $SL_2(\mathbb{Q})$ is an arithmetic subgroup (see 3.4) but not necessarily a congruence subgroup.

Every congruence subgroup is an arithmetic subgroup, and for a simply connected group the converse is often, but not always, true. For a survey of what is known about the relation of congruence subgroups to arithmetic groups (the congruence subgroup problem), see [49].

Aside 3.16. Let H be a connected adjoint real Lie group without compact factors. The pairs (G, φ) consisting of a simply connected algebraic group over $\mathbb{Q}$ and a surjective homomorphism $\varphi\colon G(\mathbb{R}) \to H$ with compact kernel have been classified (this requires class field theory). Therefore the arithmetic subgroups of H have been classified up to commensurability. When all arithmetic subgroups are congruence, there is even a classification of the groups themselves in terms of congruence conditions or, equivalently, in terms of compact open subgroups of $G(\mathbb{A}_f)$.

4. Locally symmetric varieties

To obtain an algebraic variety from a hermitian symmetric domain, we need to pass to the quotient by an arithmetic group.

Quotients of hermitian symmetric domains

Let D be a hermitian symmetric domain, and let Γ be a discrete subgroup of $\mathrm{Hol}(D)^+$. If Γ is torsion free, then Γ acts freely on D, and there is a unique complex structure on $\Gamma\backslash D$ for which the quotient map $\pi\colon D \to \Gamma\backslash D$ is a local isomorphism. Relative to this structure, a map φ from $\Gamma\backslash D$ to a second complex manifold is holomorphic if and only if $\varphi \circ \pi$ is holomorphic.

When Γ is torsion free, we often write $D(\Gamma)$ for $\Gamma\backslash D$ regarded as a complex manifold. In this case, D is the universal covering space of $D(\Gamma)$ and Γ is the group of covering transformations. The choice of a point $p \in D$ determines an isomorphism of Γ with the fundamental group $\pi_1(D(\Gamma), \pi p)$.[14]

The complex manifold $D(\Gamma)$ is locally symmetric in the sense that, for each $p \in D(\Gamma)$, there is an involution s_p defined on a neighbourhood of p having p as an isolated fixed point.

The algebraic structure on the quotient

Recall that X^{an} denotes the complex manifold attached to a smooth complex algebraic variety X. The functor $X \rightsquigarrow X^{an}$ is faithful, but it is far from being surjective on arrows or on objects. For example, $(\mathbb{A}^1)^{an} = \mathbb{C}$ and the exponential function is a nonpolynomial holomorphic map $\mathbb{C} \to \mathbb{C}$. A Riemann surface arises from an

[14]Let $\gamma \in \Gamma$, and choose a path from p to γp; the image of this in $\Gamma\backslash D$ is a loop whose homotopy class does not depend on the choice of the path.

algebraic curve if and only if it can be compactified by adding a finite number of points. In particular, if a Riemann surface is an algebraic curve, then every bounded function on it is constant, and so the complex upper half plane is not an algebraic curve (the function $\frac{z-i}{z+i}$ is bounded).

Chow's theorem An algebraic variety (resp. complex manifold) is *projective* if it can be realized as a closed subvariety of $\mathbb{P}^n$ for some n (resp. closed submanifold of $(\mathbb{P}^n)^{an}$).

Theorem 4.1 (Chow 1949 [11]). *The functor $X \rightsquigarrow X^{an}$ from smooth projective complex algebraic varieties to projective complex manifolds is an equivalence of categories.*

In other words, a projective complex manifold has a unique structure of a smooth projective algebraic variety, and every holomorphic map of projective complex manifolds is regular for these structures. See [63], 13.6, for the proof.

Chow's theorem remains true when singularities are allowed and "complex manifold" is replaced by "complex space".

The Baily-Borel theorem

Theorem 4.2 (Baily-Borel 1966 [3]). *Every quotient $D(\Gamma)$ of a hermitian symmetric domain D by a torsion-free arithmetic subgroup Γ of* $\mathrm{Hol}(D)^+$ *has a canonical structure of an algebraic variety.*

More precisely, let G be the algebraic group over $\mathbb{Q}$ attached to (D, Γ) in Theorem 3.13, and assume, for simplicity, that G has no normal algebraic subgroup of dimension 3. Let A_n be the vector space of automorphic forms on D for the nth power of the canonical automorphy factor. Then $A = \bigoplus_{n \geqslant 0} A_n$ is a finitely generated graded $\mathbb{C}$-algebra, and the canonical map

$$D(\Gamma) \to D(\Gamma)^* \stackrel{\text{def}}{=} \mathrm{Proj}(A)$$

realizes $D(\Gamma)$ as a Zariski-open subvariety of the projective algebraic variety $D(\Gamma)^*$ ([3], §10).

Borel's theorem

Theorem 4.3 (Borel 1972 [5]). *Let $D(\Gamma)$ be the quotient $\Gamma \backslash D$ in (4.2) endowed with its canonical algebraic structure, and let V be a smooth complex algebraic variety. Every holomorphic map $f\colon V^{an} \to D(\Gamma)^{an}$ is regular.*

In the proof of Proposition 1.2, we saw that for curves this theorem follows from the big Picard theorem. Recall that this says that every holomorphic map from a punctured disk to $\mathbb{P}^1(\mathbb{C}) \smallsetminus \{\text{three points}\}$ extends to a holomorphic map from the whole disk to $\mathbb{P}^1(\mathbb{C})$. Following earlier work of Kwack and others, Borel generalized the big Picard theorem in two respects: the punctured disk is replaced by a product of punctured disks and disks, and the target space is allowed to be any quotient of a hermitian symmetric domain by a torsion-free arithmetic group.

Resolution of singularities ([28]) shows that every smooth quasi-projective algebraic variety V can be embedded in a smooth projective variety $\bar{V}$ as the complement of a divisor with normal crossings. This condition means that $\bar{V}^{an} \smallsetminus V^{an}$ is locally a product of disks and punctured disks. Therefore $f|V^{an}$ extends to a holomorphic map $\bar{V}^{an} \to D(\Gamma)^*$ (by Borel) and so is a regular map (by Chow).

Locally symmetric varieties

A *locally symmetric variety* is a smooth algebraic variety X over $\mathbb{C}$ such that X^{an} is isomorphic to $\Gamma\backslash D$ for some hermitian symmetric domain D and torsion-free subgroup Γ of Hol(D).[15] In other words, X is a locally symmetric variety if the universal covering space D of X^{an} is a hermitian symmetric domain and the group of covering transformations of D over X^{an} is a torsion-free subgroup Γ of Hol(D). When Γ is an arithmetic subgroup of $\mathrm{Hol}(D)^+$, X is called an *arithmetic locally symmetric variety*. The group Γ is automatically a lattice, and so the Margulis arithmeticity theorem (3.12) shows that nonarithmetic locally symmetric varieties can occur only when there are factors of low dimension.

A nonsingular projective curve over $\mathbb{C}$ has a model over $\mathbb{Q}^{al}$ if and only if it contains an arithmetic locally symmetric curve as the complement of a finite set (Belyi; see [55], p. 71). This suggests that there are too many arithmetic locally symmetric varieties for us to be able to say much about their arithmetic.

Let $D(\Gamma)$ be an arithmetic locally symmetric variety. Recall that Γ is arithmetic if there is a simply connected algebraic group G over $\mathbb{Q}$ and a surjective homomorphism $\varphi\colon G(\mathbb{R}) \to \mathrm{Hol}(D)^+$ with compact kernel such that Γ is commensurable with $\varphi(G(\mathbb{Z}))$. If there exists a *congruence subgroup* Γ_0 of $G(\mathbb{Z})$ such that Γ contains $\varphi(\Gamma_0)$ as a subgroup of finite index, then we call $D(\Gamma)$ a *connected Shimura variety*. Only for Shimura varieties do we have a rich arithmetic theory (see [14], [19], and the many articles of Shimura, especially, [57, 58, 59, 60, 61]).

Example: Siegel modular varieties

For an abelian variety A over $\mathbb{C}$, the exponential map defines an exact sequence

$$0 \longrightarrow \Lambda \longrightarrow T_0(A^{an}) \xrightarrow{\exp} A^{an} \longrightarrow 0$$

with $T_0(A^{an})$ a complex vector space and Λ a lattice in $T_0(A^{an})$ canonically isomorphic to $H_1(A^{an}, \mathbb{Z})$.

[15] As Hol(D) has only finitely many components, $\Gamma \cap \mathrm{Hol}(D)^+$ has finite index in Γ. Sometimes we only allow discrete subgroups of Hol(D) contained in $\mathrm{Hol}(D)^+$. In the theory of Shimura varieties, we generally consider only "sufficiently small" discrete subgroups, and we regard the remainder as "noise". Algebraic geometers do the opposite.

Theorem 4.4 (Riemann's Theorem). *The functor* $A \rightsquigarrow (T_0(A), \Lambda)$ *is an equivalence from the category of abelian varieties over* $\mathbb{C}$ *to the category of pairs consisting of a* $\mathbb{C}$*-vector space* V *and a lattice* Λ *in* V *that admits a Riemann form.*

Proof. See, for example, [47], Chapter I. □

A Riemann form for a pair (V, Λ) is an alternating form $\psi\colon \Lambda \times \Lambda \to \mathbb{Z}$ such that the pairing $(x, y) \mapsto \psi_{\mathbb{R}}(x, \sqrt{-1}y)\colon V \times V \to \mathbb{R}$ is symmetric and positive definite. Here $\psi_{\mathbb{R}}$ denotes the linear extension of ψ to $\mathbb{R} \otimes_{\mathbb{Z}} \Lambda \simeq V$. A principal polarization on an abelian variety A over $\mathbb{C}$ is Riemann form for $(T_0(A), \Lambda)$ whose discriminant is ± 1. A level-N structure on an abelian variety over $\mathbb{C}$ is defined similarly to an elliptic curve (see §1; we require it to be compatible with the Weil pairing).

Let (V, ψ) be a symplectic space over $\mathbb{R}$, and let Λ be a lattice in V such that $\psi(\Lambda, \Lambda) \subset \mathbb{Z}$ and $\psi|_{\Lambda \times \Lambda}$ has discriminant ± 1. The points of the corresponding Siegel upper half space D are the complex structures J on V such that ψ_J is Riemann form (see §2). The map $J \mapsto (V, J)/\Lambda$ is a bijection from D to the set of isomorphism classes of principally polarized abelian varieties over $\mathbb{C}$ equipped with an isomorphism $\Lambda \to H_1(A, \mathbb{Z})$. On passing to the quotient by the principal congruence subgroup $\Gamma(N)$, we get a bijection from $D_N \stackrel{\text{def}}{=} \Gamma(N)\backslash D$ to the set of isomorphism classes of principally polarized abelian over $\mathbb{C}$ equipped with a level-N structure.

Proposition 4.5. *Let* $f\colon A \to S$ *be a family of principally polarized abelian varieties on a smooth algebraic variety* S *over* $\mathbb{C}$*, and let* η *be a level-*N *structure on* A/S*. The map* $\gamma\colon S(\mathbb{C}) \to D_N(\mathbb{C})$ *sending* $s \in S(\mathbb{C})$ *to the point of* $\Gamma(N)\backslash D$ *corresponding to* (A_s, η_s) *is regular.*

Proof. The holomorphicity of γ can be proved by the same argument as in the proof of Proposition 1.2. Its algebraicity then follows from Borel's theorem 4.3. □

Let $\mathcal{F}$ be the functor sending a scheme S of finite type over $\mathbb{C}$ to the set of isomorphism classes of pairs consisting of a family of principally polarized abelian varieties $f\colon A \to S$ over S and a level-N structure on A. When $N \geqslant 3$, $\mathcal{F}$ is representable by a smooth algebraic variety S_N over $\mathbb{C}$ ([44], Chapter 7). This means that there exists a (universal) family of principally polarized abelian varieties A/S_N and a level-N structure η on A/S_N such that, for any similar pair $(A'/S, \eta')$ over a scheme S, there exists a unique morphism $\alpha\colon S \to S_N$ for which $\alpha^*(A/S_N, \eta) \approx (A'/S', \eta')$.

Theorem 4.6. *There is a canonical isomorphism* $\gamma\colon S_N \to D_N$.

Proof. The proof is the same as that of Theorem 1.4. □

Corollary 4.7. *The universal family of complex tori on* D_N *is algebraic.*

5. Variations of Hodge structures

We review the definitions.

The Deligne torus

The *Deligne torus* is the algebraic torus $\mathbb{S}$ over $\mathbb{R}$ obtained from $\mathbb{G}_m$ over $\mathbb{C}$ by restriction of the base field; thus

$$\mathbb{S}(\mathbb{R}) = \mathbb{C}^\times, \quad \mathbb{S}_\mathbb{C} \simeq \mathbb{G}_m \times \mathbb{G}_m.$$

The map $\mathbb{S}(\mathbb{R}) \to \mathbb{S}(\mathbb{C})$ induced by $\mathbb{R} \to \mathbb{C}$ is $z \mapsto (z, \bar{z})$. There are homomorphisms

$$\mathbb{G}_m \xrightarrow{w} \mathbb{S} \xrightarrow{t} \mathbb{G}_m, \quad t \circ w = -2,$$
$$\mathbb{R}^\times \xrightarrow{a \mapsto a^{-1}} \mathbb{C}^\times \xrightarrow{z \mapsto z\bar{z}} \mathbb{R}^\times.$$

The kernel of t is $\mathbb{S}^1$. A homomorphism $h\colon \mathbb{S} \to G$ of real algebraic groups gives rise to cocharacters

$$\mu_h\colon \mathbb{G}_m \to G_\mathbb{C},\ z \mapsto h_\mathbb{C}(z,1),\ z \in \mathbb{G}_m(\mathbb{C}) = \mathbb{C}^\times,$$
$$w_h\colon \mathbb{G}_m \to G, \quad w_h = h \circ w \quad (\textit{weight homomorphism}).$$

The following formulas are useful ($\mu = \mu_h$):

(5.1) $$h_\mathbb{C}(z_1, z_2) = \mu(z_1) \cdot \bar{\mu}(z_2); \quad h(z) = \mu(z) \cdot \overline{\mu(z)}$$

(5.2) $$h(i) = \mu(-1) \cdot w_h(i).$$

Real Hodge structures

A *real Hodge structure* is a representation $h\colon \mathbb{S} \to GL_V$ of $\mathbb{S}$ on a real vector space V. Equivalently, it is a real vector space V together with a *Hodge decomposition*,

$$V_\mathbb{C} = \bigoplus\nolimits_{p,q \in \mathbb{Z}} V^{p,q} \text{ such that } \overline{V^{p,q}} = V^{q,p} \text{ for all } p, q.$$

To pass from one description to the other, use the rule ([16, 19]):

$$v \in V^{p,q} \iff h(z)v = z^{-p}\bar{z}^{-q}v, \text{ all } z \in \mathbb{C}^\times.$$

The integers $h^{p,q} \stackrel{\text{def}}{=} \dim_\mathbb{C} V^{p,q}$ are called the *Hodge numbers* of the Hodge structure. A real Hodge structure defines a (weight) gradation on V,

$$V = \bigoplus\nolimits_{m \in \mathbb{Z}} V_m, \quad V_m = V \cap \left(\bigoplus\nolimits_{p+q=m} V^{p,q}\right),$$

and a descending *Hodge filtration*,

$$V_\mathbb{C} \supset \cdots \supset F^p \supset F^{p+1} \supset \cdots \supset 0, \quad F^p = \bigoplus\nolimits_{p' \geqslant p} V^{p',q'}.$$

The weight gradation and Hodge filtration together determine the Hodge structure because

$$V^{p,q} = (V_{p+q})_\mathbb{C} \cap F^p \cap \overline{F^q}.$$

Note that the weight gradation is defined by w_h. A filtration F on $V_\mathbb{C}$ arises from a Hodge structure of weight m on V if and only if

$$V = F^p \oplus \overline{F^q} \text{ whenever } p + q = m + 1.$$

The $\mathbb{R}$-linear map $C = h(i)$ is called the *Weil operator*. It acts as i^{q-p} on $V^{p,q}$, and C^2 acts as $(-1)^m$ on V_m.

Thus a Hodge structure on a real vector space V can be regarded as a homomorphism $h\colon \mathbb{S} \to GL_V$, a Hodge decomposition of V, or a Hodge filtration together with a weight gradation of V. We use the three descriptions interchangeably.

5.3. Let V be a real vector space. To give a Hodge structure h on V of type $\{(-1,0), (0,-1)\}$ is the same as giving a complex structure on V: given h, let J act as $C = h(i)$; given a complex structure, let $h(z)$ act as multiplication by z. The Hodge decomposition $V_{\mathbb{C}} = V^{-1,0} \oplus V^{0,-1}$ corresponds to the decomposition $V_{\mathbb{C}} = V^{+} \oplus V^{-}$ of $V_{\mathbb{C}}$ into its J-eigenspaces.

Rational Hodge structures

A *rational Hodge structure* is a $\mathbb{Q}$-vector space V together with a real Hodge structure on $V_{\mathbb{R}}$ such that the weight gradation is defined over $\mathbb{Q}$. Thus to give a rational Hodge structure on V is the same as giving

- a gradation $V = \bigoplus_m V_m$ on V together with a real Hodge structure of weight m on $V_{m\mathbb{R}}$ for each m, *or*
- a homomorphism $h\colon \mathbb{S} \to GL_{V_{\mathbb{R}}}$ such that $w_h\colon \mathbb{G}_m \to GL_{V_{\mathbb{R}}}$ is defined over $\mathbb{Q}$.

The *Tate Hodge structure* $\mathbb{Q}(m)$ is defined to be the $\mathbb{Q}$-subspace $(2\pi i)^m\mathbb{Q}$ of $\mathbb{C}$ with $h(z)$ acting as multiplication by $\mathrm{Norm}_{\mathbb{C}/\mathbb{R}}(z)^m = (z\bar{z})^m$. It has weight $-2m$ and type $(-m,-m)$.

Polarizations

A *polarization* of a real Hodge structure (V,h) of weight m is a morphism of Hodge structures

$$\psi\colon V \otimes V \to \mathbb{R}(-m), \quad m \in \mathbb{Z}, \tag{5.4}$$

such that

$$(x,y) \mapsto (2\pi i)^m \psi(x, Cy)\colon V \times V \to \mathbb{R} \tag{5.5}$$

is symmetric and positive definite. The condition (5.5) means that ψ is symmetric if m is even and skew-symmetric if it is odd, and that $(2\pi i)^m \cdot i^{p-q}\psi_{\mathbb{C}}(x,\bar{x}) > 0$ for $x \in V^{p,q}$.

A *polarization* of a rational Hodge structure (V,h) of weight m is a morphism of rational Hodge structures $\psi\colon V \otimes V \to \mathbb{Q}(-m)$ such that $\psi_{\mathbb{R}}$ is a polarization of $(V_{\mathbb{R}},h)$. A rational Hodge structure (V,h) is polarizable if and only if $(V_{\mathbb{R}},h)$ is polarizable (cf. 2.2).

Local systems and vector sheaves with connection

Let S be a complex manifold. A *connection* on a vector sheaf $\mathcal{V}$ on S is a $\mathbb{C}$-linear homomorphism $\nabla\colon \mathcal{V} \to \Omega^1_S \otimes \mathcal{V}$ satisfying the Leibniz condition

$$\nabla(fv) = df \otimes v + f \cdot \nabla v$$

for all local sections f of $\mathcal{O}_S$ and v of $\mathcal{V}$. The *curvature* of ∇ is the composite of ∇ with the map

$$\nabla_1\colon \Omega^1_S \otimes \mathcal{V} \to \Omega^2_S \otimes \mathcal{V}$$
$$\omega \otimes v \mapsto d\omega \otimes v - \omega \wedge \nabla(v).$$

A connection ∇ is said to be *flat* if its curvature is zero. In this case, the kernel $\mathcal{V}^\nabla$ of ∇ is a local system of complex vector spaces on S such that $\mathcal{O}_S \otimes \mathcal{V}^\nabla \simeq \mathcal{V}$.

Conversely, let V be a local system of complex vector spaces on S. The vector sheaf $\mathcal{V} = \mathcal{O}_S \otimes V$ has a canonical connection ∇: on any open set where V is trivial, say $V \approx \mathbb{C}^n$, the connection is the map $(f_i) \mapsto (df_i)\colon (\mathcal{O}_S)^n \to (\Omega^1_S)^n$. This connection is flat because $d \circ d = 0$. Obviously for this connection, $\mathcal{V}^\nabla \simeq V$.

In this way, we obtain an equivalence between the category of vector sheaves on S equipped with a flat connection and the category of local systems of complex vector spaces.

Variations of Hodge structures

Let S be a complex manifold. By a *family of real Hodge structures on* S we mean a holomorphic family. For example, a family of real Hodge structures on S of weight m is a local system V of $\mathbb{R}$-vector spaces on S together with a filtration F on $\mathcal{V} \overset{\text{def}}{=} \mathcal{O}_S \otimes_{\mathbb{R}} V$ by holomorphic vector subsheaves that gives a Hodge filtration at each point, i.e., such that

$$F^p\mathcal{V}_s \oplus \overline{F^{m+1-p}\mathcal{V}_s} \simeq \mathcal{V}_s, \quad \text{all } s \in S,\ p \in \mathbb{Z}.$$

For the notion of a *family of rational Hodge structures*, replace $\mathbb{R}$ with $\mathbb{Q}$.

A *polarization* of a family of real Hodge structures of weight m is a bilinear pairing of local systems

$$\psi\colon V \times V \to \mathbb{R}(-m)$$

that gives a polarization at each point s of S. For rational Hodge structures, replace $\mathbb{R}$ with $\mathbb{Q}$.

Let ∇ be connection on a vector sheaf $\mathcal{V}$. A holomorphic vector field Z on S is a map $\Omega^1_S \to \mathcal{O}_S$, and it defines a map $\nabla_Z\colon \mathcal{V} \to \mathcal{V}$. A family of rational Hodge structures V on S is a *variation* of rational Hodge structures on S if it satisfies the following axiom (*Griffiths transversality*):

$$\nabla_Z(F^p\mathcal{V}) \subset F^{p-1}\mathcal{V} \text{ for all } p \text{ and } Z.$$

Equivalently,

$$\nabla(F^p\mathcal{V}) \subset \Omega^1_S \otimes F^{p-1}\mathcal{V} \text{ for all } p.$$

Here ∇ is the flat connection on $\mathcal{V} \stackrel{\text{def}}{=} \mathcal{O}_S \otimes_{\mathbb{Q}} V$ defined by V.

These definitions are motivated by the following theorem.

Theorem 5.6 (Griffiths 1968 [24]). *Let $f\colon X \to S$ be a smooth projective map of smooth algebraic varieties over $\mathbb{C}$. For each m, the local system $R^m f_* \mathbb{Q}$ of $\mathbb{Q}$-vector spaces on S^{an} together with the de Rham filtration on $\mathcal{O}_S \otimes_{\mathbb{Q}} Rf_*\mathbb{Q} \simeq Rf_*(\Omega^\bullet_{X/\mathbb{C}})$ is a polarizable variation of rational Hodge structures of weight m on S^{an}.*

This theorem suggests that the first step in realizing an algebraic variety as a moduli variety should be to show that it carries a polarized variation of rational Hodge structures.

6. Mumford-Tate groups and their variation in families

We define Mumford-Tate groups, and we study their variation in families. Throughout this section, "Hodge structure" means "rational Hodge structure".

The conditions (SV)

We list some conditions on a homomorphism $h\colon \mathbb{S} \to G$ of real connected algebraic groups:

SV1: the Hodge structure on the Lie algebra of G defined by $\mathrm{Ad} \circ h\colon \mathbb{S} \to GL_{\mathrm{Lie}(G)}$ is of type $\{(1,-1),(0,0),(-1,1)\}$;

SV2: $\mathrm{inn}(h(i))$ is a Cartan involution of G^{ad}.

In particular, (SV2) says that the Cartan involutions of G^{ad} are inner, and so G^{ad} is an inner form of its compact form. This implies that the simple factors of G^{ad} are geometrically simple (see footnote 9, p. 479).

Condition (SV1) implies that the Hodge structure on $\mathrm{Lie}(G)$ defined by h has weight 0, and so $w_h(\mathbb{G}_m) \subset Z(G)$. In the presence of this condition, we sometimes need to consider a stronger form of (SV2):

SV2*: $\mathrm{inn}(h(i))$ is a Cartan involution of $G/w_h(\mathbb{G}_m)$.

Note that (SV2*) implies that G is reductive.

Let G be an algebraic group over $\mathbb{Q}$, and let h be a homomorphism $\mathbb{S} \to G_{\mathbb{R}}$. We say that (G, h) satisfies the condition (SV1) or (SV2) if $(G_{\mathbb{R}}, h)$ does. When w_h is defined over $\mathbb{Q}$, we say that (G, h) satisfies (SV2*) if $(G_{\mathbb{R}}, h)$ does. We shall also need to consider the condition:

SV3: G^{ad} has no $\mathbb{Q}$-factor on which the projection of h is trivial.

In the presence of (SV1,2), the condition (SV3) is equivalent to G^{ad} being of non-compact type (apply Lemma 4.7 of [40]).

Each condition holds for a homomorphism h if and only if it holds for a conjugate of h by an element of $G(\mathbb{R})$.

Let G be a reductive group over $\mathbb{Q}$. Let h be a homomorphism $\mathbb{S} \to G_{\mathbb{R}}$, and let $\bar{h}\colon \mathbb{S} \to G^{\mathrm{ad}}_{\mathbb{R}}$ be $\mathrm{ad} \circ h$. Then (G, h) satisfies (SV1,2,3) if and only if $(G^{\mathrm{ad}}, \bar{h})$ satisfies the same conditions.[16]

Remark 6.1. Let H be a real algebraic group. The map $z \mapsto z/\bar{z}$ defines an isomorphism $\mathbb{S}/w(\mathbb{G}_m) \simeq \mathbb{S}^1$, and so the formula

$$h(z) = u(z/\bar{z}) \tag{6.2}$$

defines a one-to-one correspondence between the homomorphisms $h\colon \mathbb{S} \to H$ trivial on $w(\mathbb{G}_m)$ and the homomorphisms $u\colon \mathbb{S}^1 \to H$. When H has trivial centre, h satisfies SV1 (resp. SV2) if and only if u satisfies SU1 (resp. SU2).

Notes. Conditions (SV1), (SV2), and (SV3) are respectively the conditions (2.1.1.1), (2.1.1.2), and (2.1.1.3) of [19], and (SV2*) is the condition (2.1.1.5).

Definition of Mumford-Tate groups

Let (V, h) be a rational Hodge structure. Following [15], 7.1, we define the *Mumford-Tate group* of (V, h) to be the smallest algebraic subgroup G of GL_V such that $G_{\mathbb{R}} \supset h(\mathbb{S})$. It is also the smallest algebraic subgroup G of GL_V such that $G_{\mathbb{C}} \supset \mu_h(\mathbb{G}_m)$ (apply (5.1), p. 490). We usually regard the Mumford-Tate group as a pair (G, h), and we sometimes denote it by MT_V. Note that G is connected, because otherwise we could replace it with its identity component. The weight map $w_h\colon \mathbb{G}_m \to G_{\mathbb{R}}$ is defined over $\mathbb{Q}$ and maps into the centre of G.[17]

Let (V, h) be a polarizable rational Hodge structure, and let $T^{m,n}$ denote the Hodge structure $V^{\otimes m} \otimes V^{\vee \otimes n}$ $(m, n \in \mathbb{N})$. By a *Hodge class* of V, we mean an element of V of type (0,0), i.e., an element of $V \cap V^{0,0}$, and by a *Hodge tensor* of V, we mean a Hodge class of some $T^{m,n}$. The elements of $T^{m,n}$ fixed by the Mumford-Tate group of V are exactly the Hodge tensors, and MT_V is the largest algebraic subgroup of GL_V fixing all the Hodge tensors of V (cf. [20], 3.4).

The real Hodge structures form a semisimple tannakian category[18] over $\mathbb{R}$; the group attached to the category and the forgetful fibre functor is $\mathbb{S}$. The rational Hodge structures form a tannakian category over $\mathbb{Q}$, and the polarizable rational Hodge structures form a *semisimple* tannakian category, which we denote $\mathrm{Hdg}_{\mathbb{Q}}$. Let (V, h) be a rational Hodge structure, and let $\langle V, h\rangle^{\otimes}$ be the tannakian subcategory generated by (V, h). The Mumford-Tate group of (V, h) is the algebraic group attached $\langle V, h\rangle^{\otimes}$ and the forgetful fibre functor.

---

[16] For (SV1), note that $\mathrm{Ad}(h(z))\colon \mathrm{Lie}(G) \to \mathrm{Lie}(G)$ is the derivative of $\mathrm{ad}(h(z))\colon G \to G$. The latter is trivial on $Z(G)$, and so the former is trivial on $\mathrm{Lie}(Z(G))$.

[17] Let $Z(w_h)$ be the centralizer of w_h in G. For any $a \in \mathbb{R}^{\times}$, $w_h(a)\colon V_{\mathbb{R}} \to V_{\mathbb{R}}$ is a morphism of real Hodge structures, and so it commutes with the action of $h(\mathbb{S})$. Hence $h(\mathbb{S}) \subset Z(w_h)_{\mathbb{R}}$. As h generates G, this implies that $Z(w_h) = G$.

[18] For the theory of tannakian categories, we refer the reader to [21]. In fact, we shall only need to use the elementary part of the theory (ibid. §§1,2).

Let G and G^e respectively denote the Mumford-Tate groups of V and $V \oplus \mathbb{Q}(1)$. The action of G^e on V defines a homomorphism $G^e \to G$, which is an isogeny unless V has weight 0, in which case $G^e \simeq G \times \mathbb{G}_m$. The action of G^e on $\mathbb{Q}(1)$ defines a homomorphism $G^e \to GL_{\mathbb{Q}(1)}$ whose kernel we denote G^1 and call the *special Mumford-Tate group* of V. Thus $G^1 \subset GL_V$, and it is the smallest algebraic subgroup of GL_V such that $G^1_{\mathbb{R}} \supset h(\mathbb{S}^1)$. Clearly $G^1 \subset G$ and $G = G^1 \cdot w_h(\mathbb{G}_m)$.

Proposition 6.3. *Let G be a connected algebraic group over $\mathbb{Q}$, and let h be a homomorphism $\mathbb{S} \to G_{\mathbb{R}}$. The pair (G, h) is the Mumford-Tate group of a Hodge structure if and only if the weight homomorphism $w_h \colon \mathbb{G}_m \to G_{\mathbb{R}}$ is defined over $\mathbb{Q}$ and G is generated by h (i.e., any algebraic subgroup H of G such that $h(\mathbb{S}) \subset H_{\mathbb{R}}$ equals G).*

Proof. If (G, h) is the Mumford-Tate group of a Hodge structure (V, h), then certainly h generates G. The weight homomorphism w_h is defined over $\mathbb{Q}$ because (V, h) is a rational Hodge structure.

Conversely, suppose that (G, h) satisfy the conditions. For any faithful representation $\rho \colon G \to GL_V$ of G, the pair $(V, h \circ \rho)$ is a rational Hodge structure, and (G, h) is its Mumford-Tate group. □

Proposition 6.4. *Let (G, h) be the Mumford-Tate group of a Hodge structure (V, h). Then (V, h) is polarizable if and only if (G, h) satisfies (SV2*).*

Proof. Let $C = h(i)$. For notational convenience, assume that (V, h) has a single weight m. Let G^1 be the special Mumford-Tate group of (V, h). Then $C \in G^1(\mathbb{R})$, and a pairing $\psi \colon V \times V \to \mathbb{Q}(-m)$ is a polarization of the Hodge structure (V, h) if and only if $(2\pi i)^m \psi$ is a C-polarization of V for G^1 in the sense of §2. It follows from (2.1) and (2.2) that a polarization ψ for (V, h) exists if and only if $\mathrm{inn}(C)$ is a Cartan involution of $G^1_{\mathbb{R}}$. Now $G^1 \subset G$ and the quotient map $G^1 \to G/w_h(\mathbb{G}_m)$ is an isogeny, and so $\mathrm{inn}(C)$ is a Cartan involution of G^1 if and only if it is a Cartan involution of $G/w_h(\mathbb{G}_m)$. □

Corollary 6.5. *The Mumford-Tate group of a polarizable Hodge structure is reductive.*

Proof. An algebraic group G over $\mathbb{Q}$ is reductive if and only if $G_{\mathbb{R}}$ is reductive, and we have already observed that (SV2*) implies that $G_{\mathbb{R}}$ is reductive. Alternatively, polarizable Hodge structures are semisimple, and an algebraic group in characteristic zero is reductive if its representations are semisimple (e.g., [21], 2.23). □

Remark 6.6. Note that (6.4) implies the following statement: let (V, h) be a Hodge structure; if there exists an algebraic group $G \subset GL_V$ such that $h(\mathbb{S}) \subset G_{\mathbb{R}}$ and (G, h) satisfies (SV2*), then (V, h) is polarizable.

Notes. The Mumford-Tate group of a complex abelian variety A is defined to be the Mumford-Tate group of the Hodge structure $H_1(A^{an}, \mathbb{Q})$. In this context, special Mumford-Tate groups were first introduced in the talk of Mumford [45] (which is "partly joint work with J. Tate").

Special Hodge structures

A rational Hodge structure is *special*[19] if its Mumford-Tate group satisfies (SV1, 2*) or, equivalently, if it is polarizable and its Mumford-Tate group satisfies (SV1).

Proposition 6.7. *The special Hodge structures form a tannakian subcategory of* $\mathsf{Hdg}_{\mathbb{Q}}$.

Proof. Let (V,h) be a special Hodge structure. The Mumford-Tate group of any object in the tannakian subcategory of $\mathsf{Hdg}_{\mathbb{Q}}$ generated by (V,h) is a quotient of MT_V, and hence satisfies (SV1,2*). □

Recall that the *level* of a Hodge structure (V,h) is the maximum value of $|p-q|$ as (p,q) runs over the pairs (p,q) with $V^{p,q} \neq 0$. It has the same parity as the weight of (V,h).

Example 6.8. Let $V_n(a_1,\ldots,a_d)$ denote a complete intersection of d smooth hypersurfaces of degrees $a_1,\ldots,a_d$ in general position in $\mathbb{P}^{n+d}$ over $\mathbb{C}$. Then $H^n(V_n,\mathbb{Q})$ has level $\leqslant 1$ only for the varieties $V_n(2)$, $V_n(2,2)$, $V_2(3)$, $V_n(2,2,2)$ (n odd), $V_3(3)$, $V_3(2,3)$, $V_5(3)$, $V_3(4)$ ([50]).

Proposition 6.9. *Every polarizable Hodge structure of level* $\leqslant 1$ *is special.*

Proof. A Hodge structure of level 0 is direct sum of copies of $\mathbb{Q}(m)$ for some m, and so its Mumford-Tate group is $\mathbb{G}_m$. A Hodge structure (V,h) of level 1 is of type $\{(p,p+1),(p+1,p)\}$ for some p. Then

$$\mathrm{Lie}(MT_V) \subset \mathrm{End}(V) = V^\vee \otimes V,$$

which is of type $\{(-1,1),(0,0),(1,-1)\}$. □

Example 6.10. Let A be an abelian variety over $\mathbb{C}$. The Hodge structures $H^n_B(A)$ are special for all n. To see this, note that $H^1_B(A)$ is of level 1, and hence is special by (6.9), and that

$$H^n_B(A) \simeq \bigwedge\nolimits^n H^1_B(A) \subset H^1_B(A)^{\otimes n},$$

and hence $H^n_B(A)$ is special by (6.7).

It follows that a nonspecial Hodge structure does not lie in the tannakian subcategory of $\mathsf{Hdg}_{\mathbb{Q}}$ generated by the cohomology groups of abelian varieties.

Proposition 6.11. *A pair* (G,h) *is the Mumford-Tate group of a special Hodge structure if and only if* h *satisfies (SV1,2*), the weight* w_h *is defined over* $\mathbb{Q}$, *and* G *is generated by* h.

Proof. Immediate consequence of Proposition 6.3, and of the definition of a special Hodge structure. □

Note that, because h generates G, it also satisfies (SV3).

---

[19] Poor choice of name, since "special" is overused and special points on Shimura varieties don't correspond to special Hodge structures, but I can't think of a better one. Perhaps an "SV Hodge structure"?

Example 6.12. Let $f\colon X \to S$ be the universal family of smooth hypersurfaces of a fixed degree δ and of a fixed odd dimension n. For s outside a meagre subset of S, the Mumford-Tate group of $H^n(X_s, \mathbb{Q})$ is the full group of symplectic similitudes (see 6.23 below). This implies that $H^n(X_s, \mathbb{Q})$ is not special unless it has level $\leqslant 1$. According to (6.8), this rarely happens.

The generic Mumford-Tate group

Throughout this subsection, (V, F) is a family of Hodge structures on a connected complex manifold S. Recall that "family" means "holomorphic family".

Lemma 6.13. *For any* $t \in \Gamma(S, V)$, *the set*

$$Z(t) = \{s \in S \mid t_s \textit{ is of type } (0,0) \textit{ in } V_s\}$$

is an analytic subset of S.

Proof. An element of V_s is of type $(0,0)$ if and only if it lies in F^0V_s. On S, we have an exact sequence

$$0 \to F^0\mathcal{V} \to \mathcal{V} \to \mathcal{Q} \to 0$$

of locally free sheaves of $\mathcal{O}_S$-modules. Let U be an open subset of S such that $\mathcal{Q}$ is free over U. Choose an isomorphism $\mathcal{Q} \simeq \mathcal{O}_U^r$, and let $t|U$ map to $(t_1, \ldots, t_r)$ in $\mathcal{O}_U^r$. Then

$$Z(t) \cap U = \{s \in U \mid t_1(s) = \cdots = t_r(s) = 0\}.$$

□

For $m, n \in \mathbb{N}$, let $T^{m,n} = T^{m,n}V$ be the family of Hodge structures $V^{\otimes m} \otimes V^{\vee \otimes n}$ on S. Let $\pi\colon \tilde{S} \to S$ be a universal covering space of S, and define

$$\mathring{S} = S \smallsetminus \bigcup\nolimits_t \pi_*(Z(t)) \tag{6.14}$$

where t runs over the global sections of the local systems $\pi^*T^{m,n}$ $(m, n \in \mathbb{N})$ such that $\pi_*(Z(t)) \neq S$. Thus $\mathring{S}$ is the complement in S of a countable union of proper analytic subvarieties — we shall call such a subset *meagre*.

Example 6.15. For a "general" abelian variety of dimension g over $\mathbb{C}$, it is known that the $\mathbb{Q}$-algebra of Hodge classes is generated by the class of an ample divisor class ([12], [34]). It follows that the same is true for all abelian varieties in the subset $\mathring{S}$ of the moduli space S. The Hodge conjecture obviously holds for these abelian varieties.

Let t be a section of $T^{m,n}$ over an open subset U of $\mathring{S}$; if t is a Hodge class in $T^{m,n}_s$ for one $s \in U$, then it is Hodge tensor for every $s \in U$. Thus, there exists a local system of $\mathbb{Q}$-subspaces $HT^{m,n}$ on $\mathring{S}$ such that $(HT^{m,n})_s$ is the space of Hodge classes in $T^{m,n}_s$ for each s. Since the Mumford-Tate group of (V_s, F_s) is the largest algebraic subgroup of GL_{V_s} fixing the Hodge tensors in the spaces $T^{m,n}_s$, we have the following result.

Proposition 6.16. *Let* G_s *be the Mumford-Tate group of* (V_s, F_s). *Then* G_s *is locally constant on* $\mathring{S}$.

More precisely:

> Let U be an open subset of S on which V is constant, say, $V = V_U$; identify the stalk V_s $(s \in U)$ with V, so that G_s is a subgroup of GL_V; then G_s is constant for $s \in U \cap \mathring{S}$, say $G_s = G$, and $G \supset G_s$ for all $s \in U \smallsetminus (U \cap \mathring{S})$.

6.17. We say that G_s is *generic* if $s \in \mathring{S}$. Suppose that V is constant, say $V = V_S$, and let $G = G_{s_0} \subset GL_V$ be generic. By definition, G is the smallest algebraic subgroup of GL_V such that $G_{\mathbb{R}}$ contains $h_{s_0}(\mathbb{S})$. As $G \supset G_s$ for all $s \in S$, the generic Mumford-Tate group of (V, F) is the smallest algebraic subgroup G of GL_V such that $G_{\mathbb{R}}$ contains $h_s(\mathbb{S})$ for *all* $s \in S$.

Let $\pi\colon \tilde{S} \to S$ be a universal covering of S, and fix a trivialization $\pi^*V \simeq V_S$ of V. Then, for each $s \in S$, there are given isomorphisms

$$V \simeq (\pi^*V)_s \simeq V_{\pi s}. \tag{6.18}$$

There is an algebraic subgroup G of GL_V such that, for each $s \in \pi^{-1}(\mathring{S})$, G maps isomorphically onto G_s under the isomorphism $GL_V \simeq GL_{V_{\pi s}}$ defined by (6.18). It is the smallest algebraic subgroup of GL_V such that $G_{\mathbb{R}}$ contains the image of $h_s\colon \mathbb{S} \to GL_{V_{\mathbb{R}}}$ for all $s \in \tilde{S}$.

Aside 6.19. For a polarizable integral variation of Hodge structures on a smooth algebraic variety S, Cattani, Deligne, and Kaplan ([8], Corollary 1.3) show that the sets $\pi_*(Z(t))$ in (6.14) are algebraic subvarieties of S. This answered a question of Weil [65].

Variation of Mumford-Tate groups in families

Definition 6.20. Let (V, F) be a family of Hodge structures on a connected complex manifold S.

(a) An *integral structure* on (V, F) is a local system of $\mathbb{Z}$-modules $\Lambda \subset V$ such that $\mathbb{Q} \otimes_{\mathbb{Z}} \Lambda \simeq V$.

(b) The family (V, F) is said to *satisfy the theorem of the fixed part* if, for every finite covering $a\colon S' \to S$ of S, there is a Hodge structure on the $\mathbb{Q}$-vector space $\Gamma(S', a^*V)$ such that, for all $s \in S'$, the canonical map $\Gamma(S', a^*V) \to a^*V_s$ is a morphism of Hodge structures, or, in other words, if the largest constant local subsystem V^f of a^*V is a constant family of Hodge substructures of a^*V.

(c) The *algebraic monodromy group* at point $s \in S$ is the smallest algebraic subgroup of GL_{V_s} containing the image of the monodromy homomorphism $\pi_1(S, s) \to GL(V_s)$. Its identity connected component is called the *connected monodromy group* M_s at s. In other words, M_s is the smallest connected algebraic subgroup of GL_{V_s} such that $M_s(\mathbb{Q})$ contains the image of a subgroup of $\pi_1(S, s)$ of finite index.

6.21. Let $\pi\colon \tilde{S} \to S$ be the universal covering of S, and let Γ be the group of covering transformations of $\tilde{S}/S$. The choice of a point $s \in \tilde{S}$ determines an isomorphism $\Gamma \simeq \pi_1(S, \pi s)$. Now choose a trivialization $\pi^* V \approx V_{\tilde{S}}$. The choice of a point $s \in \tilde{S}$ determines an isomorphism $V \simeq V_{\pi(s)}$. There is an action of Γ on V such that, for each $s \in \tilde{S}$, the diagram

$$\begin{array}{ccccc}
\Gamma & \times & V & \longrightarrow & V \\
\downarrow{\simeq} & & \downarrow{\simeq} & & \downarrow{\simeq} \\
\pi_1(S, \pi s) & \times & V_s & \longrightarrow & V_s
\end{array}$$

commutes. Let M be the smallest connected algebraic subgroup of GL_V such $M(\mathbb{Q})$ contains a subgroup of Γ of finite index; in other words,

$$M = \bigcap \{H \subset GL_V \mid H \text{ connected}, (\Gamma : H(\mathbb{Q}) \cap \Gamma) < \infty\}.$$

Under the isomorphism $V \simeq V_{\pi s}$ defined by $s \in S$, M maps isomorphically onto M_s.

Theorem 6.22. *Let* (V, F) *be a polarizable family of Hodge structures on a connected complex manifold* S, *and assume that* (V, F) *admits an integral structure. Let* G_s *(resp.* M_s*) denote the Mumford-Tate (resp. the connected monodromy group) at* $s \in S$.

(a) *For all* $s \in \mathring{S}$, $M_s \subset G_s^{der}$.

(b) *If* $T^{m,n}$ *satisfies the theorem of the fixed part for all* m, n, *then* M_s *is normal in* G_s^{der} *for all* $s \in \mathring{S}$; *moreover, if* $G_{s'}$ *is commutative for some* $s' \in S$, *then* $M_s = G_s^{der}$ *for all* $s \in S$.

The theorem was proved by Deligne (see [15], 7.5; [66], 7.3) except for the second statement of (b), which is Proposition 2 of [1]. The proof of the theorem will occupy the rest of this subsection.

Example 6.23. Let $f\colon X \to \mathbb{P}^1$ be a Lefschetz pencil over $\mathbb{C}$ of hypersurfaces of fixed degree and odd dimension n, and let S be the open subset of $\mathbb{P}^1$ where X_s is smooth. Let (V, F) be the variation of Hodge structures $R^n f_* \mathbb{Q}$ on S. The action of $\pi_1(S, s)$ on $V_s = H^n(X_s^{an}, \mathbb{Q})$ preserves the cup-product form on V_s, and a theorem of Kazhdan and Margulis ([17], 5.10) says that the image of $\pi_1(S, s)$ is Zariski-dense in the symplectic group. It follows that the generic Mumford-Tate group G_s is the full group of symplectic similitudes. This implies that, for $s \in \mathring{S}$, the Hodge structure V_s is not special unless it has level $\leqslant 1$.

Proof of (a) **of Theorem 6.22** We first show that $M_s \subset G_s$ for $s \in \mathring{S}$. Recall that on $\mathring{S}$ there is a local system of $\mathbb{Q}$-vector spaces $HT^{m,n} \subset T^{m,n}$ such that $HT^{m,n}_s$ is the space of Hodge tensors in $T^{m,n}_s$. The fundamental group $\pi_1(S, s)$ acts on $HT^{m,n}_s$ through a discrete subgroup of $GL(HT^{m,n}_s)$ (because it preserves a lattice in $T^{m,n}_s$), and it preserves a positive definite quadratic form on $HT^{m,n}_s$. It therefore acts on $HT^{m,n}_s$ through a finite quotient. As G_s is the algebraic subgroup of GL_{V_s} fixing the Hodge tensors in some finite direct sum of spaces $T^{m,n}_s$, this shows that the image of some finite index subgroup of $\pi_1(S, s)$ is contained in $G_s(\mathbb{Q})$. Hence $M_s \subset G_s$.

We next show that M_s is contained in the special Mumford-Tate group G^1_s at s. Consider the family of Hodge structures $V \oplus \mathbb{Q}(1)$, and let G^e_s be its Mumford-Tate group at s. As $V \oplus \mathbb{Q}(1)$ is polarizable and admits an integral structure, its connected monodromy group M^e_s at s is contained in G^e_s. As $\mathbb{Q}(1)$ is a constant family, $M^e_s \subset \mathrm{Ker}(G^e_s \to GL_{\mathbb{Q}(1)}) = G^1_s$. Therefore $M_s = M^e_s \subset G^1_s$.

There exists an object W in $\mathsf{Rep}_{\mathbb{Q}} G_s \simeq \langle V_s \rangle^{\otimes} \subset \mathsf{Hdg}_{\mathbb{Q}}$ such that $G^{der}_s \cdot w_{h_s}(\mathbb{G}_m)$ is the kernel of $G_s \to GL_W$. The Hodge structure W admits an integral structure, and its Mumford-Tate group is $G' \simeq G_s / \left(G^{der}_s \cdot w_{h_s}(\mathbb{G}_m)\right)$. As W has weight 0 and G' is commutative, we find from (6.4) that $G'(\mathbb{R})$ is compact. As the action of $\pi_1(S, s)$ on W preserves a lattice, its image in $G'(\mathbb{R})$ must be discrete, and hence finite. This shows that

$$M_s \subset \left(G^{der}_s \cdot w_{h_s}(\mathbb{G}_m)\right) \cap G^1_s = G^{der}_s.$$

Proof of the first statement of (b) **of Theorem 6.22** We first prove two lemmas.

Lemma 6.24. *Let V be a $\mathbb{Q}$-vector space, and let $H \subset G$ be algebraic subgroups of GL_V. Assume:*

(a) *the action of H on any H-stable line in a finite direct sum of spaces $T^{m,n}$ is trivial;*
(b) *$(T^{m,n})^H$ is G-stable for all $m, n \in \mathbb{N}$.*

Then H is normal in G.

Proof. There exists a line L in some finite direct sum T of spaces $T^{m,n}$ such that H is the stabilizer of L in GL_V (Chevalley's theorem, [20], 3.1a,b). According to (a), H acts trivially on L. Let W be the intersection of the G-stable subspaces of T containing L. Then $W \subset T^H$ because T^H is G-stable by (b). Let φ be the homomorphism $G \to GL_{W^\vee \otimes W}$ defined by the action of G on W. As H acts trivially on W, it is contained in the kernel of φ. On the other hand, the elements of the kernel of φ act as scalars on W, and so stabilize L. Therefore $H = \mathrm{Ker}(\varphi)$, which is normal in G. □

Lemma 6.25. *Let (V, F) be a polarizable family of Hodge structures on a connected complex manifold S. Let L be a local system of $\mathbb{Q}$-vector spaces on S contained in a finite direct sum of local systems $T^{m,n}$. If (V, F) admits an integral structure and L has dimension 1, then M_s acts trivially on L_s.*

Proof. The hypotheses imply that L also admits an integral structure, and so $\pi_1(S, s)$ acts through the finite subgroup $\{\pm 1\}$ of GL_{L_s}. This implies that M_s acts trivially on L_s. □

We now prove the first part of (b) of the theorem. Let $s \in \mathring{S}$; we shall apply Lemma 6.24 to $M_s \subset G_s \subset GL_{V_s}$. After passing to a finite covering of S, we may suppose that $\pi_1(S, s) \subset M_s(\mathbb{Q})$. Any M_s-stable line in $\bigoplus_{m,n} T_s^{m,n}$ is of the form L_s for a local subsystem L of $\bigoplus_{m,n} T_s^{m,n}$, and so hypothesis (a) of Lemma 6.24 follows from (6.25). It remains to show $(T_s^{m,n})^{M_s}$ is stable under G_s. Let H be the stabilizer of $(T_s^{m,n})^{M_s}$ in $GL_{T_s^{m,n}}$. Because $T^{m,n}$ satisfies the theorem of the fixed part, $(T_s^{m,n})^{M_s}$ is a Hodge substructure of $T_s^{m,n}$, and so $(T_s^{m,n})_{\mathbb{R}}^{M_s}$ is stable under $h(\mathbb{S})$. Therefore $h(\mathbb{S}) \subset H_{\mathbb{R}}$, and this implies that $G_s \subset H$.

Proof of the second statement of (b) of Theorem 6.22 We first prove a lemma.

Lemma 6.26. *Let* (V, F) *be a variation of polarizable Hodge structures on a connected complex manifold* S. *Assume:*

(a) M_s *is normal in* G_s *for all* $s \in \mathring{S}$;
(b) $\pi_1(S, s) \subset M_s(\mathbb{Q})$ *for one (hence every)* $s \in S$;
(c) (V, F) *satisfies the theorem of the fixed part.*

Then the subspace $\Gamma(S, V)$ *of* V_s *is stable under* G_s, *and the image of* G_s *in* $GL_{\Gamma(S,V)}$ *is independent of* $s \in S$.

In fact, (c) implies that $\Gamma(S, V)$ has a well-defined Hodge structure, and we shall show that the image of G_s in $GL_{\Gamma(S,V)}$ is the Mumford-Tate group of $\Gamma(S, V)$.

Proof. We begin with observation: let G be the affine group scheme attached to the tannakian category $\mathsf{Hdg}_{\mathbb{Q}}$ and the forgetful fibre functor; for any (V, h_V) in $\mathsf{Hdg}_{\mathbb{Q}}$, G acts on V through a surjective homomorphism $G \to MT_V$; therefore, for any (W, h_W) in $\langle V, h_V\rangle^{\otimes}$, MT_V acts on W through a surjective homomorphism $MT_V \to MT_W$.

For every $s \in S$,

$$\Gamma(S, V) = \Gamma(S, V^f) = (V^f)_s = V_s^{\pi_1(S,s)} \overset{(b)}{=} V_s^{M_s}.$$

The subspace $V_s^{M_s}$ of V_s is stable under G_s when $s \in \mathring{S}$ because then M_s is normal in G_s, and it is stable under G_s when $s \notin \mathring{S}$ because then G_s is contained in some generic Mumford-Tate group. Because (V, F) satisfies the theorem of the fixed part, $\Gamma(S, V)$ has a Hodge structure (independent of s) for which the inclusion $\Gamma(S, V) \to V_s$ is a morphism of Hodge structures. From the observation, we see that the image of G_s in $GL_{\Gamma(S,V)}$ is the Mumford-Tate group of $\Gamma(S, V)$, which does not depend on s. □

We now prove that $M_s = G_s^{der}$ when some Mumford-Tate group $G_{s'}$ is commutative. We know that M_s is a normal subgroup of G_s^{der} for $s \in \mathring{S}$, and so it remains to show that G_s/M_s is commutative for $s \in \mathring{S}$ under the hypothesis.

We begin with a remark. Let N be a normal algebraic subgroup of an algebraic group G. The category of representations of G/N can be identified with the category of representations of G on which N acts trivially. Therefore, to show that G/N is commutative, it suffices to show that G acts through a commutative quotient on every V on which N acts trivially. If G is reductive and we are in characteristic zero, then it suffices to show that, for one faithful representation V of G, the group G acts through a commutative quotient on $(T^{m,n})^N$ for all $m, n \in \mathbb{N}$.

Let $T = T^{m,n}$. According to the remark, it suffices to show that, for $s \in \mathring{S}$, G_s acts on $T_s^{M_s}$ through a commutative quotient. This will follow from the hypothesis, once we check that T satisfies the hypotheses of Lemma 6.26. Certainly, M_s is a normal subgroup of G_s for $s \in \mathring{S}$, and $\pi_1(S, s)$ will be contained in M_s once we have passed to a finite cover. Finally, we are assuming that T satisfies the theorem of the fixed part.

Variation of Mumford-Tate groups in algebraic families

When the underlying manifold is an algebraic variety, we have the following theorem.

Theorem 6.27 (Griffiths, Schmid). *A variation of Hodge structures on a smooth algebraic variety over $\mathbb{C}$ satisfies the theorem of the fixed part if it is polarizable and admits an integral structure.*

Proof. When the variation of Hodge structures arises from a projective smooth map $X \to S$ of algebraic varieties and S is complete, this is the original theorem of the fixed part ([25], §7). In the general case it is proved in [54], 7.22. See also [13], 4.1.2 and the footnote on p. 45. □

Theorem 6.28. *Let (V, F) be a variation of Hodge structures on a connected smooth complex algebraic variety S. If (V, F) is polarizable and admits an integral structure, then M_s is a normal subgroup of G_s^{der} for all $s \in \mathring{S}$, and the two groups are equal if G_s is commutative for some $s \in S$.*

Proof. If (V, F) is polarizable and admits an integral structure, then $T^{m,n}$ is polarizable and admits an integral structure, and so it satisfies the theorem of the fixed part (Theorem 6.27). Now the theorem follows from Theorem 6.22. □

7. Period subdomains

We define the notion of a period subdomain, and we show that the hermitian symmetric domains are exactly the period subdomains on which the universal family of Hodge structures is a *variation* of Hodge structures.

Flag manifolds

Let V be a complex vector space and let $\mathbf{d} = (d_1, \ldots, d_r)$ be a sequence of integers with $\dim V > d_1 > \cdots > d_r > 0$. The *flag manifold* $\mathrm{Gr}_{\mathbf{d}}(V)$ has as points the filtrations

$$V \supset F^1V \supset \cdots \supset F^rV \supset 0, \qquad \dim F^iV = d_i.$$

It is a projective complex manifold, and the tangent space to $\mathrm{Gr}_{\mathbf{d}}(V)$ at the point corresponding to a filtration F is

$$T_F(\mathrm{Gr}_{\mathbf{d}}(V)) \simeq \mathrm{End}(V)/F^0\,\mathrm{End}(V)$$

where

$$F^j\,\mathrm{End}(V) = \{\alpha \in \mathrm{End}(V) \mid \alpha(F^iV) \subset F^{i+j}V \text{ for all } i\}.$$

Theorem 7.1. *Let V_S be the constant sheaf on a connected complex manifold S defined by a real vector space V, and let (V_S, F) be a family of Hodge structures on S. Let $\mathbf{d}$ be the sequence of ranks of the subsheaves in F.*

(a) *The map $\varphi\colon S \to \mathrm{Gr}_{\mathbf{d}}(V_{\mathbb{C}})$ sending a point s of S to the point of $\mathrm{Gr}_{\mathbf{d}}(V_{\mathbb{C}})$ corresponding to the filtration F_s on V is holomorphic.*

(b) *The family (V_S, F) satisfies Griffiths transversality if and only if the image of the map*

$$(d\varphi)_s\colon T_sS \to T_{\varphi(s)}\,\mathrm{Gr}_{\mathbf{d}}(V_{\mathbb{C}})$$

lies in the subspace $F_s^{-1}\,\mathrm{End}(V_{\mathbb{C}})/F_s^0\,\mathrm{End}(V_{\mathbb{C}})$ of $\mathrm{End}(V_{\mathbb{C}})/F_s^0\,\mathrm{End}(V_{\mathbb{C}})$ for all $s \in S$.

Proof. Statement (a) simply says that the filtration is holomorphic, and (b) restates the definition of Griffiths transversality. □

Period domains

We now fix a real vector space V, a Hodge filtration F_0 on V of weight m, and a polarization $t_0\colon V \times V \to \mathbb{R}(m)$ of the Hodge structure (V, F_0).

Let $D = D(V, F_0, t_0)$ be the set of Hodge filtrations F on V of weight m with the same Hodge numbers as (V, F_0) for which t_0 is a polarization. Thus D is the set of descending filtrations

$$V_{\mathbb{C}} \supset \cdots \supset F^p \supset F^{p+1} \supset \cdots \supset 0$$

on $V_{\mathbb{C}}$ such that

(a) $\dim_{\mathbb{C}} F^p = \dim_{\mathbb{C}} F_0^p$ for all p,
(b) $V_{\mathbb{C}} = F^p \oplus \overline{F^q}$ whenever $p + q = m + 1$,
(c) $t_0(F^p, F^q) = 0$ whenever $p + q = m + 1$, and
(d) $(2\pi i)^m\, t_{0\mathbb{C}}(v, C\bar{v}) > 0$ for all nonzero elements v of $V_{\mathbb{C}}$.

Condition (b) requires that F be a Hodge filtration of weight m, condition (a) requires that (V, F) have the same Hodge numbers as (V, F_0), and the conditions (c) and (d) require that t_0 be a polarization.

Let $D^\vee = D^\vee(V, F_0, t_0)$ be the set of filtrations on $V_\mathbb{C}$ satisfying (a) and (c).

Theorem 7.2. *The set* $D^\vee$ *is a compact complex submanifold of* $\mathrm{Gr}_d(V)$, *and* D *is an open submanifold of* $D^\vee$.

Proof. We first remark that, in the presence of (a), condition (c) requires that F^{m+1-p} be the orthogonal complement of F^p for all p. In particular, each of F^p and F^{m+1-p} determines the other.

When m is odd, t_0 is alternating, and the remark shows that $D^\vee$ can be identified with the set of filtrations

$$V_\mathbb{C} \supset F^{(m+1)/2} \supset F^{(m+3)/2} \supset \cdots \supset 0$$

satisfying (a) and such that $F^{(m+1)/2}$ is totally isotropic for t_0. Let S be the symplectic group for t_0. Then $S(\mathbb{C})$ acts transitively on these filtrations, and the stabilizer P of the filtration F_0 is a parabolic subgroup of S. Therefore $S(\mathbb{C})/P(\mathbb{C})$ is a compact complex manifold, and the bijection $S(\mathbb{C})/P(\mathbb{C}) \simeq D^\vee$ is holomorphic. The proof when m is even is similar.

The submanifold D of $D^\vee$ is open because the conditions (b) and (d) are open. □

The complex manifold $D = D(V, F_0, t_0)$ is the (Griffiths) *period domain* defined by (V, F_0, t_0).

Theorem 7.3. *Let* (V, F, t) *be a polarized family of Hodge structures on a complex manifold* S. *Let* U *be an open connected subset of* S *on which the local system* V *is trivial, and choose an isomorphism* $V|U \simeq V_U$ *and a point* $o \in U$. *The map* $\mathcal{P}\colon U \to D(V, F_o, t_o)$ *sending a point* $s \in U$ *to the point* (V_s, F_s, t_s) *is holomorphic.*

Proof. The map $s \mapsto F_s \colon U \to \mathrm{Gr}_d(V)$ is holomorphic by (7.1) and it takes values in D. As D is a complex submanifold of $\mathrm{Gr}_d(V)$ this implies that the map $U \to D$ is holomorphic ([23], 4.3.3). □

The map $\mathcal{P}$ is called the *period map*.

The constant local system of real vector spaces V_D on D becomes a polarized family of Hodge structures on D in an obvious way (called the *universal family*)

Theorem 7.4. *If the universal family of Hodge structures on* $D = D(V, F_0, t_0)$ *satisfies Griffiths transversality, then* D *is a hermitian symmetric domain.*

Proof. Let $h_0 \colon \mathbb{S} \to GL_V$ be the homomorphism corresponding to the Hodge filtration F_0, and let G be the algebraic subgroup of GL_V whose elements fix t_0 up to scalar. Then h_0 maps into G, and $h_0 \circ w$ maps into its centre (recall that V has a single weight m). Therefore (see 6.1), there exists a homomorphism $u_0\colon \mathbb{S}^1 \to G^{\mathrm{ad}}$ such that $h_0(z) = u_0(z/\bar{z}) \bmod Z(G)(\mathbb{R})$.

Let o be the point F_0 of D, and let $\mathfrak{g}$ denote Lie G with the Hodge structure provided by $\mathrm{Ad} \circ h_0$. Then

$$\mathfrak{g}_{\mathbb{C}}/\mathfrak{g}^{00} \simeq T_o(D) \subset T_o(\mathrm{Gr}_d(V)) \simeq \mathrm{End}(V)/F^0\,\mathrm{End}(V).$$

If the universal family of Hodge structures satisfies Griffiths transversality, then $\mathfrak{g}_{\mathbb{C}} = F^{-1}\mathfrak{g}_{\mathbb{C}}$ (by 7.1b). As $\mathfrak{g}$ is of weight 0, it must be of type $\{(1,-1),(0,0),(-1,1)$, and so h_0 satisfies the condition SV1. Hence u_0 satisfies condition SU1 of Theorem 2.5.

Let G^1 be the subgroup of G of elements fixing t_0. As t_0 is a polarization of the Hodge structure, $(2\pi i)^m t_0$ is a C-polarization of V relative to G^1, and so $\mathrm{inn}(C)$ is a Cartan involution of G^1 (Theorem 2.1). Now $C = h_0(i) = u_0(-1)$, and so u_0 satisfies condition SU2 of Theorem 2.5. The set D is a connected component of the space of homomorphisms $u\colon \mathbb{S}^1 \to (G^1)^{\mathrm{ad}}$, and so it is equal to the set of conjugates of u_0 by elements of $(G^1)^{\mathrm{ad}}(\mathbb{R})^+$ (apply 7.6 below with $\mathbb{S}$ replaced by $\mathbb{S}^1$). Any compact factors of $(G^1)^{\mathrm{ad}}$ can be discarded, and so Theorem 2.5 shows that D is a hermitian symmetric domain. □

Remark 7.5. The universal family of Hodge structures on the period domain $D(V,h,t_0)$ satisfies Griffiths transversality only if (a) (V,h) is of type $\{(-1,0),(0,-1)\}$, or (b) (V,h) of type $\{(-1,1),(0,0),(1,-1)\}$ and $h^{-1,1} \leqslant 1$, or (c) (V,h) is a Tate twist of one of these Hodge structures.

Period subdomains

7.6. We shall need the following statement ([19], 1.1.12.). Let G be a real algebraic group, and let X be a (topological) connected component of the space of homomorphisms $\mathbb{S} \to G$. Let G_1 be the smallest algebraic subgroup of G through which all the $h \in X$ factor. Then X is again a connected component of the space of homomorphisms of $\mathbb{S}$ into G_1. Since $\mathbb{S}$ is a torus, any two elements of X are conjugate, and so the space X is a $G_1(\mathbb{R})^+$-conjugacy class of morphisms from $\mathbb{S}$ into G. It is also a $G(\mathbb{R})^+$-conjugacy class, and G_1 is a normal subgroup of the identity component of G.

Let (V,F_0) be a real Hodge structure of weight m. A tensor $t\colon V^{\otimes 2r} \to \mathbb{R}(-mr)$ of V is a *Hodge tensor* of (V,F_0) if it is a morphism of Hodge structures. Concretely, this means that t is of type $(0,0)$ for the natural Hodge structure on

$$\mathrm{Hom}(V^{\otimes 2r},\mathbb{R}(-mr)) \simeq \left(V^{\vee}\right)^{\otimes 2r}(-mr),$$

or that it lies in $F^0\left(\mathrm{Hom}(V^{\otimes 2r},\mathbb{R}(-mr))\right)$.

We now fix a real Hodge structure (V,F_0) of weight m and a family $t = (t_i)_{i\in I}$ of Hodge tensors of (V,F_0). We assume that I contains an element 0 such that t_0 is a polarization of (V,F_0). Let $D(V,F_0,t)$ be a connected component of the set of Hodge filtrations F in $D(V,F_0,t_0)$ for which every t_i is a Hodge tensor. Thus, $D(V,F_0,t)$ is a connected component of the space of Hodge structures on V for which every t_i is a Hodge tensor and t_0 is a polarization.

Let G be the algebraic subgroup of $GL_V \times GL_{\mathbb{Q}(1)}$ fixing the t_i. Then $G(\mathbb{R})$ consists of the pairs (g, c) such that

$$t_i(gv_1, \ldots, gv_{2r}) = c^{rm} t_i(v_1, \ldots, v_{2r})$$

for $i \in I$. Let h be a homomorphism $\mathbb{S} \to GL_V$. The t_i are Hodge tensors for (V, h) if and only if the homomorphism

$$z \mapsto (h(z), z\bar{z}) \colon \mathbb{S} \to GL_V \times \mathbb{G}_m$$

factors through G. Thus, to give a Hodge structure on V for which all the t_i are Hodge tensors is the same as giving a homomorphism $h \colon \mathbb{S} \to G$, and so D is a connected component of the space of homomorphisms $\mathbb{S} \to G$.

Let G_1 be the smallest algebraic subgroup of G through which all the h in D factor. According to (7.6), D is a $G_1(\mathbb{R})^+$-conjugacy class of homomorphisms $\mathbb{S} \to G_1$. The group $G_1(\mathbb{C})$ acts on $D^\vee(V, F_0, t_0)$, and we let $D^\vee(V, F_0, t)$ denote the orbit of F_0.

Theorem 7.7. *The set* $D^\vee(V, F_0, t)$ *is a compact complex submanifold of* $D^\vee(V, F_0, t_0)$, *and* D *is an open complex submanifold of* $D^\vee$.

Proof. In fact, $D^\vee(V, F_0, t_0)$ is a smooth projective algebraic variety. The stabilizer P of F_0 in the algebraic group $G_{1\mathbb{C}}$ is parabolic, and so the orbit of F_0 in the algebraic variety $D^\vee(V, F_0, t_0)$ is smooth projective variety. Thus, its complex points form a compact complex submanifold. As

$$D(V, h_0, t_0) = D(V, h_0, t_0) \cap D^\vee(V, h_0, t_0),$$

it is an open complex submanifold of $D^\vee(V, h_0, t_0)$. □

We call $D = D(V, F_0, t)$ the *period subdomain* defined by (V, F_0, t).

Theorem 7.8. *Let* (V, F) *be a family of Hodge structures on a complex manifold* S, *and let* $t = (t_i)_{i \in I}$ *be a family of Hodge tensors of* V. *Assume that* I *contains an element* 0 *such that* t_0 *is a polarization. Let* U *be a connected open subset of* S *on which the local system* V *is trivial, and choose an isomorphism* $\mathsf{V}|U \xrightarrow{\approx} V_U$ *and a point* $o \in U$. *The map* $\mathcal{P} \colon U \to D(V, F_o, t_o)$ *sending a point* $s \in U$ *to the point* (V_s, F_s, t_s) *is holomorphic.*

Proof. Same as that of Theorem 7.3. □

Theorem 7.9. *If the universal family of Hodge structures on* D *satisfies Griffiths transversality, then* D *is a hermitian symmetric domain.*

Proof. Essentially the same as that of Theorem 7.4. □

Theorem 7.10. *Every hermitian symmetric domain arises as a period subdomain.*

Proof. Let D be a hermitian symmetric domain, and let $o \in D$. Let H be the real adjoint algebraic group such that $H(\mathbb{R})^+ = \mathrm{Hol}(D)^+$, and let $u: \mathbb{S}^1 \to H$ be the homomorphism such that $u(z)$ fixes o and acts on $T_0(D)$ as multiplication by z (see §2). Let $h: \mathbb{S} \to H$ be the homomorphism such that $h(z) = u_o(z/\bar{z})$ for $z \in \mathbb{C}^\times = \mathbb{S}(\mathbb{R})$. Choose a faithful representation $\rho: H \to GL_V$ of G. Because u satisfies (2.5, SU2), the Hodge structure $(V, \rho \circ h)$ is polarizable. Choose a polarization and include it in a family t of tensors for V such that H is the subgroup of $GL_V \times GL_{\mathbb{Q}(1)}$ fixing the elements of t. Then $D \simeq D(V, h, t)$. □

Notes. The interpretation of hermitian symmetric domains as moduli spaces for Hodge structures with tensors is taken from [19], 1.1.17.

Why moduli varieties are (sometimes) locally symmetric

Fix a base field k. A *moduli problem* over k is a contravariant functor $\mathcal{F}$ from the category of (some class of) schemes over k to the category of sets. A variety S over k together with a natural isomorphism $\phi: \mathcal{F} \to \mathrm{Hom}_k(-, S)$ is called a *fine solution to the moduli problem*. A variety that arises in this way is called a *moduli variety*.

Clearly, this definition is too general: every variety S represents the functor $h_S = \mathrm{Hom}_k(-, S)$. In practice, we only consider functors for which $\mathcal{F}(T)$ is the set of isomorphism classes of some algebro-geometric objects over T, for example, families of algebraic varieties with additional structure.

If S represents such a functor, then there is an object $\alpha \in \mathcal{F}(S)$ that is universal in the sense that, for any $\alpha' \in \mathcal{F}(T)$, there is a unique morphism $a: T \to S$ such that $\mathcal{F}(a)(\alpha) = \alpha'$. Suppose that α is, in fact, a smooth projective map $f: X \to S$ of smooth varieties over $\mathbb{C}$. Then $R^m f_* \mathbb{Q}$ is a polarizable variation of Hodge structures on S admitting an integral structure (Theorem 5.6). A polarization of X/S defines a polarization of $R^m f_* \mathbb{Q}$ and a family of algebraic classes on X/S of codimension m defines a family of global sections of $R^{2m} f_* \mathbb{Q}(m)$. Let D be the universal covering space of S^{an}. The pull-back of $R^m f_* \mathbb{Q}$ to D is a variation of Hodge structures whose underlying locally constant sheaf of $\mathbb{Q}$-vector spaces is constant, say, equal to V_S; thus we have a variation of Hodge structures (V_S, F) on D. We suppose that the additional structure on X/S defines a family $t = (t_i)_{i \in I}$ of Hodge tensors of V_S with t_0 a polarization. We also suppose that the family of Hodge structures on D is universal[20], i.e., that $D = D(V, F_0, t)$. Because (V_S, F) is a variation of Hodge structures, D is a hermitian symmetric domain (by 7.9). The Margulis arithmeticity theorem (3.12) shows that Γ is an arithmetic subgroup of $G(D)$ except possibly when $G(D)$ has factors of small dimension. Thus, when looking at moduli varieties, we are naturally led to consider arithmetic locally symmetric varieties.

Remark 7.11. In fact it is unusual for a moduli problem to lead to a locally symmetric variety. The above argument will usually break down where we assumed that the

[20]This happens rarely!

variation of Hodge structures is universal. Essentially, this will happen only when a "general" member of the family has a Hodge structure that is special in the sense of §6. Even for smooth hypersurfaces of a fixed degree, this is rarely happens (see 6.8 and 6.12). Thus, in the whole universe of moduli varieties, locally symmetric varieties form only a small, but important, class.

Application: Riemann's theorem in families

Let A be an abelian variety over $\mathbb{C}$. The exponential map defines an exact sequence

$$0 \to H_1(A^{an}, \mathbb{Z}) \to T_0(A^{an}) \xrightarrow{\exp} A^{an} \to 0.$$

From the first map in this sequence, we get an exact sequence

$$0 \to \mathrm{Ker}(\alpha) \to H_1(A^{an}, \mathbb{Z})_{\mathbb{C}} \xrightarrow{\alpha} T_0(A^{an}) \to 0.$$

The $\mathbb{Z}$-module $H_1(A^{an}, \mathbb{Z})$ is an integral Hodge structure with Hodge filtration

$$F^{-1} = H_1(A^{an}, \mathbb{Z})_{\mathbb{C}} \supset F^0 = \mathrm{Ker}(\alpha) \supset 0.$$

Let ψ be a Riemann form for A. Then $2\pi i\psi$ is a polarization for the Hodge structure $H_1(A^{an}, \mathbb{Z})$.

Theorem 7.12. *The functor* $A \rightsquigarrow H_1(A^{an}, \mathbb{Z})$ *is an equivalence from the category of abelian varieties over* $\mathbb{C}$ *to the category of polarizable integral Hodge structures of type* $\{(-1,0), (0,-1)\}$.

Proof. In view of the correspondence between complex structures and Hodge structures of type $\{(-1,0), (0,-1)\}$ (see 5.3), this is simply a restatement of Theorem 4.4. □

Theorem 7.13. *Let S be a smooth algebraic variety over* $\mathbb{C}$. *The functor*

$$(A \xrightarrow{f} S) \rightsquigarrow R_1 f_* \mathbb{Z}$$

is an equivalence from the category of families of abelian varieties over S to the category of polarizable integral variations of Hodge structures of type $\{(-1,0), (0,-1)\}$.

Proof. Let $f^A \colon A \to S$ be a family of abelian varieties over S. The exponential defines an exact sequence of sheaves on S^{an},

$$0 \to R_1 f_*^A \mathbb{Z} \to \mathcal{T}_0(A^{an}) \to A^{an} \to 0.$$

From this one sees that the map $\mathrm{Hom}(A^{an}, B^{an}) \to \mathrm{Hom}(R_1 f_*^A \mathbb{Z}, R_1 f_*^B \mathbb{Z})$ is an isomorphism. The S-scheme $\mathcal{H}om_S(A, B)$ is unramified over S, and so its algebraic sections coincide with its holomorphic sections (cf. [13], 4.4.3). Hence the functor is fully faithful. In particular, a family of abelian varieties is uniquely determined by its variation of Hodge structures up to a unique isomorphism. This allows us to construct the family of abelian varieties attached to a variation of Hodge structures locally. Thus, we may suppose that the underlying local system of $\mathbb{Z}$-modules is trivial. Assume initially that the variation of Hodge structures on S has a principal polarization, and endow it with a level-N structure. According Proposition 4.5, the variation of Hodge structures on S is the pull-back of the canonical variation of Hodge structures on D_N by a regular map $\alpha\colon S \to D_N$. Since the latter variation arises from a family of abelian varieties (Theorem 4.6), so does the former.

In fact, the argument still applies when the variation of Hodge structures is not *principally* polarized, since [44], Chapter 7, hence Theorem 4.6, applies also to nonprincipally polarized abelian varieties. Alternatively, Zarhin's trick (cf. [36], 16.12) can be used to show that (locally) the fourth multiple of the variation of Hodge structures is principally polarized. □

8. Variations of Hodge structures on locally symmetric varieties

In this section, we explain how to classify variations of Hodge structures on arithmetic locally symmetric varieties in terms of certain auxiliary reductive groups. Throughout, we write "family of integral Hodge structures" to mean "family of rational Hodge structures that admits an integral structure".

Existence of Hodge structures of CM-type in a family

Proposition 8.1. *Let* G *be a reductive group over* $\mathbb{Q}$, *and let* $h\colon \mathbb{S} \to G_{\mathbb{R}}$ *be a homomorphism. There exists a* $G(\mathbb{R})^+$*-conjugate* h_0 *of* h *such that* $h_0(\mathbb{S}) \subset T_{0\mathbb{R}}$ *for some maximal torus* T_0 *of* G.

Proof. (Mumford 1969 [46, p. 348]) Let K be the centralizer of h in $G_{\mathbb{R}}$, and let T be the centralizer in $G_{\mathbb{R}}$ of some regular element of Lie K; it is a maximal torus in K. Because $h(\mathbb{S})$ centralizes T, $h(\mathbb{S}) \cdot T$ is a torus in K, and so $h(\mathbb{S}) \subset T$. If T' is a torus in $G_{\mathbb{R}}$ containing T, then T' centralizes h, and so $T' \subset K$; therefore $T = T'$, and so T is maximal in $G_{\mathbb{R}}$. For a regular element λ of $\mathrm{Lie}(T)$, T is the centralizer of λ. Choose a $\lambda_0 \in \mathrm{Lie}(G)$ that is close to λ in $\mathrm{Lie}(G)_{\mathbb{R}}$, and let T_0 be its centralizer in G. Then T_0 is a maximal torus of G (over $\mathbb{Q}$). Because $T_{0\mathbb{R}}$ and $T_{\mathbb{R}}$ are close, they are conjugate: $T_{0\mathbb{R}} = gTg^{-1}$ for some $g \in G(\mathbb{R})^+$. Now $h_0 \overset{\text{def}}{=} \mathrm{inn}(g) \circ h$ factors through $T_{0\mathbb{R}}$. □

A rational Hodge structure is said to be of *CM-type* if it is polarizable and its Mumford-Tate group is commutative (hence a torus by 6.5).

Proposition 8.2. *Let* (V, F_0) *be a rational Hodge structure of some weight* m, *and let* $t = (t_i)_{i \in I}$ *be a family of tensors of* (V, F_0) *including a polarization. Then the period subdomain defined by* $(V, F_0, t)_{\mathbb{R}}$ *includes a Hodge structure of CM-type.*

Proof. We are given a $\mathbb{Q}$-vector space V, a homomorphism $h_0\colon \mathbb{S} \to GL_{V_{\mathbb{R}}}$, and a family of Hodge tensors $V^{\otimes 2r} \to \mathbb{Q}(-mr)$ including a polarization. Let G be the algebraic subgroup of $GL_V \times GL_{\mathbb{Q}(1)}$ fixing the t_i. Then G is a reductive group because $\mathrm{inn}(h_0(i))$ is a Cartan involution. The period subdomain D is the connected component containing h_0 of the space of homomorphisms $h\colon \mathbb{S} \to G_{\mathbb{R}}$ (see §7).

This contains the $G(\mathbb{R})^+$-conjugacy class of h_0, and so the statement follows from Proposition 8.1. □

Description of the variations of Hodge structures on $D(\Gamma)$

Consider an arithmetic locally symmetric variety $D(\Gamma)$. Recall that this means that $D(\Gamma)$ is an algebraic variety whose universal covering space is a hermitian symmetric domain D and that the group of covering transformations Γ is an arithmetic subgroup of the real Lie group $\mathrm{Hol}(D)^+$; moreover, $D(\Gamma)^{\mathrm{an}} = \Gamma\backslash D$.

According to Theorem 3.2, D decomposes into a product $D = D_1 \times \cdots \times D_r$ of hermitian symmetric domains with the property that each group $\Gamma_i \overset{\text{def}}{=} \Gamma \cap \mathrm{Hol}(D_i)^+$ is an irreducible arithmetic subgroup of $\mathrm{Hol}(D_i)^+$ and the map

$$D_1(\Gamma_1) \times \cdots \times D_r(\Gamma_r) \to D(\Gamma)$$

is finite covering. In order to be able to apply the theorems of Margulis we assume that

$$\text{(8.3)} \qquad \mathrm{rank}(\mathrm{Hol}(D_i)) \geqslant 2 \text{ for each } i$$

in the remainder of this subsection. We also fix a point $o \in D$.

Recall (2.3) that there exists a unique homomorphism $u\colon U^1 \to \mathrm{Hol}(D)$ such that $u(z)$ fixes o and acts as multiplication by z on $T_o(D)$. That Γ is arithmetic means that there exists a simply connected algebraic group H over $\mathbb{Q}$ and a surjective homomorphism $\varphi\colon H(\mathbb{R}) \to \mathrm{Hol}(D)^+$ with compact kernel such that Γ is commensurable with $\varphi(H(\mathbb{Z}))$. The Margulis superrigidity theorem implies that the pair (H, φ) is unique up to a unique isomorphism (see 3.13).

Let

$$H^{\mathrm{ad}}_{\mathbb{R}} = H_c \times H_{nc}$$

where H_c (resp. H_{nc}) is the product of the compact (resp. noncompact) simple factors of $H^{\mathrm{ad}}_{\mathbb{R}}$. The homomorphism $\varphi(\mathbb{R})\colon H(\mathbb{R}) \to \mathrm{Hol}(D)^+$ factors through $H_{nc}(\mathbb{R})^+$, and defines an isomorphism of Lie groups $H_{nc}(\mathbb{R})^+ \to \mathrm{Hol}(D)^+$. Let $\bar{h}$ denote the homomorphism $\mathbb{S}/\mathbb{G}_m \to H^{\mathrm{ad}}_{\mathbb{R}}$ whose projection into H_c is trivial and whose projection into H_{nc} corresponds to u as in (6.1). In other words,

$$\text{(8.4)} \qquad \bar{h}(z) = (h_c(z), h_{nc}(z)) \in H_c(\mathbb{R}) \times H_{nc}(\mathbb{R})$$

where $h_c(z) = 1$ and $h_{nc}(z) = u(z/\bar{z})$ in $H_{nc}(\mathbb{R})^+ \simeq \mathrm{Hol}(D)^+$. The map ${}^{g}h \mapsto go$ identifies D with the set of $H^{\mathrm{ad}}(\mathbb{R})^+$-conjugates of $\bar{h}$ (Theorem 2.5).

Let (V, F) be a polarizable variation of integral Hodge structures on $D(\Gamma)$, and let $V = V_{\pi(o)}$. Then $\pi^*V \simeq V_D$ where $\pi\colon D \to \Gamma\backslash D$ is the quotient map. Let $G \subset GL_V$ be the generic Mumford-Tate group of (V, F) (see p. 498), and let t be a family of tensors of V (in the sense of §7), including a polarization t_0, such that G is the subgroup of $GL_V \times GL_{\mathbb{Q}(1)}$ fixing the elements of t. As G contains the Mumford-Tate group at each point of D, t is a family of Hodge tensors of (V_D, F). The period map $\mathcal{P}\colon D \to D(V, h_o, t)$ is holomorphic (Theorem 7.8).

We now assume that the monodromy map $\varphi' \colon \Gamma \to \mathrm{GL}(V)$ has finite kernel, and we pass to a finite covering, so that $\Gamma \subset G(\mathbb{Q})$. Now the elements of t are Hodge tensors of (V, F).

There exists an arithmetic subgroup Γ' of $H(\mathbb{Q})$ such that $\varphi(\Gamma') \subset \Gamma$. The Margulis superrigidity theorem 3.10, shows that there is a (unique) homomorphism $\varphi'' \colon H \to G$ of algebraic groups that agrees with $\varphi' \circ \varphi$ on a subgroup of finite index in Γ',

$$\begin{array}{ccccc} H(\mathbb{Q})^+ & \xrightarrow{\varphi} & \mathrm{Hol}(D)^+ & & \\ \cup & & \cup & & \\ \Gamma' & \xrightarrow{\varphi|\Gamma'} & \Gamma & \xrightarrow{\varphi'} & G(\mathbb{Q}) \end{array} \qquad\qquad H \xrightarrow{\varphi''} G$$

It follows from the Borel density theorem 3.11 that $\varphi''(H)$ is the connected monodromy group at each point of $D(\Gamma)$. Hence $H \subset G^{\mathrm{der}}$, and the two groups are equal if the Mumford-Tate group at some point of $D(\Gamma)$ is commutative (Theorem 6.22). When we assume that, the homomorphism $\varphi'' \colon H \to G$ induces an isogeny $H \to G^{\mathrm{der}}$, and hence[21] an isomorphism $H^{\mathrm{ad}} \to G^{\mathrm{ad}}$. Let $(V, h_o) = (V, F)_o$. Then

$$\mathrm{ad} \circ h_o \colon \mathbb{S} \to G^{\mathrm{ad}}_{\mathbb{R}} \simeq H^{\mathrm{ad}}$$

equals $\bar{h}$. Thus, we have a commutative diagram

$$\begin{array}{ccccc} H & & & & \\ \downarrow & \searrow & & & \\ (H^{\mathrm{ad}}, \bar{h}) & \longleftarrow & (G, h) & \stackrel{\rho}{\hookrightarrow} & \mathrm{GL}_V \end{array} \tag{8.5}$$

in which G is a reductive group, the homomorphism $H \to G$ has image G^{der}, w_h is defined over $\mathbb{Q}$, and h satisfies (SV2*).

Conversely, suppose that we are given such a diagram (8.5). Choose a family t of tensors for V, including a polarization, such that G is the subgroup of $\mathrm{GL}_V \times \mathbb{G}_{\mathbb{Q}(1)}$ fixing the tensors. Then we get a period subdomain $D(V, h, t)$ and a canonical variation of Hodge structures (V, F) on it. Pull this back to D using the period isomorphism, and descend it to a variation of Hodge structures on $D(\Gamma)$. The monodromy representation is injective, and some fibre is of CM-type by Proposition 8.2.

Summary 8.6. *Let $D(\Gamma)$ be an arithmetic locally symmetric domain satisfying the condition (8.3) and fix a point $o \in D$. To give*

---

[21] Let G be a reductive group. The algebraic subgroup $Z(G) \cdot G^{\mathrm{der}}$ is normal, and the quotient $G/\left(Z(G)^{\circ} \cdot G^{\mathrm{der}}\right)$ is both semisimple and commutative, and hence is trivial. Therefore $G = Z(G)^{\circ} \cdot G^{\mathrm{der}}$, from which it follows that $Z(G^{\mathrm{der}}) = Z(G) \cap G^{\mathrm{der}}$. For any isogeny $H \to G^{\mathrm{der}}$, the map $H^{\mathrm{ad}} \to (G^{\mathrm{der}})^{\mathrm{ad}}$ is certainly an isomorphism, and we have just shown that $(G^{\mathrm{der}})^{\mathrm{ad}} \to G^{\mathrm{ad}}$ is an isomorphism. Therefore $H^{\mathrm{ad}} \to G^{\mathrm{ad}}$ is an isomorphism.

a polarizable variation of integral Hodge structures on $D(\Gamma)$ *such that some fibre is of CM-type and the monodromy representation has finite kernel*

is the same as giving

a diagram (8.5) in which G *is a reductive group, the homomorphism* $H \to G$ *has image* G^{der}, w_h *is defined over* $\mathbb{Q}$, *and* h *satisfies (SV2*).*

Fundamental Question 8.7. For which arithmetic locally symmetric varieties $D(\Gamma)$ is it possible to find a diagram (8.5) with the property that the corresponding variation of Hodge structures underlies a family of algebraic varieties? or, more generally, a family of motives?

In §§10,11, we shall answer Question 8.7 completely when "algebraic variety" and "motive" are replaced with "abelian variety" and "abelian motive".

Existence of variations of Hodge structures

In this subsection, we show that, for every arithmetic locally symmetric variety, there exists a diagram (8.5), and hence a variation of polarizable integral Hodge structures on the variety.

Proposition 8.8. *Let* H *be a semisimple algebraic group over* $\mathbb{Q}$, *and let* $\bar{h}\colon \mathbb{S} \to H^{ad}$ *be a homomorphism satisfying (SV1,2,3). Then there exists a reductive algebraic group* G *over* $\mathbb{Q}$ *and a homomorphism* $h\colon \mathbb{S} \to G_{\mathbb{R}}$ *such that*

(a) $G^{der} = H$ *and* $\bar{h} = \mathrm{ad} \circ h$,

(b) *the weight* w_h *is defined over* $\mathbb{Q}$, *and*

(c) *the centre of* G *is split by a CM field (i.e., a totally imaginary quadratic extension of a totally real number field).*

Proof. We shall need the following statement:

Let G be a reductive group over a field k (of characteristic zero), and let L be a finite Galois extension of k splitting G. Let $G' \to G^{der}$ be a covering of the derived group of G. Then there exists a central extension

$$1 \to N \to G_1 \to G \to 1$$

such that G_1 is a reductive group, N is a product of copies of $(\mathbb{G}_m)_{L/k}$, and

$$(G_1^{der} \to G^{der}) = (G' \to G^{der}).$$

See [41], 3.1.

A number field L is CM if and only if it admits a nontrivial involution ι_L such that $\sigma \circ \iota_L = \iota \circ \sigma$ for every homomorphism $\sigma\colon L \to \mathbb{C}$. We may replace $\bar{h}$ with an $H^{ad}(\mathbb{R})^+$-conjugate, and so assume (by Proposition 8.1) that there exists a maximal torus $\bar{T}$ of H^{ad} such that $\bar{h}$ factors through $\bar{T}_{\mathbb{R}}$. Then $\bar{T}_{\mathbb{R}}$ is anisotropic (by (SV2)), and so ι acts as -1 on $X^*(\bar{T})$. It follows that, for any $\sigma \in \mathrm{Aut}(\mathbb{C})$, $\sigma\iota$ and $\iota\sigma$ have the same action on $X^*(\bar{T})$, and so $\bar{T}$ splits over a CM-field L, which can be chosen to be Galois over $\mathbb{Q}$. From the statement, there exists a reductive group G and a central extension

$$1 \to N \to G \to H^{ad} \to 1$$

such that $G^{der} = H$ and N is a product of copies of $(\mathbb{G}_m)_{L/\mathbb{Q}}$. The inverse image T of $\bar{T}$ in G is a maximal torus, and the kernel of $T \twoheadrightarrow \bar{T}$ is N. Because N is connected, there exists a $\mu \in X_*(T)$ lifting $\mu_{\bar{h}} \in X_*(\bar{T})$.[22] The weight $w = -\mu - \iota\mu$ of μ lies in $X_*(Z)$, where $Z = Z(G) = N$. Clearly $\iota w = w$ and so, as the Tate cohomology group[23] $H^0_T(\mathbb{R}, X_*(Z)) = 0$, there exists a $\mu_0 \in X_*(Z)$ such that $(\iota+1)\mu_0 = w$. When we replace μ with $\mu - \mu_0$, we find that $w = 0$; in particular, w is defined over $\mathbb{Q}$. Let $h\colon \mathbb{S} \to G_{\mathbb{R}}$ correspond to μ as in (5.1), p. 490. Then (G, h) fulfils the requirements. □

Corollary 8.9. *For any semisimple algebraic group* H *over* $\mathbb{Q}$ *and homomorphism* $\bar{h}$: $\mathbb{S}/\mathbb{G}_m \to H^{ad}_{\mathbb{R}}$ *satisfying (SV1,2,3), there exists a reductive group* G *with* $G^{der} = H$ *and a homomorphism* $h\colon \mathbb{S} \to G_{\mathbb{R}}$ *lifting* $\bar{h}$ *and satisfying (SV1,2*,3).*

Proof. Let (G, h) be as in the proposition. Then G/G^{der} is a torus, and we let T be the smallest subtorus of it such that $T_{\mathbb{R}}$ contains the image of h. Then $T_{\mathbb{R}}$ is anisotropic, and when we replace G with the inverse image of T, we obtain a pair (G, h) satisfying (SV1,2*,3). □

Let G be a reductive group over $\mathbb{Q}$, and let $h\colon \mathbb{S} \to G_{\mathbb{R}}$ be a homomorphism satisfying (SV1,2,3). The homomorphism h is said to be *special* if $h(\mathbb{S}) \subset T_{\mathbb{R}}$ for some torus $T \subset G$.[24] In this case, there is a smallest such T, and when (T, h) is the Mumford-Tate group of a CM Hodge structure we say that h is *CM*.

Proposition 8.10. *Let* $h\colon \mathbb{S} \to G_{\mathbb{R}}$ *be special. Then* h *is CM if*

(a) w_h *is defined over* $\mathbb{Q}$, *and*
(b) *the connected centre of* G *is split by a CM-field.*

Proof. It is known that a special h is CM if and only if it satisfies the Serre condition:

$$(\tau - 1)(\iota + 1)\mu_h = 0 = (\iota + 1)(\tau - 1)\mu_h \text{ for all } \tau \in \mathrm{Gal}(\mathbb{Q}^{al}/\mathbb{Q}).$$

As $w_h = (\iota + 1)\mu_h$, the first condition says that

$$(\tau - 1)(\iota + 1)\mu_h = 0 \text{ for all } \tau \in \mathrm{Aut}(\mathbb{C}),$$

and the second condition implies that

$$\tau\iota\mu_h = \iota\tau\mu_h \text{ for all } \tau \in \mathrm{Aut}(\mathbb{C}).$$

[22]The functor X^* is exact, and so $0 \to X^*(\bar{T}) \to X^*(T) \to X^*(N) \to 0$ is exact. In fact, it is split-exact (as a sequence of $\mathbb{Z}$-modules) because $X^*(N)$ is torsion-free. On applying $\mathrm{Hom}(-, \mathbb{Z})$ to it, we get the exact sequence $\cdots \to X_*(T) \to X_*(\bar{T}) \to 0$.

[23]Let $g = \mathrm{Gal}(\mathbb{C}/\mathbb{R})$. The g-module $X_*(Z)$ is induced, and so the Tate cohomology group $H^0_T(g, X_*(Z)) = 0$. By definition, $H^0_T(g, X_*(Z)) = X_*(Z)^g/(\iota + 1)X_*(Z)$.

[24]Of course, $h(\mathbb{S})$ is always contained in a subtorus of $G_{\mathbb{R}}$, even a maximal subtorus; the point is that there should exist such a torus defined over $\mathbb{Q}$.

Let $T \subset G$ be a maximal torus such that $h(\mathbb{S}) \subset T_{\mathbb{R}}$. The argument in the proof of (8.8) shows that $\tau\iota\mu = \iota\tau\mu$ for $\mu \in X_*(T)$, and since

$$X_*(T)_{\mathbb{Q}} = X_*(Z)_{\mathbb{Q}} \oplus X_*(T/Z)_{\mathbb{Q}}$$

we see that the same equation holds for $\mu \in X_*(T)$. Therefore $(\iota+1)(\tau-1)\mu = (\tau-1)(\iota+1)\mu$, and we have already observed that this is zero. □

9. Absolute Hodge classes and motives

In order to be able to realize all but a handful of Shimura varieties as moduli varieties, we shall need to replace algebraic varieties and algebraic classes by more general objects, namely, by motives and absolute Hodge classes.

The standard cohomology theories

Let X be a smooth complete[25] algebraic variety over an algebraically closed field k (of characteristic zero as always).

For each prime number ℓ, the étale cohomology groups[26] $H^r_\ell(X)(m) \stackrel{\text{def}}{=} H^r_\ell(X_{\text{et}}, \mathbb{Q}_\ell(m))$ are finite dimensional $\mathbb{Q}_\ell$-vector spaces. For any homomorphism $\sigma\colon k \to k'$ of algebraically closed fields, there is a canonical base change isomorphism

$$(9.1) \qquad H^r_\ell(X)(m) \xrightarrow{\sigma} H^r_\ell(\sigma X)(m), \quad \sigma X \stackrel{\text{def}}{=} X \otimes_{k,\sigma} k'.$$

When $k = \mathbb{C}$, there is a canonical comparison isomorphism

$$(9.2) \qquad \mathbb{Q}_\ell \otimes_{\mathbb{Q}} H^r_B(X)(m) \to H^r_\ell(X)(m).$$

Here $H^r_B(X)$ denotes the Betti cohomology group $H^r(X^{an}, \mathbb{Q})$.

The de Rham cohomology groups $H^r_{dR}(X)(m) \stackrel{\text{def}}{=} \mathbb{H}^r(X_{\text{Zar}}, \Omega^\bullet_{X/k})(m)$ are finite dimensional k-vector spaces. For any homomorphism $\sigma\colon k \to k'$ of fields, there is a canonical base change isomorphism

$$(9.3) \qquad k' \otimes_k H^r_{dR}(X)(m) \xrightarrow{\sigma} H^r_{dR}(\sigma X)(m).$$

When $k = \mathbb{C}$, there is a canonical comparison isomorphism

$$(9.4) \qquad \mathbb{C} \otimes_{\mathbb{Q}} H^r_B(X)(m) \to H^r_{dR}(X)(m).$$

We let $H^r_{k\times\mathbb{A}_f}(X)(m)$ denote the product of $H^r_{dR}(X)(m)$ with the restricted product of the topological spaces $H^r_\ell(X)(m)$ relative to their subspaces $H^r(X_{\text{et}}, \mathbb{Z}_\ell)(m)$. This is a finitely generated free module over the ring $k \times \mathbb{A}_f$. For any homomorphism

[25] Many statements hold without this hypothesis, but we shall need to consider only this case.

[26] The "(m)" denotes a Tate twist. Specifically, for Betti cohomology it denotes the tensor product with the Tate Hodge structure $\mathbb{Q}(m)$, for de Rham cohomology it denotes a shift in the numbering of the filtration, and for étale cohomology it denotes a change in Galois action by a multiple of the cyclotomic character.

$\sigma\colon k \to k'$ of algebraically closed fields, the maps (9.1) and (9.3) give a base change homomorphism

$$(9.5) \qquad H^r_{k\times\mathbb{A}_f}(X)(m) \xrightarrow{\sigma} H^r_{k'\times\mathbb{A}_f}(\sigma X)(m).$$

When $k = \mathbb{C}$, the maps (9.2) and (9.4) give a comparison isomorphism

$$(9.6) \qquad (\mathbb{C} \times \mathbb{A}_f) \otimes_{\mathbb{Q}} H^r_B(X)(m) \to H^r_{\mathbb{C}\times\mathbb{A}_f}(X)(m).$$

Notes. For more details and references, see [20], §1.

Absolute Hodge classes

Let X be a smooth complete algebraic variety over $\mathbb{C}$. The cohomology group $H^{2r}_B(X)(r)$ has a Hodge structure of weight 0, and an element of type $(0,0)$ in it is called a *Hodge class of codimension* r on X.[27] We wish to extend this notion to all base fields of characteristic zero. Of course, given a variety X over a field k, we can choose a homomorphism $\sigma\colon k \to \mathbb{C}$ and define a Hodge class on X to be a Hodge class on σX, but this notion depends on the choice of the embedding. Deligne's idea for avoiding this problem is to use all embeddings ([18], 0.7).

$$\begin{array}{ccc} H^{2r}_B(\sigma X)(r) \cap H^{0,0} & \xrightarrow{(9.6)} & H^{2r}_{\mathbb{C}\times\mathbb{A}_f}(\sigma X)(r) \\ \uparrow & & \uparrow {\scriptstyle (9.5)\ \sigma} \\ AH^r(X) & \hookrightarrow & H^{2r}_{k\times\mathbb{A}_f}(X)(r) \end{array}$$

Let X be a smooth complete algebraic variety over an algebraically closed field k of characteristic zero, and let σ be a homomorphism $k \to \mathbb{C}$. An element γ of $H^{2r}_{k\times\mathbb{A}_f}(X)(r)$ is a σ-*Hodge class of codimension* r if $\sigma\gamma$ lies in the subspace $H^{2r}_B(\sigma X)(r) \cap H^{0,0}$ of $H^{2r}_{\mathbb{C}\times\mathbb{A}_f}(\sigma X)(r)$. When k has finite transcendence degree over $\mathbb{Q}$, an element γ of $H^{2r}_{k\times\mathbb{A}}(X)(r)$ is an *absolute Hodge class* if it is σ-Hodge for all homomorphisms $\sigma\colon k \to \mathbb{C}$. The absolute Hodge classes of codimension r on X form a $\mathbb{Q}$-subspace $AH^r(X)$ of $H^{2r}_{k\times\mathbb{A}_f}(X)(r)$.

We list the basic properties of absolute Hodge classes.

9.7. The inclusion $AH^r(X) \subset H^{2r}_{k\times\mathbb{A}_f}(X)(r)$ induces an injective map

$$(k \times \mathbb{A}_f) \otimes_{\mathbb{Q}} AH^r(X) \to H^{2r}_{k\times\mathbb{A}_f}(X)(r);$$

in particular $AH^r(X)$ is a finite dimensional $\mathbb{Q}$-vector space.

This follows from (9.6) because $AH^r(X)$ is isomorphic to a $\mathbb{Q}$-subspace of $H^{2r}_B(\sigma X)(r)$ (each σ).

9.8. For any homomorphism $\sigma\colon k \to k'$ of algebraically closed fields of finite transcendence degree over $\mathbb{Q}$, the map (9.5) induces an isomorphism $AH^r(X) \to AH^r(\sigma X)$ ([20], 2.9a).

[27] As $H^{2r}_B(X)(r) \simeq H^{2r}_B(X) \otimes \mathbb{Q}(r)$, this is essentially the same as an element of $H^{2r}_B(X)$ of type (r,r).

This allows us to define $AH^r(X)$ for a smooth complete variety over an arbitrary algebraically closed field k of characteristic zero: choose a model X_0 of X over an algebraically closed subfield k_0 of k of finite transcendence degree over $\mathbb{Q}$, and define $AH^r(X)$ to be the image of $AH^r(X_0)$ under the map $H^{2r}_{k_0\times\mathbb{A}_f}(X_0)(r) \to H^{2r}_{k\times\mathbb{A}_f}(X)(r)$. With this definition, (9.8) holds for all homomorphisms of algebraically closed fields k of characteristic zero. Moreover, if k admits an embedding in $\mathbb{C}$, then a cohomology class is absolutely Hodge if and only if it is σ-Hodge for every such embedding.

9.9. The cohomology class of an algebraic cycle on X is absolutely Hodge; thus, the algebraic cohomology classes of codimension r on X form a $\mathbb{Q}$-subspace $A^r(X)$ of $AH^r(X)$ ([20], 2.1a).

9.10. The Künneth components of the diagonal are absolute Hodge classes (ibid., 2.1b).

9.11. Let X_0 be a model of X over a subfield k_0 of k such that k is algebraic over k_0; then $\mathrm{Gal}(k/k_0)$ acts on $AH^r(X)$ through a finite discrete quotient (ibid. 2.9b).

9.12. Let

$$AH^*(X) = \bigoplus_{r\geqslant 0} AH^r(X);$$

then $AH^*(X)$ is a $\mathbb{Q}$-subalgebra of $\bigoplus H^{2r}_{k\times\mathbb{A}_f}(X)(r)$. For any regular map $\alpha\colon Y \to X$ of complete smooth varieties, the maps α_* and α^* send absolute Hodge classes to absolute Hodge classes. (This follows easily from the definitions.)

Theorem 9.13 (Deligne 1982 [20], 2.12, 2.14). *Let S be a smooth connected algebraic variety over $\mathbb{C}$, and let $\pi\colon X \to S$ be a smooth proper morphism. Let $\gamma \in \Gamma(S, R^{2r}\pi_*\mathbb{Q}(r))$, and let γ_s be the image of γ in $H^{2r}_B(X_s)(r)$ ($s \in S(\mathbb{C})$).*

(a) *If γ_s is a Hodge class for one $s \in S(\mathbb{C})$, then it is a Hodge class for every $s \in S(\mathbb{C})$.*

(b) *If γ_s is an absolute Hodge class for one $s \in S(\mathbb{C})$, then it is an absolute Hodge class for every $s \in S(\mathbb{C})$.*

Proof. Let $\bar{X}$ be a smooth compactification of X whose boundary $\bar{X} \smallsetminus X$ is a union of smooth divisors with normal crossings, and let $s \in S(\mathbb{C})$. According to [14], 4.1.1, 4.1.2, there are maps

$$H^{2r}_B(\bar{X})(r) \xrightarrow{\text{onto}} \Gamma(S, R^{2r}\pi_*\mathbb{Q}(r)) \xrightarrow{\text{injective}} H^{2r}_B(X_s)(r)$$

whose composite $H^{2r}_B(\bar{X})(r) \to H^{2r}_B(X_s)(r)$ is defined by the inclusion $X_s \hookrightarrow \bar{X}$; moreover $\Gamma(S, R^{2r}\pi_*\mathbb{Q}(r))$ has a Hodge structure (independent of s) for which the injective maps are morphisms of Hodge structures (theorem of the fixed part).

Let $\gamma \in \Gamma(S, R^{2r}\pi_*\mathbb{Q}(r))$. If γ_s is of type $(0,0)$ for one s, then so also is γ; then γ_s is of type $(0,0)$ for all s. This proves (a).

Let σ be an automorphism of $\mathbb{C}$ (as an abstract field). It suffices to prove (b) with "absolute Hodge" replaced with "σ-Hodge". We shall use the commutative diagram ($\mathbb{A} = \mathbb{C} \times \mathbb{A}_f$):

$$\begin{array}{ccccc}
H_B^{2r}(\bar{X})(r) & \xrightarrow{\text{onto}} & \Gamma(S, R^{2r}\pi_*\mathbb{Q}(r)) & \xrightarrow{\text{injective}} & H_B^{2r}(X_s)(r) \\
\downarrow{\scriptstyle ?\mapsto ?_{\mathbb{A}}} & & \downarrow & & \downarrow \\
H_{\mathbb{A}}^{2r}(\bar{X})(r) & \xrightarrow{\text{onto}} & \Gamma(S, R^{2r}\pi_*\mathbb{A}(r)) & \xrightarrow{\text{injective}} & H_{\mathbb{A}}^{2r}(X_s)(r) \\
\downarrow{\scriptstyle \sigma} & & \downarrow{\scriptstyle \sigma} & & \downarrow{\scriptstyle \sigma} \\
H_{\mathbb{A}}^{2r}(\sigma\bar{X})(r) & \xrightarrow{\text{onto}} & \Gamma(\sigma S, R^{2r}(\sigma\pi)_*\mathbb{A}(r)) & \xrightarrow{\text{injective}} & H_{\mathbb{A}}^{2r}(\sigma X_s)(r) \\
\uparrow & & \uparrow & & \uparrow \\
H_B^{2r}(\sigma\bar{X})(r) & \xrightarrow{\text{onto}} & \Gamma(\sigma S, R^{2r}(\sigma\pi)_*\mathbb{Q}(r)) & \xrightarrow{\text{injective}} & H_B^{2r}(\sigma X_s)(r).
\end{array}$$

The middle map σ uses a relative version of the base change map (9.5). The other maps σ are the base change isomorphisms and the remaining vertical maps are essential tensoring with $\mathbb{A}$ (and are denoted $? \mapsto ?_{\mathbb{A}}$).

Let γ be an element of $\Gamma(S, R^{2r}\pi_*\mathbb{Q}(r))$ such that γ_s is σ-Hodge for one s. Recall that this means that there is a $\gamma_s^\sigma \in H_B^{2r}(\sigma X_s)(r)$ of type (0,0) such that $(\gamma_s^\sigma)_{\mathbb{A}} = \sigma(\gamma_s)_{\mathbb{A}}$ in $H_{\mathbb{A}}^{2r}(\sigma X_s)(r)$. As γ_s is in the image of

$$H_B^{2r}(\bar{X})(r) \to H_B^{2r}(X_s)(r),$$

$\sigma(\gamma_s)_{\mathbb{A}}$ is in the image of

$$H_{\mathbb{A}}^{2r}(\sigma\bar{X})(r) \to H_{\mathbb{A}}^{2r}(\sigma X_s)(r).$$

Therefore $(\gamma_s^\sigma)_{\mathbb{A}}$ is also, which implies (by linear algebra[28]) that γ_s^σ is in the image of

$$H_B^{2r}(\sigma\bar{X})(r) \to H_B^{2r}(\sigma X_s)(r).$$

Let $\tilde{\gamma}^\sigma$ be a pre-image of γ_s^σ in $H_B^{2r}(\sigma\bar{X})(r)$.

Let s' be a second point of S, and let $\tilde{\gamma}_{s'}^\sigma$ be the image of $\tilde{\gamma}^\sigma$ in $H_B^{2r}(\sigma X_{s'})(r)$. By construction, $(\tilde{\gamma}^\sigma)_{\mathbb{A}}$ maps to $\sigma\gamma_{\mathbb{A}}$ in $\Gamma(\sigma S, R^{2r}(\sigma\pi)_*\mathbb{A}(r))$, and so $(\tilde{\gamma}_{s'}^\sigma)_{\mathbb{A}} = \sigma(\gamma_{s'})_{\mathbb{A}}$ in $H_{\mathbb{A}}^{2r}(\sigma X_{s'})(r)$, which demonstrates that $\gamma_{s'}$ is σ-Hodge. □

Conjecture 9.14 (Deligne [18], 0.10). *Every σ-Hodge class on a smooth complete variety over an algebraically closed field of characteristic zero is absolutely Hodge, i.e.,*

$$\sigma\text{-Hodge (for one } \sigma) \implies \text{absolutely Hodge.}$$

Theorem 9.15 (Deligne 1982 [20], 2.11). *Conjecture 9.14 is true for abelian varieties.*

---

[28] Apply the following elementary statement:

> Let E, W, and V be vector spaces, and let $\alpha\colon W \to V$ be a linear map; let $v \in V$; if $e \otimes v$ is in the image of $1 \otimes \alpha\colon E \otimes W \to E \otimes V$ for some nonzero $e \in E$, then v is in the image of α.

To prove the statement, choose an $f \in E^\vee$ such that $f(e) = 1$. If $\sum e_i \otimes \alpha(w_i) = e \otimes v$, then $\sum f(e_i) w_i = v$.

To prove the theorem, it suffices to show that every Hodge class on an abelian variety over $\mathbb{C}$ is absolutely Hodge.[29] We defer the proof of the theorem to the next subsection.

Aside 9.16. Let $X_{\mathbb{C}}$ be a smooth complete algebraic variety over $\mathbb{C}$. Then $X_{\mathbb{C}}$ has a model X_0 over a subfield k_0 of $\mathbb{C}$ finitely generated over $\mathbb{Q}$. Let k be the algebraic closure of k_0 in $\mathbb{C}$, and let $X = X_{0k}$. For a prime number ℓ, let

$$\mathcal{T}_\ell^r(X) = \bigcup_U H_\ell^{2r}(X)(r)^U \qquad \text{(space of Tate classes)}$$

where U runs over the open subgroups of $\mathrm{Gal}(k/k_0)$ — as the notation suggests, $\mathcal{T}_\ell^r(X)$ depends only on X/k. The Tate conjecture ([62], Conjecture 1) says that the $\mathbb{Q}_\ell$-vector space $\mathcal{T}_\ell^r(X)$ is spanned by algebraic classes. Statement 9.11 implies that $AH^r(X)$ projects into $\mathcal{T}_\ell^r(X)$, and (9.7) implies that the map $\mathbb{Q}_\ell \otimes_{\mathbb{Q}} AH^r(X) \to \mathcal{T}_\ell^r(X)$ is injective. Therefore the Tate conjecture implies that $A^r(X) = AH^r(X)$, and so the Tate conjecture for X and one ℓ implies that all absolute Hodge classes on $X_{\mathbb{C}}$ are algebraic. Thus, in the presence of Conjecture 9.14, the Tate conjecture implies the Hodge conjecture. In particular, Theorem 9.15 shows that, for an abelian variety, the Tate conjecture implies the Hodge conjecture.

Proof of Deligne's theorem

It is convenient to prove Theorem 9.15 in the following more abstract form.

Theorem 9.17. *Suppose that for each abelian variety A over $\mathbb{C}$ we have a $\mathbb{Q}$-subspace $C^r(A)$ of the Hodge classes of codimension r on A. Assume:*

(a) *$C^r(A)$ contains all algebraic classes of codimension r on A;*
(b) *pull-back by a homomorphism $\alpha\colon A \to B$ of abelian varieties maps $C^r(B)$ into $C^r(A)$;*
(c) *let $\pi\colon \mathcal{A} \to S$ be an abelian scheme over a connected smooth complex algebraic variety S, and let $t \in \Gamma(S, R^{2r}\pi_*\mathbb{Q}(r))$; if t_s lies in $C^r(A_s)$ for one $s \in S(\mathbb{C})$, then it lies in $C^r(A_s)$ for all s.*

Then $C^r(A)$ contains all the Hodge classes of codimension r on A.

Corollary 9.18. *If hypothesis (c) of the theorem holds for algebraic classes on abelian varieties, then the Hodge conjecture holds for abelian varieties. (In other words, for abelian varieties, the variational Hodge conjecture implies the Hodge conjecture.)*

Proof. Immediate consequence of the theorem, because the algebraic classes satisfy (a) and (b). □

The proof of Theorem 9.17 requires four steps.

---

[29] Let A be an abelian variety over k, and suppose that γ is σ_0-Hodge for some homomorphism $\sigma_0 : k \to \mathbb{C}$. We have to show that it is σ-Hodge for every $\sigma : k \to \mathbb{C}$. But, using the Zorn's lemma, one can show that there exists a homomorphism $\sigma' : \mathbb{C} \to \mathbb{C}$ such that $\sigma = \sigma' \circ \sigma_0$. Now γ is σ-Hodge if and only if $\sigma_0\gamma$ is σ'-Hodge.

Step 1: The Hodge conjecture holds for powers of an elliptic curve As Tate observed ([62], p. 19), the $\mathbb{Q}$-algebra of Hodge classes on a power of an elliptic curve is generated by those of type $(1,1)$.[30] These are algebraic by a theorem of Lefschetz.

Step 2: Split Weil classes lie in C Let A be a complex abelian variety, and let ν be a homomorphism from a CM-field E into $\mathrm{End}(A)_{\mathbb{Q}}$. The pair (A,ν) is said to be of *Weil type* if the tangent space $T_0(A)$ is a free $E\otimes_{\mathbb{Q}}\mathbb{C}$-module. In this case, $d \stackrel{\text{def}}{=} \dim_E H^1_B(A)$ is even and the subspace $\bigwedge^d_E H^1_B(A)(\frac{d}{2})$ of $H^d_B(A)(\frac{d}{2})$ consists of Hodge classes ([20], 4.4). When E is quadratic over $\mathbb{Q}$, these Hodge classes were studied by Weil [65], and for this reason are called *Weil classes*. A *polarization* of (A,ν) is a polarization λ of A whose whose Rosati involution acts on $\nu(E)$ as complex conjugation. The Riemann form of such a polarization can be written

$$(x,y)\mapsto \mathrm{Tr}_{E/\mathbb{Q}}(f\phi(x,y))$$

for some totally imaginary element f of E and E-hermitian form ϕ on $H_1(A,\mathbb{Q})$. If λ can be chosen so that ϕ is split (i.e., admits a totally isotropic subspace of dimension $d/2$), then the Weil classes are said to be *split*.

Lemma 9.19. *All split Weil classes of codimension r on an abelian variety A lie in $C^r(A)$.*

Proof. Let (A,ν,λ) be a polarized abelian variety of split Weil type. Let $V=H_1(A,\mathbb{Q})$, and let ψ be the Riemann form of λ. The Hodge structures on V for which the elements of E act as morphisms and ψ is a polarization are parametrized by a period subdomain, which is hermitian symmetric domain (cf. 7.9). On dividing by a suitable arithmetic subgroup, we get a smooth proper map $\pi\colon \mathcal{A}\to S$ of smooth algebraic varieties whose fibres are abelian varieties with an action of E (Theorem 7.13). There is a $\mathbb{Q}$-subspace W of $\Gamma(S, R^d\pi_*\mathbb{Q}(\frac{d}{2}))$ whose fibre at every point s is the space of Weil classes on A_s. One fibre of π is (A,ν) and another is a power of an elliptic curve. Therefore the lemma follows from Step 1 and hypotheses (a,c). (See [20], 4.8, for more details.) □

Step 3: Theorem 9.17 for abelian varieties of CM-type A simple abelian variety A is of *CM-type* if $\mathrm{End}(A)_{\mathbb{Q}}$ is a field of degree $2\dim A$ over $\mathbb{Q}$, and a general abelian variety is of *CM-type* if every simple isogeny factor of it is of CM-type. Equivalently, it is of CM-type if the Hodge structure $H_1(A^{an},\mathbb{Q})$ is of CM-type. According to [2]:

> For any complex abelian variety A of CM-type, there exist complex abelian varieties B_J of CM-type and homomorphisms $A\to B_J$ such that every Hodge class on A is a linear combination of the pull-backs of split Weil classes on the B_J.

Thus Theorem 9.17 for abelian varieties of CM-type follows from Step 2 and hypothesis (b). (See [20], §5, for the original proof of this step.)

---

[30] This is most conveniently proved by applying the criterion [39], 4.8.

Step 4: Completion of the proof of Theorem 9.17 Let t be a Hodge class on a complex abelian variety A. Choose a polarization λ for A. Let $V = H_1(A, \mathbb{Q})$ and let h_A be the homomorphism defining the Hodge structure on $H_1(A, \mathbb{Q})$. Both t and the Riemann form t_0 of λ can be regarded as Hodge tensors for V. The period subdomain $D = D(V, h_A, \{t, t_0\})$ is a hermitian symmetric domain (see 7.9). On dividing by a suitable arithmetic subgroup, we get a smooth proper map $\pi: \mathcal{A} \to S$ of smooth algebraic varieties whose fibres are abelian varieties (Theorem 7.13) and a section t of $R^{2r}\pi_*\mathbb{Q}(r)$. For one $s \in S$, the fibre $(\mathcal{A}, t)_s = (A, t)$, and another fibre is an abelian variety of CM-type (apply 8.1), and so the theorem follows from Step 3 and hypothesis (c). (See [20], §6, for more details.)

Motives for absolute Hodge classes

We fix a base field k of characteristic zero; "variety" will mean "smooth projective variety over k".

For varieties X and Y with X connected, we let

$$C^r(X, Y) = AH^{\dim X + r}(X \times Y)$$

(correspondences of degree r from X to Y). When X has connected components X_i, $i \in I$, we let

$$C^r(X, Y) = \bigoplus_{i \in I} C^r(X_i, Y).$$

For varieties X, Y, Z, there is a bilinear pairing

$$f, g \mapsto g \circ f: C^r(X, Y) \times C^s(Y, Z) \to C^{r+s}(X, Z)$$

with

$$g \circ f \stackrel{\text{def}}{=} (p_{XZ})_*(p_{XY}^* f \cdot p_{YZ}^* g).$$

Here the p's are projection maps from $X \times Y \times Z$. These pairings are associative and so we get a "category of correspondences", which has one object hX for every variety over k, and whose Homs are defined by

$$\mathrm{Hom}(hX, hY) = C^0(X, Y).$$

Let $f: Y \to X$ be a regular map of varieties. The transpose of the graph of f is an element of $C^0(X, Y)$, and so $X \rightsquigarrow hX$ is a contravariant functor.

The category of correspondences is additive, but not abelian, and so we enlarge it by adding the images of idempotents. More precisely, we define a "category of effective motives", which has one object $h(X, e)$ for each variety X and idempotent e in the ring $\mathrm{End}(hX) = AH^{\dim X}(X \times X)$, and whose Homs are defined by

$$\mathrm{Hom}(h(X, e), h(Y, f)) = f \circ C^0(X, Y) \circ e.$$

This contains the old category by $hX \leftrightarrow h(X, \mathrm{id})$, and $h(X, e)$ is the image of $hX \xrightarrow{e} hX$.

The category of effective motives is abelian, but objects need not have duals. In the enlarged category, the motive $h\mathbb{P}^1$ decomposes into $h\mathbb{P}^1 = h^0\mathbb{P}^1 \oplus h^2\mathbb{P}^1$, and it turns out that, to obtain duals for all objects, we only have to "invert" the motive $h^2\mathbb{P}^1$. This is most conveniently done by defining a "category of motives" which has one object $h(X, e, m)$ for each pair (X, e) as before and integer m, and whose Homs are defined by

$$\mathrm{Hom}(h(X, e, m), h(Y, f, n)) = f \circ C^{n-m}(X, Y) \circ e.$$

This contains the old category by $h(X, e) \leftrightarrow h(X, e, 0)$.

We now list some properties of the category $\mathsf{Mot}(k)$ of motives.

9.20. The Hom's in $\mathsf{Mot}(k)$ are finite dimensional $\mathbb{Q}$-vector spaces, and $\mathsf{Mot}(k)$ is a semisimple abelian category.

9.21. Define a tensor product on $\mathsf{Mot}(k)$ by

$$h(X, e, m) \otimes h(X, f, n) = h(X \times Y, e \times f, m + n).$$

With the obvious associativity constraint and a suitable[31] commutativity constraint, $\mathsf{Mot}(k)$ becomes a tannakian category.

9.22. The standard cohomology functors factor through $\mathsf{Mot}(k)$. For example, define

$$\omega_\ell(h(X, e, m)) = e\left(\bigoplus\nolimits_i H^i_\ell(X)(m)\right)$$

(image of e acting on $\bigoplus_i H^i_\ell(X)(m)$). Then ω_ℓ is an exact faithful functor $\mathsf{Mot}(k) \to \mathsf{Vec}_{\mathbb{Q}_\ell}$ commuting with tensor products. Similarly, de Rham cohomology defines an exact tensor functor $\omega_{dR}\colon \mathsf{Mot}(k) \to \mathsf{Vec}_k$, and, when $k = \mathbb{C}$, Betti cohomology defines an exact tensor functor $\mathsf{Mot}(k) \to \mathsf{Vec}_{\mathbb{Q}}$. The functors ω_ℓ, ω_{dR}, and ω_B are called the ℓ-adic, de Rham, and Betti fibre functors, and they send a motive to its ℓ-adic, de Rham, or Betti *realization*.

The Betti fibre functor on $\mathsf{Mot}(\mathbb{C})$ takes values in $\mathsf{Hdg}_{\mathbb{Q}}$, and is faithful (almost by definition). Deligne's conjecture 9.14 is equivalent to saying that it is full.

Abelian motives

Definition 9.23. A motive is *abelian* if it lies in the tannakian subcategory $\mathsf{Mot}^{ab}(k)$ of $\mathsf{Mot}(k)$ generated by the motives of abelian varieties.

The Tate motive, being isomorphic to $\bigwedge^2 h_1 E$ for any elliptic curve E, is an abelian motive. It is known that $h(X)$ is an abelian motive if X is a curve, a unirational variety of dimension $\leqslant 3$, a Fermat hypersurface, or a K3 surface.

Deligne's theorem 9.15 implies that $\omega_B\colon \mathsf{Mot}^{ab}(\mathbb{C}) \to \mathsf{Hdg}_{\mathbb{Q}}$ is fully faithful.

---

[31] Not the obvious one! It is necessary to change some signs.

CM motives

Definition 9.24. A motive over $\mathbb{C}$ is of *CM-type* if its Hodge realization is of CM-type.

Lemma 9.25. *Every Hodge structure of CM-type is the Betti realization of an abelian motive.*

Proof. Elementary (see, for example, [37], 4.6). □

Therefore ω_B defines an equivalence from the category of abelian motives of CM-type to the category of Hodge structures of CM-type.

Proposition 9.26. *Let G_{Hdg} (resp. G_{Mab}) be the affine group scheme attached to $\mathsf{Hdg}_{\mathbb{Q}}$ and its forgetful fibre functor (resp. $\mathsf{Mot}^{ab}(\mathbb{C})$ and its Betti fibre functor). The kernel of the homomorphism $G_{\mathrm{Hdg}} \to G_{\mathrm{Mab}}$ defined by the tensor functor $\omega_B\colon \mathsf{Mot}^{ab}(\mathbb{C}) \to \mathsf{Hdg}_{\mathbb{Q}}$ is contained in $(G_{\mathrm{Hdg}})^{\mathrm{der}}$.*

Proof. Let S be the affine group scheme attached to the category $\mathsf{Hdg}_{\mathbb{Q}}^{\mathrm{cm}}$ of Hodge structures of CM-type and its forgetful fibre functor. The lemma shows that the functor $\mathsf{Hdg}_{\mathbb{Q}}^{\mathrm{cm}} \hookrightarrow \mathsf{Hdg}_{\mathbb{Q}}$ factors through $\mathsf{Mot}^{\mathrm{ab}}(\mathbb{C}) \hookrightarrow \mathsf{Hdg}_{\mathbb{Q}}$, and so $G_{\mathrm{Hdg}} \to S$ factors through $G_{\mathrm{Hdg}} \to G_{\mathrm{Mab}}$:

$$G_{\mathrm{Hdg}} \to G_{\mathrm{Mab}} \twoheadrightarrow S.$$

Hence

$$\mathrm{Ker}(G_{\mathrm{Hdg}} \to G_{\mathrm{Mab}}) \subset \mathrm{Ker}(G_{\mathrm{Hdg}} \twoheadrightarrow S) = \left(G_{\mathrm{Hdg}}\right)^{\mathrm{der}}.$$

□

Special motives

Definition 9.27. A motive over $\mathbb{C}$ is *special* if its Hodge realization is special (see p. 496).

It follows from (6.7) that the special motives form a tannakian subcategory of $\mathsf{Mot}(k)$, which includes the abelian motives (see 6.10).

Question 9.28. Is every special Hodge structure the Betti realization of a motive? (Cf. [19], p. 248; [32], p. 216; [56], 8.7.)

More explicitly: for each simple special Hodge structure (V, h), does there exist an algebraic variety X over $\mathbb{C}$ and an integer m such that (V, h) is a direct factor of $\bigoplus_{r \geqslant 0} H_B^r(X)(m)$ and the projection $\bigoplus_{r\geqslant 0} H_B^r(X)(m) \to V \subset \bigoplus_{r\geqslant 0} H_B^r(X)(m)$ is an absolute Hodge class on X.

A positive answer to (9.28) would imply that all connected Shimura varieties are moduli varieties for motives (see §11). Apparently, no special motive is known that is not abelian.

Families of abelian motives For an abelian variety A over k, let

$$\omega_f(A) = \varprojlim A_N(k^{al}), \quad A_N(k^{al}) = \mathrm{Ker}(N\colon A(k^{al}) \to A(k^{al})).$$

This is a free $\mathbb{A}_f$-module of rank $2\dim A$ with a continuous action of $\mathrm{Gal}(k^{al}/k)$.

Let S be a smooth connected variety over k, and let $k(S)$ be its function field. Fix an algebraic closure $k(S)^{al}$ of $k(S)$, and let $k(S)^{un}$ be the union of the subfields L of $k(S)^{al}$ such that the normalization of S in L is unramified over S. We say that an action of $\mathrm{Gal}(k(S)^{al}/k(S))$ on a module is *unramified* if it factors through $\mathrm{Gal}(k(S)^{un}/k(S))$.

Theorem 9.29. *Let S be a smooth connected variety over* k. *The functor* $A \rightsquigarrow A_\eta \stackrel{\text{def}}{=} A_{k(S)}$ *is a fully faithful functor from the category of families of abelian varieties over* S *to the category of abelian varieties over* $k(S)$, *with essential image the abelian varieties* B *over* $k(S)$ *such that* $\omega_f(B)$ *is unramified.*

Proof. When S has dimension 1, this follows from the theory of Néron models. In general, this theory shows that an abelian variety (or a morphism of abelian varieties) extends to an open subvariety U of S such that $S \smallsetminus U$ has codimension at least 2. Now we can apply[32] [10], I 2.7, V 6.8. □

The functor ω_f extends to a functor on abelian motives such that $\omega_f(h_1 A) = \omega_f(A)$ if A is an abelian variety.

Definition 9.30. Let S be a smooth connected variety over k. A *family* M *of abelian motives* over S is an abelian motive M_η over $k(S)$ such that $\omega_f(M_\eta)$ is unramified.

Let M be a family of motives over a smooth connected variety S, and let $\bar{\eta} = \mathrm{Spec}(k(S)^{al})$. The fundamental group $\pi_1(S, \bar{\eta}) = \mathrm{Gal}(k(S)^{un}/k(S))$, and so the representation of $\pi_1(S, \bar{\eta})$ on $\omega_f(M_\eta)$ defines a local system of $\mathbb{A}_f$-modules $\omega_f(M)$. Less obvious is that, when the ground field is $\mathbb{C}$, M defines a polarizable variation of Hodge structures on S, $\mathcal{H}_B(M/S)$. When M can be represented in the form (A, p, m) on S, this is obvious. However, M can always be represented in this fashion on an open subset of S, and the underlying local system of $\mathbb{Q}$-vector spaces extends to the whole of S because the monodromy representation is unramified. Now it is possible to show that the variation of Hodge structures itself extends (uniquely) to the whole of S, by using results from [54], [9], and [26]. See [38], 2.40, for the details.

Theorem 9.31. *Let S be a smooth connected variety over* $\mathbb{C}$. *The functor sending a family* M *of abelian motives over* S *to its associated polarizable Hodge structure is fully faithful, with essential image the variations of Hodge structures* (V, F) *such that there exists a dense open subset* U *of* S, *an integer* m, *and a family of abelian varieties* $f\colon A \to S$ *such that* (V, F) *is a direct summand of* $Rf_*\mathbb{Q}$.

Proof. This follows from the similar statement (7.13) for families of abelian varieties (see [38], 2.42). □

[32] Recall that we are assuming that the base field has characteristic zero — the theorem is false without that condition.

10. Symplectic Representations

In this subsection, we classify the symplectic representations of groups. These were studied by Satake in a series of papers (see especially [51, 52, 53]). Our exposition follows that of Deligne [19].

In §8 we proved that there exists a correspondence between variations of Hodge structures on locally symmetric varieties and certain commutative diagrams

$$\begin{array}{ccccc} H & & & & \\ \downarrow & \searrow & & & \\ (H^{ad},\bar{h}) & \longleftarrow & (G,h) & \xhookrightarrow{\rho} & GL_V \end{array} \tag{10.1}$$

In this section, we study whether there exists such a diagram and a nondegenerate alternating form ψ on V such that $\rho(G) \subset G(\psi)$ and $\rho_{\mathbb{R}} \circ h \in D(\psi)$. Here $G(\psi)$ is the group of *symplectic similitudes* (algebraic subgroup of GL_V whose elements fix ψ up to a scalar) and $D(\psi)$ is the Siegel upper half space (set of Hodge structures h on V of type $\{(-1,0),(0,-1)\}$ for which $2\pi i\psi$ is a polarization[33]). Note that $G(\psi)$ is a reductive group whose derived group is the symplectic group $S(\psi)$.

Preliminaries

10.2. The *universal covering torus* $\tilde{T}$ of a torus T is the projective system $(T_n, T_{nm} \xrightarrow{m} T_n)$ in which $T_n = T$ for all n and the indexing set is $\mathbb{N} \smallsetminus \{0\}$ ordered by divisibility. For any algebraic group G,

$$\mathrm{Hom}(\tilde{T}, G) = \varinjlim_{n \geqslant 1} \mathrm{Hom}(T_n, G).$$

Concretely, a homomorphism $\tilde{T} \to G$ is represented by a pair (f, n) in which f is a homomorphism $T \to G$ and $n \in \mathbb{N} \smallsetminus \{0\}$; two pairs (f,n) and (g,m) represent the same homomorphism $\tilde{T} \to G$ if and only if $f \circ m = g \circ n$. A homomorphism $f\colon \tilde{T} \to G$ factors through T if and only if it is represented by a pair $(f,1)$. A homomorphism $\tilde{\mathbb{G}}_m \to GL_V$ represented by (μ, n) defines a gradation $V = \bigoplus V_r$, $r \in \frac{1}{n}\mathbb{Z}$; here $V_{\frac{a}{n}} = \{v \in V \mid \mu(t)v = t^a v\}$; the r for which $V_r \neq 0$ are called the *weights* the representation of $\tilde{\mathbb{G}}_m$ on V. Similarly, a homomorphism $\tilde{\mathbb{S}} \to GL_V$ represented by (h, n) defines a fractional Hodge decomposition $V_{\mathbb{C}} = \bigoplus V^{p,q}$ with $p, q \in \frac{1}{n}\mathbb{Z}$.

The real case

Throughout this subsection, H is a simply connected real algebraic group without compact factors, and $\bar{h}$ is a homomorphism $\mathbb{S}/\mathbb{G}_m \to H^{ad}$ satisfying the conditions (SV1,2), p. 493, and whose projection on each simple factor of H^{ad} is nontrivial.

---

[33] This description agrees with that in §2 because of the correspondence in (5.3).

Definition 10.3. A homomorphism $H \to GL_V$ with finite kernel is a *symplectic representation* of $(H, \bar{h})$ if there exists a commutative diagram

$$\begin{array}{ccccc} H & & & & \\ \downarrow & \searrow & & & \\ (H^{ad}, \bar{h}) & \longleftarrow & (G, h) & \longrightarrow & (G(\psi), D(\psi)), \end{array}$$

in which ψ is a nondegenerate alternating form on V, G is a reductive group, and h is a homomorphism $\mathbb{S} \to G$; the homomorphism $H \to G$ is required to have image G^{der}.

In other words, there exists a real reductive group G, a nondegenerate alternating form ψ on V, and a factorization

$$H \xrightarrow{a} G \xrightarrow{b} GL_V$$

of $H \to GL_V$ such that $a(H) = G^{der}$, $b(G) \subset G(\psi)$, and $b \circ h \in D(\psi)$; the isogeny $H \to G^{der}$ induces an isomorphism $H^{ad} \xrightarrow{c} G^{ad}$ (see footnote 21, p. 511), and it is required that $\bar{h} = c^{-1} \circ ad \circ h$.

We shall determine the complex representations of H that occur in the complexification of a symplectic representation (and we shall omit "the complexification of").

Proposition 10.4. *A homomorphism* $H \to GL_V$ *with finite kernel is a symplectic representation of* $(H, \bar{h})$ *if there exists a commutative diagram*

$$\begin{array}{ccccc} H & & & & \\ \downarrow & \searrow & & & \\ (H^{ad}, \bar{h}) & \longleftarrow & (G, h) & \xrightarrow{\rho} & GL_V, \end{array}$$

in which G *is a reductive group, the homomorphism* $H \to G$ *has image* G^{der}, *and* $(V, \rho \circ h)$ *has type* $\{(-1,0), (0,-1)\}$.

Proof. Let G' be the algebraic subgroup of G generated by G^{der} and $h(\mathbb{S})$. After replacing G with G', we may suppose that G itself is generated by G^{der} and $h(\mathbb{S})$. Then (G, h) satisfies (SV2*), and it follows from Theorem 2.1 that there exists a polarization ψ of $(V, \rho \circ h)$ such that G maps into $G(\psi)$ (cf. the proof of 6.4). $\square$

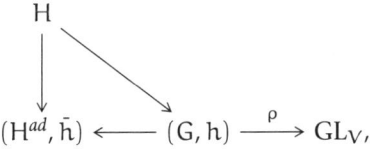

Let $(H, \bar{h})$ be as before. The cocharacter $\mu_{\bar{h}}$ of $H^{ad}_{\mathbb{C}}$ lifts to a fractional cocharacter $\tilde{\mu}$ of $H_{\mathbb{C}}$:

$$\begin{array}{ccc} \tilde{\mathbb{G}}_m & \xrightarrow{\tilde{\mu}} & H_{\mathbb{C}} \\ \downarrow & & \downarrow \text{ad} \\ \mathbb{G}_m & \xrightarrow[\mu_{\bar{h}}]{} & H^{ad}_{\mathbb{C}}. \end{array}$$

Lemma 10.5. *If an irreducible complex representation W of H occurs in a symplectic representation, then $\tilde{\mu}$ has at most two weights a and $a+1$ on W.*

Proof. Let $H \xrightarrow{\varphi} (G, h) \longrightarrow GL_V$ be a symplectic representation of $(H, \bar{h})$, and let W be an irreducible direct summand of $V_{\mathbb{C}}$. The homomorphisms $\varphi_{\mathbb{C}} \circ \tilde{\mu} \colon \tilde{\mathbb{G}}_m \to G_{\mathbb{C}}$ coincides with μ_h when composed with $G_{\mathbb{C}} \to G^{ad}_{\mathbb{C}}$, and so $\varphi_{\mathbb{C}} \circ \tilde{\mu} = \mu_h \cdot \nu$ with ν central. On V, μ_h has weights $0,1$. If a is the unique weight of ν on W, then the only weights of $\tilde{\mu}$ on W are a and $a+1$. □

Lemma 10.6. *Assume that H is almost simple. A nontrivial irreducible complex representation W of H occurs in a symplectic representation if and only if $\tilde{\mu}$ has exactly two weights a and $a+1$ on W.*

Proof. $\Rightarrow$: Let (μ, n) represent $\tilde{\mu}$. As $H_{\mathbb{C}}$ is almost simple and W nontrivial, the homomorphism $\mathbb{G}_m \to GL_W$ defined by μ is nontrivial, therefore noncentral, and the two weights a and $a+1$ occur.

$\Leftarrow$: Let (W, r) be an irreducible complex representation of H with weights $a, a+1$, and let V be the real vector space underlying W. Define G to be the subgroup of GL_V generated by the image of H and the homotheties: $G = r(H) \cdot \mathbb{G}_m$. Let $\tilde{h}$ be a fractional lifting of $\bar{h}$ to $\tilde{H}$:

$$\begin{array}{ccc} \tilde{\mathbb{S}} & \xrightarrow{\tilde{h}} & H_{\mathbb{C}} \\ \downarrow & & \downarrow \text{ad} \\ \mathbb{S} & \xrightarrow[\bar{h}]{} & H^{ad}_{\mathbb{C}}. \end{array}$$

Let W_a and W_{a+1} be the subspaces of weight a and $a+1$ of W. Then $\tilde{h}(z)$ acts on W_a as $(z/\bar{z})^a$ and on W_{a+1} as $(z/\bar{z})^{a+1}$, and so $h(z) \stackrel{\text{def}}{=} \tilde{h}(z) z^{-a} \bar{z}^{1+a}$ acts on these spaces as $\bar{z}$ and z respectively. Therefore h is a true homomorphism $\mathbb{S} \to G$, projecting to $\bar{h}$ on H^{ad}, and V is of type $\{(-1,0), (0,-1)\}$ relative to h. We may now apply Lemma 10.4. □

We interpret the condition in Lemma 10.6 in terms of roots and weights. Let $\tilde{\mu} = \mu_{\bar{h}}$. Fix a maximal torus T in $H_{\mathbb{C}}$, and let $R = R(H, T) \subset X^*(T)_{\mathbb{Q}}$ be the corresponding root system. Choose a base S for R such that $\langle \alpha, \tilde{\mu} \rangle \geqslant 0$ for all $\alpha \in S$ (cf. §2).

Recall that, for each $\alpha \in R$, there exists a unique $\alpha^\vee \in X_*(T)_\mathbb{Q}$ such that $\langle \alpha, \alpha^\vee \rangle = 2$ and the symmetry $s_\alpha \colon x \mapsto x - \langle x, \alpha^\vee \rangle \alpha$ preserves R; moreover, for all $\alpha \in R$, $\langle R, \alpha^\vee \rangle \subset \mathbb{Z}$. The lattice of weights is

$$P(R) = \{\varpi \in X^*(T)_\mathbb{Q} \mid \langle \varpi, \alpha^\vee \rangle \in \mathbb{Z} \text{ all } \alpha \in R\},$$

the fundamental weights are the elements of the dual basis $\{\varpi_1, \ldots, \varpi_n\}$ to $\{\alpha_1^\vee, \ldots, \alpha_n^\vee\}$, and that the dominant weights are the elements $\sum n_i \varpi_i$, $n_i \in \mathbb{N}$. The quotient $P(R)/Q(R)$ of $P(R)$ by the lattice $Q(R)$ generated by R is the character group of $Z(H)$:

$$P(R)/Q(R) \simeq X^*(Z(H)).$$

The irreducible complex representations of H are classified by the dominant weights. We shall determine the dominant weights of the irreducible complex representations such that $\tilde{\mu}$ has exactly two weights a and $a+1$.

There is a unique permutation τ of the simple roots, called the *opposition involution*, such that the $\tau^2 = 1$ and the map $\alpha \mapsto -\tau(\alpha)$ extends to an action of the Weyl group. Its action on the Dynkin diagram is determined by the following rules: it preserves each connected component; on a connected component of type A_n, D_n (n odd), or E_6, it acts as the unique nontrivial involution, and on all other connected components, it acts trivially ([64], 1.5.1). Thus:

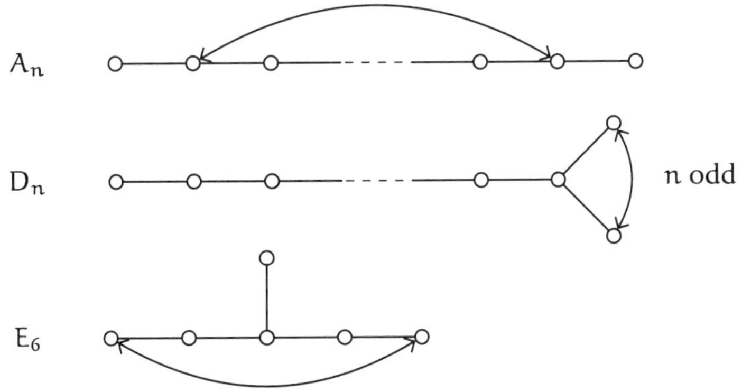

Proposition 10.7. *Let W be an irreducible complex representation of* H, *and let* ϖ *be its highest weight. The representation W occurs in a symplectic representation if and only if*

$$\langle \varpi + \tau\varpi, \bar{\mu} \rangle = 1. \tag{10.8}$$

Proof. The lowest weight of W is $-\tau(\varpi)$. The weights β of W are of the form

$$\beta = \varpi + \sum_{\alpha \in R} m_\alpha \alpha, \quad m_\alpha \in \mathbb{Z},$$

and

$$\langle \beta, \bar{\mu} \rangle \in \mathbb{Z}.$$

Thus, $\langle\beta,\bar{\mu}\rangle$ takes only two values $a, a+1$ if and only if

$$\langle -\tau(\varpi),\bar{\mu}\rangle = \langle\varpi,\bar{\mu}\rangle - 1,$$

i.e., if and only if (10.8) holds. □

Corollary 10.9. *If* W *is symplectic, then* ϖ *is a fundamental weight. Therefore the representation factors through an almost simple quotient of* H.

Proof. For every dominant weight ϖ, $\langle\varpi+\tau\varpi,\bar{\mu}\rangle \in \mathbb{Z}$ because $\varpi+\tau\varpi\in Q(R)$. If $\varpi\neq 0$, then $\langle\varpi+\tau\varpi,\bar{\mu}\rangle > 0$ unless $\bar{\mu}$ kills all the weights of the representation corresponding to ϖ. Hence a dominant weight satisfying (10.8) can not be a sum of two dominant weights. □

The corollary allows us to assume that H is almost simple. Recall from §2 that there is a unique special simple root α_s such that, for $\alpha\in S$,

$$\langle\alpha,\bar{\mu}\rangle = \begin{cases} 1 \text{ if } \alpha=\alpha_s \\ 0 \text{ otherwise.}\end{cases}$$

When a weight ϖ is expressed as a $\mathbb{Q}$-linear combination of the simple roots, $\langle\varpi,\bar{\mu}\rangle$ is the coefficient of α_s. For the fundamental weights, these coefficients can be found in the tables in [7], VI. A fundamental weight ϖ satisfies (10.8) if and only if

(10.10) $$(\text{coefficient of } \alpha_s \text{ in } \varpi+\tau\varpi) = 1.$$

In the following, we write $\alpha_1,\ldots,\alpha_n$ for the simple roots and $\varpi_1,\ldots,\varpi_n$ for the fundamental weights with the usual numbering. In the diagrams, the solid node is the special node corresponding to α_s, and the nodes ✿ correspond to symplectic representations (and we call them *symplectic nodes*).

Type A_n. The opposition involution τ switches the nodes i and $n+1-i$. According to the tables in Bourbaki, for $1\leqslant i\leqslant (n+1)/2$,

$$\varpi_i = \tfrac{n+1-i}{n+1}\alpha_1 + \tfrac{2(n+1-i)}{n+1}\alpha_2 + \cdots + \tfrac{i(n+1-i)}{n+1}\alpha_i + \cdots + \tfrac{2i}{n+1}\alpha_{n-1} + \tfrac{i}{n+1}\alpha_n.$$

Replacing i with $n+1-i$ reflects the coefficients, and so

$$\tau\varpi_i = \varpi_{n+1-i} = \tfrac{i}{n+1}\alpha_1 + \tfrac{2i}{n+1}\alpha_2 + \cdots + \tfrac{2(n+1-i)}{n+1}\alpha_{n-1} + \tfrac{n+1-i}{n+1}\alpha_n.$$

Therefore,

$$\varpi_i+\tau\varpi_i = \alpha_1+2\alpha_2+\cdots+i\alpha_i+i\alpha_{i+1}+\cdots+i\alpha_{n+1-(i+1)}+i\alpha_{n+1-i}+\cdots+2\alpha_{n-1}+\alpha_n,$$

i.e., the sequence of coefficients is

$$(1,2,\ldots,i,i,\ldots,i,i,\ldots,2,1).$$

Let $\alpha_s=\alpha_1$ or α_n. Then every fundamental weight satisfies (10.10):[34]

[34] [19], Table 1.3.9, overlooks this possibility.

$A_n(1)$ 

$A_n(n)$ 

Let $\alpha_s = \alpha_j$, with $1 < j < n$. Then only the fundamental weights ϖ_1 and ϖ_n satisfy (10.10):

$A_n(j)$ 

As P/Q is generated by ϖ_1, the symplectic representations form a faithful family.

Type B_n. In this case, $\alpha_s = \alpha_1$ and the opposition involution acts trivially on the Dynkin diagram, and so we seek a fundamental weight ϖ_i such that $\varpi_i = \frac{1}{2}\alpha_1 + \cdots$. According to the tables in Bourbaki,

$$\varpi_i = \alpha_1 + 2\alpha_2 + \cdots + (i-1)\alpha_{i-1} + i(\alpha_i + \alpha_{i+1} + \cdots + \alpha_n) \quad (1 \leqslant i < n)$$
$$\varpi_n = \tfrac{1}{2}(\alpha_1 + 2\alpha_2 + \cdots + n\alpha_n),$$

and so only ϖ_n satisfies (10.10):

$B_n(1)$ 

As P/Q is generated by ϖ_n, the symplectic representations form a faithful family.

Type C_n. In this case $\alpha_s = \alpha_n$ and the opposition involution acts trivially on the Dynkin diagram, and so we seek a fundamental weight ϖ_i such that $\varpi_i = \cdots + \frac{1}{2}\alpha_n$. According to the tables in Bourbaki,

$$\varpi_i = \alpha_1 + 2\alpha_2 + \cdots + (i-1)\alpha_{i-1} + i(\alpha_i + \alpha_{i+1} + \cdots + \alpha_{n-1} + \tfrac{1}{2}\alpha_n),$$

and so only ϖ_1 satisfies (10.10):

C_n 

As P/Q is generated by ϖ_1, the symplectic representations form a faithful family.

Type D_n. The opposition involution acts trivially if n is even, and switches α_{n-1} and α_n if n is odd. According to the tables in Bourbaki,

$$\varpi_i = \alpha_1 + 2\alpha_2 + \cdots + (i-1)\alpha_{i-1} + i(\alpha_i + \cdots + \alpha_{n-2}) + \tfrac{i}{2}(\alpha_{n-1} + \alpha_n),$$
$$1 \leqslant i \leqslant n-2$$
$$\varpi_{n-1} = \tfrac{1}{2}\left(\alpha_1 + 2\alpha_2 + \cdots + (n-2)\alpha_{n-2} + \tfrac{1}{2}n\alpha_{n-1} + \tfrac{1}{2}(n-2)\alpha_n\right)$$
$$\varpi_n = \tfrac{1}{2}\left(\alpha_1 + 2\alpha_2 + \cdots + (n-2)\alpha_{n-2} + \tfrac{1}{2}(n-2)\alpha_{n-1} + \tfrac{1}{2}n\alpha_n\right)$$

Let $\alpha_s = \alpha_1$. As α_1 is fixed by the opposition involution, we seek a fundamental weight ϖ_i such that $\varpi_i = \frac{1}{2}\alpha_1 + \cdots$. Both ϖ_{n-1} and ϖ_n give rise to symplectic representations:

$D_n(1)$ 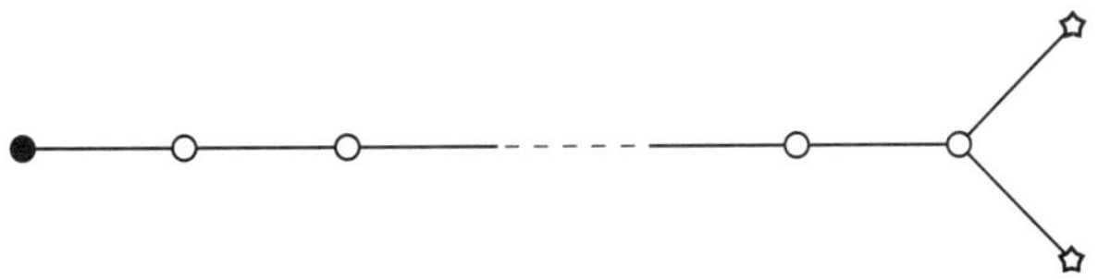

When n is odd, ϖ_{n-1} and ϖ_n each generates P/Q, and when n is even ϖ_{n-1} and ϖ_n together generate P/Q. Therefore, in both cases, the symplectic representations form a faithful family.

Let $\alpha_s = \alpha_{n-1}$ or α_n and let $n = 4$. The nodes α_1, α_3, and α_4 are permuted by automorphisms of the Dynkin diagram (hence by outer automorphisms of the corresponding group), and so this case is the same as the case $\alpha_s = \alpha_1$:

$D_4(1)$ 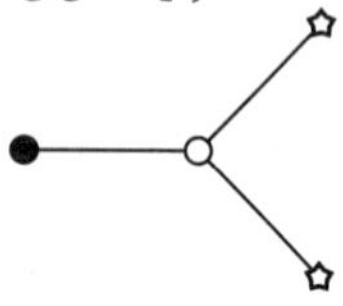 $D_4(3)$ 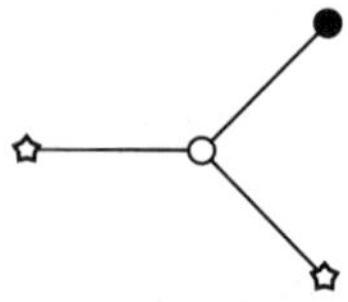 $D_4(4)$ 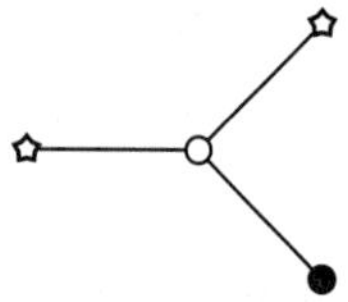

The symplectic representations form a faithful family.

Let $\alpha_s = \alpha_{n-1}$ or α_n and let $n \geqslant 5$. When n is odd, τ interchanges α_{n-1} and α_n, and so we seek a fundamental weight ϖ_i such that $\varpi_i = \cdots + a\alpha_{n-1} + b\alpha_n$ with $a + b = 1$; when n is even, τ is trivial, and we seek a fundamental weight ϖ_i such that $\varpi_i = \cdots + \frac{1}{2}\alpha_{n-1} + \cdots$ or $\cdots + \frac{1}{2}\alpha_n$. In each case, only ϖ_1 gives rise to a symplectic representation:

$D_n(n-1)$ 

$D_n(n)$ 

The weight ϖ_1 generates a subgroup of order 2 (and index 2) in P/Q. Let $C \subset Z(H)$ be the kernel of ϖ_1 regarded as a character of $Z(H)$. Then every symplectic representation factors through H/C, and the symplectic representations form a faithful family of representations of H/C.

Type E_6. In this case, $\alpha_s = \alpha_1$ or α_6, and the opposition involution interchanges α_1 and α_6. Therefore, we seek a fundamental weight ϖ_i such that $\varpi_i = a\alpha_1 + \cdots + b\alpha_6$ with $a+b = 1$. In the following diagram, we list the value $a+b$ for each fundamental weight ϖ_i:

$E_6(1)$ 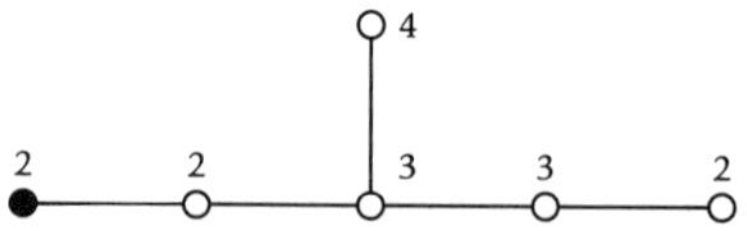

As no value equals 1, there are no symplectic representations.

Type E_7. In this case, $\alpha_s = \alpha_7$, and the opposition involution is trivial. Therefore, we seek a fundamental weight ϖ_i such that $\varpi_i = \cdots + \frac{1}{2}\alpha_7$. In the following diagram, we list the coefficient of α_7 for each fundamental weight ϖ_i:

$E_7(7)$ 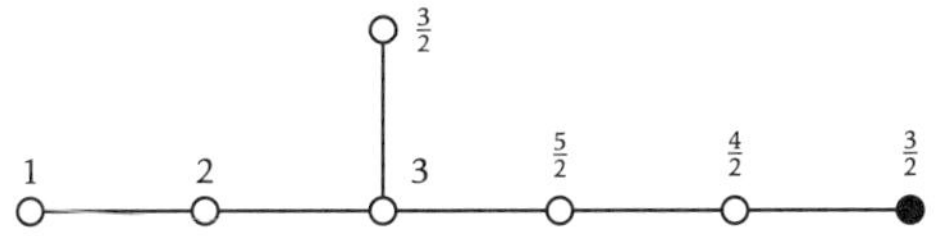

As no value is $\frac{1}{2}$, there are no symplectic representations.

Following [19], 1.3.9, we write $D^{\mathbb{R}}$ for the case $D_n(1)$ and $D^{\mathbb{H}}$ for the cases $D_n(n-1)$ and $D_n(n)$.

Summary 10.11. *Let H be a simply connected almost simple group over $\mathbb{R}$, and let $\bar{h}\colon \mathbb{S}/\mathbb{G}_m \to H^{ad}$ be a nontrivial homomorphism satisfying (SV1,2). There exists a symplectic representation of $(H, \bar{h})$ if and only if it is of type A, B, C, or D. Except when $(H, \bar{h})$ is of type $D_n^{\mathbb{H}}$, $n \geqslant 5$, the symplectic representations form a faithful family of representations of H; when $(H, \bar{h})$ is of type $D_n^{\mathbb{H}}$, $n \geqslant 5$, they form a faithful family of representations of the quotient of the simply connected group by the kernel of ϖ_1.*

The rational case

Now let H be a semisimple algebraic group over $\mathbb{Q}$, and let $\bar{h}$ be a homomorphism $\mathbb{S}/\mathbb{G}_m \to H^{ad}_{\mathbb{R}}$ satisfying (SV1,2) and generating H^{ad}.

Definition 10.12. A homomorphism $H \to GL_V$ with finite kernel is a *symplectic representation* of $(H, \bar{h})$ if there exists a commutative diagram

$$\begin{array}{ccccc} H & & & & \\ \downarrow & \searrow & & & \\ (H^{ad}, \bar{h}) & \longleftarrow & (G, h) & \xrightarrow{\rho} & (G(\psi), D(\psi)), \end{array} \tag{10.13}$$

in which ψ is a nondegenerate alternating form on V, G is a reductive group (over $\mathbb{Q}$), and h is a homomorphism $\mathbb{S} \to G_{\mathbb{R}}$; the homomorphism $H \to G$ is required to have image G^{der},

Given a diagram (10.13), we may replace G with its image in GL_V and so assume that the representation ρ is faithful.

We now assume that H is simply connected and almost simple. Then $H = (H^s)_{F/\mathbb{Q}}$ for some geometrically almost simple algebraic group H^s over a number field F. Because $H_{\mathbb{R}}$ is an inner form of its compact form, the field F is totally real (see the proof of 3.13). Let $I = \mathrm{Hom}(F, \mathbb{R})$. Then,

$$H_{\mathbb{R}} = \prod_{v \in I} H_v, \quad H_v = H^s \otimes_{F,v} \mathbb{R}.$$

The Dynkin diagram D of $H_{\mathbb{C}}$ is a disjoint union of the Dynkin diagrams D_v of the group $H_{v\mathbb{C}}$. The Galois group $\mathrm{Gal}(\mathbb{Q}^{\mathrm{al}}/\mathbb{Q})$ acts on it in a manner consistent with its projection to I. In particular, it acts transitively on D and so all the factors H_v of $H_{\mathbb{R}}$ are of the same type. We let I_c (resp. I_{nc}) denote the subset of I of v for which H_v is compact (resp. not compact), and we let $H_c = \prod_{v\in I_c} H_v$ and $H_{nc} = \prod_{v\in I_{nc}} H_v$. Because $\bar{h}$ generates H^{ad}, I_{nc} is nonempty.

Proposition 10.14. *Let F be a totally real number field. Suppose that for each real prime v of F, we are given a pair $(H_v, \bar{h}_v)$ in which H_v is a simply connected algebraic group over $\mathbb{R}$ of a fixed type, and $\bar{h}_v$ is a homomorphism $\mathbb{S}/\mathbb{G}_m \to H_v^{\mathrm{ad}}$ satisfying (SV1,2) (possibly trivial). Then there exists an algebraic group H over $\mathbb{Q}$ such that $H \otimes_{F,v} \mathbb{R} \approx H_v$ for all v.*

Proof. There exists an algebraic group H over F such that $H \otimes_{F,v} \mathbb{R}$ is an inner form of its compact form for all real primes v of F. For each such v, H_v is an inner form of $H \otimes_{F,v} \mathbb{R}$, and so defines a cohomology class in $H^1(F_v, H^{\mathrm{ad}})$. The proposition now follows from the surjectivity of the map

$$H^1(F, H^{\mathrm{ad}}) \to \prod_{v \text{ real}} H^1(F_v, H^{\mathrm{ad}})$$

([48], Proposition 1). □

Pairs $(H, \bar{h})$ for which there do not exist symplectic representations

H is of exceptional type Assume that H is of exceptional type. If there exists an $\bar{h}$ satisfying (SV1,2), then H is of type E_6 or E_7 (see §2). A symplectic representation of $(H, \bar{h})$ over $\mathbb{Q}$ gives rise to a symplectic representation of $(H_{\mathbb{R}}, \bar{h})$ over $\mathbb{R}$, but we have seen (10.11) that no such representations exist.

$(H, \bar{h})$ is of mixed type D. By this we mean that H is of type D_n with $n \geqslant 5$ and that at least one factor $(H_v, \bar{h}_v)$ is of type $D_n^{\mathbb{R}}$ and one of type $D_n^{\mathbb{H}}$. Such pairs $(H, \bar{h})$ exist by Proposition 10.14. The Dynkin diagram of $H_{\mathbb{R}}$ contains connected components

$D_n(1)$

and

$D_n(n)$

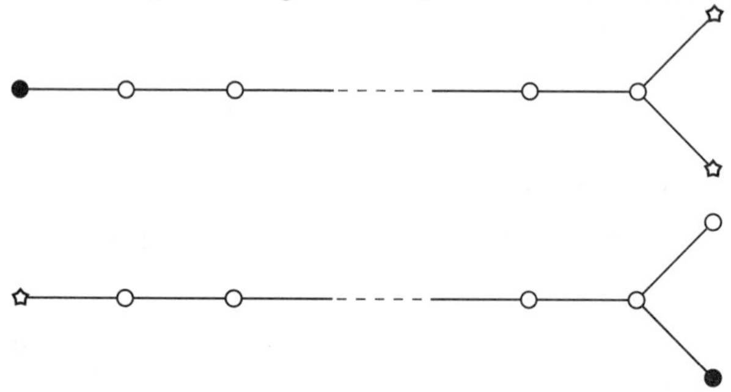

or $D_n(n-1)$. To give a symplectic representation for $H_{\mathbb{R}}$, we have to choose a symplectic node for each real prime v such that H_v is noncompact. In order for the representation to be rational, the collection of symplectic nodes must be stable under $\mathrm{Gal}(\mathbb{Q}^{\mathrm{al}}/\mathbb{Q})$, but this is impossible, because there is no automorphism of the Dynkin diagram of type D_n, $n \geqslant 5$, carrying the node 1 into either the node $n-1$ or the node n.

Pairs $(H, \bar{h})$ for which there exist symplectic representations

Lemma 10.15. *Let* G *be a reductive group over* $\mathbb{Q}$ *and let* h *be a homomorphism* $\mathbb{S} \to G_{\mathbb{R}}$ *satisfying (SV1,2*) and generating* G*. For any representation* (V, ρ) *of* G *such that* $(V, \rho \circ h)$ *is of type* $\{(-1,0), (0,-1)\}$*, there exists an alternating form* ψ *on* V *such that* ρ *induces a homomorphism* $(G, h) \to (G(\psi), D(\psi))$.

Proof. The pair $(\rho G, \rho \circ h)$ is the Mumford-Tate group of $(V, \rho \circ h)$ and satisfies (SV2*). The proof of Proposition 6.4 constructs a polarization ψ for $(V, \rho \circ h)$ such that $\rho G \subset G(\psi)$. □

Proposition 10.16. *A homomorphism* $H \to GL_V$ *is a symplectic representation of* $(H, \bar{h})$ *if there exists a commutative diagram*

$$\begin{array}{ccccc} H & & & & \\ \downarrow & \searrow & & & \\ (H^{ad}, \bar{h}) & \longleftarrow & (G, h) & \stackrel{\rho}{\hookrightarrow} & GL_V, \end{array}$$

in which G *is a reductive group whose connected centre splits over a CM-field, the homomorphism* $H \to G$ *has image* G^{der}*, the weight* w_h *is defined over* $\mathbb{Q}$*, and the Hodge structure* $(V, \rho \circ h)$ *is of type* $\{(-1,0), (0,-1)\}$.

Proof. The hypothesis on the connected centre Z° says that the largest compact subtorus of $Z^{\circ}_{\mathbb{R}}$ is defined over $\mathbb{Q}$. Take G' to be the subgroup of G generated by this torus, G^{der}, and the image of w_h. Now (G', h) satisfies (SV2*), and we can apply 10.15. □

We classify the symplectic representations of $(H, \bar{h})$ with ρ faithful. Note that the quotient of H acting faithfully on V is isomorphic to G^{der}.

Let (V, r) be a symplectic representation of $(H, \bar{h})$. The restriction of the representation to H_{nc} is a real symplectic representation of H_{nc}, and so, according to Corollary 10.9, every nontrivial irreducible direct summand of $r_{\mathbb{C}}|H_{nc}$ factors through H_v for some $v \in I_{nc}$ and corresponds to a symplectic node of the Dynkin diagram D_v of H_v.

Let W be an irreducible direct summand of $V_{\mathbb{C}}$. Then

$$W \approx \bigotimes\nolimits_{v \in T} W_v$$

$$\begin{array}{ccc} D & & \\ \downarrow & \nwarrow^{s} & \\ I & \hookleftarrow & T \end{array}$$

for some irreducible symplectic representations W_v of $H_{v\mathbb{C}}$ indexed by a subset T of I. The irreducible representation W_v corresponds to a symplectic node $s(v)$ of D_v. Because r is defined over $\mathbb{Q}$, the set $s(T)$ is stable under the action of $\mathrm{Gal}(\mathbb{Q}^{al}/\mathbb{Q})$. For $v \in I_{nc}$, the set $s(T) \cap D_v$ consists of a single symplectic node.

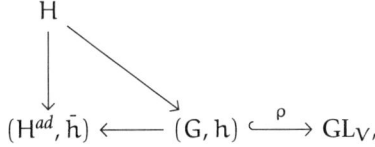

Given a diagram (10.13), we let $\mathcal{S}(V)$ denote the set of subsets $s(T)$ of the nodes of D as W runs over the irreducible direct summands of V. The set $\mathcal{S}(V)$ satisfies the following conditions:

(10.17a) for $S \in \mathcal{S}(V)$, $S \cap D_{nc}$ is either empty or consists of a single symplectic node of D_v for some $v \in I_{nc}$;

(10.17b) $\mathcal{S}$ is stable under $\mathrm{Gal}(\mathbb{Q}^{al}/\mathbb{Q})$ and contains a nonempty subset.

Given such a set $\mathcal{S}$, let $H(\mathcal{S})_{\mathbb{C}}$ be the quotient of $H_{\mathbb{C}}$ that acts faithfully on the representation defined by $\mathcal{S}$. The condition (10.17b) ensures that $H(\mathcal{S})$ is defined over $\mathbb{Q}$. According to Galois theory (in the sense of Grothendieck), there exists an étale $\mathbb{Q}$-algebra $K_{\mathcal{S}}$ such that

$$\mathrm{Hom}(K_{\mathcal{S}}, \mathbb{Q}^{al}) \simeq \mathcal{S} \qquad \text{(as sets with an action of } \mathrm{Gal}(\mathbb{Q}^{al}/\mathbb{Q})).$$

Theorem 10.18. *For any set $\mathcal{S}$ satisfying the conditions (10.17), there exists a diagram (10.13) such that the quotient of H acting faithfully on V is $H(\mathcal{S})$.*

Proof. We prove this only in the case that $\mathcal{S}$ consists of one-point sets. For an $\mathcal{S}$ as in the theorem, the set $\mathcal{S}'$ of $\{s\}$ for $s \in S \in \mathcal{S}$ satisfies (10.17) and $H(\mathcal{S})$ is a quotient of $H(\mathcal{S}')$.

Recall that $H = (H^s)_{F/\mathbb{Q}}$ for some totally real field F. We choose a totally imaginary quadratic extension E of F and, for each real embedding v of F in I_c, we choose an extension σ of v to a complex embedding of E. Let T denote the set of σ's. Thus

$$\begin{array}{ccc} E & \xrightarrow{\sigma} & \mathbb{C} \\ \cup & & \cup \\ F & \xrightarrow{v} & \mathbb{R} \end{array} \qquad T = \{\sigma \mid v \in I_c\}.$$

We regard E as a $\mathbb{Q}$-vector space, and define a Hodge structure h_T on it as follows: $E \otimes_{\mathbb{Q}} \mathbb{C} \simeq \mathbb{C}^{\mathrm{Hom}(E,\mathbb{C})}$ and the factor with index σ is of type $(-1,0)$ if $\sigma \in T$, type $(0,-1)$ if $\bar{\sigma} \in T$, and of type (0,0) if σ lies above I_{nc}. Thus ($\mathbb{C}_\sigma = \mathbb{C}$):

$$\begin{array}{ccccccc} E \otimes_{\mathbb{Q}} \mathbb{C} = & \bigoplus_{\sigma \in T} \mathbb{C}_\sigma & \oplus & \bigoplus_{\bar{\sigma} \in T} \mathbb{C}_\sigma & \oplus & \bigoplus_{\sigma \notin T \cup \bar{T}} \mathbb{C}_\sigma. \\ h_T(z) & z & & \bar{z} & & 1 \end{array}$$

Because the elements of $\mathcal{S}$ are one-point subsets of D, we can identify them with elements of D, and so regard $\mathcal{S}$ as a subset of D. It has the properties:

(a) if $s \in \mathcal{S} \cap D_{nc}$, then s is a symplectic node;

(b) $\mathcal{S}$ is stable under $\mathrm{Gal}(\mathbb{Q}^{al}/\mathbb{Q})$ and is nonempty.

Let K_D be the smallest subfield of $\mathbb{Q}^{al}$ such that $\mathrm{Gal}(\mathbb{Q}^{al}/K_D)$ acts trivially on D. Then K_D is a Galois extension of $\mathbb{Q}$ in $\mathbb{Q}^{al}$ such that $\mathrm{Gal}(K_D/K)$ acts faithfully on D. Complex conjugation acts as the opposition involution on D, which lies in the centre of Aut(D); therefore K_D is either totally real or CM.

The $\mathbb{Q}$-algebra $K_{\mathcal{S}}$ can be taken to be a product of subfields of K_D. In particular, $K_{\mathcal{S}}$ is a product of totally real fields and CM fields. The projection $\mathcal{S} \to I$ corresponds to a homomorphism $F \to K_{\mathcal{S}}$.

For $s \in \mathcal{S}$, let $V(s)$ be a complex representation of $H_{\mathbb{C}}$ with dominant weight the fundamental weight corresponding to s. The isomorphism class of the representation $\bigoplus_{s\in\mathcal{S}} V(s)$ is defined over $\mathbb{Q}$. The obstruction to the representation itself being defined over $\mathbb{Q}$ lies in the Brauer group of $\mathbb{Q}$, which is torsion, and so some multiple of the representation is defined over $\mathbb{Q}$. Let V be a representation of H over $\mathbb{Q}$ such that $V_{\mathbb{C}} \approx \bigoplus_{s\in\mathcal{S}} nV(s)$ for some integer n, and let V_s denote the direct summand of $V_{\mathbb{C}}$ isomorphic to $nV(s)$. These summands are permuted by $\mathrm{Gal}(\mathbb{Q}^{\mathrm{al}}/\mathbb{Q})$ in a fashion compatible with the action of $\mathrm{Gal}(\mathbb{Q}^{\mathrm{al}}/\mathbb{Q})$ on $\mathcal{S}$, and the decomposition $V_{\mathbb{C}} = \bigoplus_{s\in\mathcal{S}} V_s$ corresponds therefore to a structure of a $K_{\mathcal{S}}$-module on V: let $s'\colon K_{\mathcal{S}} \to \mathbb{Q}^{\mathrm{al}}$ be the homomorphism corresponding to $s \in \mathcal{S}$; then $a \in K_{\mathcal{S}}$ acts on V_s as multiplication by $s'(a)$.

Let H' denote the quotient of H that acts faithfully on V. Then $H'_{\mathbb{R}}$ is the quotient of $H_{\mathbb{R}}$ described in (10.11).

A lifting of $\bar{h}$ to a fractional morphism of $\mathbb{S}$ into $H'_{\mathbb{R}}$ defines a fractional Hodge structure on V of weight 0, which can be described as follows. Let $s \in \mathcal{S}$, and let v be its image in I; if $v \in I_c$, then V_s is of type $(0,0)$; if $v \in I_{nc}$, then V_s is of type $\{(r,-r),(r-1,1-r)\}$ where $r = \langle \varpi_s, \bar{\mu}\rangle$ (notations as in 10.7). We renumber this Hodge structure to obtain a new Hodge structure on V:

	old	new
$V_s,\ v \in I_c$	$(0,0)$	$(0,0)$
$V_s,\ v \in I_{nc}$	$(r,-r)$	$(0,-1)$
$V_s,\ v \in I_{nc}$	$(r-1,1-r)$	$(-1,0)$.

We endow the $\mathbb{Q}$-vector space $E \otimes_F V$ with the tensor product Hodge structure. The decomposition

$$(E \otimes_F V) \otimes_{\mathbb{Q}} \mathbb{R} = \bigoplus\nolimits_{v\in I} (E \otimes_{F,v} \mathbb{R}) \otimes_{\mathbb{R}} (V \otimes_{F,v} \mathbb{R}),$$

is compatible with the Hodge structures. The type of the Hodge structure on each direct summand is given by the following table:

	$E \otimes_{F,v} \mathbb{R}$	$V \otimes_{F,v} \mathbb{R}$
$v \in I_c$	$\{(-1,0),(0,-1)\}$	$\{(0,0)\}$
$v \in I_{nc}$	$\{(0,0)\}$	$\{(-1,0),(0,-1)\}$.

Therefore, $E \otimes_F V$ has type $\{(-1,0),(0,-1)\}$. Let G be the algebraic subgroup of $GL_{E\otimes_F V}$ generated by $E^\times$ and H'. The homomorphism $h\colon \mathbb{S} \to (GL_{E\otimes_F V})_{\mathbb{R}}$ corresponding to the Hodge structure factors through $G_{\mathbb{R}}$, and the derived group of G is H'. Now apply (10.16). □

Aside 10.19. The trick of using a quadratic imaginary extension E of F in order to obtain a Hodge structure of type $\{(-1,0),(0,-1)\}$ from one of type $\{(-1,0),(0,0),(0,-1)\}$ in essence goes back to Shimura (cf. [14], §6).

Conclusion Now let H be a semisimple algebraic group over $\mathbb{Q}$, and let $\bar{h}$ be a homomorphism $\mathbb{S} \to H^{\mathrm{ad}}_{\mathbb{R}}$ satisfying (SV1,2) and generating H.

Definition 10.20. The pair $(H,\bar{h})$ is of *Hodge type* if it admits a faithful family of symplectic representations.

Theorem 10.21. *A pair $(H,\bar{h})$ is of Hodge type if it is a product of pairs $(H_i,\bar{h}_i)$ such that either*

(a) *$(H_i,\bar{h}_i)$ is of type A, B, C, or $D^{\mathbb{R}}$, and H is simply connected, or*
(b) *$(H_i,\bar{h}_i)$ is of type $D^{\mathbb{H}}_n$ $(n \geqslant 5)$ and equals $(H^s)_{F/\mathbb{Q}}$ for the quotient H^s of the simply connected group of type $D^{\mathbb{H}}_n$ by the kernel of ϖ_1 (cf. 10.11).*

Conversely, if $(H,\bar{h})$ is a Hodge type, then it is a quotient of a product of pairs satisfying (a) or (b).

Proof. Suppose that $(H,\bar{h})$ is a product of pairs satisfying (a) and (b), and let $(H',\bar{h}')$ be one of these factors with H' almost simple. Let $\tilde{H}'$ be the simply connected covering group of H. Then (10.11) allows us to choose a set $\mathcal{S}$ satisfying (10.17) and such that $H' = H(\mathcal{S})$. Now Theorem 10.18 shows that $(H',\bar{h}')$ admits a faithful symplectic representation. A product of pairs of Hodge type is clearly of Hodge type.

Conversely, suppose that $(H,\bar{h})$ is of Hodge type, let $\tilde{H}$ be the simply connected covering group of H, and let $(H',\bar{h}')$ be an almost simple factor of $(\tilde{H},\bar{h})$. Then $(H',\bar{h}')$ admits a symplectic representation with finite kernel, and so $(H',\bar{h}')$ is not of type E_6, E_7, or mixed type D (see p. 532). Moreover, if $(H',\bar{h}')$ is of type $D^{\mathbb{H}}_n$, $n \geqslant 5$, then (10.11) shows that it factors through the quotient described in (b). □

Notice that we haven't completely classified the pairs $(H,\bar{h})$ of Hodge type because we haven't determined exactly which quotients of products of pairs satisfying (a) or (b) occur as $H(\mathcal{S})$ for some set $\mathcal{S}$ satisfying (10.17).

11. Moduli

In this section, we determine (a) the pairs (G,h) that arise as the Mumford-Tate group of an abelian variety (or an abelian motive); (b) the arithmetic locally symmetric varieties that carry a faithful family of abelian varieties (or abelian motives); (c) the Shimura varieties that arise as moduli varieties for polarized abelian varieties (or motives) with Hodge class and level structure.

Mumford-Tate groups

Theorem 11.1. *Let G be an algebraic group over $\mathbb{Q}$, and let $h\colon \mathbb{S} \to G_{\mathbb{R}}$ be a homomorphism that generates G and whose weight is rational. The pair (G,h) is the Mumford-Tate* *group of an abelian variety if and only if* h *satisfies (SV2*) and there exists a faithful representation* $\rho\colon G \to GL_V$ *such that* $(V, \rho \circ h)$ *is of type* $\{(-1,0),(0,-1)\}$

Proof. The necessity is obvious (apply (6.4) to see that (G,h) satisfies (SV2*)). For the sufficiency, note that (G,h) is the Mumford-Tate group of $(V, \rho\circ h)$ because h generates G. The Hodge structure is polarizable because (G,h) satisfies (SV2*) (apply 6.4), and so it is the Hodge structure $H_1(A^{\mathrm{an}},\mathbb{Q})$ of an abelian variety A by Riemann's theorem 4.4. □

The Mumford-Tate group of a motive is defined to be the Mumford-Tate group of its Betti realization.

Theorem 11.2. *Let* (G,h) *be an algebraic group over* $\mathbb{Q}$, *and let* $h\colon \mathbb{S} \to G_{\mathbb{R}}$ *be a homomorphism satisfying (SV1,2*) and generating* G. *Assume that* w_h *is defined over* $\mathbb{Q}$. *The pair* (G,h) *is the Mumford-Tate group of an abelian motive if and only if* $(G^{\mathrm{der}},\bar{h})$ *is a quotient of a product of pairs satisfying (a) and (b) of (10.21).*

The proof will occupy the rest of this subsection. Recall that G_{Hdg} is the affine group scheme attached to the tannakian category $\mathrm{Hdg}_{\mathbb{Q}}$ of polarizable rational Hodge structures and the forgetful fibre functor (see 9.26). It is equipped with a homomorphism $h_{\mathrm{Hdg}}\colon \mathbb{S} \to (G_{\mathrm{Hdg}})_{\mathbb{R}}$. If (G,h) is the Mumford-Tate group of a polarizable Hodge structure, then there is a unique homomorphism $\rho(h)\colon G_{\mathrm{Hdg}} \to G$ such that $h = \rho(h)_{\mathbb{R}} \circ h_{\mathrm{Hdg}}$. Moreover, $(G_{\mathrm{Hdg}}, h_{\mathrm{Hdg}}) = \varprojlim (G,h)$.

Lemma 11.3. *Let* H *be a semisimple algebraic group over* $\mathbb{Q}$, *and let* $\bar{h}\colon \mathbb{S}/\mathbb{G}_m \to H^{\mathrm{ad}}_{\mathbb{R}}$ *be a homomorphism satisfying (SV1,2,3). There exists a unique homomorphism*

$$\rho(H,\bar{h})\colon \left(G_{\mathrm{Hdg}}\right)^{\mathrm{der}} \to H$$

such that the following diagram commutes:

$$\begin{array}{ccc} (G_{\mathrm{Hdg}})^{\mathrm{der}} & \xrightarrow{\rho(H,\bar{h})} & H \\ \downarrow & & \downarrow \\ G_{\mathrm{Hdg}} & \xrightarrow[\rho(\bar{h})]{} & H^{\mathrm{ad}}. \end{array}$$

Proof. Two such homomorphisms $\rho(H,\bar{h})$ would differ by a map into $Z(H)$. Because $(G_{\mathrm{Hdg}})^{\mathrm{der}}$ is connected, any such map is constant, and so the homomorphisms are equal.

For the existence, choose a pair (G,h) as in (8.9). Then (G,h) is the Mumford-Tate group of a polarizable Hodge structure, and we can take $\rho(H,\bar{h}) = \rho(h)|(G_{\mathrm{Hdg}})^{\mathrm{der}}$. □

Lemma 11.4. *The assignment* $(H,\bar{h}) \mapsto \rho(H,\bar{h})$ *is functorial: if* $\alpha\colon H \to H'$ *is a homomorphism taking* $Z(H)$ *into* $Z(H')$ *and carrying* $\bar{h}$ *to* $\bar{h}'$, *then* $\rho(H',\bar{h}') = \alpha\circ\rho(H,\bar{h})$.

Proof. The homomorphism $\bar{h}'$ generates H'^{ad} (by SV3), and so the homomorphism α is surjective. Choose a pair (G, h) for $(H, \bar{h})$ as in (8.9), and let $G' = G/\operatorname{Ker}(\alpha)$. Write α again for the projection $G \to G'$ and let $h' = \alpha_{\mathbb{R}} \circ h$. This equality implies that

$$\rho(h') = \alpha \circ \rho(h).$$

On restricting this to $(G_{Hdg})^{der}$, we obtain the equality

$$\rho(H', \bar{h}') = \alpha \circ \rho(H, \bar{h}).$$

□

Recall that G_{Mab} is the affine group scheme attached to the category of abelian motives over $\mathbb{C}$ and the Betti fibre functor. The functor $\mathsf{Mot}^{ab}(\mathbb{C}) \to \mathsf{Hdg}_{\mathbb{Q}}$ is fully faithful by Deligne's theorem (9.15), and so it induces a surjective map $G_{Hdg} \to G_{Mab}$.

Lemma 11.5. *If* (H, h) *is of Hodge type, then* $\rho(H, \bar{h})$ *factors through* $(G_{Mab})^{der}$.

Proof. Let (G, h) be as in the definition (10.12), and replace G with the algebraic subgroup generated by h. Then (G, h) is the Mumford-Tate group of an abelian variety (Riemann's theorem 4.4), and so $\rho(h)\colon G_{Hdg} \to G$ factors through $G_{Hdg} \to G_{Mab}$. Therefore $\rho(H, \bar{h})$ maps the kernel of $\left(G_{Hdg}\right)^{der} \to (G_{Mab})^{der}$ into the kernel of $H \to G$. By assumption, the intersection of these kernels is trivial. □

Lemma 11.6. *The homomorphism* $\rho(H, \bar{h})$ *factors through* $(G_{Mab})^{der}$ *if and only if* $(H, \bar{h})$ *has a finite covering by a pair of Hodge type.*

Proof. Suppose that there is a finite covering $\alpha\colon H' \to H$ such that $(H', \bar{h})$ is of Hodge type. By Lemma 11.5, $\rho(H', \bar{h})$ factors through $(G_{Mab})^{der}$, and therefore so also does $\rho(H, \bar{h}) = \alpha \circ \rho(H', \bar{h})$.

Conversely, suppose that $\rho(H, \bar{h})$ factors through $(G_{Mab})^{der}$. There will be an algebraic quotient (G, h) of (G_{Mab}, h_{Mab}) such that $(H, \bar{h})$ is a quotient of $(G^{der}, \operatorname{ad} \circ h)$. Consider the category of abelian motives M such that the action of G_{Mab} on $\omega_B(M)$ factors through G. By definition, this category is contained in the tensor category generated by $h_1(A)$ for some abelian variety A. We can replace G with the Mumford-Tate group of A. Then $(G^{der}, \operatorname{ad} \circ h)$ has a faithful symplectic embedding, and so it is of Hodge type. □

We can now complete the proof of the Theorem 11.2. From (9.26), we know that $\rho(h)$ factors through G_{Mab} if and only if $\rho(G^{der}, \operatorname{ad} \circ h)$ factors through $(G_{Mab})^{der}$, and from (11.6) we know that this is true if and only if $(G^{der}, \operatorname{ad} \circ h)$ has a finite covering by a pair of Hodge type.

Aside 11.7. Let G be an algebraic group over $\mathbb{Q}$ and let h be a homomorphism $\mathbb{S} \to G_{\mathbb{R}}$. If (G, h) is the Mumford-Tate group of a motive, then h generates G, w_h is defined over $\mathbb{Q}$, and h satisfies (SV2*). Assume that (G, h) satisfies these conditions. A positive answer to Question 9.28 would imply that (G, h) is the Mumford-Tate group of a motive if h satisfies (SV1). If G^{der} is of type E_8, F_4, or G_2, then there does not exist an h satisfying (SV1) (apply §2 to $h|\mathbb{S}^1$). Nevertheless, it has recently been shown that there exist motives whose Mumford-Tate group is of type G_2 ([22]).

Notes. This subsection follows §1 of [38].

Families of abelian varieties and motives

Let S be a connected smooth algebraic variety over $\mathbb{C}$, and let $o \in S(\mathbb{C})$. A family $f\colon A \to S$ of abelian varieties over S defines a local system $V = R_1 f_* \mathbb{Z}$ of $\mathbb{Z}$-modules on S^{an}. We say that the family is *faithful* if the monodromy representation $\pi_1(S^{an}, o) \to GL(V_o)$ is injective.

Let $D(\Gamma) = \Gamma \backslash D$ be an arithmetic locally symmetric variety, and let $o \in D$. By definition, there exists a simply connected algebraic group H over $\mathbb{Q}$ and a surjective homomorphism $\varphi\colon H(\mathbb{R}) \to \mathrm{Hol}(D)^+$ with compact kernel such that $\varphi(H(\mathbb{Z}))$ is commensurable with Γ. Moreover, with a mild condition on the ranks, the pair (H, φ) is uniquely determined up to a unique isomorphism (see 3.13). Let $\bar{h}\colon \mathbb{S} \to H^{ad}$ be the homomorphism whose projection into a compact factor of H^{ad} is trivial and is such that $\varphi(\bar{h}(z))$ fixes o and acts on $T_o(D)$ as multiplication by $z/\bar{z}$ (cf. (8.4), p. 510).

Theorem 11.8. *There exists a faithful family of abelian varieties on $D(\Gamma)$ having a fibre of CM-type if and only if $(H, \bar{h})$ admits a symplectic representation (10.12).*

Proof. Let $f\colon A \to D(\Gamma)$ be a faithful family of abelian varieties on $D(\Gamma)$, and let (V, F) be the variation of Hodge structures $R_1 f_* \mathbb{Q}$. Choose a trivialization $\pi^* V \approx V_D$, and let $G \subset GL_V$ be the generic Mumford-Tate group (see 6.17). As in (§8), we get a commutative diagram

$$\begin{array}{ccccc} H & & & & \\ \downarrow & \searrow & & & \\ (H^{ad}, \bar{h}) & \longleftarrow & (G, h) & \xrightarrow{\ \rho\ } & GL_V \end{array} \tag{11.9}$$

in which the image of $H \to G$ is G^{der}. Because the family is faithful, the map $H \to G^{der}$ is an isogeny, and so $(H, \bar{h})$ admits a symplectic representation.

Conversely, a symplectic representation of $(H, \bar{h})$ defines a variation of Hodge structures (8.6), which arises from a family of abelian varieties by Theorem 7.13 (Riemann's theorem in families). □

Theorem 11.10. *There exists a faithful family of abelian motives on $D(\Gamma)$ having a fibre of CM-type if and only if $(H, \bar{h})$ has finite covering by a pair of Hodge type.*

Proof. The proof is essentially the same as that of Theorem 11.8. The points are the determination of the Mumford-Tate groups of abelian motives in (11.2) and Theorem 9.31, which replaces Riemann's theorem in families. □

Shimura varieties

In the above, we have always considered connected varieties. As Deligne [14] observed, it is often more convenient to consider nonconnected varieties.

Definition 11.11. A *Shimura datum* is a pair (G, X) consisting of a reductive group G over $\mathbb{Q}$ and a $G(\mathbb{R})^+$-conjugacy class of homomorphisms $\mathbb{S} \to G_{\mathbb{R}}$ satisfying (SV1,2,3).[35]

Example 11.12. Let (V, ψ) be a symplectic space over $\mathbb{Q}$. The group $G(\psi)$ of symplectic similitudes together with the space $X(\psi)$ of all complex structures J on $V_{\mathbb{R}}$ such that $(x, y) \mapsto \psi(x, Jy)$ is positive definite is a Shimura datum.

Let (G, X) be a Shimura datum. The map $h \mapsto \bar{h} \overset{\text{def}}{=} \mathrm{ad} \circ h$ identifies X with a $G^{\mathrm{ad}}(\mathbb{R})^+$-conjugacy class of homomorphisms $\bar{h}\colon \mathbb{S}/\mathbb{G}_m \to G^{\mathrm{ad}}_{\mathbb{R}}$ satisfying (SV1,2,3). Thus X is a hermitian symmetric domain (2.5, 6.1). More canonically, the set X has a unique structure of a complex manifold such that, for every representation $\rho_{\mathbb{R}}\colon G_{\mathbb{R}} \to GL_V$, $(V_X, \rho \circ h)_{h \in X}$ is a holomorphic family of Hodge structures. For this complex structure, $(V_X, \rho \circ h)_{h \in X}$ is a variation of Hodge structures, and so X is a hermitian symmetric domain.

The Shimura variety attached to (G, X) and the choice of a compact open subgroup K of $G(\mathbb{A}_f)$ is[36]

$$\mathrm{Sh}_K(G, X) = G(\mathbb{Q})_+ \backslash X \times G(\mathbb{A}_f)/K$$

where $G(\mathbb{Q})_+ = G(\mathbb{Q}) \cap G(\mathbb{R})^+$. In this quotient, $G(\mathbb{Q})_+$ acts on both X (by conjugation) and $G(\mathbb{A}_f)$, and K acts on $G(\mathbb{A}_f)$. Let $\mathcal{C}$ be a set of representatives for the (finite) double coset space $G(\mathbb{Q})_+ \backslash G(\mathbb{A}_f)/K$; then

$$G(\mathbb{Q})_+ \backslash X \times G(\mathbb{A}_f)/K \simeq \bigsqcup\nolimits_{g \in \mathcal{C}} \Gamma_g \backslash X, \quad \Gamma_g = gKg^{-1} \cap G(\mathbb{Q})_+.$$

Because Γ_g is a congruence subgroup of $G(\mathbb{Q})$, its image in $G^{\mathrm{ad}}(\mathbb{Q})$ is arithmetic (3.4), and so $\mathrm{Sh}_K(G, X)$ is a finite disjoint union of connected Shimura varieties. It therefore has a unique structure of an algebraic variety. As K varies, these varieties form a projective system.

We make this more explicit in the case that G^{der} is simply connected. Let $\nu\colon G \to T$ be the quotient of G by G^{der}, and let Z be the centre of G. Then ν defines an isogeny $Z \to T$, and we let

$$T(\mathbb{R})^{\dagger} = \mathrm{Im}(Z(\mathbb{R}) \to T(\mathbb{R})),$$
$$T(\mathbb{Q})^{\dagger} = T(\mathbb{Q}) \cap T(\mathbb{R})^{\dagger}.$$

---

[35] In the usual definition, X is taken to be a $G(\mathbb{R})$-conjugacy class. For our purposes, it is convenient to choose a connected component of X.

[36] This agrees with the usual definition because of [40], 5.11.

The set $T(\mathbb{Q})^{\dagger}\backslash T(\mathbb{A}_f)/\nu(K)$ is finite and discrete. For K sufficiently small, the map

$$[x, a] \mapsto [\nu(a)]: G(\mathbb{Q})\backslash X \times G(\mathbb{A}_f)/K \to T(\mathbb{Q})^{\dagger}\backslash T(\mathbb{A}_f)/\nu(K) \tag{11.13}$$

is surjective, and each fibre is isomorphic to $\Gamma\backslash X$ for some congruence subgroup Γ of $G^{der}(\mathbb{Q})$. For the fibre over [1], the congruence subgroup Γ is contained in $K \cap G^{der}(\mathbb{Q})$, and equals it if $Z(G^{der})$ satisfies the Hasse principal for H^1, for example, if G^{der} has no factors of type A.

Example 11.14. Let $G = GL_2$. Then

$$(G \xrightarrow{\nu} T) = (GL_2 \xrightarrow{\det} \mathbb{G}_m)$$
$$(Z \xrightarrow{\nu} T) = (\mathbb{G}_m \xrightarrow{2} \mathbb{G}_m),$$

and therefore

$$T(\mathbb{Q})^{\dagger}\backslash T(\mathbb{A}_f)/\nu(K) = \mathbb{Q}^{>0}\backslash \mathbb{A}_f^{\times} / \det(K).$$

Note that $\mathbb{A}_f^{\times} = \mathbb{Q}^{>0} \cdot \hat{\mathbb{Z}}^{\times}$ (direct product) where $\hat{\mathbb{Z}} = \varprojlim_n \mathbb{Z}/n\mathbb{Z} \simeq \prod_{\ell} \mathbb{Z}_{\ell}$. For

$$K = K(N) \stackrel{\text{def}}{=} \{a \in \hat{\mathbb{Z}}^{\times} \mid a \equiv 1, \bmod N\},$$

we find that

$$T(\mathbb{Q})^{\dagger}\backslash T(\mathbb{A}_f)/\nu(K) \simeq (\mathbb{Z}/N\mathbb{Z})^{\times}.$$

Definition 11.15. A Shimura datum (G, X) is of *Hodge type* if there exists an injective homomorphism $G \to G(\psi)$ sending X into $X(\psi)$ for some symplectic pair (V, ψ) over $\mathbb{Q}$.

Definition 11.16. A Shimura datum (G, X) is of *abelian type* if, for one (hence all) $h \in X$, the pair $(G^{der}, \mathrm{ad} \circ h)$ is a quotient of a product of pairs satisfying (a) or (b) of (10.21).

A Shimura variety $\mathrm{Sh}(G, X)$ is said to be of Hodge or abelian type if (G, X) is.

Notes. See [40], §5, for proofs of the statements in this subsection. For the structure of the Shimura variety when G^{der} is not simply connected, see [19], 2.1.16.

Shimura varieties as moduli varieties

Throughout this subsection, (G, X) is a Shimura datum such that

(a) w_X is defined over $\mathbb{Q}$ and the connected centre of G is split by a CM-field, and

(b) there exists a homomorphism $\nu: G \to \mathbb{G}_m \simeq GL_{\mathbb{Q}(1)}$ such that $\nu \circ w_X = -2$.

Fix a faithful representation $\rho: G \to GL_V$. Assume that there exists a pairing $t_0: V \times V \to \mathbb{Q}(m)$ such that (i) $gt_0 = \nu(g)^m t_0$ for all $g \in G$ and (ii) t_0 is a polarization of $(V, \rho_{\mathbb{R}} \circ h)$ for all $h \in X$. Then there exist homomorphisms $t_i: V^{\otimes r_i} \to \mathbb{Q}(\frac{mr_i}{2})$, $1 \leqslant i \leqslant n$, such that G is the subgroup of GL_V whose elements fix $t_0, t_1, \ldots, t_n$. When (G, X) is of Hodge type, we choose ρ to be a symplectic representation.

Let K be a compact open subgroup of $G(\mathbb{A}_f)$. Define $\mathcal{H}_K(\mathbb{C})$ to be the set of triples

$$(W, (s_i)_{0 \leqslant i \leqslant n}, \eta K)$$

in which

- $W = (W, h_W)$ is a rational Hodge structure,
- each s_i is a morphism of Hodge structures $W^{\otimes r_i} \to \mathbb{Q}(\frac{mr_i}{2})$ and s_0 is a polarization of W,
- ηK is a K-orbit of $\mathbb{A}_f$-linear isomorphisms $V_{\mathbb{A}_f} \to W_{\mathbb{A}_f}$ sending each t_i to s_i,

satisfying the following condition:

(*) there exists an isomorphism $\gamma: W \to V$ sending each s_i to t_i and h_W onto an element of X.

Lemma 11.17. *For $(W, \ldots)$ in $\mathcal{H}_K(\mathbb{C})$, choose an isomorphism γ as in (*), let h be the image of h_W in X, and let $a \in G(\mathbb{A}_f)$ be the composite $V_{\mathbb{A}_f} \xrightarrow{\eta} W_{\mathbb{A}_f} \xrightarrow{\gamma} V_{\mathbb{A}_f}$. The class $[h, a]$ of the pair (h, a) in $G(\mathbb{Q})_+ \backslash X \times G(\mathbb{A}_f)/K$ is independent of all choices, and the map*

$$(W, \ldots) \mapsto [h, a]: \mathcal{H}_K(\mathbb{C}) \to \mathrm{Sh}_K(G, X)(\mathbb{C})$$

is surjective with fibres equal to the isomorphism classes.

Proof. The proof involves only routine checking. □

For a smooth algebraic variety S over $\mathbb{C}$, let $\mathcal{F}_K(S)$ be the set of isomorphism classes of triples $(A, (s_i)_{0 \leqslant i \leqslant n}, \eta K)$ in which

- A is a family of abelian motives over S,
- each s_i is a morphism of abelian motives $A^{\otimes r_i} \to \mathbb{Q}(\frac{mr_i}{2})$, and
- ηK is a K-orbit of $\mathbb{A}_f$-linear isomorphisms $V_S \to \omega_f(A/S)$ sending each t_i to s_i,[37]

satisfying the following condition:

(**) for each $s \in S(\mathbb{C})$, the Betti realization of $(A, (s_i), \eta K)_s$ lies in $\mathcal{H}_K(\mathbb{C})$.

With the obvious notion of pullback, $\mathcal{F}_K$ becomes a functor from smooth complex algebraic varieties to sets. There is a well-defined injective map $\mathcal{F}_K(\mathbb{C}) \to \mathcal{H}_K(\mathbb{C})/\approx$, which is surjective when (G, X) is of abelian type. Hence, in this case, we get an isomorphism $\alpha: \mathcal{F}_K(\mathbb{C}) \to \mathrm{Sh}_K(\mathbb{C})$.

Theorem 11.18. *Assume that (G, X) is of abelian type. The map α realizes Sh_K as a coarse moduli variety for $\mathcal{F}_K$, and even a fine moduli variety when $Z(\mathbb{Q})$ is discrete in $Z(\mathbb{R})$ (here $Z = Z(G)$).*

Proof. To say that (Sh_K, α) is coarse moduli variety means the following:

---

[37] The isomorphism η is defined only on the universal covering space of S^{an}, but the family ηK is stable under $\pi_1(S, o)$, and so is "defined" on S.

(a) for any smooth algebraic variety S over $\mathbb{C}$, and $\xi \in \mathcal{F}(S)$, the map $s \mapsto \alpha(\xi_s)\colon S(\mathbb{C}) \to \mathrm{Sh}_K(\mathbb{C})$ is regular;
(b) (Sh_K, α) is universal among pairs satisfying (a) .

To prove (a), we use that ξ defines a variation of Hodge structures on S (see p. 523). Now the universal property of hermitian symmetric domains (7.8) shows that the map $s \mapsto \alpha(\xi_s)$ is holomorphic (on the universal covering space, and hence on the variety), and Borel's theorem 4.3 shows that it is regular.

Next assume that $Z(\mathbb{Q})$ is discrete in $Z(\mathbb{R})$. Then the representation ρ defines a variation of Hodge structures on Sh_K itself (not just its universal covering space), which arises from a family of abelian motives. This family is universal, and so Sh_K is a fine moduli variety.

We now prove (b). Let S' be a smooth algebraic variety over $\mathbb{C}$ and let $\alpha'\colon \mathcal{F}_K(\mathbb{C}) \to S'(\mathbb{C})$ be a map with the following property: for any smooth algebraic variety S over $\mathbb{C}$ and $\xi \in \mathcal{F}(S)$, the map $s \mapsto \alpha'(\xi_s)\colon S(\mathbb{C}) \to S'(\mathbb{C})$ is regular. We have to show that the map $s \mapsto \alpha'\alpha^{-1}(s)\colon \mathrm{Sh}_K(\mathbb{C}) \to S'(\mathbb{C})$ is regular. When $Z(\mathbb{Q})$ is discrete in $Z(\mathbb{R})$, the map is that defined by α' and the universal family of abelian motives on Sh_K, and so it is regular by definition. In the general case, we let G' be the smallest algebraic subgroup of G such that $h(\mathbb{S}) \subset G'_{\mathbb{R}}$ for all $h \in X$. Then (G', X) is a Shimura datum (cf. 7.6), which now is such that $Z(\mathbb{Q})$ is discrete in $Z(\mathbb{R})$; moreover, $\mathrm{Sh}_{K\cap G'(\mathbb{A}_f)}(G', X)$ consists of a certain number of connected components of $\mathrm{Sh}_K(G, X)$. As the map is regular on $\mathrm{Sh}_{K\cap G'(\mathbb{A}_f)}(G', X)$, and $\mathrm{Sh}_K(G, X)$ is a union of translates of $\mathrm{Sh}_{K\cap G'(\mathbb{A}_f)}(G', X)$, this shows that the map is regular on $\mathrm{Sh}_K(G, X)$. □

Remarks

11.19. When (G, X) is of Hodge type in Theorem 11.18, the Shimura variety is a moduli variety for abelian *varieties* with additional structure. In this case, the moduli problem can be defined for all schemes algebraic over $\mathbb{C}$ (not necessarily smooth), and Mumford's theorem can be used to prove that the Shimura variety is moduli variety for the expanded functor.

11.20. It is possible to describe the structure ηK by passing only to a finite covering, rather than the full universal covering. This means that it can be described purely algebraically.

11.21. For certain compact open groups K, the structure ηK can be interpreted as a level-N structure in the usual sense.

11.22. Consider a pair $(H, \bar{h})$ having a finite covering of Hodge type. Then there exists a Shimura datum (G, X) of abelian type such that $(G^{\mathrm{der}}, \mathrm{ad} \circ h) = (H, \bar{h})$ for some $h \in X$. The choice of a faithful representation ρ for G gives a realization of the connected Shimura variety defined by any (sufficiently small) congruence subgroup of $H(\mathbb{Q})$ as a fine moduli variety for abelian motives with additional structure. For example, when H is simply connected, there is a map $\mathcal{H}_K(\mathbb{C}) \to T(\mathbb{Q})^{\dagger}\backslash T(\mathbb{A}_f)/\nu(K)$

(see (11.13), p. 541), and the moduli problem is obtained from $\mathcal{F}_K$ by replacing $\mathcal{H}_K(\mathbb{C})$ with its fibre over [1]. Note that the realization involves many choices.

11.23. For each Shimura variety, there is a well-defined number field $E(G, X)$, called the reflex field. When the Shimura variety is a moduli variety, it is possible choose the moduli problem so that it is defined over $E(G, X)$. Then an elementary descent argument shows that the Shimura variety itself has a model over $E(G, X)$. A priori, it may appear that this model depends on the choice of the moduli problem. However, the theory of complex multiplication shows that the model satisfies a certain reciprocity law at the special points, which characterize it.

11.24. The (unique) model of a Shimura variety over the reflex field $E(G, X)$ satisfying (Shimura's) reciprocity law at the special points is called the *canonical model*. As we have just noted, when a Shimura variety can be realized as a moduli variety, it has a canonical model. More generally, when the associated connected Shimura variety is a moduli variety, then $Sh(G, X)$ has a canonical model ([61], [19]). Otherwise, the Shimura variety can be embedded in a larger Shimura variety that contains many Shimura subvarieties of type A_1, and this can be used to prove that the Shimura variety has a canonical model ([35]).

Notes. For more details on this subsection, see [38].

References

[1] Y. André. Mumford-Tate groups of mixed Hodge structures and the theorem of the fixed part. *Compositio Math.*, **82**, 1–24, 1992a. ← 499

[2] Y. André. Une remarque à propos des cycles de Hodge de type CM, in: Séminaire de Théorie des Nombres, Paris, 1989–1990, 1–7. *Progr. Math.*, **102** , Birkhäuser Boston, Boston, MA, 1992b. ← 519

[3] Jr. W. L. Baily and A. Borel. Compactification of arithmetic quotients of bounded symmetric domains. *Ann. of Math. (2)*, **84**, 442–528, 1966. ← 487

[4] A. Borel. Introduction aux groupes arithmétiques, in: Publications de l'Institut de Mathématique de l'Université de Strasbourg, XV. *Actualités Scientifiques et Industrielles*, **1341**, Hermann, Paris, 1969. ← 482, 483

[5] A. Borel. Some metric properties of arithmetic quotients of symmetric spaces and an extension theorem. *J. Differential Geometry*, **6**, 543–560, 1972. ← 487

[6] A. Borel and Harish-Chandra. Arithmetic subgroups of algebraic groups. *Ann. of Math. (2)*, **75**, 485–535, 1962. ← 482

[7] N. Bourbaki. Groupes et Algèbres de Lie. Éléments de mathématique. Hermann; Masson, Paris. Chap. I, Hermann 1960; Chap. II,III, Hermann 1972; Chap. IV,V,VI, Masson 1981;Chap. VII,VIII, Masson 1975; Chap. IX, Masson 1982 (English translation available from Springer). ← 480, 528

[8] E. Cattani, P. Deligne, and A. Kaplan. On the locus of Hodge classes. *J. Amer. Math. Soc.*, **8**, 483–506, 1995. ← 498

[9] E. Cattani, A. Kaplan, and W. Schmid. Degeneration of Hodge structures. *Ann. of Math. (2)*, **123**, 457–535, 1986. ← 523

[10] C.-L. Chai and G. Faltings. Degeneration of abelian varieties, volume 22 of *Ergebnisse der Mathematik und ihrer Grenzgebiete (3)*. Springer-Verlag, Berlin, 1990. ← 523

[11] W.-L. Chow. On compact complex analytic varieties. *Amer. J. Math.*, **71**, 893–914, 1949. ← 487

[12] A. Comessatti. Sugl'indici di singolarita a più dimensioni della varieta abeliane. *Rendiconti del Seminario Matematico della Università di Padova*, 50–79, 1938. ← 497

[13] P. Deligne. Théorie de Hodge. II. *Inst. Hautes Études Sci. Publ. Math.*, 5–57, 1971a. ← 502, 508

[14] P. Deligne. Travaux de Shimura, in: Séminaire Bourbaki, 23ème année (1970/71), Exp. No. 389, 123–165.*Lecture Notes in Math.*, **244**, Springer, Berlin, 1971b. ← 488, 516, 536, 540

[15] P. Deligne. La conjecture de Weil pour les surfaces K3. *Invent. Math.*, **15**, 206–226, 1972. ← 494, 499

[16] P. Deligne. Les constantes des équations fonctionnelles des fonctions L, in: Modular functions of one variable, II (Proc. Internat. Summer School, Univ. Antwerp, Antwerp, 1972), 501–597. *Lecture Notes in Math.*, **349**, Springer, Berlin, 1973. ← 490

[17] P. Deligne. La conjecture de Weil. I. *Inst. Hautes Études Sci. Publ. Math.*, 273–307, 1974. ← 499

[18] P. Deligne. Valeurs de fonctions L et périodes d'intégrales, in: Automorphic forms, representations and L-functions (Proc. Sympos. Pure Math., Oregon State Univ., Corvallis, Ore., 1977), Part 2, 313–346. *Proc. Sympos. Pure Math.*,**XXXIII**, Amer. Math. Soc., Providence, R.I., 1979a. ← 515, 517

[19] P. Deligne. Variétés de Shimura: interprétation modulaire, et techniques de construction de modèles canoniques, in: Automorphic forms, representations and L-functions (Proc. Sympos. Pure Math., Oregon State Univ., Corvallis, Ore., 1977), Part 2, 247–289. *Proc. Sympos. Pure Math.*,**XXXIII**, Amer. Math. Soc., Providence, R.I., 1979b. ← 480, 488, 490, 494, 505, 507, 522, 524, 528, 531, 541, 544

[20] P. Deligne. Hodge cycles on abelian varieties (notes by J.S. Milne), in : Hodge cycles, motives, and Shimura varieties, 9–100. *Lecture Notes in Mathematics*, Springer-Verlag, Berlin, 1982. A corrected version is available at `jmilne.org`. ← 494, 500, 515, 516, 517, 519, 520

[21] P. Deligne and J. S. Milne. Tannakian categories, in: Hodge cycles, motives, and Shimura varieties, 101–228. *Lecture Notes in Mathematics*, **900**, Springer-Verlag, Berlin, 1982. ← 494, 495

[22] M. Dettweiler and S. Reiter. Rigid local systems and motives of type G_2. *Compos. Math.*, **146**, 929–963, 2010. With an appendix by Michael Dettweiler and Nicholas M. Katz. ← 539

[23] H. Grauert and R. Remmert. Coherent analytic sheaves, *Grundlehren der Mathematischen Wissenschaften*, **265**, Springer-Verlag, Berlin, 1984. ← 504

[24] P. A. Griffiths. Periods of integrals on algebraic manifolds. I, II. *Amer. J. Math.*, **90**, 568–626, 805–865, 1968. ← 493

[25] P. A. Griffiths. Periods of integrals on algebraic manifolds: Summary of main results and discussion of open problems. *Bull. Amer. Math. Soc.*, **76**, 228–296, 1970. ← 502

[26] P. A. Griffiths and W. Schmid. Locally homogeneous complex manifolds. *Acta Math.*, **123**, 253–302, 1969. ← 523

[27] S. Helgason. Differential geometry, Lie groups, and symmetric spaces. *Pure and Applied Mathematics*, **80**, Academic Press Inc., New York, 1978. ← 477

[28] H. Hironaka. Resolution of singularities of an algebraic variety over a field of characteristic zero. I, II. *Ann. of Math.*, **79**, 109–326, 1964. ← 488

[29] J. E. Humphreys. Linear algebraic groups. Springer-Verlag, New York, 1975. ← 479

[30] N. M. Katz and B. Mazur. Arithmetic moduli of elliptic curves, *Annals of Mathematics Studies*, **108**, Princeton University Press, Princeton, NJ, 1985. ← 475

[31] S. Kobayashi. Geometry of bounded domains. *Trans. Amer. Math. Soc.*, **92**, 267–290, 1959. ← 477

[32] R. P. Langlands. Automorphic representations, Shimura varieties, and motives. Ein Märchen, in: Automorphic forms, representations and L-functions (Proc. Sympos. Pure Math., Oregon State Univ., Corvallis, Ore., 1977), Part 2, 205–246. *Proc. Sympos. Pure Math.*, **XXXIII**, Amer. Math. Soc., Providence, R.I., 1979. ← 522

[33] G. A. Margulis. Discrete subgroups of semisimple Lie groups, *Ergebnisse der Mathematik und ihrer Grenzgebiete (3)*, **17**, Springer-Verlag, Berlin, 1991. ← 484

[34] A. Mattuck. Cycles on abelian varieties. *Proc. Amer. Math. Soc.*, **9**, 88–98, 1958. ← 497

[35] J. S. Milne. The action of an automorphism of **C** on a Shimura variety and its special points, in: Arithmetic and geometry, Vol. I, 239–265. *Progr. Math.*, Birkhäuser Boston, Boston, MA, 1983. ← 544

[36] J. S. Milne. Abelian varieties, in: Arithmetic geometry (Storrs, Conn., 1984), 103–150. Springer, New York, 1986. ← 509

[37] J. S. Milne. Motives over finite fields, in: Motives (Seattle, WA, 1991), 401–459. *Proc. Sympos. Pure Math.*, **55**, Amer. Math. Soc., Providence, RI, 1994a. ← 522

[38] J. S. Milne. Shimura varieties and motives, in: Motives (Seattle, WA, 1991), 447–523. *Proc. Sympos. Pure Math.*, Amer. Math. Soc., Providence, RI, 1994b. ← 523, 539, 544

[39] J. S. Milne. Lefschetz classes on abelian varieties. *Duke Math. J.*, **96**, 639–675, 1999. ← 519

[40] J. S. Milne. Introduction to Shimura varieties, in: J. Arthur and R. Kottwitz (eds.), Harmonic analysis, the trace formula, and Shimura varieties, 265–378. *Clay Math. Proc.*, **4**, Amer. Math. Soc., Providence, RI, 2005. Available at: `http://www.claymath.org/library/`. ← 478, 480, 493, 540, 541

[41] J. S. Milne and K.-Y. Shih. Conjugates of Shimura varieties, in: Hodge cycles, motives, and Shimura varieties, 280–356. *Lecture Notes in Mathematics*, Springer-Verlag, Berlin, 1982. ← 512

[42] D. Morris. Introduction to arithmetic groups, 2008. `arXive:math/0106063v3` ← 481, 484

[43] G. D. Mostow and T. Tamagawa. On the compactness of arithmetically defined homogeneous spaces. *Ann. of Math. (2)*, **76**, 446–463, 1962. ← 482

[44] D. Mumford. Geometric invariant theory. *Ergebnisse der Mathematik und ihrer Grenzgebiete, Neue Folge*, Band **34**, Springer-Verlag, Berlin, 1965. ← 475, 489, 509

[45] D. Mumford. Families of abelian varieties, in: Algebraic Groups and Discontinuous Subgroups (Proc. Sympos. Pure Math., Boulder, Colo., 1965), 347–351. Amer. Math. Soc., Providence, R.I., 1966. ← 495

[46] D. Mumford. A note of Shimura's paper "Discontinuous groups and abelian varieties". *Math. Ann.*, **181**, 345–351. 1969. ← 509

[47] D. Mumford. Abelian varieties. *Tata Institute of Fundamental Research Studies in Mathematics*, **5**. Published for the Tata Institute of Fundamental Research, Bombay, 1970. ← 489

[48] G. Prasad and A. S. Rapinchuk. On the existence of isotropic forms of semisimple algebraic groups over number fields with prescribed local behavior. *Adv. Math.*, **207**, 646–660, 2006. ← 532

[49] G. Prasad and A. S. Rapinchuk. Developments on the congruence subgroup problem after the work of Bass, Milnor and Serre, 2008. `(arXiv:0809.1622)` ← 486

[50] M. Rapoport. Complément à l'article de P. Deligne "La conjecture de Weil pour les surfaces K3". *Invent. Math.*, **15**, 227–236, 1972. ← 496

[51] I. Satake. Holomorphic imbeddings of symmetric domains into a Siegel space. *Amer. J. Math.*, **87**, 425–461, 1965. ← 524

[52] I. Satake. Symplectic representations of algebraic groups satisfying a certain analyticity condition. *Acta Math.*, **117**, 215–279, 1967. ← 524

[53] I. Satake. Algebraic structures of symmetric domains, *Kanô Memorial Lectures*, **4**, Iwanami Shoten, Tokyo, 1980. ← 476, 477, 478, 524

[54] W. Schmid. Variation of Hodge structure: the singularities of the period mapping. *Invent. Math.*, **22**, 211–319, 1973. ← 502, 523

[55] J.-P. Serre. Lectures on the Mordell-Weil theorem, 2nd ed. Aspects of Mathematics, E15. Friedr. Vieweg & Sohn, Braunschweig. Translated from the French and edited by Martin Brown from notes by Michel Waldschmidt, 1990. ← 488

[56] J.-P. Serre. Propriétés conjecturales des groupes de Galois motiviques et des représentations l-adiques, in: Motives (Seattle, WA, 1991), 377–400. *Proc. Sympos. Pure Math.*, **55**, Amer. Math. Soc., Providence, RI, 1994. ← 522

[57] G. Shimura. On the field of definition for a field of automorphic functions. *Ann. of Math. (2)*, **80**, 160–189. 1964. ← 488

[58] G. Shimura. Moduli and fibre systems of abelian varieties. *Ann. of Math. (2)*, **83**, 294–338, 1966. ← 488

[59] G. Shimura. Algebraic number fields and symplectic discontinuous groups. *Ann. of Math. (2)*, **86**, 503–592, 1967a. ← 488

[60] G. Shimura. Construction of class fields and zeta functions of algebraic curves. *Ann. of Math. (2)*, **85**, 58–159, 1967b. ← 488

[61] G. Shimura. On canonical models of arithmetic quotients of bounded symmetric domains. *Ann. of Math. (2)*, **91**, 144–222; II ibid. **92**, 528–549, 1970. ← 488, 544

[62] J. T. Tate. Algebraic cohomology classes, in: Lecture Notes Prepared in Connection with Seminars held at the Summer Institute on Algebraic Geometry, Woods Hole, MA, July 6 - July 31, 1964. American Mathematical Society, 1964. Reprinted as: Algebraic cycles and poles of zeta functions, Arithmetical Algebraic Geometry, Harper & Row, (1965), 93–110. ← 518, 519

[63] J. L. Taylor. Several complex variables with connections to algebraic geometry and Lie groups, *Graduate Studies in Mathematics*, **46**, American Mathematical Society, Providence, RI, 2002. ← 487

[64] J. Tits. Classification of algebraic semisimple groups, in: Algebraic Groups and Discontinuous Subgroups (Proc. Sympos. Pure Math., Boulder, Colo., 1965), 33–62. Amer. Math. Soc., Providence, R.I., 1966. ← 527

[65] A. Weil. Abelian varieties and the Hodge ring. Talk at a conference held at Harvard in honor of Lars Ahlfors (Oevres Scientifiques 1977c, 421–429), 1977. ← 498, 519

[66] Y. G. Zarhin. Weights of simple Lie algebras in the cohomology of algebraic varieties. *Izv. Akad. Nauk SSSR Ser. Mat.*, **48**, 264–304, 1984. ← 499

[67] R. J. Zimmer. Ergodic theory and semisimple groups, *Monographs in Mathematics*, **81**, Birkhäuser Verlag, Basel, 1984. ← 484

Ann Arbor, MI 48104 USA
E-mail address: `jmilne@umich.edu`

The Torelli locus and special subvarieties

Ben Moonen and Frans Oort

Abstract. We study the Torelli locus T_g in the moduli space A_g of abelian varieties. We consider special subvarieties (Shimura subvarieties) contained in the Torelli locus. We review the construction of some non-trivial examples, and we discuss some conjectures, techniques and recent progress.

Contents

2000 *Mathematics Subject Classification.* Primary 14G35, 14H, 14K; Secondary: 11G.
Key words and phrases. Curves and their Jacobians, Torelli locus, Shimura varieties.

Introduction

The Torelli morphism

An algebraic curve C has a Jacobian $J_C = \mathrm{Pic}^0(C)$, which comes equipped with a canonical principal polarization $\lambda\colon J_C \xrightarrow{\sim} J_C^t$. The dimension of J_C equals the genus g of the curve. One should like to understand J_C knowing certain properties of C, and, conversely, we like to obtain information about C from properties of J_C or the pair (J_C, λ).

As an important tool for this we have the moduli spaces of algebraic curves and of polarized abelian varieties. The construction that associates to C the principally polarized abelian variety (J_C, λ) defines a morphism

$$j\colon \mathsf{M}_g \to \mathsf{A}_g ,$$

the *Torelli morphism*, which by a famous theorem of Torelli is injective on geometric points. (Here we work with the coarse moduli schemes.) The image $\mathsf{T}_g^\circ := j(\mathsf{M}_g)$ is called the (open) Torelli locus, also called the Jacobi locus. Its Zariski closure T_g inside A_g is called the *Torelli locus*. We should like to study the interplay between curves and their Jacobians using the geometry of M_g and of A_g, and relating these via the inclusion $\mathsf{T}_g \subset \mathsf{A}_g$.

Special subvarieties

Although the moduli spaces M_g and A_g have been studied extensively, it turns out that there are many basic questions that are still open. The moduli space A_g locally has a group-like structure, in a sense that can be made precise. As a complex manifold, $\mathsf{A}_g(\mathbb{C})$ is a quotient of the Siegel space, which is a principally homogeneous space under the group $\mathrm{CSp}_{2g}(\mathbb{R})$. Hence it is natural to study subvarieties that appear as images of orbits of an algebraic subgroup; such a subvariety is called a *special subvariety*. (We refer to Section 3 for a more precise definition of this notion.)

The zero dimensional special subvarieties are precisely the CM points, corresponding to abelian varieties where $\mathrm{End}(A)\otimes\mathbb{Q}$ contains a commutative semi-simple algebra of rank $2\cdot\dim(A)$. These are fairly well understood. The arithmetical properties of CM points form a rich and beautiful subject, that plays an important role in the study of the moduli space. The presence of a dense set of CM points enabled Shimura to prove for a large class of Shimura varieties that they admit a model over a number field. In its modern form this theory owes much to Deligne's presentation in [19] and [21], and the existence of canonical models for all Shimura varieties was established by Borovoi and Milne.

For the problems we want to discuss, the special points again play a key role. It is not hard to show that on any special subvariety of A_g over $\mathbb{C}$ the special points are dense, even for the analytic topology. A conjecture by Y. André and F. Oort says that, conversely, an algebraic subvariety with a Zariski dense set of CM points is in fact a special variety. See 3.14. There is a nice analogy between this conjecture and the Manin-Mumford conjecture, proven by Raynaud, which says that an algebraic subvariety of an abelian variety that contains a Zariski dense set of torsion points is the translate of an abelian subvariety under a torsion point.

Klingler and Yafaev, using work of Ullmo and Yafaev, have announced a proof of the André-Oort conjecture under assumption of the Generalized Riemann Hypothesis for CM fields.

Coleman's Conjecture and special subvarieties in the Torelli locus

In view of the above, it is natural to investigate the subvariety $T_g \subset A_g$ from the perspective of special subvarieties of A_g. A conjecture by Coleman, again based on the analogy with the Manin-Mumford conjecture, says that for $g > 3$ there are only finitely many complex curves of genus g such that J_C is an abelian variety with CM. See 4.1.

In terms of moduli spaces, this says that for $g > 3$ the number of special points on T_g° should be finite. Combining this with the André-Oort conjecture, it suggests that for large g there are no special subvarieties $Z \subset A_g$ of positive dimension that are contained inside T_g and meet T_g°; see 4.2. We should like to point out that there is no clear evidence for this expectation, although there are several partial results in support of it. See for instance Theorem 4.6 below, which gives a weaker version of Coleman's conjecture.

As we shall explain in Section 5, Coleman's Conjecture is not true for $g \in \{4,5,6,7\}$. The reason is that for these genera we can find explicit families of curves such that the corresponding Jacobians trace out a special subvariety of positive dimension in A_g. All known examples of this kind, for $g \geqslant 4$, arise from families of abelian covers of $\mathbb{P}^1$, where we fix the Galois group and the local monodromies, and we vary the branch points. For cyclic covers it can be shown that, beyond the known examples, this construction gives no further families that give rise to a special subvariety in A_g; see Theorem 5.13.

In this paper, we give a survey of what is known about special subvarieties in the Torelli locus, and we try to give the reader a feeling for the difficulties one encounters in studying these. Going further, one could ask for a complete classification of such special subvarieties $Z \subset T_g$ for every g. We should like to point out that already for $g = 4$ we do not have such a classification, in any sense.

Acknowledgments

We thank Dick Hain for patiently explaining to us some details related to his paper [32].

0.1. Notation

If G is a group scheme of finite type over a field k, we denote its identity component by G^0. In case k is a subfield of $\mathbb{R}$ we write $G(\mathbb{R})^+$ for the identity component of $G(\mathbb{R})$ with regard to the analytic topology. If G is reductive, G^{ad} denotes the adjoint group.

The group of symplectic similitudes. Let a positive integer g be given. Write $V := \mathbb{Z}^{2g}$, and let $\phi\colon V \times V \to \mathbb{Z}$ be the standard symplectic pairing, so that on the standard ordered basis $\{e_i\}_{i=1,\dots,2g}$ we have $\phi(e_i, e_j) = 0$ if $i + j \neq 2g + 1$ and $\phi(e_i, e_{2g+1-i}) = 1$ if $i \leqslant g$. The group $\mathrm{CSp}(V, \phi)$ is the reductive group over $\mathbb{Z}$ of symplectic similitudes of V with regard to the form ϕ. If there is no risk of confusion we simply write CSp_{2g} for this group. We denote by $\nu\colon \mathrm{CSp}_{2g} \to \mathbb{G}_m$ the multiplier character. The kernel of ν is the symplectic group Sp_{2g} of transformations of V that preserve the form ϕ.

1. The Torelli morphism

1.1. Moduli of abelian varieties

Fix an integer $g > 0$. If S is a base scheme and A is an abelian scheme over S, we write A^t for the dual abelian scheme. There is a canonical homomorphism $\kappa_A\colon A \to A^{tt}$, which Cartier and Nishi proved to be an isomorphism; see [9], [10], [56]; also see [61], Theorem 20.2.

On $A \times_S A^t$ we have a Poincaré bundle $\mathcal{P}$. A *polarization* of A is a symmetric isogeny $\lambda\colon A \to A^t$ with the property that the pull-back of $\mathcal{P}$ via the morphism $(\mathrm{id}, \lambda)\colon A \to A \times_S A^t$ is a relatively ample bundle on A over S. The polarization λ is said to be principal if in addition it is an isomorphism of abelian schemes over S. A pair (A, λ) consisting of an abelian scheme over S together with a principal polarization is called a *principally polarized abelian scheme*. We abbreviate "principally polarized abelian scheme(s)/variety/varieties" to "ppav" (sic!).

We denote by $\mathcal{A}_g$ the moduli stack over $\mathrm{Spec}(\mathbb{Z})$ of g-dimensional ppav. It is a connected Deligne-Mumford stack that is quasi-projective and smooth over $\mathbb{Z}$, of relative dimension $g(g+1)/2$. The characteristic zero fiber $\mathcal{A}_{g,\mathbb{Q}}$ can be shown to be geometrically irreducible by transcendental methods; cf. subsection 3.2. Using this, Chai and Faltings have proved that the characteristic p fibers are geometrically irreducible, too; see [27], [12], and [28] Chap. IV, Corollary 6.8. A pure characteristic p proof can be obtained using the results of [64].

Note that later in this article we shall use the same notation $\mathcal{A}_g$ for the moduli stack over some given base field.

For many purposes it is relevant to consider ppav together with a level structure. We introduce some notation and review some facts. Let again (A, λ) be a ppav of relative dimension g over some base scheme S. Given $m \in \mathbb{Z}$ we have a "multiplication by m" endomorphism $[m]\colon A \to A$. We denote by $A[m]$ the kernel of this endomorphism, in the scheme-theoretic sense. Assume that $m \neq 0$. Then $A[m]$ is a finite locally free group scheme of rank m^{2g}. There is a naturally defined bilinear pairing

$$e_m^\lambda\colon A[m] \times A[m] \to \mu_m\,,$$

called the Weil pairing, that is non-degenerate and that is symplectic in the sense that $e_m^\lambda(x, x) = 1$ for all S-schemes T and all T-valued points $x \in A[m](T)$. If m is invertible in $\Gamma(S, O_S)$ then the group scheme $A[m]$ is étale over S.

Fix an integer $m \geqslant 1$. Write $\mathbb{Z}[\zeta_m]$ for $\mathbb{Z}[t]/(\Phi_m)$, where $\Phi_m \in \mathbb{Z}[t]$ is the mth cyclotomic polynomial. We shall first define a Deligne-Mumford stack $\mathcal{A}'_{g,[m]}$ over the spectrum of the ring $R_m := \mathbb{Z}[\zeta_m, 1/m]$. The moduli stack $\mathcal{A}_{g,[m]}$ is then defined to be the same stack, but now viewed as a stack over $\mathbb{Z}[1/m]$, with as structural morphism the composition

$$\mathcal{A}'_{g,[m]} \to \operatorname{Spec}(R_m) \to \operatorname{Spec}\big(\mathbb{Z}[1/m]\big)\,.$$

The reason that we want to view $\mathcal{A}_{g,[m]}$ as an algebraic stack over $\mathbb{Z}[1/m]$ is that this is the stack whose characteristic zero fiber has a natural interpretation as a Shimura variety; see the discussion in subsection 3.2 below.

Let S be a scheme over R_m. Note that this just means that S is a scheme such that m is invertible in $\Gamma(S, O_S)$, and that we are given a primitive mth root of unity $\zeta \in \Gamma(S, O_S)$. We may view this ζ as an isomorphism of group schemes $\zeta\colon (\mathbb{Z}/m\mathbb{Z}) \xrightarrow{\sim} \mu_m$. Let V and ϕ be as defined in Subsection 0.1. Given a ppav (A, λ) of relative dimension g over S, by a *(symplectic) level* m *structure* on (A, λ) we then mean an isomorphism of group schemes $\alpha\colon (V/mV) \xrightarrow{\sim} A[m]$ such that the diagram

$$\begin{array}{ccc} (V/mV) \times (V/mV) & \xrightarrow{\ \phi\ } & (\mathbb{Z}/m\mathbb{Z}) \\ {\scriptstyle \alpha\times\alpha}\downarrow & & \downarrow{\scriptstyle \zeta} \\ A[m] \times A[m] & \xrightarrow{\ e_m^\lambda\ } & \mu_m \end{array}$$

is commutative.

We write $\mathcal{A}'_{g,[m]}$ for the moduli stack over R_m of ppav with a level m structure. Again this is a quasi-projective smooth Deligne-Mumford stack with geometrically irreducible fibers. If $m \geqslant 3$, it is a scheme. The resulting stack

$$\mathcal{A}_{g,[m]} := \Big(\mathcal{A}'_{g,[m]} \to \operatorname{Spec}(R_m) \to \operatorname{Spec}\big(\mathbb{Z}[1/m]\big)\Big)$$

is a quasi-projective smooth Deligne-Mumford stack over $\mathbb{Z}[1/m]$ whose geometric fibers have $\varphi(m)$ irreducible components.

We now assume that $m \geqslant 3$. The finite group $\mathcal{G} := \mathrm{CSp}_{2g}(\mathbb{Z}/m\mathbb{Z})$ acts on the stack $\mathcal{A}'_{g,[m]}$ via its action on the level structures. Note that this is not an action over R_m; only the subgroup $\mathrm{Sp}_{2g}(\mathbb{Z}/m\mathbb{Z})$ acts by automorphisms over R_m. We do, however, obtain an induced action of $\mathcal{G}$ on $\mathcal{A}_{g,[m]}$ over $\mathbb{Z}[1/m]$. The stack quotient of $\mathcal{A}_{g,[m]}$ modulo $\mathcal{G}$ is just $\mathcal{A}_g \otimes \mathbb{Z}[1/m]$. The scheme quotient of $\mathcal{A}_{g,[m]}$ modulo $\mathcal{G}$ is the coarse moduli scheme $\mathsf{A}_g \otimes \mathbb{Z}[1/m]$. For $m_1, m_2 \geqslant 3$ these coarse moduli schemes agree over $\mathbb{Z}[1/(m_1 m_2)]$; hence we can glue them to obtain a coarse moduli scheme A_g over $\mathbb{Z}$.

The stacks $\mathcal{A}_{g,[m]}$ are not complete. There are various ways to compactify them. The basic reference for this is the book [28]. There are toroidal compactifications that depend on the choice of some cone decomposition Σ (see [28] for details) and that map down to the Baily-Borel (or minimal, or Satake) compactification.

1.2. Moduli of curves

Fix an integer $g \geqslant 2$. We denote by $\mathcal{M}_g$ the moduli stack of curves of genus g over $\mathrm{Spec}(\mathbb{Z})$. It is a quasi-projective smooth Deligne-Mumford stack over $\mathbb{Z}$ of relative dimension $3g-3$ for $g \geqslant 2$, with geometrically irreducible fibers.

We write $\overline{\mathcal{M}}_g$ for the Deligne-Mumford compactification of $\mathcal{M}_g$; it is the stack of stable curves of genus g. See [24]. The boundary $\overline{\mathcal{M}}_g \setminus \mathcal{M}_g$ is a divisor with normal crossings; it has irreducible components Δ_i for $0 \leqslant i \leqslant \lfloor g/2 \rfloor$. There is a non-empty open part of Δ_0 that parametrizes irreducible curves with a single node. Similarly, for $i > 0$ there is a non-empty open part of Δ_i parametrizing curves with precisely two irreducible components, of genera i and $g-i$. The complement of Δ_0 is the open substack $\mathcal{M}_g^{\mathrm{ct}} \subset \overline{\mathcal{M}}_g$ of curves of compact type. Here we recall that a stable curve C over a field k is said to be *of compact type* if the connected components of its Picard scheme are proper over k. If k is algebraically closed this is equivalent to the condition that the dual graph of the curve has trivial homology; so the irreducible components of C are regular curves, the graph of components is a tree, and the sum of the genera of the irreducible components equals g. Note that $\mathcal{M}_g \subset \mathcal{M}_g^{\mathrm{ct}}$.

For the corresponding coarse moduli schemes we use the notation M_g and $\mathsf{M}_g^{\mathrm{ct}}$.

1.3. The Torelli morphism and the Torelli locus

Let C be a stable curve of compact type over a base scheme S, with fibers of genus $g \geqslant 2$. The Picard scheme $\mathrm{Pic}_{C/S}$ is a smooth separated S-group scheme whose components are of finite type and proper over S. In particular, the identity component $J_C := \mathrm{Pic}^0_{C/S}$ is an abelian scheme over S. It has relative dimension g and comes equipped with a canonical principal polarization λ.

The functor that sends C to the pair (J_C, λ) defines a representable morphism of algebraic stacks

$$j \colon \mathcal{M}_g^{\mathrm{ct}} \to \mathcal{A}_g ,$$

called the *Torelli morphism*. It is known that this morphism is proper.

Torelli's celebrated theorem says that for an algebraically closed field k the restricted morphism $j\colon \mathcal{M}_g \to \mathcal{A}_g$ is injective on k-valued points. In other words, if C_1 and C_2 are smooth curves over k such that (J_{C_1}, λ_1) and (J_{C_2}, λ_2) are isomorphic then C_1 and C_2 are isomorphic. This injectivity property of the Torelli morphism does not hold over the boundary; for instance, if C is a stable curve of compact type with two irreducible components then the Jacobian of C does not "see" at which points the two components are glued. For a detailed study of what happens over the boundary, see [8].

Let us now work over a field k. On coarse moduli schemes the Torelli morphism gives rise to a morphism $j\colon M_g^{ct} \to A_g$. We define the *Torelli locus* $T_g \subset A_g$ to be the image of this morphism. It is a reduced closed subscheme of A_g. By the *open Torelli locus* $T_g^\circ \subset T_g$ we mean the image of M_g, i.e., the subscheme of T_g whose geometric points correspond to Jacobians of nonsingular curves. Note that $T_g^\circ \subset T_g$ is open and dense.

The boundary $T_g \setminus T_g^\circ$ is denoted by T_g^{dec}. The notation refers to the fact that a point $s \in T_g(\bar{k})$ is in T_g^{dec} if and only if the corresponding ppav (A_s, λ_s) is decomposable, meaning that we have ppav (B_1, μ_1) and (B_2, λ_2) of smaller dimension such that $(A_s, \lambda_s) \cong (B_1, \mu_1) \times (B_2, \mu_2)$ as ppav. In one direction this is clear, for if C is a reducible curve of compact type, (J_C, λ) is the product, as a ppav, of the Jacobians of the irreducible components of C. Conversely, if (A_s, λ_s) is decomposable, it has a symmetric theta divisor that is reducible, which implies it cannot be the Jacobian of an irreducible (smooth and proper) curve.

Remark 1.1. Let $g > 2$. The Torelli morphism $j\colon \mathcal{M}_g \to \mathcal{A}_g$ is ramified at the hyperelliptic locus. Outside the hyperelliptic locus it is an immersion. The picture is different for the Torelli morphism $j\colon M_g \to A_g$ on coarse moduli schemes. The morphism $j\colon M_{g,\mathbb{Q}} \to A_{g,\mathbb{Q}}$ on the characteristic zero fibers is an immersion; however, in positive characteristic this is not true in general. See [68], Corollaries 2.8, 3.2 and 5.3.

2. Some abstract Hodge theory

In this section we review some basic notions from abstract Hodge theory. We shall focus on examples related to abelian varieties.

2.1. Basic definitions

We start by recalling some basic definitions. For a more comprehensive treatment we refer to [18] and [69].

Let $\mathbb{S} := \mathrm{Res}_{\mathbb{C}/\mathbb{R}}(\mathbb{G}_m)$, which is an algebraic torus over $\mathbb{R}$ of rank 2, called the Deligne torus. Its character group is isomorphic to $\mathbb{Z}^2$, on which complex conjugation, the non-trivial element of $\mathrm{Gal}(\mathbb{C}/\mathbb{R})$, acts by $(m_1, m_2) \mapsto (m_2, m_1)$. We have $\mathbb{S}(\mathbb{R}) = \mathbb{R}^*$ and $\mathbb{S}(\mathbb{C}) = \mathbb{C}^* \times \mathbb{C}^*$. Let $w\colon \mathbb{G}_{m,\mathbb{R}} \to \mathbb{S}$ be the cocharacter given on real points by the natural embedding $\mathbb{R}^* \hookrightarrow \mathbb{C}^*$; it is called the weight cocharacter.

Let $\mathrm{Nm}\colon \mathbb{S} \to \mathbb{G}_{m,\mathbb{R}}$ be the homomorphism given on real points by $z \mapsto z\bar{z}$; it is called the norm character.

A *$\mathbb{Z}$-Hodge structure of weight* k is a torsion-free $\mathbb{Z}$-module H of finite rank, together with a homomorphism of real algebraic groups $h\colon \mathbb{S} \to \mathrm{GL}(H)_{\mathbb{R}}$ such that $h \circ w\colon \mathbb{G}_m \to \mathrm{GL}(H)_{\mathbb{R}}$ is the homomorphism given by $z \mapsto z^{-k} \cdot \mathrm{id}_H$. The Hodge decomposition

$$H_{\mathbb{C}} = \oplus_{p+q=k} H_{\mathbb{C}}^{p,q}$$

is obtained by taking $H_{\mathbb{C}}^{p,q}$ to be the subspace of $H_{\mathbb{C}}$ on which $(z_1, z_2) \in \mathbb{C}^* \times \mathbb{C}^* = \mathbb{S}(\mathbb{C})$ acts as multiplication by $z_1^{-p} z_2^{-q}$. (We follow the sign convention of [22]; see loc. cit., Remark 3.3.)

As an example, the Tate structure $\mathbb{Z}(n)$ is the Hodge structure of weight $-2n$ with underlying $\mathbb{Z}$-module $\mathbb{Z}(n) = (2\pi i)^n \cdot \mathbb{Z}$ given by the homomorphism $\mathrm{Nm}^n\colon \mathbb{S} \to \mathbb{G}_{m,\mathbb{R}} = \mathrm{GL}(\mathbb{Z}(n))_{\mathbb{R}}$. If H is any Hodge structure, we write $H(n)$ for $H \otimes_{\mathbb{Z}} \mathbb{Z}(n)$.

The endomorphism $C = h(i)$ of $H_{\mathbb{R}}$ is known as the Weil operator. If H is a $\mathbb{Z}$-Hodge structure of weight k, a *polarization* of H is a homomorphism of Hodge structures $\psi\colon H \otimes H \to \mathbb{Z}(-k)$ such that the bilinear form

$$H_{\mathbb{R}} \times H_{\mathbb{R}} \to \mathbb{R}$$

given by $(x, y) \mapsto (2\pi i)^k \cdot \psi_{\mathbb{R}}(x \otimes C(y))$ is symmetric and positive definite. (Here $\psi_{\mathbb{R}}$ is obtained from ψ by extension of scalars to $\mathbb{R}$.) This condition implies that ψ is symmetric if k is even and alternating if k is odd.

Instead of working with integral coefficients, we may also consider $\mathbb{Q}$-Hodge structures. The above definitions carry over verbatim.

Example 2.1. Let (A, λ) be a polarized complex abelian variety. The first homology group $H = H_1(A, \mathbb{Z})$ carries a canonical Hodge structure of type $(-1,0) + (0,-1)$. (By this we mean that $H^{p,q} = (0)$ for all pairs (p, q) different from $(-1,0)$ and $(0,-1)$.) The polarization $\lambda\colon A \to A^t$ (in the sense of the theory of abelian varieties) induces a polarization (in the sense of Hodge theory) $\psi\colon H \otimes H \to \mathbb{Z}(1)$. Let $\mathrm{CSp}(H, \psi) \subset \mathrm{GL}(H)$ be the algebraic group of symplectic similitudes of H with respect to the symplectic form ψ. The homomorphism $h\colon \mathbb{S} \to \mathrm{GL}(H)_{\mathbb{R}}$ that gives the Hodge structure factors through $\mathrm{CSp}(H, \psi)_{\mathbb{R}}$.

A crucial fact for much of the theory we want to discuss is that the map $(A, \lambda) \mapsto (H, \psi)$ gives an equivalence of categories

$$\left\{ \begin{matrix} \text{polarized complex} \\ \text{abelian varieties} \end{matrix} \right\} \xrightarrow{\text{eq}} \left\{ \begin{matrix} \text{polarized Hodge structures} \\ \text{of type } (-1,0) + (0,-1) \end{matrix} \right\}. \tag{2.2}$$

In subsection 3.2 below we shall explain how this fact gives rise to a modular interpretation of certain Shimura varieties. As a variant, the category of abelian varieties is equivalent to the category of polarizable $\mathbb{Z}$-Hodge structures. Another variant is obtained by working with $\mathbb{Q}$-coefficients; we get that the category of abelian varieties up to isogeny is equivalent to the category of polarizable $\mathbb{Q}$-Hodge structures of type $(-1,0) + (0,-1)$.

2.2. Hodge classes and Mumford-Tate groups

Let H be a $\mathbb{Q}$-Hodge structure of weight k. By a *Hodge class* in H we mean an element of

$$H \cap H_{\mathbb{C}}^{0,0};$$

so, a *rational* class that is purely of type (0,0) in the Hodge decomposition. Clearly, a non-zero Hodge class can exist only if $k = 0$. If the weight is 0 and $\mathrm{Fil}^{\bullet}H_{\mathbb{C}}$ is the Hodge filtration, the space of Hodge classes is also equal to

$$H \cap \mathrm{Fil}^0 H_{\mathbb{C}},$$

where again the intersection is taken inside $H_{\mathbb{C}}$.

As an example, we note that a $\mathbb{Q}$-linear map $H \to H$, viewed as an element of $\mathrm{End}_{\mathbb{Q}}(H) = H^{\vee} \otimes H$, is a homomorphism of Hodge structures if and only it is a Hodge class for the induced Hodge structure on $\mathrm{End}_{\mathbb{Q}}(H)$. (Note that $H^{\vee} \otimes H$ indeed has weight 0.) Similarly, a polarization $H \otimes H \to \mathbb{Q}(-k)$ gives rise to a Hodge class in $(H^{\vee})^{\otimes 2}(-k)$.

Let $h\colon \mathbb{S} \to \mathrm{GL}(H)_{\mathbb{R}}$ be the homomorphism that defines the Hodge structure on H. Consider the homomorphism $(h, \mathrm{Nm})\colon \mathbb{S} \to \mathrm{GL}(H)_{\mathbb{R}} \times \mathbb{G}_{m,\mathbb{R}}$. The *Mumford-Tate group of* H, notation MT(H), is the smallest algebraic subgroup $M \subset \mathrm{GL}(H) \times \mathbb{G}_m$ over $\mathbb{Q}$ such that (h, Nm) factors through $M_{\mathbb{R}}$. As we shall see in Example 2.3 below, in some important cases the definition takes a somewhat simpler form.

It is known that MT(H) is a connected algebraic group. If the Hodge structure H is polarizable, MT(H) is a reductive group; this plays an important role in many applications.

The crucial property of the Mumford-Tate group is that it allows us to calculate spaces of Hodge classes, as we shall now explain. For simplicity we shall assume the Hodge structure H to be polarizable.

We consider the Hodge structures we can build from H by taking direct sums, duals, tensor products and Tate twists. Concretely, given a triple $m = (m_1, m_2, m_3)$ with $m_1, m_2 \in \mathbb{Z}_{\geqslant 0}$ and $m_3 \in \mathbb{Z}$, define

$$T(m) := H^{\otimes m_1} \otimes (H^{\vee})^{\otimes m_2} \otimes \mathbb{Q}(m_3).$$

The group $\mathrm{GL}(H) \times \mathbb{G}_m$ naturally acts on H via the projection onto its first factor; hence it also acts on $H^{\vee}$. We let it act on $\mathbb{Q}(1)$ through the second projection. This gives us an induced action of $\mathrm{GL}(H) \times \mathbb{G}_m$ on $T(m)$. The crucial property of the Mumford-Tate group $\mathrm{MT}(H) \subset \mathrm{GL}(H) \times \mathbb{G}_m$ is that for any such tensor construction $T = T(m)$, the subspace of Hodge classes $T \cap T_{\mathbb{C}}^{0,0}$ is precisely the subspace of MT(H)-invariants. Note that the MT(H)-invariants are just the invariants in T under the action of the group $\mathrm{MT}(H)(\mathbb{Q})$ of $\mathbb{Q}$-rational points of the Mumford-Tate group. As the $\mathbb{Q}$-points of MT(H) are Zariski dense in the $\mathbb{C}$-group $\mathrm{MT}(H)_{\mathbb{C}}$, the subspace of $\mathrm{MT}(H)(\mathbb{C})$-invariants in $T \otimes \mathbb{C}$ equals

$$(T \cap T_{\mathbb{C}}^{0,0}) \otimes_{\mathbb{Q}} \mathbb{C}.$$

The above property of the Mumford-Tate group characterizes it uniquely. In other words, if $M \subset GL(H) \times \mathbb{G}_m$ is an algebraic subgroup with the property that for any $T = T(m)$ the space of Hodge classes $T \cap T_{\mathbb{C}}^{0,0}$ equals the subspace of M-invariants in T, we have $M = MT(H)$. (See [22], Section 3.) In this way we have a direct coupling between the Mumford-Tate group and the various spaces of Hodge classes in tensor constructions obtained from H.

Example 2.3. Consider a polarized abelian variety (A, λ) over $\mathbb{C}$. Let $H_{\mathbb{Q}} = H_1(A, \mathbb{Q})$ with polarization form $\psi\colon H_{\mathbb{Q}} \otimes H_{\mathbb{Q}} \to \mathbb{Q}(1)$ be the associated $\mathbb{Q}$-Hodge structure, and recall that the Hodge structure is given by a homomorphism $h\colon \mathbb{S} \to CSp(H_{\mathbb{Q}}, \psi)_{\mathbb{R}}$. Define the Mumford-Tate group of A, notation $MT(A)$, to be the smallest algebraic subgroup $M \subset CSp(H_{\mathbb{Q}}, \psi)$ over $\mathbb{Q}$ with the property that h factors through $M_{\mathbb{R}}$.

The Mumford-Tate group as defined here is really the same as the Mumford-Tate group of the Hodge structure $H_{\mathbb{Q}}$. To be precise, $MT(H_{\mathbb{Q}}) \subset GL(H_{\mathbb{Q}}) \times \mathbb{G}_m$ as defined above is the graph of the multiplier character $\nu\colon CSp(H_{\mathbb{Q}}, \psi) \to \mathbb{G}_m$ restricted to $MT(A)$. When considering Hodge classes in tensor constructions T as above, we let $MT(A)$ act on the Tate structure $\mathbb{Q}(1)$ through the multiplier character.

If we want to understand how the Mumford-Tate group is used, the simplest non-trivial examples are obtained by looking at endomorphisms. As customary we write $\mathrm{End}^0(A)$ for $\mathrm{End}(A) \otimes \mathbb{Q}$. By the equivalence of categories of (2.2), in the version with $\mathbb{Q}$-coefficients, we have

$$\mathrm{End}^0(A) \xrightarrow{\sim} \mathrm{End}_{\mathbb{Q}\text{-HS}}(H_{\mathbb{Q}}) = \mathrm{End}_{\mathbb{Q}}(H_{\mathbb{Q}})^{MT(A)}. \tag{2.4}$$

Once we know the Mumford-Tate group, this allows us to calculate the endomorphism algebra of A. In practice we often use this the other way around, in that we assume $\mathrm{End}^0(A)$ known and we use (2.4) to obtain information about $MT(A)$.

In general Hodge classes on A are not so easy to interpret geometrically; see for instance the example in [54], § 4. (In this example, A is a simple abelian fourfold and there are Hodge classes in $H^4(A^2, \mathbb{Q})$ that are not in the algebra generated by divisor classes; see [50] and [51], (2.5).) According to the Hodge conjecture, all Hodge classes should arise from algebraic cycles on A but even for abelian varieties this is far from being proven.

2.3. Hodge loci

We study the behavior of Mumford-Tate groups in families. The setup here is that we consider a polarizable $\mathbb{Q}$-VHS (variation of Hodge structure) of weight k over a complex manifold S. Let $\mathcal{H}$ denote the underlying $\mathbb{Q}$-local system. For each $s \in S(\mathbb{C})$ we have a Mumford-Tate group $MT_s \subset GL(\mathcal{H}(s)) \times \mathbb{G}_m$, and we should like to understand how this group varies with s. This is best described by passing to a universal cover $\pi\colon \tilde{S} \to S$. Write $\tilde{\mathcal{H}} := \pi^*\mathcal{H}$, and let $H := \Gamma(\tilde{S}, \tilde{\mathcal{H}})$. We have a trivialization $\tilde{\mathcal{H}} \xrightarrow{\sim} H \times \tilde{S}$. Using this we may view the Mumford-Tate group MT_x of a point $x \in \tilde{S}$ (by which we really mean the Mumford-Tate group of the image of x

in S) as an algebraic subgroup of $\mathrm{GL}(H) \times \mathbb{G}_m$. So, passing to the universal cover has the advantage that we can describe the VHS as a varying family of Hodge structures on some given space H and that we may describe the Mumford-Tate groups MT_x as subgroups of one and the same group.

Given a triple $\mathfrak{m} = (m_1, m_2, m_3)$ as in subsection 2.2, consider the space

$$T(\mathfrak{m}) := H^{\otimes m_1} \otimes (H^\vee)^{\otimes m_2} \otimes \mathbb{Q}(m_3)\,,$$

on which we have a natural action of $\mathrm{GL}(H) \times \mathbb{G}_m$. We may view an element $t \in T(\mathfrak{m})$ as a global section of the local system

$$\tilde{\mathcal{T}}(\mathfrak{m}) := \tilde{\mathcal{H}}^{\otimes m_1} \otimes (\tilde{\mathcal{H}}^\vee)^{\otimes m_2} \otimes \mathbb{Q}(m_3)_{\tilde{S}}\,,$$

which underlies a $\mathbb{Q}$-VHS of weight $(m_1 - m_2)k - 2m_3$ on $\tilde{S}$. (The Tate twist $\mathbb{Q}(m_3)_{\tilde{S}}$ is a constant local system; it only has an effect on how we index the Hodge filtration on $\tilde{\mathcal{T}}(\mathfrak{m})_{\mathbb{C}}$.) In particular, it makes sense to consider the set $Y(t) \subset \tilde{S}$ of points $x \in \tilde{S}$ for which the value t_x of t at x is a Hodge class. As remarked earlier, we can only have nonzero Hodge classes if the weight is zero, so we restrict our attention to triples $\mathfrak{m} = (m_1, m_2, m_3)$ with $(m_1 - m_2)k = 2m_3$. With this assumption on $\mathfrak{m}$, define

$$Y(t) := \left\{x \in \tilde{S} \;\middle|\; t_x \in \mathrm{Fil}^0\tilde{\mathcal{T}}(\mathfrak{m})_{\mathbb{C}}(x)\right\}.$$

As $\mathrm{Fil}^0\tilde{\mathcal{T}}(\mathfrak{m})_{\mathbb{C}} \subset \tilde{\mathcal{T}}(\mathfrak{m})_{\mathbb{C}}$ is a holomorphic subbundle, $Y(t) \subset \tilde{S}$ is a countable union of closed irreducible analytic subsets of $\tilde{S}$.

The first thing we deduce from this is that there is a countable union of proper analytic subsets $\Sigma \subset S$ such that the Mumford-Tate group MT_s is constant (in a suitable sense) on $S \setminus \Sigma$, and gets smaller if we specialize to a point in Σ. Let us make this precise. Define $\tilde{\Sigma} \subset \tilde{S}$ by

$$\tilde{\Sigma} := \cup\, Y(t)$$

where, with notation as introduced above, the union is taken over all triples $\mathfrak{m}$ with $(m_1 - m_2)k = 2m_3$, and all $t \in \Gamma(\tilde{S}, \tilde{\mathcal{T}}(\mathfrak{m}))$ such that $Y(t)$ is *not* the whole $\tilde{S}$.

The main point of this definition of $\tilde{\Sigma}$ is the following. If $x \in \tilde{S} \setminus \tilde{\Sigma}$ and t_x is a Hodge class in the fiber of $\tilde{\mathcal{T}}(\mathfrak{m})$ at the point x, we can extend t_x in a unique way to a global section t of $\tilde{\mathcal{T}}(\mathfrak{m})$ over $\tilde{S}$, and by definition of $\tilde{\Sigma}$ this global section is a Hodge class in every fiber. By contrast, if $x \in \tilde{\Sigma}$ then there is a tensor construction $\tilde{\mathcal{T}}(\mathfrak{m})$ and a Hodge class t_x whose horizontal extension t is not a Hodge class in every fiber.

It follows that for $x \in \tilde{S} \setminus \tilde{\Sigma}$ the Mumford-Tate group $\mathrm{MT}_x \subset \mathrm{GL}(H) \times \mathbb{G}_m$ is independent of the choice of x. We call the subgroup of $\mathrm{GL}(H) \times \mathbb{G}_m$ thus obtained the *generic Mumford-Tate group*. Let us denote it by $\mathrm{MT}^{\mathrm{gen}}$. Further, for *any* $x \in \tilde{S}$ we have an inclusion $\mathrm{MT}_x \subseteq \mathrm{MT}^{\mathrm{gen}}$.

The subset $\tilde{\Sigma} \subset \tilde{S}$ is stable under the action of the covering group of $\tilde{S}/S$ and therefore defines a subset $\Sigma \subset S$. It follows from the construction that Σ is a countable union of closed analytic subspaces of S. We say that a point $s \in S(\mathbb{C})$ is *Hodge generic* if $s \notin \Sigma$. In somewhat informal terms we may restate the constancy of the Mumford-Tate group over $\tilde{S} \setminus \tilde{\Sigma}$ by saying that over Hodge generic locus the Mumford-Tate group is constant.

We can also draw conclusions pertaining to the loci in S where we have some given collection of Hodge classes. Start with a point $s_0 \in S(\mathbb{C})$, let $x_0 \in \tilde{S}$ be a point above s_0, and consider a finite collection of nonzero classes $t^{(i)}$, for $i = 1, \ldots, r$, in tensor spaces $T(m^{(i)}) = \Gamma(\tilde{S}, \tilde{\mathcal{T}}(m^{(i)}))$ that are Hodge classes for the Hodge structure at the point x_0. With notation as above, the locus of points in $\tilde{S}$ where all classes $t^{(i)}$ are again Hodge classes is $Y(t^{(1)}) \cap \cdots \cap Y(t^{(r)})$. The image of this locus in S is a countable union of closed irreducible analytic subspaces. These components are called the Hodge loci of the given VHS.

Definition 2.5. Let $\mathcal{H}$ be a polarizable $\mathbb{Q}$-VHS over a complex manifold S. A closed irreducible analytic subspace $Z \subseteq S$ is called a *Hodge locus* of $\mathcal{H}$ if there exist nonzero classes $t^{(1)}, \ldots, t^{(r)}$ in tensor spaces $T(m^{(i)}) = \Gamma(\tilde{S}, \tilde{\mathcal{T}}(m^{(i)}))$ such that Z is an irreducible component of the image of $Y(t^{(1)}) \cap \cdots \cap Y(t^{(r)})$ in S.

Remark 2.6. If we start with a polarizable $\mathbb{Z}$-VHS over a nonsingular complex algebraic variety S then by a theorem of Cattani, Deligne and Kaplan, see Corollary 1.3 in [11], the Hodge loci are algebraic subvarieties of S. See also [81].

We note that in the definition of Hodge loci there is no loss of generality to consider only a finite collection of Hodge classes. The requirement that a (possibly infinite) collection of classes $t^{(i)}$ are all Hodge classes translates into the condition that the homomorphism $h\colon \mathbb{S} \to GL(H)_{\mathbb{R}}$ that defines the Hodge structure factors through $M_{\mathbb{R}}$, where $M \subset GL(H)$ is the common stabilizer of the given classes. Among the classes $t^{(i)}$ we can then find a finite subcollection $t^{(i_1)}, \ldots, t^{(i_n)}$ that have M as their common stabilizer; so the Hodge locus we are considering is defined by these classes.

3. Special subvarieties of A_g, and the André-Oort conjecture

In this section we discuss the notion of a special subvariety in a given Shimura variety. The abstract formalism of Shimura varieties provides a good framework for this but it has the disadvantage that it requires a lot of machinery. For this reason we shall give several equivalent definitions and we focus on concrete examples related to moduli of abelian varieties. In particular the version given in Definition 3.7 can be understood without any prior knowledge of Shimura varieties.

Further we state the André-Oort conjecture. A proof of this conjecture, under the assumption of the Generalized Riemann Hypothesis (GRH), has been announced by Klingler, Ullmo and Yafaev.

3.1. Shimura varieties

In this paper, by a *Shimura datum* we mean a pair (G, X) consisting of an algebraic group G over $\mathbb{Q}$ together with a $G(\mathbb{R})$-conjugacy class X of homomorphisms $\mathbb{S} \to G_{\mathbb{R}}$, such that the conditions (2.1.1.1–3) of [21] are satisfied. The space X is the disjoint union of finitely many connected components. These components have the structure of a hermitian symmetric domain of non-compact type.

Associated with such a datum is a subfield $E(G, X) \subset \mathbb{C}$ of finite degree over $\mathbb{Q}$, called the reflex field. Given a compact open subgroup $K \subset G(\mathbb{A}_{\mathrm{fin}})$ we have a Shimura variety $\mathrm{Sh}_K(G, X)$, which is a scheme of finite type over $E(G, X)$, and for which we have

$$\mathrm{Sh}_K(G, X)(\mathbb{C}) = G(\mathbb{Q})\backslash\big(X \times G(\mathbb{A}_{\mathrm{fin}})/K\big)\,.$$

For $x \in X$ and $\gamma K \in G(\mathbb{A}_{\mathrm{fin}})/K$ we denote by $[x, \gamma K] \in \mathrm{Sh}_K(G, X)(\mathbb{C})$ the class of $(x, \gamma K)$.

If $K' \subset K$ is another compact open subgroup of $G(\mathbb{A}_{\mathrm{fin}})$ we have a natural morphism $\mathrm{Sh}_{K',K}\colon \mathrm{Sh}_{K'}(G, X) \to \mathrm{Sh}_K(G, X)$.

Let $\gamma \in G(\mathbb{A}_{\mathrm{fin}})$. Given compact open subgroups K, $K' \subset G(\mathbb{A}_{\mathrm{fin}})$ with $K' \subset \gamma K \gamma^{-1}$ we have a morphism $T_\gamma = [\cdot\gamma]\colon \mathrm{Sh}_{K'}(G, X) \to \mathrm{Sh}_K(G, X)$ that is given on $\mathbb{C}$-valued points by $T_\gamma[x, aK'] = [x, a\gamma K]$. (For $\gamma = 1$ we recover the morphism $\mathrm{Sh}_{K',K}$.) This induces a right action of the group $G(\mathbb{A}_{\mathrm{fin}})$ on the projective limit

$$\mathrm{Sh}(G, X) := \varprojlim_K \mathrm{Sh}_K(G, X)\,,$$

and $\mathrm{Sh}_K(G, X)$ can be recovered from $\mathrm{Sh}(G, X)$ as the quotient modulo K. More generally, for compact open subgroups K_1, $K_2 \subset G(\mathbb{A}_{\mathrm{fin}})$ and $\gamma \in G(\mathbb{A}_{\mathrm{fin}})$ we have a Hecke correspondence T_γ from $\mathrm{Sh}_{K_1}(G, X)$ to $\mathrm{Sh}_{K_2}(G, X)$, given by the diagram

$$\mathrm{Sh}_{K_1}(G, X) \xleftarrow{\ \mathrm{Sh}_{K',K_1}\ } \mathrm{Sh}_{K'}(G, X) \xrightarrow{\ [\cdot\gamma]\ } \mathrm{Sh}_{K_2}(G, X)\,, \tag{3.1}$$

where $K' := K_1 \cap \gamma K_2 \gamma^{-1}$.

3.2. The description of $\mathcal{A}_g$ as a Shimura variety

With notation as in Subsection 0.1, let $\mathfrak{H}_g$ denote the space of homomorphisms $h\colon \mathbb{S} \to \mathrm{CSp}_{2g,\mathbb{R}}$ that define a Hodge structure of type $(-1,0)+(0,-1)$ on V for which $\pm(2\pi i)\cdot\phi\colon V \times V \to \mathbb{Z}(1)$ is a polarization. The group $\mathrm{CSp}_{2g}(\mathbb{R})$ acts transitively on $\mathfrak{H}_g$ by conjugation, and the pair $(\mathrm{CSp}_{2g,\mathbb{Q}}, \mathfrak{H}_g)$ is an example of a Shimura datum. The reflex field of this datum is $\mathbb{Q}$.

The associated Shimura variety is known as the *Siegel modular variety* and may be identified with the moduli space of principally polarized abelian varieties with a level structure. To explain this in detail, we define, for a positive integer m, a compact open subgroup $K_m \subset G(\mathbb{A}_{\mathrm{fin}})$ by

$$K_m = \big\{\gamma \in \mathrm{CSp}(V \otimes \hat{\mathbb{Z}}, \phi) \mid \gamma \equiv \mathrm{id} \pmod{m}\big\}\,.$$

For $m \geqslant 3$ we then have an isomorphism

$$\beta\colon \mathcal{A}_{g,[m],\mathbb{Q}} \xrightarrow{\sim} \mathrm{Sh}_{K_m}(\mathrm{CSp}_{2g,\mathbb{Q}}, \mathfrak{H}_g)\,.$$

On $\mathbb{C}$-valued points β is given as follows. To begin with, a $\mathbb{C}$-valued point of $\mathcal{A}_{g,[m]}$ is a triple (A, λ, α) consisting of a complex ppav of dimension g together with a level m structure that is symplectic for some choice of a primitive mth root of unity $\zeta \in \mathbb{C}$. Write $H := H_1(A, \mathbb{Z})$ for the singular homology group in degree 1 of (the complex manifold associated with) A, and let $\psi\colon H \times H \to \mathbb{Z}(1)$ be the polarization associated with λ; see (2.2). Choose a symplectic similitude $s\colon H \xrightarrow{\sim} V$. (Even though ψ takes values in $\mathbb{Z}(1)$ and ϕ takes values in $\mathbb{Z}$, the notion of a symplectic similitude makes sense.) Via s, the natural Hodge structure on H corresponds to an element $x \in \mathfrak{H}_g$. Further, we have an identification $A[m] \xrightarrow{\sim} H/mH$, such that $e^{\lambda}(P, Q) = \exp(\psi(y, z)/m)$ if $P, Q \in A[m](\mathbb{C})$ correspond to $y, z \in H/mH$, respectively. (Note that $\exp(\psi(y, z)/m)$ is well-defined.) Hence the given level structure α can be viewed, via s, as an element $\gamma K_m \in \mathrm{CSp}(V \otimes (\mathbb{Z}/m\mathbb{Z}), \phi) = \mathrm{CSp}(V \otimes \hat{\mathbb{Z}}, \phi)/K_m$. The isomorphism β then sends (A, λ) to the point $[x, \gamma K_m]$.

For $m \geqslant 3$, the group $\mathrm{CSp}_{2g}(\mathbb{Q})$ acts properly discontinuously on the product $\mathfrak{H}_g \times \mathrm{CSp}_{2g}(\mathbb{A}_{\mathrm{fin}})/K_m$. The space $\mathfrak{H}_g$ has two connected components, and if $\mathfrak{H}_g^+$ is one of these, the $\varphi(m)$ components of $\mathcal{A}_{g,[m],\mathbb{C}}$ all have the form $\Gamma\backslash\mathfrak{H}_g^+$ for an arithmetic subgroup $\Gamma \subset \mathrm{CSp}_{2g}(\mathbb{Q})$. For $m = 1$ it is no longer true that $\mathrm{CSp}_{2g}(\mathbb{Q})$ acts properly discontinuously on $\mathfrak{H}_g \times \mathrm{CSp}_{2g}(\mathbb{A}_{\mathrm{fin}})/K_1$, and we should interpret the quotient space $\mathrm{CSp}_{2g}(\mathbb{Q})\backslash(\mathfrak{H}_g \times \mathrm{CSp}_{2g}(\mathbb{A}_{\mathrm{fin}})/K_1)$ as an orbifold. Alternatively, we may take the actual quotient space; this gives us an isomorphism

$$\mathrm{CSp}_{2g}(\mathbb{Q})\backslash\big(\mathfrak{H}_g \times \mathrm{CSp}_{2g}(\mathbb{A}_{\mathrm{fin}})/K_1\big) \xrightarrow{\sim} A_g(\mathbb{C})$$

between $\mathrm{Sh}_{K_1}(\mathrm{CSp}_{2g,\mathbb{Q}}, \mathfrak{H}_g)(\mathbb{C})$ and the set of $\mathbb{C}$-valued points of the coarse moduli space.

Remark 3.2. Let $\mathcal{L} = \mathcal{L}(V_{\mathbb{Q}}, \phi)$ denote the symplectic Grassmanian of Lagrangian (i.e., maximal isotropic) subspaces of $V_{\mathbb{Q}}$ with respect to the form ϕ. The map $\mathfrak{H}_g \to \mathcal{L}(\mathbb{C})$ that sends a Hodge structure $y \in \mathfrak{H}_g$ to the corresponding Hodge filtration $\mathrm{Fil}^0 \subset V_{\mathbb{C}}$ is an open immersion, known as the Borel embedding. For an arbitrary Shimura datum (G, X) we have, in a similar manner, a Borel embedding of X into the $\mathbb{C}$-points of a homogeneous projective variety.

Because $\mathrm{Fil}^0 \subset V_{\mathbb{C}}$ is maximal isotropic, the form ϕ induces a perfect pairing $\bar{\phi}\colon \mathrm{Fil}^0 \times V_{\mathbb{C}}/\mathrm{Fil}^0 \to \mathbb{C}$. Via the Borel embedding, the tangent space of $\mathfrak{H}_g$ at the point y maps isomorphically to the tangent space of $\mathcal{L}$ at the point Fil^0, which gives us

$$\begin{aligned} T_y(\mathfrak{H}_g) &\xrightarrow{\sim} \mathrm{Hom}^{\mathrm{sym}}(\mathrm{Fil}^0, V_{\mathbb{C}}/\mathrm{Fil}^0) \\ &:= \big\{\beta\colon \mathrm{Fil}^0 \to V_{\mathbb{C}}/\mathrm{Fil}^0 \mid \bar{\phi}(v, \beta(v')) = \bar{\phi}(v', \beta(v)) \text{ for all } v, v' \in \mathrm{Fil}^0\big\}. \end{aligned} \tag{3.3}$$

The condition that $\bar{\phi}(v, \beta(v')) = \bar{\phi}(v', \beta(v))$ means that β is its own dual, via the isomorphisms $\mathrm{Fil}^0 \xrightarrow{\sim} (V_{\mathbb{C}}/\mathrm{Fil}^0)^\vee$ and $V_{\mathbb{C}}/\mathrm{Fil}^0 \xrightarrow{\sim} (\mathrm{Fil}^0)^\vee$ induced by $\bar{\phi}$; whence the notation $\mathrm{Hom}^{\mathrm{sym}}$.

3.3. Special subvarieties

There are several possible approaches to the notion of a special subvariety of a given Shimura variety. Let us start with the construction that gives the quickest definition. We use the language of Shimura varieties; the example to keep in mind is the Siegel modular variety $\mathrm{Sh}(\mathrm{CSp}_{2g,\mathbb{Q}}, \mathfrak{H}_g)$. After this we shall give some alternative definitions that lean less heavily on the abstract formalism of Shimura varieties. The reader who prefers to avoid the language of Shimura varieties is encouraged to skip ahead to Definition 3.7.

We fix a Shimura datum (G, X) and a compact open subgroup $K \subset G(\mathbb{A}_{\mathrm{fin}})$. If (M, Y) is a second Shimura datum, by a morphism $f\colon (M, Y) \to (G, X)$ we mean a homomorphism of algebraic groups $f\colon M \to G$ such that for any $y\colon \mathbb{S} \to M_{\mathbb{R}}$ in Y the composite $f \circ y\colon \mathbb{S} \to G_{\mathbb{R}}$ is an element of X. The existence of such a morphism of Shimura data implies that $E(M, Y)$ contains $E(G, X)$, and f gives rise to a morphism of schemes

$$\mathrm{Sh}(f)\colon \mathrm{Sh}(M, Y) \to \mathrm{Sh}(G, X)_{E(M,Y)}\,.$$

Let $\gamma \in G(\mathbb{A}_{\mathrm{fin}})$. The image of the composite morphism

$$\mathrm{Sh}(M, Y)_{\mathbb{C}} \xrightarrow{\mathrm{Sh}(f)} \mathrm{Sh}(G, X)_{\mathbb{C}} \xrightarrow{[\cdot\gamma]} \mathrm{Sh}(G, X)_{\mathbb{C}} \longrightarrow \mathrm{Sh}_K(G, X)_{\mathbb{C}} \tag{3.4}$$

is a reduced closed subscheme of $\mathrm{Sh}_K(G, X)_{\mathbb{C}}$.

Definition 3.5 (Version 1). A closed subvariety $Z \subset \mathrm{Sh}_K(G, X)_{\mathbb{C}}$ is called a *special subvariety* if there exists a morphism of Shimura varieties $f\colon (M, Y) \to (G, X)$ and an element $\gamma \in G(\mathbb{A}_{\mathrm{fin}})$ such that Z is an irreducible component of the image of the morphism (3.4).

For a second approach we fix an algebraic subgroup $M \subset G$ over $\mathbb{Q}$. Define $Y_M \subset X$ as the set of all $x\colon \mathbb{S} \to G_{\mathbb{R}}$ in X that factor through $M_{\mathbb{R}} \subset G_{\mathbb{R}}$. We give Y_M the topology induced by the natural topology on X. The group $M(\mathbb{R})$ acts on Y_M by conjugation. It can be shown (see [45], Section I.3, or [46], 2.4) that Y_M is a finite union of orbits under $M(\mathbb{R})$. We remark that the condition that Y_M is nonempty imposes strong restrictions on M; it implies, for instance, that M is reductive.

Definition 3.6 (Version 2). A closed subvariety $Z \subset \mathrm{Sh}_K(G, X)_{\mathbb{C}}$ is called a *special subvariety* if there exists an algebraic subgroup $M \subset G$ over $\mathbb{Q}$, a connected component $Y^+ \subset Y_M$, and an element $\gamma \in G(\mathbb{A}_{\mathrm{fin}})$ such that $Z(\mathbb{C}) \subset \mathrm{Sh}_K(G, X)(\mathbb{C})$ is the image of $Y^+ \times \{\gamma K\} \subset X \times G(\mathbb{A}_{\mathrm{fin}})/K$ under the natural map to $\mathrm{Sh}_K(G, X)(\mathbb{C}) = G(\mathbb{Q})\backslash(X \times G(\mathbb{A}_{\mathrm{fin}})/K)$.

For the equivalence of this definition with Version 1 we refer to [45], Proposition 3.12 or [46], Remark 2.6.

Our third version of the definition, which in some sense is the most conceptual one, describes special subvarieties of $\mathrm{Sh}_K(G, X)$ as the Hodge loci of certain natural VHS associated with representations of the group G. As we wish to highlight the

Siegel modular case, we only state this version of the definition for special subvarieties of $\mathcal{A}_{g,[m],\mathbb{C}}$ for some $m \geqslant 3$. (See, however, Remark 3.8.)

In order to talk about Hodge loci we need some VHS. The most natural $\mathbb{Q}$-VHS on $\mathcal{A}_{g,[m],\mathbb{C}}$ to consider is the variation $\mathcal{H}$ whose fiber at a point (A, λ, α) is given by $H_1(A, \mathbb{Q})$. (One could also work with the $\mathbb{Q}$-VHS given by the first cohomology groups; for what we want to explain it makes no difference.) The Hodge loci of this VHS are precisely the special subvarieties.

Definition 3.7 (Version 3). A closed subvariety $Z \subset \mathcal{A}_{g,[m],\mathbb{C}}$ is called a *special subvariety* if it is a Hodge locus of the $\mathbb{Q}$-VHS $\mathcal{H}$ over $\mathcal{A}_{g,[m],\mathbb{C}}$ whose fiber at a point (A, λ, α) is $H_1(A, \mathbb{Q})$.

Concretely this means that the special subvarieties are "defined by" the existence of certain Hodge classes, i.e., they are the maximal closed irreducible subvarieties of $\mathcal{A}_{g,[m],\mathbb{C}}$ on which certain given classes are Hodge classes. In order to make this precise, one has to pass to the universal cover, where the underlying local system can be trivialized, as explained in subsection 2.3. Note that in this case the Hodge loci are algebraic subvarieties of $\mathcal{A}_{g,[m],\mathbb{C}}$; this can be shown by proving that this notion of a special subvariety agrees with the one given by Version 1 of the definition but it also follows from the theorem of Cattani, Deligne and Kaplan mentioned in Remark 2.6.

Remark 3.8. Version 3 of the definition of a special subvariety, in terms of Hodge loci, can be extended without much difficulty to arbitrary Shimura varieties $\mathrm{Sh}_K(G, X)$. The variations of Hodge structure one considers are those associated with representations of the group G. See for instance [46], Proposition 2.8, which also proves the equivalence with the earlier definitions. We note that, even if one is only interested in statements about special subvarieties of $\mathcal{A}_g$, there are good reasons to extend this notion to more general Shimura varieties. The formalism of Shimura varieties enables one to perform some useful constructions, such as the passage to an adjoint Shimura datum, or the reduction of a problem to the case of a simple Shimura datum.

Example 3.9. Let (A, λ, α) be a complex ppav of dimension g with a symplectic level m structure, for some $m \geqslant 3$. Let $D := \mathrm{End}^0(A)$ be its endomorphism algebra. We should like to describe the largest closed (irreducible) subvariety $Z \subset \mathcal{A}_{g,[m],\mathbb{C}}$ that contains the moduli point $[A, \lambda, \alpha] \in \mathcal{A}_{g,[m]}(\mathbb{C})$, and such that all endomorphisms of A extend to endomorphisms of the universal abelian scheme over Z. In terms of Hodge classes this means that all elements of D, viewed as Hodge classes in $\mathrm{End}_{\mathbb{Q}}(H_1(A, \mathbb{Q}))$, should extend to Hodge classes in the whole $\mathbb{Q}$-VHS over Z given by the endomorphisms of the first homology groups. (Note that we may pass to $\mathbb{Q}$-coefficients, for if $f \in \mathrm{End}(A)$ is an endomorphism such that some positive multiple nf extends to an endomorphism of the whole family, f itself extends.)

As before, choose a symplectic similitude $s\colon H_1(A,\mathbb{Z}) \xrightarrow{\sim} V$ with respect to the polarization forms on both sides. This gives $V_\mathbb{Q}$ the structure of a left module over the algebra D. Let $h_0 \in \mathfrak{H}_g$ be the point given by the Hodge structure on $H_1(A,\mathbb{Z})$, viewed as a Hodge structure on V via s. Further, let $M \subset \mathrm{CSp}_{2g,\mathbb{Q}} = \mathrm{CSp}(V_\mathbb{Q},\phi)$ be the algebraic subgroup given by

$$M = \mathrm{CSp}(V_\mathbb{Q},\phi) \cap \mathrm{GL}_D(V_\mathbb{Q})\,.$$

The homomorphism $h_0\colon \mathbb{S} \to \mathrm{CSp}_{2g,\mathbb{R}}$ factors through $M_\mathbb{R}$. Conversely, if $h \in \mathfrak{H}_g$ factors through $M_\mathbb{R}$ then D acts by endomorphisms on the corresponding abelian variety. This means that $Z \subset \mathrm{Sh}_{K_m}(\mathrm{CSp}_{2g,\mathbb{Q}},\mathfrak{H}_g)_\mathbb{C} = \mathcal{A}_{g,[m],\mathbb{C}}$ is the special subvariety that is obtained, with notation as in Definition 3.6, as the image of $Y_M^+ \times \{\gamma K_m\}$ in $\mathcal{A}_{g,[m],\mathbb{C}}$, were $Y_M^+ \subset Y_M$ is the connected component containing h_0 and $\gamma K_m \in \mathrm{CSp}_{2g}(\mathbb{A}_{\mathrm{fin}})/K_m$ is the class corresponding (via s) with the given level structure α.

We conclude that the closed subvarieties $Z \subset \mathcal{A}_{g,[m],\mathbb{C}}$ "defined by" the existence of endomorphisms, as made precise above, are examples of special subvarieties.

Remark 3.10. The special subvarieties considered in Example 3.9 are referred to as special subvarieties *of PEL type*; the name comes from the fact that they have a modular interpretation in terms of abelian varieties with a *p*olarization, given *e*ndomorphisms and a *l*evel structure.

In the above example our focus is on the concrete modular interpretation of these special subvarieties. For many purposes it is relevant to also have a good description of these examples in the language of Shimura varieties; see also [19], Section 4. Here one starts with data $(D,*,V,\phi)$ where D is a finite dimensional semisimple $\mathbb{Q}$-algebra, $*$ is a positive involution, V is a faithful (left) D-module of finite type, and $\phi\colon V\times V \to \mathbb{Q}$ is an alternating form such that

$$\phi(dv,v') = \phi(v,d^*v') \qquad \text{for all } v, v' \in V \text{ and } d \in D. \tag{3.11}$$

Let $G = \mathrm{CSp}(V,\phi)\cap \mathrm{GL}_D(V)$ be the group of D-linear symplectic similitudes of V. Next one considers a $G^0(\mathbb{R})$-conjugacy class X of homomorphisms $\mathbb{S} \to G_\mathbb{R}$ defining on V a Hodge structure of type $(-1,0)+(0,-1)$ such that $\pm 2\pi i \cdot \phi$ is a polarization. The pair (G^0,X) is then a Shimura datum.

In Example 3.9, the data (D,V,ϕ) are given, and we take for $*$ the involution on D defined by (3.11). For X we take the conjugacy class of the homomorphism h_0. By construction, the inclusion $G \hookrightarrow \mathrm{CSp}_{2g}$ defines a morphism of Shimura data $f\colon (G^0,X) \to (\mathrm{CSp}_{2g},\mathfrak{H}_g)$, and the special subvariety $Z \subset \mathcal{A}_{g,[m],\mathbb{C}}$ of 3.9 is an irreducible component of a Hecke translate of the image of $\mathrm{Sh}(G^0,X)$ in $\mathrm{Sh}_{K_m}(\mathrm{CSp}_{2g,\mathbb{Q}},\mathfrak{H}_g)_\mathbb{C} = \mathcal{A}_{g,[m],\mathbb{C}}$.

Example 3.12. Let (G,X) be a Shimura datum. If $x\colon \mathbb{S} \to G_\mathbb{R}$ is an element of X, we define the Mumford-Tate group of x to be the smallest algebraic subgroup $\mathrm{MT}_x \subset G$ over $\mathbb{Q}$ such that x factors through $\mathrm{MT}_{x,\mathbb{R}}$.

Let $s = [x, \gamma K] \in \mathrm{Sh}_K(G,X)(\mathbb{C})$. Up to conjugation by an element of $G(\mathbb{Q})$, the Mumford-Tate group $\mathrm{MT}_x \subset G$ is independent of the choice of the chosen representative $(x, \gamma K)$ for s. Hence we may define the Mumford-Tate group of s to be $\mathrm{MT}_s := \mathrm{MT}_x$.

A point $x \in X$ is called a *special point* if MT_x is a torus. A point $s \in \mathrm{Sh}_K(G,X)(\mathbb{C})$ is a special subvariety of $\mathrm{Sh}_K(G,X)_{\mathbb{C}}$ if and only if MT_s is a torus, i.e., if for some (equivalently, any) representative $[x, \gamma K]$ the point $x \in X$ is special.

In the Siegel modular variety A_g over $\mathbb{C}$, the special points are precisely the CM points, i.e., the points corresponding to ppav (A, λ) such that A is an abelian variety of CM type; see Mumford [54], § 2.

3.4. Basic properties of special subvarieties

We list a number of elementary properties.

(a) *Hecke images of special subvarieties are again special.* More precisely, consider a Hecke correspondence T_γ as given by (3.1). Let $Z \subset \mathrm{Sh}_{K_1}(G,X)_{\mathbb{C}}$ be a special subvariety. Write $T_\gamma(Z)$ for the image of $\mathrm{Sh}_{K',K_1}^{-1}(Z)$ under the map $[\cdot\gamma]$. Then all irreducible components of $T_\gamma(Z)$ are special subvarieties of $\mathrm{Sh}_{K_2}(G,X)_{\mathbb{C}}$.

As a particular case, if we have compact open subgroups $K' \subset K \subset G(\mathbb{A}_{\mathrm{fin}})$ and if $Y \subset \mathrm{Sh}_{K'}(G,X)_{\mathbb{C}}$ is a special subvariety, the image of Y in $\mathrm{Sh}_K(G,X)_{\mathbb{C}}$ is again a special subvariety. Conversely, if $Z \subset \mathrm{Sh}_K(G,X)_{\mathbb{C}}$ is a special subvariety, the irreducible components of $\mathrm{Sh}_{K',K}^{-1}(Z)$ are special subvarieties of $\mathrm{Sh}_{K'}(G,X)_{\mathbb{C}}$. In the study of special subvarieties, these remarks often allow us to choose the level subgroup K as small as needed.

(b) *The special points in a special subvariety are dense.* If $Z \subset \mathrm{Sh}_K(G,X)_{\mathbb{C}}$ is a special subvariety then the special points in Z are dense for the analytic topology on $Z(\mathbb{C})$; in particular they are Zariski dense. In order to prove this, the essential point is to show that Z contains at least *one* special point. The density of special points then follows from the fact that, with notation as in Def. 3.6, the set of special points in Y_M is stable under the action of $M(\mathbb{Q})$, and that $M(\mathbb{Q})$ is analytically dense in $M(\mathbb{R})$.

In view of the importance of special points, let us sketch a proof of the existence of at least one special point, following [54], § 3. The argument uses the fact from the theory of reductive groups that, given a maximal torus $T \subset G_{\mathbb{R}}$, the $G(\mathbb{R})$-conjugacy class of T contains a maximal torus that is defined over $\mathbb{Q}$.

Start with any $x\colon \mathbb{S} \to G_{\mathbb{R}}$ in X. Let $C \subset G_{\mathbb{R}}$ be the centralizer of $x(\mathbb{S})$, which is a connected reductive subgroup of $G_{\mathbb{R}}$. (See [78], Lemma 15.3.2.) Choose a maximal torus $T \subset C$. Because $x(\mathbb{S})$ is contained in the center of C, we have $x(\mathbb{S}) \subseteq T$. Hence, if $T' \subset G_{\mathbb{R}}$ is any torus that contains T then T' centralizes $x(\mathbb{S})$ and therefore $T' \subset C$. It follows that T is also a maximal torus of $G_{\mathbb{R}}$. By the general fact stated above, there exists an element $g \in G(\mathbb{R})$ and a maximal torus $S \subset G$ (over $\mathbb{Q}$) such that $gTg^{-1} = S_{\mathbb{R}}$. Then $gx = \mathrm{Inn}(g) \circ x\colon \mathbb{S} \to G_{\mathbb{R}}$ factors through $S_{\mathbb{R}}$; hence $gx \in X$ is a special point.

A more refined version of the existence of special points plays a role in the theory of canonical models of Shimura varieties. See [19], Theorem 5.1.

(c) *Intersections of special subvarieties are again special.* This is easily seen using Version 2 of the definition.

(d) *After passage to an appropriate level cover, special subvarieties are locally symmetric.* Suppose $Z \subset \mathrm{Sh}_K(G, X)$ is a special subvariety. We may then find a subgroup $K' \subset K$ of finite index and an irreducible component $Z' \subset \mathrm{Sh}_{K'}(G, X)$ of the inverse image of Z such that Z' is a locally symmetric variety. (We shall recall the definition of this notion in subsection 4.2 below.)

Remark 3.13. A situation we often encounter is that we have an abelian scheme $A \to T$ over some complex algebraic variety T (assumed to be irreducible), with a principal polarization $\lambda\colon A \to A^t$ and a level m structure α. This gives us a morphism $\tau\colon T \to \mathcal{A}_{g,[m],\mathbb{C}}$, and we should like to know if the scheme-theoretic image of τ is a special subvariety. (In this situation, the scheme-theoretic image is the reduced closed subscheme of $\mathcal{A}_{g,[m]}$ that has as underlying set the Zariski closure of the topological image of τ.)

We remark that the answer to this question only depends on A up to isogeny, and is independent of the polarization and the level structure.

Of course, if we change the polarization or the level structure, or if we replace A/T by an isogenous abelian scheme, the morphism τ is replaced by another morphism $\tau'\colon T \to \mathcal{A}_{g,[m]}$, but if the image of τ is special, so is the image of τ'. (For simplicity of exposition, we here assume the new polarization is again principal, but even this assumption can be dropped.) The reason that this is true is that the Mumford-Tate group of an abelian variety A only depends on A up to isogeny, and not on any additional structures.

The conclusion, then, is that one may set up the situation as one finds it convenient. We could equip (A, λ) with a level structure (possibly after replacing T with a cover), allowing us to work with a fine moduli scheme $\mathcal{A}_{g,[m]}$. Alternatively, one could forget about the level structure and consider the morphism $T \to \mathcal{A}_g$ to the moduli stack; in this case we should talk about special substacks of $\mathcal{A}_g$, which makes perfectly good sense. Yet another option is to consider the morphism $T \to A_g$ to the coarse moduli scheme. For the question whether the closed image of T is special, it does not matter which version we consider.

Further, since we are only interested in the closure of $\tau(T)$, we may replace T by an open subset, and in fact it suffices to know the generic fiber of A/T.

Conjecture 3.14 (André-Oort, [2], [62], [63]). *Let* (G, X) *be a Shimura datum. Let* $K \subset G(\mathbb{A}_{\mathrm{fin}})$ *be a compact open subgroup and let* $Z \subset \mathrm{Sh}_K(G, X)_{\mathbb{C}}$ *be an irreducible closed algebraic subvariety such that the special points on Z are dense for the Zariski topology. Then Z is a special subvariety.*

An alternative way of stating the conjecture is that, given a set of special points $\mathfrak{S} \subset \mathrm{Sh}_K(G, X)(\mathbb{C})$, the irreducible components of the Zariski closure of $\mathfrak{S}$ are special subvarieties.

We refer to Noot's Bourbaki lecture [60] for an excellent overview of what was known about the conjecture in 2004.

Klingler and Yafaev [38], using the work of Ullmo and Yafaev [79], have announced a proof of the André-Oort conjecture, assuming the Generalized Riemann hypothesis (GRH) for CM fields.

Theorem 3.15 (Klingler-Yafaev). *Assume the GRH holds for all CM fields. Then the André-Oort conjecture is true.*

There are some special cases of the André-Oort conjecture that are known to hold without any assumptions on the GRH. See for instance [47], Theorem 5.7, later refined by Yafaev in [82], Theorem 1.2 of [26], and Theorem 1.2.1 (with condition (2)) of [38]. In all these cases, however, further assumptions are needed on the set of special points. Apart from trivial cases, the only completely unconditional case of the André-Oort conjecture that we know of, is the main result of André's paper [3], which proves the conjecture for subvarieties of a product of two modular curves, and the more recent extension of this by Pila [70] to arbitrary products of modular curves.

Remark 3.16. As in Remark 3.13, consider a principally polarized abelian scheme (A, λ) over some complex algebraic variety T. Again we consider the question whether the closed image of the morphism $\tau\colon T \to A_g$ is a special subvariety. Suppose that A/T is isogenous to a product $A_1 \times A_2$ with $\dim(A_i/T) = g_i$. After choosing polarizations (principal, say) on the factors A_i we get morphisms $\tau_i\colon T \to A_{g_i}$. In this situation, if the closure of $\tau(T)$ in A_g is a special subvariety, the closure of $\tau_i(T)$ in A_{g_i} is special, too. If one believes the André-Oort conjecture this is clear, and it is in fact not very difficult to show this using our definitions of a special subvariety.

The converse implication does not hold, in general. So, if the closures of $\tau_1(T)$ and $\tau_2(T)$ are special, this does not imply that the closure of $\tau(T)$ is special. Looking at it from the perspective of the André-Oort conjecture, suppose that for each of the two factors A_i/T separately, we have a Zariski dense collection of points in $T(\mathbb{C})$ at which the fiber of A_i/T is of CM type. Then it is not true, in general, that there is a dense set of points at which the fibers are *simultaneously* of CM type. (For a concrete example, take T to be an open part of $\mathbb{A}^1 \setminus \{0,1\}$, let A_1/T be a family of elliptic curves that is not isotrivial, and take for A_2 the same family with a suitable shift in the parameter, i.e., such that $A_{2,t} = A_{1,t+\epsilon}$ for some fixed $\epsilon \in \mathbb{C}$.)

4. Special subvarieties in the Torelli locus

Throughout this section we work over $\mathbb{C}$.

Conjecture 4.1 (Coleman, [16], Conjecture 6). *Given $g \geqslant 4$ there are only finitely many non-singular projective curves C over $\mathbb{C}$, up to isomorphism, of genus g and the Jacobian J_C is a CM abelian variety.*

As we shall discuss below, the conjecture is known to be false for $g \leqslant 7$, so a corrected version of the conjecture should have as an assumption that $g \geqslant 8$.

Both Coleman's conjecture 4.1 and the André-Oort conjecture 3.14 were inspired by the analogy with the Manin-Mumford conjecture, now a theorem of Raynaud. For proofs of the Manin-Mumford conjecture, see [72], [73]; for an easy proof see [71]. The analogy between this conjecture and the André-Oort conjecture is discussed for instance in [48], Section 6 and in [60], Section 3. As we shall discuss next, the André-Oort conjecture also has important implications for Coleman's conjecture.

Expectation 4.2 (Oort, [63], § 5). *For large g (in any case $g \geqslant 8$), there does not exist a special subvariety $Z \subset A_g$ with $\dim(Z) \geqslant 1$ such that $Z \subseteq T_g$ and $Z \cap T_g^\circ$ is nonempty.*

Note that the assumption that $Z \cap T_g^\circ$ is nonempty implies that this intersection is open and dense in Z.

Remark 4.3. The condition that Z meets T_g° is important. Indeed, it is easy to see that for any $g \geqslant 2$ there exist special subvarieties $Z \subset A_g$ of positive dimension with $Z \subset T_g$. For $g \leqslant 3$ this is clear, as $T_g = A_g$ is special. Assume $g > 3$. Choose a base variety T and a stable curve $C \to T$ of genus 3, such that the closure of the image of T in A_3 is a special subvariety of positive dimension d. Let $J \to T$ be the Jacobian, λ its canonical principal polarization. Next take an elliptic curve E with complex multiplication, and let μ be its principal polarization. Then for every $t \in T(\mathbb{C})$ the moduli point $\xi_t \in A_g(\mathbb{C})$ of the ppav $(J_t, \lambda) \times (E, \mu)^{g-3}$ lies in the closed Torelli locus T_g, as it is the Jacobian of the curve that is obtained from C_t by attaching to it a tail of $g-3$ copies of E. Moreover, it is not hard to see that the closure of the set of points ξ_t is a d-dimensional special subvariety of A_g. In this way we can produce many positive dimensional special subvarieties in T_g for any $g \geqslant 2$.

Remark 4.4. Assume we have an integer g for which Expectation 4.2 holds. Assume furthermore that the André-Oort Conjecture 3.14 is true. Then Coleman's Conjecture 4.1 is true in genus g. In fact, let $CM(T_g^\circ) \subset T(\mathbb{C})$ be the set of all CM Jacobians of dimension g. If this set is infinite, its Zariski closure in A_g has at least one irreducible component Z of positive dimension, which by 3.14 is special, contradicting 4.2. Hence 4.2 and 3.14 together imply 4.1.

Remark 4.5. For $g > 3$ we know that T_g itself is not a special subvariety, and in fact there is no special subvariety $S \subsetneq A_g$ that contains T_g. (Note that $T_g \neq A_g$ for $g > 3$.) The simplest argument we know for this is to use information about the geometric monodromy. For convenience, let us pass to moduli spaces with a level m structure, for some $m \geqslant 3$. Write $\mathcal{T}_{g,[m]} \subset \mathcal{A}_{g,[m]}$ for the Torelli locus in $\mathcal{A}_{g,[m]}$. Over $\mathcal{A}_{g,[m]}$

we have the natural $\mathbb{Q}$-VHS $\mathcal{H}$ considered before (see just before Definition 3.7), and the assertion that T_g is not contained in any special subvariety $S \subsetneq A_g$ is implied by the fact that the generic Mumford-Tate group of the restriction of $\mathcal{H}$ to $\mathcal{T}_{g,[m]}$ is the full group $\mathrm{CSp}_{2g,\mathbb{Q}}$.

Write $\mathcal{H}'$ for the restriction of $\mathcal{H}$ to $\mathcal{T}_{g,[m]}$. Choose a Hodge generic base point $b \in \mathcal{T}_{g,[m]}$, let H be the fiber of $\mathcal{H}'$ at b, and let ψ be the polarization form on H. Further, let $M \subset \mathrm{CSp}_{2g,\mathbb{Q}} = \mathrm{CSp}(H,\psi)$ be the Mumford-Tate group at b, which is the generic Mumford-Tate group of $\mathcal{H}'$. Then M contains $\mathbb{G}_m \cdot 1$. On the other hand we have a monodromy representation

$$\rho\colon \pi_1(\mathcal{T}_{g,[m]}, b) \to \mathrm{GL}(H)\,,$$

and since we have passed to a level cover, ρ factors through $\mathrm{Sp}(H,\psi)$. By a result of Deligne, see [20], Proposition 7.5, the image of ρ has a subgroup of finite index that is contained in $M(\mathbb{C})$.

So we are done if we can show that the image of ρ is Zariski dense in $\mathrm{Sp}(H,\psi)$. This can be seen, for instance, using transcendental methods. In fact, the homomorphism $\rho\colon \pi_1(\mathcal{T}_{g,[m]}, b) \to \mathrm{Sp}(H,\psi)$ may be identified with the natural homomorphism $\Gamma_g \to \mathrm{Sp}(H,\psi)$ from the mapping class group in genus g to the symplectic group, and it is a classical result that this homomorphism is surjective. (See for instance [5], Chap. 15, especially § 3.)

The argument given here is entirely based on information about the geometric monodromy, and there are other loci in T_g to which the same reasoning applies. As an example, if $H_g \subset T_g$ is the hyperelliptic locus, the image of the geometric monodromy representation on H_g is again dense in the symplectic group $\mathrm{Sp}(H,\psi)$; see [1], Theorem 1. By the above argument, it follows that there is no special subvariety $S \subsetneq A_g$ that contains H_g.

4.1. A modified version of Coleman's Conjecture

In [13] we find a modified version of 4.1 which does hold, at least conditionally, for every $g > 3$. A g-dimensional abelian variety A, say over $\mathbb{C}$, is said to be a *Weyl CM abelian variety* if $L := \mathrm{End}^0(A)$ is a field of degree $2g$ over $\mathbb{Q}$ whose Galois closure has degree $2^g \cdot g!$ over $\mathbb{Q}$. It can be shown that, in a suitable sense, most CM abelian varieties are of this type.

Theorem 4.6 (Chai and Oort). *Assume the André-Oort Conjecture* 3.14 *to be true. Then for every* $g > 3$ *the number of Weyl CM points in* T_g° *is finite.*

See [13], 3.7. We remark that for $g \geqslant 4$ we do not know any example of a Jacobian of Weyl CM type. In connection with this, note that for a non-hyperelliptic curve C of genus $g(C) > 1$ with $\mathrm{Aut}(C) \neq \{\mathrm{id}\}$ the Jacobian J_C does not give a Weyl CM point.

4.2. Results of Hain

In his paper [32], Hain proved some results inspired by 4.2. Though the results are conditional, they point in an interesting direction, as they suggest that ball quotients should play a special role. (The open unit ball in $\mathbb{C}^n$ is the symmetric space $SU(n,1)/U(n)$.) It should be pointed out that all non-trivial examples known to us, in genus at least 4, are indeed ball quotients; see the next sections.

In order to state Hain's results, let us first recall that a complex algebraic variety S is called a *locally symmetric variety* if there is a semisimple algebraic group G over $\mathbb{Q}$, a maximal compact subgroup $K \subset G(\mathbb{R})$, and an arithmetic subgroup $\Gamma \subset G(\mathbb{Q})$, such that the symmetric space $G(\mathbb{R})/K$ is hermitian and $S = \Gamma\backslash G(\mathbb{R})/K$. As we have seen in subsection 3.4, special subvarieties are (at least after passage to a level cover) locally symmetric.

If $S_1 = \Gamma_1\backslash G_1(\mathbb{R})/K_1$ and $S_2 = \Gamma_2\backslash G_2(\mathbb{R})/K_2$ are locally symmetric varieties, a morphism $f\colon S_1 \to S_2$ is called a map of locally symmetric varieties if it is induced by a homomorphism of algebraic groups $G_1 \to G_2$ over $\mathbb{Q}$.

Hain calls a locally symmetric variety S *good* if it has no locally symmetric divisors. Further, he mostly restricts his attention to locally symmetric varieties S for which the corresponding $\mathbb{Q}$-group G is almost simple (i.e., G^{ad} is simple); for the problems that interest us this is no loss of generality. In case G is almost simple, it can be shown that S is not good only if the $\mathbb{Q}$-rank of G is $\leqslant 2$ and if the non-compact factors of $G_{\mathbb{R}}$ are all of the form $SO(n,2)$ or $SU(n,1)$, up to isogeny.

Define a *locally symmetric family of abelian varieties* to be a principally polarized abelian scheme $X \to S$ such that S is a locally symmetric variety and the corresponding morphism $S \to A_g$ is a map of locally symmetric varieties. By a *locally symmetric family of curves* we mean a curve $C \to S$ of compact type such that the corresponding relative Jacobian $J \to S$ is a locally symmetric family of abelian varieties.

Theorem 4.7 (Hain). *Let $\pi\colon C \to S$ be a locally symmetric family of curves that is not isotrivial. Assume the $\mathbb{Q}$-group that gives S is almost simple. Assume further that either π is smooth, or S is good and the generic fiber of π is smooth. Then S is a quotient of a complex n-ball.*

See [32], Theorem 1.

It should be realized that for applications to Coleman's conjecture, the assumptions of the theorem are too strong. The main problem is that a special subvariety Z as in 4.2 only gives us (possibly after passing to a level cover) a locally symmetric family of abelian varieties $J \to Z$ such that the geometric generic fiber is a Jacobian. We may find a dominant morphism $Z' \to Z$ such that the pullback $J' \to Z'$ is the relative Jacobian of a smooth curve $C \to Z'$ but in general it is not possible to do this with Z' an open part of a locally symmetric variety. See [32], in particular Proposition 8.3.

Under weaker assumptions, Hain proves a second result.

Theorem 4.8 (Hain). *Let $A \to S$ be a locally symmetric family of abelian varieties such that the morphism $p\colon S \to A_g$ is not constant and factors through the Torelli locus T_g. Write*

$$\begin{aligned} S^{dec} &:= \{s \in S(\mathbb{C}) \mid A_s \text{ is the Jacobian of a singular curve,}\} \\ S^* &:= S(\mathbb{C}) \setminus S^{dec}, \\ S^{he} &:= \{s \in S(\mathbb{C}) \mid A_s \text{ is the Jacobian of a hyperelliptic curve.}\} \end{aligned}$$

We assume the $\mathbb{Q}$-group that gives S is almost simple, S is good, and $S^* \neq \emptyset$. Then either S is a ball quotient, or else $g \geqslant 3$, each component of S^{dec} has complex codimension $\geqslant 2$ in S, the family does not lift to a locally symmetric family of curves, and $S^* \cap S^{he}$ is a non-empty divisor in S^*, which moreover for $g > 3$ is nonsingular.

See [32], Theorem 2.

The proofs of Hain's results rely on a rigidity property of mapping class groups, which is a special case of the theorem of Farb and Masur in [29].

De Jong and Zhang [36] have pushed Hain's results further, based on the observation that the hyperelliptic locus in $A_g \setminus A_g^{dec}$ is affine.

Theorem 4.9 (de Jong and Zhang). *With assumptions as in Theorem 4.8, either S is a ball quotient, or S^{dec} has codimension $\leqslant 2$ in S, or the Baily-Borel compactification of S has a boundary of codimension $\leqslant 2$.*

Corollary 4.10. *Let $S \subset A_g$, with $g \geqslant 4$, be a Hecke translate of a Hilbert modular subvariety of A_g, i.e., S is a special subvariety of PEL type obtained from a totally real field F of degree g. Then S is not contained in T_g.*

The Corollary was proved in [36], except when $g = 4$ and F contains a quadratic subfield. That case was settled by Bainbridge and Möller in [6].

In addition to the results discussed here, there are several other results inspired by 4.2. Among these are papers of Möller, Viehweg and Zuo; see [43], [44] and [80], and the recent paper [4] by F. Andreatta. These results support the Expectation 4.2.

4.3. Results of Ciliberto, van der Geer and Teixidor i Bigas

In [15] and [14], Ciliberto, van der Geer and Teixidor i Bigas have obtained some interesting results about the number of moduli of curves whose Jacobians have nontrivial endomorphisms. As we shall discuss below, when combined with the results of Hain and de Jong-Zhang, these results can be used to obtain some restrictions on the special subvarieties of PEL type that are contained in the Torelli locus.

As always in this section we work over $\mathbb{C}$.

Theorem 4.11 (Ciliberto, van der Geer and Teixidor i Bigas, [15]). *Let $Z \subset M_g$, for $g \geqslant 2$, be an irreducible closed subvariety. Let Ω be an algebraic closure of the function field $\mathbb{C}(Z)$, let C/Ω be the curve corresponding to the geometric generic point of Z, and* *assume that the Jacobian* J_C/Ω *has the property that* $\mathbb{Z} \subsetneq \mathrm{End}(J_C)$. *Then* $\dim(Z) \leqslant 2g-2$, *and the intersection of* Z *with the hyperelliptic locus in* M_g *has dimension at most* g.

Note that in this theorem it is not assumed that the image of Z in A_g is a special subvariety.

In addition to the result as quoted here, the authors have some finer results about the case when $\dim(Z) \geqslant 2g-3$. For instance, they show that if $\dim(Z) = 2g-2$ then either C is a cover of a non-constant elliptic curve E over Ω, or $g = 2$ and $\mathrm{End}^0(J_C)$ is a real quadratic field. (The case $\dim(Z) = 2g-3$ is analyzed by Ciliberto and van der Geer in [14].)

4.4. Excluding certain special subvarieties of PEL type

One may use the results we have discussed to obtain restrictions on the special subvarieties $Z \subset A_g$ of PEL type that can be contained in T_g with $Z \cap T_g^\circ \neq \emptyset$. (We do not, however, see a way to apply such arguments to arbitrary special subvarieties.) Our result is as follows.

Theorem 4.12. *Consider a a special subvariety* $S \subset A_g$ *of PEL type, arising from PEL data* $(D, *, V, \phi)$ *as in* Remark 3.10, *with* $\dim_{\mathbb{Q}}(V) = 2g$.

(i) *Suppose* $D = F$ *is a totally real field. (Albert Type I.) If* $g > 4$ *then* S *is not contained in* T_g.

(ii) *Suppose* D *is a quaternion algebra over a totally real field* F *that splits at all infinite places of* F. *(Albert Type II.) If* $g > 8$ *then* S *is not contained in* T_g.

For Albert's classification of the possible endomorphism algebras of abelian varieties we refer to [55], Section 21.

Proof. In both cases it follows from the results of [76] that D equals the endomorphism algebra of the abelian variety corresponding to the geometric generic point of S. Hence the generic abelian variety in our family is geometrically simple; in particular, if $S \subset T_g$ then $S \cap T_g^\circ \neq \emptyset$.

First suppose $D = F$ is a totally real field. Let $e = [F : \mathbb{Q}]$, which is an integer dividing g. Then S arises as a quotient of a product of e copies of the Siegel space $\mathfrak{H}_h$ with $h = g/e$. This gives

$$\dim(S) = e \cdot \frac{h(h+1)}{2} = \frac{g(g+e)}{2e}.$$

Theorem 4.11 gives the inequality $g(g+e)/2e \leqslant 2g-2$, so we find

$$e \geqslant \frac{g^2}{3g-4} > \frac{g}{3}.$$

As e divides g, either g is even and $e = g/2$ or $e = g$. However, for $g > 4$ the case $e = g$ is excluded by Corollary 4.10. Assume then $g > 4$ is even and $e = g/2$; in this case $\dim(S) = 3g/2$. The assumptions in Hain's theorem 4.8 are satisfied and S is not a ball quotient. Hence we obtain that the intersection with the hyperelliptic locus is a nonempty divisor and therefore has dimension $(3g/2)-1$. This contradicts Theorem 4.11.

Next we consider the case where D is a quaternion algebra over a totally real field F such that $D \otimes_{\mathbb{Q}} \mathbb{R}$ is isomorphic to a product of $e = [F : \mathbb{Q}]$ copies of $M_2(\mathbb{R})$. In this case $2e$ divides g and S is a quotient of a product of e copies of $\mathfrak{H}_h$ with $h = g/2e$; so $\dim(S) = e \cdot h(h+1)/2$ and S is not a ball quotient. The boundary in the Baily-Borel compactification has codimension $g/2 > 2$. By Theorem 4.9 it therefore suffices to show that S^{dec} cannot have codimension $\leqslant 2$.

There are two possibilities for the geometric generic point of a component of S^{dec}. Either the corresponding abelian variety is isogenous to a product $X_1 \times X_2$, where X_1 and X_2 both have an action by an order in D. Straightforward calculation gives that in this case the codimension is at least $e \cdot (h-1)$. Using the relation $g = 2eh$ we find that for $g > 8$ the codimension is > 2. The other possibility is that there is a subfield $F' \subset F$ and a quaternion algebra D' with center F' such that $D \cong F \otimes_{F'} D'$. With $e' = [F' : \mathbb{Q}]$ the codimension in this case equals $(e-e') \cdot h(h+1)/2$. Straightforward checking of the possibilities shows that for $g > 8$ this is greater than 2. □

It seems that similar arguments will also work for the Albert Type III, and even for Type IV we expect that we can obtain some non-trivial conclusions. We have not yet pursued this.

4.5. The Schottky problem

Expectation 4.2 is of course intimately related to the Schottky problem, which is the problem of characterizing which ppav (A, λ) are Jacobians of curves. There are several solutions or conjectural solutions of this problem. We refer to the overview papers [7], [17], [31] for an introduction to this beautiful topic. It seems that none of the (conjectural) solutions discussed in these papers can be directly applied to problems such as Conjecture 4.1 or Expectation 4.2. Though for both sides—algebraic curves and their moduli on the one hand, abelian varieties and special subvarieties in the moduli space on the other hand—we have many techniques and results at our disposal, it is difficult to find a language, or a set of techniques, using which both sides can be described simultaneously. This is a difficulty that we believe lies at the heart of the matter.

5. Examples of special subvarieties in the Torelli locus

In this section we discuss examples of special subvarieties $S \subset T_g$ of positive dimension, with $g \geqslant 4$ and $S \cap T_g^{\circ} \neq \emptyset$. The examples we consider arise from families of cyclic covers of $\mathbb{P}^1$. Throughout this section we work over $\mathbb{C}$.

We first look at a concrete example, following [35]. (The example was already given in [77].)

Example 5.1. Consider the family of curves given by

$$y^5 = x(x-1)(x-t),$$

where $t \in \mathbb{C} \setminus \{0,1\}$ is a parameter. The complete and regular model C_t of this curve is a cyclic cover of $\mathbb{P}^1$ with group μ_5, with total ramification above 0, 1, t and ∞. The Hurwitz formula gives $\chi = 5 \cdot -2 + 4 \cdot (5-1) = 6$ so $g(C_t) = 4$. In the moduli space A_4 the corresponding family of Jacobians gives a 1-dimensional subvariety, whose closure we call $Z \subset A_4$. Clearly these Jacobians admit an action by $\mathbb{Z}[\zeta_5]$, the ring of integers of the cyclotomic field $F := \mathbb{Q}[\zeta_5]$.

Consider the special subvariety $S \subset A_4$ containing Z that is defined by this action of $\mathbb{Z}[\zeta_5]$. More formally, fix a base point $b \in \mathbb{C} \setminus \{0,1\}$ and choose a symplectic similitude $s\colon H_1(C_b, \mathbb{Z}) \xrightarrow{\sim} V$ as in subsection 3.2. Via s, the Hodge structure on $H_1(C_b, \mathbb{Q})$ corresponds to a point $y \in \mathfrak{H}_4$ and we obtain the structure of an F-vector space on $V_\mathbb{Q}$. Let $M \subset G_\mathbb{Q} = \mathrm{CSp}(V, \phi)_\mathbb{Q}$ be the algebraic subgroup obtained as the intersection of G with $\mathrm{GL}_F(V_\mathbb{Q})$. With notation as in Def. 3.6, y lies in Y_M. If we choose the base point b such that the Jacobian J_b is not of CM type (which is certainly possible) then there is a unique connected component $Y^+ \subset Y_M$ containing y, and $S \subset A_4$ is the special subvariety obtained as the image of this Y^+ under the quotient map $\mathfrak{H}_4 \to \mathrm{CSp}(V, \phi)\backslash\mathfrak{H}_4 \xrightarrow{\sim} A_4(\mathbb{C})$.

As we shall show, $\dim(S) = 1$. Assuming we know this, we conclude that Z, which is also 1-dimensional and is contained in S, contains an open dense subset of S, in which case it follows from property (b) in subsection 3.4 that there are infinitely many values of t such that the Jacobian of C_t is of CM type.

In order to calculate the dimension of S, we need to know how F acts on the tangent space of the Jacobian J_t at the origin. More precisely, $T_0(J_t)$ has the structure of a module over the ring $F \otimes_\mathbb{Q} \mathbb{C} = \prod_{\sigma\colon F \to \mathbb{C}} \mathbb{C}$. Hence we obtain a direct sum decomposition $T_0(J_t) = \oplus_\sigma T(\sigma)$. Let n_σ denote the $\mathbb{C}$-dimension of $T(\sigma)$. These multiplicities n_σ do not depend on t. By using the polarization, one can show that $n_\sigma + n_{\bar\sigma} = 2g/\varphi(5) = 2$ for all σ, where $\bar\sigma$ denotes the complex conjugate of σ. With this notation we have $H_\mathbb{R}^{\mathrm{der}} \cong \prod \mathrm{SU}(n_\sigma, n_{\bar\sigma})$ and $\dim(S) = \sum n_\sigma n_{\bar\sigma}$, where the sum is taken over a set of representatives of the complex embeddings of F modulo complex conjugation. See [76], Theorem 5, and see below for further details.

As $T_0(J_t)$ is canonically dual to $H^0(C_t, \omega)$, we can calculate the multiplicities n_σ by writing down a basis of regular differentials on C_t. (In fact, as we shall explain below the Chevalley-Weil formula gives a much quicker method.) In the example at hand, if P_i, for $i \in \{0, 1, t, \infty\}$, is the unique point above i, we have

$$\mathrm{div}(x) = 5P_0 - 5P_\infty, \quad \mathrm{div}(y) = P_0 + P_1 + P_t - 3P_\infty,$$

and

$$\mathrm{div}(dx) = 4P_0 + 4P_1 + 4P_t - 6P_\infty.$$

As a basis of regular differentials we find

$$\frac{dx}{y^2}, \quad \frac{dx}{y^3}, \quad \frac{dx}{y^4}, \quad \frac{x\,dx}{y^4}.$$

The weights of the first two are dual, whereas the last two forms have the same weight. The conclusion, then, is that there is one pair $(\sigma, \bar{\sigma})$ with multiplicities $(2,0)$, and one pair where the multiplicities are $(1,1)$. Hence $\dim(S) = 1 \cdot 1 + 2 \cdot 0 = 1$, and we conclude that $S = Z$ is a special subvariety contained in T_4 with $S \cap T_4^\circ \neq \emptyset$. In particular this proves that Coleman's conjecture 4.1 does not hold for $g = 4$.

Remark 5.2. In the above example, we do not know a method to decide for which values of the parameter t the Jacobian of C_t is a CM abelian variety. The same remark applies to the examples we shall discuss next.

5.1. Families of cyclic covers

To obtain further such examples, the idea is to fix an integer $m \geqslant 2$, an integer $N \geqslant 4$, and monodromy elements $a_1, \ldots, a_N$ in $\mathbb{Z}/m\mathbb{Z}$; then we consider cyclic covers of $\mathbb{P}^1$ with group μ_m, branch points $t_1, \ldots, t_N$ in $\mathbb{P}^1$ and local monodromy $\exp(2\pi i a_j/m) \in \mu_m$ about t_j. If the branch points are all in $\mathbb{A}^1$, this cover is given by the affine equation

$$y^m = (x - t_1)^{a_1}(x - t_2)^{a_2} \cdots (x - t_N)^{a_N}, \tag{5.3}$$

with $\zeta \in \mu_m$ acting by $(x, y) \mapsto (x, \zeta \cdot y)$. Varying the branch points t_i gives us a family of curves, and it turns out that for certain choices of the data involved, the corresponding family of Jacobians traces out a special subvariety in A_g.

Write $a = (a_1, \ldots, a_N)$. The triple (m, N, a) that serves as input for our construction has to satisfy some conditions. As already indicated, we want $m \geqslant 2$, as taking $m = 1$ is clearly of no interest. Next, we require that $N \geqslant 4$. The reason is that from our family we obtain an $(N-3)$-dimensional subvariety in the moduli space (see below), and we are interested in special subvarieties of positive dimension. We assume that $a_i \not\equiv 0 \pmod{m}$ for all i, and that $a_1 + \cdots + a_N \equiv 0 \pmod{m}$. This means that the t_i are branch points and that there are no further branch points. Further we need to assume that the elements a_i generate the group $\mathbb{Z}/m\mathbb{Z}$ for otherwise the (smooth projective) curves given by (5.3) are reducible.

We can find an open subscheme $T \subset (\mathbb{P}^1)^N$, disjoint from the big diagonals, and a smooth proper curve $f\colon C \to T$ such that the fiber of f at a point $t = (t_1, \ldots, t_N)$ in $T(\mathbb{C})$ with all t_i in $\mathbb{A}^1(\mathbb{C})$ is the complete regular curve given by (5.3). The genus of these curves is given by

$$g = 1 + \frac{(N-2)m - \sum_{i=1}^{N} \gcd(a_i, m)}{2}. \tag{5.4}$$

The relative Jacobian $J \to T$ then defines a morphism $\tau\colon T \to A_g$, and we define $Z(m, N, a) \subset A_g$ as the closure of the image of τ. Note that since we are only interested in the closure of $\tau(T)$ there is no need to specify exactly which open subscheme $T \subset (\mathbb{A}^1)^N$ we choose; see Remark 3.13.

We call two triples (m, N, a) and (m', N', a') equivalent if $m = m'$ and $N = N'$ and if the classes of a and a' in $(\mathbb{Z}/m\mathbb{Z})^N$ are in the same orbit under $(\mathbb{Z}/m\mathbb{Z})^* \times \mathfrak{S}_N$. Here $(\mathbb{Z}/m\mathbb{Z})^*$ acts diagonally on $(\mathbb{Z}/m\mathbb{Z})^N$ by multiplication, and the symmetric group $\mathfrak{S}_N$ acts by permutation of the indices. The closed subvariety $Z(m, N, a) \subset A_g$ only depends on the equivalence class of the triple (m, N, a).

The morphism $\tau\colon T \to A_g$ factors through the quotient of T modulo the action of the group $PGL_2(\mathbb{C}) \times \mathfrak{S}_N$, with $PGL_2(\mathbb{C}) = \mathrm{Aut}(\mathbb{P}^1)$ acting diagonally on $(\mathbb{P}^1)^N$ and $\mathfrak{S}_N$ acting by permutation of the diagonals. (Without loss of generality we may assume $T \subset (\mathbb{P}^1)^N$ is stable under $PGL_2(\mathbb{C}) \times \mathfrak{S}_N$.) The subvariety $Z(m, N, a) \subset A_g$ has dimension $N - 3$. Note that we may also fix three of the branch points to lie at 0, 1 and ∞, as we did in Example 5.1; this has the effect of replacing T with a closed subvariety of dimension $N - 3$ on which the morphism ϕ is generically finite. For instance, the example considered in 5.1 corresponds to the triple $(m, N, a) = (5,4,(1,1,1,2))$.

The Jacobians J_t in our family come equipped with an action of the group ring $\mathbb{Z}[\mu_m]$. Let $S(\mu_m) \subset A_g$ be the special subvariety containing $Z(m, N, a)$ that is defined by this action. (In order to make this precise we follow the recipe given in Example 5.1.) In all cases we have the inequality

$$N - 3 \leqslant \dim S(\mu_m), \tag{5.5}$$

and if we have $N - 3 = \dim S(\mu_m)$ then the conclusion is that $Z(m, N, a) = S(\mu_m)$ is a special subvariety in the Torelli locus that meets the open Torelli locus. (If $N - 3 < \dim S(\mu_m)$ then a priori we know nothing; see subsection 5.4 below.) The question is therefore how to calculate the dimension of $S(\mu_m)$. Before we discuss this in general, let us look at another example.

Example 5.6. Consider the family of curves given by

$$y^9 = x(x-1)(x-t).$$

With notation as above we have $m = 9$ and $N = 4$, with local monodromy about the branch points given by $a = (1,1,1,6)$.

The complete regular model C_t is a cyclic cover of $\mathbb{P}^1$ with group μ_9. The points 0, 1 and t are totally ramified, so we have unique points P_0, P_1, and P_t above them. There are three points $P_\infty^{(1)}$, $P_\infty^{(2)}$ and $P_\infty^{(3)}$ above ∞, each with ramification index 3. The Hurwitz formula gives $\chi = 9 \cdot -2 + 3 \cdot 8 + 3 \cdot 2 = 12$ so $g = 7$, which agrees with (5.4).

The Jacobian J_t contains as an isogeny factor the Jacobian of the curve C'_t given by $u^3 = x(x-1)(x-t)$. By a similar calculation, C'_t has genus 1. Its Jacobian J'_t is an elliptic curve with complex multiplication by $\mathbb{Z}[\zeta_3]$, and the family of Jacobians J'_t is isotrivial over $\mathbb{P}^1 \setminus \{0,1\}$. (Note that in the given equation for C'_t there is no ramification over ∞, so C'_t is geometrically isomorphic to the curve given by $s^3 = v(v-1)$. More explicitly, let $\tilde{T} \to \mathbb{P}^1 \setminus \{0,1\}$ be the cyclic cover of degree 3 given by the equation $\mu^3 = t(t-1)$. After base change to $\tilde{T}$ the family of curves C'_t becomes isomorphic to the constant curve defined by $s^3 = v(v-1)$ via the isomorphism given by $s = \mu u/(x-t)$ and $v = (1-t)x/(x-t)$.)

Let J_t^{new} (the "new part") be the quotient of J_t modulo J'_t; this is an abelian variety of dimension 6 on which we have an action by $\mathbb{Z}[\zeta_9]$. In order to determine the dimension of the special subvariety in A_6 given by this action, we again calculate the multiplicities of the action on the tangent space. In this case, we already know in advance that there will be one non-primitive character of μ_9 occurring in $H^0(C_t, \omega)$. We have

$$\operatorname{div}(x) = 9P_0 - 3P_\infty^{(1)} - 3P_\infty^{(2)} - 3P_\infty^{(2)}\,, \quad \operatorname{div}(y) = P_0 + P_1 + P_t - P_\infty^{(1)} - P_\infty^{(2)} - P_\infty^{(3)}\,,$$

and

$$\operatorname{div}(dx) = 8P_0 + 8P_1 + 8P_t - 4P_\infty^{(1)} - 4P_\infty^{(2)} - 4P_\infty^{(3)}\,.$$

As a basis of regular differentials we find

$$\frac{dx}{y^4}\,, \quad \frac{dx}{y^5}\,, \quad \frac{dx}{y^6}\,, \quad \frac{dx}{y^7}\,, \quad \frac{dx}{y^8}\,, \quad \frac{x\,dx}{y^7}\,, \quad \frac{x\,dx}{y^8}\,.$$

As predicted, this gives one non-primitive character of μ_9 (corresponding to dx/y^6); for the rest we find one complex conjugate pair with multiplicities (1,1), and two complex conjugate pairs for which the multiplicities are (2,0). The special subvariety in A_6 given by the action of $\mathbb{Q}[\zeta_9]$ with this collection of multiplicities has dimension $1 \cdot 1 + 2 \cdot 0 + 2 \cdot 0 = 1$. It follows that the J_t^{new} trace out a dense open subset of this special subvariety. As the original Jacobians J_t are isogenous to $J'_t \times J_t^{\mathrm{new}}$ with J'_t an isotrivial family with fibers of CM type, we conclude that the family of J_t traces out in A_7 a (dense subset of a) special subvariety. As before, this implies that there are infinitely many values of t for which J_t is of CM type. (The t for which this happens are even analytically dense in $\mathbb{P}^1(\mathbb{C})$.)

Remark 5.7. Let (M, Y) be a Shimura datum. The dimension of the associated Shimura varieties $\mathrm{Sh}_K(M, Y)$ only depends on the structure of the real adjoint group $M_{\mathbb{R}}^{\mathrm{ad}}$. Indeed, the dimension of the Shimura variety equals the complex dimension of the space Y, and the components of Y can be described as hermitian symmetric domains associated with the connected Lie group $M^{\mathrm{ad}}(\mathbb{R})^+$, where the superscript "+" denotes the identity component for the analytic topology. In particular, if $M_{\mathbb{R}}^{\mathrm{ad}} \cong Q_1 \times \cdots \times Q_l$ is the decomposition of $M_{\mathbb{R}}^{\mathrm{ad}}$ as a product of $\mathbb{R}$-simple groups, the dimension of $\mathrm{Sh}_K(M, Y)$ can be calculated as a sum $\delta(Q_1) + \cdots + \delta(Q_l)$, where the contribution $\delta(Q_i)$ of the factor Q_i can be looked up in [33], Table V. (Caution: the dimensions given there are the real dimensions.) The cases most relevant for our discussion are the following.

(1) $\delta(Q_i) = 0$ if Q_i is anisotropic, i.e., if $Q_i(\mathbb{R})$ is compact;

(2) $\delta(Q_i) = h(h+1)/2$ if $Q_i \cong \mathrm{PSp}_{2h,\mathbb{R}}$;
(3) $\delta(Q_i) = pq$ if $Q_i \cong \mathrm{PSU}(p,q)$.

5.2. Calculating the dimension of $S(\mu_m)$

We return to the general situation of a family of curves $f\colon C \to T$ associated with the data (m, N, a). Our next goal is to calculate the dimension of the special subvariety $S(\mu_m) \subset A_g$ that contains $Z = Z(m, N, a)$. Choose a Hodge-generic base point $b \in T(\mathbb{C})$, choose a symplectic similitude $s\colon H_1(C_b, \mathbb{Z}) \xrightarrow{\sim} V$ as in subsection 3.2, and let $y \in \mathfrak{H}_g$ be the point corresponding to the Hodge structure on $H_1(C_b, \mathbb{Z})$, viewed as a Hodge structure on V via s.

The Jacobian $J \to T$ comes equipped with an action of the group ring $\mathbb{Z}[\mu_m]$. The Hodge classes we want to have over $S(\mu_m)$ are these endomorphisms, which means we are in the situation of Example 3.9. The algebraic group $M \subset \mathrm{CSp}(V_\mathbb{Q}, \phi)$ we need to consider is given by

$$M := \mathrm{CSp}(V_\mathbb{Q}, \phi) \cap \mathrm{GL}_{\mathbb{Q}[\mu_m]}(V_\mathbb{Q})\,.$$

We can calculate the dimension of $S(\mu_m)$ via a deformation argument. In other words, if $Y_M \subset \mathfrak{H}_g$ is as defined just before Definition 3.6, we calculate the dimension of the tangent space of Y_M at te point y.

The Hodge structure on $V_\mathbb{Q}$ is of type $(-1,0) + (0,-1)$, and the polarization gives us a symplectic form $\phi\colon V_\mathbb{Q} \times V_\mathbb{Q} \to \mathbb{Q}(1)$. As discussed in Remark 3.2, the Hodge filtration $\mathrm{Fil}^0 = V_\mathbb{C}^{0,-1} \subset V_\mathbb{C}$ is a Lagrangian subspace, so we have an induced isomorphism $\bar{\phi}\colon \mathrm{Fil}^0 \xrightarrow{\sim} V_\mathbb{C}/\mathrm{Fil}^0$. The tangent space of $\mathfrak{H}_g$ at the point y is given by (3.3).

The extra structure we now have is an action of the algebra $D := \mathbb{Q}[\mu_m]$ on $V_\mathbb{Q}$. Let $d \mapsto \bar{d}$ denote the involution of D induced by the inversion in the group μ_m. The form ϕ has the property that $\phi(dv, v') = \phi(v, \bar{d}v')$ for all $d \in D$ and $v, v' \in V_\mathbb{Q}$.

The D-action on $V_\mathbb{Q}$ induces on $V_\mathbb{C}^{-1,0}$ and $V_\mathbb{C}^{0,-1}$ the structure of a module over the ring

$$D \otimes_\mathbb{Q} \mathbb{C} = \prod_{n \in \mathbb{Z}/m\mathbb{Z}} \mathbb{C}\,.$$

Correspondingly, we have decompositions

$$V_\mathbb{C}/\mathrm{Fil}^0 = V_\mathbb{C}^{-1,0} = \bigoplus_{n \in \mathbb{Z}/m\mathbb{Z}} V_{\mathbb{C},(n)}^{-1,0} \quad \text{and} \quad \mathrm{Fil}^0 = V_\mathbb{C}^{0,-1} = \bigoplus_{n \in \mathbb{Z}/m\mathbb{Z}} V_{\mathbb{C},(n)}^{0,-1}\,.$$

The involution of $D_\mathbb{C}$ obtained by linear extension of the involution $d \mapsto \bar{d}$ exchanges the factors $\mathbb{C}$ indexed by the classes n and $-n$. It follows that the perfect pairing $\bar{\phi}\colon \mathrm{Fil}^0 \times (V_\mathbb{C}/\mathrm{Fil}^0) \to \mathbb{C}$ restricts to perfect pairings

$$\bar{\phi}_n\colon V_{\mathbb{C},(n)}^{0,-1} \times V_{\mathbb{C},(-n)}^{-1,0} \to \mathbb{C}\,.$$

For $n \in \mathbb{Z}/m\mathbb{Z}$, define

$$d_n := \dim_\mathbb{C} V_{\mathbb{C},(n)}^{0,-1}\,.$$

Note that $d_{(0 \bmod m)} = 0$, i.e., $V_{\mathbb{C},(0 \bmod m)} = (0)$, as the μ_m-invariant subspace in V is the first homology of the base curve $\mathbb{P}^1$, which is zero. We shall see in Proposition 5.9 below how to calculate the d_n in terms of the given data (m, N, a).

The tangent space $T_y(Y_M) \subset T_y(\mathfrak{H}_g)$ is the $\mathbb{C}$-subspace consisting of those elements $\beta \in \mathrm{Hom}^{\mathrm{sym}}(\mathrm{Fil}^0, V_{\mathbb{C}}/\mathrm{Fil}^0)$ that are $D_{\mathbb{C}}$-linear. Any such β can be written as $\beta = \sum \beta_n$, where the $\beta_n\colon V^{0,-1}_{\mathbb{C},(n)} \to V^{-1,0}_{\mathbb{C},(n)}$ are $\mathbb{C}$-linear maps that satisfy

$$\bar{\phi}_n(v, \beta_{-n}(v')) = \bar{\phi}_{-n}(v', \beta_n(v)) \quad \text{for all } v \in V^{0,-1}_{\mathbb{C},(n)} \text{ and } v' \in V^{0,-1}_{\mathbb{C},(-n)}. \tag{5.8}$$

If $n \not\equiv -n \pmod m$, this last condition gives a duality between β_{-n} and β_n; this means the linear map β_n can be chosen arbitrarily and β_{-n} is determined by β_n. The situation is different if $n \equiv -n \pmod m$. Of course, this only occurs (with $n \not\equiv 0$) if $m = 2k$ is even. In this case we have a perfect pairing $\bar{\phi}_k\colon V^{0,-1}_{\mathbb{C},(k)} \times V^{-1,0}_{\mathbb{C},(k)} \to \mathbb{C}$, and (5.8) gives

$$\begin{aligned}\beta_k \in \mathrm{Hom}^{\mathrm{sym}}&\left(V^{0,-1}_{\mathbb{C},(k)}, V^{-1,0}_{\mathbb{C},(k)}\right)\\ &:= \left\{\beta\colon V^{0,-1}_{\mathbb{C},(k)} \to V^{-1,0}_{\mathbb{C},(k)} \mid \bar{\phi}_k(v, \beta_k(v')) = \bar{\phi}_k(v', \beta_k(v)) \text{ for all } v, v' \in V^{0,-1}_{\mathbb{C},(k)}\right\}.\end{aligned}$$

(Cf. (3.3).) We find that the dimension of $T_y(Y_M)$ equals

$$\sum_{\substack{\pm n \in (\mathbb{Z}/m\mathbb{Z})/\{\pm 1\} \\ 2n \not\equiv 0 \pmod m}} d_{-n} d_n + \begin{cases} \frac{d_k \cdot (d_k+1)}{2} & \text{if } m = 2k \text{ is even;} \\ 0 & \text{if } m \text{ is odd.} \end{cases}$$

As the final step in the calculation we need to calculate the dimensions d_n in terms of the given triple (m, N, a). The result is as follows.

Proposition 5.9 (Hurwitz, Chevalley-Weil). *The dimensions* $d_n := \dim_{\mathbb{C}} V^{0,-1}_{\mathbb{C},(n)}$ *are given by* $d_n = 0$ *if* $n \equiv 0 \pmod m$ *and*

$$d_n = -1 + \sum_{i=1}^{N} \left\langle \frac{-n a_i}{m} \right\rangle \quad \text{if } n \not\equiv 0 \pmod m, \tag{5.10}$$

where $\langle x \rangle = x - \lfloor x \rfloor$ *denotes the fractional part of a number* x.

We note that $V^{0,-1}_{\mathbb{C}}$ is naturally isomorphic to the space $H^0(C_b, \Omega^1)$ of global differentials on the curve. The given formula is then a special case of a classical result by Chevalley and Weil about the structure of $H^0(C_b, \Omega^1)$ as a representation of the Galois group μ_m, which in the cyclic case is already due to Hurwitz. See [52] Section 3 for a modern proof. Another proof, using the holomorphic Lefschetz formula, can be found in [75], Lemma 1.6b.

Putting everything together, the dimension of $S(\mu_m)$ is given by the following result.

Proposition 5.11. *Consider a triple* (m, N, a) *as in subsection 5.1. Then the dimension of the special subvariety* $S(\mu_m)$ *that contains* $Z(m, N, a)$ *is given by*

$$\dim S(\mu_m) = \sum d_{-n} d_n + \begin{cases} \frac{d_k \cdot (d_k+1)}{2} & \text{if } m = 2k \text{ is even;} \\ 0 & \text{if } m \text{ is odd,} \end{cases}$$

where the sum runs over the pairs $\pm n$ *in* $\mathbb{Z}/m\mathbb{Z}$ *with* $2n \not\equiv 0 \pmod{m}$, *and where the* d_n *are given by* (5.10).

Remark 5.12. Instead of using a tangent space computation, we may calculate the dimension of $S(\mu_m)$ by analyzing the real adjoint group $M_{\mathbb{R}}$; see Remark 5.7 and [49]. It can be shown that

$$M_{\mathbb{R}}^{\mathrm{ad}} \cong \prod \mathrm{PSU}(d_n, d_{-n}) \times \begin{cases} \mathrm{PSp}_{2d_k,\mathbb{R}} & \text{if } m = 2k \text{ is even;} \\ \{1\} & \text{if } m \text{ is odd,} \end{cases}$$

where the first product runs over the pairs $\pm n$ in $\mathbb{Z}/m\mathbb{Z}$ with $2n \not\equiv 0 \pmod{m}$.

5.3. An inventory of known special subvarieties in the Torelli locus

Now that we have an explicit formula for the dimension of $S(\mu_m)$ in terms of the given data, it is easy to check, in each given example, if in the inequality (5.5) we have an equality. Recall that if $N - 3 = \dim S(\mu_m)$, we conclude that $Z(m, N, a)$ is dense in the special subvariety $S(\mu_m) \subset A_g$, in which case it follows that the family of Jacobians $J \to T$ (as in subsection 5.1) contains infinitely many different Jacobians of CM type.

In order to illustrate how well this works, let us redo Example 5.1. We have $(m, N, a) = (5, 4, (1, 1, 1, 2))$. The Hurwitz-Chevalley-Weil formula gives

$$d_1 = -1 + 3 \cdot \left\langle \frac{-1}{5} \right\rangle + \left\langle \frac{-2}{5} \right\rangle = 2,$$

and in a similar way we find

$$d_2 = 1, \quad d_3 = 1, \quad d_4 = 0,$$

which agrees with the basis of regular differentials we have found. Proposition 5.11 gives $\dim S(\mu_m) = 1 = N - 3$, and as before we conclude that $Z(m, N, a)$ is a special subvariety.

Using a computer it is not hard to do a systematic search for triples (m, N, a) with $N - 3 = \dim S(\mu_m)$. Up to equivalence (as defined in Section 5.1), one finds twenty such triples. They are listed in Table 1. Moreover, it was proven by Rohde in [74] that these are the only triples, up to equivalence, for which $N - 3 = \dim S(\mu_m)$.

5.4. Excluding further examples

In the discussion so far, the argument is based on the fact that $Z(m, N, a) \subset A_g$ is visibly contained in the special subvariety $S(\mu_m)$, and we look for examples where the two have the same dimension. If in some given case we find that $\dim Z(m, N, a) =$

Table 1. Examples of special subvarieties in the Torelli locus

	genus	m	N	a
(1)	1	2	4	(1,1,1,1)
(2)	2	2	6	(1,1,1,1,1,1)
(3)	2	3	4	(1,1,2,2)
(4)	2	4	4	(1,2,2,3)
(5)	2	6	4	(2,3,3,4)
(6)	3	3	5	(1,1,1,1,2)
(7)	3	4	4	(1,1,1,1)
(8)	3	4	5	(1,1,2,2,2)
(9)	3	6	4	(1,3,4,4)
(10)	4	3	6	(1,1,1,1,1,1)

	genus	m	N	a
(11)	4	5	4	(1,3,3,3)
(12)	4	6	4	(1,1,1,3)
(13)	4	6	4	(1,1,2,2)
(14)	4	6	5	(2,2,2,3,3)
(15)	5	8	4	(2,4,5,5)
(16)	6	5	5	(2,2,2,2,2)
(17)	6	7	4	(2,4,4,4)
(18)	6	10	4	(3,5,6,6)
(19)	7	9	4	(3,5,5,5)
(20)	7	12	4	(4,6,7,7)

$N - 3 < \dim S(\mu_m)$, this does not, a priori, imply that $Z(m, N, a)$ is not special. Put differently, in addition to the endomorphisms in $\mathbb{Z}[\mu_m]$, there might be Hodge classes in our family of Jacobians $J \to T$ that we just happen not to see. Even in individual examples, it is usually not so easy to exclude this.

One method to do this was given by de Jong and Noot in [35], Section 5; they prove there that for $m > 7$ not divisible by 3 the family of curves given by $y^m = x(x-1)(x-t)$, which in our language corresponds to the triple $(m, N, a) = (m, 4, (1, 1, 1, m-3))$, does *not* give a special subvariety. The method is based on results of Dwork and Ogus in [25].

Extending this to arbitrary families, it was proven in [49] that the twenty examples we have found are the only ones such that the image $Z(m, N, a) \subset A_g$ is a special subvariety.

Theorem 5.13. *Consider data* (m, N, a) *as in subsection* 5.1. *Then* $Z(m, N, a) \subset A_g$ *is a special subvariety if and only if* (m, N, a) *is equivalent to one of the twenty triples listed in* Table 1.

Remark 5.14. Examples (6), (8), (10), (11), (16) and (17) in Table 1 were given by Shimura in [77]. Examples (10), (11) and (17) were given, in a language that is closer to the present paper, by de Jong and Noot in [35]; they also explained the relevance of such examples for Coleman's conjecture. (The fact that these examples already occurred in [77] was recognized only later. In connection with this, note that [77] appeared more than twenty years before Coleman stated his conjecture.)

In retrospect, it is surprising that the complete list of examples was obtained only recently. Example (19) was found by one of us in 2003; see [65]. All twenty examples were found by Rohde in [74] (Example (2) is somewhat hidden there, but see op. cit. Corollary 5.5.2) and independently by one of us, with the help of a small computer program. See 6.7 below for some further examples, coming from families of covers with a non-cyclic Galois group.

It should be mentioned that there is some connection between these examples and the theory of Deligne and Mostow in [23], [53]; see also Looijenga's overview paper [40]. The relation is that, in the setting of subsection 5.1, the family of Jacobians $J \to T$ has a "new part" (cf. Example 5.6), and that the monodromy of the corresponding VHS can be described as the monodromy on a space of hypergeometric functions. In this context there is an easy criterion for the arithmeticity of the monodromy group; see [23], Proposition 12.7 or [40], Theorem 4.3. As far as we know there is, however, no easy way to obtain from this, and the resulting tables in [23] and [53], a classification result such as in Theorem 5.13. Apart from the fact that the Deligne-Mostow arithmeticity criterion only applies under some condition on the local monodromy elements (condition (INT) of [23], p. 25) that in our situation is not always satisfied, it only gives us information about certain direct summands of the VHS that we want to study, and as already explained in Remark 3.16 we in general need more.

6. Some questions

In this section, with the exception of subsections 6.2 and 6.3, the base field is $\mathbb{C}$. Most questions below could also be formulated over an algebraic closure of $\mathbb{Q}$.

Question 6.1. *How do we construct curves for which the Jacobian is a CM abelian variety?* If we have a special subvariety in T_g that intersects T_g°, the existence of CM Jacobians follows from property (b) in subsection 3.4. Note that in such cases we typically have no control over which fibers in our family are the CM fibers; cf. Remark 5.2.

On the other hand, we could look for curves that have many automorphisms. For instance, the Fermat curves $X^n + Y^n + Z^n = 0$ in $\mathbb{P}^2$ have Jacobians of CM type. Similarly, cyclic covers of $\mathbb{P}^1$ with 3 branch points have CM Jacobians. For a given $g \geqslant 2$ there are only finitely many such covers of genus g, however. Further such examples are given in [66], 5.15.

In [66], a curve C is called a curve *with many automorphisms* if the deformation functor of the pair $(C, \mathrm{Aut}(C))$ is a 0-dimensional scheme. In many cases such a curve has a CM Jacobian.

These are the only methods known to us to construct, or to prove the existence of, CM Jacobians.

Question 6.2. *Do we know the existence of, or can we construct, a curve* C *of genus at least* 4 *such that* $\mathrm{Aut}(C) = \{\mathrm{id}\}$ *and such that the Jacobian of* C *is an abelian variety of CM type?*

Question 6.3. *Does there exist* $g > 3$ *and a special subvariety* $Z \subset \mathcal{A}_g$ *contained in the Torelli locus* T_g *such that the geometric generic fiber over* T *gives an abelian variety with endomorphism ring equal to* $\mathbb{Z}$ *?* We do not know a single example.

Remark 6.4. In 4.2 we have seen an expectation about the (non-)existence of special subvarieties in the Torelli locus. In order to make the question more precise, consider, for $g \in \mathbb{Z}_{>0}$ the set

$$\mathcal{ST}(g) := \{\text{special subvarieties } Z \subset T_g \text{ with } \dim(Z) > 0 \text{ and } Z \cap T_g^\circ \neq \emptyset\}$$

of special subvarieties of positive dimension, contained in the Torelli locus, and not fully contained in the boundary of T_g. The expectation is that for $g \gg 0$ we have $\mathcal{ST}(g) = \emptyset$; see 4.2.

We would like to classify all pairs (g, Z) with $Z \in \mathcal{ST}(g)$. For $g = 2$ and $g = 3$ we have $T_g = A_g$ and in this case every special subvariety of A_g is of PEL type; see [51]. Hence in this case we can classify all pairs (g, Z), up to Hecke translation, by listing all possible endomorphism algebras. We know that $\mathcal{ST}(g) \neq \emptyset$ for all $g < 8$. However, already for $g = 4$ we do not have a good description of $\mathcal{ST}(4)$. It seems very difficult to describe $\mathcal{ST}(g)$ for arbitrary g.

6.1. Non-PEL Shimura curves for $g = 4$

In [54], § 4, Mumford shows there exist 1-dimensional special subvarieties $Z \subset A_4$ that are not of PEL type. The abelian variety corresponding to the geometric generic fiber of Z has endomorphism algebra $\mathbb{Z}$. The curves Z are complete. Note that $T_4 \subset A_4$ is a closed subvariety of codimension one, which by a result of Igusa [34] is ample as a divisor. Hence we see that $Z \cap T_4 \neq \emptyset$.

Question 6.5. *Is there a "Mumford curve"* $Z \subset A_4$ *that is contained in* T_4 *?*

If $Z \subset A_4$ is a 1-dimensional special subvariety of the type constructed by Mumford, the abelian variety corresponding to the geometric generic point of Z has endomorphism ring $\mathbb{Z}$; hence if $Z \subset T_4$ then Z meets T_4°.

The examples constructed by Mumford, and generalizations thereof, have been studied in detail by Noot in [58] and [59]. In particular, [59], Section 3 contains a detailed analysis of the possible CM points on the special curves $Z \subset A_4$ constructed by Mumford. In particular, it is shown there that the (geometric) CM fibers are either absolutely simple or are isogenous to a product $E \times Y$ with E an elliptic curve, Y an abelian threefold, and $\mathrm{End}^0(E)$ isomorphic to a subfield of $\mathrm{End}^0(Y)$. It might be possible to use this to get some non-trivial information related to the above question. In connection with what was discussed in subsection 4.1, let us note that there are no Weyl type CM fibers in these families.

Question 6.6. *For which* $g \geqslant 2$ *does there exist a positive dimensional subvariety* $Z \subset T_g$ *with* $Z \cap T_g^\circ \neq \emptyset$*, such that the abelian variety corresponding with the geometric generic point of* Z *is isogenous to a product of elliptic curves?*

This question was stimulated by results in [39], which say that under more restrictive conditions such a family does not exist for large g.

Question 6.7. In Section 5 we have seen examples of special subvarieties $Z \subset T_g$ with $Z \cap T_g^\circ \neq \emptyset$ arising from families of cyclic covers of $\mathbb{P}^1$. *Can one obtain further such examples by taking non-cyclic covers, or from a family of covers of another base curve?*

To make this more precise, consider a (complete, nonsingular) curve B over $\mathbb{C}$ of genus h, and let $\mathcal{G}$ be a finite group. Define a group $\Pi = \Pi(h, N)$ by

$$\Pi := \left\langle \alpha_1, \ldots, \alpha_h, \beta_1, \ldots, \beta_h, \gamma_1, \ldots, \gamma_N \;\middle|\; [\alpha_1, \beta_1] \cdots [\alpha_h, \beta_h] \cdot \gamma_1 \cdots \gamma_N = 1 \right\rangle .$$

Given $t_1, \ldots, t_N$ in B, fix a presentation

$$\pi_1\big(B \setminus \{t_1, \ldots, t_N\}\big) \xrightarrow{\sim} \Pi ,$$

and fix a surjective homomorphism $\psi\colon \Pi \twoheadrightarrow \mathcal{G}$ such that $\psi(\gamma_i) \neq 1$ for all $i = 1, \ldots, N$. Correspondingly, we have a Galois cover $C_t \to B$ with group $\mathcal{G}$, branch points $t_1, \ldots, t_N$ in B, and with local monodromy about t_i given by the element $\psi(\gamma_i)$. Varying the branch points, we get a family of curves $C \to T$, for some open $T \subset B^N$. The corresponding family of Jacobians gives a moduli map $T \to A_g$; denote the image by Z°, with Zariski closure $Z \subset T_g \subset A_g$. Having set the scene in this way we can ask for which choices of the data involved Z is a special subvariety of positive dimension.

(a) *Are there examples with non-cyclic Galois group such that $Z \subset A_g$ is a special subvariety of positive dimension?*

(b) *Are there examples where B is not a rational curve, and such that $Z \subset A_g$ is a special subvariety of positive dimension?* Note that if Z is special, the Jacobian of B is a CM abelian variety.

As for question (a), we do have some examples that are obtained from families of covers of $\mathbb{P}^1$ with a non-cyclic abelian Galois groups. The examples presently known to us are listed in Table 2. In these examples we consider families of covers of $\mathbb{P}^1$ with Galois group of the form $A = (\mathbb{Z}/m_1\mathbb{Z}) \times (\mathbb{Z}/m_2\mathbb{Z})$ with $m_1|m_2$, with $N \geqslant 4$ branch points, and with local monodromy about the branch points t_i given by an N-tuple $a = (a_1, \ldots, a_N)$ in A^N. This gives rise to an $(N-3)$-dimensional closed irreducible subvariety $Z(m, N, a)$, where now $m = (m_1, m_2)$. For the Jacobians J_t in our family we have a (generally non-injective) homomorphism $\mathbb{Q}[A] \to \mathrm{End}^0(J_t)$; this defines a special subvariety $S(m) \subset A_g$ of PEL type with $Z(m, N, a) \subseteq S(m)$. We calculate $\dim\big(S(m)\big)$, and if this dimension equals $N-3$ then we conclude that $Z(m, N, a)$ is special.

Let us give some further details concerning Example (24). In this case the covers $C_t \to \mathbb{P}^1$ are branched over 4 points $t_1, \ldots, t_4$. There are 6 points above t_1, each with ramification index $e = 2$. Likewise, there are 2 points above t_2, both with $e = 6$, there are 4 points above t_3, each with $e = 3$, and finally there are 6 points above t_4, each

Table 2. Examples of special subvarieties in the Torelli locus (continued)

	genus	group	N	a
(21)	1	$(\mathbb{Z}/2\mathbb{Z}) \times (\mathbb{Z}/2\mathbb{Z})$	4	$((1,0),(1,0),(0,1),(0,1))$
(22)	3	$(\mathbb{Z}/2\mathbb{Z}) \times (\mathbb{Z}/4\mathbb{Z})$	4	$((1,0),(1,1),(0,1),(0,2))$
(23)	3	$(\mathbb{Z}/2\mathbb{Z}) \times (\mathbb{Z}/4\mathbb{Z})$	4	$((1,0),(1,2),(0,1),(0,1))$
(24)	4	$(\mathbb{Z}/2\mathbb{Z}) \times (\mathbb{Z}/6\mathbb{Z})$	4	$((1,0),(1,1),(0,2),(0,3))$
(25)	4	$(\mathbb{Z}/3\mathbb{Z}) \times (\mathbb{Z}/3\mathbb{Z})$	4	$((1,0),(1,0),(1,2),(0,1))$
(26)	2	$(\mathbb{Z}/2\mathbb{Z}) \times (\mathbb{Z}/2\mathbb{Z})$	5	$((1,0),(1,0),(1,0),(1,1),(0,1))$
(27)	3	$(\mathbb{Z}/2\mathbb{Z}) \times (\mathbb{Z}/2\mathbb{Z})$	6	$((1,0),(1,0),(1,1),(1,1),(0,1),(0,1))$

with $e = 2$. The Hurwitz formula gives $\chi = -24+6\cdot 1+2\cdot 5+4\cdot 2+6\cdot 1 = 6$, so $g = 4$. The group ring Q[A], with $A = (\mathbb{Z}/2\mathbb{Z}) \times (\mathbb{Z}/6\mathbb{Z})$, is isomorphic to $\mathbb{Q}^4 \times \mathbb{Q}(\zeta_3)^4$. Accordingly, the Jacobians J_t split, up to isogeny, as a product of eight factors. The simple factors of $\mathbb{Q}[A]$ correspond with the $\mathrm{Gal}(\overline{\mathbb{Q}}/\mathbb{Q})$-orbits in $\mathrm{Hom}(A, \overline{\mathbb{Q}}^*)$, which in this example may be identified with the set of complex characters $\rho \in \mathrm{Hom}(A, \mathbb{C}^*)$ taken modulo complex conjugation. Given a complex character $\rho\colon A \to \mathbb{C}^*$, the corresponding factor $J_{\rho,t}$ of J_t can be described as follows. We consider the cover $\pi_{\rho,t}\colon C_{\rho,t} \to \mathbb{P}^1$ with group $\rho(A)$, branched only above the points t_i, with local monodromy $\rho(a_i)$ about t_i. Note that $\rho(a_i)$ may be the identity element of $\rho(A)$, in which case t_i is not a branch point of $\pi_{\rho,t}$. Also note that $\pi_{\rho,t}$ is a cyclic cover, which brings us back to the situation considered in Section 4. Then $J_{\rho,t}$ is the new part of the Jacobian of $C_{\rho,t}$. With this description, it is easy to verify that:

- there are five pairs $(\rho, \overline{\rho})$ for which $J_{\rho,t} = 0$; this includes the four real characters, for which $\rho = \overline{\rho}$,
- there are two pairs $(\rho, \overline{\rho})$ for which $J_{\rho,t}$ is an elliptic curve with CM by an order in $\mathbb{Q}(\zeta_3)$,
- there is one pair $(\rho, \overline{\rho})$ for which $J_{\rho,t}$ is 2-dimensional, carrying an action by an order in $\mathbb{Q}(\zeta_3)$; varying t these give a family that is isogenous to the family of Jacobians of Example (5) in Table 1.

It follows from this description that the special subvariety $S(m)$ containing $Z(m, N, a)$ is 1-dimensional, and therefore $Z(m, N, a) = S(m)$ is special.

It should be noted that Example (24) is a sub-family of the family given in Example (14). To see this, let D_t be the quotient of C_t modulo the action of $\{1\} \times (\mathbb{Z}/6\mathbb{Z})$, and factor $\pi_t\colon C_t \to \mathbb{P}^1$ as

$$C_t \xrightarrow{q_t} D_t \xrightarrow{r_t} \mathbb{P}^1 .$$

We have $D_t \cong \mathbb{P}^1$. The cover q_t has group $(\mathbb{Z}/6\mathbb{Z})$ and is branched above five points, namely the unique point $\tilde{t}_2$ of D_t above t_2, the two points $\tilde{t}_{3,1}$ and $\tilde{t}_{3,2}$ above t_3, and the two points $\tilde{t}_{4,1}$ and $\tilde{t}_{4,2}$ above t_4. The local monodromy about these points is given by the 5-tuple (2,2,2,3,3) in $(\mathbb{Z}/6\mathbb{Z})^5$, which gives exactly the data of Example (14). The sub-family considered here is given by the constraint that the five branch points on $D_t \cong \mathbb{P}^1$ do not move freely but form three orbits under the action of $\mathbb{Z}/2\mathbb{Z}$ on D_t.

6.2. The Serre-Tate formal group structure and linearity properties

Let $m \geqslant 3$. On $\mathcal{A}_{g,[m]}(\mathbb{C})$ we have a metric, obtained from the uniformization by the Siegel space $\mathfrak{H}_g$; see subsection 3.2. In [46] it is proven that an irreducible algebraic subvariety $Z \subset \mathcal{A}_{g,[m]}$ is a special subvariety if and only if Z is totally geodesic and contains at least one special point. This may be viewed as a characterization of special subvarieties in terms of *linearity properties.*

This characterization has a nice arithmetic analogue, by which it was in fact inspired. Recall that if k is a perfect field of characteristic $p > 0$ and $x \in \mathcal{A}_g(k)$ corresponds to a ppav (A, λ) over k such that A is *ordinary*, the formal completion $\mathfrak{A}_x$ of $\mathcal{A}_{g,[m]}$ at the point x has a canonical structure of a formal torus over the ring of Witt vectors $W(k)$; see [37] or [41], Chap. 5. Using this we can again give meaning to the notion of "linearity". Let us elucidate this. Consider a closed irreducible algebraic subvariety $Z \subset \mathcal{A}_{g,[m],F}$, where F is a number field. Let $\mathcal{Z}$ denote the Zariski closure of Z inside $\mathcal{A}_{g,[m]}$ over O_F. If some ordinary point x as above lies in $\mathcal{Z}(k)$, the formal completion of $\mathcal{Z}$ at x gives a formal subscheme $\mathfrak{Z}_x \subset \mathfrak{A}_x$. It was shown by Noot in [57] that if Z is a special subvariety, the components of the formal subscheme $\mathfrak{Z}_x \subset \mathfrak{A}_x$ over $W(\bar{k})$ are translates of formal subtori of $\mathfrak{A}_x$ over torsion points. At the cost of excluding finitely many primes of O_F this may be sharpened to the conclusion that the components of the formal subschemes $\mathfrak{Z}_x$ are formal subtori; see also [47], Theorem 4.2(ii).

The converse of Noot's result was proven in [47]; the result is that if $Z \subset \mathcal{A}_{g,[m],F}$ is a closed irreducible algebraic subvariety such that for some ordinary point $x \in \mathcal{Z}(k)$ some component of $\mathfrak{Z}_x \subset \mathfrak{A}_x$ is a translate of a formal subtorus of $\mathfrak{A}_x$ over a torsion point, Z is a special subvariety. These results again give a characterization of special subvarieties in terms of linearity properties. It was shown in [46] that the result over $\mathbb{C}$ may reformulated in terms of formal group structures, in a way that makes the analogy between the two situations even clearer.

These results lead us to investigate the structure of the Torelli locus $T_g \subset A_g$ (or its analogue with a level structure) locally near an ordinary point x in characteristic p. The identity section of the formal torus $\mathfrak{A}_x$ gives a lifting of the ppav (A, λ) over k to a polarized abelian scheme $(A^{\mathrm{can}}, \lambda^{\mathrm{can}})$ over the ring of Witt vectors $W(k)$. This lifting is called the *Serre-Tate canonical lifting* of (A, λ). Note, however, that if (A, λ) is the Jacobian of a curve, the canonical lifting $(A^{\mathrm{can}}, \lambda^{\mathrm{can}})$ over $W(k)$ need not be a Jacobian; see [25], [67]. This leads to another view on Expectation 4.2. Indeed, suppose we have a special subvariety $Z \subset T_g$ with $Z \cap T_g^\circ \neq \emptyset$. Then Z is defined over some number field F, and, with notation as above, we can find ordinary points $x \in \mathcal{Z}(k)$, for finite fields k, such that the corresponding ppav (A, λ) is the Jacobian of a curve C/k, and such that the formal completion $\mathfrak{Z}_x \subset \mathfrak{A}_x$ is a union of formal subtori (or even just a formal subtorus). In this situation, the canonical lifting $(A^{\mathrm{can}}, \lambda^{\mathrm{can}})$ is again the Jacobian of a curve, which is a highly nontrivial fact. Indeed, the main results of Dwork and Ogus in [25] are based on the observation that in general, already the first-order canonical lifting, over the ring $W_2(k)$ of Witt vectors of length 2, is no longer a Jacobian.

For the next two questions, let $m \geqslant 3$ be an integer, and fix a genus g that is large enough, at least $g > 7$.

Question 6.8. Let $\mathcal{T}_{g,[m],\mathbb{Z}} \subset \mathcal{A}_{g,[m],\mathbb{Z}}$ denote the scheme-theoretic image of the Torelli morphism $\mathcal{M}_{g,[m]} \to \mathcal{A}_{g,[m]}$ over $\mathrm{Spec}(\mathbb{Z})$. Let k be a perfect field of characteristic $p > 0$, and suppose $x \in \mathcal{T}_{g,[m],\mathbb{Z}}(k)$ is an ordinary point, i.e., the corresponding abelian variety is ordinary. *Is it true that the formal completion $\mathfrak{T}_x \subset \mathfrak{A}_x$ of $\mathcal{T}_{g,[m],\mathbb{Z}}$ at the point x does not contain a formal subscheme $\mathfrak{Z}$, flat over $W(k)$, of positive dimension, such that $\mathfrak{Z}$ is a formal subtorus of $\mathfrak{A}_x$?*

A positive answer to this question would confirm Expectation 4.2. It should be noted, however, that what we ask here is stronger than what we need for 4.2. The difference is that in 6.8 we do not require the formal subscheme $\mathfrak{Z}$ to be algebraic, i.e., to be the formal completion of an algebraic subvariety of $\mathcal{A}_{g,[m]}$ passing through the point x. The interesting point, however, is that Question 6.8 only depends on the formal completion of the Torelli locus at a single ordinary point. One might expect that, in terms of the "linear structure" provided by the Serre-Tate structure of a formal torus on $\mathfrak{A}_x$, the Torelli locus $\mathfrak{T}_x \subset \mathfrak{A}_x$ should be highly non-linear.

Over $\mathbb{C}$ we have a question that is similar in spirit.

Question 6.9. *Does the Torelli locus $\mathcal{T}_{g,[m],\mathbb{C}} \subset \mathcal{A}_{g,[m],\mathbb{C}}$ contain any totally geodesic subvarieties of positive dimension?*

Again this question is stronger than what is needed for Expectation 4.2, as we do not require the totally geodesic subvariety to be algebraic. (In addition we should ask that the subvariety contains at least one CM point.)

Let us mention the paper [42], in which Möller investigates algebraic curves in $\mathcal{M}_g$ over $\mathbb{C}$ that are totally geodesic with respect to the Teichmüller metric (these are called Teichmüller curves), such that the image of this curve in $\mathcal{A}_g$ is a special subvariety. It is shown in [42] that for $g = 2$ and $g \geqslant 6$ there are no such curves. For $g = 3$ and $g = 4$ there is precisely one example, corresponding to Examples (7) and (12) in Table 1.

6.3. An analogy

We can view the boundary of $\mathcal{A}_g$ in a compactification as the locus of degenerating abelian varieties, but one can also view, working over $\mathbb{Z}_p$, say, the space $\mathcal{A}_g \otimes \mathbb{F}_p$ as a boundary of $\mathcal{A}_g \otimes \mathbb{Z}_p$. In this last case the abelian variety does not degenerate, but the p-structure does change. These two points of view have striking similarities. Often geometric questions are settled by studying properties at the boundary, and then to lift back to the interior of the moduli space considered.

Here we want to draw attention to the analogy between the results in [25] and the results of [30] and [4]. In the first case Dwork and Ogus study ordinary Jacobians in positive characteristic and their Serre-Tate canonical lifts to characteristic zero. In analogy with this, Fresnel and van der Put [30] and Andreatta [4] study liftings of Jacobians of degenerate curves. In both cases the authors show that, for $g > 3$, these liftings in general do not lie in the Torelli locus. In particular, Andreatta explains in [4] that special subvarieties in $\mathcal{A}_g$ have a linear structure at the boundary in a toroidal compactification of $\mathcal{A}_g$, and he shows that the closure of the Torelli locus is not linear at the boundary.

References

[1] N. A'Campo. Tresses, monodromie et le groupe symplectique. *Comment. Math. Helv.*, **54**, 318–327, 1979. ← 570

[2] Y. André. G-functions and geometry. *Aspects of Math.*, **E13**, Vieweg, Braunschweig, 1989. ← 567

[3] Y. André. Finitude des couples d'invariants modulaires singuliers sur une courbe algébrique plane non modulaire. *J. reine angew. Math.*, **505**, 203–208, 1998. ← 568

[4] F. Andreatta. Coleman-Oort's conjecture for degenerate irreducible curves. *Israel J. Math.* **187**, 231–285, 2012. ← 572, 589

[5] E. Arbarello, M. Cornalba, and Ph. Griffiths. Geometry of algebraic curves, Vol. II. *Grundlehren der math. Wissenschaften*, **268**, Springer, 2011. ← 570

[6] M. Bainbridge and M. Möller. The locus of real multiplication and the Schottky locus. To appear in the *J. Reine Angew. Math.* `http://dx.doi.org/10.1515/crelle-2012-0019`. ← 572

[7] A. Beauville. Le problème de Schottky et la conjecture de Novikov, in: Séminaire Bourbaki, année 1986–87, Exposé 675. *Astérisque*, **152–153**, 101–112, 1987. ← 574

[8] L. Caporaso and F. Viviani. Torelli theorem for stable curves. *J. European Math. Soc.*, **13**, 1289–1329, 2011. ← 555

[9] P. Cartier. Dualité des variétés abéliennes. *Séminaire Bourbaki, année 1957–58*, Exposé **164**. ← 552

[10] P. Cartier. Isogenies and duality of abelian varieties. *Ann. of Math. (2)*, **71**, 315–351, 1960. ← 552

[11] E. Cattani, P. Deligne, and A. Kaplan. On the locus of Hodge classes. *J. American Math. Soc.*, **8**, 483–506, 1995. ← 560

[12] C.-L. Chai. Compactification of Siegel moduli schemes. *London Math. Soc. Lecture Note Series*, **107**, Cambridge University Press, Cambridge, 1985. ← 552

[13] C.-L. Chai and F. Oort. Abelian varieties isogenous to a Jacobian. *Ann. of Math.* **176**, 589–635, 2012. ← 570

[14] C. Ciliberto and G. van der Geer. Subvarieties of the moduli space of curves parametrizing Jacobians with nontrivial endomorphisms. *Amer. J. Math.*, **114**, 551–570, 1992. ← 572, 573

[15] C. Ciliberto, G. van der Geer, and G. M. Teixidor i Bigas. On the number of parameters of curves whose Jacobians possess nontrivial endomorphisms. *J. Algebraic Geom.*, **1**, 215–229, 1992. ← 572

[16] R. Coleman. Torsion points on curves, in: Galois representations and arithmetic algebraic geometry, 235-247. *(Y. Ihara, ed.) Adv. Studies Pure Math.*, **12**, North-Holland, Amsterdam, 1987. ← 569

[17] O. Debarre. The Schottky problem: an update, in: Current topics in complex algebraic geometry (Berkeley, CA, 1992/93), 57–64. *(H. Clemens and J. Kollár, eds.) Math. Sci. Res. Inst. Publ.*, **28**, Cambridge Univ. Press, Cambridge, 1995. ← 574

[18] P. Deligne. Théorie de Hodge. II. *Inst. Hautes Études Sci. Publ. Math.*, **40**, 5–57, 1971. ← 555

[19] P. Deligne. Travaux de Shimura, in: Séminaire Bourbaki, année 1970–71, Exposé 389, 123–165. *Lecture Notes in Math.*, **244**, Springer, Berlin, 1971. ← 551, 565, 567

[20] P. Deligne. La conjecture de Weil pour les surfaces K3. *Inventiones Math.*, **15**, 206–226, 1972. ← 570

[21] P. Deligne. Variétés de Shimura: interprétation modulaire, et techniques de construction de modèles canoniques, in: Automorphic forms, representations, and L-functions; Part 2, 247–289. *(A. Borel, W. Casselman, eds.) Proc. of Symposia in Pure Math.*, **33**, American Math. Soc., Providence, R.I., 1979. ← 551, 561

[22] P. Deligne. Hodge cycles on abelian varieties. (Notes by J. Milne.) in: Hodge cycles, motives, and Shimura varieties, 9–100. *Lecture Notes in Math.*, **900**, Springer, Berlin-New York, 1982. ← 556, 558

[23] P. Deligne and G. Mostow. Monodromy of hypergeometric functions and nonlattice integral monodromy. *Inst. Hautes Études Sci. Publ. Math.*, **63**, 5–89, 1986. ← 583

[24] P. Deligne and D. Mumford. The irreducibility of the space of curves of given genus. *Inst. Hautes Études Sci. Publ. Math.*, **36**, 75–109, 1969. ← 554

[25] B. Dwork and A. Ogus. Canonical liftings of Jacobians. *Compositio Math.*, **58**, 111–131, 1986. ← 582, 588, 589

[26] B. Edixhoven and A. Yafaev. Subvarieties of Shimura varieties. *Ann. of Math. (2)*, **157**, 621–645, 2003. ← 568

[27] G. Faltings. Arithmetische Kompaktifizierung des Modulraums der abelschen Varietäten, in: Arbeitstagung Bonn 1984, 321–383. *(F. Hirzebruch et al., eds.) Lecture Notes in Math.* 1111, Springer, Berlin, 1985. ← 552

[28] G. Faltings and C.-L. Chai. Degeneration of abelian varieties. *Ergebnisse der Math. und ihrer Grenzgebiete (3)*, **22**, Springer, Berlin, 1990. ← 552, 554

[29] B. Farb and H. Masur. Superrigidity and mapping class groups. *Topology*, **37**, 1169–1176, 1998. ← 572

[30] J. Fresnel and M. van der Put. Uniformisation de variétés de Jacobi et déformations de courbes. *Ann. Fac. Sci. Toulouse Math. (6)*, **3**, 363–386, 1994. ← 589

[31] G. van der Geer. The Schottky problem, in: Arbeitstagung Bonn 1984; pp. 385–406. *(F. Hirzebruch et al., eds.) Lecture Notes in Math.*, **1111**, Springer, Berlin, 1985. ← 574

[32] R. Hain. Locally symmetric families of curves and Jacobians, in: Moduli of curves and abelian varieties, 91–108. *(C. Faber, E. Looijenga, eds.) Aspects of Math.*, **E33**, Vieweg, Braunschweig, 1999. ← 552, 571, 572

[33] S. Helgason.Differential geometry, Lie groups, and symmetric spaces. *Pure and Applied Mathematics*, **80**, Academic Press, Inc., New York-London, 1978. ← 578

[34] J. Igusa. On the irreducibility of Schottky's divisor. *J. Fac. Sci. Univ. Tokyo Sect. IA Math.*, **28**, 531–545, 1981. ← 584

[35] A. de Jong and R. Noot. Jacobians with complex multiplication, in: Arithmetic algebraic geometry (Texel, 1989), 177–192. *(G. van der Geer et al., eds.) Progr. Math.*, **89**, Birkhäuser Boston, Boston, MA, 1991. ← 574, 582

[36] A. de Jong and S. Zhang. Generic abelian varieties with real multiplication are not Jacobians, in: Diophantine Geometry, 165–172. *(U. Zannier, ed.) CRM Series*, **4**, Scuola Normale Pisa, 2007. ← 572

[37] N. Katz. Serre-Tate local moduli, in: Algebraic surfaces; pp. 138–202. *Lecture Notes in Math.*, **868**, Springer, Berlin, 1981. ← 587

[38] B. Klingler and A. Yafaev. The André-Oort conjecture. Preprint 2006. Available at `http://www.math.jussieu.fr/~klingler/papers.html`. ← 568

[39] S. Kukulies. On Shimura curves in the Schottky locus. *J. Algebraic Geom.*, **19**, 371–397, 2010. ← 585

[40] E. Looijenga. Uniformization by Lauricella functions—an overview of the theory of Deligne-Mostow, in: Arithmetic and geometry around hypergeometric functions, 207–244. *(R.-P. Holzapfel et al., eds.) Progr. Math.*, **260**, Birkhäuser, Basel, 2007. ← 583

[41] W. Messing. The crystals associated to Barsotti-Tate groups: with applications to abelian schemes. *Lecture Notes in Math.*, **264**, Springer, 1972. ← 587

[42] M. Möller. Shimura- and Teichmüller curves. *J. modern dynamics*, **5**, 1–32, 2011. ← 588

[43] M. Möller, E. Viehweg, and K. Zuo. Special families of curves, of abelian varieties, and of certain minimal manifolds over curves, in: Global aspects of complex geometry; pp. 417–450. (F. Catanese et al., eds.) Springer, Berlin, 2006. ← 572

[44] M. Möller, E. Viehweg, and K. Zuo. Stability of Hodge bundles and a numerical characterization of Shimura varieties. Preprint, `arXiv: /0706.3462`. ← 572

[45] B. Moonen. Special points and linearity properties of Shimura varieties. Ph.D. Thesis, University of Utrecht, 1995. ← 563

[46] B. Moonen. Linearity properties of Shimura varieties. I. *J. Algebraic Geom.*, **7**, 539–567, 1998. ← 563, 564, 587

[47] B. Moonen. Linearity properties of Shimura varieties. II. *Compositio Math.*, **114**, 3–35, 1998. ← 568, 587

[48] B. Moonen. Models of Shimura varieties in mixed characteristics, in: Galois representations in arithmetic algebraic geometry, 267–350. *(A. Scholl, R. Taylor, eds.) London Math. Soc. Lecture Note Series*, **254**, Cambridge Univ. Press, Cambridge, 1998. ← 569

[49] B. Moonen. Special subvarieties arising from families of cyclic covers of the projective line. *Documenta Math.*, **15**, 793–819, 2010. ← 581, 582

[50] B. Moonen and Yu. Zarhin. Hodge classes and Tate classes on simple abelian fourfolds. *Duke Math. J.*, **77**, 553–581, 1995. ← 558

[51] B. Moonen and Yu. Zarhin. Hodge classes on abelian varieties of low dimension. *Math. Ann.*, **315**, 711–733, 1999. ← 558, 584

[52] I. Morrison and H. Pinkham. Galois Weierstrass points and Hurwitz characters. *Ann. of Math. (2)*, **124**, 591–625, 1986. ← 580

[53] G. Mostow. On discontinuous action of monodromy groups on the complex n-ball. *J. American Math. Soc.*, **1**, 555–586, 1988. ← 583

[54] D. Mumford. A note of Shimura's paper "Discontinuous groups and abelian varieties". *Math. Ann.*, **181**, 345–351, 1969. ← 558, 566, 584

[55] D. Mumford. Abelian varieties. *Tata Institute of Fundamental Research Studies in Math.*, **5**, Oxford University Press, London, 1970. ← 573

[56] M. Nishi. The Frobenius theorem and the duality theorem on an abelian variety. *Mem. Coll. Sc. Kyoto*, **32**, 333–350, 1959. ← 552

[57] R. Noot. Models of Shimura varieties in mixed characteristic. *J. Algebraic Geom.*, **5**, 187–207, 1996. ← 587

[58] R. Noot. Abelian varieties with ℓ-adic Galois representation of Mumford's type. *J. reine angew. Math.*, **519**, 155–169, 2000. ← 584

[59] R. Noot. On Mumford's families of abelian varieties. *J. Pure Appl. Alg.*, **157**, 87–106, 2001. ← 584

[60] R. Noot. Correspondances de Hecke, action de Galois et la conjecture de André-Oort (d'après Edixhoven et Yafaev), in: Séminaire Bourbaki, année 2004-2005, Exposé 942. *Astérisque*, **307**, 165–197, 2006. ← 568, 569

[61] F. Oort. Commutative group schemes. *Lecture Notes in Math.*, **15**, Springer, Berlin-New York, 1966. ← 552

[62] F. Oort. Some questions in algebraic geometry. Unpublished manuscript, June 1995. Available at `http://www.math.uu.nl/people/oort/`. ← 567

[63] F. Oort. Canonical liftings and dense sets of CM-points, in: Arithmetic Geometry; pp. 228–234. *(F. Catanese, ed.) Symposia Mathematica*, **37**, Cambridge University Press, Cambridge, 1997. ← 567, 569

[64] F. Oort. A stratification of a moduli space of polarized abelian varieties, in: Moduli of abelian varieties, 345–416. *(C. Faber, G. van der Geer, F. Oort, eds.). Progr. Math.*, **195**, Birkhäuser, Basel, 2001. ← 552

[65] F. Oort. Special points in Shimura varieties, an introduction. Informal notes. Available at `http://www.math.uu.nl/people/oort/`. ← 582

[66] F. Oort. Moduli of abelian varieties in mixed and in positive characteristic, in: Handbook of moduli, Vol. II (Gavril Farkas and Ian Morrison eds.), 75–134. *Advanced Lectures in Mathematics*, **25**, Higher Education Press & International Press, Beijing-Boston, 2012. ← 583

[67] F. Oort and T. Sekiguchi. The canonical lifting of an ordinary Jacobian variety need not be a Jacobian variety. *J. Math. Soc. Japan*, **38**, 427–437, 1986. ← 588

[68] F. Oort and J. Steenbrink. The local Torelli problem for algebraic curves, in: Journées de Géométrie Algébrique d'Angers; pp. 157–204. *(A. Beauville, ed.) Sijthoff & Noordhoff*, Alphen aan den Rijn–Germantown, Md., 1980. ← 555

[69] C. Peters and J. Steenbrink. Mixed Hodge structures. *Ergebnisse der Mathematik und ihrer Grenzgebiete*, 3. Folge, 52, Springer, Berlin, 2008. ← 555

[70] J. Pila. O-minimality and the André-Oort conjecture for $\mathbb{C}^n$. *Ann. of Math. (2)*, **173**, 1779–1840, 2011. ← 568

[71] R. Pink and D. Rössler. On Hrushovski's proof of the Manin-Mumford conjecture, in: Proceedings of the International Congress of Mathematicians (Beijing, 2002); Vol. I, 539–546. (Tatsien Li, ed.), Higher Education Press, Beijing, 2002. ← 569

[72] M. Raynaud. Courbes sur une variété abélienne et points de torsion. *Inventiones Math.*, **71**, 207–233, 1983. ← 569

[73] M. Raynaud. Sous-variétés d'une variété abélienne et points de torsion, in: Arithmetic and geometry; Vol. I, 327–352. *(M. Artin, J. Tate, eds.) Progr. Math.*, **35**, Birkhäuser Boston, Boston, MA, 1983. ← 569

[74] J. Rohde. Cyclic coverings, Calabi-Yau manifolds and complex multiplication. *Lecture Notes in Math.*, **1975**, Springer, Berlin, 2009. ← 581, 583

[75] C. Schoen. Hodge classes on self-products of a variety with an automorphism. *Compositio Math.*, **65**, 3–32, 1988. ← 580

[76] G. Shimura. On analytic families of polarized abelian varieties and automorphic functions. *Ann. of Math. (2)*, **78**, 149–193, 1963. ← 573, 575

[77] G. Shimura. On purely transcendental fields automorphic functions of several variable. *Osaka J. Math.*, **1**, 1–14, 1964. ← 574, 582

[78] T. Springer. Linear algebraic groups. Second edition. *Progr. Math.*, **9**, Birkhäuser Boston, Boston, MA, 1998. ← 566

[79] E. Ullmo and A. Yafaev. Galois orbits and equidistribution of special subvarieties: towards the André-Oort conjecture. Preprint 2006. `arXiv: 1209.0934`. ← 568

[80] E. Viehweg and K. Zuo. A characterization of certain Shimura curves in the moduli stack of abelian varieties. *J. Differential Geom.*, **66**, 233–287, 2004. ← 572

[81] C. Voisin. Hodge loci, in: Handbook of moduli, Vol. III (Gavril Farkas and Ian Morrison eds.), 507–546. *Advanced Lectures in Mathematics*, **26**, Higher Education Press & International Press, Beijing-Boston, 2012. ← 560

[82] A. Yafaev. On a result of Moonen on the moduli space of principally polarised abelian varieties. *Compositio Math.*, **141**, 1103–1108, 2005. ← 568

University of Amsterdam
E-mail address: `bmoonen@uva.nl`

University of Utrecht
E-mail address: `f.oort@uu.nl`

图书在版编目（CIP）数据

模手册．2 = Handbook of Moduli. Vol. Ⅱ：英文 /（德）法卡斯（Farkas, G.），（美）莫里森（Morrison, I.）编．—北京：高等教育出版社，2013.1
ISBN 978-7-04-035168-2

Ⅰ.①模… Ⅱ.①法… ②莫… Ⅲ.①代数几何-文集-英文 Ⅳ.①O187-53

中国版本图书馆CIP数据核字（2012）第242458号

策划编辑 王丽萍　　责任编辑 王丽萍　　封面设计 张申申　　责任印制 朱学忠

出版发行	高等教育出版社	咨询电话	400-810-0598
社　　址	北京市西城区德外大街4号	网　　址	http://www.hep.edu.cn
邮政编码	100120		http://www.hep.com.cn
印　　刷	涿州市星河印刷有限公司	网上订购	http://www.landraco.com
开　　本	787mm × 1092mm　1/16		http://www.landraco.com.cn
印　　张	38	版　　次	2013年1月第1版
字　　数	720千字	印　　次	2013年1月第1次印刷
购书热线	010-58581118	定　　价	128.00元

本书如有缺页、倒页、脱页等质量问题，请到所购图书销售部门联系调换

物 料 号　35168-00